ENVIRONMENTAL FOREST SCIENCE

FORESTRY SCIENCES

Volume 54

The titles published in this series are listed at the end of this volume.

Environmental
Forest Science

*Proceedings of the IUFRO Division 8 Conference
Environmental Forest Science,
held 19–23 October 1998, Kyoto University, Japan*

Edited by

KYOJI SASSA

*Disaster Prevention Research Institute,
Kyoto University,
Kyoto, Japan*

SPRINGER-SCIENCE+BUSINESS MEDIA, B.V.

A C.I.P. Catalogue record for this book is available from the Library of Congress.

ISBN 978-94-010-6237-4   ISBN 978-94-011-5324-9 (eBook)
DOI 10.1007/978-94-011-5324-9

Printed on acid-free paper

TABLE OF CONTENTS

Preface   xi

IUFRO: History and Role in the 21st Century
---- Risto SEPPÄLÄ   1

Toward the New Stage of Environmental Forest Science Research in the 21st Century
-IUFRO Status and Perspective
---- Heinrich SCHMUTZENHOFER   7

Chapter 1 *FORESTS AS ENVIRONMENT*

Environment and Production -
Two Sides of the Same Coin or Incompatible Goals ?
---- Lisa SENNERBY- Forsse   15

Integrating Environmental Research into Forest Ecosystems Planning
---- Klaus von GADOW   29

Forests in Sustainable Mountain Development
---- Martin F. PRICE   39

Tactical Sediment Yield Control to Mitigating Terrestrial and/or Aquatic Environmental
Impacts of Forest Road Construction
---- Masami SHIBA   51

The Environment in Planning a Forest Road Network
---- Igor POTOCNIK   67

Forest Disturbance and Efficiency of Vehicle Logging Operations through Selective
Cutting Harvest in a Natural Forest
---- Toshio NITAMI, Shin'ich IWAMOTO, Kenji FUKUSHI
       and Hisatomi KASAHARA   75

Status of Waste Generation, Disposal Ways and Poplar Species as a Landfill Cover
in South Korea
---- Don Koo LEE and Su-Young WOO   83

Environmentally Friendly Resource Use
- Taiwan's Experience in Pulp and Paper Production
---- Hsiu Hwa WANG and Sheau Horng LIN   93

Forests as Ecosystems Within a Changing Environment
---- John L. INNES   107

Carbon Absorption by Temperate Forest Ecosystems: Problems and Responses to a
Changing Environment
---- Giuseppe SCARASCIA-MUGNOZZA, Paolo DE ANGELIS,
     Giorgio MATTEUCCI and Riccardo VALENTINI   119

Economic Value of the Carbon Sink Services of Costa Rica's Forestry Plantations
---- Octavio A. RAMIREZ and Manuel GOMEZ   129

Trees and Woodland in a Cultural Landscape: the History of Woods in England
---- Oliver RACKHAM   139

Review of Forest Culture Research in Japan: Toward a New Paradigm of Forest Culture
---- Taiichi ITO   149

Forest Landscapes of the Kanto Region, Japan in the 1880's and Human Impact on
Them
---- Jun-ichi OGURA   157

Japanese Attitudes Towards Forests According to Comparative Opinion Surveys
---- Shigejiro YOSHIDA, Masaaki IMANAGA, Koji MATSUSHITA
     and Okihiro OHSAKA   165

Chapter 2 FOREST ECOSYSTEMS AND BIODIVERSITY

History, Condition and Management of Floodplain Forest Ecosystems in Europe
---- Emil KLIMO   173

Ecosystem Management as an Approach for Sustaining Forests and Their Biodiversity
---- Robert C. SZARO and William T. SEXTON   187

Decomposition Processes of Litter Along a Latitudinal Gradient
---- Hiroshi TAKEDA   197

Size Distribution and Carbon-to-Nitrogen Ratios of Size-Fractionated Organic Matter in
the Forest Floor of Coniferous and Broadleaved Stands
---- Masamichi TAKAHASHI   207

Nitrogen Dynamics of Decomposing Japanese Cedar and Japanese Cypress Litter in
Plantation Forests
---- Tamon YAMASHITA, Hiroyuki TOBIT and Hiroshi TAKEDA   215

Decomposition Process of Leaf Litter in a Coniferous Forest
---- Xing-jun TIAN and Hiroshi TAKEDA   223

Changes in Nitrogen and Carbon Fractions of the Decomposing Litters of Bamboo
---- Nimfa K. TORRETA, Hiroshi TAKEDA
and Jun-Ichi AZUMA   231

Gross Soil N Transformations in a Coniferous Forest in Japan
---- Naoko TOKUCHI, Muneto HIROBE and Keisuke KOBA   239

Topographic Differences in Soil N Transformation Patterns Along a Forest Slope
---- Muneto HIROBE, Naoko TOKUCHI and Goro IWATSUBO   245

Dynamics of Soil Microbial Activities in Different Vegetation Types of the Seasonally
Dry Tropics
---- Pitayakon LIMTONG, Kazuhiro ISHIZUKA, Masamichi TAKAHASHI,
Vanlada SUNANTAPONGSUK and Prasode TUMMAKATE   253

The Role of Soil Microbial Biomass in Burned Japanese Red Pine Forest
---- Takahiro TATEISHI   261

Allelopathic Interactions in Forestry Systems
---- Ravinder Kumar KOHLI   269

Comparative Vegetation Analysis under Multipurpose Plantations
---- Ravinder Kumar KOHLI   285

Allelopathic Effects of Exotic Tree Species on Microorganisms and Plants in Galicia
(Spain)
---- Manuel J. REIGOSA, Xosé C. SOUTO, Luis GONZÁLEZ
and Juan C. BOLAÑO   293

Soil Microorganisms and Phenolics : Their Implication in Spruce Natural Regeneration
Failure
---- X. Carlos SOUTO, Geneviève CHIAPUSIO and François PELLISSIER   301

Allelopathy and Competition in Coniferous Forests
---- A. U. MALLIK   309

Role of Allelopathy in Regulating the Understorey Vegetation of Casuarina
Equisetifolia
---- Daizy R. BATISH and Harminder Pal SINGH   317

Long-term Response of Radiata Pine to Phosphate Fertiliser on a Strongly P-Fixing Soil
--- Peter HOPMANS and David W. FLINN   325

Monitoring Population Viability in Declining Tree Species Using Indicators of
Genetic Diversity and Reproductive Success
---- Alexander MOSSELER and O. P. RAJORA   333

Using GIS to Review Wildlife Habitat Information in the Kii Peninsula, Japan
---- Tsuyoshi YOSHIDA, Manabu ISHIZAKI and Kazuhiro TANAKA   345

Analysis of Species Hyper-diversity in the Tropical Rain Forests of Indonesia : the Problem of Non-observance
---- Keith RENNOLLS and Yves LAUMONIER   355

Mixed Stands Between Description and Modelling
---- Alain FRANC   363

Quantifying Biodiversity
The Effect of Sampling Method and Intensity on Diversity Indices
---- Dieter R. PELZ and Paul LUEBBERS   373

Photosynthetic Light Environment of Tropical Lowland Forest and Growth Response of *Shorea Leprosula*
---- Muhamad AWANG, Ahmad Makmom ABDULLAH and Akio FURUKAWA   379

Growth Performance of Malaysian Tropical Trees under Different Light Regimes
---- Toshihiro YAMADA, Taketo YOKOTA, Akio FURUKAWA,
Makmon ABUDULLA, Samusddin JOHAN and Muhamad AWANG   387

Chapter 3 *FOREST HYDROLOGY*

Hydrological Effects of Afforestation and Pasture Improvement in Montane Grasslands, South Island, New Zealand
---- Barry FAHEY, Rick JACKSON and Lindsay ROWE   395

Water Balance Modelling on Small Forested Catchments
---- Pavel KOVAR   405

Evaluation of Forest Canopy Shape from Standpoint of Thermal Exchange above Forest
---- Koji TAMAI, Shigeaki HATTORI and Yoshiaki GOTO   411

Estimating Rates of Nutrient Recovery Following Timber Harvesting in a Second Growth Forest of Peninsular Malaysia
---- Yusop ZULKIFLI, Kasran BAHARUDDIN and Nik ABDUL RAHIM   419

Hydrological Variations of Discharge, Soil Loss and Recession Coefficient in Three Small Forested Catchments
---- Kyongha KIM and Yongho JEONG   431

Evaluating the Effectiveness of Forest Crop to Mitigate Erosion Using a Sediment
Delivery Distributed Model
---- Vito FERRO, Paolo PORTO, Giovanni CALLEGARI, Francesco IOVINO,
Vittoria MENDICINO and Antonella VELTRI 439

A Concept for Runoff Processes on a Steep Forested Hillslope
---- Makoto TANI 455

Flow Pathways on Steep Forested Hillslopes: the Tracer, Tensiometer and Trough
Approach
---- Jeffrey MCDONNELL, Dean BRAMMER, Carol KENDALL,
Niclas HJERDT, Lindsay ROWE, Mike STEWART and Ross WOODS 463

Intrastorm Fluctuations of Piezometric Head and Soil Temperature within a Steep
Forested Hollow
---- Yoshio TSUBOYAMA, Shoji NOGUCHI, Toshio SHIMIZU,
Roy C. SIDLE and Ikuhiro HOSODA 475

Progress Towards Understanding Stormflow Generation in Headwater Catchments
---- Roy C. SIDLE, Yoshio TSUBOYAMA, Shoji NOGUCHI,
Ikuhiro HOSODA, Motohisa FUJIEDA and Toshio SHIMIZU 483

Chapter 4 *NATURAL DISASTERS IN MOUNTAINS*

Mechanisms of Landslide Triggered Debris Flows
---- Kyoji SASSA 499

Assessment of Failure and Success of Preventing Damages of Debris Flows Caused by
Landslide in Laogan Ravine, Yunnan, China
---- Tianchi LI 519

The Estimation of the Hazard Potency of Debris Flows and the Step to Step Method
---- Gernot FIEBIGER 529

Statistical Analysis on the Planimetry of Debris Flow Fans
---- Xilin LIU 541

The Largest Debris Flow in the World, Seimareh Landslide, Western Iran
---- Zieaoddin SHOAEI and Jafar GHAYOUMIAN 553

Assessment of Hazard Potential of Debris Flows in Relation to the Reclamation of
Forests on the Foot of Volcanoes
---- Kazuo OKUNISHI and Hiroshi SUWA 563

Motion and Fluidization of a Hariharagawa Landslide, South Japan
---- Hiromu MORIWAKI and Teruko SATO   569

Cyclic-Loading Ring-Shear Tests to Study High-Mobility of Earthquake-Induced-Landslides
---- Fawu WANG, Kyoji SASSA and Hiroshi FUKUOKA   575

Initiation of Rapid and Slow Landslides in Experimental Model
---- Tayoko KUBOTA and Yasuo TAKEDA   583

An Experimental Study on the Rainfall-Induced-Flowslides
---- Gonghui WANG and Kyoji SASSA   591

Computed Powder Avalanche Impact Pressures on a Tunnel-bridge in Au²erfern-Tirol
---- Lambert RAMMER, Horst SCHAFFHAUSER and Peter SAMPL   599

Evaluation of Slope-failure-debris Mass Using a Digital Elevation Model with Stereo Pair Aerophotographs
---- Satoshi TSUCHIYA   607

Energy Approach to Evaluation of Grain Crushing
---- Dimitri A. VANKOV and Kyoji SASSA   615

Comparison of Shear Behavior of Sandy Soils by Ring-Shear Test with Conventional Shear Tests
---- Yasuhiko OKADA, Kyoji SASSA and Hiroshi FUKUOKA   623

Soil Bioengineering - an Environmental Alternative for Erosion and Torrent Control
---- Christoph GERSTGRASER   633

Forest Fire Prevention Through Prescribed Burning in Acacia Mangium Plantation in South Sumatra, Indonesia
---- Bambang Hero SAHARJO and Hiroyuki WATANABE   641

Demands on a Nature Orientated Flood Control
---- Hansjoerg HUFNAGL   649

Preface

This proceedings volume has been edited from sixty-nine full text papers of the 132 papers presented to the IUFRO (International Union of Forestry Research Organizations) Conference on Environmental Forest Science, which was jointly organized by IUFRO Division 8, "Forest Environment", and Kyoto University in Kyoto, Japan, on 19-23 October 1998.

The International Union of Forestry Research Organizations (IUFRO) is one of the oldest scientific societies. It was founded in 1892 to foster cooperation of research units on forestry. IUFRO consists of 650 research organizations from 100 countries. IUFRO Division 8 is the latest division, founded at the 20th World Congress in 1995 by subdividing the previous Division 1, "Forest Environment and Silviculture". The objective of this first general Conference of Division 8 is to consider research needs in the 21st century for forest environment, and the integration of related fields of sciences to a new concept of environmental forest science.

The 1995 Kobe earthquake killed about 5,400 people in Japan. The disaster was much magnified by the dense regional development and land use of mankind, which were not in harmony with the natural environment. Population growth and economical development are necessarily accompanied by regional development and expansion of the area used by mankind. This causes changes in natural ground and vegetation, filling valleys and ponds, reclaiming seashores, changing slope geometries, cutting forests, and destroying farmlands. These changes result in flooding, liquefaction, landslides, debris flows, slope failures, rock falls, and forest fires resulting in many fatalities and destruction of buildings and earth structures. The people of Japan have been reminded that a safe environment is a fundamental factor for society and its industries. To protect residential areas from disasters that occur in mountainous areas, and to reserve places of recreation and relaxation, so called "green belts" (forest zones) are being planned between urban areas and the mountains. The Japanese Government is continuing to purchase very expensive lands at the margins of urban areas to return them to forest use as green belts.

In the Kobe area, a large population and industrial activities are confined to a narrow zone between the Rokko Mountains and the Sea in the Kobe area. It is a cause of big disaster. This situation is a typical one, which many countries are encountering or are likely to encounter in the next century. The photo on the front cover shows a landslide-debris flow in Japan, which occurred in the outskirts of a small city in 1997. The event killed 21 inhabitants of orange farms on the debris fan.

Apart from the aforementioned direct disasters, a less hazardous environment that is in harmony with Nature and the Earth is critically important for mankind. We have to (1) understand ecosystems, biodiversity, and the cycle of water in forests, (2) study engineering and industrial works for forests and forestry so as not to deteriorate the environment, and (3) develop forest culture as the interaction between mankind and forests. The present stage is only the beginning of Environmental Forest Science, thus this volume does not include all aspects of the science, but at least presents the direction and early efforts. It includes four chapters: Chapter 1, Forests as Environment; Chapter 2, Ecosystems and Biodiversity; Chapter 3, Forest Hydrology; Chapter 4, Natural Disasters in Mountains. It would be a great pleasure to me if this book could be effective in tempting forest and environmental engineers and scientists to join the activities to develop Environmental Forest Science.

The Editor acknowledges the help of Associate Professor Hiroshi Fukuoka, Dr. Gen Furuya, and other colleagues of the Landslide Section, Disaster Prevention Research Institute, Kyoto University in editing this book. My thanks are also due to the organizing committee members of this conference, research group coordinators of Division 8, and members of IUFRO Executive Board and IUFRO-Japan.

Kyoji SASSA
Coordinator
IUFRO Division 8 "Forest Environment"

IUFRO: History and Role in the 21st Century

Risto SEPPÄLÄ

IUFRO Vice President for Programme
Finnish Forest Research Institute
Metsätutkimuslaitos (Metla), Unioninkatu 40 A, FIN-00170 Helsinki
Finland

ABSTRACT From its small beginnings in Central Europe, the IUFRO has grown into a major global organization. The Union has gained reputation for facilitating research cooperation, maintaining high scientific standards and promoting the exchange of information. These will remain a focus of the IUFRO also in the 21st century. To survive and to prosper in the next millennium, the Union must, however, reconsider its organizational and financial structure, develop further its network, put more emphasis on problem-oriented research, and act as a bridge between the scientific community and policy makers.
Keywords: Forest science, IUFRO

1. THE IUFRO'S HISTORY

1.1 From a Central European Core to a Global Union

In 1890, during the Congress of Agriculture and Forestry in Vienna, a proposal was made to establish a "central organ" for applied forest research in the European countries. Soon after this, statutes were outlined, and in 1892 the Union of Forest Experiment Organizations was founded in Eberswalde, Germany.

Originally only three governments, Austria, Germany and Switzerland, agreed that their forest experiment stations would join the Union. Until 1903 when Japan came along, the members were only from Europe. By the First World War the number of member countries had already risen to 22, including also the USA and Canada outside Europe.

Between the two world wars the IUFRO gradually lost its Central European character. For the first time, representatives from Africa, Asia and South America were able to bring forward and discuss their forest research on an international forum. Not only experiment stations, but also universities, forestry education centres, and other forestry institutions gathered together in the IUFRO that also changed its name to the present form.

After the Second World War the IUFRO's geographical focus shifted. The centre of gravity in forest research was partly displaced from Europe to North America where the emerging new disciplines were booming. Forest research institutions in the developing countries joined the IUFRO. The union became truly global.

1. 2 The Birth of Today's IUFRO

The new world order after the Second World War had an impact on the IUFRO, too. FAO had been founded and efforts were made to incorporate the IUFRO into

1

it. However, the Union objected strongly to such a centralization and the plan was rejected. Finally, an agreement was signed with the FAO whereby the IUFRO retained its independent status but had a special consultative role with the FAO. After the war the IUFRO Secretariat was even housed within the FAO in Rome from where it moved to Vienna in 1973.

Before the Second World War the IUFRO was very much a collaboration network between the leading persons of its member institutions. The first Congress after the war, in 1948, established a new structure based on research sections which the staff of the member institutions could join freely. This carried the international exchange of knowledge and experience in forest research from the level of directors to the level of active scientists. Consequently, while the emphasis previously was on harmonizing and standardizing forest research methods, it was then put more on the exchange of experience. Thus, the new structure formed a basis for the IUFRO's current network of scientists.

The rapid economic and political development after the Second World War brought substantial changes also in forest science that developed from empirical descriptions to the explanation of causal connections. New methods and scientific disciplines were needed and developed, many of them outside the traditional forest research.

The multiplication of tasks and members, as well as new disciplines made a substantial reorganization of the Union necessary. A new structure with divisions and research groups, created in 1971, remains the basis of the IUFRO's structure today. The constitution and the number of Divisions have been slightly modified but practically the only visible change since 1971 is the creation of Special Programmes and Task Forces.

1.3 Activities and Research Topics

IUFRO has played an essential role in establishing standards and harmonizing field investigations in forest research. In 1903 the Union initiated international forestry bibliographies that did not exist earlier. Gradually the development led to the well-known Oxford System of Decimal Classification (ODC) for Forestry in the 1950s. The ODC marked the beginning of a new form of scientific cooperation within the IUFRO.

Although the IUFRO traditionally does not carry out its own research, collaboration through the IUFRO in experimental research between the member institutions has been rather common. Examples include studies of exotic tree species, international seed exchange, thinning and incremental felling trials, as well as testing of the technical properties of wood.

The interaction between forests and water was one of the early subjects in which the member institutes established extensive experiments to collect data within the IUFRO. The influence of forests on the water balance was perhaps the most important single topic of the Union before the First World War. Since then, the subject was on the programme of some IUFRO groups but a new impetus was given in 1997 when a special Task Force on forests and water was set up.

Another example of the cyclical variation in research preferences comes from forest genetics. Genetic research has been an important part of IUFRO cooperation since the very beginning of the organization, and international provenance trials have been one of the Union's trade marks in its history. Although genetics continued to be on IUFRO's programme, in recent years it has received more attention than for decades. In 1995 genetic research (together with physiology) was assigned its own Division and in 1997 a new Task Force was allotted for gene resources.

Until the 1960s forest was usually understood to be just a field stocked with trees. Gradually the idea gained ground that it is also a complex ecosystem which needs to be protected and maintained for future generations. This change, together with the problems caused by air pollution, has had a great impact on the IUFRO's work since the 1970s. Sustainability, an old forestry principle formulated almost 300 years ago, has become a commonly used slogan that now also in forest research means both economic and ecological and social sustainability.

2. IUFRO IN THE 21ST CENTURY

From its small beginnings in Central Europe, IUFRO has now grown into a major global organization with 15,000 cooperating scientists in 650 member institutions in 100 countries. The Union has gained a reputation for facilitating research cooperation, maintaining high scientific standards and promoting the exchange of information. These will remain a focus of the IUFRO also in the 21st century. But this is not enough: the Union must also develop and open up new avenues. In the following some ideas relating mainly to the structure of the Union are presented.

2.1 Discipline or Problem Oriented?

Since the establishment of the IUFRO most of the forestry problems have become more and more complex and can only be solved through interdisciplinary cooperation. The IUFRO's divisional structure is, however, based on disciplines. Collaboration between the disciplines through the IUFRO divisions has not been very intensive. Consequently, problem-oriented Task Forces have begun to play an increasingly important role in the 1990s.

In the period 1990 - 95 there was only one Task Force but now IUFRO has already six of them, and their number may increase. This raises a question about the relation between the Divisions and the Task Forces, and perhaps also about the existence of Divisions.

Although practical problems are studied and solved in projects in most research organizations, science itself often progresses on a disciplinary basis. A long-term science-oriented activity usually supports short-term interdisciplinary projects. Therefore, disciplinary-oriented and project-oriented researches are not exclusive but rather complementary. This means that the IUFRO will have room for both Divisions and an increasing number of Task Forces.

2.2 Union of Organizations or Individuals?

The IUFRO's basic funding comes from its member institutions. The present trend of the governments to reduce public expenditure has led to severe budget cuts in many of these organizations. Partly as a consequence of this development, an increasing number of the IUFRO's members are no longer able or willing to pay their membership fees. If this trend persists and new financial sources are not found, the IUFRO's economic base gradually begins to deteriorate.

Recently the IUFRO has cleaned its files and dropped tens of members who have not paid for years. In principle, it is fair to get rid of freeriders but, as for individuals, this has led in some cases to a difficult situation if active IUFRO scientists have worked in the former member organizations. Although the IUFRO has an Associate Member category for individuals, the fee is so high that most scientists, especially those coming from developing or transition countries, may find it economically impossible to pay the fee. Consequently, the number of Associate Members has remained very low.

The fact that the IUFRO is an organization of organizations does not guarantee that individual scientists working in the member institutions know about IUFRO or get information about it. For scientists working in non-member organizations, the information flow has been practically non-existent. The Internet has obviously improved the situation and will do so also in the future because it makes the distribution of IUFRO information easy and fast.

The time may be ripe to review IUFRO's membership. One possibility is to combine organizational and individual memberships so that the member organizations will identify and list those of their scientists who want to be involved in the IUFRO's work. These people then become individual members of the Union and their parent organizations pay the fees on their behalf. If a parent organization is not able or willing to be a member of the IUFRO, individual scientists may become individual members like the current Associate Members. The fee must, however, be much lower than today in order to encourage people to join in. On the other hand, it must be in a proper proportion to the organizational fee so that, when an organization has enough individual IUFRO members, it becomes more profitable if the organization joins and collects the fees from the scientists, or preferably pays them on their behalf.

2.3 The Role of IUFRO Congresses and Meetings

IUFRO convenes a World Congress every five years. These congresses have grown bigger and bigger. Practically every time the previous record is broken for the number of delegates, sessions and papers. The size of the World Congress begins to reach a limit which cannot be exceeded without changing its basic nature. This change is needed and has in fact commenced: the scientific programme of the next Congress in Kuala Lumpur, the first in the 21st century, will be different compared with its predecessors.

So far the IUFRO World Congresses have been relatively technical when evaluated on their average papers. In the future the Congress should concentrate more on general topics and above all, provide a forum for synthesis presentations.

After the Congress the whole world should know about the state of the knowledge in all areas of forest research. We have not yet fully utilized the enormous media potential of the IUFRO World Congresses.

The possibly diminishing role of the technical presentations in the World Congress does not mean that technical topics are less important than before. Instead of being presented in the Congress, technical papers should be more often delivered in smaller meetings of experts to specialized audiences. In fact, the number of small and medium-sized meetings between the World Congresses is increasing. Practically all Divisions and also some Task Forces have now their mid-term meetings. This is a trend that should go stronger in the future.

2.4 The IUFRO Network

Throughout its history the IUFRO has been a network of forest scientists. This network supports the exchange of scientific and technical information, and ideas among the researchers worldwide. In addition, by mobilizing its network, the IUFRO can provide the different international forums with information about the implications of scientific findings or about status of science on a particular issue.

By establishing in 1995 a special Task Force for the Internet resources IUFRO can guarantee that it develops its network with the most up-to-date knowledge and advanced technical tools. When the IUFRO Net is ready, all of IUFRO's almost 300 units form a huge information base. If in the future also individual scientists and their special expertise can be listed and included in the IUFRO Net, both the scientific community and the policy-makers and other user communities have at their disposal an instrument which is one of the most revolutionary single inventions in the history of the Union. This means that the network will be the IUFRO's asset also in the 21st century.

2.5 IUFRO as a Bridge Builder between Scientists and Policy Makers

The world is growing more and more complex. Consequently, research will gain increasing importance in the policy-making process. The demand for sound scientific information also for forest policy deliberations has grown significantly in recent years. The problem is that the communications and interactions between the research community and the users of the research results is not yet good enough.

In collaboration with FAO, CIFOR and some other international organizations, the IUFRO is developing mechanisms to improve the interface between researchers and policy makers. With its international network IUFRO can play a key role in this process.

REFERENCES

1892-1992. 100 Years of IUFRO, International Union of Forestry Research Organizations. 1992. 35p.

IUFRO - the Global Network for Forest Science Cooperation. International Union of Forestry Research Organizations. 1997. 16p.

Toward the New Stage of Environmental Forest Science Research in the 21st Century

IUFRO Status and Perspective

Heinrich SCHMUTZENHOFER

IUFRO Secretary
Forstliche Bundesversuchsanstalt
Seckendorff-Gudent-Weg 8, A-1131 Wien
Austria

ABSTRACT   IUFRO developed from a union of forest experiment organizations represented exclusively by their directors to a Union of Forestry Research Organizations involving individual scientists as networking members from diverse fields of research. Research conducted in IUFRO's member organizations was always geared to and dependent on the needs of the respective time, society and its problems. IUFRO follows a mission, vision and guiding principles and defined goals and objectives. A strategy was adopted to define the orientation needed for the future. Basic statistical figures of the Union are presented.
Keywords: Environment, forest science, research, IUFRO

1. IUFRO'S DEVELOPMENT

1.1 Foundation and historic development until 1920

International cooperation in forestry began in the 19th century after the great floods that haunted Central Europe in those times. The need for international forest protection laws was evident. Applied forest research was undertaken in Central Europe and rules for harmonization and standardization of research methods were sought. In addition, a forum was needed where researchers could discuss their plans, methods, results and instruments and where they could advise each other. In 1892, in Eberswalde, Germany, this situation led to the foundation of the International Union of Forest Experiment Organizations, the first name of what is today IUFRO, the International Union of Forestry Research Organizations.

In this first stage, the directors, and only the directors, of the Central European research, or at that time, experiment centres met periodically at congresses to harmonize their respective research programmes and to set standards, such as the Breast Height Diameter – BHD-1.3m for measurements or the "Guide for Thinning and Incremental Trials". It might be interesting to know that an experimental plot under these guidelines was established in Japan in 1913. The topics of main importance were international agreement on seed material, provenance trials, introduction of exotic tree species conversion of pure stands to mixed stands, increment and weather, technical properties of wood and forestry bibliographies. We see already that soon countries outside Central Europe joined this Union, among them the USA, Canada, Japan and Russia. Unfortunately, this development came to a halt with the First World War.

1.2 IUFRO's Development between 1929 and 1948

The next stage of the development of the Union was initiated in 1929 at the Congress in Stockholm where a permanent Secretariat with its seat in Stockholm was set up. Forest research institutions such as universities in Europe joined the Union, now called International Union of Forestry Research Organizations, IUFRO. The exclusive participation of directors ended, senior researchers started to be involved in the process and in defined projects. International interest grew and led research centres from some 40 countries throughout the world to join the Union.

In addition to the permanent Secretariat in Stockholm, a Permanent Working Committee and technical working parties were established. These bodies dealt with general forest issues, forest ecology, forest soil sciences and entomology. In 1932, at the Congress of Nancy, tropical and Mediterranean forest issues were included – Dehra Dun Station was entrusted with projects in this field -, other new topics were harvest operations, forest protection, reforestation, site ecology and, in 1936, forest management plannings. 1936, with the Congress in Hungary, was a decisive year for the Union: great progress was made there in the fields mentioned above on an international scale, e.g. agreement was achieved on three major subjects: tropical forestry, yield and increment studies and site descriptions for cool-temperate regions. Again it was a war, World War II, that seriously hampered IUFRO's activities for some time. Yet, international provenience trials were started in 1938/39 for Picea abies and Pinus sylvestris in Europe and the USA. In 1944, the famous Larix decidua, leptolepis and sibirica provenance trials were started, again in Europe and the USA.

1.3 IUFRO after 1948

After World War II, a new – and third – stage began: for the first time, individual researchers were given the opportunity to establish close personal relations with colleagues from other countries worldwide, for the benefit of national research programmes through the exchange of ideas, experiences and results in the entire range of forest and wood research. In the first decades, common interest and emphasis were directed mainly towards a uniform terminology (Ford-Robertson), bibliography (Oxford Decimal Classification) and methods for forestry research such as evaluation of damage by air pollution, torrent and erosion research, properties of forest products - wood technology. It could be said that this was the beginning of networking as we know it today.

In this stage, IUFRO was to be integrated into the newly established UN organization FAO, but finally remained independent. A contract was concluded between FAO and IUFRO on cooperation and a Secretariat of IUFRO was established at FAO Headquarters in Rome and remained there until 1956. Membership was growing continually and in particular so in Asia and the Austral-Pacific region and later also in Latin America and Africa.

In the period from 1956 to 1971, there was no official permanent Secretariat. In 1973, finally the headquarters of the Union were established in Vienna at the Federal Forest Research Centre at Schoenbrunn.

In the initial phase, the IUFRO working language was German. French was dominant to a certain extent in the second phase and English replaced German and French in the late 1930s. English continued to be the general working language and in 1990, Spanish became the fourth official IUFRO language.

1.4 IUFRO's Structure in the 1960s and Development after 1971

During the 1960, IUFRO was structured in five Research Units which were subdivided into so-called Sections:

General: Bibliography and Terminology, History of Forestry
General Forest Influences: Forest Influences and Watershed Management
Forest Production, Establishment and Maintenance of Forests: Research on Site Factors, Study on Forest Plans, Means for Amelioration of Forest Production (Silviculture), Forest Protection, Study of Growth and Yield and of Forest Management
Forest Economics – Operational Efficiency: Forest Economics, Operational Efficiency
Forest Products: Forest Products

At the 1971 Congress at Gainesville the foundations for the present IUFRO Structure were laid. First, the whole range of research was dealt with in six technical Divisions: D1 - Forest Environment and Silviculture; D2 - Forest Plans and Forest Protection; D3 - Forest Operations and Techniques; D4 - Inventory, Growth, Yield, Quantitative and Management Sciences; D5 - Forest Products; D6 - Social, Information and Policy Sciences. These Divisions were divided into Project and Subject Groups as well as Working Parties that were established as needed.

In 1983, the Special Programme for Developing Countries (SPDC) was founded as a result of the 1981 Kyoto Congress Resolution.

In 1995, the present Structure was adopted: the former Divisions 1 and 2 were split into four Divisions: Division 1 – Silviculture; Division 2 – Physiology and Genetics; Divisions 3-6 remained unchanged; Division 7 – Forest Health; Division 8 – Forest Environment. Subject and Project Groups were combined in Research Groups; the Working Parties remained. Since the mid-1980s, Task Forces according to topical and political requirements have been established.

Also in 1995, IUFRO started its terminology project SylvaVoc.

1.5 Main Research Topics in Defined Periods

1.5.1 Reconstruction after WWII

During WWII, forests in Europe had suffered considerably both because of direct war damage and because of the absence of active forest management. Immediately after the war, climatic stress and draughts caused mass outbreaks of forest pests. The main aim of research and forestry was therefore the establishment of modern pest management methods (introduction of organic pesticides such DDT), national forest inventories, harvesting operations and the establishment of adequate infrastructure.

1.5.2 Period of Unlimited Growth

During the 1950s, the general industrial development and its principles had a direct impact on forestry and forest research. Increased productivity and production through the use of agrochemicals, fertilizers and pesticides were the main objectives at the time. These were to be achieved through pest control and fertilization by means of aerial application, large-scale clear cuts and the development of a fully mechanized harvesting technology. Large-scale reforestation and afforestation programmes for soils not needed anymore for agricultural crops were launched. The development of herbicides and studies into their side effects and that of other agrochemicals were becoming more and more important.

At the same time, forest laws that had been passed in the 19th century, had to be adapted to the new situation. It was necessary to define pest management conditions as well as the prerequisites for road building and modern logging methods. New fields of research arose, among them ergonomics, molecular genetics and evaluation of forest damage by air pollutants. Provenience trials, plant breeding activities, growth and yield research and silviculture developments were started again.

1.5.3 Consideration of the Environment

Rachel Carson's book "The Silent Spring" brought about a change in thinking. Unlimited growth started to be seen in a negative light, while an increased consideration of the environment was demanded. Forestry research had to reconsider its aspirations and its direction. Energy crisis and plantation forestry as well as deforestation in the Tropics became main keywords. IUFRO, too, took account of this new development in its 1986 Ljubljana Congress Resolution and, based on research results published by one of its Working Parties, defined SO_2 and HF concentration limits which were later used in national legislation in Europe to protect the environment and to fight forest decline. This clearly shows that IUFRO became a serious source of advice, consultancy and information for governments and society as a whole. Subject-related research lost some of its importance and dominance, while project-oriented research increased.

2. IUFRO AT PRESENT

2.1 Mission, Vision and Guiding Principles

IUFRO is a non-profit, non-governmental international network of forest scientists. Its objectives are to promote international cooperation in forestry research and related sciences. IUFRO's activities are organized primarily through its 270 specialized Units in 8 technical Divisions.

IUFRO aspires to bring together scientific knowledge about all aspects of trees and forests through the cooperative efforts of its worldwide member research organizations and scientists. Through this means it seeks to promote the sustainable use of forest ecosystems to provide multiple benefits for local people and for society as a whole.

IUFRO's statutes require that it remains independent and non-political which in practice means it must be non-discriminatory. All actions of the Union must conform with these guiding principles.

2.2 Goals and Objectives

During the last years IUFRO has developed a strategic plan to cope better with today's requirements. IUFRO's goals and objectives are permanently under review.

Objectives:
- Enhance cooperation between forestry research organizations and between individual scientists;
- Promote the dissemination and application of research results and the standardization of research terminology and techniques;
- Address issues of regional or global significance which require inter-agency or inter-disciplinary action;
- Publicize the outcome of the above activities and make awards for outstanding work which contributes to the advancement for forest science;
- Assist developing or disadvantaged countries to strengthen their research knowledge and capability.

Goals:
- Expand the coverage of cooperative activities;
- Increase the number of research institutes;
- Improve communications between scientists;
- Strengthen contacts with other international organizations;
- Increase the number of IUFRO conferences and seminars of all types;
- Extend the coverage of IUFRO publications on research methods, techniques and terminology;
- Expand the number of regional workshops, seminars and other activities;
- Set up inter-disciplinary project groups or task forces to deal with global issues;
- Increase the number of publications on inter-disciplinary, regional or global issues;
- Make appropriate awards to recognize scientific excellence;
- Improve public activities;
- Expand programme support to developing or disadvantaged countries;
- Seek to ensure adequate funding for above.

2.3 IUFRO's Statistics

IUFRO's importance and scope can be illustrated by some key figures:
- Membership: 657 Member Organizations in 100 countries in 9 geographic regions worldwide; the Member Organizations are mainly forest research institutes, universities, faculties of forestry and forest schools. A small number are private enterprises conducting forest research, NGOs and inter-

governmental organizations. Membership is still concentrated in Europe, Northern America and parts of Asia.

- 270 IUFRO Units: 8 Divisions, 64 Research Groups, 220 Working Parties.
- 716 officers (both Coordinators and Deputy Coordinators of IUFRO Units): these come from 69 countries. Unfortunately, some of the IUFRO Regions are still underrepresented, these are the Mediterranean (38 officers), Latin America (38 officers), Africa south of Sahara (25 officers – only 3 of those are Coordinators).
- Meetings: 1996 – 62, 1997 – 64, 1998 – 66. From 1991 to 1995, the average number of meetings held was 57 per year. This shows an increased interest and need for meetings, conferences and workshops.
- Publications: more than 50 proceedings are published annually and presently, more than 1,100 publications are available in IUFRO's Reference Library at the Secretariat. In addition, there is the IUFRO World Series (started in 1990; 8 volumes) and the IUFRO Occasional Paper Series (started in 1995; 11 volumes). Other publications: quarterly IUFRO News, Annual Report and brochures.

2.4 IUFRO on the Internet

In 1994 IUFRO launched its own website at http://iufro.boku.ac.at/. At present, this website is mirrored daily to servers at IUFRO Member Organizations in Finland, USA, Chile, Costa Rica, Australia, South Africa and Japan. One part of the webpages contain the organizational information on the Union, including LIBERO, a database for IUFRO's reference library; the second part contains the IUFRO Network which is a forum where IUFRO Units share information about their own activities and about other issues of interest to member scientists and people associated with forest research. IUFRO Network involves hundreds of persons in creation of the network and the participation is steadily increasing. More than 50% of all IUFRO Units maintain their own homepage and are actively involved in networking. Support is given by IUFRO's Task Force on Internet Resources. There is a marked trend towards the increased use of Internet which is illustrated by the number of hits at the server in Vienna: daily maxima have already passed 3,200 hits. A remarkable part of activity is dedicated to the preparation of meetings, conferences and congresses.

3. IUFRO'S PERSPECTIVE

The basis for the perspective of an organization such as IUFRO is naturally the ad hoc status and the adopted strategy. The strategy has to be oriented towards medium- and long-term requirements deriving from the development of our world and the arising challenges. Forestry is now considered as a part of the landscape and does not stand alone as a science anymore. Networking among scientists in **different** fields, such as ecology, biology, zoology, agriculture, social and economic sciences, is imperative. A decentralization process, i.e. networking in regions, has recently been started. This trend is furthered by the use of modern communication techniques such as the Internet.

3.1 IUFRO's Strategy Background

The condition of the world's forests and science to support sustainable forest resources has changed dramatically since the establishment of the International Union of Forestry Research Organizations (IUFRO) more than 100 years ago. IUFRO, an organization envisioned as the world network for forestry research and often presented as the conscience of forestry, has changed as well.

Today, environmental stresses and increased human demands on natural resources require knowledge and decisions that address social, economic and environmental needs. The scope of these needs requires more inter-disciplinary science and stronger linkages among all stakeholders.

As demands on science and technology have increased, so too have the expectations and opportunities for IUFRO. IUFRO was established primarily to "...promote international cooperation in scientific studies embracing the entire field of research related to forests." In the past, this mission was redeemed through activities to standardize methods and terminology within the field of forestry, and to enhance information exchange among forest scientists.

Although the Union affirms the value of its traditional goals, it is prudent to evaluate new opportunities and challenges. This strategy reflects the realization that Union membership has changed, along with the changing nature of forestry science and multiple-use management.

Today, IUFRO activities are addressing a wider array of research questions, many of which are linked to key policy issues. Similarly, membership has broadened dramatically since the beginning of the Union to include many scientific disciplines beyond those of traditional forestry and forest products. This increase in diversity is likely to continue.

The mode of organizational operations is also changing as members employ the Internet and other technological advances for internal and external communications. IUFRO has initiated Task Forces to accelerate the integration of multi-disciplinary information, and IUFRO seeks to contribute to intergovernmental policy dialogues related to sustainable resource management, forestry and forest science.

3.2 IUFRO's Strategic Goals for the New Millennium

The Union's vision of success for the 21st century is the realization of sustainable forestry worldwide, through the scientific and educational contributions of its members. To achieve this over-arching goal, IUFRO will address several objectives:

- IUFRO will continue to be a Union of organizations while seeking new ways to engage its individual members - especially those in non-forestry disciplines. This objective reflects the reality that IUFRO is no longer primarily an organization of forestry research directors; rather, it is an organization of individuals working on issues in, or associated with, forests.
- IUFRO will be an advocate for forest science, helping to assess needs for forest science infrastructure and state-of-knowledge reports for significant global issues. Moreover, IUFRO Divisions, Research Groups and Working Parties are

encouraged to utilize multi-institutional efforts for networked research projects. This approach draws upon IUFRO's role as a facilitator and convenor.

- IUFRO will use Task Forces (Forests in Sustainable Mountain Development, Sustainable Forestry, Environmental Change, Water and Forests, Management and Conservation of Forest Gene Resources, Internet Resources) to accelerate the synthesis of scientific and technical information, for the benefit of member organizations and policy makers. These reports will draw upon and integrate information from IUFRO's Divisions.

- IUFRO will monitor international policy fora and identify issues for which, in partnership with other organizations, the Union can make scientific contributions. It will be the goal of the Union to provide scientific information for consideration by policy makers, but not as a statement of a Union "position". Rather, IUFRO will seek to present accurately the range of scientific thought on key global forestry issues.

- IUFRO affirms the importance of the following resolutions accepted in its 1995 World Congress: a) to maintain and enhance well targeted forestry and forest products research; b) to expand research capacity, especially in developing countries; c) to enhance partnerships that strengthen research and improve communications between scientific and general communities; and, d) to increase policy and problem-oriented research in economic and social sciences.

REFERENCES

1892-1992. 100 Years of IUFRO, International Union of Forestry Research Organizations. 1992. 35p.

IUFRO – The Global Network of Forest Science Cooperation. Union of Forestry Research Organizations. 1997. 16p.

Forest Research for the New Millenium – IUFRO's Strategy for Entering the 21st Century. 1997. 10p.

Environment and Production –
Two Sides of the Same Coin or Incompatible Goals ?

Lisa SENNERBY- FORSSE

The Swedish Forestry Research Institute, SkogForsk
The Uppsala Science Center
S-751 83 Uppsala
Sweden

1. INTRODUCTION

Today there is common agreement that forests are natural resources of vital importance and that they have to be managed, restored and created with great care in order to sustain healthy and productive ecosystems for future generations. Important global agreements, such as the United Nations Conference on Environment and Development held in Rio de Janeiro in June 1992, the Montreal Process 1993 and the Helsinki Process 1993 (FAO 1997) have been ratified by many countries, however we have a long way to go before words are put into action. Still, forest operations around the world are too often examples of short-sighted, profit-oriented business. Also, the poverty situation in many countries is such that the struggle for survival overwhelms considerations for environment and future generations. We therefore need to develop methods and techniques for silviculture as well as tools for communication that will allow economical forest management at the same time as high biodiversity and good health conditions in our forests can be maintained.

Most of the world's forests have been affected by man for thousands of years. Unmanaged, old forests are therefore very rare. According to a rough estimation, about 35% of the old forests in the world are believed to remain in a close-to-original state. These old forests are mainly found in Canada, the Russian Federation and in some tropical rainforest areas. In Europe, the few remaining areas of old forest are found in northern Scandinavia. Less than 6 million hectares, or 5% of the forest land of Europe are considered to be relatively undisturbed (i.e. retaining much of the natural character) (Anon. 1995). A rapid decrease of forested areas is in progress, which is especially obvious in tropical regions. FAO (1997) reported that forest areas in developing countries are disappearing at a rate of approximately 13 million hectares per year (0.7%). A contradictory trend, although much smaller, is seen in industrialised countries, where ongoing re- and afforestation programmes are leading to a slight increase, approximately about 2.5% per year.

Forests differ in many respects in different parts of the world. However, the conflict between environmental consideration and production is similar in most forest administrations. In Sweden, forest policy has focussed on preserving biodiversity in managed forests rather than setting aside large areas as nature reserves. The prerequisites for this policy and the present forest management systems will be discussed in this paper.

2. Swedish Forests

The forests in Sweden as in all of Scandinavia largely belong to the taiga area, i.e. the northern soft-wood belt covering large parts of North America, the Russian Federation and northern Europe. The soils vary from fine textured sedimentary soils, mostly used for agriculture, to rocky, morainic till soils. In Sweden the forests contain large areas of wetlands, which are rich biotopes for flora and fauna but often regarded as trouble spots for silviculture. There are several vegetation and forest types in Sweden, with the dominating tree species being *Picea abies* (45%), *Pinus silvestris* (39%) and *Betula alba* + other deciduous trees (16%) (Fig. 1). In the southernmost part of the country, the nemoral forests with *Fagus silvatica* and *Quercus robur* as dominating species, make up a very special biotope with a high biodiversity level. Between the latitudes 56° and 58° mixed forest with deciduous trees and spruce dominates the forests while north of 58° latitude the real boreal forests with pine and spruce and some birch, alder and aspen take over. Towards the northernmost part of the country, there are the mountaineous birch forests which gradually becomes tree-less land (Anon. 1997a).

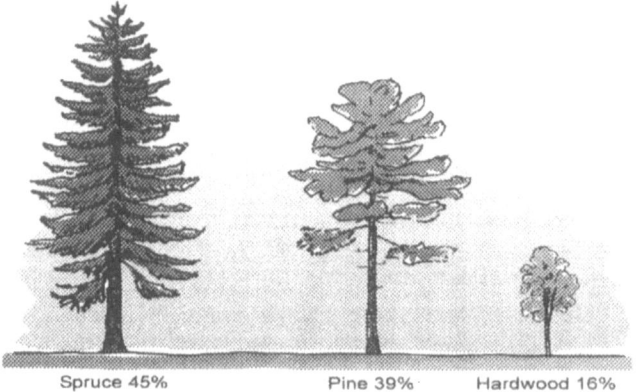

Spruce 45% Pine 39% Hardwood 16%

Figure 1. The most common forest tree species are spruce, pine and birch (Anon. 1997a).

Sweden has a land area of 41 million ha, stretching from latitude 55° to 69° N (Anon. 1997b). (Fig. 2). Forests cover more than half of the land area and the population is at present almost 9 million inhabitants. Forests have always played a major role in the life of most Swedes, not only as a part of the natural landscape but also as a place for recreation, hunting, berry picking, skiing etc. An ancient right entitles Swedes to enter freely and enjoy any area of forest land, regardless of ownership.

Figure 2. Sweden has a land area of 41 million ha, stretching from latitude 55° to latitude 69° (Anon. 1997a).

However, historically forest were taken for granted and during the 19th century the Swedish forests were being over-exploited and no forest regeneration measures were taken. In the southern part of the country forests were transformed into grazing land. At the same time forests were cut down all over the country for various industrial purposes and exported to Europe where there was a great demand for forest products from Sweden. The population pressure at this time caused all possible land to be put under plough and during 50 years (1850–1900) more than 1 million people emigrated from Sweden, mainly to North America. During this time a fast growing and successful expansion of the forestry industry started in the northern part of the country and the large rivers were used for the transport of timber from the interior. Sawmills and pulp industries grew up along the coast at a rapid rate. Towards the end of the 19th century there was a broad opinion in favour of the need for political measures to control forestry and in 1903 Sweden got her first Forestry Act. Since then it has been several Acts, and the one now valid was implemented in 1994 (Ekelund and Dahlin 1997).

2.1 Production

The National Forest Survey is responsible for regularly monitoring the standing stock, increment and cut. The annual increment varies markedly from north to south. The national average is 4.3 m³/ha which is equivalent to a total national increment of about 100 million m³ per year. The annual cut, which is about 70 million m³ has been lower than the annual increment for many years. As a consequence, the volume of the standing stock is steadily increasing and now averages 120 m³/ha although in mature stands the volume can be more than double this figure (Anon. 1997a). In connection with felling, most of the material is removed in the form of sawlogs and pulpwood. The removal of fuelwood amounted to 16% of the total volume during the first halt of the 1990s and is expected to increase (Fig. 3).

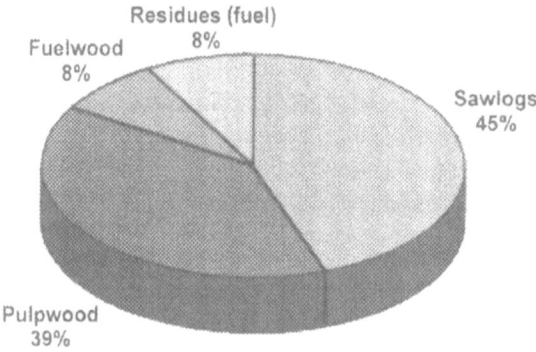

Figure 3. Distribution of assortments removed at felling 1991–1995 (Anon.1997a).

The length of a typical stand rotation generally varies between 70–120 years. After establishment, the stand is cleaned, thinned one to three times, and sometimes fertilised (Fig. 4). In contrast to natural forests, stands under long-term management contain large-diameter, even-aged trees, relatively little understorey and few dead trees, standing or toppled.

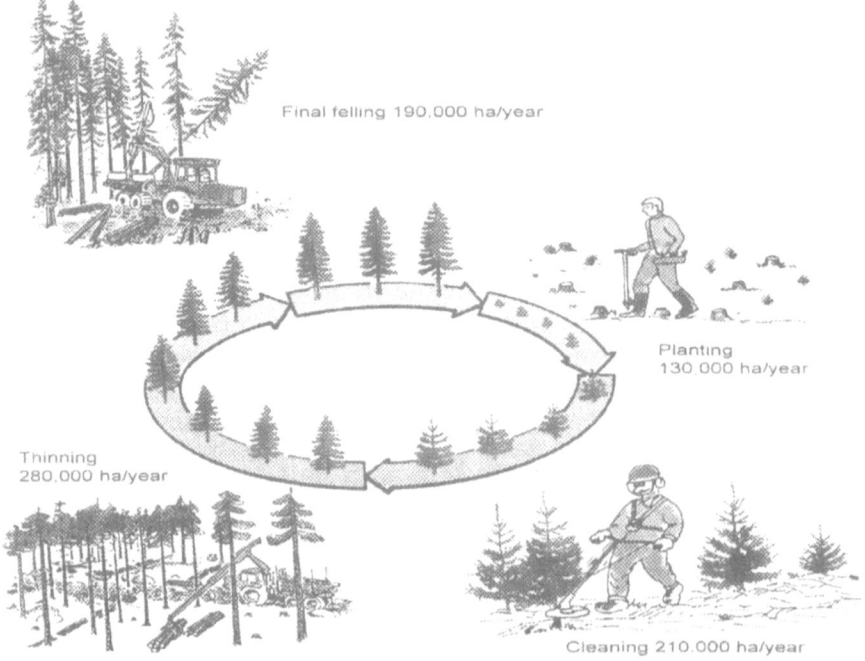

Figure 4. Operations during one rotation in the typical Swedish forest stand (Anon. 1997a).

About 13% of the forest land is owned by the public sector, and another 37% belongs to forest companies. The remaining 50% is distributed among some 250 000 private woodlot holdings (Fig. 5). These holdings have an average area of 46 ha and are often spread over a number of individual woodlots. Individual stands, the treatment units, are frequently no larger than a hectare. Obviously, the methods used in these stands often have to differ from those used in large-scale forestry, in which treatment areas smaller than 10 ha are avoided whenever possible.

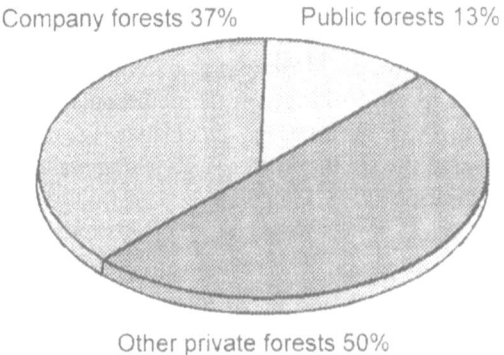

Figure 5. Half of the forest area is privately owned (Anon. 1997a).

2.2 Forest Health

The present forests are facing several environmental problems created by modern industrial society, including acidification, eutrophication and, possibly, greenhouse warming. Swedish forestry is affected by these environmental changes but can also contribute to increasing or decreasing the strength of the changes.

2.3 Acidification

The forest soils have undergone rapid acidification during the last century. Soil acidification results in a lowering of pH, an enrichment of free aluminium and a depletion of plant-available base cations, i.e. calcium, potassium and magnesium. These changes may threaten forest nutrition and health.

Besides air pollution from industry, traffic and husbandry, forest harvesting can contribute to acidification as a result of the export of base cations from the site. This contribution may be especially significant in cases where whole-tree harvesting is applied. This may happen since interest in using forest residues as a renewable energy source is rapidly increasing in Sweden. The acidifying effect of removing tips, branches and needles from the felling sites may be counteracted by recycling the wood ash remaining after combustion (Olsson et al., 1993; Ohno 1992). Nitrogen fertilisers may be needed on poorer sites to prevent losses in fiber production remaining from whole-tree harvesting. Ash recycling has yet to be implemented in practical forestry and studies are ongoing concerning the quality, environmental risks and economy of this practice.

Surface waters and coastal seas have been eutrophied by a large influx of nutrients from society. The main sources of emissions on national levels are agriculture, municipal waters and direct depositions. Most of the leaching from the forest consists of contributions of smaller size, but forest operations such as clear-felling and fertilisation may cause some elevation (Nohrstedt et al., 1994). Simultaneously, forest ecosystems act as filters, removing most of the atmospherically deposited nitrogen. The deposition of nitrogeneous pollutants has caused vegetation changes (Kellner and Redbo-Torstensson 1995) and is probably enhancing tree growth in Swedish forests, as also in large parts of Europe (Canell 1995). The low current level of nitrogen leaching associated with forestry could probably be reduced even further by increasing the use of shelterwood management systems and by leaving buffer strips along waterways in connection with fellings and fertilisation.

2.4 Greenhouse Gases

The concentration of greenhouse gases is increasing in the atmosphere. The effect of this increase is believed to cause climatic changes, which in the most generally accepted hypothesis will cause a warmer and wetter climate in the Nordic countries. The most important greenhouse gases are carbon dioxide, methane and nitrous oxide. The main reason behind the net increase is the burning of fossil fuels and conversion of rainforest into farmland. Growing forests are generally sinks for CO_2 and the carbon accumulates in both trees and soil over time (Linder and Murrey 1997). At present, we do not know how fast carbon is accumulating in the soil. However, the retention of carbon in the biomass will not solve the problem created by global warming. A possible countermeasure to this development is to replenish fossil fuels with biomass fuels and other renewable energy sources (Börjesson et al., 1997). Forest operations that may increase the flux of carbon dioxide from soil to the atmosphere are ditching of peatland, liming and ash recycling. The flux from the soil has to be weighed against any possible increased fixation by trees. Forest ecosystems also affect, and are affected by, methane and nitrous oxide net fluxes to the atmosphere. The ongoing eutrophication induced by nitrogen deposition may increase this net flux.

3. BIODIVERSITY

Forests are species-rich ecosystems containing a wide variety of taxa from both the plant and the animal kingdom. Furthermore, biodiversity has many different aspects. Most of the knowledge we have today is on the species level. Other aspects, such as genetical biodiversity, i.e. the gene pools of different species, are largely hidden. Also Franklin (1989) refer to structural and process diversity as important components of biodiversity. However, most commonly, when referring to biodiversity it is the species richness that is addressed. Among the factors influencing species richness are the need for specific home-range areas for different species, population dynamics over time and space, interactions between different species, the sequences of disturbance, etc.

Sweden has about 19 500 – 24 000 known species that depend on forest biotopes and of these 1 500 are on the red-list of endangered species (Berg et al., 1994). The invertebrates, fungi, lichens and mosses are taxa where the largest number of threatened species can be found in Sweden. Thirty-three different species of birds are red-listed or have even disappeared during the last 20 years (Fig. 6). The strategies to secure forest biodiversity in Sweden can be separated into two major ways. One is through legal protection of forests regulated mainly by the National Forest Act. The other way is voluntary, i.e. land-owners decide to set aside productive forest land from production in order to save key biotopes and areas of importance for fauna and flora.

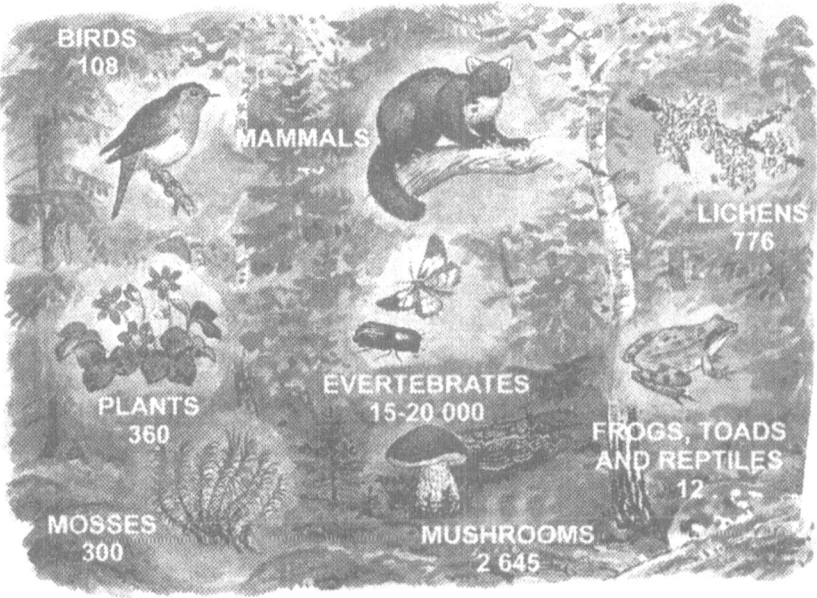

Figure 6. Redlisted species in Swedish forest.

3.1 Legal Protection

There are several categories of protected forest land in Sweden. One is when large-scale forestry operations, such as thinning, final felling, drainage and fertilisation are prohibited in non-productive forests, i.e. stands that produce less than 1 m^3 of wood per ha and year. However, single trees may be felled in such stands as long as the felling does not substantially alter the character of the stand. Non-productive forests account for about 1.7 million ha or 7 % of the forest land in Sweden. Another 0.8 million ha, or 3%, of the country's productive forest is set aside in national parks or nature reserves. Most reserves are situated in mountainous forest areas in the north. There are also areas that receive limited protection. These so-called nature management areas, wildlife sanctuaries and semi-protected nature reserves constitute about 1% of the forest land in total (Anon. 1997b).

3.2 Voluntary Protection

Aspects of protection are routinely considered in forestry planning and practical forestry operations. An example of this is the recommendations in the Forestry Act which have led to ca 5% of the stand area being left for conservation purposes after final felling operations. Forest stands expected to contain many red-listed species, the so-called "woodland key-habitats" are being inventoried by the National Board of Forestry on all privately owned forest land. Although certain key habitats can be protected by the Nature Conservancy Act, most of these valuable areas will have to be saved or restored on a voluntary basis. Most forest companies are currently conducting surveys in search of key habitats and aim to protect these habitats in their ecological landscape plans.

4. SILVICULTURE AND ENVIRONMENT

4.1 The Swedish Forestry Act

In the Forestry Act of 1994, production and the environment are given equal priority (Fig. 7). This means that the need for nature conservation and the impacts of various forestry practises on the environment must be considered on a daily basis in practical forestry. Great effort is required to achieve the preservation goals formulated for a large number of Swedish plant and animal species associated with natural forests which should enable them to survive in managed systems. Naturally, in this case, nature reserves play an important role, but we must also introduce some of the characteristics of natural forest into our managed forest stands. This requires that nature conservation be routinely considered by foresters.

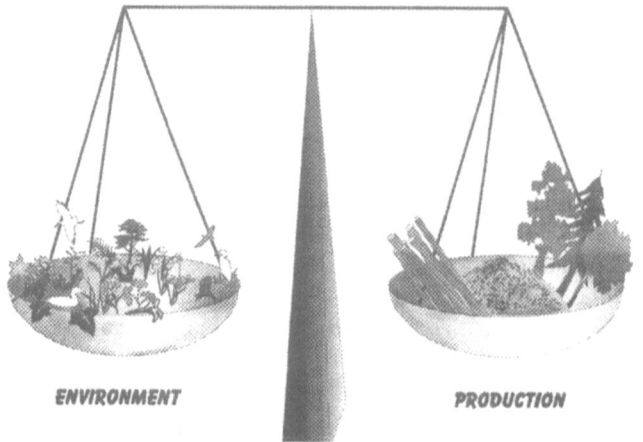

ENVIRONMENT   **PRODUCTION**

Figure 7. The balance between Environment and Production

4.2 Silvicultural Methods

The concept of "Sustainable forestry" is widely used and sometimes the definitions of the expression are somewhat diffuse, to say the least. The following criteria have been proposed to qualify for a sustainable forest management:

- All species should be able to persist within a landscape over a period.

- Critical habitats across the full range of spatial scales from individual structures to large areas of natural forest (reserves) exist within the landscape.

- Retention–harvest systems are used, retaining individual structures in the new stand (old, decadent trees, snags, fallen logs).

- There is significant difference in the degree of resilience and resistance shown by populations and communities when exposed to natural disturbance or human-caused disturbance (Anon. 1997c)

In this definition the focus is on biodiversity, however there are other aspects on forest ecosystems that have to be considered in order to achieve sustainability. Little is known about effects of air pollution on forests, or the biodiversity and biochemical balances in forest soils and water systems. All these factors being important for forest health, nutrient balance and water relations, must be included in ecological studies in order to increase the present knowledge and understanding of forest ecosystems.

4.3 Forest Practises

Forest management has certain implications for stand structure. For example, man-made forests create younger, even-aged and homogeneous stands with more roads and fewer fires compared with natural forests. In the northern hemisphere, in the pine and spruce dominated forests, the final felling was commonly done as a clear-cut over a larger area without leaving sufficient old trees, logs, residues, etc. on the site. The forests we have today are the result of forest practises focussed on lucrative short-term wood production. R&D involved in those days was characterised largely by forest engineering aspects on production rather than on forest biology and species occurrence.

4.4 Ecosystem Management

The new attitudes towards forests and forestry has led to many different models for sustainable management. One is the concept of ecosystem management strategy which tries to mimic the structures and dynamics of natural forests although keeping the production level and the efficiency level of forest operations high and economical. The ecosystem approach can be used to create a situation closer to natural conditions. However, various scientists in this area have raised objections and warnings about the uncritical use of this term (Simberloff 1997), especially since there is no consensus definition of the term. In ecosystem management the focus is on ecological processes rather than individual species (Meffe and Carrol 1994). This has alarmed some conservation biologists, arguing that many ecosystem processes can be maintained even if component species are lost (Tracy and Brussard 1994). Usually the definition of ecosystem management includes the consideration of biodiversity in management regimes, i.e. to protect and keep all the species needed to make the ecosystem function. However, not all species are needed all the time at every place in order to get a functional ecosystem and thus we must ensure that species are maintained even though their importance might not be of the highest dignity according to the knowledge available.

In addition to new methods and systems developed to sustain a productive and healthy forest, another important condition for conserving biodiversity is the variation in management practices. This we know from regions and areas where the ownership structure consists of many small forest owners. With few big owners, all having a similar way of managing the forests, the risk of biodiversity decline increases.

4.5 Landscape Planning

The overall strategy in the planning of forest management is to adjust forest operations to natural conditions, landscape properties and local socio-economic aspects in combination. Different scales, time aspects, levels such as landscape, stand and individual trees or structures must be considered when deciding upon which regimes to use. Forest planning at the landscape level and the inclusion of ecology-related information in annual reports are now part of the standard routine used at larger forest companies. Also the land owners with smaller properties are looking for ways of cooperation in order to be able to adapt a landscape perspective for their forest management planning.

In accordance with current laws stipulating that the environmental impacts of forestry practises be reduced, which may lead to decreases in yield or profitability, it has been necessary to apply a wider variety of management types. The use of silvicultural methods tailored to a given site increase the possibility to achieve environmentally and economically sound sustainable production (Berg et al., 1995). This view provides the basis for FSC certification, which is carried out by the forest companies. Stand management practises, such as cleaning and thinning, must be carried out in ways that meet the demands placed on both production and nature conservation. Here, as in final felling, aspects of biological diversity need to be considered. For example, nowadays clear-cut size is varied, wetlands are not ditched any longer, and isolated living trees as well as some snags and logs remain after the machines have left the site. On suitable soils, the use of shelterwoods, which enhance the survival of natural regeneration, has increased. Almost all forest land in Sweden has been affected by wildfires and controlled burning is now increasingly used as a tool to promote forest regeneration and species dependent on fire. The point of view is that in such a landscape it is possible to preserve the biological diversity while producing trees that can be used to manufacture paper and wood products.

4.6 Tools for improved forest practises

The goal for the forest industry in Sweden is to combine environment and production. Provided that the environmental aspects, including biodiversity, are fully implemented from the initial steps of planning to the final cutting and regeneration of forest stands, we might have a chance to realign today's common practises towards a more sustainable form of management. The practical implementation of new knowledge is a major obstacle to achieve the goal. The tools that are available today, i.e. different certification schemes, are still in their initial stages and need to be adjusted and improved as our knowledge increases. However, as a starting point for restructuring of forest practises and operations they are useful. In Sweden, the Private Farmers Federations are working towards implementing the ISO 14001 (International Standardisation Organisation) system into their forest operations. The big forest owners, i.e. the forest companies, on the other hand, have decided to certify their forest management according to the FSC (Forest Stewardship Council). Certification is being promoted with the two objectives to improve the management of forests in order to achieve sustainability and to provide market access for the products from these forests. The two main components to certification is certifying the standard of forest management and certifying the products that are made from wood from these forests. At present, most of the focus is on determining whether products are derived from sustainably managed forests.

In addition to objective evaluation of forest owners and organisations in order to ensure that environmental concern is a real part of forest management, we also need to improve ways of calculating the economic outcome of these approaches. The valuation of forests needs better, objective methods, that enable values to be quantified and compared. This is especially true for the values and products that are non-marketed and less tangible (Kengen 1996). If the market demands are such that consumers ask for certified wood and wood products, the system run on its own. However, certification and similar systems cannot be relied upon as the only means of achieving the goal of sustainable forestry. A change of attitudes is necessary and may be reached with the help of improved methods for valuation, which would improve the estimates of forestry in national accounts.

5. CONCLUSIONS

To accomplish the goal of preserving biodiversity a wealth of scientific research is needed in order to gain a better understanding of the forest ecosystems. We do not know whether it is possible to maintain species richness in the forest and at the same time conduct economic wood production. Therefore, virgin and unmanaged forest reserves are necessary complements to different nature-mimicking approaches in forestry management systems. The question whether environment and production are possible to combine can not yet be answered, but by working towards greater acceptance of the view of the forest as a community of species, rather than a wood factory, the chances should increase.

In the efforts to develop sustainable management methods, the role of science should be to:

1. monitor extensively and intensively, in natural and managed forests, trends and dynamics of species richness and populations, biological processes and structural changes;

2. carry out well-designed experiments on key problems;

3. provide information and education to foresters, and

4. develop tools needed for implementation of results into practical forest management.

REFERENCES

Anon, (1995): The Swedish National Encyclopedia, Stockholm, Sweden.

Anon, (1997a): Forestry in Sweden, SkogForsk, (The Forestry Institute of Sweden), Uppsala, Sweden, ISBN 91-7614-088-1

Anon, (1997b): Statistical Yearbook of Forestry 1997, National Board of Forestry, Jönköping, Sweden, ISBN 91-88462-33-1.

Anon, (1997c): The Swedish Biodiversity Group, Concepts and Solutions, Presented at the Third Meeting of the Subsidiary Body on Scientific, UNEP/CBD/SBSTTA/3/Inf 29, Discussion paper.

Berg, Å., B. Ehnström, L. Gustafsson, T. Hallingbäck, M. Jonsell, and J. Weslien (1994): Threatened forest plants, animals and fungus species in Swedish forests distribution and habitat associations, Conservation Biology, Vol 8, pp 718-731.

Berg, Å., B. Ehnström, L. Gustafsson, T. Hallingbäck, M. Jonsell, and J. Weslien, (1995): Threat levels and threats to red-listed species in Swedish forests, Conservation Biology, Vol 9, pp 1629–1633.

Börjesson, P., L. Gustavsson, L. Christersson and S. Linder, (1997): Future production and utilization of biomass in Sweden; Potentials and CO2 mitigation, Biomass and Bioenergy. (In press).

Cannell, M.G.R. (1995): Forests and the global carbon cycle in the past, present and future, European Forest Institute, Report No 2, 66 p.

Ekelund, H. and C-G. Dahlin (1997): Development of the Swedish Forests and Forest Policy during the last 100 years, National Board of Forestry, No 0518, Jönköping, Sweden.

FAO (1997): State of the Worlds Forests, Words and Publications, Oxford, ISBN 92-5-103977-1.

Franklin, J.F. (1989): Toward a new forestry, American Forests, Nov/Dec. 1989, pp 1-8.

Kellner, O. and P. Redbo-Torstensson (1995): Effects of elevated nitrogen deposition on the field-layer vegetation in coniferous forests, Ecological Bulletins Vol 44, pp 227-237.

Kengen, S. (1996): Forestry valuation – purpose, context and process, André Mayer Research Fellowship Draft Report, FAO, Rome.

Linder, S. and M. Murrey (1997): Do elevated CO2 concentrations and nutrients interact? In: P.G., Jarvis (Ed.), The Likely Impact of Rising CO2 and Temperature on European Forests, Cambridge University Press, (In press).

Meffe, G.K. and C.R. Carrol (1994): Principles of Conservation Biology. Sinauer, Sunderland, Massachusetts, 600 p.

Nohrstedt, H-Ö., E. Ring, L. Klemendtsson and Å. Nilsson (1994): Nitrogen losses and soil water acidity after clear-felling of fertilized experimental plots in a *Pinus silvestris* stand, Forest Ecology and Management Vol 66, pp 69-86.

Ohno, T. (1992): Neutralization of soil acidity and release of phosphorous and potassium by wood ash, Journal of Environmental Quality Vol 21, pp 433-438.

Olsson, M., K. Rosén and P-A. Melkerud (1993): Regional modelling of base cation losses from Swedish forest soils due to whole-tree harvesting, Applied Geochemistry Supplement, No. 2, pp 189-194.

Simberloff, D. (1997): Flagships, umbrellas, and keystones: Is single-species management passe in the era of ecosystem management and landscapes? Biological Conservation, (In press).

Tracy, C.R. and P.F. Brussard (1994): Preserving biodiversity: species in landscapes. Ecological Applications, Vol 42, pp 205-207. 1994

Integrating Environmental Research into Forest Ecosystems Planning

Klaus von GADOW

Institute of Forest Management, Faculty of Forest Sciences and Woodland Ecology, Georg-August-University Göttingen

ABSTRACT One of the basic challenges facing forest ecosystems planning is to generate information required by management for solving a variety of problems, generally with a limited budget. This cannot be achieved merely by gathering data about the forest resource. It is also necessary to transform the data into useful information, based on new research findings. One of the challenges facing forestry research is the quest for more effective, varied and flexible forms of cooperation among the disciplines. Suitable networking models for integrated forestry research projects are the top-down approach involving interdisciplinary groups, the bottom-up approach involving bargaining groups and the systems networking approach complemented by sporadic communication through the Internet. **Forest ecosystems planning** is in a position to apply existing discipline-specific knowledge from various domains towards a common goal, thus enabling effective, socially acceptable planning and control of forestry operations. A principal basis for medium-term planning is the **silvicultural action space** defined by normative indicators. Another practical approach which aims to apply discipline-specific knowledge with a common focus, is a technique known as **thinning inventory**.

Keywords: research networking, systems networking, thinning inventory, silvicultural action space.

1. NEW CHALLENGES FOR FOREST ECOSYSTEMS PLANNING

Although the contribution of the forest sector to the gross national product has declined to insignificant levels in many industrialized countries, forests are usually considered a *Central Resource* which is essential for the existence of human life and culture (Volz, 1995). Woodland conservation and recreation are given higher ratings than timber production and a growing emphasis on *biocentric* functions, such as species, habitat and process conservation, has been observed, as well as a subtle shift from *multiple-use management* to *ecosystems management* (Sedjo, 1995). The new ecological perspective of multi-functional forest management is based on the principles of ecosystem diversity, stability and elasticity, and the dynamic equilibrium of primary and secondary production

(Ulrich, 1987; 1993; Beese, 1996). Forests are a renewable resource, producing essential raw materials with minimum waste and energy use (*production* function). Multi-aged, mixed forests are rich in habitat and species diversity and contribute to increased ecosystem stability (*habitat* function). Forests can absorb the effects of unwanted deposition and other disturbances. They protect neighbouring ecosystems by maintaining stable nutrient and energy cycles and by preventing soil degradation and erosion (*regulating* function). Forestry provides much-needed recreation and contributes to stabilizing rural communities by offering job opportunities (*social and cultural* function).

The large-scale application of new silvicultural systems has become a political reality in many parts of the world. This involves a gradual transformation of traditional silvicultural practice towards *Continuous Cover Forestry,* favouring mixed uneven-aged stands, site-adapted tree species and selective harvesting (Griesel & Gadow, 1995).

One of the basic challenges facing forest ecosystems planning is to generate information required by management for solving a variety of problems, generally with a limited budget. The information is gathered in the field using different scales of time and space, and applying new technology, such as remote sensing and new terrestrial sampling strategies for biodiversity assessment. This is not an easy task as large amounts of data are becoming available at spatial scales ranging from individual sites to the whole biosphere. There are some important challenges facing forest planning today:

a) *New variables*. Greater emphasis will be placed on variables representing forest spatial information, biodiversity and details about the quality of timber products available in the forest. Growing stock volumes and diameter distributions are not considered sufficiently informative to satisfy the increasing information needs. Thus, new methods of forest sampling will have to be developed for assessing the new variables (Lund, 1998).

b) *Criteria and indicators for forest management*. Forest management planning and forest research need to assist in defining suitable criteria and indicators for resource monitoring and operational control. Public opinion has become an important factor influencing forest management, and the demands of society need to be carefully considered.

c) *Cost reduction*. The need for useful information will continue to increase, including the *new* ecosystems variables and the *old* timber variables. At the same time, there will be continuing demands to reduce costs, especially regarding the methods of forest resource assessment. The need for cost reduction will influence the methods of forest ecosystems planning.

d) *Better timing of resource assessment activities*. One of the major disadvantages of current sampling techniques is the fact that assessment activities, involving temporary or permanent plots, are scheduled to take place at fixed intervals. More useful information may be obtained if resource assessments coincide with forestry operations, thus informing not only about the system state but also about management-induced system changes (Gadow and Stüber, 1993).

e) *Improved use of multiple data sources.* Effective use of existing data requires new appoaches in forest inventory involving the use of prior information and mixed data modelling (Song, 1991; Dees, 1996; Puumalainen, 1998).

f) *Market-orientation of information generating efforts.* New *Just-In-Time* production systems und reduced storage time require accurate information about standing timber products. Estimating details about future removals in uneven-aged *Continuous Cover Forests* has become a central element of research (Albert, 1997; Staupendahl and Puumalainen, 1997; Daume et al., 1998).

2. NEW CHALLENGES FOR FOREST ECOSYSTEMS RESEARCH

Forest research depends on private or public project funding. For a private investor, it is worthwhile to fund a research project if cost savings can be achieved resulting from new research findings, or if greater product quantities can be sold at lower cost (Dasgupta and Stiglitz, 1980; Hyde et al., 1992; Hellström et al., 1995). A peculiarity of forest research is the fact that it may involve both applied and basic aspects, addressing immediate concrete problems as well as long-term basic issues which do not necessarily produce tangible benefits. The transition between applied and basic research is not always clearly defined.

There are many examples of **industry-funded** forest research (FAO, 1995). Commercially oriented research activities are rather common in Australia, South Africa and New Zealand, where state-owned forest research organizations have been privatized, at least partially. In New Zealand, it will be increasingly difficult in the future to distinguish between private and public research activities (Leslie, 1995). Industry-sponsored forest research is also fairly common in certain regions within the United States[1], and in the Scandinavian countries, but rare in Central Europe.

State-funded forest research is fairly common in most countries. Forest ecosystems research is often conducted within large integrated projects, involving several applied and basic disciplines and concentrating on a few *flagship-experiments*. Examples are the integrated *Sustainable Forest Management* project in Southern Sweden and the *Forest Ecosystems Research Project* in Göttingen, Germany (Bredemeyer and Wiedey, 1994). Large ecosystems projects can produce very good synergy effects, on condition that a

---

[1] Examples are the *Research Cooperatives* in the South-eastern US, e.g. the *Plantation Management Research Cooperative*, at the University of Georgia, or the *Loblolly Pine Growth and Yield Research Cooperative* at the *Virginia Polytechnic Institute and State University*.

suitable *project champion* is available, a person who is capable of providing the necessary leadership and coherence.

Scientists tend to be individualists, pursuing their goals with energy and a sense of vision. A certain amount of curiosity and sportsman-like ambition are the prerequisites for success in science. The creative contributions of individuals, though they may not always seem to fit into a logical plan, are essential for scientific development. Equally important in an applied science such as forestry, are efforts to bring together and coordinate the work of individuals (Kollmann, 1952; Luhmann, 1983). Thus, a central task of forestry research is the quest for varied and flexible forms of cooperation. Among the most suitable networking models for integrated forestry research projects are the *top-down* approach involving interdisciplinary groups, the *bottom-up* approach involving bargaining groups, the *systems networking* approach and informal collaboration *through the Internet* (c.f. Rossini and Porter, 1978; Krott, 1994).

In group networking, the participating scientists cooperate as a multi-disciplinary group under a common objective. According to Krott (1994) each of the individual disciplines will strive to secure as much funding as possible for conducting their own specialized research, acting as independently as possible. They are willing to cooperate with other disciplines when such cooperation is seen to bring advantages, albeit for a limited period of time.

A second approach involves bargaining between individuals or small groups belonging to different disciplines. The participants must be prepared to meet regularly and to spend time discussing strategy and getting to know each other's potential for contributing to a common goal. The bargaining process may be time consuming, involving a lot of human interaction, but the capacity for improved understanding of complex issues is excellent. The discussions concentrate on the interfaces between ecological *compartments* and management *windows* (root/soil solution; leaf/atmosphere; forest sampling/timber utilization). Again, an important precondition for success is the availability of a capable *project champion*. Communication provided by the *Internet* facilitates rapid exchange of data and publications, and the organisation of meetings.

A third method for promoting integrated forestry research involves *systems networking*. The approach is most suitable for large complex projects generating a high data load. In systems networking, the factors which are most essential for system performance are identified to maintain clarity, even with increasing complexity of the sub-problems. A useful basis for systems networking in large projects are planning systems or geo-information systems (Rose and McDill, 1996).

3. INTEGRATING RESEARCH RESULTS INTO FOREST ECOSYSTEMS PLANNING

Forest ecosystems planning is in a position to apply existing discipline-specific knowledge towards a common goal, thus enabling effective planning and control of forestry operations. A principal basis for medium-term planning is the *silvicultural action space* defined by normative indicators. Another practical approach which aims to apply discipline-specific experience for monitoring management activity, is a technique known as *thinning inventory*.

3.1 Planning Management Activity: Generating Multiple Scenarios of Forest Development

There is no unanimity about optimum silviculture, and the principle of multiple forest developments implies that a managed forest may develop in a variety of ways. The number of possible development scenarios for a given forest may be very high. Finding the optimum one is a problem of multi-objective analysis (Saaty, 1980; Haedrich et al., 1986; Lillich, 1992; Steinmeyer u. Gadow, 1994; Pukkala et al., 1998). Numerous practical tools for generating alternative scenarios of forest development, not only in even-aged but also in uneven-aged mixed forests, are already being used (Ek and Monserud, 1974; Wykoff et al., 1984; Pretzsch, 1992; Nagel, 1996). A key element in a scenario of forest development, is the analysis of thinning operations and the definition of the permissible silvicultural action space. A thinning reduces the amount of saleable products. It modifies the spatial distribution of the temperature and radiation regime and thus influences the process of mineralization (Gemesi et al., 1995) and the composition of the ground flora. Thus, a principal basis for medium-term forest ecosystems planning is the silvicultural *action space* defined by normative indicators (Fig. 1).

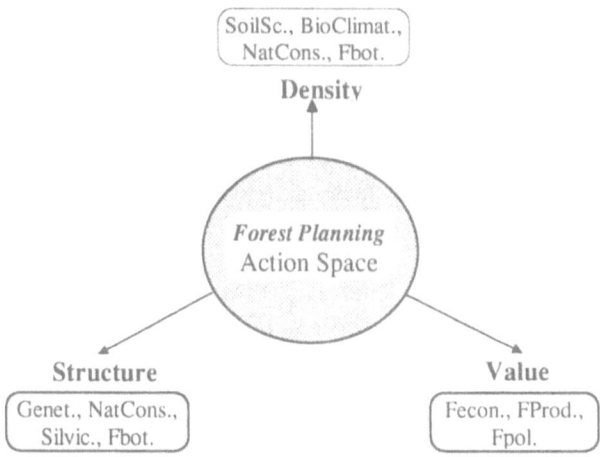

Figure 1. The silvicultural action space is defined by normative indicators, specifying allowable modifications of forest density, structure and value. The indicators are defined by the various disciplines (Forest Policy,

*Planning and Economics, Soil Science, Bioclimatology, Genetics, Botany,
Nature Conservation, Forest Products etc.).*

All the different disciplines, including less typical forestry disciplines such as
biology, mathematics and computer science, as well as third parties such as
Sierra Club, Greenpeace, NABU etc., are involved in defining a *silvicultural
action space*, which includes the range of permissible silvicultural activities in a
scenario of forest development. Each one of the different thinnings which are
scheduled along the path of forest development will determine the long-term
cash flow and risk, requiring input from Forest Economics and Forest Products
research and the spatial structure of a forest ecosystem, which need to be
evaluated by disciplines such as Forest Genetics, Forest Botany and Soil Science.

3.2 Monitoring Forest Management Activity: Periodic Inventories and Thinning Inventories

Forest planning requires data about the current state of the resource. Such data
are gathered during stock-taking activities, involving either temporary or
permanent sample plots. Most of the classical stock-taking activities, such as
compartment sampling or systematic strata sampling, are scheduled to take place
at periodic time intervals.

Virtually all the current methods of forest sampling were designed for
situations where the modification of stand structure following a thinning is
assumed to be known. Stand inventories are carried out at regular intervals
because the changes caused by thinning operations between successive
inventories are assumed predictable. This was a plausible approach, as long as
the actual silvicultural operations agreed with the standard ones prescribed by a
yield table.

In reality, the data obtained in a periodic inventory are often very short-
lived on account of the changes resulting from intermediate thinning operations.
It is not in the sampling design which is deficient, but the timing of the
enumeration activities, the most popular timing pattern being the regular,
periodic one. In the traditional forest sampling schemes, periodic inventories are
carried out in all stands at regular intervals. However, the values of the state
variables assessed during an inventory change as a result of natural growth, but
even more so as a result of thinnings, which are often hard to predict, especially
in uneven-aged mixed forests. The information gathered during a strictly
periodic stock-taking will be worthless after the first thinning. Thus, despite the
phenomenal progress which has been achieved in the theory of forest sampling,
there is no doubt that the effectiveness of forest inventory schemes may still be
considerably enhanced by taking account of the right timing of stock-taking
activities.

An alternative timing strategy for stock-taking is applied in an approach
known as *thinning inventory* (Gadow and Stüber, 1993). A *thinning inventory*
captures stand data immediately after marking the trees, but before the marked

trees are removed (Fig. 2). Using the same effort as that applied in a periodic inventory, a thinning inventory is capable of simultaneously producing information about three different states: a) all the trees before a thinning, b) the removed trees and c) the trees remaining after the thinning.

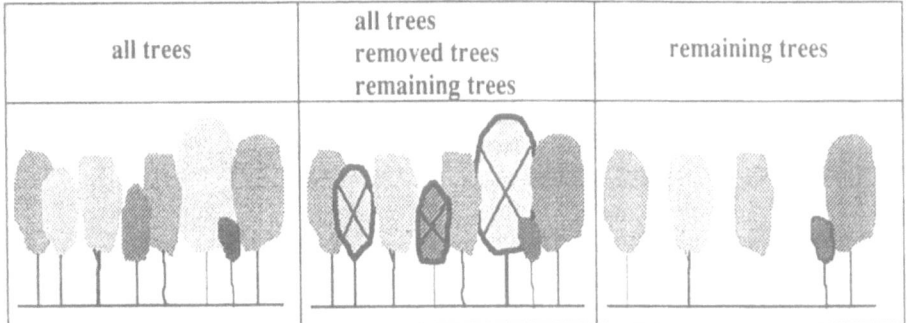

all trees	all trees removed trees remaining trees	remaining trees

Figure 2. Most resource inventories assess the entire growing stock (which may soon change due to a thinning operation, left) or the growing stock remaining after a thinning, (e.g. to evaluate the damage done by harvesting operations, right). The information obtained in a thinning inventory includes three data sets: the before-, removed- and after-data (centre).

A thinning inventory captures information about thinning effects, thus providing the data for modelling thinning effects and thinning behaviour of individual foresters. A thinning inventory permits continuous evaluation of all silvicultural operations, considering the effects of management on the modification of the ecosystem: the density and spatial structure of the remaining trees and the value of the residual timber growing stock.

REFERENCES

Albert, M. (1997): Bericht zum Forschungsprojekt Erfassung und Beschreibung von Strukturveränderungen durch forstliche Eingriffe in Buchen-Edellaubholz-Mischbeständen. Report, German Research Foundation, 26 pp.

Bäumler, R., Blessing, U. and Zech, W. (1995): Untersuchungen zur Stoffdynamik zweier bewaldeter Kleineinzugsgebiete im Flysch - Auswirkungen eines geregelten forstlichen Eingriffs, FW Cbl 114, pp. 261-271.

Beese, F. (1996): Indikatoren für eine multifunktionelle Waldnutzung, Forstw. Cbl. 115, pp. 65-79.

Bredemeier, M. and Wiedey, G.A. (1994): Forschungsantrag an das BMFT zum Verbundprojekt Veränderungsdynamik von Waldökosystemen, Teil I. Forschungszentrum Waldökosysteme, Göttingen: 183 S.

Dasgupta, P. and Stiglitz, J.E. (1980): Uncertainty, industrial structure and the speed of R&D. Bell Journal of Economics, 11, pp.1-28.

Daume, S., Füldner, K. and Gadow, K. v. (1998): Zur Modellierung personenspezifischer Durchforstungen in ungleichaltrigen Mischbeständen, AFJZ 169 (2), pp. 21-26.

Dees, H., (1996): Effektivitätssteigerung bei forstlichen Grossrauminventuren durch Hilfsinformationen. Proc. Ecology Group, German Region of the International Biometric Society, No. 7, pp.134-145.

Ek, A.R. and Monserud, R.A. (1974): FOREST - a computer model for simulating the growth and reproduction of mixed-species forest stands. Univ. Wisc. School Nat. Resources Res. Rep. R2635.

FAO (1995): The role of the private sector in forestry research, Rome. 186 p.

Gadow, K. v. and Stüber, V. (1993): Die Inventuren der Forsteinrichtung. Forst u. Holz 49 (5), pp. 129-131.

Gemesi, O., Skambracks, D. and Topp, W. (1995): Einfluß eines geregelten forstlichen Eingriffs auf die Besiedelungsdichte und den Streuabbau der Regenwürmer im Flysch der Tegernseer Berge, FWCbl 114, pp.272-281.

Griesel, F. and Gadow, K. v. (1995): Naturgemäßer - naturnaher - ökologischer Waldbau - Begriffsbestimmungen, Konzepte und Richtlinien, Anwendungsbeispiele, Arbeitsbericht, Institut für Forsteinrichtung und Ertragskunde, Univ. of Göttingen, 20 pages.

Haedrich, G., Kuss, A. and Kreilkamp, E. (1986): Der Analytic Hierarchy Process - Ein neues Hilfsmittel zur Analyse und Entwicklung von Unternehmens- und Marktstrategien, WiSt, pp. 120-126.

Hellström, E., Palo, M. and Solberg, B. (1995): Privatization of forest sector research - theory and European empirical findings, Paper presented at XX IUFRO World Congress in Tampere, 19 pages.

Hyde, W.F., Newman, D.H. and Seldon, B.J. (1992): The economic benefits of forestry research, Iowa State Univ. Press, Ames, 249 pages.

Kollmann, F. (1952): Integrierte Forschung - ein Ausweg aus der Krise der Wissenschaft. Selbstverlag der Univ. Hamburg, 24 pages.

Krott, M. (1994): Management vernetzter Umweltforschung - Wissenschafts-politisches Lehrstück Waldsterben, Böhlau Verlag, Wien - Köln - Graz.

Leslie, A.. (1995): Private research in forestry in New Zealand and Australia, In: FAO, 1995, pp.141-160.

Lillich, L. (1992): Nutzwertverfahren. Schriften zur quantitativen Betriebs-wirtschaftslehre, Bd.3, Heidelberg, Physica-Verlag, 196 pages.

Luhmann, N. (1983): Interdisziplinäre Theoriebildung in den Sozial-wissenschaften, In: Schneider, pp.155-159.

Lund, G. (1998): Multiple resource inventory, IUFRO World Series No. 8.

Nagel, J. (1996): Anwendungsprogramm zur Bestandesbewertung und zur Prognose der Bestandesentwicklung, Forst u. Holz 51 (3), pp.76-78.

Pretzsch, H. (1992): Konzeption und Konstruktion von Wuchsmodellen für Rein- und Mischbestände, Schriftenreihe d. Forstw. Fak. Univ. München, No. 115, pp.332 p.

Pukkala, T., Miina, J. and Rautiainen, O. (1998): Dependence of stand management on management goal, In: Pukkala, T. and Eerikäinen, K. (eds), Tree seedling production and management of plantation forests, University of Joensuu, Res. Notes 68, pp.165-180.

Puumalainen, J. (1998): Marktorientierte Vorratsschätzung und terrestrische Vorinformationen in kleinräumigen Waldinventuren, Diss., Fakultät für Forstwissenschaften und Waldökologie, Georg-August-Universität Göttingen.

Rose, D.W. and McDill, M. (1996): Incorporating non-commodity considerations in strategic planning of natural resources, Report, Univ. of Minnesota, 13 pages.

Rossini, F.A. u. Porter, L.A. (1978): The Management of interdisciplinary, policy-related research, In: Sutherland u. Legasto, A. (ed): Management Handbook of Public Administration, New York.

Saaty, T.L. (1980): The Analytic Hierarchy Process.McGraw - Hill.

Sedjo, R.A. (1995): Ecosystem Management - an uncharted Path for Public Forests. Resources, (Resources for the Future) No. 121, pp.10-20.

Song, X. (1991): Die Bayes-Schätzung in der forstlichen Betriebsinventur. Mitt. D. Abt. f. Forstl. Biometrie, Albert-Ludwigs-Univ. Freiburg, 153 p.

Staupendahl, K. and Puumalainen, J. (1997): Beschreibung von Durchforstungen im Forstamt Paderborn. Project Report, Institute of Forest Management, Univ. of Göttingen, 27 pages.

Steinmeyer. A. and Gadow, K. v. (1994): Saaty's AHP - dargestellt am Beispiel der Waldbiotopkartierung, Centralblatt f. d. ges. Forstwesen 112 (1), pp.53-65.

Ulrich, B. (1987): Stabilität, Elastizität und Resilienz von Waldökosystemen unter dem Einfluß saurer Depositionen, Forstarchiv, 58, pp.232-239.

Ulrich, B. (1993): Prozeßhierarchie in Waldökosystemen, Biologie in Unserer Zeit 23 (5), pp.322-329.

Volz, K.-R. (1995): Zur ordnungspolitischen Diskussion über die nachhaltige Nutzung der Zentralressource Wald, Forst u. Holz, 50 (6), pp.163-170.

Wykoff, W. R., Crookston, N. L. and Stage, A. R. (1982): User's guide to the stand prognosis model, USDA For. Serv., Gen. Tech. Rep. INT-133.

Forests in Sustainable Mountain Development

Martin F. PRICE

Coordinator, IUFRO Task Force on Forests in Sustainable Mountain Development,
Mountain Regions Programme, Environmental Change Unit
University of Oxford, Oxford, UK

ABSTRACT Mountain regions have global significance. They occupy a fifth of the Earth's land surface. About one tenth of humankind lives in them, and they affect the lives of over half the world's population. The values of mountain forests to mountain people, those living downstream, and much of the remainder of humanity include their roles as centres of biodiversity; the storage and supply of water; the provision of fuelwood and many other wood and non-wood products; essential complements to agriculture; sites for recreation; and protection against natural hazards and soil erosion.   These values are underlined in documents resulting from regional meetings around the world and emphasised in many recent and ongoing projects, and are the focus of a IUFRO Task Force.

Keywords: mountains, forests, sustainable development, IUFRO.

1. MOUNTAINS ON THE GLOBAL AGENDA

Mountains and uplands occupy approximately a fifth of the Earth's land surface. About one tenth of humankind lives in them, and they affect the lives of more than half of the world's population (Ives 1992). Building on a number of initiatives over the past three decades (Ives and Messerli 1990; Stone 1992), their global significance was specifically recognised at the UN Conference on Environment and Development (UNCED), or 'Earth Summit', in Rio de Janeiro in June 1992 by the inclusion of a chapter in "Agenda 21", a plan for action into the 21st century which was endorsed by the heads of state or government from most of the world's countries.   The inclusion of Chapter 13 of "Agenda 21" – "Managing Fragile Ecosystems: Sustainable Mountain Development" – places mountain regions on a equal footing with climate change, tropical deforestation, desertification, and similar issues of global change (Turner et al. 1990) in the global debate on environment and development.

In September 1993, the UN Inter-Agency Committee of Sustainable Development gave the Food and Agriculture Organization of the United Nations (FAO) responsibility for acting as Task Manager for Chapter 13, which includes two 'programme areas':

A) generating and strengthening knowledge about the ecology and sustainable development of mountain ecosystems;

B) promoting integrated watershed development and alternative livelihood opportunities.

In this role, FAO has fostered a number of global and regional initiatives; and many countries have also taken steps towards implementing various elements of Chapter 13 (Chipeta and Michaelsen 1995; Sène and McGuire 1997; Price 1998).

This paper discusses the roles of mountain forests in relation to the major issues of sustainable mountain development, and reviews initiatives that have taken place since 1992.

2. THE ROLES OF MOUNTAIN FORESTS IN SUSTAINABLE DEVELOPMENT

Chapter 13 begins "Mountains are an important source of water, energy, and biological diversity." Forests play crucial roles in fulfilling all of these functions. While global statistics on the proportion of the world's mountains covered by forests are not available - in fact, there is no globally-accepted definition of mountains (Ives et al. 1997) - this proportion is large, encompassing a very wide range of ecosystems at all latitudes. In any mountain range, the diversity of ecosystems reflects the range of altitudes, slopes, aspects, geological substrates, and soils; as well as the effects of human activities. All of the centres of greatest vascular plant species diversity are in, or include, mountains: Costa Rica, the tropical eastern Andes, the Atlantic forest of Brazil, the eastern Himalayan-Yunnan region, northern Borneo, and Papua New Guinea. Secondary centres are found in Mediterranean and arid mountains, the Rocky Mountains, and Central Asia (Barthlott et al. 1996). In Europe, a very high proportion of centres of plant diversity is in mountains (Davis et al. 1994). Species diversity typically decreases with altitude; thus, forests generally have greater diversity than the non-forested zones above (Jenik 1997).

Mountains are of vital importance to the globe as sources of water, typically having higher rates of precipitation than surrounding lowlands. Forests play a number of important roles in the water cycle. These include the capture of atmospheric moisture, particularly by cloud forests, which are widespread but of particular importance in tropical areas (Hamilton et al. 1994); storage of snow and water in and beneath canopies; and release to watercourses, which is strongly affected by the degree of disturbance by natural and human forces (Hamilton and Bruijnzeel 1997). In semi-arid and arid regions, over 90 percent of river flow comes from the mountains. Even in temperate Europe, the importance of the Alps is shown by the fact that, while the Alps occupy only 11 percent of the area of the Rhine basin, they supply 31 percent of the annual flow – and more than 50 percent in summer (Bandyopadhyay et al. 1997). Much of the water flowing through and from the mountains is used to produce hydro-electric power. This is largely used on the plains below; though water power has also long been important in mountain economies, especially for grinding grain.

Wood is the primary source of fuel for those living in the mountains of the developing world, and also many industrialised countries, where it has only been supplanted by electricity and fossil fuels in recent decades, if at all. Mountain forests are also the primary sources of fuel for nearby settlements in the foothills and plains, so that the collection of wood for fuel or the production of charcoal is a major factor in deforestation in many regions where mountain and nearby populations – both rural and urban – are increasing. Globally, more than half of estimated wood consumption is for fuelwood and charcoal, largely in developing countries (FAO 1992). The highest rate of deforestation in any biome is for tropical upland forests: 1.1 percent per year. Rates of clearing are particularly high in Central America, East and Central Africa, Southeast Asia, and the Andes (FAO 1993a).

The second sentence of Chapter 13 notes that mountains are also a source of forest products, agricultural products, and recreation. Mountain forests are important sources of timber in many countries, though the costs of harvesting are typically higher than on low-angle land. However, in many parts of the world, such land, even if once forested, provides a greater economic yield for agriculture or settlement; a process of conversion that took place centuries ago in Europe and has spread to many tropical countries in recent decades. Thus, the pressures on mountain forests increase. Until recently, the concept of sustained yield from forests focussed primarily on the production of wood for industrial and other uses. New models of sustainable forest management (e.g., FAO 1993b; Maser 1994) give greater emphasis to the need to manage forests as ecosystems, rather than 'mines' of exploitable resources. One critical element of this is to ensure sustainable yields of non-wood forest products - valuable for food, nutrition, medicine, construction materials, and household and cultural uses - which contribute to the subsistence and market economies of mountain people; and, in some cases, national economies (Taylor 1996). Such products are vital for many in developing countries, but also provide supplementary sources of food and income in industrialized countries (Hansis 1996; Richards and Creasy 1996), and have recently been of increasing importance in the uncertain conditions being experienced in the former communist countries of Central and Eastern Europe.

In many mountain regions of the world, it is inappropriate to speak of 'foresters' and 'farmers' (Price and Thompson 1997). Agriculture and forestry are inseparable activities within a complex mountain landscape, with forests providing essential resources to ensure the viability of agriculture. These resources depend greatly on local ecological and social conditions. In many parts of the world, grazing animals rely on forage within forests, fodder brought from them, and the shade they provide during hot summer days. The expansion of grazing land by clearing and/or burning at the upper timberline has led to its depression in many mountains; by as much as 300 m in the Alps (Langenegger 1984). Restoration of forests in this dynamic zone is often difficult or impossible because of changes in microclimate and soil conditions following clearing.

In many tropical countries, slash-and-burn agriculture represents a successful means of utilising forested ecosystems when human population densities are low and government policies do not limit movement. However, settlement and/or high population densities often lead to a level of degradation from which forests can not recover; and clearance in order to grow illegal crops has affected forests in many less-accessible mountain regions. Agroforestry - through planting, natural regeneration, or controlled use of remnant forests, often combined with intercropping of food crops - can often permit continued use of land by large human populations (Nair 1993). However, this requires that those who make the necessary investments of time and labour have security of tenure and, in many cases, that extension services and other organisations provide new tree or crop species in addition to those of local origin in order to provide the desirable mixture of food security, income generation, and slope stability.

The longest-established uses of forests for recreation are for hunting and fishing for pleasure: activities that also remain important for subsistence in many regions. The more widespread use of forests for recreation derives from a series of linked factors; particularly, increases in accessibility, leisure time, and income in both industrialised, but also many developing, countries. This phenomenon began in the early 19th century in western Europe and spread to the mountains of South Asia, eastern North America, and Latin America over the next century (Price 1992). However, until after the Second World War, it was limited to relatively few locations. Mass recreation now takes place in mountain forests around the world, with a vast range of activities: both active (e.g., skiing, mountain-biking, hunting, walking) and passive (e.g., bird-watching, painting, sightseeing). A critical issue is how to manage such rapidly-growing activities to minimise environmental and societal impacts, and to maximise benefits to local communities (Price et al. 1997).

Following the mention of the vital resources provided by mountains - notably, as discussed above, by their forests - the first paragraph of Chapter 13 notes the rapid changes that are taking place in mountains: soil erosion, landslides, and rapid loss of habitat and genetic diversity. The loss of diversity can result from many of the processes discussed above, such as deforestation, excessive harvesting of forest products, hunting, and recreation. The protective function of forests, in stabilizing slopes and limiting damage from extreme events, such as avalanches and floods, has long been known by mountain people with, for instance, local regulation in the Swiss Alps since the 13th century, and national legislation from the late 19th century (Price 1990). Nevertheless, it should be recognised that this legislation followed major floods after decades of unheeded warnings of the risks of excessive utilisation of the forests. A more recent example is in Thailand, where landslides and floods followed widespread replacement of forests by agriculture and plantations (Hamilton 1992). The major issue for minimising risks from 'natural hazards' is therefore how to ensure stable forest cover, particularly when this has been lost through human action, extreme events, or a complex mixture of these.

3. ACTIONS TOWARDS SUSTAINABLE MOUNTAIN DEVELOPMENT

The fact that forests are implicated in the majority of the issues addressed in the introductory paragraph of Chapter 13 underlines their central importance to sustainable mountain development. Forests are also specifically mentioned in a number of the sub-paragraphs describing activities that governments are encouraged to implement within the two programme areas of Chapter 13:

- 13.6(f) "integrate all forest, rangeland and wildlife activities in such a way that specific mountain ecosystems are maintained";
- 13.7(b) "build an inventory of different forms of soils, forests, water use, and crop, plant, and animal genetic resources, giving priority to those under threat of extinction. Genetic resources should be protected in situ ...";
- 13.10 "Strengthen scientific research and technological development programmes, including diffusion through national and regional institutions, particularly in meteorology, hydrology, forestry, soil sciences and plant sciences";
- 13.16(b) "establish task forces or watershed development committees, complementing existing institutions, to coordinate integrated services to support local initiatives in animal husbandry, forestry, horticulture and rural development at all administrative levels".

Under programme area B (paragraphs 13.13-22), there are also a number of sub-paragraphs that propose activities related more generally to various aspects of forests: pilot projects combining environmental protection, development and environmental management practices/systems (13.21a); participatory generation of technology for watershed management (13.21b); technology, training, and dissemination of knowledge relating to agroforestry and other aspects of land and water management (13.21c, 22a); and national centres for watershed management (13.23).

The importance of forests is also underlined in the documents resulting from the regional inter-governmental consultations on sustainable mountain development which have taken place since UNCED, in Africa (International Livestock Research Institute 1997), Asia (Banskota and Karki 1995), Europe (Backmeroff et al. 1997), and Latin America (Mujica and Rueda 1996). All of these documents, endorsed by representatives of a total of 62 countries and the European Union, stress three major issues of direct relevance to forests. First, integrated approaches to land use, involving all stakeholders, are necessary, but require investment. Second, land use practices should be appropriate to land capability and community needs and capabilities. Third, there are strong linkages between land management and the conservation of both water and biodiversity; mechanisms to ensure the security of these important values of forests include the protection of areas with minimum human impacts and the development of ecological corridors and transboundary protected areas.

Many of these cross-cutting issues can be identified in activities which have begun in different countries since UNCED. However, it is rarely possible to say that such activities result directly from Chapter 13. Even when they are specifically in the context of a national plan or strategy for Agenda 21, conservation, or biodiversity (as, for example, in Bulgaria, Japan, Pakistan, Papua New Guinea, or Philippines), such activities may have been planned or underway before 1992 (Price 1998). The following paragraphs illustrate a few of these activities, recognising that similar examples could probably be cited for other countries. However, few national governments specifically identified such activities in the 'country statements' submitted to the fifth session of the UN Commission on Sustainable Development in April 1997.

In Japan, the Council of Ministers for Global Environmental Conservation approved the National Action Plan for Agenda 21 in December 1993. Chapter 13 of this plan places a very strong emphasis on the management of forests for protection, production, recreation, and conservation. The links between the latter two 'outputs' and economic development is clearly identified in relation to the development of eco-tourism. Nevertheless, while the maintenance and encouragement of mountain economies is a clear goal of the chapter and a law passed in the same year, this is within the constraints imposed by the need to minimise health and environmental impacts. The Basic Plan for Forest Resources, passed in 1996, emphasizes five main functions of forests: 1) conservation of water resources; 2) disaster prevention; 3) environmental conservation; 4) timber production; and 5) health and cultural activities.

One activity that clearly notes its antecedence in Chapter 13 is the Inter-regional Project for Participatory Upland Conservation and Development, which the FAO/Italy Cooperative Programme began in 1992 in Bolivia, Burundi, Nepal, Pakistan, and Tunisia. Its aim is to identify and field test "strategies, methods and techniques for the promotion and consolidation of people's participation in the conservation and development of upland watersheds" (d'Ostiani and Warren 1996: 11). A number of preliminary conclusions have emerged. First, the inter-regional institutional framework and the flexibility in planning and implementation are both important. Second, participatory methods are necessary but not sufficient; teams must also address technical elements of watershed management and underlying environmental, socio-economic and organisational issues. Third, participatory integrated watershed management involves both process-based activities and physical interventions. Finally, while limited resources can be very valuable for catalysing development investments, most actions take longer than anticipated.

Many of the projects funded by the World Bank and the Global Environment Facility (GEF) in recent years also emphasize participatory approaches and multi-functional economies (Global Environment Facility 1996; World Bank Environment Department 1996). Examples of World Bank projects include those in Bhutan, Indonesia, Pakistan, Tunisia, Turkey, and Yemen, which focus on watershed management, improving and restoring forest productivity, and institutional strengthening. Many GEF projects, which focus on conserving biodiversity in national parks and other protected areas, also emphasize cooperative forest management within these areas and in their buffer zones, recognising the importance of ensuring sustainable livelihoods and joint decision-making as necessary elements of biodiversity conservation. These include projects in Bolivia, Cameroon, China, Costa Rica, Ecuador, Indonesia, Mexico, Nepal, and Uganda, as well as Slovakia and Ukraine; thus, such approaches are valid not only in developing but also industrialised countries, as recognised by the Pan-European Biological and Landscape Diversity Strategy, endorsed by Ministers of Environment from 55 European states in 1995 (Council of Europe/UNEP/ECNC 1996).

In June 1997, the UN General Assembly held a Special Session (UNGASS) to review the implementation of all of "Agenda 21". The final document of this meeting specifically mentions mountains in relation to four issues:

- continued deterioration of mountain ecosystems, resulting in diminishing biological diversity (para. 9);
- the need to formulate and implement policies and programmes for integrated watershed management (para. 34);
- the need for ecosystem approaches to combat or reverse soil degradation, recognising the multiple functions of agriculture (para. 62);
- the need for national policy development and implementation to ensure sustainable patterns of consumption and production in tourism (para 68).

These concerns relate very closely to many of the values of mountain regions explored in the first section of this paper. According to the UNGASS final document, all countries are expected to have prepared national strategies for sustainable development by 2002, involving all interested parties and integrating economic, social, and environmental objectives. There is considerable scope, and need, for the large number of countries with mountain forests to consider how these should be considered in such strategies.

4. THE ROLE OF IUFRO

In recent decades and years, there has been a widespread shift in the science and practice of forestry, from emphasis on the production of wood towards management based on recognition that forests serve multiple functions and produce a wide range of outputs (Hamilton et al. 1997). This multi-functionality has long been present in most mountain forests, even when they have primarily been managed for sustained harvests of timber. A useful summary of these two different perspectives is given in Table 1 (Jodha 1992).

Recognising the general shift in forestry, as well as the changing expectations of populations around the world regarding mountain forests, and the rapid rates of change in the cover and uses of mountain forest ecosystems, the International Union of Forestry Research Organisations (IUFRO) has established a Task Force on Forests in Sustainable Mountain Development for the period 1996-2000.

The Task Force has two main purposes with respect to the roles of forests in the sustainable development of mountain areas:

- to provide a framework for developing and strengthening linkages within IUFRO and between IUFRO members and other relevant organisations and initiatives;
- to prepare an assessment of major issues for mountain forests at the beginning of the 21st century.

Table 1: Dominant features of conventional and mountain perspective based approaches to forest management (from Jodha 1992)

CONVENTIONAL APPROACH	MOUNTAIN PERSPECTIVE BASED APPROACH
Primary focus and concern	
Forest treated as an isolated, revenue-generating sector of a region; focus on yield of selected key products (e.g., timber)	Forest as an integral component of ecosystem; inseparability of sustainability of the two; emphasis on both 'service' and 'product functions
Dominant products and usage system	
Timber and other high value products; market-directed over-extractions; insensitivity to negative side effect; isolated sectoral activity run through legal and administrative superstructures	Diversified biomass-based, interlinked activity patterns (e.g., farm-forest linkages); compatibility with ecosystem needs, people sustenance strategies, and user perspectives
Valuation norms/yardsticks	
Market-based narrow yardsticks for pricing products, compensating for extractions, and determining investment and subsidies; unequal terms of exchange (compensations); insensitivity to local concerns	Focus on health and stability of total system and interlinked activities with concern for multiple externalities; compensation mechanisms also involve biophysical components
Research and development approach	
'Extraction'-oriented approach with focus on monocultures of selected attributes (e.g., high-value timber); with little concern for folk knowledge, local needs	Focus on sustained biodiversity and linkages, regeneration and conservation, and people-centred possibilities; effective use of folk knowledge (folk agronomy, ethnoecology, etc.)
Sustainability prospects	
Emergence of indicators of unsustainability (i.e., persistent negative changes in forest health, productivity, usage patterns)	Possibility of restoring sustainability by sensitising forest interventions to mountain perspective (mountain specificities)

It is recognised that trends in the cover and/or density of mountain forests in different parts of the world are different, with two broad types which will often have to be considered separately in the assessment. The first is in the temperate zone (industrialised countries), where cover and/or density are generally stabilised or increasing, from land abandonment and re- or afforestation. In certain regions, significant impacts of tourism and air pollution occur. The second is in tropical, semi-arid, and arid countries, where there are significant decrease in cover and/or density, especially through clearance for agriculture. Serious erosion and desertification, exacerbated by human activities, are often major problems.

However, there have been some successes in reforestation, especially through community-based projects.

A number of means are being used to fulfil the purposes of the Task Force. One of the first activities has been to create a 'mountain forests network' on the IUFRO website, listing active individuals and organisations both within and outside IUFRO. Linkages have been developed with other relevant initiatives at different scales, notably regional (e.g., European Cooperation in the Field of Scientific and Technical Research [COST] Action E3, FAO/European Forestry Commission Working Party on Management of Mountain Watersheds; European Observatory of Mountain Forests; FAO Latin American Technical Cooperation Network) and global (e.g., Inter-governmental Forum on Forests, the Mountain Forum). A side meeting was held at the World Forestry Congress in Antalya, Turkey, in October 1997, in order to begin to define further work. In addition to the issues raised throughout this paper, the participants suggested that the Task Force should work on the following areas:

- advocacy of the commonalities of mountain regions, including the promotion of North-South dialogue;
- valuation of the benefits of mountain forests, which requires the close involvement of social scientists from diverse disciplines, with a focus on the service-related functions of mountain forests;
- contributions of science to awareness-raising and policy-making, through strong linkages between social and natural scientists, practitioners, and decision-makers.

The IUFRO Inter-divisional Conference on Forest Ecosystem and Land Use in Mountainous Areas in Seoul, Korea and the IUFRO Division 8 Conference on Environmental Forest Science in Kyoto, Japan, both in October 1998, are key events in the activities of the Task Force. It is anticipated that the experts participating in these meetings will identify and discuss the key issues facing the world's mountain forests and those who depend on them. From this basis, experts – both within and outside IUFRO – will be invited to contribute to a state-of-knowledge assessment to be presented at the IUFRO Congress 2000 in Malaysia. This will be IUFRO's contribution to ensuring that the manifold roles of forests in sustainable mountain development are realised into the next millennium.

REFERENCES

Backmeroff, C., Chemini, C. and P. La Spada (eds.) (1997) : European Intergovernmental Consultation on Sustainable Mountain Development: Proceedings of the Final Trento Session. Government of the Autonomous Province of Trento, Trento.

Bandyopadhyay, J., J.C. Rodda, R. Kattelmann, D. Kraemer and Z.W. Kundzewicz (1997) : Highland Waters - A Resource of Global Significance In J.D. Ives and B. Messerli (eds.) Mountains of the World: A Global Priority. Parthenon, Carnforth, pp. 131-155.

Banskota, M. and Karki, A.S. (eds.) (1995) : Sustainable Development of Fragile Mountain Areas of Asia. ICIMOD, Kathmandu.

Barthlott, W., W. Lauer and A. Placke (1996) : Global Distribution of Species Diversity in Vascular Plants: towards a World Map of Phytodiversity. Erdkunde, Vol. 50, pp. 317-327.

Chipeta, M. and T. Michaelsen (1995) : Post-UNCED Forestry and Mountain Development: New Challenges for FAO. Unasylva, Vol. 46, No. 3, pp.16-24.

Council of Europe/United Nations Environment Programme(UNEP)/European Centre for Nature Conservation (ECNC) (1996) : The Pan-European Biological and Landscape Diversity Strategy. Council of Europe, Strasbourg.

Davis, S.D., V.H. Heywood and A.C. Hamilton (eds.) (1994) : Centres of Plant Diversity: A Guide and Strategy for their Conservation, Vol. 1, Europe, Africa, South West Asia and the Middle East. IUCN, Cambridge.

d'Ostiani, L.F. and P. Warren (eds.) (1996) : Steps towards a Participatory and Integrated and Integrated Approach to Watershed Management. Coordination Unit, FAO/Italy Cooperative Programme Inter-regional Project for Participatory Upland Conservation and Development, Tunis.

FAO (1992) : FAO Yearbook of Forest Products 1990. FAO, Rome.

FAO (1993a) : Forest Resources Assessment 1990 - Tropical Countries. FAO, Rome.

FAO (1993b) : The Challenge of Sustainable Forest Management. FAO, Rome.

Global Environment Facility (1996) : Quarterly Operational Report, November 1996. GEF Secretariat, Washington DC.

Hamilton, L.S. (1992) : The Wrong Villain? Journal of Forestry, Vol. 90, No. 2, p. 7.

Hamilton, L.S. and L.A. Bruijnzeel (1997) : Mountain Watersheds - Integrating Water, Soils, Gravity, Vegetation, and People In J.D. Ives and B. Messerli (eds.) Mountains of the World: A Global Priority. Parthenon, Carnforth, pp. 337-370.

Hamilton, L.S., D.A. Gilmour and D.S. Cassells (1997) : Montane Forests and Forestry In J.D. Ives and B. Messerli (eds.) Mountains of the World: A Global Priority. Parthenon, Carnforth, pp. 281-311.

Hamilton, L.S., J.O. Juvik and F.N. Scatena (eds.) (1994) : Tropical Montane Cloud Forests. Springer, New York.

Hansis, R. (1996) : The Harvesting of Special Forest Products by Latinos and Southeast Asians in the Pacific Northwest: Preliminary Observations. Society and Natural Resources, Vol. 9, pp. 611-615.

International Livestock Research Institute (1997) : Proceedings, African Inter-governmental Consultation on Sustainable Mountain Development, June 3-7 1996. International Livestock Research Institute, Addis Ababa.

Ives, J.D. and B. Messerli (1990) : Progress in Theoretical and Applied Mountain Research, 1973-1989, and Major Future Needs. Mountain Research and Development, Vol. 10, pp. 101-127.

Ives, J.D., B. Messerli, and E. Spiess (1997) : Mountains of the World - A Global Priority In J.D. Ives and B. Messerli (eds.) Mountains of the World: A Global

Priority. Parthenon, Carnforth, pp. 1-15.

Jenik, J. (1997) : The Diversity of Mountain Life In J.D. Ives and B. Messerli (eds.) Mountains of the World: A Global Priority. Parthenon, Carnforth, pp. 199-235.

Jodha, N.S. (1992) : Sustainability of Himalayan Forests: Some Perspectives In A. Agarwal (ed.) The Price of Forests. Centre for Science and Environment, New Delhi, pp. 285-290.

Langenegger, H. (1984) : Mountain Forests: Dynamics and Stability In E.A. Brugger, et al. (eds.) The Transformation of Swiss Mountain Regions. Haupt, Berne, pp. 361-372.

Maser, C. (1994) : Sustainable Forestry: Philosophy, Science and Economics. St. Lucie Press, Delray Beach.

Mujica, E. and J.L. Rueda (eds.) (1996) : El Desarollo Sostenible de Montañas en América Latina. CONDESAN/CIP.FAO, Lima.

Nair, P.K.R. (1993) : An Introduction to Agroforestry. Kluwer, Dordrecht.

Price, M.F. (1990) : Mountain Forests as Common-property Resources: Management Policies and their Outcomes in the Colorado Rockies and the Swiss Alps. Forstwissenchaftliche Beiträge 9, ETH Zürich.

Price, M.F. (1992) : Patterns of the Development of Tourism in Mountain Environments. GeoJournal, Vol. 27, pp. 87-96.

Price, M.F. (1998) : Chapter 13 in Action 1992-97 – A Task Manager's Report. FAO, Rome.

Price, M.F., L.A.G. Moss and P.W. Williams (1997) : Tourism and Amenity Migration In J.D. Ives and B. Messerli (eds.) Mountains of the World: A Global Priority. Parthenon, Carnforth, pp. 249-280.

Price, M.F. and Thompson, M. (1997) : The Complex Life: Human Land Uses in Mountain Ecosystems. Global Ecology and Biogeography Letters, Vol. 6, pp. 77-90.

Richards, R.T. and Creasy, M. (1996) : Ethnic Diversity, Resources Values, and Ecosystem Management: Matsutake Mushroom Harvesting in the Klamath Bioregion. Society and Natural Resources, Vol. 6, pp. 359-374.

Sène, E.H. and D. McGuire (1997) : Sustainable Mountain Development - Chapter 13 in Action In J.D. Ives and B. Messerli (eds.) Mountains of the World: A Global Priority. Parthenon, Carnforth, pp. 447-453.

Stone, P.B. (ed.) (1992) : The State of the World's Mountains: A Global Report. Zed Books, London.

Taylor, D.A. (1996) : Income Generation from Non-wood Forest Products in Upland Conservation. FAO Conservation Guide 30, FAO, Rome.

Turner, B.L. et al. (1990) : Two Types of Global Environmental Change: Definitional and Spatial-scale Issues in their Human Dimensions. Global Environmental Change, Vol. 1, pp. 14-22.

World Bank Environment Department (1996) : World Bank Environmental Projects, July 1986 - July 1996. World Bank, Washington DC.

Tactical Sediment Yield Control to Mitigating Terrestrial and/or Aquatic Environmental Impacts of Forest Road Construction

Masami SHIBA

University Forest, Faculty of Agriculture, Kyoto University, Kyoto, Japan

ABSTRACT   Many studies have been identified forest roads as major non-point source impact contributors of sediment resulting from forest activities in mountain environment.   Thus, sediment deposits on slopes below roads make an important environmental issue of management approaches to riparian zones because of the numerous ecological linkages between terrestrial and aquatic ecosystems.   This paper demonstrates a PC-based imagery data processing system using digital orthophoto through the application to estimate the granitic sediment deposits eroded from roads constructed on watersheds in the mountains of Shiga Prefecture. The results showed the same time trends in sediment yield as earlier studies that documented large areas of sediment deposits during the earlier years after road construction followed by a rapid reduction in deposits in subsequent years.   Also, a relationship was developed to describe the area percent of the sediment accumulated on the slope in relation to the travel distance.
Key words: forest road, sediment, environmental impact, mitigation, imagery data processing system

1. INTRODUCTION

Traditional forest practices in mountainous regions in Japan were based on principles of simplification and homogenization of the forest resources at the tree, stand, and landscape levels in order to achieved economical timber production. Consequently, foresters developed a management paradigm for Japanese cedar (*Cryptomeria japonica*) and cypress (*Chamaecyparis obtusa*) in the 1960s based on clearcutting and planting of conifer monocultures.   These approaches also reflected societal emphasis on timber production in the 1960s.

As public forest land use pressure increases, circumstances have remarkably changed.   As partly outlined in the above-mentioned, our understanding of forest ecosystem and landscapes has improved dramatically.   We are learning that what enhances efficient production and harvesting of timber in the short term does not always enhance other forest values and long-term productivity.   Society has become concerned with many other forest values, including the maintenance of biological diversity.   With expanded knowledge and objectives, the goal becomes one of using ecological knowledge to create alternative forest management systems.   Management approaches must be developed which better integrate the maintenance of ecological values with production of commodities

Road development is one of the most environmentally damaging activities in forested areas and often leads to serious and unnecessary degradation of natural resources. In seep mountainous regions, road construction and the sequent maintenance practices cause greater soil disturbance. Where roads are located near streams, excavated material that is cast aside on steep hillsides may reach the stream. Redistribution of drainage water and deposition of soil on slopes can cause soil instability and accelerated soil erosion. Most of mid-slope roads which traverse steep slopes and cross tributary streams cause the erosion and deterioration of water quality. Thus, the sediments on slopes below roads make an important environmental issue of management approaches to riparian zones because of the numerous ecological linkages between terrestrial and aquatic ecosystems (Brown et al. 1971, Cline et al. 1983, Franklin 1992, Fredriksen 1970, LaFayette et al. 1992).

This paper presents the results of a case study from sensitive sites in granitic watersheds and provides information necessary to develop road design criteria and evaluate risks and trade-offs. Objectives of the study were to:

· Quantify the dispersion and aggregation of sediment deposits;
· Evaluate the distribution of sediment deposits in relation to travel distance;
· Investigate time trends in sediment deposits;
· Demonstrate a PC-based imagery data processing system through the application to sediment deposit measure.

2. DESCRIPTION OF THE STUDY AREA

2.1 Overall site characteristics

The study site is in the headwater of the Tanakami drainage (276 ha), a tributary to the Ukawa river flowing into Lake Biwa (673 km^2) in Shiga Prefecture (Figure 1). The study was conducted on roads constructed across three study watersheds within the Tanakami drainage area. Elevations range from 249 to 731 m (average: 460 m) and slopes are generally moderately steep, averaging approximately 35 %. Bedrock in the area, characteristic of middle elevation in this section of the western part of Shiga, is primarily coarse-grained quartz monzonite and moderate-to-well weathered. Soil are weakly developed with a horizons ranging from 5 to 30 cm thick overlaying moderately weathered granitic parent material. Soil textures are loamy sands to sandy loams. Annual precipitation averages about 2,000 mm with most of the precipitation occurring during the rainy seasons. Streamflows from study area are mainly dominated by a long spell of heavy rain occurring during June and July, and also a springtime melt of the snowpack that accumulates from about mid-December to mid-April. Stand composition is mainly second-growth Japanese red pine (*Pinus densiflora*) accompanied by substantial amount of white oak (*Quercus serrata*) in a variety of age classes.

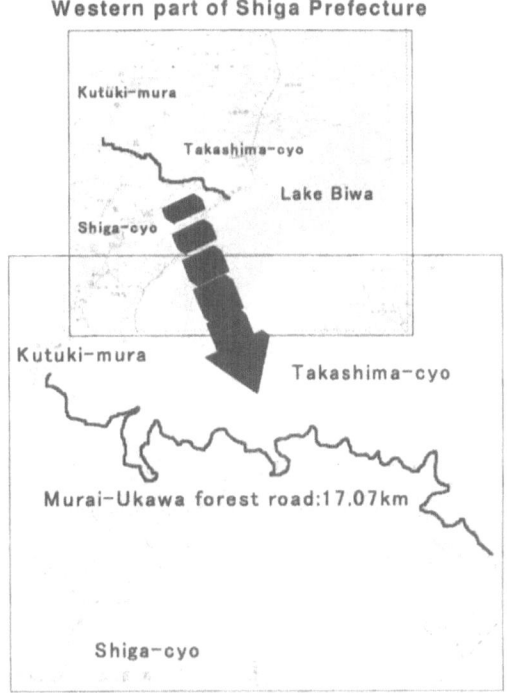

Figure 1. Schematic map of the Tanakami study area in Shiga Prefecture.

2.2 Road construction

A total of 17.07 km of forest roads was constructed in this area. Construction began in 1973 and was completed in 1995. About 3.5 km of roads run through the study sites were built between 1980 and 1995. Roads in this area were design for year-round and/or partially seasonal access to support rural traffic facilities, as well as forest management activities. The design 5 m wide running surface was widened for curves, ditches, turnouts, and shoulders to much as 7 m. Approximately 14 km of the main road section improved in subsequent year, were design for maximum erosion control and used; partial asphalt pavement on the running surface and crushed rock surfacing on the unpaved remainder; grass seed, plastic netting, and transplanted shrubs and trees on fillslopes; grass seed and hydromulch on cut slopes. While road designs varied on study sites. Road design features were representative of typical local road design features for steep slope area and included; sharper curvatures, rolling grades that minimize excavation, a native material road surface, native material berms to protect fillslopes from direct runoff on outsloped sections, and small rock placed at culvert outlets. An about 1.6 km of these dirt road sections was selected to assess the effect of sediment deposits along the area of construction (Figure 2).

54

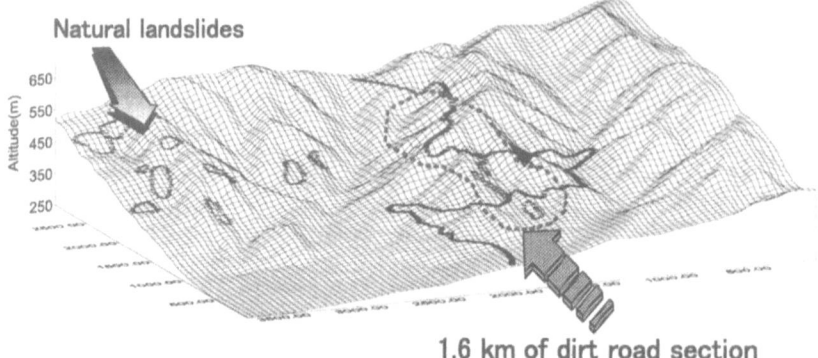

1.6 km of dirt road section

Figure 2. A 1:5,000 based-scale DEM of a portion of the study sites showing topographic characteristics, road location and study site boundaries.

2.3 Three study sites

Ground surveys and as well as aerial photo interpretation of all sediment deposits on slopes below the roads constructed on the study sites were made in 1996 and 1997. Consequently, 520 m of road segment on each site were selected to measure the sediment deposits originated from road fillslopes. According to an overall site characteristic and construction dates, three sites were named as follows:

- Site A: a relatively small amount of sediment deposit, elevation range from 275 to 600 m and 28.6 ha in size, hillslope range from 11 to 31 degrees, 2 stream channels in proximity to the site, and road construction in 1986;
- Site B: higher sediment deposits during the earlier years following road construction in 1987, elevation range form 300 to 663 m and 50.6 ha in size, hillslope range from 22 to 39 degrees, and 1 relatively large stream;
- Site C: a relatively constant of higher sediment yield following construction in 1982/83, elevation range from 300 to 600 m and 33.2 ha in size, hillslope range from 11 to 31 degrees, and 3 small stream channels (Figure 3).

Figure 3. Photomosaic representation of overall site situations including natural landslide location, buffer zones (corridors), road segment boundaries and settlement area, respectively.

3. DATA SOURCES AND ANALYSIS

3.1 Aerial photo as source data

Aerial photo has developed into a standardized, technically mature tool for photo interpretation, reconnaissance and inventory. Traditional procedures of photogrammetric data gathering characterized by interdisciplinary methods, however, may be insufficient to address all of the today's environmental protection issues concerned with the road planning phase, which claims details, site-specific, carefully timed quantitative and qualitative information. Aerial photo as a type of analogous memory media may not represent meaningful information; it is comparatively difficult to discern the specified anomalies or site-specific target conditions in relation to the complex interactions between different landscape components of ecological and hydrological characteristics on photo image. New scanning sensors and automated imagery data processing software to identify site investigation data in a detailed spatial context have recently become affordable.

Aerial photo can be scanned electronically and the data converted from analog to digital. In the digital form, aerial photo can be analyzed in the same manner as digital satellite images. A digital image is a matrix of rows and columns whose smallest units a pixcel. The pixcel is defined by the resolution of the scanner and corresponds spatially to a specific area on the ground. Each pixcel has a digital number which represents some portion of the electromagnetic spectrum and is the radiance value emanating from the ground at a particular point. Since these images are digital, they can be subjected to a variety of computer operations. These operations emphasize certain characteristics of the image that aid in the classification of objects on the basis of their multispectral radiance values (Burrough 1986, Falkner 1994).

3.2 Image analysis with a digital orthophoto

Photo-based data sources were used for this research: photo interpretation data, sample point interpretation (pixcel count), and digital imagery classifications. The photo interpretation data were used to demonstrate the spatial accuracy of the digital imagery data, while the pixcel count data for sediment areas and travel distance measures. The digital imagery interpretation and pixcel count estimates were derived from the same orthophotos, which were rectified aerial photos used as the photo interpretation base for the inventory.

Both black and white panchromatic (BW) and color aerial photos of a standard 9 x 9 inch format taken during the spring between 1968 (pre-road construction) and 1995, and these enlarged orthophotos (50 x 75 cm; scale 1: 5,000) were utilized to obtain the required information, such as sediment deposit sites associated with road construction, and planimetric features represented in their geographic positions. The interval analysis of digital orthophotos between four different dates (1968, 1985, 1990 and 1995) offers the identification of landslides and sediment deposits related to natural erosion process which were already present in earlier term prior to road constructions, and provides a comparison of the pre- and post-road construction conditions.

A orthophoto image was digitized by a series of 8-bit binary image data records; original orthophoto was scanned with 200 DPI, therefore a pixcelsize corresponds to approximately 1.27 m on the ground. The radiometric image brightness values were stored as 256 RGB levels. Digital classification for the coverage of the study sites was carried out using Adobe Photoshop 4.0j, an image processing software. Consequently, it was easy to distinguished nonforested from forested land. With these groups, further color differences considered in association with textural differences permitted a very accurate delineation of nonforested land types, such as scattered natural landslides, unstable rocky slopes, scoured stream channels, and sediment deposits from specific components of the road prism: traveled way, cutslopes and fillslopes. Linear and area measures of a defined portion on digital photo image can be performed by counting the number of composite pixcel. This method was applied to measure a sediment deposit area and travel distance from road corridor. Also orthophoto-based measures with a digitizer were carried out for 'photo truthing' Consequently, the measurement results indicated that the measurement accuracy of both methods is almost of same order. Based upon a mathematical relationship between distribution and travel distance of sediment deposits (Burroughs et all. 1989, King 1989, Megaham et al. 1991, 1996), virtual buffer zones (corridors) that generally defined the spatial proximity of forests to road and major streams were generated at 10m intervals on each side of roads. These zones were automatically compiled by the computer and the area percent and travel distance of sediment deposits within each zone were estimated by pixcel count method.

4. RESULTS

4.1 Dispersion and aggregation of sediment deposits

Most studies that have measured sediment yield from fillslopes over time show that, initially, rates in this unconsolidated material are high and exponentially decrease over time (Fredriksen 1970, Megahan and Kidd 1972, Megahan 1974). Table 1 shows the area percent of sediment deposits within 50 m and 100 m corridors at each observed year in three study sites. As was typical for all sites, most of the sediments was deposited the earlier period with significant increases in area, but generally subsequent years. Data from Site A, representing a relatively small sediment area within each corridor, suggest that all sediments originated from road fillslopes are smaller than from any other site and sediment deposits gradually decrease in area as they move downslope. A reduction rate of sediment area between Site B and C, compared to Site A, ranged from 34 to 53 % within 50 m corridor and 42 to 52 % within 100 m corridor, respectively. The results indicate that fillslope erosion to be considered the primary source of sediments from road components continued with a relatively constant of higher sediment yield, and that during the earlier years following road construction, erosion control measures would be more effective to reducing sediment yield.

Table 1. Area percent of sediments deposited within 50 m and 100 m corridors on fillslope side of road segments in three study sites.

	50 m zone			100 m zone		
	Site A	Site B	Site C	Site A	Site B	Site C
1968	0.00	0.00	0.00	0.00	0.00	0.00
1985	0.00	0.00	17.81	0.00	0.00	11.30
1990	6.71	19.30	11.76	3.91	11.44	6.81
1995	2.40	9.78	6.77	1.36	5.57	3.92

58

Although the initial rate of fillslope erosion can be high compared to erosion rates on other road components as running surface, cutslopes and ditch, it is the distribution and travel distance of eroded material below the fillslopes that determine the degree that streams and its tributaries are affected by fillslope erosion. The slope distance required to prevent material from reaching a streamside is a function of many interacting site factors, making it difficult to predict with any degree of accuracy. However, both the cumulative area percent and travel distances of sediments below fillslopes provide some insight into relationships between spatial distribution and site characteristics. The cumulative area percent of sediment deposits for study sites are shown in Figure 4.

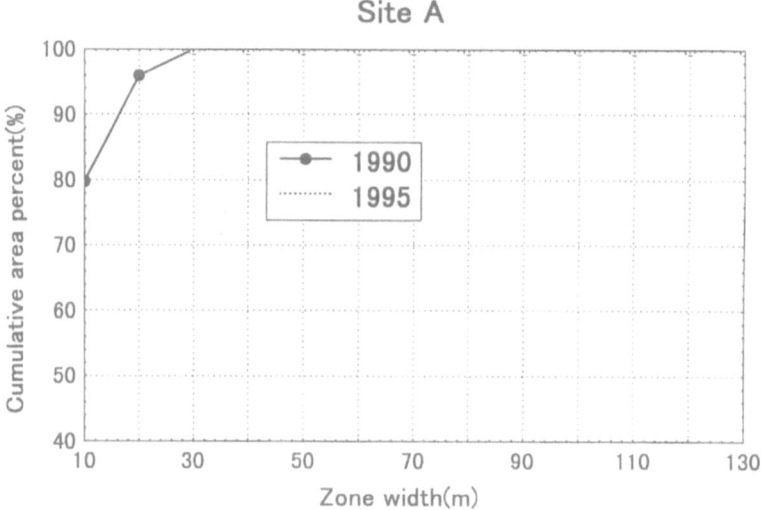

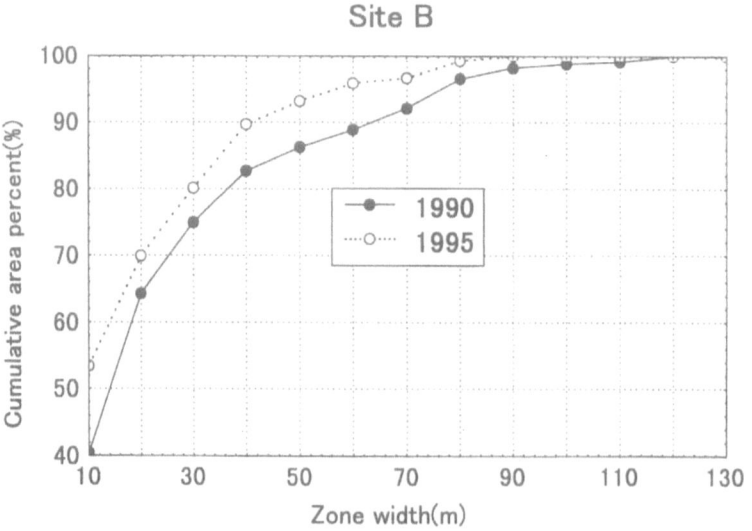

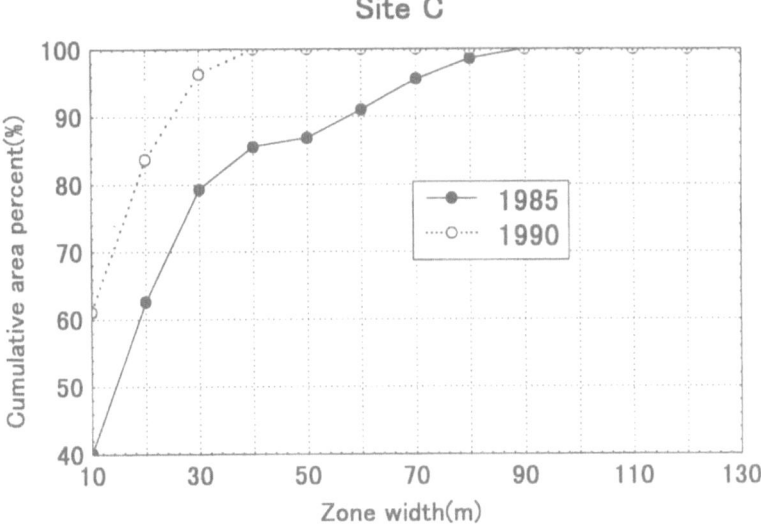

Figure 4. Cumulative area percent of sediment deposits below fillslopes.

Although the range of distribution remains similar, the time delay following road construction shifts the curve toward the shorter distances. The importance of the timing in reducing fillslope erosion for any treatment is apparent in the similar shape of the curve through the study sites. For Site B and C at early stages following construction, an about 90 % area cover would require a buffer zone of at least 60 m. The maximum distance was 120 m for Site B and 90 m for Site C, respectively. In contrast, Site A in 1990 resulted in relatively short distances to about 40 m and rapid reduced the distribution; only 4 % of the sediment area had distances of exceeding 20 m and all of the deposits dropped into 10 m zone in 1995. This relationship can be mathematically expressed but probably varies substantially from place to place. Because the scarcity of this type of data, however the information could be used to estimate leave strip width (buffer zone) below roads on sites similar to those in Tanakami drainage; management of both road corridor and streamsides.

Figure 5 illustrates the relationship between the frequency of sediment flow paths and travel distances. Of the 9 sediment flow paths that formed on the fillslopes in Site A in 1990, only 2 with few changes in length remained in 1995. Although 5 flow paths in Site B were quite same in number after 5 years, the path location and travel distances varied considerably. The maximum travel distance was about 140 m in 1990 and 90m in 1995, considerably greater than data from Site A and C. 5 flow paths were counted in Site C in 1985, and 3 and 2 flow paths joined other paths remained in 1990 and 1995, respectively. These data provide estimates of distance required between fillslopes and streams to minimizing transport of fillslope-derived sediment to the streams. However, the mean travel distances is not useful because the population of travel distances is skewed to extreme values. The cumulative area percent of sediment deposits combined with the maximum travel distance is more useful for mitigation planning.

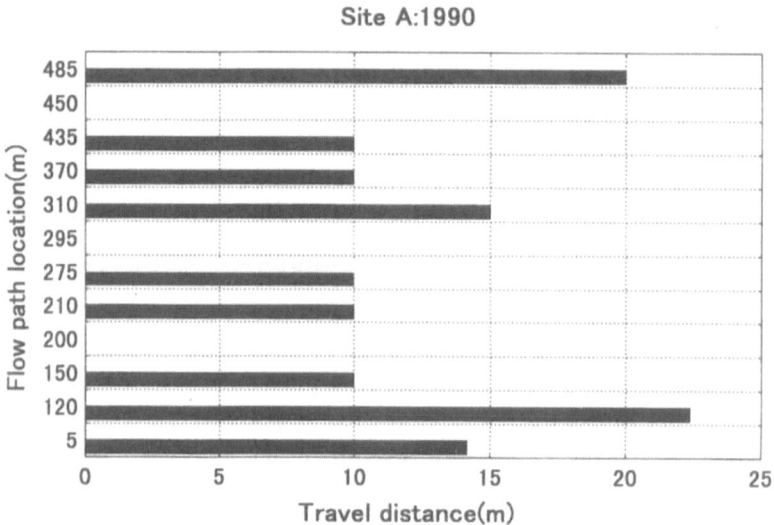

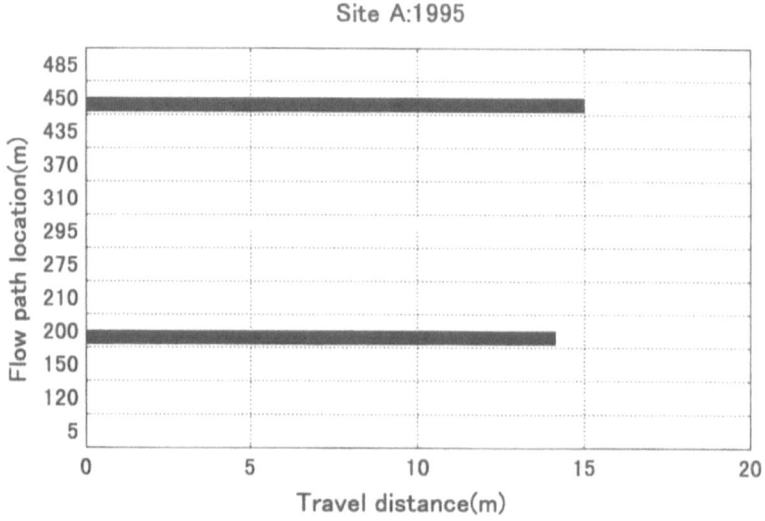

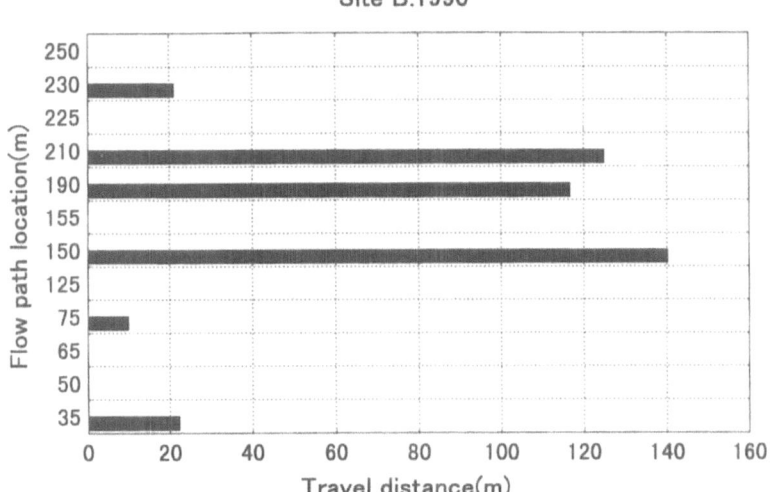

Site B:1990

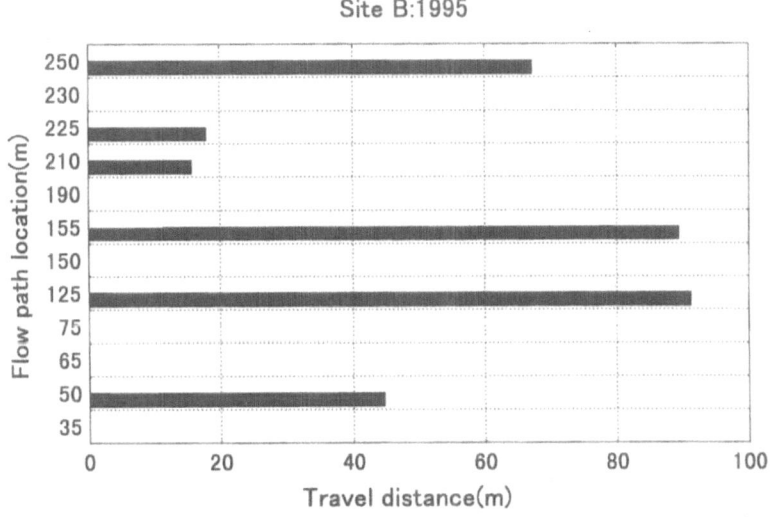

Site B:1995

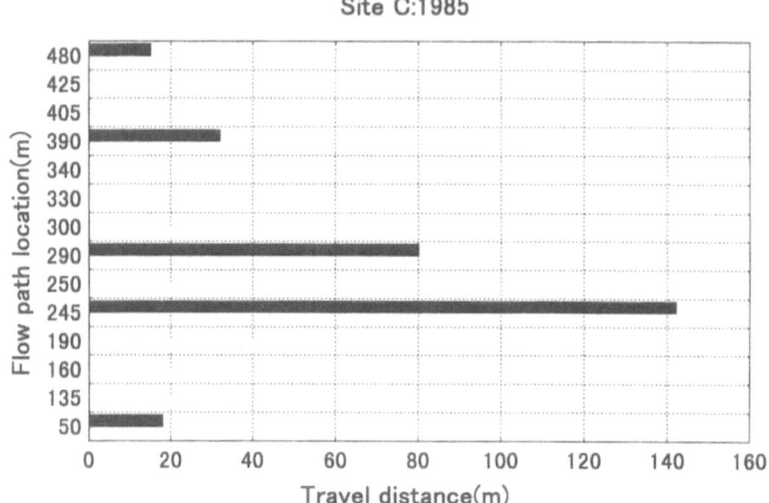

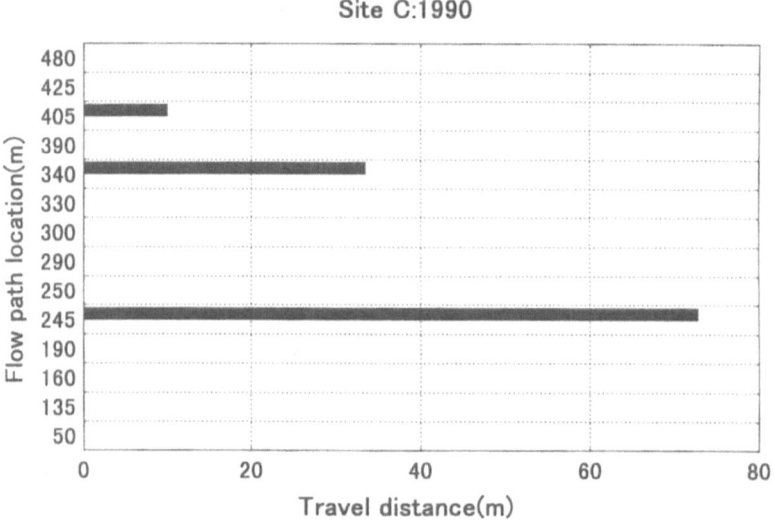

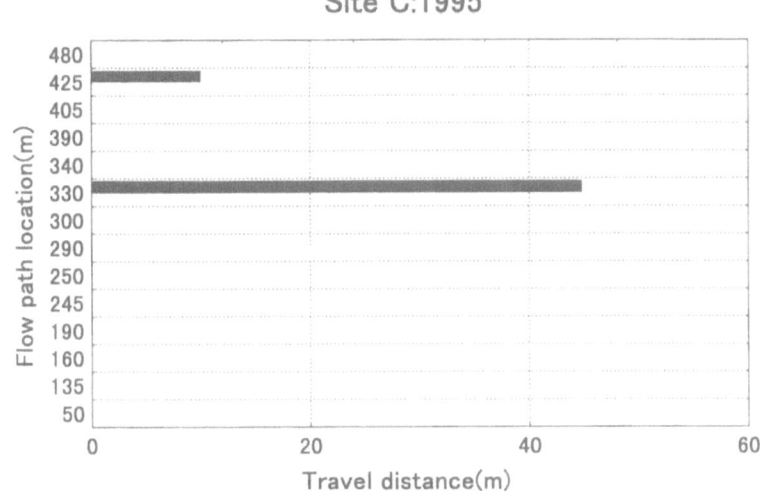

Figure 5. Relationship between the frequency of sediment flow paths and travel distances.

4.2 Sediments supplied to streams and its tributaries

Additional measurements were made of sediment deposits supplied to the streams at points where sediments extend far enough downslope to reach active streams. The cumulative area percent of stream channel aggregation as the result of sediment storage within the main stream and its tributaries for the study sites are shown in Figure 6.

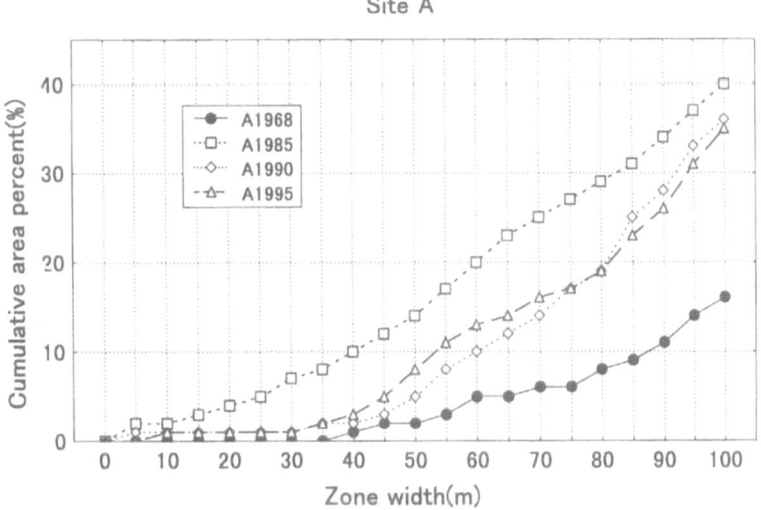

Site B

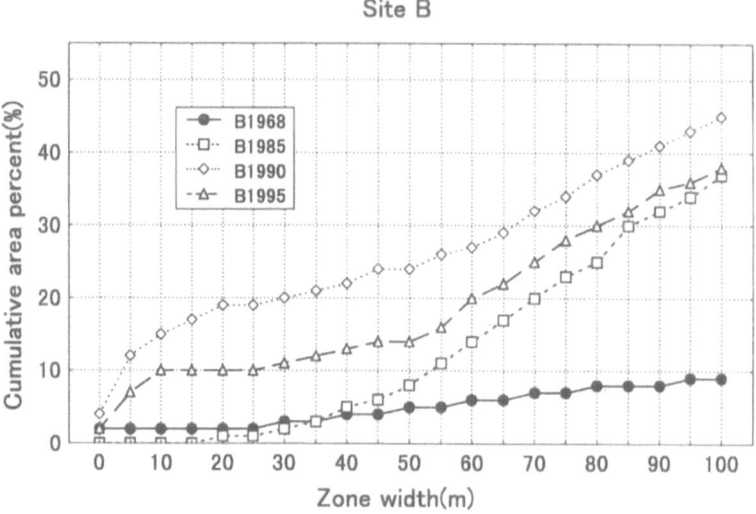

Site C

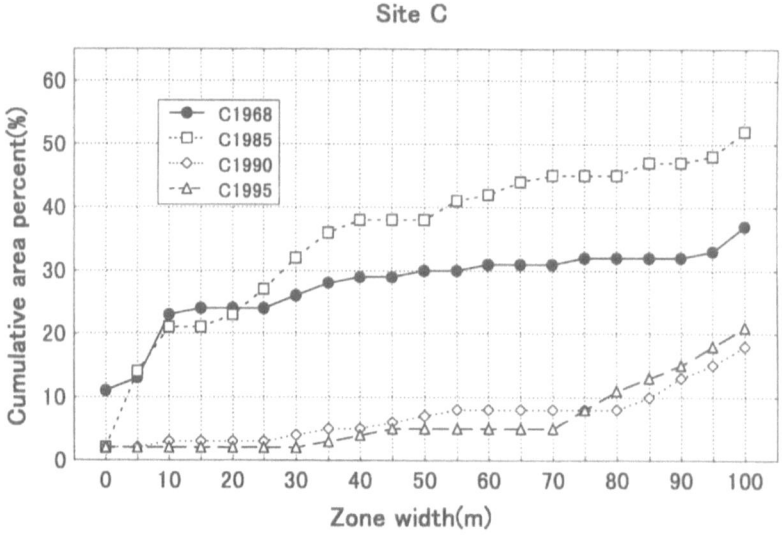

Figure 6. Cumulative area percent of stream channel aggregation as the result of sediment storage within the streams and its tributaries.

The variation in sediment storage is quite small at Site A as indicated by changing in channel bed area. This is because sediment supplies from fillslopes were lower after road construction. In contrast, for Site B and C, effect of the accelerated sediment yields created sediment greatly in excess of the transport capabilities of the stream. The result was widespread channel aggregation as shown significant increases in area percent in figure 6. One of the reasons for greater sediment storage within channel bed is the longer travel distances exceeding about 60m; the majority of the sediments reached the stream via fillslope side of the road. These results are an oversimplification because the travel distance must be influenced by many interacting site and climate factors. However, these data provide a good example of recognizing and becoming aware of new perspectives for alternative management regimes for forests and associated streams- that is, ecological linkages between terrestrial and aquatic ecosystems.

5. CONCLUSION

Many studies have identified forest roads as major non-point source impact contributors of sediment resulting from forest activities in mountain environment. This paper presents the results of a case study from sensitive sites in granitic watershed and provides basic information necessary to develop road design criteria and evaluate risks and trade-offs. Distribution of sediment travel distance from road fillslopes and cross drains provide a measure of the risk of total sediment travel distance by road source condition. Relationship between travel distance and sediment area allows field personnel to estimate the risk of sediment travel below the road to a stream, and how much sediment reaches the stream. If the risk of sediment dispersion is too high, alternative road design or location can then be planned or provision can be made to catch additional sediment below the road.

The present method of estimating the sediments and site conditions by photogrammetric manner, as well as field surveys is expensive and time consuming. Reliable estimates of sediment deposits from a PC-based imagery data processing system introduced in this paper (Shiba et al. 1992, 1996, 1998), would eliminate the need for expensive visit to each permanent and intensification plots. While this research demonstrates that the capabilities of aerial photo and computerized imagery data processing systems are not basically competitive but compensative. Consequently, they are accountable for the accuracy of their interpretation, and correct and upgrade their results with field work and on-site verification.

REFERENCES

Brown, G.W. and J.K. Krygier (1971): Clearcut logging and sediment production in the Oregon Coast Range, Water Resource Res., No.7, pp.489-498.

Burroughs, E.R. and J.G.King (1989): Reduction of Soil Erosion on Forest Roads, USDA Forest Serv., Res. Gen. Tec. Rep. INT-264, pp.1-21.

Burrough, P.A. (1986): Principles of Geographical Information Systems for Land Resources Assessment, Oxford Univ. Press Inc., New York, 194p.

Cline, L.D., Short, R.A., Ward, J.V., Carlson, C.A. and L.G. Howard (1983):

Effects of Highway Construction on Water Quality and Biota in an Adjacent Colorado Mountain Stream, USDA Forest Serv. Res. Note, RM-429, pp.1-11.

Falkner, E. (1994): Aerial Mapping, CRC Press Inc., New York, 322p.

Franklin, J.F. (1992): Scientific basis for New Perspective in Forest and Streams, pp.25-72 in J.N. Robert, editor, Watershed Management, Springer-Verlag, New York, 542p.

Fredriksen, R.L. (1970): Erosion and sedimentation following road construction and timber harvest on unstable soils in three small western Oregon watersheds, USDA Forest Serv. Res. Pap. PNW-104, pp.1-15.

King, J.G. (1989): Responses to Road Building and Harvesting: a Comparison With the Equivalent Clearcut Area Procedure, USDA Forest Serv. Res. Pap. INT-401, pp.1-13.

LaFayette, R.A., Pruitt, J.R. and W.D. Zeedyk (1992): Riparian Area Enhancement Through Road Design and Maintenance, USDA Forest Ser., Res. Gen. Tec. Rep. RM-GTR-279, pp.85-95.

Magahan, W.F. (1974): Erosion over time on severely disturbed granitic soils, USDA Forest Serv. Res. Pap. INT-156, pp.1-14

Magahan, W.F. and G.L. Ketcheson (1996): Predicting downslope travel of granitic sediments from forest roads in Idaho, Water Resource Res. Bull. No.2, pp.371-382.

Magahan, W.F. and W.J. Kidd (1972): Effect of logging roads on sediment production rates in the Idaho Batholith, USDA Forest Serv. Res. Pap. INT-123, pp.1-14.

Magahan, W.F., Monson, S.B. and M.D. Wilson (1991): Probability of sediment yields from surface erosion on granitic roadfills in Idaho, J. Environment Quality, No.20, pp.53-60.

Shiba, M. (1996): Imagery Data Processing Systems Using Aerial Photography for Sensitive Site Investigation in the Route Selection Process, J. Forest Engineering, Vol.7, No.3, pp.53-65.

Shiba, M. (1998): Quantitative assessment of forest landscape resources as an integral component of rural landscape modification, Res. Pro. Grant-in-Aid for Scientific Research: No.08660185, pp.1-87 (in Japanese with English abstracts).

Shiba, M. and H. Loeffler (1992): Anwndungsmoeglichkeiten von Nutzwertanalyse fuer Variantenvergleich bei Walderschliessungsplanung, Bull. Bio. Mie Univ., No.7, pp.1-20 (in Japanese with German abstract).

The Environment in Planning a Forest Road Network

Igor POTOCNIK

Biotechnical Faculty, Department of Forestry and Renewable Forest Sources
University of Ljubljana, Slovenia

ABSTRACT Slovenia is a small European country with varied natural characteristics. 30% of Slovenia's total surface has a slate, tertiary or flysch basis and approximately 19% of the total surface is threatened by soil erosion. When planning the forest road network, three main groups of factors are studied: natural conditions, state of forests and existing traffic communications. It is essential to find an equilibrium between the forests (nature), economic with organisation and technology. Inside the environmental impact at planning forest road network the role of topography, relief, geological conditions, climatic conditions and water conditions is described. It is concluded that the knowledge and consideration of the forest's natural laws and of the environmental factors is crucial for the planning of the forest road network.
Key words: forest, forest road, environment, planning

1. INTRODUCTION

Slovenia is a small European country (with a total surface area of 2 million ha, 2 million inhabitants, 52% of territory covered with forest). Nature has endowed Slovenia with varied characteristics, from the Adriatic coast through the Dinaric and Alpine world to the Subpannonian valley. It is therefore essential to have a sound knowledge of the soil's natural characteristics and limitations when planning and constructing forest roads. Slovenia's characteristic shapes and forms are largely the consequence of agitated tectonic and geological conditions, as well as the climatic and vegetational situation combined with the soil's stability and erosiveness. As much as 30% of Slovenia's total surface has a slate, tertiary or flysch basis, which is unstable and has a high potential for slides. Moreover, approximately 19% of the total surface is threatened by soil erosion (Anonym 1970). The average incline is approximately 40%. When planning and constructing the forest road network, the natural conditions were also respected and considered in the past (cognitive approach). All this has an impact on the planning of a forest road network which must respect both the natural conditions and meet the economic and technical requirements.

1.1. Legislation

1.1.1. Forest Act (1993)

The Forest Act stipulates that, considering the technical, economic and ecological conditions, the forest infrastructure must be designed, built and maintained in a way that would affect the forest soil, vegetation and fauna to the least possible extent. When planning forest roads, special emphasis must be given to their adaptation to the natural environment. As early as the planning stage, and later during the construction, maintenance and exploitation of the forest road system, it has to be taken into account that forest roads must not:
− endanger water sources;
− cause erosion processes;
− prevent high waters (floods) from torrents from flowing off;
− increase the danger of avalanches and landslides;
− destroy the equilibrium in unstable ground;
− affect the flowing off of rain water in a way that would jeopardise the forest and other surfaces;
− affect areas which are important for the preservation of wild animals;
− affect the natural or cultural heritage;
− jeopardise the forest's other functions or its manifold exploitation/utilisation.

1.1.2 Environmental Protection Act (1993)

For the planning of the forest road network, it is most important to first assess the environmental impact. This will show whether individual anticipated interventions in the environment are acceptable in view of their short- and long-term, as well as direct and indirect, consequences. The natural environmental conditions should be preserved to the greatest possible extent considering the required values of environmental protection. This involves intervention which could markedly influence the environment, such as forests exploitation. An assessment of the environmental impact of the planned forest road network comprises, in particular, an assessment of the total and integral potential for environmental hazards. Moreover, the Ministry of Environment must give its consent for any planned interventions into the forest concerning the forest road network.

The legal provisions covering interventions into the forests are sufficiently restrictive, and define environmental protection with adequate precision. Nevertheless, the practical control and inspection of whether works are performed in line with the plans shows some defaults which must be dealt with by the competent governmental body and concern the organisation of inspection services.

1.1.3 Planning of Forest Road Network

When planning the forest road network, three main groups of factors are studied (Anonym 1982):
– Natural Conditions - covering general information on the area being opened up, the economic and environmental situation, etc.
– State of Forests - covering information on the state of forest stocks, the forests' state of health, the forests' biological composition, ownershipand the dispersion of forest properties, etc.
– Existing traffic communications: we are interested in information on how well the forests are opened up with public traffic lines (train, public roads), and main links (primary and secondary) through forests.

When planning the forest road network, it is essential to find an equilibrium between the forests (nature), economic organisation and technology. Different interests must operate in a balanced dynamism through time and space in order to guarantee a sustainable and stable economic situation in the forests. Moreover, public interests and the system of the existing integral traffic system has an impact on the forests (Figure 1). The forest warden's or forest road network planner's duty is to harmonise the influential factors in time.

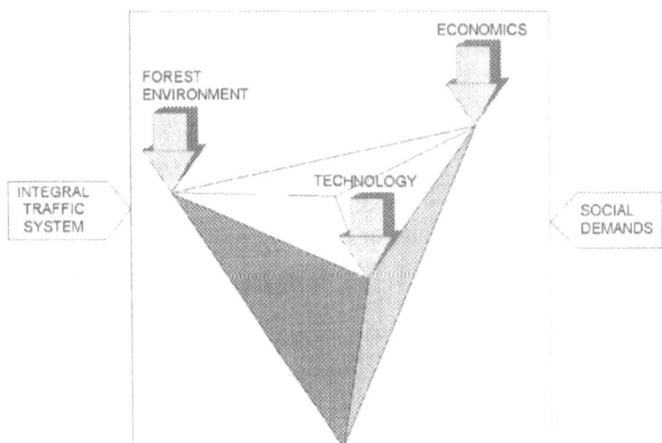

Figure 1. Dynamic equilibrium of planned forest road network

The process of planning the forest road network comprises several steps (Figure 2). The initial state gives rise to a need followed by the prospective construction plan for a forest road network. The next phase is collecting the necessary data concerning important natural conditions and limitations. On such a basis, alternative solutions for a forest road network are worked out. The key to selecting the most suitable variant is applying a multi-criteria assessment of the variants. Ultimately in a last step, the forest road network is realised. The feedback during the planning stage supplies corrective measures for enhancing the forest road network, in particular, from the perspective of the impact on the forest's stability and on the water system, as well as of the integration of the road network into the forest, and of its extension.

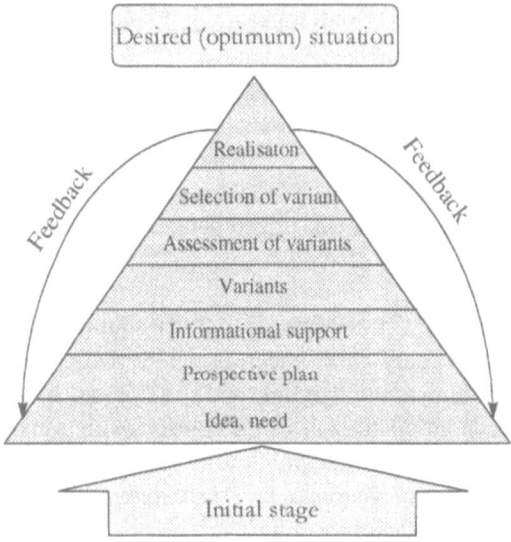

Figure 2. Planning process for forest road network

There are three possibilities of integrating the forest road network into the environment: appropriate planning of route (ground plan and elevation), designing embankments (of removed earth), and biotechnical renovation measures of the intervention in the forest. It is essential to treat the forest road network in three dimensions (space) while taking account of the forest's time component. If these principles are considered, the forest road network will represent the least possible disturbance to the forest ecosystem.

2. ENVIRONMENTAL IMPACT AT PLANNING FOREST ROAD NETWORK

2.1 Topography

We are dealing here with a description and cartographic view of the area which is the subject of development (Anonym 1970). We use classical maps and aerial photographs which help determine the current state of the area, and, particularly, help define the different uses of the forest area. Planning the forest road network is an integral process, and tries to consider not only the forests' economic exploitation, but also its social and environmental value.

2.2. Relief

The macro-relief which structures the forest surface being opened up with mountain crests, valleys and large ditches is most important as it indicates the macro-route of the future forest road. This natural situation implies individual gravitational areas for the future forest roads. They need to be adapted to the natural course of the micro-relief to the greatest possible extent. We are confronted here with types of level and undulatory micro-relief with ditches and sinkholes. Each condition has a different impact on the micro-course of the future forest road.

2.3. Geological conditions

In view of the construction of the forest road network (technological choice, utility of material for the construction, dimension and type of drainage system) and the environment's potential fragility it is important to know the geological conditions. As a rule, on less stable and softer geological grounds there are problems with ground and rain water; if these influences are not considered, this may create a centre of erosion. Moreover, the geological ground has an impact on the incline chosen for the embankments. On soft ground, the embankments are more gentle (1/1 to 1.5/1). However, this increases the width of the roadway, and, in turn, the width of the belt of felled trees (Table 1). Likewise, humid locations and unstable slopes above the embankment necessitate a larger width of the belt of felled trees for the forest road.

Table 1. Average width of felled forest belt

Incline of ground (in %)	20	30	40	50	60	> 60
Soft ground	8 m	9 m	10 m	11 m	13 m	-
Hard ground	-	7 m	7 m	8 m	9 m	11 m

In stands of conifers (spruces, firs), a 5 m - wide cut through the forest may be considered as not having a marked impact on the stand. In stands of deciduous trees (beech trees), such a cut may have a width of 8 m. In beech tree forests, a period of 20 years after the construction of a forest road is usually sufficient for the border trees to totally cover the road profile with their crowns (Potocnik 1997).

For this reason, we plan a roadway width of 2.8 to 3.0 m for the most problematic ground, which is sufficient for normal lorry traffic. Lengthways, the road incline must not be in excess of 8 % on soft surfaces. Special attention should be paid to the length- and crosswise drainage, as well as the thickness and structure of the upper roadway layer (thick supporting and wear layer, and extra impervious layer).

2.4. Climatic Conditions

Important data for the planning of the forest road network is provided by the yearly quantity of precipitation, the distribution of precipitation over the year, as well as the average minimum and maximum temperatures. In the Alpine and Dinaric regions, the distribution of precipitation over the year is crucial as most rainwater falls in showers in early summer and autumn. Showers have a very unfavourable impact on the state and passableness of forest roads. Therefore we must take into account the natural situation when designing the drainage, the upper roadway structure, as well as traffic regulations (limitation of axle weight for when the road is wet, etc.). The situation also implies regular maintenance due to water washing away the foundation (the roadway may thus become unstable) - particularly in winter.

The temperature extremes influence the erosion of the embankments and the necessary depth for the construction of foundations. Here, exposed locations must be considered with special care: southern locations are exposed to greater changes in temperature than northern locations, but the latter face more time with frost, snow and (black) ice.

2.5 Water Conditions

Water conditions depend on the geological basis. In the area of development, the water balance of inshore waters must be considered (torrents are characteristic here), and the problem of ground water in embankments tackled. Moreover, the right dimension for bridges or dips must be found. When planning the forest road network, special care must be taken in the surroundings of fresh water reservoirs.

3. CONCLUSION

The knowledge of natural conditions, of the technical possibilities, and of economic forestry is a prerogative for a successful plan for the construction and operation of the forest road network. True enough, a forest road is an intervention in nature - but economic forests cannot be efficient otherwise. This is why it is crucial at the planning stage to anticipate that the forest road will become part of the forest and its culture. The forest is such a complex natural formation that external influences can affect its balance. Each artificial infrastructure in the forest is an additional burden and causes instability. The knowledge and consideration of the forest's natural laws and of the environmental factors is crucial for the planning of the forest road network. Only in this way the forest can preserve its economic interest, and maintain all its environmental and social functions. For a successful completion, this demanding work requires forestry professionals with a broad educational horizon.

4. REFERENCES

Potocnik, I. (1997): Filling-in the Clearance of a Forest Road Cross-section. International scientific Conference Forest - Wood - Environment '97, Zvolen, Slovakia, pp. 68 - 76.

Anonym. (1970): Ceste in krajina (Roads and Landscape), Zbornik seminarja, Ljubljana, 171p.

Anonym. (1982): Smernice za projektiranje gozdnih cest (Guidelines for Forest Road Network Planning), Institut za gozdno in lesno gospodarstvo, Ljubljana, 63p.

Anonym. (1993): Forest Act, Official Gazette of the Republic of Slovenia, No. 30/93

Anonym. (1993): Environment Protection Act, Official Gazette of the Republic of Slovenia, No. 32/93, 1/96

Forest Disturbance and Efficiency of Vehicle Logging Operations through Selective Cutting Harvest in a Natural Forest

Toshio NITAMI*[1], Shin'ich IWAMOTO*[2], Kenji FUKUSHI*[2] and Hisatomi KASAHARA*[2]

*1 University Forest in Chichibu, The University of Tokyo, Saitama, Japan
*2 University Forest in Hokkaido, The University of Tokyo, Hokkaido, Japan

ABSTRACT   Forest disturbance and operational efficiency were discussed through a ground logging system in a selective cutting harvesting. Grapple skidder operation together with a winch skidder resulted in small disturbance when the grapple skidder mainly execute logging operations. Trees were injured often at slopes where gentler than ten degrees and at slopes steeper than twenty degrees. The former is because of easiness to run a skidder to make free direction long skidding path, and the latter is because that felled trees often roll down on the slope to hit trees when hauled. Decreasing branching of skidding paths enables shorten the paths length and is expected to suppress forest disturbance.
Keywords: disturbance, efficiency, vehicle, logging, selective cutting

1. INTRODUCTION

Selective cutting management has been conducting almost a century and vehicle based ground harvesting operations have been operating more than forty years in University Forest in Hokkaido, the University of Tokyo. Results through the operations compiled and discussed the forest disturbance of the mechanized ground harvesting operation system after a grapple skidder was introduced and the operational efficiency was also discussed.

2. FIELD AND SURVEY

Forest stand used for the experiment had 270 m³ in the storage over a hector with 670 trees in the average, and have been managed through selective cutting by twenty year rotation and seventeen percentage of storage volume to cut, which is classified into the second management section. The first section is low altitude and has better growth and managed through ten years rotation and sixteen percentage of the storage volume to cut. Harvesting operations through ground logging system, which follows manual mortor saw felling by a contractor in order to cut thick trees, damage residual standing trees, disturb forest floor and diminishe succession trees. The ground harvesting system in the university forest has been conducting by two D4-class skidders, a winch skidder and a grapple

75

skidder, after a grapple skidder was introduced to the experimental operations in 1996. The winch skidder is followed by a worker for choking timbers and for guide to the felled trees. Timber cross cutting is done by an excavator mounted grapple-saw and logs were piled up by it after scaling and quality classification by two workers. Additionally, three workers were introduced for timber quality evaluation, scaling and marking for data collection. These jobs were supervised by a chief staff. The grapple skidder lifts logs at their ends with its large scissors-like grapple and hauls to a landing (Figure 1). The grapple enables one-man choking and does not need manual choking work. This promotes operation productivity and offers more safety working condition by keeping clear around the machine. But, pushing felled trees to make them lie at the convenient position for the following skidding and approaching to felled trees without guidance by a choker man often leads to longer running distance in the stand. This made us worry about if it leads to more sever disturbance in the forest stand. And a chief staff gave the vehicle advice at suitable sites to establish the major skidding paths for frequent hauls and gave guide for advancing direction to the felled trees (Nitami et al, 1997).

Fig. 1 Grapple skidder used for the discussion, operating grappling and hauling in the stand.

3. OPERATIONAL EFFICIENCY AND SAFTY

The selective cutting management here needs six to seven hectors of natural forest to produce 200 m^3 of timber piled up at a landing adjacent to forest road. The average payload of a skidder is about 3 m^3. This means a skidder transports timber around 70 times along paths, which are 130 m in the average, between the landing and choking sites and results in around 18 kilometer of the total run through length. The major skidding paths on which skidder hauls timber frequently are designed to be used fixedly at each harvesting operation in the managing rotation. These major paths spread in the stands about 110 m/ha and the whole skidding paths including the other branch paths spread about 280 m/ha. The most major paths were concentrated near landings by frequent run of skidders.

The number of skidding times was eight per a day and which was the same for the grapple skidder and winch skidder. The productivity was 28.617 m^3/day for the winch skidder and 24.255 m^3/day for the grapple skidder. Regarding that the former needed two workers and the latter needed one worker, this shows the grapple skidder had 1.7 times productivity per a worker-day (Nitami, et al, 1997).

The operational safety of the grapple skidder is better than that of the winch skidder because of the one man operation excludes risky man-machine co-work such as choking and sign to escape from the vehicle. But the operator needs to poses skill to know and evaluate surrounding conditions of the vehicle in the stand to make approach vehicle smooth to the felled trees by judging felled direction, ground inclination and tree spaces against the vehicle body and among them for avoiding injure trees.

The grapple skidder took such subsidiary works as skidding path reopening and landing reshaping, because it had hydraulic maneuver system for dozing blade and run faster and clime slope more easily than a winch skidder. Skidding productivity will increase when they were excluded. Additionally, logs were hauled by picked up at the end keeping the end section clear without mud. This also makes crosscutting and scaling operation easy and effective at the landing.

4. DISTURBANCE OF STANDS

Observed areas were choose to have 50 m square samples in the meshed field by longitude and latitude along nineteen coordinates by 50 m interval to have area of 3 % of the whole area in the 1996 operated field. Nine samples were used through sixteen areas in the harvested field 79.42 ha, in which we could compare the stand situation before and after the operation.

The inclination and the change of number of young trees are on the Figure 2. Here, young tree means small DBH tree not larger than 6 cm. This classification has been used in the selective cutting harvesting of the stand based natural forest management in the University Forest in Hokkaido, The University of Tokyo. The survival ratio of young trees was 0.66 – 0.98, 0.85 in the average. The log skidding results in 1995 by winch skidders showed number of diminished young trees was 245 /ha in the average and survived young trees was 199 /ha in the average and was less than the diminished trees. The average number of diminished young trees decreased to 88 /ha in 1996 when the grapple skidder was introduced. In spite that much young trees were stood in the 1996 than that in 1995, small number of young trees were disappeared and so many as 723 young

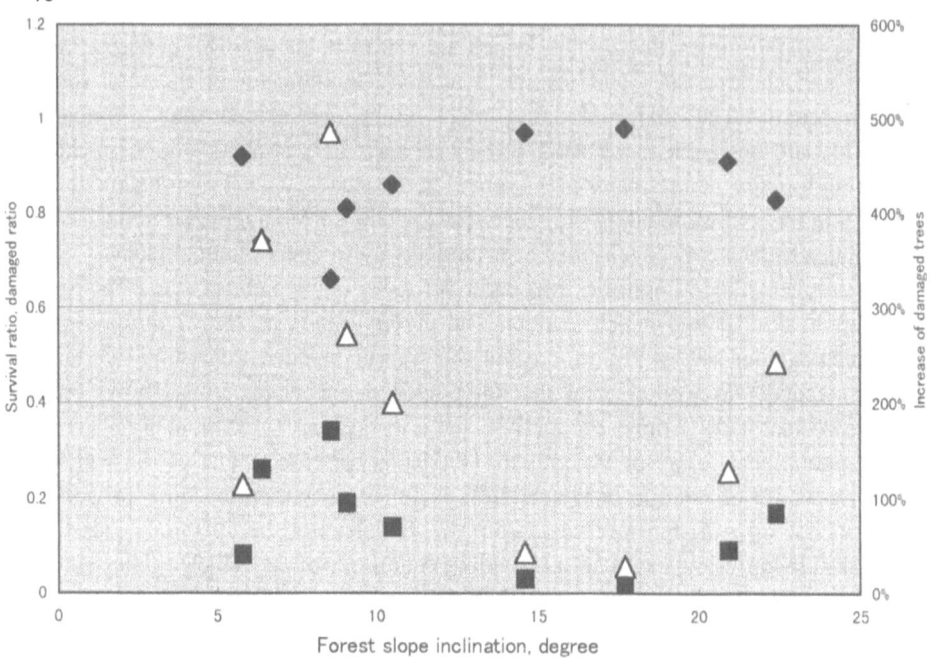

Fig. 2 Forest slope inclination and damage of young trees

◆ survival ratio ■ damaged ratio △ increase of damaged trees

trees were remained.

Slope inclination and survival ratio of young trees after the harvesting operations showed varieties among sample plots. The varieties were thought that they were from if skidding paths are in the area or not, size of logged timber, felled direction, hauled direction and the amounts of logged timber. With considerations on such factors, as shown on the Figure 2, many young trees were lost at the area of rather gentle area less than ten degrees and at the area steeper than twenty degrees (Nitami et al, 1997).

5. MODEL FOR FOREST DISTURBANCE BY VEHICLE SKIDDING

The skidder logging through selective cutting operation here spreads 280 m/ha of skidding paths by a 2.5 m width skidder in the average. This makes the run-over area, skidding paths, to occupy seven percentage of the cutting block area. The young trees are damaged or diminished in this area. The rates of increase of damaged or diminished trees by comparing to the area of skid run-over are superimposed on the Figure 2. This figure also shows that the damaged rate is high in the gentle slope less than ten degrees and in the steep slope more than twenty degrees. When slope inclination comes to steep, the skidder manages to find gentle plots to run and are not always approach close to felled trees but often extend winch wire to pull them close to the vehicle. This operation avoids disturbance of forest stands. When the slope comes to steeper, the skidder advances opening up the paths and leads to alteration of stand condition and most young trees disappear. Additionally, winched trees are easy to slip downward on slopes and damage young trees. The mechanism of the skidding operation to damage and diminish young trees is thought as the above.

A skidder can run through spaces among standing trees, when the spaces are bigger than the vehicle horizontal width on the slope, when slipping downward of the vehicle is ignored. A vehicle of 2.5 m width can run through a stand of 600 – 700 trees per hector on a slope gentler than ten degrees as shown on Figure 3. The distances among trees are multiplied 1.5 times to that of geometrically estimated ones, because forest here is natural stand (Kamiizawa, 1985). The inclination of slope, ten degrees could be considered the upper limit of slope inclination where a skidder can operate without winching operation.

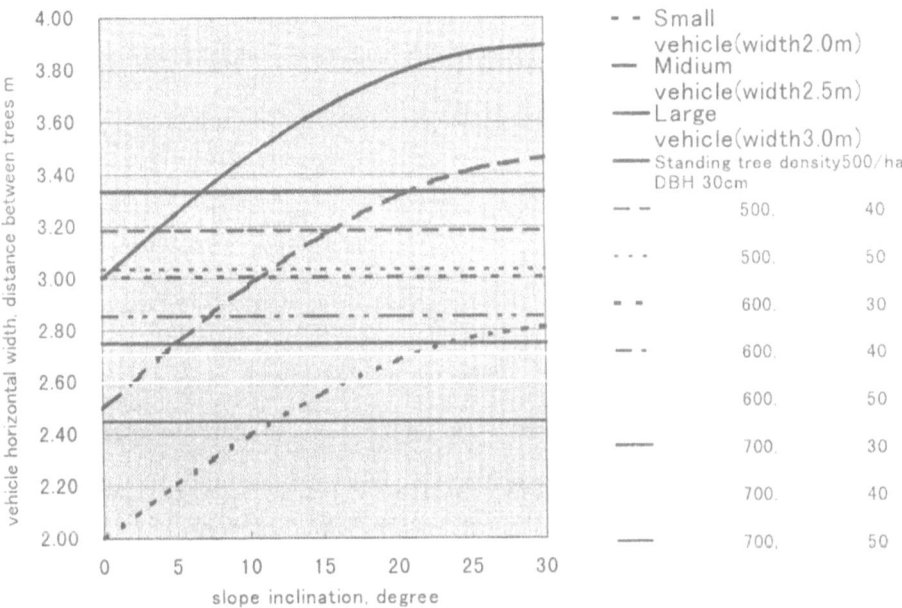

Fig. 3 Forest slope inclination and vehicle horizontal width, distance between trees

6. POSSIBILITY FOR SUPPRESS DISTURBANCES BY VEHICLE OPERATIONS

In order to have environmentally favorable results by the ground harvesting operation system, to suppress forest disturbance at gentle slope area has an importance. The skidding paths spread out from a landing as if they forms a stream system. The highest order of streams at landings were five at almost blocks, when they were ordered by method of basin management. Many of the streams were narrow and low order branched frequent and not obvious major skidding paths were found. Harvesting operation above showed that the supervisor regarded a half of skidding paths were major ones, but not much timber was transported through the end sections of the paths.

A skidder advances or extends winch 5 m after it branched and run last time. Number of felled trees is 30 /ha and area of a cutting block was 7 ha in the average. The total length of the skidding paths in the block is represented by the Horton's second law concerning length of stream system as;

$$\log \sum_{\omega=1}^{\Omega} L_\omega = a + b\omega.$$

Here, Ω is the highest order of the skidding paths and ω is order of skidding paths, a, b are constants, and $\log^{-1} b = R_1 - R_b$. R_1 is stream (path) length ratio (Takayama, 1974). By substituting total paths length, $\omega = 1$ and $5 \times 30 \times 7$ m for the first order and $\omega = 5$ and whole paths length 280×7 m for fifth order (Nitami, 1997), we obtain,

Log1050=a+b,

Log1960=a+50b.

These give us a=2.954, b=0.0668. When we change branch ratio R_b, the total length of skidding paths ΣL_ω varies as on Figure 4 and it comes to large when branching ratio exceeds 3.5. The branching ratio of the operated field showed 3.45 by applying the figure to the site condition. And, when we reduce the branching ratio to 3, the total length of skidding paths comes to 950 m, which is a half amount of the operated field situation. It can be interpreted that a skidding vehicle should be controlled to branch and run less than the probability p,

$$p = 2 / R_b.$$

The p is 0.67 when R_b is 3. This means two thirds of paths would branch from major (high order) paths. This also means that one should lay major skidding paths and should make a skidder less branch and run to reach choking points. The major skidding paths posses high importance and the length can be reduced to a half by adequate layouts and leads to decrease of forest disturbances.

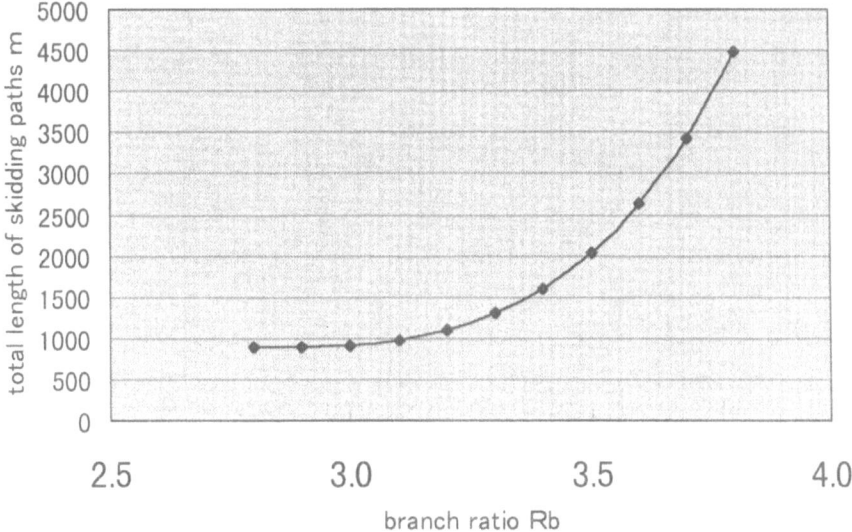

Fig. 4 Branch ratio and total length of skidding paths

7. CONCLUSION

Forest stand damages through ground based selective cutting operation could be decreased when suppress branching to run skidder from skidding paths. Especially at the gentle inclination sites, when branching suppressed to $R_b=3$, the total length of skidding paths would be reduced to a half of today's situation.

REFERENCES

Kamiizaka, M. (1985): Optimizing for Operational Function of a Small Forestry Vehicle to Avoid Obstacles in Forest Stands, Report for Research Grant Ministry of Education, pp.11-18.

Nitami,T., Iwamoto, S., Fukushi, K., Kasahara, H., Sakaguti, T. and Ihara, S. (1997): Logging Operation by a Grapple Skidder, J. Forestry Mechanization, 520, pp. 54-67.

Nitami, T. (1997): Development of Selective Harvesting System due to the Forest Road Network in a Natural Forest, IUFRO, Furano, pp.367-373.

TAKAYAMA, S., (1974): Watershed Terrain, 304pp, Kyoritu Press, Tokyo.

Status of Waste Generation, Disposal Ways and Poplar Species as a Landfill Cover in South Korea

Don Koo LEE*[1] and Su-Young WOO*[2]

*1 Department of Forest Resources, College of Agriculture and Life Sciences,
  Seoul National University, Suwon, 411-744, Republic of Korea
*2 Department of Forest Resources, College of Agriculture,
  Sangju National University, Sangju, 742-711, Republic of Korea

ABSTRACT The objectives of this paper were 1) to overview present waste generation and disposal ways in Korea and 2) to discuss the possible phytoremediation using poplar species in landfill.

In Korea, municipal waste occupies a large portion of total waste generation, even though the amount of briquet ashes in municipal waste decreases. Furthermore, industrial waste has gradually increased during the last couples of years. Landfill area in Korea has been also enlarged due to increased municipal and industrial wastes.

Leachate and heavy metals in landfills can be environmentally problematic. In Korea, several studies have suggested that poplar trees could be suitable species for phytoremediation in landfills because these species transpire leachate and take up heavy metals. Especially, leachate absorption capacity of *Populus alba* x *P. gladulosa* and *P. euramericana* was better than that of other poplars. The growth of *Populus euramericana* in certain landfills was better than that in mountain areas. In addition, poplar plantation in landfills can provide an environmentally friendly ecosystem between landfill and its neighbors.

Keywords: Waste, landfill, phytoremediation with poplars, leachate and heavy metal

1. INTRODUCTION

Waste in the world has enormously increased during the industrial periods in the 20th century with geometrical explosive increase of population. In Korea, industrialization and increase of population in urban area also gave rise to serious waste problems. Briquet ashes had occupied most municipal waste in the past couples of decades but dramatically decreased due to changes of heating systems in houses and industrial facilities. Even though the ratio of briquet ashes in municipal waste decreases, municipal waste in the metro cities occupies a large portion of total waste generation. Furthermore, industrial waste such as synthetic materials has gradually increased during the last decades(Ministry of Environment, 1997).

84

Landfill area near the big cities has been increased to remove municipal and industrial waste in Korea. Serious environmental problems in landfill area could be classified into three general categories: 1) harmful and smelling gases such as CH_4 2) leachate and 3) soil contamination due to industrial waste which contains heavy metals(Nutter and Red, 1986). Especially, leachate which contains high NH_3 and NO_3 seeps into near river or ocean and gives negative impacts on ground water. In addition, heavy metals in contaminated soil can alter tree growth. Generally, municipal and industrial waste landfill can induce many negative influences on the ecosystems.

To absolve gases, leachate and heavy metals, planting of trees in landfill has been often considered. Trees have the potential to remediate or prevent both soil and ground water pollution in waste landfill. Trees, especially fast growing species such as *Populus* and *Salix* can take up the toxic materials through their roots and transport them to stems or leaves.

The objectives of this paper were 1) to overview present waste generation and disposal ways in Korea and 2) to discuss the possible phytoremediation using poplar species in landfill.

2. CURRENT STATUS OF WASTE AND DISPOSAL WAYS IN KOREA

2.1 Classification of waste

Waste can be classified into two big categories: municipal and industrial wastes(Figure 1). Briquet ashes, food and paper occupy most portion of municipal waste. Industrial waste can be divided into non-hazardous and hazardous wastes. Non-hazardous waste includes flyashes, waste of construction, waste lime, lemnant of livingthing and waste gypsum. Especially, Korean Government has classified and managed hazardous industrial wastes as specific waste. Waste alkali, acid, organic solvent, synthetic rubber, dust, sludge, slag, paint and resin are important components of specific waste(Ministry of Environment, 1997).

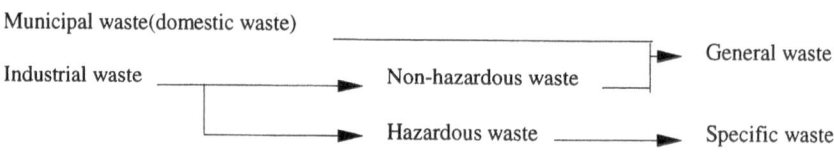

Figure 1. Classification of waste in Korea

2.2 Generation of waste

Total waste generation per day in Korea gradually increased during the last decade(Table 1). In 1996, total generation of waste was 180, 573 tons per day. Interestingly, domestic waste has gradually decreased during the last 5 years. Changes of heating system turn to decrease briquet ashes and contribute to decrease domestic waste(Table 2). But domestic waste contains much moisture, and can create leachate from landfill after dispose. In contrast, industrial waste has dramatically increased(Table 1). Fast industrialization in Korea produced much industrial waste, which could be a potential source of soil contamination for the near future.

Table 1. Changes of waste in Korea during the last 5 years (Unit: tons/day)

Year	Total	General waste	Specific waste	Domestic waste	Industrial waste
1992	144,535	123,154	21,381	75,096	69,439
1993	141,383	118,909	22,474	62,940	78,443
1994	147,049	143,347	3,702	58,118	88,931
1995	148,041	143,597	4,444	47,774	100,267
1996	180,573	175,334	5,239	49,925	130,648

(Ministry of Environment, 1997)

In 1995, Korean Government promulgated the Municipal Waste Regulation Act, and this Act contributes to reduce the generation of domestic waste(Table 2). Decrease of domestic waste in Korea was caused by these two reasons; reduced briquet ashes and municipal waste regulation act. However, generation of other waste, such as food, paper and metal(+ glass) have been still produced as much as the amount of prior years(Table 2). In 1996, waste generation per person in a day was 1.1kg.

Specific waste in Korea has gradually increased after 1994(Ministry of Environment, 1997). Since 1994, waste acid, alkali, oil, organic solvent and dust occupied most portion of specific waste. Among them, dust produced as a byproduct of the combustion process can be deposited with rain. So, these wastes seem to be potential sources of air and soil pollutions.

Table 2. Changes of domestic waste in Korea (Unit: tons/day)

Year	Total	Briquet ashes	Food	Paper	Metal (+glass)	Wood	Others	Amounts (Kg/day/ person)
1992	75,096	17,750	21,807	13,125	4,957	3,077	14,380	1.8
1993	62,940	9,780	19,764	11,546	3,732	2,822	15,296	1.5
1994	58,118	5,534	18,055	12,468	3,264	2,443	16,354	1.3
1995	47,774	3,235	15,075	11,203	3,639	1,938	12,684	1.1
1996	49,925	1,853	14,532	13,327	5,262	1,857	13,094	1.1

(Ministry of Environment, 1997)

2.3 Waste disposal

Disposal ways of waste in Korea have been classified into recycling, combustion and landfill(Table 3). Recycling rate has been increasing during last 4 years. In 1996, 54.7% of total waste was recycled, thus it would help prevent environmental pollution. Even though the landfill rate showed decreasing trends on every year, it occupied more than 38% of total waste in 1996.

Table 3. Disposal ways of waste in Korea (Unit: tons/day)

Year	Total	Recycling	Combustion	Landfill	Others
1993	141,383	55,894	5,822	76,449	3,218
	(100)*	(39.5)	(4.1)	(54.1)	(2.3)
1994	147,049	62,940	6,512	76,565	1,032
	(100)	(42.8)	(4.4)	(52.1)	(0.7)
1995	148,041	72,375	8,303	65,986	1,377
	(100)	(48.8)	(5.6)	(44.5)	(0.9)
1996	180,573	98,692	9,933	70,225	1,723
	(100)	(54.7)	(5.5)	(38.9)	(1.0)

(Ministry of Environment, 1997)
* The numbers in parentheses denote percentage

Similarly, the recycling rate of municipal waste in Korea has gradually increased, but the landfill has decreased(Table 4). However, the great part of municipal wastes has been disposed by sanitary landfill method. Actually, 68.3% of municipal waste was disposed by this method in 1996.

Table 4. Disposal ways of municipal waste in Korea (Unit: tons/day)

Year	Total	Recycling	Combustion	Landfill	Others
1992	75,096	5,912	1,132	66,965	1,087
	(100)*	(7.9)	(1.5)	(89.2)	(1.4)
1993	62,940	7,233	1,480	54,227	-
	(100)	(11.4)	(2.4)	(86.2)	(0.0)
1994	58,118	8,927	2,025	47,166	-
	(100)	(15.3)	(3.5)	(81.2)	(0.0)
1995	47,774	11,306	1,922	34,564	-
	(100)	(23.7)	(4.0)	(72.3)	(0.0)
1996	49,925	13,084	2,725	34,116	-
	(100)	(26.2)	(5.5)	(68.3)	(0.0)

(Ministry of Environment, 1997)
* The numbers in parentheses denote percentage

Table 5. Disposal ways of specific waste in Korea (Unit: tons/day)

Year	Total	Recycling	Combustion	Landfill	Others
1992	21,381	10,142	2,748	6,203	2,288
	(100)*	(47.4)	(12.9)	(29.0)	(10.7)
1993	22,474	11,310	3,297	4,649	3,218
	(100)	(50.3)	(14.7)	(20.7)	(14.3)
1994	3,702	1,805	575	290	1,032
	(100)	(48.8)	(15.5)	(7.9)	(27.8)
1995	4,444	2,140	690	219	1,395
	(100)	(48.2)	(15.5)	(4.9)	(31.4)
1996	5,238	2,432	705	379	1,722
	(100)	(46.4)	(13.5)	(7.2)	(32.9)

(Ministry of Environment, 1997)
* The numbers in parentheses denote percentage

Disposal ways of specific waste in Korea are shown in Table 5. Recycling, combustion and landfill rate of disposal ways remained at steady condition after 1994. One year later waste classification system has been changed. During three years from 1994 to 1996, percentage of recycling, combustion and landfill showed approximately 47, 15 and 7%, respectively.

2.4 Landfill condition

The number of landfill in Korea has been gradually decreased during the last 5 years, but total landfill area has been increased(Table 6). The size of landfill becomes larger than before. For example, the area of Kimpo municipal waste landfill near Seoul, is more than 2090ha(Ministry of Environment, 1998). Probably, the landfill area will increase with increasing population. Bigger landfill could make latent environmental problems such as leachate and heavy metals

Table 6. Landfill sites in Korea

Year	No. of landfill	Total landfill area(m^2)	Total landfill capacity ($1000m^3$)	Already landfilled capacity ($1000m^3$)	Future landfill capacity ($1000m^3$)
1992	618	13,841	286,224	161,268	125,955
1993	622	37,190	467,575	95,286	372,289
1994	586	35,114	448,555	91,987	356,568
1995	537	34,678	456,484	101,827	355,070
1996	529	36,468	485,087	112,094	369,337

(Ministry of Environment, 1997)

3. POSSIBLE PHYTOREMEDIATION IN KOREA

Phytoremediation is an approach to clean up contaminated soils using plant species. Many countries have been interested in phytoremediation because it appears to be cheaper than chemical and engineering-oriented methods and may also offer a restoration and long-term environmental benefits around contaminated area such as landfills, mining sites, farmland and nuclear waste dumps(Litch and Madision, 1995). In Korea, poplar species seems to be an outstanding candidate for phytoremediation because of robust growth rate, rapid establishment of many leaves and bulky root systems.

3.1 Growth of poplar in landfill

Nanjido landfill is a municipal waste deposit area near Seoul, the biggest city in Korea. To test the suitability of poplars for landfill reclamation, *Populus euramericana* was planted on Nanjido landfill in 1994 and 1995. The growth of *Populus euramericana* in Nanjido landfill was better than that in mountain area(Table 7). The 3-year-growth of poplar planted in 1994, showed average height and DBH, 5.9m and 6.2cm, respectively. These impressive growth rates were 1.6 and 1.8 times higher than those in mountain area. Poplar planted at Nanjido landfill in 1995 also showed higher growth than that in mountain area.

High NH_3 and NO_3 concentration in landfill area might help improve tree growth(Nutter and Cole, 1986). The leachate from landfill or sludge contains high concentration of nutrients. Today there are numerous examples of the application and treatment of both leachate and sludge on forest land and agricultural crop land areas not only to dispose the waste but also to improve plant growth. In the U. S. A., the growth of Douglas-fir from sludge application was twice as great as when this species was fertilized with urea(Cole and Henry, 1986).

Table 7. *Populus euramericana* growth in Nanjido landfill and mountain area

Planting year	Nanjido landfill		Mountain area		Growth ratio	
	Height (m)	Diameter (cm)	Height (m)	Diameter (cm)	Height	Diameter
1995	4.0	3.4	3.0	1.9	1.3 times	1.8 times
1994	5.9	6.2	3.8	3.5	1.6 times	1.8 times

(Koo, 1996)

3.2 Absorption of heavy metals by poplar

The heavy metal concentrations of both soil and poplar wood grown in Nanjido landfill were higher than those in other natural sites(Tables 8 and 9). Especially, Zn, Cd, Pb and Cu concentrations in Nanjido landfill were more than twice higher than those in non-polluted area. It is not clear whether waste or soil brought in to cover the waste was the source of the high heavy metal concentrations in soils.

However, increased industrial wastes could contaminate soil with a high level of heavy metals(Table 1). Those heavy metals mentioned above, showed higher concentration in *Populus euramericana* wood grown in Nanjido landfill than those in non-polluted area. Many tree species naturally absorb metals from soil and store them in their tissues(Balsberg, 1989; Kukaszewiski et al., 1993). Woody plant species seem to have a lower degree of immobilization to heavy metals than grasses and other herbaceous species. But, the genus *Betula* including several pioneer species are often found on soils contaminated with large amounts of various heavy metals(Eltrop et al., 1991).

Trees also can immobilize toxic compounds. Organic compounds can be degraded by enzymes expressed in the membranes of poplar trees. These plants may also stimulate the growth of chemical-degrading bacteria around their roots(Harrison, 1996). Harrison(1996) has reported in laboratory experiments using poplar that monooxygenase enzymes break down trichloroethylene.

Table 8. Heavy metal concentrations of soil and leachate in Nanjido landfill area in Korea
(Unit: ppm)

	Zn	Cd	Pb	Cu
Nanjido landfill	9.45	0.032	5.01	8.09
Leachate	7.00	0.000	0.00	4.00
Non-polluted soil	4.30	0.014	1.14	4.91

(Koo *et al.*, 1997)

Table 9. Heavy metal concentrations in *Populus euramericana* grown in Nanjido landfill
(Unit: ppm)

	Grown in Nanjido landfill				Grown in non-polluted area			
	Zn	Cd	Pb	Cu	Zn	Cd	Pb	Cu
Trunk	0.0	0.0	0.3	1.7	0.0	0.0	1.3	1.7
Stem	0.7	1.7	6.5	3.3	0.0	0.0	2.7	0.7
Branch	1.0	1.0	6.3	2.7	0.0	0.0	0.8	2.3
Leaf	2.7	0.0	1.0	1.7	0.7	0.0	0.2	1.8

(Koo *et al.*, 1997)

3.3 Absorption of leachate by poplar

Leachate absorption capacity of *Populus alba* x *Populus gladulosa* and *P. euramericana* was better than that of *P. nigra* x *P. maximowiczii* grown in Kimpo landfill(Table 10). Average amounts of transpiration in *Populus alba* x *P. gladulosa* and *P. euramericana* were 409 and 391ml per day, respectively. These two species showed higher transpiration rate than those of *P. nigra* x *P. maximowiczii*. Hybrid poplar trees in the U. S. A. can remove 600 liters of water from the soil for every kilogram of stem dry matter growth(Stomp *et al.*, 1994; Licht and Madision, 1995).

Rainwater, surface runoff and seeping through landfill waste created a leachate of water and dissolved waste. Poplar can immobilize leachate that waste landfill produces. Since the fast growing trees utilize a large amount of water in the respiration processes, the moisture is extracted from the soil before it percolates beyond the root zone. Poplars grow dense, root into the landfill cover soil, thus acting as a pump that transpires the soil water back to atmosphere(Schnoor *et al.*, 1992). The plant uptakes, removes water and harmful compounds from root zones.

Table 10. Leachate absorption capacity of two-year-old poplar trees (ml/tree/day)

	Leachate	Tab water
Average	361	429
Populus alba x *P. glandulosa*	409	470
P. nigra x *P. maximowiczii*	283	332
P. euramericana	391	484

(Koo *et al.*, 1998)

Poplar also immobilizes toxic leachate(Table 11). Leachate goes through and contaminates near river. Poplar trees reduced BOD and NH$_3$

concentration which could contaminate ground water. Poplars are being used to effectively remove toxic compounds and other contaminants from leachate.

Even though the concentration of heavy metals in leachate was not seriously high in Kimpo landfill(lower than legal standard), heavy metals in

Table 11. Changes of leachate collected from Kimpo landfill ingredient before and after irrigation

(Unit: ppm)

	Legal Standard	Before irrigation		After irrigation		
		Tab water	Leachate solution	P. alba x P. glandulosa	P. nigra x P.maximowiczii	P. euramericana
PH	8.6	6.9	7.9	7.7	7.8	7.7
BOD	100	0.0	4050	99	92	95
NH₃	NA	0.03	1256	860	760	820
Cr	2.0	0.0	0.19	0.058	0.058	0.061
Fe	37.0	0.018	7.10	0.528	0.78	0.0625
Zn	5.0	0.009	0.67	0.384	0.326	0.383
Hg	0.005	ND	0.0007	0.0009	ND	0.0006
As	0.5	ND	0.01	ND	ND	ND
Total-P	NA	-	7.89	3.9	ND	3.6

-: could not be detected due to lack of sample
NA: no data available
ND: could not be detected
(Koo et al., 1998)

leachate such as Cr, Fe, Zn and As were also removed by poplars after irrigation. As mentioned previously, these high NH_3 and heavy metals might be removed and accumulated in the poplar tissues(Table 9). This fact suggested that poplar might be a good candidate for immobilizing contaminated leachate.

4. SUMMARY

Municipal waste in Korea occupies a large portion of total waste generation, even though the amount of briquet ashes in municipal waste becomes decreased. The industrial waste has been gradually increased during the last decades. To dispose municipal and industrial wastes in Korea, landfill area has been also increased in the past decades.

Leachate and heavy metals in landfills can be environmentally problematic. Several studies in Korea suggested that poplars could be a suitable species for phytoremediation in landfills because these take up leachate and heavy metals. Especially, leachate absorption capacity of *Populus alba* x *Populus gladulosa* and *P. euramericana* was better than that of *P. nigra* x *P. maximowiczii*. Furthermore, the growth of *Populus euramericana* in landfill was better than that in mountain area. In addition, poplar plantation in landfill can provide an environmentally-friendly ecosystem between landfill and its neighbors.

5. REFERENCES

Balsberg, A-M. (1989): Toxicity of heavy metals (Zn, Cu, Cd, Pb) to vascular plants Water, Air and Soil Pollution 47:287-319

Cole D. W. and C. L. Henry(1986): Future directions; Forest sludge application. In: Cole D. W., C. L. Henry and W.L. Nutter (Eds.). The forest alternative for treatment and utilization of municipal and industrial waste. P62-69 Univ. of Washington , Seattle

Eltrop, L., G. Brown, O. Joachim and K. Brinkmann(1991): Lead tolerance of *Betula* and *Salix* in the mining area of mechernich/Germany. Plants and Soil 131:275-285

Harrison, R. B. (1996): Snoqualmie Pass Sewer District Wastewater Treatment System Monitoring Project. Web site. http://weber.u.wasgington.edu/~robh

Koo Y. B. (1996): Let us make plantation for poplar. Poplar 13: 34-38

Koo Y. B., S. K. Lee, P.G. Kim, K. O. Byun and S.Y. Woo(1997): Growth and absorption capacity of heavy metals of *Populus euramericana* at Nangido landfill. Poplar 14: 25-32

Koo Y. B., E. R. Noh, S. Y. Woo and S. K. Lee(1998): Using poplar trees as a landfill cover and leachate treatment. Poplar 15: 19-29

Kukaszewiski, Z., R, Siwecki, J. Opydo and W. Zembrzuski(1993): The effects of industrial pollution on copper, lead, zinc and cadmium concentration in xylem rings of resistant (*Populus marilanica*) and sensitive (*P. balsamifera*) species of poplar. Trees 7:169-174

Litch, L.A. and M. Madision(1995): Using Poplar Trees as a Landfill Cover: Experiences with the Ecolotree Cap. SWANA 11th Annual Northwest Regional Soil Waste Symposium, Portland Oregon April 12-14

Ministry of Environment(1997): Environmental Statistics Yearbook. Seoul p543

Ministry of Environment(1998): Waste landfill facilities. Web site. http://www.moenv.go. kr/pae/pae.htm

Nutter W. and J. T. Red(1986): Future directions: Forest wastewater application. In: Cole D. W., C. L. Henry and W.L. Nutter(Eds.). The forest alternative for treatment and utilization of municipal and industrial waste. p55-61. Univ. of Washington , Seattle

Schnoor, J. L., L.A. Licht, S. C. McCutcheon, N. Lee and L.H. Carreira(1992): Phytoremediation of organic and nutrient contaminants. Environmental Science and Tech. 29(7): 318-323

Stomp, A-N, K. H. Han, S. Wilbert, M. P. Gordon and S. D. Cunningham(1994): Genetic strategies for enhancing Phytoremediation. Reprint of Recombinant DNA Technology II Vol. 721 of the Annals of the New York Academy of Sciences

Environmentally Friendly Resource Use
— Taiwan's Experience in Pulp and Paper Production

Hsiu Hwa WANG and Sheau Horng LIN

Department of Forest Products, National Pingtung University of Science and
Technology, Taiwan

ABSTRACT   Fiber supply, waste reduction and pollution control have long
been key challenges facing the pulp and paper industry in Taiwan. Water usage,
energy consumption and capital effectiveness are also high on the list. An average
of 5,213 hectares of plantation was established each year on the island in the past
decade to provide a wood volume of approximately 26 thousand cubic meters
annually. To complement the supply of increasingly expensive virgin fiber, the
industry in 1997 re-utilized about 4 million tons of waste paper, of which nearly
70% was collected domestically.   This secondary fiber input for paper and board
production eliminated both the construction of ten incinerators and 3.65 million
tons of CO_2 emission.

The industry is in a better position than most other industries to institute
sustainable development in its operating practices. Innovative techniques and
substantial capital investments have been implemented to accomplish a very
impressive degree of environmental protection·7 2.5% of the mills meet the
environmental regulations, water usage per ton of product has been reduced from
$50m^3$ to 10-12 m^3, chemical oxygen demand (COD) and suspension solid (SS) in
the effluent have been minimized to conform to the most stringent regulations
issued for 1998.All of this has resulted from an average investment in facilities of
2.72 millions U.S. dollars per annum during the last four years. As for energy
reduction, many co-generation systems have been in operation and those under
construction will double the capacity by the year 2002.

With all these endeavors, it is hoped that the industry will become a
sustainable and healthy one that is globally competitive; environmentally
compatible and a key contributor to the global economy and society.

1. INTRODUCTION

In order to support the world's second highest population density, Taiwan
has witnessed various kinds of mills established along the very narrow shores of
her 3,591,500 ha island since world war II. With double-digit growth of her
economy in the sixties and seventies, there was little concern for non-economical
issues. Forests were heavily exploited to support and maintain the economic
development in that period. The rapid development of information technology in

recent years has, however, broadened the vision of the people on the island. An environmental protection consciousness emerged in Taiwan. The public recognized the increasing importance of maintaining the critical balance between economic growth, social and environmental responsibility.

Realizing that economic development should go hand in hand with environmental protection, the government has begun to issue more and more stringent environmental regulations. Various factions of the society have swiftly accepted the concept of sustainable development.

Paper industry utilizes plant fibers as the main raw material and uses water as carrier of chemicals and fibers to obtain uniformly distributed webs during the formation of paper. From the toxicity characteristics leaching procedure (TCLP) test[1], it is evident that sludge from treatment of pulp and paper mill effluents has much less toxic substances than the permitted levels. Nevertheless, the effluent discharge standards in Taiwan have become more and more tightened in the past decade.

In order to meet increasingly severe waste discharge limitations, the pulp and paper industry has implemented five stages of pollution control since 1970.First emphasis was on end-of-pipe treatment. Second was an array of four projects to deal with wastewater treatment, removal of foul gases, solid waste cleaning, and noise abatement. The third stage emphasized pollution reduction in processing, because success there reduces cost of end-of-pipe treatment. The fourth stage focused on waste minimization and pollution control, moving toward minimum impact manufacturing. This also reduced reliance on end-of-pipe treatment while keeping costs in line to maintain profitability. The goal of the fifth stage is to conserve resources and ecosystems to benefit people both now and in the future. Efforts to meet Taiwan's 4R (reduce, recover, reuse, recycle) policy have gained valuable experience that will benefit the conservation of global forest resources.

2. CURRENT STATUS OF FOREST RESOURCES AND UTILIZATION

2.1 Resources

Taiwan's geographic location makes her a vital link in the Pacific Rim and a bridgehead to Southeast Asia. With economical and cultural affinities, Taiwan has played a very important role in the enhancement of regional development in this part of the world. Almost 60%of the island's 3.6 million hectares are forested. This is 6.5% more than 20 years ago when the second forest resource inventory was conducted[2]. Timber production has played an important role in support of the economic development in Taiwan. However, unlimited exploitation of forest resources has been severely limited since the late eighties. The steep topography in most areas of the island contributes to this limitation as well. Table 1 shows the decreasing annual cut of timber, while Table 2 indicates that 34.3% of the area of forest reserves in Taiwan is on a slope steeper than 35: These areas include nearly 40% of the total timber volume. Timber production plays less and less important role in support of the economic development in Taiwan. Recent forest policy has emphasized conservation, sustainable management and multiple uses of forest resources.

Table1.Annual cut in Taiwan forest

Year	Area(ha)	Volume(m^3)	Volume %
1987	5,546	670,410	100.0
1988	5,208	426,483	63.6
1989	2,493	264,492	39.5
1990	1,917	203,213	30.3
1991	1,046	126,059	18.8
1992	1,036	118,323	17.6
1993	575	71,735	10.7
1994	439	56,128	8.4
1995	625	63,176	9.4
1996	500	56,374	8.4

According to the 3rd survey of forest resources and land utilization started in 1990[2], 72.7% of the total forest is natural forest, the other 422,600 hectares is 51% coniferous plantation and the rest is hardwoods and mixture of both categories.

As to species distribution on the total 2,102,400 hectares sites, 20.86% is coniferous, 53.29% deciduous, 18.61% mixture of the two, while 7.24% is bamboo plantation. A total volume of 0.36 billion m^3 was recorded in the third resources inventory. Changes of forest resource volume in the past three decades are shown in Fig 1.

96

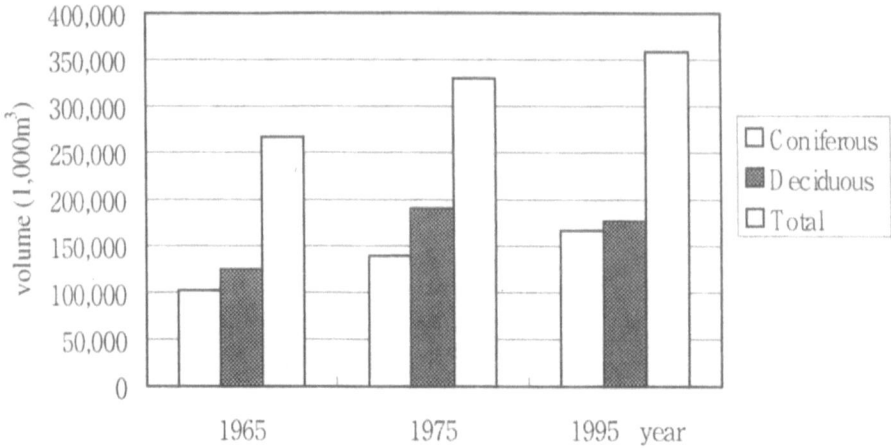

Fig 1. Changes of forest resources in Taiwan

There are approximately 200 tree species with economic potential, which indicates high diversity in the forest ecosystem in Taiwan.

Table 2.Forest reserves distribution by slope class

Slope class	Area %	Volume stock %	Vol./ha M³
<5 ·	2.19	0.98	76
5-15 ·	8.68	5.45	107
15-25 ·	20.72	19.15	158
25-30 ·	34.15	35.61	178
35-45 ·	26.52	31.75	204
>45 ·	7.77	7.05	155
Total	100	100	171

2.2 Utilization

Forest products produced in 1996 were valued at 5,210 million US dollars, representing 1.9% of the total GDP [3]. This figure, 0.4% less than that in the previous year, indicates that this traditional industry has experienced some recession, though it is still fairly close to that of some advanced developed countries like the United States. However, the industry has been experiencing operational difficulties, like several other traditional industries, due to the lack of raw material, high labor cost, more and more stringent environmental regulations, competition from the other side of the Strait, as well as totally changing career planning of the X or Y generation.

Domestic timber consumption has declined from 20.2% of the total in 1975 to less than 1 % twenty years later. The exact data are shown in Table 3[4]. Pulp and paper industry in Taiwan used 1,420 thousand tons of fast grown pulpwood, 3,340 thousand tons of wastepaper pulp and 863 thousand tons of imported pulp to produce 4,507 thousand tons of paper and board in 1997.

Table 3.The supply and consumption of timber

Year	Domestic supply($1,000/m^3$)	Total consumption ($1,000/m^3$)	% of self supply
1975	1,027	5,323	19.3
1980	680	6,734	10.1
1985	574	6,105	9.4
1990	145	6,918	2.1
1995	40	6,528	0.6

3. PULP AND PAPER INDUSTRY

3.1 Production statistics

Two pulp mills and 145 papermaking companies are members of the Taiwan Paper Industry Association (TPIA). There are also some minor non-member companies. The products of paper mills range from printing and writing, packaging, household, newsprint, and joss paper to specialty products such as insoles and handmade Chinese calligraphy paper manufactured mainly from bast fibers.

In 1997, 4.50 million tons of paper and board were produced, comprising 1.25 millions tons of paper and 3.25 millions tons of paperboard. Increase in production fluctuated from -1.8 % to 7.5 % during the past five years, with an average increase of 2.58 % per year. Most of the mills are of small to medium scale compared to those in North America, Scandinavia and Japan. The production of the first 13 biggest companies (with annual business volume exceeding US$50 million) accounts for about 60% of the total of all companies combined. Fig.2 shows production and per capital GNP[6] and paper consumption in the past decade. Wastepaper contributes about 73.4% of total raw materials required [5]. 4.09 million tons of wastepaper were consumed in 1997 with 2.79 million tons collected domestically and 1.30 million tons imported. The other 26.2% came mostly from exotic pulps. In 1997, 862,600 tons were purchased mainly from USA, Canada, Chile and South Africa, which accounted for more than three-quarters of the total imported. Fig.3 shows material supply scenario in 1997.

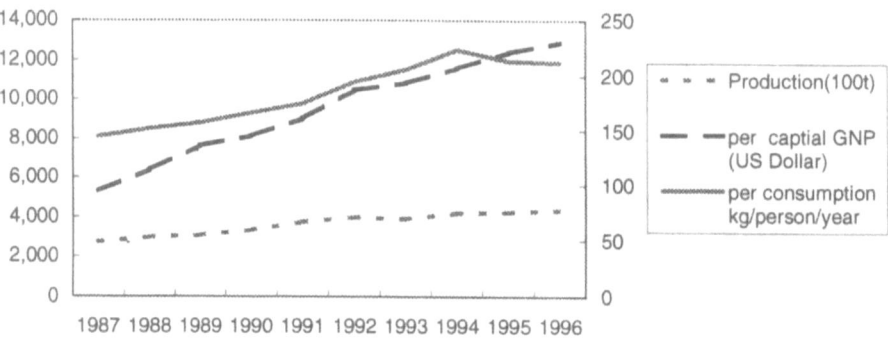

Fig.2 Paper and board production and per capital GNP& per consumption in Taiwan

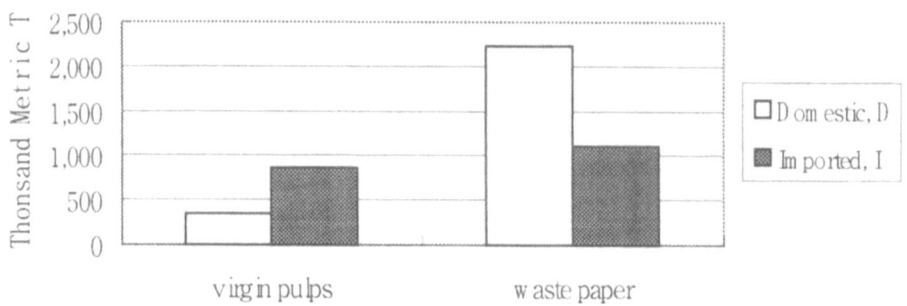

Fig.3 Raw materials supplied for paper industry in Taiwan in 1997

3.2 Management of fiber resources in paper industry

The double-digit growth rate enjoyed by the industry is unlikely to continue unless market conditions change dramatically for the better. For the industry to maintain healthy growth and simultaneously meet environmental regulations, raw material, particularly waste paper, must be secured in an increasingly tight international market.

The forest sector has worked to cultivate a healthy plantation resource to support the desperately hungry industry. Table 4 shows the changes in plantation area in the past ten years.

Table 4. Plantation and cutting areas in Taiwan in the last 10 years (ha)

Year	1988	1989	1990	1991	1992	1993	1994	1995	1996	1997	Ave.
plantation	9,676	9,394	3,883	4,656	5,081	4,783	4,622	3,558	5,230	5,246.3	5,212.9
cutting	5,208	2,493	1,917	1,046	1,036	575	439	625	500	448.2	1,428.7

In addition to the work completed by the Forest Bureau, the paper industry has also invested in developing forest plantations for their own use. It is reported that 61,120 hectares have been planted on the island, while 172,300 hectares are planted abroad. The most popular genera introduced are Leucaena and Eucalyptus.

3.3 Utilization of wastepaper

Taiwan's paper industry relies heavily on the utilization of wastepaper owing to its very limited fiber resources. The domestic recovery in 1992 was 2.30 million metric tons, which reached 2.79 million in 1997 with an average annual rate of increase of slightly more than 4%. The rate of increase in wastepaper recovery declined significantly in recent years compared to the average annual increase of 11.0% from 1982 to 1991, indicating that people on the island have almost reached a limit in recycling reusable resources. Fig.4 shows the paper and paperboard   consumption trend and the increase in its recovery.

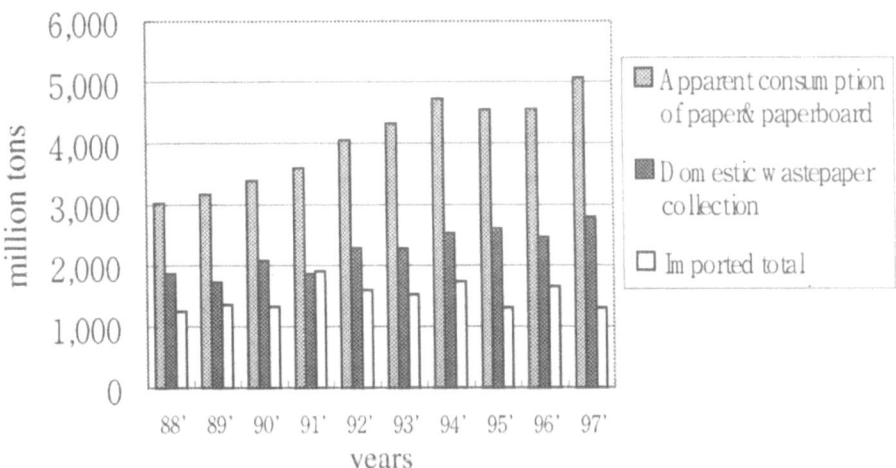

Fig.4 Paper and board consumption trend and the its increase in its recovery

It is clear from the chart that paper and board consumption has increased steadily at an average rate of 3.1% in the last decade, with some minor fluctuations. Two indices, the average wastepaper recovery and utilization rates, highlight the achievement of people in Taiwan in the past ten years. The figures are 56% and 75.3% respectively; both ranked the very top in the whole world. Table 5 shows the comparison of wastepaper recovery rates among four countries [7,8]

Table 5.The wastepaper recovery rates of four countries

	US	Canada	Japan	Taiwan
1986	28.2	23.4	50.0	47.3
1990	33.8	31.7	49.7	61.5
1995	45.6	41.1	52.2	57.3

Wastepaper recovery is expected to increase over the next ten years, hoping to reach even about 70% of total paper and board consumption, due to economic and regulatory incentives for the enhancement of raw material supply and reduction of solid wastes.

3.4 Sustainable uses of water and energy resources in the paper industry

The paper industry consumes a large quantity of water and heat. The manufacturing processes use, however, a minimum of hazardous chemicals. It appears technically feasible to totally close the water systems on most of the manufacturing operations, to achieve environmentally sound water recycling.

Effluent from the pulp and paper industry has received a great deal of public attention in recent years, because water contamination before treatment is quite visible. There are many reasons for reducing water consumption and effluent generation: regulatory compliance, production increase, economics as well as market forces.

The major sources of effluent pollution in a pulp and paper mill complex are the following:

(1) Water used in wood handling/barking and chip washing,

(2) Digester and evaporator condensates,

(3) White waters from screening, cleaning and thickening,

(4) Bleach plant washer filtrates,

(5) Paper machine white water,

(6) Fiber and liquor spills from all sections.

Important objectives in modern pulp and paper mill operation are to treat mill effluents effectively and to minimize process losses such that their impact on the environment is minimal and essentially non-polluting.

To implement environmentally friendly programs within the pulp and paper industry will certainly entail significant capital and operating expenses. These costs will usually be partly offset by improved energy conservation or by better retention of fiber in the process. Improved fiber retention is, in return, an effective means of reducing or controlling mill discharges.

The most effective stratagem for reducing in-plant losses has been to recycle and reuse mill process waters, which reduces mill effluents. In recent years, the amount of water consumed by pulp and paper mills has been dramatically reduced. The pulp and paper industry in Taiwan has mitigated its water consumption per ton of product significantly during the last 20 years while production has increased ten times. The lessons learned by the industry are twofold. First, changing production systems to cut down on the generation of all sorts of environmental wastes and pollutants is technically feasible. Second, waste reduction usually makes economic sense. TPPC (Taiwan Pulp & Paper Corporation), one of the three pulp mills on the island concentrated its efforts in the early eighties on end-of-pipe treatment. This required expensive technology, and resulted in huge operational costs and great opportunity to be penalized.

The company suffered an unforgettable fine of one million NT dollars a month for not meeting the discharge standards and for producing foggy surroundings. The first wastewater operation system was consequently established in 1983, using secondary biological treatments to reduce water pollution. Water color and chemical oxygen demand (COD) were somewhat improved. However, the effluent did not meet the regulatory standard due to the lack of heat control in the system. In 1987, an oxygen delignification process was added between the cooking and conventional bleaching sequences and the product was promoted as elemental-chlorine-free pulp. Closed type pure oxygen aeration treatment equipment was installed to increase the pollutant removing efficiency of the previously installed activated sludge secondary treatment system. Since then, the treated effluent has never failed to pass the most stringent 1998 standards (Table 6).

Table 6 Effluent discharge standards for pulp and paper industry in Taiwan

Industry	Year	COD(mg/l)	SS(mg/l)	Visibility(cm)
Pulp	1993	200	100	>15
	1998	150	50	>15
Paper	1993	200	50	>15
	1998	100	50	>15

TPPC has benefited from that improvement through: substantial reduction in size of the aeration tank from $30,000m^3$ to $4000m^3$, decreased retention time from 24 to 4 hours, reduced unit operation cost of the wastewater (currently about 0.184 dollars per ton, compared to 0.232 dollars in the year 1991) even with huge capital input. More than 270 thousand US dollars are meanwhile saved per year through the installation of pure oxygen production.

The pulp and paper industry consumes a high quantity of steam in various unit operations such as cooking, washing and web drying. Efficient energy conservation leads not only to increasing the competitiveness of the company but also to a better use of limited human resources. Taiwan's paper industry works to improve process efficiency, invests in new facilities to raise the consistency of various unit operations, controls power loading with system computers, and minimizes power consumption especially by adjusting operations to better utilize off-peak power supply. In additional, the industry has been encouraged to implement co-generation units to enhance energy efficiency. This effort not only reduces operating cost but also minimizes CO2 emissions. The total installed capacity has reached 283,900KW by 1996, equivalent to 12% of the self-generation capacity of all industries in Taiwan. Many new facilities have been proposed. The projected capacity in the year 2002 is 521,000KW, implying an 83.5% increase within 4 years.

3.5 Upstream minimization of the waste

The large complex of environmental regulation related to pollution control has led to high and steadily increasing costs and liabilities that make upstream waste reduction in production, a matter of self-interest for industry.

The most significant source of solid waste pollution for the pulp and paper industry is the sludge from effluent treatment. Persistent efforts in waste reduction have been made to solve this problem. One of the abatement processes is to install secondary sludge dewatering units to reduce the moisture content from 70% to below 55%. This not only concentrates solid wastes needed to be disposed of, but also enables self-combustion in fluidized bed incinerators.

There are already mills utilizing treated sludge as soil amelioration media, or blending them with bark and shives to facilitate fermentation by biochemical agents to produce organic fertilizer for agricultural use. Other uses for recovered sludge are mold casting for fruit plates, heat insulation enhancement, and additives in cement and heat generation.

The largest paper company in Taiwan, ranking 76th out of the top 150 paper mills globally in 1996[9], Yuen-Foong-Yu (YFY) has made impressive strides in reducing water use, and increasing energy self-sufficiency and raising the use of recovered fiber. Its environment strategies have been through careful planning. Understanding that upstream waste reduction in production would not only reduce the cost in end-of-pipe treatment but also improve competitiveness and increase profits, YFY has carried out continuous capital investment since 1988. Table 7 shows the results of its efforts in upstream waste reduction [10] as well as in end-of-pipe treatment.

Table 7. The improvement of waste reduction and pollution control in the largest paper company in Taiwan (YFY).

	1992	1993	1994	1995	1996	1997
Total effluent m³/day	97,657	82,840	78,065	78,387	74,444	68,690
Solid waste t/Mon.	20,074	12,864	12,199	12,689	10,568	7,399
SS kg/day	4,100	2,979	2,107	2,366	1,581	2,145
COD kg/day	14,028	10,447	9,142	9,308	9,710	9,468

Under the banner of improving Taiwan's industrial competitiveness internationally, the government has encouraged and implemented policies of refuse classification and resource recovery. To fully utilize the secondary fibers recovered from various categories of waste paper, YFY has successfully manufactured printing and writing paper out of this material. Table 8 indicates how that fiber resource has been utilized in an environmentally friendly manner.

Table 8. Sales of writing and printing paper from recovered fibers in YFY (ton)

Year	1991	1992	1993	1994	1995
Amount (ton)	3,188	7,716	9,211	9,290	11,200

3.6 Capital input for pollution control

It is now widely accepted that a healthy environment costs a lot of money.

Sustainable development has been viewed by the pulp and paper industry as an opportunity to create more international competitiveness rather than as an additional environmental burden. Cutting costs and liabilities can result eventually in reduced prices that improve competitiveness and increase profits, which, in turn, can provide capital for mill modernization, expansion or R &D.

The high investments for pollution control that occurred in the years 1992 and 1993 (Table 9[11]) are explained by a rather stringent regulation issued for all mills to follow by the end of 92. The even stricter standards established to be met with by 1998 (Table 6) provide incentives for the increasing investment in facilities for pollution prevention in the year 1995.

Table 9.Capital investment for pollution control in pulp and paper industry

Item year	Waste water	Foul gas	Solid waste	Noise	Total
1992	31,487	5,169	675	389	37,720
1993	23,095	6,188	1,148	412	30,843
1994	14,239	1,651	1,078	148	17,116
1995	17,168	2,748	1,732	1,471	23,119

The pulp and paper industry in Taiwan, due to its ceaseless efforts in the investment and construction of effective pollution abatement facilities, which vastly improved the effluent and air qualities, has generally been granted effluent and gases discharge permits by the government. Table 10 shows the improvement of water consumption and effluent quality in TPPC through the installation of oxygen delignification unit and a new effluent treatment system.

Table 10. Influence of oxygen delignification on effluent improvement in TPPC

| Bleaching sequence | Brightness % ISO | Viscosity dm^3/kg | Water consumption m/t | Effluent/t characteristics of pulp | | | | |
				Quantity m^3	Color kg	COD kg	Organic chlorine kg	Bleaching cost US/t
Conventional C-E-D-E-D	90	960	45	45	135	80	9	29.5
O-D-E-D	90	860	20	10	12	25	1	31.4

4. GENERAL

It is obvious that Taiwan has entered the middle stage of what can be termed an advanced technical civilization. This achievement has been supported by a number of factors: remarkable planning at many levels, a well-designed educational system, sound financial practices and fifty years of very hardworking and sacrifice inhabitants.The experiences some have accumulated by learning and living in other more advanced countries for a certain period should be added to the above mentioned. The strides our society has managed to accomplish economically have inevitably led to problems. Environmental pollution is listed on top.

Highly competitive market conditions coupled with tariff reduction and trade liberalization in recent years will definitely constrain the increasing growth rate enjoyed by the forest products industries, including pulp and paper. For the industry to maintain healthy growth, as well as to meet the increasingly stringent environmental restrictions, secure renewable resources, including improvement of wastepaper utilization and pollution control, are essential.

Conservation, sustainable management and multiple uses of forest resources have been emphasized in Taiwan since the major change of forest policy in the eighties. Continuous capital investment in pollution abatement in the pulp and paper industry has enabled most of the mills to obtain discharge licenses. Through effective collection and use of wastepaper, the industry has helped in reducing 230 million tons of municipal waste yearly, meaning a saving of 306.4 million US dollars per year. The reuse of 4.10 million tons of wastepaper implies also an abatement of 3.65 million tons of CO_2 emission.

Cooperation in joint sourcing and pooling of R & D information are critical in the long run, both to increase resource supply and to advancing technology of pollution prevention. Both of which are essential to the survival of the industry and improvement of living standard on the island.

5. REFERENCES

1. Yuen-Foong-Yu Paper Company (1998) : Personal correspondence.

2. The Third Forest Resources and Land Use Inventory in Taiwan 1995 Taiwan Forest Bureau.

3. Lai, Chien Hsing (1998) : The role of Forest Products Industry in the pursuing of higher National Competitiveness. Proceedings of 1998 Annul meeting for the Forest Products Association of R.O.C.

4. Lien, Ching Chang (1997) : Analysis on the supply-demand and self-sufficiency ratio of timber in Taiwan. Industries Note 27(1):44-64

5. The Statistics of Taiwan paper industry (1997) : Taiwan paper Industry Association.

6. DGBAS, Executive Yuan 1996 Statistical Abstract of National Income in Taiwan Area R.O.C., Directorate General of Budget, Accounting and Statistics.

7. Takeyama, S. and H. Otsuka (1994) : Waste paper utilization in Japan and its effect on the environment. International Environment Conference p.329-335.

8. Pulp and Paper 1997 North American Factbook

9. Pulp and Paper International 1997 September

10. Huang, Shou Tsu (1997) : Strategy for Sustainable Development of Pulp and Paper Industry. Proceedings of Japan Annual Meeting (50th Anniversary) pp.93-102

11. Report on Industrial statistics Investigation, Ministry of Economy, R.O.C. 1992-1995.

Forests as Ecosystems Within a Changing Environment

John L. INNES

Swiss Federal Institute for Forest, Snow and Landscape Research
CH-8903 Birmensdorf, Switzerland

ABSTRACT   Environmental change is a major issue facing forest managers today. Both the physico-chemical and socio-economic environments are changing, with major implications for the sustainable management of forest ecosystems. Traditionally, only air pollution and climate change have been included in assessments of climate change. However, there is now a need to examine the impacts of changes in the ways that forests are viewed, changes in technology, and changes in economic forces, such as trading patterns. Forest scientists have a responsibility to ensure that the implications of such changes are brought to the attention of those responsible for the management of forest ecosystems.
Key words: Environmental change, climate change, sustainable forestry, socio-economic changes

1. INTRODUCTION

Forests today are growing in a rapidly changing environment. Much concern has been expressed about the possible impacts of climate change on forests, but the environment is also changing in many other ways. For example, air pollution remains a problem in many countries, despite international attempts to control pollutants such as sulphur dioxide and the nitrogen oxides through the Convention on Long-Range Transboundary Air Pollution. While some of the most important point sources of pollution have been eliminated, air pollution is increasing as a problem in many rapidly industrializing countries. In addition, ozone remains a potential threat to forests in many developed countries. At the same time as these changes in the physical and chemical environment, there has been a shift in the way that many forests are viewed. While timber production remains the most important aim of forestry in many areas, there is an increasing trend to view forests as ecosystems which provide a range of different benefits. To achieve sustainability, these ecosystems need to be managed in a holistic fashion and across a variety of scales. For example, fire suppression policy in the USA has changed markedly in the last 20 years (e.g. Agee 1993; Bancroft et al. 1985) in response to the realisation that fire is an integral feature of many forest ecosystems. Similarly there is increasing interest in managing forests as ecosystems (e.g. Boyce and Haney 1997): a socio-economic change that is likely to have major and long-lasting impacts on forest environments.

The consequences of these changes in the forest environment are very poorly understood, especially for ecosystem components other than the trees. Often, the impacts of specific changes have been analysed separately without paying sufficient attention to potential interactions. In Europe in the 1980s, much research was conducted on the impacts of air pollution without taking into account other changes occurring in the forests (such as the recovery of soils from former litter-raking practices). The research initially concentrated on forest trees, later being extended to forest soils. Very little attention was paid to the forest fauna, and interactions within the ecosystem, such as the impacts of forest food webs, remain virtually unknown. Much the same can be said about studies of carbon dioxide impacts. These are often conducted without taking account of changes in other atmospheric components, particularly ozone. Consequently, it is important to recognise the deficiencies in our knowledge of forest ecosystems if they are to be sustainably managed in the future.

Increasingly, modelling is being used to develop scenarios of future forest conditions (c.f. Shugart 1998). Models are being used to develop climate scenarios given specific greenhouse gas emission projections (Houghton et al. 1995), they are being used to examine likely levels of pollution in the future (e.g. Posch et al. 1997) and they are being used to derive estimates of the likely responses of ecosystems to such changes (e.g. CCIRG 1996). While such models provide invaluable information in an area of major uncertainty, they need to be supported by real data collected in forests. The collection of these data represents a major challenge for forest ecologists and requires the development of inter- and multi-disciplinary research projects which simultaneously investigate different aspects of the forest ecosystem. Such data can then be used for the calibration and, eventually, the checking of the models. The application of the models to forest management practices is also surrounded by uncertainty, leading to the development of adaptive forest management practices (Walters 1986) and other techniques for making decisions when faced with uncertainty (e.g. Smith 1997).

2. CLIMATE

Climate is a major determinant for forest ecosystems. The Intergovernmental Panel on Climate Change (IPCC) has identified that the global average surface temperature is about 15°C, if Antarctica is excluded (Houghton et al. 1995). Since the late 19th century, global surface temperatures have increased by between 0.3°C and 0.6°C, with the increase over the last 40 years being between 0.2°C and 0.3°C. The warming has not been distributed evenly, with the greatest increases being found over the continents between 40°N and 70°N. This represents a substantial part of the temperate forests of the World and the majority of its boreal forests. Modelling efforts suggest that an average warming at these latitudes of 1–3.5°C may occur by the year 2100. This is equivalent to a pole-ward shift in the appropriate isotherms of about 150–550 km, or an altitude shift of about 150–550 m. There is much speculation as to whether forests will be able to adapt to these changes. Historical information suggests that they may not, but it is difficult to relate what has happened in the past to what might happen in the future.

The changes in temperature that have already occurred have been accompanied by changes in other climatic parameters, including precipitation, cloud cover and the magnitude–frequency relations of extreme climatic events (e.g. droughts, storms and frosts). In Switzerland, an analysis by Rebetez and Beniston (1998) has indicated that changes in mean temperatures have been accompanied by changes in the distributions of extreme values. Specifically, diurnal temperature ranges have decreased as has the variability of winter temperatures, both concurrent with an increase in mean temperatures. Such changes are very important, as the role of extreme climatic events in controlling the dynamics of species within forest ecosystems appears to have been underestimated in the past (Innes 1998). However, although it has been suggested that there may be changes in the magnitude and frequency of extreme events such as tornadoes, there is still considerable uncertainty as to the nature of such changes.

The latest report produced by the IPCC (Watson et al. 1998) examines the regional impacts of climate change, and emphasises that these are likely to vary regionally. Model simulations of vegetation distribution suggest that there may be substantial changes in the vegetation boundaries at higher elevations and latitudes, although the reliability of such simulations remains uncertain. The possible warming of 1–2°C that may occur in Africa over the next 50 years is likely to affect the distribution of forests, but changes in precipitation patterns may be equally or even more important. In tropical Asia, any impacts on forests are likely to be extremely complex and difficult to predict. The IPCC report suggests that the area of tropical forest in Thailand could increase from 45% to 80% of the total forest cover whereas in Sri Lanka, a decrease in wet forest and an increase in dry forest might occur. The report suggests that because of the differential responses of individual species to climate change, some forest types may disappear and that new ones may be formed.

3. SOILS

There is increasing evidence that the chemistry of soils in many areas is changing. For example, in Sweden, soils sampled in 1927 were re-sampled in the 1980s and a general decrease in pH of between 0.3 and 0.9 units was found (Tamm and Hallbäcken 1986). Comparisons of the rates of acidification between areas with high and low levels of acidic deposition indicated that acidification had been greater in areas with higher levels of acidic deposition (Tamm and Hallbäcken 1988). There was no evidence of confounding factors: differences in forest history, climate and soil mineralogy did not explain the observed trends. Acidification as a result of acidic deposition appears to have been particularly important in some European countries, and there is also evidence of acidification of some forest soils in eastern North America (Johnson et al. 1991). The extent of the acidification however is unclear: the presence of an acidic soil does not mean that recent soil acidification has occurred. Mapping of critical loads for sulphur and nitrogen suggests that current deposition rates will still result in soil acidification in many areas (Posch et al. 1997). The extent of the acidification is such that adverse effects on sensitive forest ecosystems can be expected.

Soil acidification resulting from acidic deposition should not be confused with acidification attributable to other processes. For example, studies in Australia (Webb 1954) and western North America (Bormann and DeBell 1981, Binckley et al. 1984) indicate that rapid soil acidification can occur as a result of internal ecosystem processes, such as the uptake of base cations by trees and nitrification associated with excessive nitrogen fixation. In cases where the nitrogen fixation exceeds the capacity of the forest ecosystem to accumulate nitrogen, nitrification is known to occur (van Miegroet and Cole 1984), resulting in the introduction of nitric acid into the soil solution.

Short-term changes associated with, for example, droughts, can cause temporary disturbances to forest nutrition. However, it is the longer term changes that are of greatest interest. Many areas are experiencing acidification, but eutrophication is also a problem, and the two process may actually occur simultaneously (Thimonier et al. 1994). Nitrogen saturation has been of particular concern in some ecosystems (e.g. Aber et al. 1989, Fenn et al. 1996, Ohrui and Mitchell 1997), but there are controversies over the way in which it is determined and the nature of its impacts on the ecosystem (e.g. Emmett and Reynolds 1996). For example, although widely used in Europe, the absence of seasonal trends in NO_3^- in surface waters does not appear to be a useful indicator of nitrogen saturation in Japanese forests, when nitrogen retention in the forests can simultaneously be very high (Mitchell et al. 1997). Alternative methods for the identification of nitrogen saturation are currently being developed, with biochemical analysis of tree foliage appearing to offer considerable potential (Näsholm et al. 1997).

Tree growth itself results in changes in soil chemistry, especially if the internal cycling is disturbed by for example timber removals or by fire. Other aspects of the mineral cycling within forests are also changing, with climatic change potentially altering the rates at which some soil reactions occur and factors such as air pollution resulting in the deposition of acids and other substances to forest soils. Some models, such as LINKAGES (Pastor and Post 1986), take these changes into account; many others do not.

In addition to changes in soil acidity, there may be changes in other chemical characteristics of the soil and in soil structure. Many soils have been damaged by inappropriate management practices in the past. For example, litter-raking resulted in the severe depletion of nutrients in some forests in central Europe (Emanuelsson 1988), and whole-tree harvesting may still represent an important drain on the nutrient status of forest soils where it is practised (Nykvist and Rosén 1985), particularly in short-rotation tropical forest plantations. Soils may also be disturbed during harvesting operations (e.g. Dyrness 1965; Bockheim et al. 1975), sometimes irreversibly. Unlike many potential environmental changes, damage to the soil during harvesting operations can usually be reduced or even avoided by the selection of suitable harvesting times (e.g. during the winter when the soil is frozen) or techniques (e.g. the use of sky-lines).

4. DISTURBANCE

Another major aspect of the forest environment is the nature, frequency and extent of disturbances, particularly those associated with anthropogenic activities. White and Pickett (1985, p.7) define a disturbance as "any relatively discrete event in time that disrupts ecosystem, community, or population structure and changes resources, substrate availability, or the physical environment". Forman (1995) simplifies this to "any event that significantly alters the pattern of variation in the structure or function of a system". As such, disturbances include both destructive events and environmental fluctuations although in most discussions of the subject, it is the destructive events that are emphasized. Disturbance is being increasingly recognized as a key biological process in forests. All natural forests are subject to disturbance, and attempts by man to prevent these (e.g. through fire suppression policies) are now seen as having adverse effects on the forest ecosystem. Forman (1995) actually argues that regular disturbances, such as low-intensity fires in pine forest ecosystems should be considered as a stress rather than a disturbance, whereas a fire suppression policy should be viewed as the disturbance. This argument could also be extended to drought: the regular, seasonal droughts that occur in some climates should be seen as a stress, whereas a drought occurring in an ecosystem that has no adaptation to such a phenomenon should be seen as a disturbance. Consequently, a disturbance is generally considered as a normal, but infrequent, event within a system (Forman 1995).

Although traditional ideas about forest ecosystems have concentrated on the role that that external factors play in creating them, internal factors may be as important or even more important. Both natural and anthropogenic disturbances play a crucial role in determining the functioning, structure and composition of forest ecosystems at any point in time. This is becoming increasingly apparent as forests previously believed to be undisturbed are found to have evidence of major disturbances. A particularly important aspect of disturbance studies is the magnitude–frequency relations of the disturbances. Within any given forest, it is possible to relate the magnitude of a disturbance to its frequency. Generally, the larger a disturbance, the more infrequent it is. Thus, any particular forest will be subject to a variety of disturbances of differing magnitude. A third characteristic of a disturbance is its intensity, which is a measure of the proportion of biomass destroyed by the disturbance within a given area (Crawley 1997).

One aspect of disturbance that deserves particular attention is the invasion of forest ecosystems by exotic species. These can include both deliberately-introduced plant species, such as the invasion of southern European forests by *Robinia pseudoacacia* and *Prunus serotina*, and accidentally-introduced pathogens, such as *Adelges piceae* and *Cryphonectria parasitica* in North America. Introduced pathogens may have particularly dramatic effects on the composition of forest ecosystems. The American Chestnut (*Castanea dentata*) was an important component of the transition forests of eastern North America, but has now been eliminated as a canopy species by chestnut blight

(*Cryphonectria parasitica*). The incidence of such invasions has been facilitated by the ease with which biological material can be carried around the world. In addition, changing climatic conditions may favour the spread of certain pathogens beyond their current ranges (c.f. Speight and Wainhouse 1989).

5. SOCIO-ECONOMIC CHANGES

In many parts of the world, there have been major changes in the way in which forestry is approached. The increasing emphasis being placed on the sustainable management of forest ecosystems has led to quite substantial changes in the ways in which forests are viewed and managed. These changes are particularly significant for the non-timber aspects of forest ecosystems, such as species diversity. Forests are no longer being seen only as a timber resource: today they must fulfil many different functions. As a result, attention is now being given to components of forest ecosystems that previously were of low priority. A variety of different approaches to forest management are being advocated, some better based on scientific knowledge than others. Unfortunately, the polarisation of the debate between proponents of ecoforestry (c.f. Drengson and Taylor 1997) and those proposing a modification of traditional practices better to fulfil ecological needs (c.f. Kohm and Franklin 1997) has resulted in considerable confusion over the best practices to be adopted. This is shown by the difficulties and controversy surrounding the certification of forests as being sustainably managed.

The extent to which these changes will affect forest ecosystems is a matter of much speculation. Kimmins (1997) usefully divides current approaches to the issue into traditional (silvicultural), ecological and environmental viewpoints. He illustrates these different viewpoints in relation to different perceptions of stand-level forest health (Table 1).

These differences in approach to the identification of stand-level forest health illustrate why it is often very difficult to obtain agreement over concepts in modern forestry, even one so widely discussed as forest health.

At the same time as these changes in the way in which forests are viewed, the technology used to exploit forests is changing. This are also likely to have an impact on forest ecosystems. For example, a trend toward whole-tree harvesting would result in lower amounts of slash being left. Similarly, changes in the use of fire following clear-cutting would have major impacts on the ecosystem (Kimmins 1997). The utilisation of smaller trees will change the dynamics of the forest ecosystem and might result in early successional species being favoured at the expense of late-successional species.

6. IMPLICATIONS FOR THE MANAGEMENT OF FOREST ECOSYSTEMS

Forest ecosystems are highly dynamic entities. The concept of a stable, climax forest today has little credibility (Shugart 1998). Instead, forests are usually seen as a mosaic of different-aged patches. Changes in the physical and socio-economic forest environment mean that the nature and dynamics of these patches will change over time. Such change is not new. For example, pollen analysis has revealed that the composition of forest communities can vary over time. There are species assemblages recorded from the last 10,000 years that are no longer found today, and some of today's communities appear to have no historical precedent (Fuller et al. 1998). This poses a problem in relation to forest conservation. Should management interventions be made to try and preserve particular combinations of species that are no longer matched to their environment? To what extent should relict populations of a species be protected? Should there be management interventions in nature reserves established to protect early ecosystem seral stages?

Table 1. Stand-level forest health, as defined from different viewpoints (from Kimmins 1997).

Traditional silvicultural	Ecological	Environmental
• forest trees are vigorous and productive • there is rapid biomass accumulation and nutrient cycling • insects and diseases have few negative effects on the development of tree populations and plant communities • processes of ecological succession and stand development proceed at rates characteristic of the particular seral stage of the particular forest ecosystem • the forest ecosystem has high 'vitality' and is able to withstand stress	• there is a physical environment, biotic resources, and trophic networks to support productive forests during at least some seral stages • there is resistance to catastrophic change and/or the ability to recover from catastrophic change at the landscape level • there is a functional equilibrium between the supply and demand of essential resources for major portions of the vegetation, or a diversity of seral stages and stand structure that provides habitat for many native species and all essential ecosystem processes	• the forest has reached its peak of development, i.e. a late seral stage • the canopy is multi-layered with many large gaps, and the trees are uneven-aged • there are many large, dead, deformed, broken, and partly decayed trees and decaying logs that provide habitat for the animals that use these structures • rare and specialist species considered to be indicators of a 'healthy' forest ecosystem are present

Much research has adopted a fairly static approach to forest ecosystems. Recent developments are moving away from this trend, with increasing interest in concepts such as forest ecosystem integrity and resilience and forest quality. These require a much more dynamic view of forests than has been held in the past. For example, the recognition of the role of disturbance in maintaining forest ecosystems is a very important development that has led many managers to allow patches of forest to recover from disturbances without major management interventions. However, both the integrity and the resilience of forest ecosystems are extremely difficult to quantify, and better indicators are required for these important characteristics.

The environmental changes also require that the concept of site quality be re-assessed. Much forestry is based on the prediction of site–yield relationships. However, there is evidence that site quality is changing (Spiecker et al. 1996), and the anticipated future changes in the environment are such that the use of yield predictions based on past growth to predict future growth must be seriously questioned. In future, knowledge-based (as opposed to experience-based) methods, particularly process models, are likely to become much more important. However, to date, they have been applied only rarely in yield forecasting work. Kimmins (1997) argues that the most promising development is the use of hybrid models, which combine experience- and knowledge-based methods. Such models include JABOWA, LINKAGES, ZELIG, FORCYTE and FORECAST.

7. CONCLUSION

The changes occurring in forests extend beyond those normally seen as part of the environment, namely physical factors such as climate and soil conditions. Today it is also necessary to take into account changes in management philosophy when considering forests. The new combinations of physical, social and economic factors mean that many of the more traditional practices in forestry, such as the assessment of site quality, require radical modification. It is no longer sufficient simply to match the most suitable species to a site, as other considerations may mean that the species is unacceptable. In addition, as our knowledge of the genetic variability of forest trees increases, it is likely that increasing emphasis will be placed on the selection of the correct genotype for a site. These changes are becoming particularly apparent in relation to plantation forestry in developed countries, but are likely to increase in importance in many other countries as pressure to ensure sustainable forestry practices increases.

The forest manager seeking to adopt a more ecological approach to forest management is faced with a huge range of recommendations. Many of these recommendations are based on dogma rather than sound scientific facts. Forest scientists have a duty to help sift through the available information, to fill some of the major gaps in our knowledge of forest ecosystem processes and to present the information in a form that can be used by those responsible for managing our forests.

REFERENCES

Aber, J.D., K.J. Nadelhoffer, P. Steudler, and J. Melillo (1989) : Nitrogen saturation in northern forest ecosystems. BioScience, Vol. 39, 278–386.

Agee, J.K. (1993) : Fire Ecology of Pacific Northwest Forests. Island Press, Covelo, California, 493p.

Bancroft, L., T. Nichols, D. Parsons, D. Graber, B. Evison, and J.W. van Waglendonk (1985) : Evolution of the natural fire management program at Sequoia and Kings Canyon National Parks. In: J.E. Lotan, B.M. Kilgore, W.C. Fischer and R.W. Mutch (tech. coords.), Proceedings – Symposium and Workshop on Wilderness Fire, Missoula, Montana, November 15–18, 1983. USDA Forest Service, General Technical Report INT–182, pp. 174–180.

Binckley, D., J.D. Lousier and K. Cromack Jr. (1984) : Ecosystem effects of Sitka alder in a Douglas-fir plantation. Forest Science, Vol. 32, 26–35.

Bockheim, J.G., T.M. Ballard, and R.P. Willington (1975) : Soil disturbance associated with timber harvesting in southwestern British Columbia. Canadian Journal of Forest Research, Vol. 5, 285–290.

Bormann, B.T. and DeBell, D.S. (1981) : Nitrogen content and other soil properties related to age of red alder stands. Soil Science Society of America, Journal, Vol. 45, 428–432.

Boyce, M.S. and A. Haney (eds.) (1997) : Ecosystem Management. Applications for Sustainable Forest and Wildlife Resources. Yale University Press, New Haven, 361p.

CCIRG (Climate Change Impacts Review Group) (1996) : The Potential Effects of Climate Change in the United Kingdom. HMSO, London, 108p.

Crawley, M.J. (1997) : Life history and environment. In: Crawley, M.J. (ed.) Plant ecology. Blackwell Science, Oxford, 73–131.

Drengson, A.R. and D.M. Taylor (eds.) (1997) : Ecoforestry. The Art and Science of Sustainable Forest Use. New Society Publishers, Gabriola Island, British Columbia, 312p.

Dyrness, C.T. (1965) : Soil surface condition following tractor and high-lead logging in the Oregon Cascade. Journal of Forestry, Vol. 63, 272–275.

Emanuelsson, U. (1988) : The relationship of different agricultural systems to the forests and woodlands of Europe. In: F. Sabiano (ed.) Human Influence on Forest Ecosystems Development in Europe. Pitagora Editrice, Bologna, 169–178.

Emmett, B.A. and B. Reynolds (1996) : Nitrogen critical loads for spruce plantations in Wales: Is there too much nitrogen? Forestry, Vol. 69, No. 3, 205–214.

Fenn, M.E., M.A. Poth and D.W. Johnson (1996) : Evidence of nitrogen saturation in the San Bernardino Mountains in southern California. Forest Ecology and Management, Vol. 82, 211–230.

Forman, R.T.T. (1995) : Land Mosaics. The ecology of landscapes and regions. Cambridge University Press, Cambridge, 632 p.

Fuller, J.L., D.R. Foster, J.S. McLachlan and N. Drake (1998) : Impact of human activity on regional forest composition and dynamics in Central New England. Ecosystems, Vol. 1, No. 1, 76–95.

Houghton, J.T., L.G. Meira Filho, B.A. Callander, N. Harris, A. Kattenberg, and K. Maskell (eds) (1995) : Climate Change 1995: The Science of Climate Change. Cambridge University Press, Cambridge, 572p.

Innes, J.L. (1998) : The impact of climate extremes on forests: an introduction. In: M. Beniston and J.L. Innes (eds.), The Impacts of Climate Change on Forests, Springer Verlag, Berlin, in press.

Johnson, D.W., M.S. Cresser, S.I. Nilsson, J. Turner, B. Ulrich, D. Binkley and D.W. Cole (1991) : Soil changes in forest ecosystems: evidence for and probable causes. Proceedings of the Royal Society of Edinburgh, Vol. 97B, 81–116.

Kauppi, P., P. Anttila, and K. Kenttämies (eds.) (1990) : Acidification in Finland. Springer Verlag, Berlin, 1237p.

Kimmins, H. (1997) : Balancing Act. Environmental Issues in Forestry. UBC Press, Vancouver, 305p.

Kohm, K.A. and J.A. Franklin (eds.) (1997) : Creating a Forestry for the 21st Century. The Science of Ecosystem Management. Island Press, Washington D.C., 475p.

Mitchell, M.J., G. Iwatsubo, K. Ohrui, and Y. Nakagawa (1997) : Nitrogen saturation in Japanese forests: an evaluation. Forest Ecology and Management, Vol. 97, No. 1, 39–51.

Näsholm, T., A. Nordin, A.-B. Edfast, and P. Högberg (1997) : Identification of coniferous forests with incipient nitrogen stauration through analysis of arginine and nitrogen-15 abundance of trees. Journal of Environmental Quality, Vol. 26, No. 1, 302–309.

Nykvist, N. and K. Rosén (1985) : Effect of clear-felling and slash removal on the acidity of northern coniferous soils. Forest Ecology and Management, Vol. 11, 157–169.

Ohrui, K. and M.J. Mitchell (1997) : Nitrogen saturation in Japanese forested watersheds. Ecological Applications, Vol. 7, No. 2, 391–401.

Pastor, J. and W.M. Post (1986) : Influences of climate, soil moisture, and succession on forest carbon and nitrogen cycles. Biogeochemistry, Vol. 2, 3–27.

Posch, M., J.-P. Hettelingh, P.A.M. de Smet and R.J. Downing (eds.) (1997) Calculation and Mapping of Critical Thresholds in Europe. Status report 1997. RIVM Report 259101007. Coordination Center for Effects, National Institute of Public Health and the Environment, Bilthoven, The Netherlands, 163p.

Rebetez, M. and M. Beniston (1998) : Changes in temperature variability in relation to shifts in mean temperatures in the Swiss Alpine region this century. In: M. Beniston and J.L. Innes (eds.), The Impacts of Climate Change on Forests, Springer Verlag, Berlin, in press.

Reuss, J.O. and D.W. Johnson (1986) : Acid Deposition and the Acidification of Soil and Water. Springer-Verlag, New York, 119p.

Shugart, H.H. (1998) : Terrestrial Ecosystems in Changing Environments. Cambridge University Press, Cambridge, 537p.

Smith, G.R. (1997) : Making decisions in a complex and dynamic world. In: Kohm, K.A. and J.A. Franklin (eds.) : Creating a Forestry for the 21st Century. The Science of Ecosystem Management. Island Press, Washington D.C., 419–436.

Speight, M.R. and D. Wainhouse (1989) : Ecology and Management of Forest Insects. Clarendon Press, Oxford, 374p.

Spiecker, H., K. Mielikäinen, M. Köhl, and J.P. Skovsgaard (eds.) (1996) : Growth Trends in European Forests. Springer Verlag, Berlin, 372p.

Tamm, C.O. and L. Hallbäcken (1986) : Changes in soil pH over a 50-year period under different forest canopies in SW Sweden. Water, Air, and Soil Pollution, Vol. 31, 337–341.

Tamm, C.O. and L. Hallbäcken (1988) : Changes in soil acidity in two forest areas with different acid deposition: 1920s to 1980s. Ambio, Vol. 17, No. 1, 56–61.

Thimonier, A., J.L. Dupouey, F. Bost and M. Becker (1994) : Simultaneous eutrophication and acidification of a forest ecosystem in North-East France. New Phytologist, Vol. 126, 533–539.

Van Miegroet, H. and D.W. Cole (1984) : The impact of nitrification on soil acidification and cation leaching in a red alder forest. Journal of Environmental Quality, Vol. 13, 586–590.

Walters, C.J. (1986) : Adaptive Management of Renewable Resources. Macmillan, New York.

Watson, R.T., M.C. Zinyowera and R.H. Moss. (1998) : The Regional Impacts of Climate Change. An Assessment of Vulnerability. Cambridge University Press, Cambridge, 517p.

Webb, L.J. (1954) : Aluminium accumulation in the Australian–New Guinea flora. Australian Journal of Botany, Vol. 2, 176–196.

White, P.S. and Pickett, S.T.A. (1985) : Natural disturbance and patch dynamics: An introduction. In: Pickett, S.T.A. and White, P.S. (eds.) The ecology of natural disturbance and patch dynamics. Academic Press, Orlando, 3–13.

Carbon Absorption by Temperate Forest Ecosystems: Problems and Responses to a Changing Environment

Giuseppe SCARASCIA-MUGNOZZA, Paolo DE ANGELIS,
Giorgio MATTEUCCI and Riccardo VALENTINI

Department of Forest Environment and Resources, University of Tuscia
Viterbo, Italy

ABSTRACT Global change is becoming a decisive environmental issue of the present time; to elucidate causes and effects at the atmosphere and biosphere levels, a major effort is being devoted world-wide, by the scientific community. Forest ecosystems can contribute to mitigate these changes because of their key role played in the energy and mass exchanges between the atmosphere and the geosphere. Several, important scientific and methodological problems still need to be solved in order to fully understand mitigation potentials of temperate forests and to answer to questions about their optimal management strategies. In this paper, results of recent researches conducted on global change and temperate forests will be provided.
Keywords: Global Change, Forest Ecosystems, Mitigation, Carbon Sequestration, Elevated CO_2.

1. INTRODUCTION

Global change is becoming a decisive environmental issue of the present time; to elucidate causes and effects at the atmosphere and biosphere levels, a major effort is being devoted world-wide, by the scientific community. Obviously, the interest is not limited to the research milieu but a vast demand for more information and management options is originating from the public opinion and political institutions. Earth is a dynamic system and dramatic environmental changes have frequently occurred in the past; however, two factors make the changes that our planet is presently experiencing, so unique:
- man's activities have now the capability to determine profound effects on the biosphere and the atmosphere;
- the rate of development of the environmental changes is much faster than those which took place in the past millions of years.
Global changes can be defined as "those changes that alter the well-mixed fluid envelopes of the Earth system (the atmosphere and the oceans) and those that occur in discrete sites but are so widespread as to constitute a global change" (Vitousek, 1992). Forest ecosystems can contribute to mitigate these changes because of their key role played in the energy and mass exchanges between the atmosphere and the geosphere. In fact, forests cover slightly more than 30% of the land surface but represent about 70% of its net primary production and stock

119

more than 90% of total biomass (Whittaker and Woodwell 1971). At the same time, it is estimated that from 50 to 70% of the precipitations that fall over terrestrial ecosystems in the tropics come directly from forest evapotranspiration (Salati and Vose 1984). In particular, temperate forests are major sinks of carbon as it has been estimated, by Taylor & Lloyd (1992), that 37% of the total carbon absorbed by terrestrial ecosystems is accumulated in the temperate zone. Therefore, especially the forests of the temperate and boreal zone are thought to be responsible for the accumulation of the "missing" carbon. This search for the "missing" carbon is particularly difficult because the search is for very small net fluxes where the gross fluxes and the stocks of carbon are very large. That explains why only a detailed carbon balance study, which includes all the components of the forest ecosystems and calculates the carbon balance in a standardised way, can yield significantly new results. Several, important scientific and methodological problems still need to be solved in order to fully understand mitigation potentials of temperate forests and to answer to questions about their optimal management strategies:

- what is the effect of different species and forest types and how environmental stresses interact with carbon absorption;
- below-ground carbon accumulation and consumption may represent a major component of a forest carbon budget but little information is presently available;
- the relationship between forests and the environment is a reciprocal one: forests can be strongly affected by environmental modifications, as the increase of atmospheric CO_2 concentration; therefore, there is large uncertainty on the effect of this change "at equilibrium" on the real capacity of forests to sequester carbon.

In this paper, results of recent researches conducted on global change and temperate forests at the University of Tuscia will be provided. Researches have been conducted on three different experimental systems:

- a high elevation, beech forest stand where the yearly carbon budget was assessed;
- a natural, *Quercus ilex* high "macchia" forest stand exposed to increasing atmospheric CO_2 concentrations;
- an experimental stand of different poplars clones.

2. CARBON BUDGET OF A BEECH FOREST

The *Fagus sylvatica* L. study forest is located in Central Italy (41°52' N, 13°38' E), at an elevation of 1564 m a.s.l. on the Apennines mountains. It is an almost pure, high stand of beech trees with a mean height of 20.2 m, a mean basal area of 26.4 m^2 ha^{-1} and standing above-ground wood volume of 293 m^3 ha^{-1}. The yearly carbon budget of the forest stand was determined by different, independent techniques: CO_2 fluxes between the atmosphere and the forest canopy, woody standing biomass and wood increments, fine roots productivity, and soil and trees respiration.

Eddy fluxes of carbon dioxide, water vapour and sensible heat have been determined by the eddy covariance technique (Baldocchi et al. 1988); the fluxes of matter or energy are given by the average product of the fluctuations around the mean of the vertical wind speed and of the concentration of matter or energy. The measurements were carried out at 5 m above the beech forest canopy, by instruments installed on a scaffold 25 m tall. The instrumental set-up consisted of a 3D sonic anemometer (Gill, UK) and a LI-6262 (Licor, USA) CO_2 and H_2O infra-red gas analyser; a pump at the bottom of the line withdraws air through the analyser at a flow rate of about 6 L min^{-1}. Power supply was given by batteries charged by solar panels and by a generator for emergency situations.

Wood volume was calculated by standard forest mensuration techniques, using harvested model trees. Soil cores were collected, at 5 different dates over the vegetative season, to analyse the distribution of fine roots and their turn-over. Respiratory losses during the dormant and the growing season were estimated by measurements of stem and branch respiration and of soil respiration conducted by means of home-built cuvettes and an IRGA analyser (ADC, UK).

Combining measurements of the net carbon dioxide flux density during the growing season and the estimation of respiratory components during winter by cuvette measurements and temperature scaling (Valentini et al., 1996), we were able to assess the annual net ecosystem production (NEP = 472 g C m^{-2} yr^{-1}). The NEP is the amount of total photosynthesized carbon over the whole year (gross primary production, GPP) decreased for the autotrophic (trees and herbs) and heterotrophic (consumers and decomposers) respiration. This fixed carbon is what remains in the ecosystem at the end of the year to make up the increment of living biomass and dead organic matter. Few studies of the NEP of forest ecosystems are available worldwide; Wofsy et al (1993) found a net carbon exchange of about 220 g C m^{-2} yr^{-1}, at Harvard forest, while Greco and Baldocchi (1996) measured a net carbon gain of 525 g C m^{-2} yr^{-1}, for a mixed deciduous forest in Oak Ridge, Tennessee. Our value is intermediate between the two, but closer to the Tennessee forest which also has a similar climate.

In our beech forest, the above-ground NEP estimated by biomass sampling, under the assumption of negligible leaf litter accumulation, was 238 g C m^{-2} yr^{-1}, that is about 50% of the total NEP, indicating a significant below-ground carbon accumulation as living roots and soil organic matter. This result can be explained by the fact that this type of forest is under an active regrowing phase and represents a southern boundary for the geographic distribution of beech in Europe. For this reason the ratio of below ground to above-ground productivity of this ecosystem is about 1.4, a rather high value when compared with other central european beech ecosystems (Reichle 1981).

A significant inter-annual variability of ecosystem carbon storage was also observed at our experimental site when the measuring period was extended for more than three years. Differences among years could be as large as 30% caused by various factors, the most evident being summer water stress, and differences in phenology and leaf area development caused by climatic variation.

3. IMPACT OF ENVIRONMENTAL CHANGES ON THE MEDITERRANEAN "MACCHIA"

Mediterranean woodland communities represent slightly more than 10% of the total forest surface of the world (Walter, 1985). Yet, their importance is relevant because they make up the natural vegetation of some of the most populated and economically active areas of the globe; at the same time, the Mediterranean biome represents the intermediate vegetation between the desert zone and the temperate forests. Their carbon sequestration potential is strongly dependent on environmental stresses and on the likely effect of future climate changes.

In Mediterranean-type ecosystems, the two main factors limiting primary productivity are water and nutrient availability (Specht, 1973; Debano and Conrad, 1978). Additionally, the frequency of disturbances is high due to the occurrence of wildfires during the summer dry season, harvesting of biomass, and animal grazing.

Predictions on the effects of environmental changes, particularly increasing CO_2 concentration, on natural communities of trees and herbs have been traditionally inferred from short term studies conducted on plants raised under controlled conditions. However, according to different authors concern has arisen over the validity of extrapolations from short term, small scale experiments.

A paucity of long-term experiments at the community and natural ecosystem levels have been conducted world-wide until presently. The objective of our research has been, therefore, to examine the impact of environmental stresses and of long-term exposure to an elevated CO_2 concentration on carbon absorption of a natural Mediterranean community of high "macchia dominated by *Quercus ilex*. The site is located near Montalto di Castro (Viterbo), along the Thyrrenian coast, 100 km north-west of Rome, at the E.N.E.L. (National Agency of Electric Energy) reservation (42°22' N, 11°32' E). The vegetation is a Mediterranean evergreen "macchia" ecosystem, 4 to 6 m tall, dominated by *Quercus ilex* trees, with a dense shrub layer made up of *Phillyrea angustifolia* L., *Pistacia lentiscus* L. and *Myrtus communis* L. (*Quercetum ilicis* Br. Bl. association). Within the belt of high "macchia", total above-ground biomass is 35 Mg ha^{-1} of dry matter, vegetation covers about 80% of the ground surface and leaf area index (LAI) ranges from 3 to 4. Holm oak (*Q. ilex*) trees represent 34% of the total number of woody plants, 53% of above-ground biomass and 62% of leaf area index of the forest community; total above-ground productivity is around 2.5 Mg ha^{-1} year^{-1} with the oak contributing for more than one half. The climate in this area is typically temperate-mediterranean, with a mean annual temperature of 15 °C; the maximum temperature in summer can be greater than 35 °C and the minimum winter temperature can be less than -5 °C. The total annual rainfall is around 610 mm; its distribution during the year typically peaks in February and in late September; consequently, the dry season lasts from May until early September.

Within this "macchia" stand, six open top chambers (OTCs) have been installed to test the effect of atmospheric $[CO_2]$ enrichment on clumps of natural, Mediterranean vegetation, starting from early spring 1992. The size of the OTCs is 4 m in diameter and 6 m in height. The air flow rate of 12000 m^3 h^{-1} inside the OTCs changes the air 3-4 times per minute in order to maintain the microclimate inside the chambers similar to outside. The CO_2 concentration of the air inside the OTCs is either ambient or ambient plus 350 μmol mol^{-1}

Carbon absorption potential of this mediterranean forest, under favorable environmental conditions, was similar to carbon net fluxes measured above the beech forest. However, summer droughts are usually pronounced, causing base plant water potential to be as low as -4 to -5 Mpa; the effect on carbon fixation is relevant determining its reduction to almost 10% of its yearly maximum value.

Furthermore, carbon metabolism was also markedly affected by elevated $[CO_2]$; under optimal soil moisture conditions, net photosynthesis of mediterranean species was almost doubled. The lack of downward regulation of A in elevated $[CO_2]$, at least after the 5-year long exposure, was evident from the A/Ci response curves carried out on the three woody species present inside the OTCs, *Quercus ilex*, *Phillyrea angustifolia* and *Pistacia lentiscus*. Photosynthetic activity (A), in elevated $[CO_2]$, was increased by 122% in *Q. ilex*, by 98% in *P. angustifolia* and by only 32% in the other shrub species, *P. lentiscus*, than at ambient $[CO_2]$. On the contrary, *P. lentiscus* had the transpiration (E) and stomatal conductance (Gs) significantly reduced (43%) by the high $[CO_2]$ treatment whereas *Q. ilex* and *P. angustifolia* were less affected: the 11% reduction of E and Gs was not significant in both species. Interestingly enough, the ratio A/E, the instantaneous transpiration efficiency (ITE), increased similarly, by 130% to 150%, in the three species.

A detailed analysis of the effects of elevated $[CO_2]$ exposure on the growth of shoots was conducted on the trees of *Q. ilex*, the ecological dominant of the community, and is presently being conducted on the other woody species. The average size of all the one-year old shoots produced by ten representative branches per tree increased as a result of the exposure to elevated $[CO_2]$; shoot length, diameter and number of leaves significantly augmented by 200%, 38% and 36% respectively. Mean area of leaves, on the other hand was slightly, but not significantly, reduced. The same occurred for the total number of shoots per sampled branch. The overall result of these changes reflected in a slight reduction of the total leaf area per branch and in a large increase of cumulative shoot length and sapwood area per branch; however, these variations were statistically not significant. On the other hand, quite large and significant was the increase of cumulative shoots volume per branch as well as the increase of the efficiency of sapwood and volume production per unit of leaf area. It is also interesting to mention that the increase of total shoots volume production per branch, under elevated $[CO_2]$, was caused by the occurrence of an autumn flush in most branches of the sampled trees whereas the second flush was irregularly present on the branches of ambient $[CO_2]$ trees.

4. GENETIC VARIATION OF CARBON SEQUESTRATION IN POPLAR CLONES

The rate of biomass growth in trees has been traditionally thought not to be comparable with that of herbaceous species or agricultural crops (Jarvis and Jarvis 1964); however, recently, Cannell (1989) showed that some temperate forest tree species, namely *Populus* spp., *Eucalyptus* spp., *Salix* spp. and *Platanus* spp. are capable of high biomass production rates. According to Cannell and Smith (1980) poplar clones, in different regions of the world, can produce as much as 10 to 14 Mg ha^{-1} yr^{-1} of aboveground dry matter. Yet, it seems that the growth potential of this genus has not been adequately evaluated; in fact, Heilman and Stettler (1985) have demonstrated that productivity rates as high as 28 Mg ha^{-1} yr^{-1} can be achieved with *P. trichocarpa* x *P. deltoides* hybrids grown under intensive culture.

The present study was conducted with a subset of four poplar clones forming part of a larger integrated field experiment implemented at the University of Washington (USA) to test recently selected poplar clones (Ceulemans et al. 1992; Scarascia Mugnozza et al. 1997). This subset included one clone each of the two parental species, ILL-005 for *P. deltoides* and 1-12 for *P. trichocarpa*, and two hybrid clones (11-11 and 44-136).

The plantation was established in February 1985 in a rich, alluvial silt loam soil (pH 6) at Farm 5 of the Washington State University Research and Extension Center in Puyallup, WA (47°10' N, 122°10' W) with 25 cm unrooted hardwood cuttings that had been soaked in water for one week. Spacing was 1x1 m (10,000 trees ha^{-1}). Each clonal plot contained three groups of five adjacent trees, surrounded by border rows to reduce the "edge" effect (Cannell and Smith 1980), designed to be utilized for experimental observations and harvesting (i.e. harvest plots). Five to seven trees from each of the four intensively studied clones, were harvested at approximately monthly intervals, each year during the 4-year long experiment. The above-ground portions were separated in smaller parts, placed in large plastic bags and brought to the lab for further analysis. In the first two years the below-ground biomass was also harvested. The below-ground components were excavated within a 1 m^3 of soil delineated at the soil surface by a 1x1 m^2 quadrat around the sample tree and 1 m deep. Although the poplar root system extends beyond these borders early in the life cycle of the tree, it was reasoned that the amount of roots extending beyond the excavated cube was compensated for by the roots of adjacent trees growing into the cube of soil. Therefore, all the roots found within the mentioned volume of soil were attributed to the harvested tree. Tree biomass components were dried in large ventilated ovens at 75° C, until they reached constant weight. Dry weights of all components were measured to 0.001g with a balance.

At the end of the experiment, in the fourth growing season just prior to leaf fall, the "average tree" had a stem height of 12.5 and 11.5 m for hybrid clones 11-11 and 44-136 respectively, compared to 10.3 m for clone 1-12 and 9.3 m for ILL-005. Biomass production data can be expressed in terms of carbon sequestration potential of the study poplar clones. A large variation was observed between the different clones, with the hybrid 11-11 being capable to fix more than 1100 g C m^{-2} yr^{-1} while the least productive, *P. deltoides* clone, less than 450 g C m^{-2} yr^{-1}; the other clones were intermediate between the two with the other hybrid, 44-136, sequestering 850 g C m^{-2} yr^{-1} and the *P. trichocarpa* clone more than 700 g C m^{-2} yr^{-1}. These values might be somehow overestimated because root turn-over in winter was not considered; however, part of leaves and roots litter can accumulate in the soil as dead organic matter.

5. CONCLUSION

Carbon sequestration potential of tree stands and natural forests has been seldom determined in all its major components, that is considering also such processes as roots productivity and soil respiration. Furthermore, to gain a complete evaluation of the role of forests in the global carbon cycle, the entire life cycle of a forest should be taken into consideration as well as the various types of wood utilization. In our experiments we attempted to cover a reasonably long time interval, at least an entire year or more, and also the below-ground processes in order to obtain more realistic information on trees and forests.

Genetic variation among and within forest tree species is one of the most important factors controlling carbon absorption potential: not only because of the variability of photosynthetic characteristics but also because it influences tree phenology, amount of foliage produced and the root to shoot distribution of biomass. The variability of net ecosystem production we observed among tree species was large, ranging from 1100 g C m^{-2} yr^{-1} for very productive poplar clones to about 400 g C m^{-2} yr^{-1} for slow growing poplars as well as for other traditional forest trees as beech. The possibility to increase by a factor of 2 to 3 the average net rate of carbon sequestration by forest trees is an interesting option; obviously, in this case good soil and environmental conditions will be needed.

From the beech study we conclude that carbon accumulation in the soil is an important process which deserves to be studied more intensively. For this reason a combination of eddy covariance fluxes and biomass sampling can be an effective strategy in providing useful information on the impact of vegetation structure and management on carbon conservation in forest ecosystem. A significant inter-annual variability of ecosystem carbon storage was also observed in our experiment and represents another important factor that requires in-depth examination.

Finally, the effect of a changing environment on productivity and carbon absorption of forest stands can be dramatic even though it is presently quite difficult to correctly anticipate how these changes will act. Climatic modifications for the mediterranean environment are expected to influence mostly the rainfall regime and the air temperature but there is not much consensus about the more likely direction of the changes of these parameters. The atmospheric concentration of carbon dioxide is, however, the only globally changing parameter that is regularly monitored world-wide and that is steadily increasing over the years. Some experiments are presently undergoing all over the world to study the impact of the raising CO_2 concentration at the whole population and ecosystem levels. Unfortunately, it is still uncertain whether the CO_2-fertilization effect on photosynthesis will persist over the years; in fact it may also be possible that leaf carbon metabolism will acclimate to a double CO_2 world or that biotic and abiotic stresses will eventually compensate the initial advantage provided by the increased CO_2. Anyhow, CO_2 may also interact with other factors such as temperature or anthropogenic nitrogen fertilization reinforcing its positive effect. Basic and applied research in this domain is badly needed.

REFERENCES

Baldocchi D.D., B.B. Hicks and T.P. Meyers (1988): Measuring biosphere-atmosphere exchanges of biologically related gases with micrometeorological methods. Ecology 69: 1331-1340.

Cannell M.G.R. and R.I. Smith (1980) Yields of minirotation closely spaced hardwoods in temperate regions: review and appraisal. For. Sci. 26: 415-428.

Cannell M.G.R. (1989): Physiological basis of wood production: a review. Scand. J. For. Res. 4: 459-490.

Ceulemans R., G.E Scarascia-Mugnozza, B.M. Wiard, J.H. Braatne, T.M. Hinckley, R.F. Stettler, J.G. Isebrands and P.E. Heilman (1992): Production physiology and morphology of *Populus* species and their hybrids grown under short rotation. I. Clonal comparison of 4-year growth and phenology. Can. J. For. Res. 22: 1937-1948.

Debano L.F. and C.E. Conrad (1978): The effect of fire on nutrients in a chaparral ecosystem. Ecology 59: 489-497.

Greco S. and D.D. Baldocchi (1996): Seasonal variations and water vapour exchange rates over a temperate deciduous forest. Global Change Biology 2: 183-197.

Heilman, P.E. and R.F. Stettler (1985): Genetic variation and productivity of *Populus trichocarpa* T. & G. and its hybrids. II. Biomass production in a 4-year plantation. Can. J. For. Res. 15: 384-388.

Jarvis P.G. and M.S. Jarvis (1964): Growth rates of woody plants. Physiol. Plant. 17: 654-666.

Reichle D.E. (1981): Dynamic properties of forest ecosystems. IBP 23, Cambridge University Press, Cambridge, 683 pp.

Salati E. and P.B. Vose (1984): Amazon Basin: system in equilibrium. Science 225: 129-138.

Scarascia-Mugnozza G., R. Ceulemans, P.E. Heilman, J.G. Isebrands, R.F. Stettler and T.M. Hinckley (1997): Production physiology and morphology of *Populus* species and their hybrids grown under short rotation. II. Biomass components and harvest index of hybrid and parental species clones. Can. J. For. Res. 27: 285-294.

Specht R.L. (1973). Structure and functional responses of ecosystems in the Mediterranean climate of Australia. In "Mediterranean type ecosystems. Origin and structure" (Castri F. and Mooney H.A., eds). Chapman and Hall, London. pp.113-120.

Taylor J.A. and J. Lloyd (1992): Sources and Sinks of atmospheric CO_2. Aust. J. Bot. 40:407-418.

Valentini R., P. De Angelis, G. Matteucci, R. Monaco, S. Dore and G. Scarascia-Mugnozza (1996): Seasonal net carbon dioxide exchange of a beech forest with the atmosphere. Global Change Biology 2: 199-207.

Vitousek P.M. (1992): Global environmental change: an introduction. Annu. Rev. Ecol. Syst. 23: 1-14.

Walter H. (1985): Vegetation of the earth and ecological systems of the geo-biosphere. Springer Verlag, Berlin. 318 pp.

Whittaker R.H. and G.M. Woodwell (1971): Measurement of net primary production of forests. In: Productivity of Forest Ecosystems (P. Duvignaud, ed.). UNESCO, Paris.

Wofsy S.C., M.L. Goulden, J.W. Munger, S.M. Fan, P.S. Bakwin, B.C. Daube, S.L. Bassow and F.A. Bazzaz (1993): Net exchange of CO_2 in a mid-latitude temperate forest. Science 260: 1314-1317.

Economic Value of the Carbon Sink Services of Costa Rica's Forestry Plantations

Octavio A. RAMIREZ*[1,2] and Manuel GOMEZ*[1]

*1 Centro Agronómico Tropical de Investigación y Enseñanza (CATIE)
   Turrialba, Costa Rica
*2 Department of Agriculture and Applied Economics, Texas Tech University
   Lubbock, Texas, USA

ABSTRACT: It is estimated that the "average" hectare of plantation forestry in Costa Rica can sequester 7.7 metric tons of Carbon, or 28.2 tons of CO_2 per year. Based on this estimate, it is calculated that the 128,000 hectares of forestry plantations reported have sequestered approximately 4.4 million metric tons of carbon to date (Figure 1.). The average net amount of carbon that has remained stored in this area during the last 20 years is calculated at 750,000 metric tons, with a potential value of 7.5 to 15 million U.S. dollars in government issued Carbon Bonds. In addition, the potential value of the average storage that is likely to occur during the next 20 years, of approximately 8.5 million tons, is estimated at between 84 and 168 million U.S. dollars, as the prices paid for the bonds may vary widely.
Keywords: Plantation Forestry, Valuation of Carbon-Sink Services.

1. INTRODUCTION

Forestry plantations have become an increasingly important part of Costa Rica´s rural economic development since 1964. During the last eight years (1990-95), more than 16,000 hectares have been planted per year, reaching a total of nearly 130,000 to date. This has required of an important investment from the public and private sectors, which has been only partially accounted for.

This study proposes a methodology to quantify and value the environmental service of carbon dioxide (CO_2) sequestration rendered by forestry plantations. The yearly amount of carbon sequestered by the "average" hectare of plantation forestry in Costa Rica is estimated. The average net amount of carbon that has remained stored in these areas during the last 20 years is also calculated as well as its potential value in government issued "carbon bonds".

In addition, the potential value of the average net sequestration that is likely to occur during the next 20 years is estimated. These figures and the methodologies used to obtain them are of great interest to Costa Rica and other developing country governments, as they have begun issuing "carbon bonds" that are now being traded in the international markets.

2. FORESTRY PLANTATIONS IN COSTA RICA

Tree cropping in the form of plantation forestry started in Costa Rica, in a small scale, more than 30 years ago. In 1964, 12 planted hectares were reported for the first time (Boletín Estadístico No.5), and small additional areas continued being registered annually until 1978. During this period, only 19 hectares per year were planted, on average. The increase in deforestation during those years started signaling to an eventual depletion of the reservoirs of natural forest. It is estimated that, from 1963 to 1973, 48,800 hectares of forest were cut per year; and 31,800 per year from 1973 to 1989 (Solórzano et al., 1991).

The annual rate of deforestation in Costa Rica has ranged from 1.8% in 1963 to 1.2% in 1989; compared with the average rate for all tropical forests, that has been estimated at 0.6% per year. Nearly 850,000 ha of forest were cut in Costa Rica between 1966 and 1984, of which only 35% were turned into possibly sustainable agricultural and livestock production land uses. About 65% of that area, on the other hand, exhibited a definite vocation for sustained forestry production or conservation.

Table 1. Hectares of plantation forestry established in Costa Rica, per year, (1964-95)

Year	Area planted that year	Accumulated area planted	Year	Area planted that year	Accumulated area planted
1964-71		84	1984	1286	6222
1972	84	168	1985	2501	8723
1973	24	192	1986	4175	12898
1974	21	213	1987	5303	18201
1975	21	234	1988	4835	23036
1976	5	239	1989	5000	28036
1977	12	251	1990	13797	41833
1978	34	285	1991	15560	57393
1979	413	698	1992	15958	73351
1980	807	1505	1993	14630	87981
1981	1098	2603	1994	14628	102609
1982	1357	3960	1995	25981	128590
1983	976	4936			

Source: Dirección General Forestal (1994) and Boletín estadístico forestal No.5

The search for alternatives to reduce the pressure on the remaining natural forests and prolong the useful life of the affected ecosystems resulted in an express national interest to promote forestry plantations, through economic incentive programs that started in 1978. In 1979, 413 hectares of trees were planted and, thereafter, an average of 2,500 ha of additional plantations were established every year until 1989.

In 1987, organized groups of small and medium-sized farmers joined this reforestation movement motivated by incentive programs that targeted such groups. They planted 34,640 hectares of trees between 1988 and 1994 (Torres, Luján and Pineda, 1995), that is, almost 5,000 annually. If the projects started by large forestry companies are taken into account, it is estimated that plantation forestry areas in Costa Rica have grown at an average of 16,759 hectares per year during the last six years.

Between 1990 and 1993 the most commonly planted specie was the "melina" (*Gmelina arborea*), in about 44% of the area established during that period. Other important species in this regard included "pilón" (*Hieronyma alchorneoides*), "eucalipto" (*Eucaliptus spp.*), "teca" (*Tectona grandis*) and "laurel" (*Cordia alliadora*). Those five were planted in an area of 48,221 hectares, a full 80% of the 1990-93 total.

3. MONITORING THE CARBON-SINK SERVICE

The carbon dioxide (CO_2) emissions, due to the combustion of fossil fuels, are considered as a key cause of the "green house" effect that is leading to an increase in global temperatures. The absorption of atmospheric CO2 by tropical forestry plantations with fast growing species can thus contribute to alleviate this global warming problem (Barres, 1993). Trees absorb CO_2 from the atmosphere and produce wood. According to the USDA Forest Products Laboratory, wood from conifer trees contains between 50 and 53% carbon, while wood from broadleaf forest trees between 47 and 50% (Barres, 1993).

The carbon stored in the trees remains as an integral element of the wood until these die and decompose. If, however, they are harvested and transformed into wood products used in construction, furniture or other durable structures, the carbon remains stored until that wood decomposes, and such is gradually freed again into the atmosphere. To estimate the absorption of CO_2, and the carbon storage by Costa Rica's plantations, the following steps must be followed:

1. Estimate the mean Average Annual Increment (AAI) of trunk volume, for the five most used species in Costa Rica, mentioned previously. The AAI was calculated for each of the five species using 10 years of annual growth field data (CATIE/Madeleña Project Silvicultural Guidebooks) representative of the conditions

of forestry plantations in Costa Rica. The five-specie mean AIA was then calculated at 18.9 cubic meters per hectare per year. The application of an adjustment factor of 0.85 to account for the previously mentioned deficiencies in the current status of the plantations (Torres, Luján y Pineda, 1995) results in a final mean AAI of 16.4 cubic meters per hectare per year.

2. Estimate the mean AAI of volume for the whole three, including trunk, branches, leaves and roots. Schroeder et. al. (1993) estimate that at least 40% of a tree's biomass is on its branches and leaves. An evaluation of data from different types of tropical forests (Brown et. al., 1988) resulted on multipliers from 1.8 to 3.5. Hollinger et. al. utilizes a relation of 1.89 to transform trunk biomass to whole-tree biomass for *Pinus radiata* forestry plantations in New Zeland. The values of this multiplier in other literature reviewed (Brown and Lugo, 1984; Thoranisorn, 1991; Faeth et. al., 1994; Nabuurs et. al., 1995) range from 1.3 to 3.6 with an average of 2.5 for natural forests, and between 1.3 and 2.7 with an average of 1.9 in the case of forestry plantations. A multiplier of 2.0 to convert trunk biomass to whole-tree biomass is used in this study, resulting on a mean AAI of volume for the whole three of 32.8 m^3/ha/year.

3. Determine mean specific weight for the 5 principal species. Considering the known values of 0.37gr/cm^3 for "melina" (*Gmelina arborea*), 0.70gr/cm^3 for "teca" (*Tectona grandis*) and 0.50gr/cm^3 for "eucalipto" (*Eucaliptus deglupta*) (CATIE/Madeleña Project Silvicultural Guidebooks), their average of 0.47gr/cm^3, or ton/m^3 is used in this study.

4. Determine the carbon content of the wood. According to the Laboratory of Forest Products of the USDA, the five principal species planted in Costa Rica contain between 47 and 53% of carbon (Barres, 1993). Therefore, an average carbon content of 50% was assumed.

5. Calculate the quantity of carbon stored per ton of wood biomass. The method described by Barres (1993) was followed for this purpose. The mean specific weight of Costa Rica's plantation forestry wood (0.47 ton/m^3) is multiplied by its average content of organic carbon (0.50) yielding a conversion factor of 0.235.

6. Calculate the amount of carbon stored per hectare of plantation forestry per year. Multiplying the mean AAI of volume for the whole three of 32.8 m^3/ha per year by the conversion factor in 5. results in an estimate of 7.7 metric tons of carbon stored per ha/year.

7. Estimate the net additional quantity of carbon stored by forestry plantations in Costa Rica during each of the years of the study. For this, the accumulated number of hectares in the different years (Table 1) is multiplied by the previously calculated factor of 7.7, and the amount of carbon exiting the system every year through the harvesting of forest products is subtracted (Table 2, columns (1) and (2)). The harvested volumes reported by MINAE (1995) have to be multiplied by 2, since they refer to trunk volumes only, and by the conversion factor given in 5.

8. Estimate the net accumulated quantity of carbon stored by forestry plantations in Costa Rica for each of the years of the study. This is done by accumulating through time the net additional quantity of carbon stored yearly, as calculated in 7.

9. Estimate average net quantity of carbon that has been stored in Costa Rica's forestry plantations, per year, during the last 25 years, by summing up the net accumulated quantity of carbon stored during each of the years of the study and dividing it by 25. The equivalent amount of CO_2 sequestered can be calculated. It is known that 27.3% of the CO_2 molecular weight is carbon. Thus, each ton of carbon stored implies a sequestration of 3.66 tons of carbon dioxide.

Table 2. Net additional quantity of carbon stored by forestry plantations in Costa Rica during each of the years of the study (1); amount of carbon exiting the system every year (2); and net accumulated quantity through time (3).

Year	Ha's	(1)	(2)	(3)	Year	Ha's	(1)	(2)	(3)
1971	84	649	0	649	1984	6222	48079	0	166832
1972	168	1298	0	1947	1985	8723	67405	0	234237
1973	192	1484	0	3431	1986	12898	99666	0	333904
1974	213	1646	0	5077	1987	18201	140645	0	474547
1975	234	1808	0	6885	1988	23036	178005	0	652553
1976	239	1847	0	8732	1989	28036	216642	0	869195
1977	251	1940	0	10672	1990	41833	314212	9044	1183406
1978	285	2203	0	12874	1991	57393	409872	33620	1593277
1979	698	5394	0	18267	1992	73351	543716	23087	2136994
1980	1505	11630	0	29897	1993	87981	604730	75124	2741723
1981	2603	20115	0	50011	1994	102609	771246	21642	3512969
1982	3960	30600	0	80611	1995	128590	908828	84822	4421797
1983	4936	38142	0	118753	**Average Annual Amount Stored:**				746769

Notice that in column (1), the additional carbon stored during that year by the existing plantation forestry area is calculated, net of the amounts exiting the system in that time period given in column (2). The total amount of carbon that has been kept in storage during a given year is computed in column (3) by accumulating column (1) through time. Because of the larger areas planted in recent years, and the fact that the carbon being stored during the previous periods has been accumulating, the net total amounts of carbon kept in storage increase rapidly at the end, to 4.42 million metric tons by 1995. On average, however, only 746,769 tons of carbon (per year) have been kept in storage in Costa Rica's forestry plantations during the last 25 years, while the average for the last 20 years raises to 932,563 metric tons.

134

Obviously, if the 128,590 hectares already established remain in plantation forestry during the next 20 years, the annual average amount of carbon stored in them would be much higher. This can be estimated using the general procedure outlined above. However, the volumes to be harvested through time in the future would have to be estimated, as well.

To precisely estimate these volumes, it would be necessary to know the turn-to-harvest of the areas planted each year since 1971. MINAE (1994) reports that, in 1990, 41% of the new areas established were planted with species with short turn-to-harvest (10-14 years, 12 on average), 19% with species with medium turn-to-harvest (16-20 years, 18 on average), and 40% with species with long turn-to-harvest (22-26 years, 24 on average). During 1991, 1992 and 1993 those percentages change to (50, 11, 39), (47, 9, 44) and (75, 3, 22), resulting in a four year average of 53.25, 10.5 and 36.25%, respectively. On the other hand, for all of the areas established from 1971 to 1995, the percentages are 40, 7 and 53%, respectively.

MINAE has not published detailed yearly data on the number of hectares of each specie planted before 1990 and after 1993. Further, the estimation of the volumes to be harvested through time is greatly simplified if it can be assumed that the same relative mix of species, at least in regards to their turn-to-harvest, has been planted every year. Therefore, a constant relative mix of 45, 10 and 45% is assumed for the base-line estimations in this study.

Then, it can be assumed that 45% of the area planted in 1971 will, on the average, be harvested in 1983, and again in 1995, 2007 and so on; while 10% will be harvested in 1989, 2007, 2015 and so on; and the other 45% will be harvested in 1995, 2019, 2043 and so on. Repeating the former process for 1972 up until 1995, a hypothetical harvesting schedule can be obtained, ad infinitum, and the volumes to be harvested through time in the future, estimated.

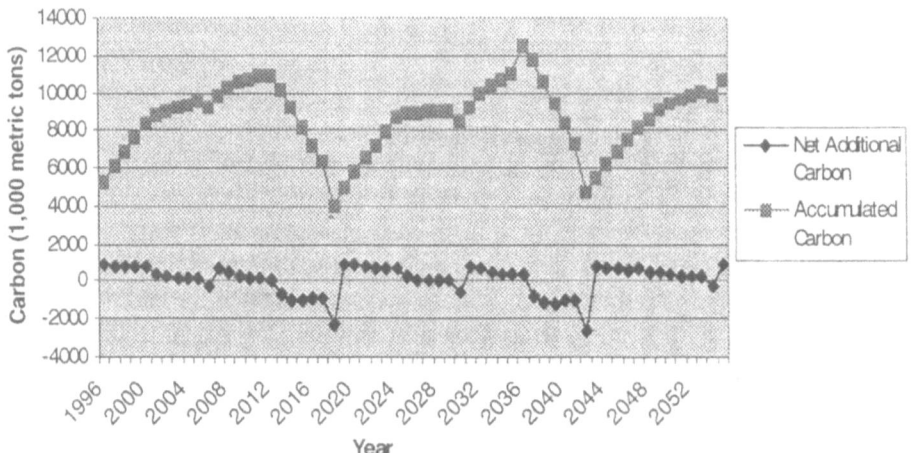

Figure 1: Carbon Flows in Costa Rica's Forestry Plantation Areas

The calculations resulting in Figure 1 indicate that, if the existing 128,590 hectares of forestry plantations area maintained through managed natural regeneration or replanting in the same or different sites after harvesting, an annual average of 9 million metric tons of carbon would remain stored in the twenty-year period from 1996 through 2015. This average is estimated at about 8.1 million tons for 2016-2035, 8 million tons for 2036-2055, and 8.3 million tons for 2056-2075 (not shown in Figure 1). Therefore, over the next 80 years, a very stable average of 8.4 million tons of carbon per year would remain stored, just in the form of forestry plantations.

To provide and idea of the potential error that could be induced by the assumption about the relative mix of species, in regards to their turn-to harvest (45, 10 and 45%), it is estimated that an equal mix (33, 33, 33) would produce a relatively similar average of 8.7 million tons. If, on the other hand, all of the species planted had either a short, medium or long turn-to-harvest, the corresponding averages would be of 5.6, 8.4 and 11.2 million tons per year. As expected, longer turns result in considerably higher average amounts of carbon remaining stored through time, amounts that are linearly proportional to the length of the turn $(12:18:24 = 5.6:8.4:11.2)$.

Actually, there is a theoretical alternative to the former method of estimation, based on the premise that forestry plantations in Costa Rica sequester an average of 7.7 metric tons of carbon per hectare, per year. A hectare of a specie with a short turn-to-harvest (12 years), thus, would reach a maximum storage of 92.4 tons by the end of the twelfth year. On average, it would have retained 46.2 tons of carbon per year. Thus, 128,000 hectares of this specie in permanent rotations should maintain a yearly average of 5.9 million tons of carbon stored.

Similarly, species with medium turn-to-harvest (18 years) would reach a maximum storage of 138.6 tons, for a mean of 69.3 tons; while species with long turn-to-harvest (24 years) would reach a maximum storage of 184.8 tons, for a mean of 92.4 tons. Then, 128,000 hectares in permanent rotations should maintain a yearly average of 8.85 and 11.8 million tons of carbon stored, respectively. The linear relationship between the length-of-turn and the yearly average quantity of carbon stored is illustrated, since $12:18:24=46.2:69.3:92.4= 5.9:8.85:11.8$

If 128,000 hectares of plantations were established during the same year, with a relative mix of species with short, medium and long turn-to-harvest of 45, 10, and 45%, respectively, and maintained ad infinitum, the yearly average amount of carbon permanently stored will be $128,000*0.45*46.2+128,000*0.10*69.3+128,000*0.45*92.4=8,870,400$, or 8.9 million tons.

The relatively small difference between the former figure and the 8.4 million tons estimate illustrated in Figure 1 is due to the fact that Costa Rica's forestry plantations were not all established during the same year, and that such estimate was obtained considering finite and relatively short time-span . Also notice that the more involved calculations behind Figure 1 would yield a more precise estimate when the former is the case. Further, they must be used if plantation forestry areas are changing over time, which is by far the most common scenario.

4. POTENTIAL ECONOMIC VALUE OF THE CARBON SEQUESTERED

Once the average amount of carbon that can be permanently stored in a given area is calculated, assuming that a relatively stable long-term productive land-use pattern will be maintained, valuation is straightforward, conceptually. In practice, however, it is complicated by the fact that a liquid, stable, open market for carbon storage does not exist yet. However, a limited number of buyers and sellers have so far managed to come together and make transactions at a wide range of per-unit-prices, as it is to be expected in this kind of situations.

In Costa Rica, for example, in the three projects already approved by the United States Initiative on Joint Implementation (USIJI), the CO_2 has been priced from U.S.$10 to U.S. $16 per metric ton; equivalent to U.S.$ 36.6 to 58.6 per metric ton of carbon permanently stored, given the previously discussed conversion factor. On the demand side, Carranza et al. (1996), cites estimations of the cost of the damage caused by a metric ton of carbon released into the atmosphere of between U.S.$ 10 and U.S.$ 20 (or U.S.$ 36.7 to U.S.$ 73.2 per ton of carbon). Fankhauser (1995) cited by Carranza et al. (1996), estimates that cost at U.S.$ 20 per ton of CO_2 (or U.S. $73.2 per ton of carbon), and projects an increase to U.S.$ 28 by the year 2,000.

On the supply side, Winjum et al. (1993), reports the cost of storing one metric ton of carbon through plantation forestry at U.S.$ 31 in Argentina, U.S.$ 5 in Australia, U.S.$ 10 in Brazil, U.S.$ 11 in Canada, U.S$ 4 in Mexico, and U.S.$ 5 in the United States. Cline (1992), on the other hand, estimates these costs at U.S.$ 5 per ton in tropical areas and U.S.$ 20 in temperate areas of developed countries. Emission control costs in industrialized countries are reported at between U.S.$10 and U.S.$20 per ton of carbon.

The former figures provide a guideline for the general range of prices per ton of carbon stored that could be assumed for the purposes of this study. At the average of the prices already negotiated by the Costa Rican Government with the USIJI, for example, the 8.4 million tons of carbon that can be permanently stored in the 128,000 hectares of plantation forestry already established in Costa Rica would have a one-time value of U.S. $400 million. This is slightly lower than the estimate obtained when the mid-point of the range of cost-avoidance figures provided by Carranza et al. (1996) (which includes Fankhauser (1995)) is used. On the supply side, the current average cost of storing one metric ton of carbon through plantation forestry of about U.S.$ 11.5 (Winjum et al., 1993, Cline, 1992) would imply a much lower value of U.S.$ 97 million. If the current marginal cost of the alternative of emissions control is used instead, this estimate increases slightly.

A last interesting aspect to consider is that the Government of Costa Rica has begun issuing "carbon bonds" (CTO's). The CTO concept enjoys the support of the World Bank and the Earth Council of the United Nations. The Norwegian government has already bought 200,000 metric tons of Costa Rican CTO's at U.S. $10 per ton, while the U.S. based Centre Financial Products Ltd. has been authorized to sell, on the Chicago stock market, as many as 4 million tons in CTO's over the next 20 years, worth as much as U.S. $80 million, or U.S.$ 20 per ton (Costa Rica's Joint Implementation Office, 1997). A unique feature of these CTO's is that they commit to maintain a given amount of carbon stored for a period of only twenty years. Under such circumstances, the estimated value of the 8.4 million tons of carbon that could, on average, be permanently stored in Costa Rica's 128,000 hectares of plantation forestry, would be of U.S.$ 168 million per 20-year period, or U.S.$ 8.4 million per year, or U.S.$65 per ha/year.

5. CONCLUSSIONS AND RECOMMENDATIONS

This study demonstrates that the monitoring and valuation of at least one key environmental service, carbon storage, is feasible in the case of plantation forestry in a developing country, using currently available data and relatively simple methods. During the last 20 years, the rapidly growing forestry plantations areas in Costa Rica, have kept an annual average of about 750,000 metric tons of carbon stored. It is estimated that the 128,000 established through 1995 could, on average, maintain 8.4 million metric tons of carbon in permanent storage. Given the terms of the current carbon bond initiative of the Costa Rican Government, these could have a value of U.S.$84 to U.S.$168 million per 20-year period, or U.S.$4.2 to U.S.$8.4 million per year, equivalent to U.S.$32.5 to U.S.$65 per ha/year.

If the current trend of growing plantation forestry areas continues in Costa Rica, the former figures would be much higher. They could be estimated using the methodology proposed in this study. In addition, the previous calculations implicitly assume that, once the plantations are harvested, the carbon stored is immediately released into the atmosphere in the form of CO_2. It is important to estimate CO_2 "release curves" that predict the % of this gas, or its carbon equivalent, that will be annually discharged back into the atmosphere after harvest. These would allow for and upward revision of the former estimates that could be substantial.

Finally, although an evaluation of Costa Rica's plantation forestry incentive programs is not one of the objectives of this study, the results raise an interesting point about them. It is shown that the average amount of carbon that can be permanently stored in a hectare of plantation forestry is directly and linearly proportional to the length of the turn-to-harvest of the specie planted. Therefore, if one of the objectives of or justifications for an incentive program is to reduce net CO_2 emissions, this should be taken into account and the value of the incentives ought to be higher for species with longer turn-to-harvest. This is not the case in Costa Rica.

6. REFERENCES

Barres, H. (1993): Carbon-Fixing and Timber Production in Tropical Klinki Pine Forest Plantations, The Klinki Pine Project.

Brown, S., A. J. Gillespie and A. E. LUGO (1989): Biomass Estimation Methods for Tropical Forests with Applications to Forest Inventory Data, Forest Science, Vol. 35, No. 4, pp. 881- 902.

Brown, S. and A. E. LUGO (1984): Biomass of Tropical Forests; A New Estimate Based on Forest Volumes, Science, No. 223, pp. 1290-1293.

Carranza, C. F. (1996): Valoración de los Servicios Ambientales de los Bosques de Costa Rica. Centro Científico Tropical/ODA/MINAE, San José, Costa Rica, 77 p.

Cline, W.R. (1992): The Economics of Global Warming, Institute for International Economics, Washington D.C.

Faeth, P., Ch. Cort and R. Livernash (1994): Evaluating the Carbon Sequestration Benefits of Forestry Projects in Developing Countries, World Resources Institute/U.S. Environmental Protection Agency, Washington D.C., U.S.A., 96 p.

MINISTERIO DE RECURSOS NATURALES, ENERGIA Y MINAS (1994): Informe Anual 1993, Programa de Desarrollo Forestal, FDF DGF-DECAFOR, San José, Costa Rica, 44 p.

MINISTERIO DE RECURSOS NATURALES, ENERGIA Y MINAS (1994): Boletín Estadístico Forestal No. 5, DGF, San José, Costa Rica.

MINISTERIO DE RECURSOS NATURALES, ENERGIA Y MINAS (1995): Estadísticas Relevantes del Sector Forestal, DGF, San José, Costa Rica.

Nabuurs, G. J. and G. M. Mohren (1995): Modeling Analysis of Potential Carbon Sequestration in Selected Forest Types, Canadian Journal of Forest Research, Vol. 25, No. 7, pp. 1157 - 1172.

Solórzano, R. (1991): La Depreciación de los Recursos Naturales en Costa Rica y su Relación con el Sistema de Cuentas Nacionales, Centro Científico Tropical/World Resource Institute, San José, Costa Rica, 139 p.

Thoranisorn, S., P. Sahunalu and K. Yoda (1991): Litterfall and Productivity of Eucalyptus Camaldulensis in Thailand, Journal of Tropical Ecology, Vol. 7, No.2, pp. 275-279.

Torres, G., R. Luján and M. Pineda (1995): Diagnóstico Técnico del Proceso de Producción Forestal en Plantaciones de Pequeña Escala en Costa Rica, Instituto Tecnológico de Costa Rica, Cartago, Costa Rica, 105 p.

Winjum, J. K., R. K. Dixon and P. E. Schroeder (1993): Forest Management and Carbon Storage; An Analysis of 12 Key Forest Nations, Water Air and Soil Pollution, No. 70, pp. 239-257.

Trees and Woodland in a Cultural Landscape : the History of Woods in England

Oliver RACKHAM

Corpus Christi College, Cambridge, England

ABSTRACT   The trees of England are deciduous species of genera such as *Quercus, Fraxinus, Corylus, Ulmus* and *Tilia.* Native conifers are insignificant. Nearly all the trees coppice or sucker when felled; this is often their main historic means of reproduction. England is historically a country with very little woodland. Most of the primaeval wildwood was destroyed more than 2000 years ago. The first inventory of land-uses, in 1086 AD, records about 15% of the area of England as woodland; this had fallen to about 6% by 1350. This remaining woodland was a valuable resource, and was managed and conserved down to the 20th century. It then fell into a period of neglect and destruction, from which it has lately been rescued by the efforts of conservationists.

Keywords: forest history, coppice, wood-pasture, savanna, England

1. INTRODUCTION

England is a land with a long history of dense population and very little forest. The land is not volcanic and not very mountainous, but has been much affected by glaciations. Our history of forest goes back some 12,000 years. Although not a forest people, we have a long tradition of maintaining small areas of forest, which we call woodland.

The trees of England are deciduous species of genera such as *Quercus, Fraxinus, Corylus, Ulmus, Tilia, Carpinus, Acer, Alnus,* and the anciently-introduced *Castanea.* These, singly or in combination, give rise to many types of woodland, even within the small proportion of forest that survives. Native conifers are insignificant. Nearly all the trees coppice or sucker (sprouting from the stump or the roots) when felled; this is their main historic means of reproduction.

Trees have long formed part of an intensively-managed cultural landscape. There are four main traditions of land-use involving trees:

(1) *Woodland,* areas of naturally-occurring trees which have been managed.

(2) *Wood-pasture,* trees combined with grazing animals.

(3) *Non-woodland trees,* forming hedges between fields or scattered within fields.

(4) *Plantations,* where the trees have been planted and are usually unrelated to the natural vegetation.

The first three traditions are ancient, going back at least as far as the earliest documents. Plantations are modern: with unimportant exceptions they begin about

139

1600 AD, but become significant only in the 19th century. Introduced trees are still largely confined to plantations.

Here I deal mainly with the woodland tradition. Woods form islands of forest, typically of from 1 to 100 ha, surrounded by farmland, heathland (dominated by Ericaceae), or other non-forest vegetation. They have traditionally been managed by *coppicing*: areas of trees

Fig.1 Coppice-wood, 1½ years since felling. The standard trees are mainly *Quercus* and the underwood mainly *Fraxinus*. Bradfield Woods, Suffolk.

would be felled every 4 to 30 years and allowed to grow again from the stools to give successive crops of poles and rods, called **underwood** (Fig.1). Underwood has been used for many purposes, but especially light construction, fuel, and charcoal. Among the stools is a scatter of trees, mainly *Quercus*, allowed to stand for several cycles of the underwood and to grow big enough to make beams and planks. (This is what academic writers call coppice-with-standards.) England, like most west and middle European countries, makes a distinction between **wood**, consisting of poles, rods, and branches, and timber, trunks big enough to saw. Woods with a long history of coppicing have a very distinctive flora and fauna, responding to the cycles of light and shade.

Coppicing is a traditional practice in west and south Europe, and still continues on a large scale in Italy. The regrowth of a felled tree is a symbol of unconquered hope, originating with the ancient Hebrews; it is read out from the Bible in every Christian church each Christmas.

Some woods are **ancient woodland**: they have existed for at least 400 years and may go back to the prehistoric wildwood. Others have arisen at various times, from prehistory onwards, on former arable or pasture land, deserted settlements, etc. Any piece of land which is not grazed or cultivated easily turns into woodland, though it will not be the same as ancient woodland.

The history of woodland is the sum of the histories of thousands of individual wood-lots. It is not the history of the things that people have said about woodland. Nor is it the history of forestry laws and policies, which in England were insignificant until the 20th century.

2. WOODLAND IN PREHISTORY

After the last ice age, England became covered with natural forests known as wildwood. We have no memory or record of wildwood, which has to be reconstructed from palynology, from trees buried in peat, and from countries where it survived late enough to be put on record.

In the Neolithic period, from about 4500 BC onwards, people began agriculture, cultivating crops and keeping domestic animals. To do this they began grubbing out wildwood to create fields and pastures; they also began to turn parts of the remaining wildwood into managed woodland. Wooden artefacts of this period have been excavated from waterlogged sites. These indicate that woodland management already existed: regular crops were produced of coppice poles and leaves for feeding animals. Farmland was extended in the Bronze and Iron Ages, until by 500 BC, at the latest, more than half the country had ceased to be forest.

In the Roman period (40-400 AD) England was a densely populated, mainly agricultural country. It had fuel-using industries such as brickmaking, iron-smelting, and the heating of baths, which implies management of the remaining woodland to yield renewable supplies of wood. Wattle — interwoven underwood rods — continued to be an immensely important constructional material.

3. EARLY HISTORIC WOODLAND

The earliest documents, from 700 AD onwards, describe a landscape not very different from that of today, with woods, wood-pastures, hedges, and non-woodland trees set in farmland. Some of the wood-lots themselves still exist. In 1086 there came the first inventory of land-uses, called Domesday Book, which records about 15% of the area of England as woodland and wood-pasture. Large areas of the country were without woodland, so that trade and transport were already significant. After this came an increase in population and in pressure on land. The area of woodland had fallen to about 6% by 1350, and much of the wood-pasture had been converted to coppice-wood. From then on the human population was held in check by epidemics, and the woodland area was stable.

From AD 1250 onwards many individual wood-lots are recorded in detail. Woods had names, owners, definite boundaries, and peculiar shapes by which they can be recognized on the map; they were managed for a sustained yield, and were permanent. Hayley Wood near Cambridge, first described in a document of 1251, is one of many that still exist today. It belonged to the Bishops of Ely: many other well-recorded woods were the property of monasteries. The emphasis was on producing underwood: the woods were coppiced, often on short cycles. Timber was also produced, although much of England's timber was imported from north and east Europe.

Written documents record the ownership and management of named wood-lots and what was done with the produce. Large-scale maps begin in about AD 1580, and can be used to trace changes in the extent of woods. Chalkney Wood, some 60 kilometres SE of Cambridge, is exactly the same today as on a map of 1595. What is not often recorded is the composition: timber trees were usually *Quercus*, but for many purposes the species of underwood did not matter.

Underwood was used for purposes such as fencing and as the wattle infill of timber-framed buildings (the form in which it still survives). The biggest use, however, was for fuel, domestic and industrial. Most of the big concentrations of woodland had some special use such as iron-smelting, tile-making, or maintaining sea-defences on low-lying coasts. The biggest concentration of all was around London, by far the biggest city.

Woodland history cannot be derived solely from documents alone, which record only those aspects that people thought worth writing down. It is essential to use other sources of evidence as well. Gamlingay Wood near Cambridge is, in my experience, the best-documented wood in England: yet fieldwork reveals features too early to have been recorded in writing.

Owners surrounded their woods with earthworks, constructed especially in the period 800-1300. These still exist and can be used to trace any alterations in a wood's boundaries (Fig.2). Where a wood has arisen on a non-wooded site, archaeological remains may survive from the previous land-use for arable land, mining, or settlement.

Woods were valuable land, and were intended to be permanent. People took conservation and sustained yield for granted, and seldom wrote about them in the abstract. The importance they attached to conservation is to be judged by the effort put into defending the boundaries. Even woods of less than 10 ha often have great earthworks round them.

Fig.2 Boundary bank of the Bradfield Woods, constructed about 1000 years ago to define the wood edge. The wood on the bank is newly felled; the interior is of one year's growth.

Another source of information is from ancient coppice stools. With repeated cutting a tree — *Fraxinus*, *Tilia*, *Quercus*, etc. — lives for many centuries and grows into a massive circular base, 2-4 m in diameter, from which the coppice poles arise (Fig.3). The presence of these stools is one way to recognize an ancient wood. There are also particular plants characteristic of woods that have certain kinds of history. *Tilia cordata*, for example, holds on

Fig.3 Giant coppice stool of *Fraxinus*, some 400 years old. Hayley Wood, Cambridgeshire.

tenaciously to its existing sites, but is reluctant to invade newly-formed woodland. Many herbaceous plants are indicative of whether a wood is ancient or recent.

England has thousands of houses, churches, barns, etc. which contain the actual timbers and underwood produced by woods. From these can be inferred the species, growth-rates, felling cycles, and the degree of competition between timber

Fig.4 Barley Barn, Cressing Temple, Essex, built by the Knights Templars about 800 years ago. The timbers are mostly small to middle-sized individuals of *Quercus*.

trees and underwood. For example the two great barns of Cressing Temple, built by the Knights Templars — soldier-monks — in c.1200 and c.1270, tell a story of large numbers of small *Quercus*, which were felled young and easily replaced (Fig.4). Specially large trees — which by modern standards would not be exceptionally big — were rare and expensive and transported hundreds of kilometres.

Relations between woods and cities or other centres of population varied. The occurrence of timber-framed buildings is not correlated with that of woodland. Some cities, such as London and Canterbury, were ringed round by extensive woodland; the roads out of these places were bordered by clearings to protect travellers against gangsters lurking in the bushes. Other places, such as Norwich, York, and Cambridge, had very little local woodland. Cambridge burnt other fuels such as peat; its timber was brought from a distance, but the study of its ancient buildings reveals that there was much recycling of timber out of demolished buildings.

4. LATER HISTORY OF WOODLAND

Woods were among the most enduring of English institutions. They outlived many changes in human society, economy, and land-use; wood-lots until recently were islands of stability in an otherwise changing landscape. From time to time woods have been grubbed out, and new woods have arisen; but as late as 1910 many woods would have been instantly recognized by someone from 1210. Even now nearly one-third of the woods of the great monastery of Bury St Edmund's are still extant, nearly 500 years after the monastery itself ceased to exist.

The demands of increasing population and industry were met by increasing use of coal and of imported timber. In the 19th century the price of oak-trees was inflated, for a few decades, by unprecedented demands from leather-tanning (using oak bark) and shipbuilding.

The woodland tradition began to decline from 1850 onwards, as railways brought cheap coal to remote rural areas and people ceased to use renewable fuels. At the same time modern 'scientific' forestry began to develop. By historical accident this was derived from Germany — not from the rival school of 'scientific' forestry in France — and was based on plantations. Forestry was not a development from indigenous historic methods of managing woodland, as it might have been had it been taken from France. At first it was concerned with plantations on non-forest sites (at which it has been moderately successful) and did not encroach on ancient woods.

Most ancient woods remained in existence, although often disused, until after World War II. It then became the fashion to regard them as vacant land which ought to be growing something. Between 1950 and 1975 very many woods were grubbed out and made into farmland. Others were taken over by modern foresters, who felled and poisoned the existing trees and tried to replace them with planted trees (a practice called **replanting**). Both practices were encouraged and subsidized by governments, and nearly half the ancient woodland fell victim to one or other.

Conservation organizations, governmental and non-governmental, went to great efforts in reaction to these crude kinds of destruction. They made it their business to acquire woods, both to preserve them from destruction and to restore the coppicing and other management to which much of the flora and fauna of English woods has become adapted.

5. WOOD-PASTURE AND ANCIENT TREES

Wood-pasture is the practice of combining trees with cattle, sheep, or deer. The shade of the trees reduces the growth of the pasture, and the animals are liable to eat the regrowth of the trees. Wood-pasture therefore involves different practices from woodland. Often the trees are not coppiced but **pollarded**, cut 2-4 m above ground so that the animals shall not be able to reach the young sprouts. Pollarding prolongs the life of a tree and also causes it to develop habitats, especially in its hollow interior, for specific invertebrates and other plants and animals.

Wood-pasture goes back to prehistory, and is well recorded in early documents. When the Normans conquered England in the 11th century they introduced new variants of it connected with keeping deer (*Cervus, Capreolus*, but especially the introduced *Dama dama*) as semi-domestic animals. Deer became an upper-class status symbol.

Wood-pasture has a less stable history than woodland. Many examples were either made into coppice-woods or destroyed altogether. The tradition, however, lives on in places like the New Forest, Epping Forest, and Hatfield Forest (Fig.5). (The English word Forest — with a capital F — means a place where the king had the right to keep deer.) Wood-pastures are often savanna-like places, with ancient pollard trees scattered among grassland. Others were felled from time to time like an ordinary wood, each felled area being fenced to keep out deer and cattle until the trees had grown again.

Wood-pastures have a special place in our culture and history. The English have usually loved their savanna-like aspect and the beauty, majesty, and oddity of their ancient trees. Particular trees are named and invested with railings; stories are told about them and their portraits are painted. Wood-pastures such as Epping

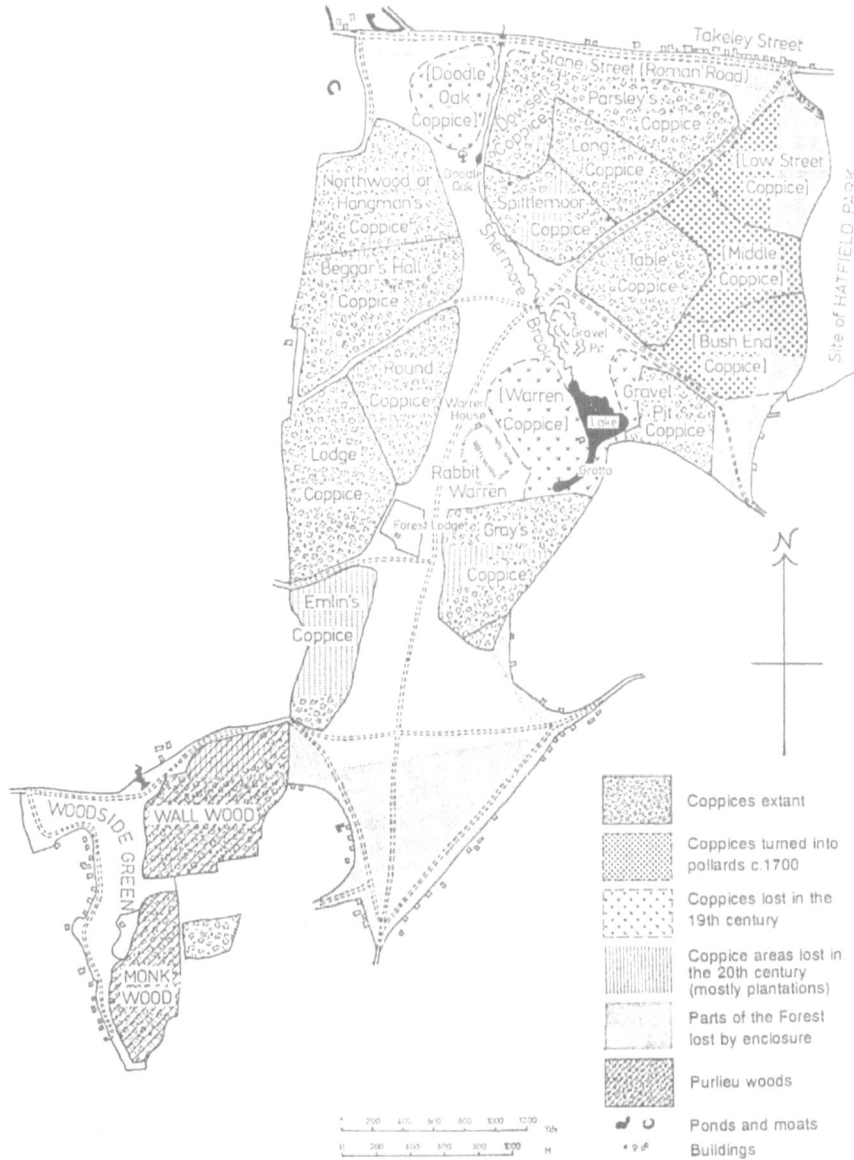

Fig.5 Complex wood-pasture: Hatfield Forest, Essex. The individual coppices were felled from time to time and then fenced to keep out deer and other animals. The plains were accessible all the time and contain pollard trees.

Forest and Burnham Beeches are traditional places of public resort. Many ancient trees from the general landscape were preserved by being incorporated into the ornamental parks round country mansions, for example Grimsthorpe Park some 80 km NW of Cambridge (Fig.6). Unfortunately the appreciation of ancient trees declined in the early 20th century, and is only now being recovered.

Fig.6 Ancient pollard *Quercus* in wood-pasture, Grimsthorpe Park, Lincolnshire. They were probably originally in a hedge, taken into the park in the 16th or 17th century.

6. NON-WOODLAND TREES

England is a country, like the United States but unlike Australia, where fences and the edges of fields quickly turn into narrow strips of woodland. There is also a strong tradition of planting hedges — hundreds of thousands of kilometres of them — to define fields and property boundaries. There are also trees round settlements and in churchyards, or (though seldom today) free-standing in fields.

Hedges have existed for at least 2000 years, and some still extant are over a thousand years old. They served as sources of timber and underwood as well as to defend boundaries.

Hedges, like woods, have passed through a period of neglect and destruction, but are now again becoming appreciated.

7. THE FUTURE

Threats of destruction have receded. Ancient woodland is now appreciated as an irreplaceable ecosystem, an antiquity, and a part of English culture. Much has been done among conservationists to restore public understanding of trees and woodland, which had sadly declined in the previous hundred years. The Woodland Trust and the county wildlife trusts are now among the biggest owners and managers of ancient woodland.

Woodland is no longer made into arable land: indeed there are proposals to make great new plantations on farmland as places of public amenity. Modern forestry has retreated from ancient woodland. Many plantations on the site of ancient woods have proved unsustainable and have fallen into neglect and decay. A major conservation task is to rehabilitate those ancient woods which had a planting episode but where the original vegetation has recovered from the poison and is now reasserting itself. Much of this work is being done by Forest Enterprise (the government forest-management body) as well as by voluntary societies.

Coppicing has recovered from a low point in the 1960s to the point where one begins to hear complaints that there is too much coppicing.

Wood-pasture is also appreciated, not only for its beauty and its historic witness, but for the many animals and lichens for which ancient, hollow trees are a specific habitat. The English love of old trees has made these a feature of European importance.

An increasing threat is the proliferation of deer. There are now more deer in England than for a thousand years: to the two native species have been added many others, including *Cervus nippon* from Japan. Too many deer result in severe browsing of young trees, coppice sprouts, and woodland herbs.

REFERENCES

Rackham, O. (1980): Ancient Woodland: its history, vegetation and uses in England, Edward Arnold, London, 402 pp.

Rackham, O. (1986): The History of the [British and Irish] Countryside, Dent, London, 445 pp.

Rackham, O. (1990): Trees and Woodland in the British Landscape (2nd edition), London, 234 pp.

Review of Forest Culture Research in Japan: Toward a New Paradigm of Forest Culture

Taiichi ITO

Institute of Agricultural and Forest Engineering, University of Tsukuba
Tsukuba 305-8572, Japan

ABSTRACT   This paper attempts to find a new paradigm for forest culture in the twenty-first century by reviewing forest culture research conducted in Japan.   Traditionally, according to folklore forest culture meant the wise use of forest resources.   However, its meaning changed radically in the 1950s as demand for forest commodities such as fuels and fertilizers virtually disappeared.   On the other hand, ecological expectations based on global environmental problems and nostalgia for lost old-fashioned but sustainable ways of life are increasing.   At the same time, recreational demand for wild lands has emerged. Reflecting these changes in perspective, interest in forest culture is surging again, but we have no experience responding to these new demands.   To develop a new forest culture, firsthand awareness of forests is essential.   Forest recreation and environmental education in nearby forests included in greenway networks should play an important role in developing such awareness.

Keywords: Forest culture, Sustainable use, Forest recreation, Environmental awareness

1. INTRODUCTION

It seems proper to define "forest culture" before discussing it.   However, the meaning of forest culture is always changing according to human demand.   Occasionally, forest culture is confused with wood culture, one of its components that focuses on timber use. Therefore, instead of defining the term, this review will try to show the changes in how forest culture is perceived by shedding light on the development of forest culture research in Japan.   Topics related to forest culture include various used of forest products, techniques of forest and watershed management, and the lifestyle of local communities. In other words, forest culture is related to both the physical and mental values of people who depend on forests.

Firstly, the nature of forest culture is discussed from an historical perspective.   Then, the history of forest culture research in Japan is traced through three steps.   Finally, I will discuss the paradigm of a new forest culture as a way to promote an understanding of forests and environmental awareness, as well as methods for the sustainable use of forests.

2. FOREST CULTURE VERSUS CIVILIZATION

Forest culture is an oxymoron, because culture has developed by cultivating forestlands to raise crops and cattle.   Cultivation implies the destruction of forests.   This becomes obvious when we look at the ruined ancient civilizations of Mesopotamia and Greece. These areas were once covered with forests, but people consumed them in a relatively short period.   Marsh (1974) wrote about this in 1864 and his book helped to initiate the conservation movement in the United States.

In Japan, Watsuji (1935) published a well-known book on the relationship between climate and culture.   Huntington (1938) published a similar work in the United States. Huntington's opinion on the correlation between civilization and climate was criticized. Still, it is obvious that climate has cultural influence.   In Japan, Watsuji's book is recognized as one of the first to stimulate popular discussion on the influence of the natural environment on human beings.

While Watsuji based his work on his musings during a voyage to Europe, Yasuda (1980, 1989) did extensive field research, employing pollen analysis as a powerful tool to reconstruct the lost vegetation in the vicinity of destroyed civilizations.   Then he asserted that civilization consumes forests by tracing changing vegetation in the case of each civilization.   These civilizations disappeared because they could not develop sustainable ways of using forest resources.

In contrast with the negative impact of civilization on forests, forest culture can be sustained once people reach a certain stage of production for survival and maintain renewable resources.   Thus, the forest culture can be nurtured by maintaining sustainable conditions for a prolonged period.   Traditionally, forests in Japan were used in this way, especially during the Edo period (1605 – 1868) when sustainable use was inevitable under the strict isolation policy.   The Japanese maintained renewable forest resources until the fuel revolution in the 1950s.   For this reason, investigating the meaning of Japan's forest culture should provide insight into how to restore sustainable relationships with forests all over the world.

3. THE DEVELOPMENT OF FOREST CULTURE RESEARCH

3.1 Approach from Folklore

A literary investigation showed that the amount of forest culture research was quite limited until 1980, or rather, we did not recognize forest culture as an independent area of research until that time.

In reality, folklorists and rural sociologists studied forest culture as a part of their field surveys of mountain villages.   In this sense, Yanagida (1875 – 1962), the founder of folklore in Japan, can also be called a pioneer of forest culture research.   He majored in jurisprudence at the University of Tokyo but he wanted to study forestry to visit the mountains (Akasaka, 1991).   He realized his dream by visiting isolated mountain villages to gather information on folklore and lifestyle, which he published in "The Life of Mountain Residents" in 1909.   Thus, the study of the traditional lifestyles of mountain villagers was recognized as an orthodox approach to the study of forest culture.

Forest culture research based on historic sources has continued to the present.   While Chiba (1956, 1973) discussed how mountains were denuded, Makino (1988) revealed how the Japanese restored these forests during the feudal period. Tsutsui (1988) investigated the development of forest conservation policies and recognized them as an aspect of sustainable forest culture.

Such research is occasionally motivated by nostalgia for vanishing objects or the traditions of isolated mountain villagers. Therefore, it tends to stress retrospective views instead of suggesting ways to create a new forest culture for the twenty-first century.

3.2 Ecological Approach

The new field of ecology gradually gained power in the 1950s. Umesao (1974), an ecologist turned anthropologist, wrote an article titled "Ecological View of Civilization" in 1957. He used ecological concepts, such as succession, as tools to investigate the development of civilization. Stimulated by Umesao's approach, Ueyama (1969) published a book on forest culture in the evergreen broadleaf forest region from an ecological perspective. These views are supported by field surveys in East Asia by ecologists and anthropologists. They have scientifically confirmed what Watsuji recognized intuitively on his way to Europe. While Ueyama evaluated the role of evergreen forests in the roots of Japanese culture, Ichikawa (1985, 1987) pointed out the importance of forest culture in the northern part of Japan, which is dominated by beech forests. The idea that the type of vegetation nourishes a culture is very persuasive in a country like Japan, where forests once dominated and the culture depended on those forests. However, as mentioned above, our dependency on forests disappeared in the 1950s because of rapid urbanization and the increased importation of oil. In other words, our culture has lost its ecological link to the forest.

With the decreased dependency on forest products, Japanese awareness of the forest has declined. Nevertheless, information on the forests in distant national parks and other protected areas is constantly supplied to urbanites by the mass media. City dwellers are aware that our forests are facing ecological problems due to mismanagement. In response to the prevailing journalistic ecological view of forests and forestry, forestry professors like Tadaki (1981, 1984, 1988) and Sugawara (1989, 1995, Yasuda and Sugawara, 1996) have published books to explain the importance of forestry in the conservation of forest ecosystem and forest culture.

The ecological approach to forest culture is persuasive. However, without a correct understanding of forest conditions in Japan, city dwellers develop a distorted ecological image of our forests.

3.3 Combined Approaches

In the 1980s, the number of books on forest culture based on either historical or ecological approaches rapidly increased. In addition, a new approach emerged that was partially influenced by landscape ecology. Combining these methods, Ogura (1992, 1996) reconstructed lost forest landscapes from historic sources such as paintings, drawings, and maps. By comparing the reconstructed landscapes with the existing ones, he investigated how such landscapes were created and maintained by local residents. His studies give us clues on how to recover lost forest landscapes in the twenty-first century.

Another new approach, the comparative studies on forest recognition by Kitamura (1981, 1993) and Shidei and Hayashi. (1984), is remarkable enough to open a new horizon in forest culture research. In 1981 Kitamura initiated an international survey of forest recognition, and revealed that the urban Japanese tend to understand forests as an image constructed by the mass media, instead of through direct experience. For example, they tend to prefer photographs of man-made forests to natural ones, while verbally they state preferences for natural forests over man-made ones. This contradiction reveals that urban Japanese prefer naturalistic images of forests to the real natural environment. According to the survey by Kitamura, the gap between reality and imagery is wider in Japan than in other countries.

The natural forests of Japan are not very accessible and there are few amenities for visitors. They are dark and humid. Furthermore, the steep topography requires demanding hiking. On the other hand, the public with illusions demands urban amenities such as hot water and flush toilets at camping sites in wild forests. In order to satisfy them and promote enjoyable outdoor activities in this environment, expensive forest recreational facilities must be constructed. The result is negative impacts to the forests.

Based on the analysis of forest recognition, Kitamura suggested that we should use urban forests to improve our environmental awareness instead of isolated forests in mountain areas. He also recognized that forests themselves are the products of culture. Although Japan imported only technical aspects of forestry from Germany in the late 1800s, it did not import the German cultural background. For this reason, the Japanese have not yet developed a recreational form of forest culture to communicate with near-by forests equivalent to the German *Wandering*.

4. DISCUSSION

In short, forest culture in Japan changed in the 1950s in two ways. Firstly, the direct tie with the use of forest products for a living was lost, and without such direct physical ties, the ecological and mental values of forests became increasingly important. However, a lack of direct knowledge of forests or forestry has lead to an unrealistic perception of forests and a longing for imagined forests deep in the mountains. Secondly, the mountain villagers who traditionally maintained forest culture are also disappearing. However, urbanites are beginning to recognize the benefits of recreational activities related to forest management.

From these two points, it is obvious that urbanites need to understand the reality of the forest environment, and shoulder responsibility for taking care of the forest and its culture as a form of recreation. In fact, there are two different movements to restore forest culture in modern Japanese society. One involves establishing a new type of mountain community in the forest; the other restores and cares for forests near urban centers.

The establishment of new forest communities is called the *Forestopia* movement. In the late 1980s, Miyazaki prefecture coined the word by shortening "forest utopia" for a group of rural development projects in its mountainous areas. Hirano (1996) also discussed developing forest cities with urban amenities especially in the national forests. The roots of this idea date back to the garden city movement in England, but the natural conditions are quite different in Japan. The white paper by the Forestry Agency (1995) supported this idea as part of a way to deal with its huge financial deficit from the national forest management.

Life in forests is appealing to urban residents. However, it is unlikely that people emigrating from cities will engage in forest management and lead a sustainable life isolated in the mountains in which most national forests are located. Constructing cities in mountains requires special consideration from the perspective of disaster prevention and also requires a considerable investment in infrastructure. Furthermore, forest life will inevitably increase the dependency on automobiles and fossil fuels. Recognizing these problems, Aoki (1996) proposed a more subsistence-oriented revival of mountain societies as ecomuseums. However, the gap between the economic power of cities and mountain villages is difficult to overcome, even with unique local products and attractions as ecomuseums. The number of people who prefer working in forested areas is on the rise, but they will remain a minority in the future because the lifestyle of the rich natural environment is gained at the sacrifice of urban amenities.

The movement to restore urban-fringe forests, which are often called *Satoyama*, is more realistic. *Satoyama* are more accessible and easier to visit, since such forests are located on flat land surrounded by agricultural fields. Already there is a nation-wide movement to protect *Satoyama* forests. Certainly, they are vulnerable to residential development or the construction of golf courses, and their preservation needs urgent measures. However, ecological motivation, such as the protection of native or endangered species, often dominates this movement, while the fact that such forests have been maintained for commodity use is often ignored.

The sustainable use of forest resources in the *Satoyama* used to guarantee that the forest succession was held at a certain stage suitable for certain species. However, without restoring a lifestyle heavily dependent on the *Satoyama*, halting the succession will not last long. How many of today's urban Japanese would depend on firewood for cooking and heating, as their ancestors did before the fuel revolution? Surely, the demand for organic agricultural products is on the rise, but urban forests grow more quickly than the present demand for products made from leaves and branches. Consequently, the material pruned to halt forest succession is sent to waste-processing plants and burnt with heavy oil.

If restoring an old-fashioned lifestyle is not realistic or sustainable, we need to reconsider the ecological approach to the *Satoyama*. As Kitamura (1993) mentioned, recognizing urban forests as cultural resources will supply a new perspective for the conservation of these forests. For instance, firewood can be utilized for small-scale local power plants, instead of being burnt with oil as garbage. Thus, we can accept a more sustainable lifestyle without sacrificing modern amenities such as electricity.

At the same time, well-managed *Satoyama* forests without undergrowth are easy and comfortable to walk through and are an excellent resource for on-site education to boost environmental awareness. Some of the protected *Satoyama* forests already have visitor centers, staffed with rangers in charge of interpretation. Nevertheless, most visitors use automobiles to visit these forests because public transportation is not convenient. A heavy dependence on automobiles is a serious dilemma common to most conservation areas (Ito, 1997). Ironically, visitors are increasing the environmental impact by enjoying and learning about these forests. Therefore, a more environmentally sound mode of transportation is required for *Satoyama* visitors. Greenway networks that connect the home with these areas by various modes of transportation are indispensable and can realistically make a true environmental contribution (Ito, 1998). Greenway networks also promote hiking as part of Japanese forest culture.

In this way, we can shift the meaning of forest culture from the sustainable use of forest resources to the use of the forest environment. The new forest culture should aim at enhancing environmental awareness by filling the gap between real forests and naturalistic images of them.

REFERENCES (J: in Japanese)

Akasaka, N. (1991): Spiritual History of the Mountains: Birth of Kunio Yanagida. Shogaku-kan, Tokyo, 350 p. (J)

Aoki, K. (1996): Japan's Culture Restored in Forests. San-ichi Shobo, Tokyo, 190 p. (J)

Chiba, T. (1956): Study of the Bold Mountains. Gakusei-sha, Tokyo, 237 p. (J)

Chiba, T. (1973): Culture of the Bold Mountains. Gakusei-sha, Tokyo, 231 p. (J)

Forestry Agency (1995): Pictorial Forestry White Paper - Aiming at Further Development of Forest Culture. Norin Tokei Kyokai, Tokyo, 175 p. (J)

Hirano, H. (1996): The Search for Forestopia. Cyuokoron-sha, Tokyo, 240 p. (J)

Huntington, E. (1938): Civilization and Climate. Iwanami Shoten, Tokyo, 400 p. (Japanese translation)

Ichikawa, T. and Saito, I. (1985): Reconsideration of the Forest Culture of Japan. Nihon Hoso Shuppan Kyokai, Tokyo, 209 p. (J)

Ichikawa, T. (1987): Beech Vegetation Belts and the Japanese. Kodan-sha, Tokyo, 204 p. (J)

Ito, T. (1997): The Dilemma of Ecotourism. Shinrin Kagaku (Forest Science), No. 21, pp. 16-22. (J)

Ito, T. (1998): The Possibility of Greenway Networks in Japan. Kokuritsu Koen (National Parks), No. 560, pp. 22-18. (J)

Kitamura, M. (1981): Forests and Culture. Toyo Keizai Shinpo-sha, Tokyo, 227 p. (J).

Kitamura, M. (1993): Forest Culture in Forests, People, and Community Design. Gakugei Shuppan, Kyoto, 301 p. (J)

Makino, K. (1988): Japanese who Restored the Forests. Nihon Hoso Shuppan Kyokai, Tokyo, 289 p. (J)

Marsh, G. P. (1974): Man and Nature. The Belknap Press of Harvard Univ. Press, USA, 472 p.

Ogura, J. (1992): History of Vegetation and Man. Yuzankaku, Tokyo 238 p. (J)

Ogura, J. (1996) Vegetation and Lives of Japanese. Yuzankaku, Tokyo, 246 p. (J)

Shidei, T. and Hayashi, C. (Eds.) (1984): Forest Images. Kyoritsu Shuppan, Tokyo, 254 p. (J)

Sugawara, S. (1989): What do forests mean to men? Kodan-sha, Tokyo, 246 p. (J)

Sugawara, S. *et al.* (1995): Isolated Woodlands and Nearby Forests - Evolution of Forest

Images and Civilization.   Aichi Shuppan, Tokyo 166 p. (J)

Tadaki, Y. (1981): The Cultural History of Forests.   Kodan-sha, Tokyo, 230 p. (J)

Tadaki, Y. (1984): The Cultural History of Forest – Human Relations.   Nihon Hoso Shuppan Kyokai, 156 p. (J)

Tadaki, Y. (1988): The Cultural History of Forests and People. Nihon Hoso Shuppan Kyokai, Tokyo, 211 p. (J)

Tsutsui, M. (1988): A Study of Forest Culture Policy.   Univ. of Tokyo Press, Tokyo, 188 p. (J)

Ueyama, S. (1969): Evergreen Forest Culture.   Cyuo-Koron, Tokyo, 208 p. (J)

Umesao, T. (1974): An Ecological View of Civilization.   Cyuokoron-sha, 290 p. (J)

Watsuji, T. (1935): Fudo or Human Investigation.   Iwanami Shoten, Tokyo, 208 p. (J)

Yasuda, Y. (1980): Introduction to Environmental Archeology.   Nihon Hoso Shuppan Kyokai, Tokyo, 270 p. (J)

Yasuda, Y. (1989): Civilization Eats Forests.   Yomiuri Shinbun-sha, Tokyo, 227 p. (J)

Yasuda, Y. and Sugawara, S., Eds. (1996): Forests and Civilization.   Asakura Shoten, Tokyo 259 p. (J)

Forest Landscapes of the Kanto Region, Japan in the 1880's and Human Impact on Them

Jun-ichi OGURA

Faculty of Humanities, Kyoto Seika University, Kyoto, Japan

ABSTRACT   One of the primary materials for the study of forest landscapes of the Kanto region in the 1880's is a set of topographic maps which were made for the first time ever in Japan to cover a wide area based on a modern survey. The other primary material is a set of records which were written as supplementary explanations of these maps. The results show that most of the forests there at that time, other than special ones such as national forests, consisted predominantly of pine trees or deciduous oak trees, and that most pine trees in the forests were less than 10 meters in height while most deciduous oak forests were less than 4 meters in height. Intensive use for fuel etc. on a rotation of 4 to 30 years was the background of such forests.
Keywords: Forest landscapes, Human impact, Coppicing cycle, Kanto region, 1880's

1. INTRODUCTION

Forest landscapes of Japan have largely changed during the post world war period through the rapid increase of plantations of Japanese cedar (*Cryptomeria japonica*) and Japanese cypress (*Chamaecyparis obtusa*) and so on, although there had been a long history of significant human impact on the forests of this country according to studies of written records, paintings, soil and so on (e.g., Chiba, 1991; Ogura, 1992). However, the conditions of forest landscapes of Japan in earlier times are still not so clear, and more studies are needed to clarify them and past human activities relating to the forests.

Here, I would like to show the forest landscapes of the Kanto region of central Japan in the 1880's, and human impact on them, as seen through the study of early topographic maps and written records of those days. The era is important as it represents the threshold of the country's modernization, and was a time of which useful new materials such as topographical maps describing land use appeared, and written records increased.

The hypothetical climax vegetation in most of the region was woodland, consisting of evergreen broad-leaved trees. The region has been greatly changed since the beginning of modernization with the expansion of cities and residential areas and so on. Tokyo, the capital city of Japan, is the largest city in this region.

2. METHODS AND THE STUDY AREA

A set of topographic maps, called the *Jinsokuzu* in Japanese, is one of the main materials in this study. Those maps, which cover most of the Kanto region with a scale of 1 to 20000, are the earliest maps in the region based on a modern survey. The survey was carried out between 1880 and 1886. The colorful drafts of the maps which are kept by the Geographical Survey Institute are more useful in studying the vegetation of the time than those printed in black and white, because they are very easy to read, the names of the main species in respective forests are included, with small illustrations on some of their margins depicting the actual landscape. The results of a study establishing the meanings of vegetation marks on these maps (Ogura, 1993) are also useful in their interpretation.

The other primary material used here is a set of records called the *Teisatsuroku*, which was written as supplementary explanation of the *Jinsokuzu* maps, by careful observation and questioning of native people and local officials (Rikugun-bunko, 1881). The

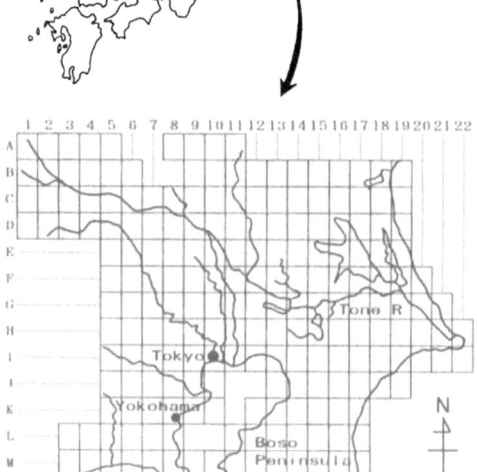

Fig.1   The study area

number of pages of these records exceeds 6000. They clarify the details of the landscapes, human lives and so on which existed more than a century ago and which would be difficult to ascertain from the maps only. They are also kept by the Geographical Survey Institute. Some other written records were used in addition in this study.

The study area is the whole area of the rectangles shown in Figure 1. It consists of the whole surveyed area of the *Jinsokuzu*, with each rectangle marking one unit of the survey. In the following text, the mark [·] is used to show a particular area. For example, [1·A] means the top left of the study area. It is often used to show an area which a record of the *Teisatsuroku* describes.

3. FOREST LANDSCAPES OF THE KANTO REGION IN THE 1880'S

Forests were commonly seen in the Kanto region in the 1880's, especially on hills and mountains. However these landscapes were usually quite different from those of the present day.

3.1 Dominant species of the forests seen in the *Jinsokuzu*

Figure 2, which is based on the *Jinsokuzu* maps, shows the dominant species of the forests in respective areas in the 1880's; each rectangle is marked to show the dominant species of the forests there. The forest area of marked dominant species is more than 2/3 of the total forest area in each rectangular unit.

As shown in Figure 2, pine trees and deciduous oak trees were usually the dominant species of the forests that existed there in those days. The main species of pines are Japanese red pine (*Pinus densiflora*) and Japanese black pine (*Pinus thunbergii*). The latter is usually seen in coastal areas and is rare in inland forests, while the former is seen all over the study area. The main species of deciduous oak trees are *konara* oak (*Quercus serrata*) and *kunugi* oak (*Quercus acutissima*).

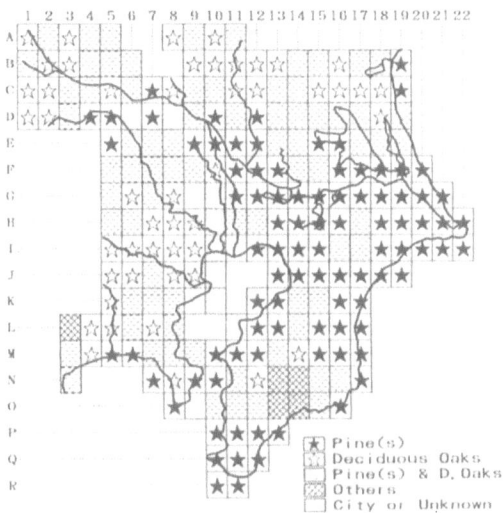

Fig.2　Dominant species of the forests

Forests dominated by pine trees were particularly common in the areas near the sea and along the Tone river and its branches, and forests dominated by deciduous oaks were most commonly seen in Musashino on the west of Tokyo, Sagamino in the west of Yokohama and the northern part of the Kanto region, although there were many areas where pine and deciduous oak forests were mixed. Main species of other forests, called 'others' in Figure 2, were fir (*Abies firma*) and deciduous oaks [13·N][14·N], fir and pines [13·O], fir and deciduous and evergreen oaks [14·O] and beech (*Fagus crenata*) and deciduous oaks [3·L].

3.2 Forest landscapes and human impact on them as seen in the *Teisatsuroku*

It is difficult to establish the details of the forest landscapes or human impact on them using the maps only. The *Teisatsuroku* provides a very useful material for this study, although descriptions of some areas are very precise while those of other areas are less comprehensive.

If the entire text is read, it is apparent that the common forests that were mainly owned by villages were different from those special forests mainly owned by the nation. So they should be described separately.

3.2.1 Common forests

3.2.1.1 Landscapes

According to the *Teisatsuroku*, most common forests which were mainly owned by villages were usually low in height in the region in those days, although a few records show that big trees or high forests existed in common forests in some areas [6·G][8·K][11·G][12·H]. Such descriptions as 'little trees' [2·B][5· I][10·C][13·P][14·J][14·K][7·H][11·H][12·F][13·F][14·G], 'shrubs' [6·G][6· I][6·J][7·H][8·H][9·I][10·H][12·G][12·H][15·K][15·M][16·M][16·L][17· B][18·C][18·D], 'very young trees' [6·F][8·H][9·I][10·M][12·H][13·I][14· I][14·L], 'the size of hedges' [5·C][10·B], 'no high forests' [5·G][6·H][8· F][12·F] or 'no big trees' [10·C][11·C][11·L][12·D][14·F][15·F] are often given in the *Teisatsuroku*. Low forests were especially common in places near Tokyo, like the Musashino, as well as in the northeastern Kanto region, some areas of the Boso peninsula and so on.

In some areas, it is possible to establish from the *Teisatsuroku* the size of those trees were: most deciduous oak trees were from 2 to 4 meters high, while some pine trees and Japanese cedar trees were relatively high, although many of them were not higher than 10 meters. There was at least one area where deciduous oak forests were less than 2.1 meters high [15·F]. Usual DBH (diameter breast height) of relatively high trees of pines or Japanese cedar was from 10 to 20 centimeters, although that of some big trees of those species was from 30 to 50 centimeters, while that of common deciduous oaks was usually less than 10 centimeters.

Mixed forests with species of deciduous oaks and pines and (or) Japanese cedar were commonly seen in at least some areas [5·F][5·H][6·G][11·G][12·H]. In those areas, relatively high pine or cedar trees, which were often sparsely grown, were conspicuous in respective forests. Figure 3, from the margin of a draft of the *Jinsokuzu* [9·N], shows such a forest landscape.

3.2.1.2 Human impact

There existed intensive human impact on the above-mentioned common forests, where landscapes were largely different from today.

Fig.3   A picture from a draft of the *Jinsokuzu*

We can find many descriptions concerning the use of the forests in the *Teisatsuroku*, although they are usually brief. According to them, the main purpose of the use of the forests in the 1880's in this region was to supply fuel or firewood and charcoal. Descriptions to show such human activities are seen almost all over the region. The fuel was often transported to the cities if it is not insufficient in an area. Tokyo and Yokohama were the main cities for that destination. On the other hand, relatively big trees like pine or cedar were used for timbering.

The period of rotation of the common forests differed according to place and species. But generally speaking, pine trees were usually cut down on a rotation of 20 to 30 years, while deciduous oak trees were cut down on a rotation of 4 to 30 years. In some areas, the deciduous oak trees or coppiced trees were cut down on a rotation of 4 to 15 years while they were cut down on a rotation of 15 to 30 years in other areas. There were also areas where the rotation ages lay between the above. Figure 4 shows the rotation ages of deciduous oak trees in respective areas. Un-marked areas mean that no description about the period exists in the *Teisatsuroku*. But that is a rough explanation. If you read it attentively, you will find the coppicing cycles were very short in some areas at that time: for example 4 to 5 years [15·D][18·D], 5 years [8·L], 5 to 6 years [9·N][14·C], 6 to 7 years [13·H][13·I][15·F]. Conversely, some pines, cedars and so on were cut down for timber on a rotation of 50 to 100 years [1·B][13·E][14·E].

According to the *Teisatsuroku*, management methods for deciduous oak forests differed according to area. In some areas, they were not managed at all after cutting. It was impossible or very difficult to enter such forests [8·M][9·M][9·N], or it was difficult to walk in them [6·J]. In other areas, deciduous oak forests were managed well: weeding was done in early spring each year [2·B], or under-branches were usually trimmed in winter [8·K].

Conversely, pine forests were usually well-managed by weeding and trimming of under-branches. It was usual for some well-growing pine trees as well as cedar trees in the forests to be left intentionally for timber use.

There is a description in the *Teisatsuroku* which shows the relation between the ease of transportation of fuel and the dominant species of forests [14·K]. It can be seen as explaining the distribution of the pine forests which were very common along the coast and the Tone river and its branches where the transportation by

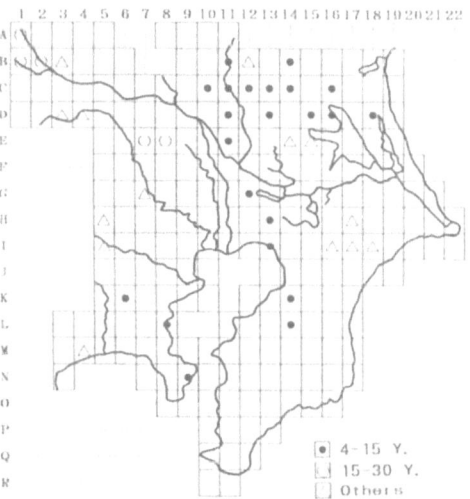

Fig.4 Coppicing cycles of deciduous oaks

ships was relatively easy (cf. Figure 2).

It was usual to plant pines and deciduous oaks, especially in the northeastern areas of Tokyo.

3.2.2 Special forests

3.2.2.1 National forests

Some national forests in the region were extensive, with areas of more than 1000 ha. They occupied large areas in some regions, although this was exceptional in Kanto region.

According to the *Teisatsuroku*, many big trees often existed in the national forests, although big trees were rare in some national forests in some areas [11·M][13·F][15·L]. The main species of the forests were usually pine(s) and Japanese cedar, while Japanese cypress was also often seen there. Other main species seen in the *Teisatsuroku* are fir (*Abies firma*) [3·L][13·C][13·N][13·O][14·N][14·O][15·N], Japanese chestnut (*Castanea crenata*) [1·D], *mizunara* oak (*Quercus mongolica*) [1·D], Japanese beech *(Fagus crenata)* [3·L][13·C], *keyaki* (*Zelkova serrata*) [5·L], Chinese nettle tree (*Celtis sinensis*) [5·L] and *mukunoki (Aphananthe aspera)* [5·L].

Some trees were very old; such expressions like 'around 100 years' [4·A][17·M], 'around 150 years' [4·C] or 'over hundreds of years' [3·L][13·N] were used to describe them. DBHs of big trees were 'between 130 and 30 centimeters' [12·N], 'more than 60 centimeters' [3·M] or 'between around 30 and 10 centimeters' [5·L] according to the records. References to the height of those trees are rare in the *Teisatsuroku*.

The management of the forests varied; weeding and trimming of under-branches were done every year in some forests, while such works were not done in other national forests. A description of planting of Japanese cypress in gaps in forests can be seen in a note for one area [17·B].

3.2.2.2 Others

There were some other special forests or those of similar status.

Shrines and temples were often recorded in the *Teisatsuroku* as having special forests which usually included big or old trees, although the area of most of them was relatively small. Main species in those forests were usually pine(s) and cedar. That was quite different from now, as evergreen broad-leaved trees are often dominant there nowadays. That suggests that most of those forests were frequently used for some purposes, although such description can not be found in the *Teisatsuroku*.

Small woods were often seen near houses in the countryside in many areas. There were usually big trees in those woods. The main purposes of the woods were for wind protection and to supply timber in many cases. The main species of those woods were Japanese cedar, *keyaki*, pines, evergreen oaks, bamboos and so on. Many villages often resembled deep forests from a distance due to their presence.

Big trees were often seen along the main roads. Pines were the most common species. They sometimes interrupted the view and also resembled forests from a distance.

4. LOW VEGETATION

Low vegetation of trees and grasses, usually lower than a man's height, was commonly seen in the study area, especially on hills and mountains. More than half of the hills and mountains was covered with such vegetation in some areas. This was quite different from today and seems to be an important aspect for the study of the forest landscapes of those days.

Fire, grazing and frequent cutting of trees and grasses for fuel, fodder etc. caused the low height of vegetation in the region, according to the *Teisatsuroku.* Grasses were dominant in the low vegetation of many areas, while shrubs were dominant in some areas.

5. CONCLUSIONS

According to the *Jinsokuzu* and the *Teisatsuroku*, the dominant species of most of the forests in the Kanto region in the 1880's were pines and deciduous oaks. Pine trees were usually less than 10 meters in height and most deciduous oaks were less than 4 meters high in common forests. Moreover, very low forests or shrubs less than about 2 meters high were commonly seen in many areas. Intensive use for fuel etc. on a short rotation period was the background of such forests.

The period of rotation differed according to place and species. Pine trees were usually cut down on a rotation of 20 to 30 years, while deciduous oak trees were cut down on a rotation of 4 to 30 years in common forests in general. In some areas, however, the deciduous oak trees were coppiced on a rotation of 4 to 6 years. This cycle is much shorter than the prevailing thought in these days, which counts a cycle as 20 to 30 years (e.g., Moriyama, 1988; Ishii et al., 1993). It is also shorter than the description in an old book which refers to the cycle as 6 to 25 years (Honda, 1908). The results of this study show that the cycle was shorter than usually practiced in some areas and suggests that no high deciduous oaks were seen in the coppiced forests there, although it might not be rare to see some high trees like pines in those forests.

Such forest landscape was a reflection of the people's lives in those days when commercialism was already well-developed and the main purpose of common forests was to supply firewood and charcoal which were vital to daily lives: coppicing was a beneficial and easy forestry method at that time, although it was apt to lead to degradation of the land (Honda, 1908), while other forestry methods seemes to have been more beneficial in some areas. The existence of big cities in the region had an important influence on the landscape, because much fuel was consumed there.

On the other hand, special forests mainly owned by the nation were often quite different from common forests as big trees were commonly seen in them, although areas covered widely with such forests were exceptional in the study area.

According to historical studies of vegetation, similar forest landscapes were also seen in many areas in Japan at that time including the Kansai region, where the big cities of Kyoto and Osaka were located (Ogura; 1992, 1996).

REFERENCES

Chiba, T. (1991): A study of bald hills (*Hageyama no kenkyu*), Soshiete, Tokyo, 349p. (in Japanese)

Honda, S. (1908): Silviculture (*Zoringaku honron*), Miura Shoten, Tokyo, 227p. (in Japanese)

Ishii, M., Ueda, K. & Shigematsu, T. (1993): Conservation of satoyama forests (*Satoyama no shizen wo mamoru*), Tsukijishokan, Tokyo, 171p. (in Japanese)

Moriyama, H. (1988): How to conserve nature (*Shizen wo mamoru towa doiu koto ka*), Nobunkyo, Tokyo, 260p. (in Japanese)

Ogura, J. (1992): History of vegetation and man (*Hito to keikan no rekishi*), Yuzankaku, Tokyo, 238p. (in Japanese)

Ogura, J. (1993): A study of the vegetation marks of the maps called *Jinsokuzu*, Journal of Kyoto Seika University, No.5, pp.40-69. (in Japanese with English summary)

Ogura, J. (1996): Vegetation and lives of Japanese (*Shokusei kara yomu nihonjin no kurashi*), Yuzankaku, Tokyo, 246p. (in Japanese)

Rikugun-bunko, (1881): Criteria for the military survey (*Heiyo sokuryo kiten*), Rikugun-bunko, Tokyo, 370p. (in Japanese)

Japanese Attitudes Towards Forests According to Comparative Opinion Surveys

Shigejiro YOSHIDA*1, Masaaki IMANAGA*2
Koji MATSUSHITA*3 and Okihiro OHSAKA*2

*1 Faculty of Agriculture, Kyusyu University, Fukuoka 812-8581, Japan
*2 Faculty of Agriculture, Shizuoka University, Shizuoka 422-8017, Japan
*3 Faculty of Agriculture, Kyoto University, Kyoto 606-8502, Japan

ABSTRACT The aim of this study was to discover what the residents of various countries think and feel about their forests. Large scale opinion surveys had been conducted throughout the world since 1979 to 1996 in various Japanese cities, four German cities, one French city, three Brazilian cities, and in three Peruvian cities. Results of the opinion surveys were analyzed, and attitudes towards forest and nature summarized as follows;
1. Many people believed that a spirit exists in natural things animism and had a quite deep emotional affinity with forest and nature except in Tokyo.
2. On the ethical behavior of forest management, positive responses were most frequent in Germany and France , followed by Brazil and Peru , and then Japan.
Key words: Attitude towards forest, Forest management, Opinion survey

1. INTRODUCTION

The problem of nature and forest conservation is today of high light priority in all parts of the world. And the methods of conservation employed are numerous and varied. This variety is caused not only by differences in the natural and forest condition of each country but also by the habits and attitudes of its residents. Studies on the opinions and feelings of residents concerning their natural surroundings, and their attitudes towards forest and forest management, are very important for all countries. In order to clarify these issues, opinion surveys were carried out in Germany, France, Japan, Brazil and Peru.

Through analysis of results from this study, not only attitudes of residents towards forest, but also rational future forest management in individual countries will become clear by analogy.

This study was carried out with the aid of funds made available by the Japanese Government.

2. METHODS

Large scale surveys were conducted by the Society before and after 1980 in six Japanese cities, four German cities (in the former Federal Republic of Germany), and one French city. Opinion surveys in one city and one town in Kagoshima Prefecture, Japan, were conducted by the authors in 1993. Three cities in Brazil in 1992 and 1993, and three cities in Peru in 1995 and 1996 were added to the program. The sample varied in size from 260 to 1,200 individuals, with valid responses numbering 190 to 410. In the Japanese and French surveys, voter registration lists were used, while the German survey used resident registration lists. The Peruvian and Brazilian surveys selected companies and individual houses.

In Japan, the survey was conducted in the following cities: (1) Miyazaki, a city of 260,000 and the prefectural capital of Miyazaki Prefecture; (2) Tokyo, the twenty-tree wards of central Tokyo, with a population of 8.51 million; (3) Kagoshima, a city of 530,000 and the prefectural capital of Kagoshima Prefecture. In Germany, the survey was held in the following four cites: (1) Freiburg, a city of 180,000 on the end of the Black Forest in south-west Germany; (2) Neuenbuerg, a small village of 7,000 at the northern end of the North Black Forest; (3) Goettingen, a college city of 260,000 near the Harz Mountains which form the border between the former East and West Germanies; (4) Hanover, a city of 540,000 and the capital of the state of Lower Saxony. In France, the survey was conducted in Nancy, provincial capital of Lorraine, with a population of 400,000. In Brazil, the survey was held in the following three cities : (1) Sao Paulo, a city of 11 million, industrial and cultural center of Brazil; (2) Curitiba, the provincial capital of Parana to the south of Sao Paulo, with a population of 1.4 million; (3) Manaus; a city of 1.2 million situated in the middle reaches of the Amazon River. In Peru, the survey was held in the following three cities : (1) Iquitos, a port city of 170,000, an upriver district of the Amazon River, market center for timber and agricultural products; (2) Lima, capital of Peru with a population of 4,160,000, a Pacific coastal city, political, industrial and cultural center of Peru and; (3) Cuzco, old capital of the Inca Empire with a population of 210,000, a cultural and tourist city, 3,500 m above sea level.

The surveys took place in Germany, France, Miyazaki and Tokyo in 1980, in Brazil in 1991 and 1992, in Kagoshima in 1993 and in Peru in 1994 and 1995. In Brazil, Kagoshima and Peru, not only responses of citizens, but also those of college and high school students were recorded.

The questionnaire consisted of a face sheet containing thirteen questions which were subsequently classified into the following seven categories;
1)Primitive and religious emotions to nature. 2)Religious and mystical feelings relating to trees and forests. 3)Affection for forests in daily life. 4)Feeling for and knowledge of trees. 5)Attitudes towards hunting. 6)Ethical behavior of forest management. 7)Preference for forest type was ascertained by paired comparison method using five pairs of photographs.

3. RESULTS

3.1 Favorite tourist destinations

Favorite tour destinations were investigated. The results are shown in Table 1. In Germany over 50% selected "forest". In France, Brazil and Peru about 10-24% made the same selection, whereas in Japan the figure was under 10%. The first choice for Japanese was "scenic mountain". In France "beach" and "forest" were the favorite destinations, and in Brazil "beach" and "lake", in Peru "scenic mountain" and "lake" were selected.

Interest towards forest in daily life is shown in Table 1. The survey in Germany yielded particularly significant results. The percentage of respondents making this selection differed very little even from city to city. It seems reasonable to conclude that Germans in general are extremely fond of forests. Only 21% of respondents from Nancy, France, by contrast, expressed the same affinity for forests. The results were somewhat lower in Brazil and somewhat higher in Peru.

It is thus apparent that Germans have an exceptional fondness for forests, followed by Peruvians, the French, Brazilians and the Japanese, in that order. These results lead to the observation that while there may be major variations in perceptions of the ideal destination on a country-to-country basis, the thinking of people of the same country displays a striking similarity, regardless of where they reside in that country.

Table 1. When you travel, which destination do you prefer ? Choose only one.

	I	L	Cz	S	C	Ma	F	Ne	G	H	N	M	K	T
Forest	23	15	24	15	10	20	55	62	56	57	21	8	7	3
Church	5	5	14	0	2	4	1	2	0	1	7	20	18	18
Beach	15	16	7	24	34	25	3	3	6	7	25	3	3	10
Meadow	5	3	4	5	3	4	9	10	7	7	12	24	23	19
Mountain	21	31	19	25	19	13	17	11	11	8	15	26	30	23
Ridge	2	3	4	1	0	1	4	6	6	4	8	0	0	1
Lake	25	24	22	22	27	31	9	5	10	13	3	16	14	22
Others	3	2	5	1	2	2	1	2	1	1	9	2	3	1

Meadow: Mountain Meadow, Mountain: Scenic Mountain, Ridge: High peak or ridge

I: Iquitos (Peru), L: Lima (Peru), Cz: Cuzco (Peru), S: Sao Paulo (Brazil), C: Curitiba (Brazil),

Ma: Manaus (Brazil), F: Freiburg(Germany), Ne: Neuenburg (Germany),

G: Goettingen (Germany), H: Hanover (Germany), N:Nancy (France), M: Miyazaki (Japan),

K: Kagoshima (Japan) ,T: Tokyo (Japan)

3.2 Attitudes towards forest and nature

The results are shown in Table 2. The percentage of positive answers was above

70 % in Brazil and Peru. In France and Japan (except in Tokyo) it was about 60%. By contrast, in Germany the number of negative answers was higher than for positive ones. It should be noted that Peruvians and Brazilians have deep primitive and religious feelings for nature.

Table 2. Do you believe in a spirit in such natural objects as mountains, valleys, streams, trees, plants, etc. ?

	I	L	Cz	S	C	Ma	F	Ne	G	H	N	M	K	T
Yes	87	78	83	79	75	72	47	43	40	44	66	57	58	24
No	13	21	16	17	24	28	52	55	60	56	34	41	40	74

3.3 Affection for forests in daily life

The results are shown in Table 3. The percentage of "enjoy" was higher than "dislike" and "indifferent" in all cities. In Germany and France over 90% of respondents selected positive answers and the percentage of "enjoy" in Japan was equal to Peru and Brazil except for Tokyo. In Tokyo only 62% of respondents gave positive answers and it was the lowest value in all cities.

Table 3. Do you enjoy walking in the forest ?

	I	L	Cz	S	C	Ma	F	Ne	G	H	N	M	K	T
Enjoy	71	70	83	88	75	71	96	98	97	96	92	79	87	62
Indifferent	22	23	15	7	18	25	4	2	3	4	7	16	11	25
Dislike	5	5	2	1	3	4	0	0	0	0	1	1	1	7

3.4 Religious or mystic feelings concerning trees and forests

The results are shown in Tables 4 and 5. Positive answers represented over 80% of responses for both questions except for Tokyo (for both questions) and for France in Table 4. The positive answers in Table 4 are equal or higher than the one in Table 5 except for France. In case of France the positive answer concerning forests is larger than the one concerning trees. It seems that they may distinguish the images between tree and forest.

Table 4. When you look at a large old tree, do you feel anything spiritual

	I	L	Cz	S	C	Ma	F	Ne	G	H	N	M	K	T
Yes	93	86	94	99	93	90	90	92	90	91	70	90	91	57
No	7	12	5	1	5	10	9	7	10	8	30	10	9	38

Table 5. When you are in a deep forest, do you have a mysterious feeling ?

	I	L	Cz	S	C	Ma	F	Ne	G	H	N	M	K	T
Yes	91	83	91	98	93	88	86	84	85	87	80	91	90	53
No	7	15	10	0	7	12	14	16	15	13	20	9	10	35

3.5 The ethical behavior of forest management and favorite forest types

The results are shown in Tables 6 and Table 7. In Table 6, about 70 % of Peruvian respondents selected 1, 61-75% in Brazil and over 80% in Germany and France. However in Japan the number of positive answers was lower than in other countries and in Kagoshima and Tokyo the percentage of negative answers was particularly higher.

In Table 7, about 65% of Peruvian respondents selected 1, 50-68% in Brazil and over 75% in Germany. On the other hand, in France and in Japan the percentage of respondents selected 2 is higher than the one of respondents selected 1, and was 71% and about 53% respectively.

Table 6. Which is your opinion ?
1. Man should manage forests to keep them beautiful.
2. Man should not manage forests at all.

	I	L	Cz	S	C	Ma	F	Ne	G	H	N	M	K	T
1	71	68	72	68	61	75	87	86	80	78	83	61	44	45
2	25	29	28	28	32	25	10	12	15	20	17	34	53	50

Table 7. Which do you prefer ?
1. Nature influenced by man, intermixing farming with meadow and forest.
2. Unspoiled nature, such as virgin forests and wilderness.

	I	L	Cz	S	C	Ma	F	Ne	G	H	N	M	K	T
1	62	66	62	50	59	68	82	83	75	78	28	47	43	42
2	35	32	37	46	40	32	16	13	24	22	71	52	55	51

The results of favorite forest type are shown in Table 8, selections having been made from a pair of picture. There were five pairs of pictures. In the fourth pair (Fig.1.), the selection was between naturally regenerated beech forest and Japanese cedar plantation. All respondents in each country selected cedar plantation in preference to naturally regenerated forest. The percentage of respondents selecting cedar plantation was higher in Germany and Japan.

From the above results, contradictory opinions were observed in the French and Japanese samples, especially in Tokyo. This seems to be the case when "man should not manage" was selected in Table 6 and "unspoiled nature" in Table 7, and yet, "man-made forest" was selected in Table 8. These contradictions should be noted.

Fig.1 The pair of picture for selection(4A is natural beech forest and 4B is Japanese cedar plantation). Source: Typical forest in Japan, Japan Forest Technical Association (1966)

Table 8. Which forest type do you prefer in the fourth pair of photographs ?

	I	L	Cz	S	C	Ma	F	Ne	G	H	N	M	K	T
4A	40	46	50	43	37	31	15	14	21	16	46	30	-	28
4B	59	53	49	46	47	67	62	66	57	62	54	67	-	64

4A: Naturally regenerated beech forest   4B: Japanese cedar plantation

4. DISCUSSION

The most salient feature of Japan's natural terrain is the predominance of forest. Forests cover 67% of Japan's total land mass. Few countries in the world are as heavily forested as Japan. Forests cover 29% of Germany, 27% of France, 66% of Brazil and 55% of Peru. However the forested area today is concentrated in the northern sections of the Amazon in Brazil and in the eastern sections of the Peruvian Amazon and in the mountainous areas in Peru.

While focusing on the forest survey, contradictions in the Japanese samples were detected. Why are they found only in Japanese samples? It may be that forests do not stand out in the Japanese consciousness but are more often viewed simply as a distant backdrop.

Naturally regenerated forests can be found in the mountainous regions of Japan's hinterlands, but these are generally surrounded by plantation. Consequently, the image projected by forests throughout Japan with the exception of some prefectures, is virtually identical. The 10 million hectares of plantation accounts for approximately 40% of Japan's forests. As most of the plantation are situated around populated mountain regions, the forest scenery to which the Japanese have become most accustomed is primarily made up of cedars and cypresses, which closely resemble each other in shape and coloring. The Japanese are not accustomed to going into forests for recreation and don't have the opportunity to come into contact with trees and forests in the course of study. People will apparently pay no particular attention to the forests and naturally not be able to contemplate whether a forest is naturally regenerated or man-made if they do not even notice it to begin with. It seems that many Japanese, under the influence of recent mass media attention, are under the impression that no special measures need be taken to protect the environment. The Japanese are said to have a deep traditional interest in and strong attachment to nature. Our surveys, however, suggest that non-Japanese are more likely than Japanese to be moved by ancient trees or dense forest, thus, throwing this commonly held belief into question.

Contemporary Japanese do not necessarily have a strong interest in nature or forests. Since the enhancement of conservation policies will require a deep understanding of nature based on actual experience, it will be necessary to provide the Japanese, especially the younger generation, with many more opportunities to come into contact with nature and forest.

The first opinion surveys were held in 1980, or about 20 years ago. In these 20 years, opinions and attitudes towards forest may have changed in the targeted countries and cities. Therefore further surveys are necessary in future.

REFERENCES

Imanaga,M.(1989) International comparisons of attitudes towards nature(XIV) An opinion survey on the forest in Kamou/Kagoshima, Bull. Kagoshima Univ. Forests, 17,1-11(in Japanese)

Imanaga,M. and Cho,M.(1993) Forest environment and people's attitudes toward forest in Brazil, Forest Economy, 540, pp.7-18(in Japanese)

Imanaga,M.(1996) The Japanese view of nature from a comparative standpoint, The Japan Foundation Newsletter, 24(No.2),pp.1-6

Shidei,T.(1981) International comparison of attitudes towards nature. Scientific Report of Toyota Foundation, I-007, pp.1-128(in Japanese)

IIistory, Condition and Management of Floodplain Forest Ecosystems in Europe

Emil KLIMO

Mendel University of Agriculture and Forestry, Brno, Czech Republic

INTRODUCTION

The European floodplain forest ecosystems represent specific forest biogeo-cenoses with a specific species composition related to ecotypes formed on Quaternary river alluvia. In spite of regional differences closely related types of production conditions developed with azonal and unique vegetation, showing the similar quality and intensity of production and decomposition processes.

At present, floodplain forests in Europe occur in different stages of their natural development in relation to site conditions which is expressed by such terms as 'soft-wooded floodplain forest' (with the predominance of poplar, alder and willow) and 'hard-wooded floodplain forest' (with the predominance of oak, ash, elm etc.).

Present main roles of floodplain forests consist in:

- high production level;
- high biodiversity based on high variability of forest sites;
- protection of watercourses against erosion and pollution;
- high number of nature reserves;
- recreational and aesthetic functions within the landscape.

Historically, but also at present, floodplain forest ecosystems throughout Europe are under heavy anthropogenic impacts which can have various results under various conditions.

Main anthropogenic impacts are as follows:

- decrease of floodplain forest area in favour of agriculture, often to the level of riparian stands;
- watercourse regulation resulting in the termination of floods and groundwater table decrease;
- dam construction (hydroelectric power stations) or building water reservoirs resulting in the destruction of large floodplain forest areas (e.g. the Dnepr river, Ukraine or the Nové Mlýny reservoirs, S. Moravia);
- interactions between floodplain forests and housing estates (particularly increased recreational use of floodplain forests, road construction, sports areas etc., e.g. the Leipzig floodplain forest);
- interactions between floodplain forests and intensively managed agricultural land in their immediate vicinity (increased input of various substances particularly through wind erosion from fields to forests); exploitation of raw materials

in alluvial plains and particularly in oxbow lakes after river regulation (sand and gravel exploitation);

- intensive game management (high game populations, establishing game preserves);
- fragmentation of floodplain forest ecosystems often below the limit of a minimum ecological range for a number of autochthonous plant communities and animal species and thus also the formation of ecotones or even barriers disturbing the integrity of ecosystems. The fragmentation of ecosystems and isolation of populations result in the fact that their subsistence requirements are not satisfied and this necessarily means increasing degeneration and population dieback. It is, therefore, very important to preserve as large as possible area of floodplain forests which are, as a matter of fact, of remnant character due to the anthropogenic activities mentioned above.

Looks at History

The development of floodplain forests in the last millennium was similar in the majority of Central European alluvial plains. In the Middle Ages, coppice forest or coppice-with standards with short rotation and high proportion of species characteristic of the 'soft-wooded floodplain forest' predominated. Rationalization of forestry in the last century resulted in the change of floodplain forests to high forests (forests of seed origin) and the proportion of species characteristic of the 'hard-wooded floodplain forest' increased. The development of biogeocenoses was continual and the biocenoses preserved substantial properties of natural ecosystems particularly high ecological stability and biodiversity. Thanks to the changes in hydrological regime, biogeocenoses related to the high level of groundwater table and periodic floods, i.e. floodplain forest biogeocenose types belonging to *Salicion albae* alliance and/or more hygric subassociations of *Querceto-Ulmetum* association characterized by the occurrence of wetland species, e.g. *Iris pseudacorus* became, however, more rare.

For example, floodplain forests of Croatia having an area of 290 000 ha, i.e. 15% of the total forested area are documented since the beginning of the 16th century. It was then that the first regulations ruling the use and protection of forests were passed ("Tripartitum opus iuris consuetudinarii inclyti regni Hungariae", 1514). The lowland part of Croatia bordering with Bosnia (the Military Border) was very sparsely populated, and forests were only utilized for military needs and for the needs of borderline families. The region abounded in vast tracts of virgin forests of authentic composition and form. Estimates based on veritable documents show that 70% of the total area of Slavonia was under forests.

According to the existing statistics, the relative area under timber in Slavonia was as follows:

in 1750   70% estimate
in 1850   60% estimate
in 1914   35%   statistics 1875-1915
in 1938   30.8% statistics 1938
in 1953   28.5% statistics 1955
in 1961   27.5% statistics 1970
in 1979   29.5% statistics 1978

Settlements progressively increased in numbers, and their inhabitants, who carried a rifle in one hand for most of their lifetime, accepted livestock breeding as the main means of maintaining their existence. In such conditions excessive grazing and feeding of their livestock on acorns interrupted natural reforestation, while older trees, having continued their growth, reached a very old age and acquired momentous dimensions. The transition from virgin forest was gradual and heterogeneous.

The forests in the vicinity of settlements were further endangered by the choice of trees for felling. Slimmer and better trees were selected for building material, for firewood also thinner trees and especially those that burnt easily. Thus, the forests became thinner, with thick and branchy oak trees.

Today, lowland forests in Croatia cover less than 300 000 ha (approximately 20% of the lowland area) and the most seriously affected area is the one along the Drava river. Between 1971 and 1988, three hydro-power stations with three storage reservoirs and canals were built there, resulting in a drop of the water table in the upper course of the Drava. The forest vegetation was devastated, and floodplain forests were completely destroyed.

Similar situation occurs in the upper course of the Sava river and along the river Danube in its Croatian part.

Also floodplain forests of Hungary were exposed to changes in land use. Inhabitants of the territory occupied further and further areas of the originally flooded region. It was enabled by the regulation of the Tisza river and other watercourses (Haraszthy 1994).

Flood control works affected not only the Tisza but almost all of the rivers in Hungary, reducing the once 2.3 million km^2 total floodplain area of the country to only 1.5 thousand km^2. The majority of floodplain forests is scattered in small clusters along rivers, and their widths often do not exceed 10 meters. Larger floodplain forests can only be found along the Danube, Dráva, Rába, Tisza, Maros, and Hármas Körös rivers with the forests of the Danube and Tisza rivers being the most prominent.

Based on the examples given above it is possible to say that one of the most important measures affecting the condition of the floodplain forests has been the period of the construction of hydroelectric power stations, water canals and river regulation. The process continued in the last fifty years when the area of alluvial forests decreased by 50% in the Rhine Alsatian forest (7500 ha as against 15 000 ha in 1930) and 25% in the case of Austrian forest of the Danube (8000 ha as against 33 000 ha in 1930) (Tendron 1981).

In the alluvium territory in the lower reach of the Morava river (Slovakia), the marked reduction of floodplain forest area occurred in favour of meadow ecosystems. Later on, the major part of the meadows was ploughed and used as arable land. The development of the proportion of floodplain forests, meadows and arable land in the territory is given in the following table:

Year	Arable land	Meadows	Forests	Urban area	Water surface
1782-1784	6.25	46.28	46.71	0.10	0.66
1882	27.26	47.90	21.84	0.10	2.90
1991	35.32	31.17	29.46	0.57	3.48

As already given above, many floodplain forests disappeared by permanent flooding of alluvial land due to reservoir construction. For example, the Dnepr river (Ukraine) can well demonstrate the destruction of floodplain forests.

Area of flooded alluvial forests connected with building water reservoirs on the Dnepr river (thousand ha)

Water reservoir	Forests	Shrubs	Total
Kievsky	10	5	15
Kanevsky	10	6	16
Kremunchugsky	30	10	40
Dnieprodzerzhinsky	12	2	14
Lenin Lake	2	-	2
Kahovsky	126	-	126
Total	**190**	**23**	**213**

At present, only small areas of the Dnepr floodplain forests are fragmentarily protected in separate locations.

In many countries of Europe, as for example in Iceland, Norway, Sweden, Finland, Denmark and Ireland, it is possible to speak only about riparian stands along water courses in relation to climatic conditions. In many other countries, the area of floodplain forests was reduced to a minimum. For example in the Netherlands, the area of floodplain forests amounts to about 2500 ha, i.e. less than 1% of the total area of forest resources in the country. Likewise in Belgium, floodplain forests along large rivers were replaced by cultivated land and meadows and there are only small areas of the forests bordering some smaller rivers (usually alder, willow or poplar stands). Floodplain forests occupy considerable areas still today in Germany, viz. floodplain forests on the alluvium of the Rhine, Danube, Main, Neckar and Elbe. Floodplain forests in the city of Leipzig represent an interesting unit fulfilling also a recreation role, further Austria (particularly on the alluvium of the Danube, Morava and Dyje rivers).

In France, considerable part of floodplain forests is situated on the alluvium of the Rhine (about 7400 ha) and other forests occur particularly along the Rhone river. Poland with 82 000 ha of floodplain forests, White Russia with large areas of wetlands and then Russia, the Ukraine and Romania (where particularly the Danube Delta is of great importance) and floodplain forests along the Danube in Bulgaria and Serbia are other countries where floodplain forests play an important role.

Changes in the forest site of floodplain forests resulted also in changes in the composition of forest stands, so called soft-wooded floodplain forest was changed to hard-wooded floodplain forest.

The floodplain forest of Leipzig had the following species composition of stands in 1870: oak 60%, elm 20%, lime 0.6%, hornbeam 13%, alder 0.7%, maple 0.4%, ash 0.4, and aspen 5.0% while in 1986, ash exhibited the highest proportion of 49.17% and oak decreased to 25%. In addition to ash, the proportion of maple increased to 13.7%.

In spite of all the consequences of human activities, in the cultivated landscape of European lowlands with the absolute predominance of agrocenoses and urbanized areas, biogeocenoses of floodplains appear to be a very important refugium of biotic diversity representing part of European natural heritage requiring a special strategy for their protection, conservation and management.

The effort aimed at the protection of floodplain forests in the alluvium of the Danube between Vienna and Heinburg resulting in the declaration of the area as the Donau Auen National Park can be evaluated as a very positive activity.

Present Condition and Processes

Considerable biodiversity of floodplain forests is given also by high heterogeneity of soil conditions when due to the variability of sedimentation processes in the course of the origin of alluvial soils of very different physical properties can be found (clayey, loamy, sandy and gravel soils). The diversification is both of vertical and horizontal character. Even small changes in topography can modify the groundwater table which results in differences in the composition of forest stands. All the factors reflect on the chemistry of the upper layer of soils.

The uneven distribution of data is caused by the heterogeneity of alluvial deposits and by the variability of stand composition. This results in different litter composition, and in various rates of decomposition and nutrient release from the forest vegetation, as well as in an uneven distribution of the shrub and herb layer in this stand.

The range of soil properties in the top layer

Soil property		Range
pH		5.5-6.8
Humus (%)		16-21
Ca	$(x\ 10^2\,mg\ g^{-1})$	57-126
K	$(x\ 10^2\,mg\ g^{-1})$	8-25
P	$(x\ 10^2\,mg\ g^{-1})$	4-17
Total N	$(x\ 10^2\,mg\ g^{-1})$	500-800
Available N	$(x\ 10^2\,mg\ g^{-1})$	14-26

Another typical characteristics of floodplain forests is the fact that the floodplain forests rank among temperate ecosystems with the highest supplies of nutrients in the stand biomass, and that in the case of some elements, these supplies are similar to those of tropical forests. The level of biodiversity and biomass production are particularly dependent on high accumulation and rapid cycling of nitrogen as given on the example of a floodplain forest in southern Moravia, Czech Republic (Klimo 1985).

Vertical distribution of nutrient reserves in the floodplain forest ecosystem (kg ha^{-1})

	N	P	K	Ca	Mg
Tree crown, branches and leaves	473	41	219	619	45
Trunks	1 044	42	358	1 326	81
Shrubs (aboveground part)	55	8	26	31	9
Herbs (shoots)	22	4	47	21	5
Total	1 594	95	650	1 997	140
Surface humus, annual mean	99	6	18	75	15
Roots: trees, shrubs, herbs	210	34	84	240	52
Nutrient reserves: rhizosphere	14 846	4 953	121 364	44 864	-
Reserves: without rhizosphere	9 093	5 085	176 532	76 321	-
Plants total	1 804	129	734	2 233	192
Soil total	24 038	10 044	297 914	121 260	-
Total	25 842	10 173	298 648	123 493	-

Heterogeneity in the composition of forest stands appears in the variety of decomposition processes as given on the example of the S. Moravian floodplain forest. The data show that the most intensive decomposition of litter occurred in a plot where ash predominated.

Plot I (*Quercus robur*)   kg ha^{-1}

Leaf reserves on	1 Dec. 1972	2 500	100%
	1 Apr. 1973	1 600	decreased by 36%
	1 July 1973	1 400	decreased by 44%
	1 Oct. 1973	1 100	decreased by 56%

Plot II (*Quercus robur* + *Cornus sanguinea* L.)

Leaf reserves on	1 Dec. 1972	4 200	100%
	1 Apr. 1973	2 000	decreased by 52%
	1 July 1973	1 700	decreased by 59%
	1 Oct. 1973	1 200	decreased by 71%

Plot III (*Fraxinus excelsior* L.)

Leaf reserves on	1 Dec. 1972	4 700	100%
	1 Apr. 1973	2 000	decreased by 57%
	1 July 1973	0.000	decreased by 100%
	1 Oct. 1973	0.000	decreased by 100%

In addition to the heterogeneity of forest sites conditioned by natural processes we can also find anthropogenic effects and their impacts on the properties of forest sites and plant communities. The effects are as follows: pollutants, forest ecosystem fragmentation and changes in the level of groundwater table.

Changes in soil chemistry due to polluted stemflow

Special attention was paid to the absence of herbs in the circle around stems of mature trees particularly in the northern margin of the southern part of the Leipzig floodplain forest. Soil samples taken from the surroundings of stems and from the area between stems were analysed. Comparison of data from the area between stems and from the surroundings of stems shows significant effect of stemflow on soil properties. It is well known that stemflow in beech is marked also from the point of view of quantity with respect to the structure of crown and smooth surface of bark, branches and stems. In the surroundings of beech stems, the greatest acidification of soil also occurred (by 1.5 pH). In general, it is possible to say that soils are more acid both in the surroundings of stems and between stems in the layer of 5-10 cm in comparison with the upper layer of 0-5 cm. It is obviously related to the higher buffer capacity of the upper layer due to the increased content of organic substances.

Some processes of affecting soils in the surroundings of tree stems (e.g. higher accumulation of organic C or N) occur generally but in the instance of our locality it is moreover a case of the effect of air and precipitation pollution. On stems of trees, dry depositions accumulate being then washed away by stemflow. The water of stemflow is, however, affected by air pollution already in the moment of fall on the surface of tree crowns. In the set of data we can see, therefore, the increase in the content of C, N and particularly SO_4 and Cl.

Although the area of land affected by stemflow amounts to about 1% of the total stand area the fact is important from the viewpoint of impacts on the heterogeneity of forest floor as well as from the viewpoint of indicating the high degree of air pollution and formation of stress situation for the harmonic development of a floodplain forest ecosystem.

Chemical analyses of soil around the stems of trees (water extract)

	Depth cm	Cl⁻ xx	SO_4^- xx	Ct %	Nt %	NH_4-N	NO_3 N	N anorg.	pH $CaCl_2$
Between stems	0-5	17.73	68.13	4.89	0.35	15.71	48.3	64.01	4.88
	5-10	16.60	67.00	3.24	0.27	10.22	21.0	31.25	4.62
Fagus	0-5	33.93	127.53	23.32	1.29	98.00	227.7	325.74	3.08
	5-10	28.70	131.97	17.32	0.94	61.89	124.6	186.49	2.95
Acer	0-5	31.27	195.23	37.49	1.97	93.24	247.3	340.57	3.56
	5-10	27.93	164.13	20.06	1.24	52.32	126.1	178.41	3.10
Quercus	0-5	27.53	358.43	15.43	0.90	33.27	170.5	203.81	3.88
	5-10	17.10	193.37	6.98	0.45	15.45	52.7	68.20	3.07
Fraxinus	0-5	29.20	170.87	16.71	0.94	26.00	98.8	152.82	3.58
	5-10	22.90	145.00	1.52	0.63	18.00	61.0	79.03	3.56

Formation of Ruderal Ecotone Associations

Ecotone associations with an important proportion or even the dominance of ruderal species and neophytes will be formed in newly created forest margins in those their segments where the planned communication will cross the forest stands. These will be non-stabilized associations enabling the penetration of ruderal species inside the forest stands and the final result of this process will be degradation of the present biodiversity. Examples of such associations can be found out in forest margins of rides for the high voltage power transmission lines. These associations, however, are already rather stabilized and began to function as a protective shield of internal forest biocenoses. Of species showing ruderal tendencies, for instance the following ones can be found: *Anthriscus sylvestris, Tussilago farfara* and *Artemisia vulgaris.* Of ruderal neophytes, this concerns the following species: *Acer negundo, Impatiens parviflora, Symphoricarpus racemosus, Solidago canadense* and *Oenothera biennis.*

A considerable length of newly formed ruderalized ecotone associations of forest margins induces a considerable fear concerning their potential negative effects on the remaining forest stands because they could represent linear sources of expansion of ruderal species and neophytes inside forest geobiocenoses.

Changes in Permanent Ecological Conditions due to Road Construction

There is no doubt that the ecological stability of forest stands remaining in the neighbourhood of newly constructed roads can be influenced by those changes in permanent ecological factors which will result from various effects of the construction and operation of various technical facilities. It is not possible to eliminate changes in the hydrological regime of soils due to disturbances in the flow of ground water, definitely there will be changes due to the thinning of new stand margins, the state of forest geobiocenoses will be adversely affected by air pollution caused by car exhausts and soil properties along the roads may be changed due to salting in the winter period. In forest geobiocenoses, all these phenomena may manifest themselves as stress factors and their synergic effects may induce symptoms of an ecological stress resulting in disturbances of ecological stability, often with irreversible consequences.

A new communication corridor will be a completely impassable barrier e.g. for myrmecochorous species and it will be also very difficult to pass for some other zoochorous species. In this context, it is also necessary to mention that some of important and typical dominating species of the floodplain forest undergrowth belong to species the diaspores of which are transported by ants (e.g. *Allium ursinum, Anemone nemorosa, Corydalis cava* and *Ficaria verna*).

This barrier effect will be manifested in the isolation of populations of a number of plant and animal species, in the disturbed connectivity of biocenoses, and all this will result in a decrease in the species diversity in isolated residual segments of the floodplain forest.

Changes in the groundwater table induced a number of changes in the floodplain forest ecosystems as well as a keen discussion between water managers and conservationists. In case of the absence of this regular source of nutrients through flooding some elements of mineral nutrition may become deficit in spite of the system which is relatively rich in nutrients. Other possible impacts of nutrient cycling following the marked change in the water regime can be as follows:

- slowing down the decomposition processes of organic residues as well as the nitrogen and other element total cycling under conditions of changes in species composition;

- reduced potassium cycling with the lower production of herbs;

- elimination of silt sedimentation and the following decrease in the entry of nitrogen and other elements into the ecosystem after flood removal;

- decrease in nutrient transport from ground water into the surface soil layers.

Changes in the soil and its microclimate after the control of flooding in the floodplain forest led to:

- the major development of aerobic heterotrophic microflora, especially bacteria, to a lesser extent micromycetes;

- a fall in the cellulose and lignin decomposers and a loss of anaerobic bacteria, including binders of atmospheric nitrogen (*Clostridium pasteurianum*);

- a fall in the rate of decomposition of cellulose and CO_2 respiration from the soil;

- a fall in the seasonal dynamics of soil respiration in the first half of the growing season compared with the time of periodic flooding but in the second half of the growing season a rise was found;

- a slight activation of ammonification and nitrification processes;

- a slight fall in the activity of soil catalase.

All the changes are closely related with changes in biodiversity (its decrease) and in the production level of a forest stand biomass. In many floodplain forests, therefore, revitalization projects appeared based on artificial flooding either by overall flooding or using canals leading water into the forest. So far, effects of the measures have not been evaluated although favourable effects are naturally expected.

Economic evaluation of floodplain forests is related to their basic roles which were given in Introduction. There is not much experience with the economic evaluation of the roles of forest ecosystems and our conclusions are based mainly upon the WWF publication "Economic Evaluation of Danube Floodplains" (1995). We quote, therefore, the introductory part of the publication here:

The wood biomass of the forest floodplains can be used for different purposes depending on the quality of the trees. Foresters expect a wood biomass production which is 20 to 30 per cent higher than average. Trees such as oaks are used for the production of furniture and trees of lesser quality for the pulp and paper industry or are used as bioenergy. The forests are also habitat for deer that are hunted and may provide a food resource. They are also regarded as valuable recreational sites, especially in spring, as the first sign of spring appears in the alluvial forest.

The forests also provide wind protection, which is important to maintain the local climate. A decline in forest cover may cause more windy conditions which, in turn, may cause higher evaporation and wind erosion.

The grasslands are used by farmers to feed their cattle. Grass is used for pasture but is also harvested for stored fodder.

A special type of orchid found in the grasslands was used as a medicinal plant for horses. The grasslands also provide food resources for species living in the surrounding forests and wetlands.

Wetlands are known for their richness in flora and fauna and provide habitats for many bird species. They provide important links in the food chain. As mentioned earlier, wetlands contribute to improved water quality by the process of denitrification.

Management in floodplain forests is considerably diversified with respect to the use of floodplain forest functions. Floodplain forests include a number of nature reserves or are subject to special management. We can find various forms of forest regeneration, from clear felling applying agroforestry principles, establishing poplar and willow plantations to method copying natural conditions.

Oak forms one of the main species of the 'hard-wooded' floodplain forest having also a fundamental importance for financial yield. Its regeneration as mentioned above requires special silvicultural/management measures. One of the methods is regeneration of oak by sowing (or planting) on prepared clear-cut areas where all organic matter is removed (including stumps, forest weed and roots) and row sowing (or planting) is carried out into the soil prepared according to the procedure. After establishing the young plantation the area between rows is frequently used for the production of agricultural crops and in further years the oak plantation is complemented by other tree species. This method of regeneration, although successful from the point of view of plantation establishment, is generally refused with respect to the marked impacts into the forest environment, particularly into the cycle and fixation of C and N, soil chemistry and soil biology, biodiversity etc. The procedure often results in the origin of pure stands of oak (or ash).

Another approach is practised by Croatian foresters which manage about 250 000 ha of floodplain forests in the watershed of the Sáva and Dráva rivers. Forest Act in Croatia does not permit to apply regeneration using clear cutting. The approach of the foresters is markedly related to the forest site and considerable attention is paid to soil conditions. Soil is protected against excessive weeding by hornbeam undergrowth and in a region where oak regeneration can be started (age of the main stand, expected seed crop) the hornbeam undergrowth is then removed and fallen seeds are completed if need be by broadcast sowing (600-800 kg.ha^{-1}).

In further years, the main stand is thinned and protection is aimed at oak seedlings (weed mowing) and care of fully non-lignified oak seedlings against fungal diseases (mould). In order to fully establish oak regeneration, final cutting of the main stand is carried out, the occurrence of other tree species is supported and silvicultural measures are conducted gradually starting at an age of about 15 years of the new stand.

In general, it is necessary to take into account the character of the site and to regulate species composition of forest stands in such a way close to nature stands to be supported.

Conclusion

Floodplain forests of Europe represent an important nature biome having exceptional importance particularly from the viewpoint of biodiversity preservation. They are highly affected by human activities being more or less of residual character. The forests often occur only in narrow strips along rivers. At present, they are markedly affected particularly by changes in the water regime or fragmentation of the forest tracts. In the region of the forests, there is a number of nature reserves and protected landscape areas including a national park situated on the Danube banks (Austria). Tendencies to preserve floodplain forests resulted in many places in revitalization projects which means artificial flooding of floodplain forests.

In general, it is possible to say that floodplain forest ecosystems in Europe should become a subject of detailed studies aimed at working up the system of their management and protection.

REFERENCES

Andreasson-Gren, M. and K. Henning Groth (1955) : Economic Evaluation of Danube Floodplain, A WWF International, Discussion Paper, 24 pp.

Haraszthy, L. (1994) : Floodplain Forests in Hungary, Proceedings of the Conference: Conservation of Forests in Central Europe, Zvolen, pp. 69-71.

Klimo, E (1985) : Cycling of Mineral Nutrients, In: M. Penka, M. Vyskot, E. Klimo, F. Vasicek: Floodplain Forest Ecosystem, Academia Praha, pp. 425-459.

Raus, D. (1994) : The Lowland Forests in Silvae Nostrae Croatiae, Zagreb, pp. 36-40.

Yon, D. and G. Tendron (1981) : Alluvial Forests of Europe, Council of Europe, 64 pp.

Attachment: 1

Floodplain forests in Croatia along the Sava river

Nature reserve in the floodplain forest of Leipzig, Germany

Attachment: 2

Hard-wooded' floodplain forest along the Morava river, Czech Republic

Ecosystem fragmentation-the Leipzig floodplain forest, Germany

Attachment: 3

'Soft-wooded' floodplain forest along the Danube river, Slovakia

Artificial flooding of the floodplain forest in southern Moravia, Czech Republic

Ecosystem Management as an Approach for Sustaining Forests and Their Biodiversity

Robert C. SZARO*[1] and William T. SEXTON*[2]

*1 IUFRO, Special Programme for Developing Countries
  Seckendorff-Gudent-Weg 8, A-1131 Vienna, Austria
*2 USDA Forest Service, P.O. BOX 96090, Washington, D.C.20090-6090, USA

ABSTRACT Ecosystem management is an approach that attempts to involve all stakeholders in defining sustainable alternatives for the interactions of people and the environments in which they live. The inadequacy of the traditional resource management paradigm to deal with multiple scales and larger areas that encompass both public and private lands coupled with growing concern over decreasing biodiversity gave rise to its development. It is based on a collaboratively developed vision of desired future ecosystem conditions that integrates ecological, economic, and social factors affecting an area that is defined by multiple boundaries including ecological and political ones. It is a goal-driven approach to restoring and sustaining healthy ecosystems and their functions and values while supporting communities and their economic base.
Keywords: Ecosystem management, biodiversity, scale factors, implementation.

1. INTRODUCTION

Conserving biodiversity involves restoring, protecting, conserving, or enhancing the variety of life in an area so that the abundances and distributions of species and communities provide for continued existence and normal ecological functioning, including adaptation and extinction (Szaro 1995). This does not mean all things must occur in all areas, but that all things must be cared for at some appropriate geographic scale. New approaches must incorporate fundamental shifts in the scale and scope of conservation practice from the more traditional single-species and stand level management approach to management of communities and ecosystems (Miller 1996).

Ecosystem management is such an approach that attempts to involve all stakeholders in defining sustainable alternatives for the interactions of people and the environments in which they live. It adopts a combination of established ecological concepts and principles that address human–environmental interactions. It is a way to better understand and manage lands and resources, their conflicting resource uses and management objectives, and the activities that impact them. This approach emphasizes place- or region-based objectives, with scopes and approaches defined appropriately for each given situation.

Ecosystem management is not a linear, highly standardized, or certain means to identify the one right way to manage resources (Szaro et al. 1998). This approach aids in the development of better options and sustainable solutions by incorporating human needs and values while recognizing that science alone has not and will not produce a single "right" answer for resource use and management. Instead, decisions will continue to be a complex blending of social, economic, political, and scientific information and interests and therefore is a strategy based on integrating ecosystem science and socioeconomic principles (Underwood 1998).

2. LANDSCAPE SCALES FOR BIODIVERSITY MANAGEMENT

Biodiversity will only be maintained successfully to the extent that land and water territories of the appropriate scales are managed as regional units (Miller 1996). Why? Five fundamental reasons can be deduced from ecological, social, and economic considerations:

2.1 Ecological Composition

Threatened, endangered, and sensitive species programs have long been the focal point of biodiversity concerns on the species level but they represent only the tip of the iceberg. The magnitude of the problem makes it clear that we do not have the resources for intense management programs for all species if only because of the enormity of the problem. Intensive programs to manage populations of particular species contribute to the welfare of those components of ecosystems, but such focused approaches can only pretend to deal with a relatively small portion of the ever-growing list of endangered species (Reid 1992). Not all species and their genetic variation will be found within established or potential networks of protected areas. The most effective approach is the maintenance of as many species as possible within landscape mosaics.

A strategy to maximize species diversity at the local level does not necessarily add to regional diversity. In fact, often in our efforts to "enhance" habitats for wildlife "edge" preferring species has been at the expense of "area" sensitive ones, and consequently regional diversity might have been decreased. It is important to realize that principles that apply at smaller scales of time and space do not necessarily apply to longer time and larger spatial scales (Crow 1989). Long-term conservation of species and their genetic variation, will require cooperative efforts across entire landscapes (Miller 1996).

2.2 Ecological Structures

Not all ecosystem types, including old growth, second growth, pioneer stages, and flood plains, are found in existing or potential protected areas (Miller 1996). The dynamics of some of these such as old growth clearly illustrates the need for a larger dynamic landscape (Spies and Franklin 1996). The maintenance of old growth within large areas is dependent the rate, size, pattern, and intensity of disturbances. If the frequency of disturbance is too high, some types of old growth, such as Douglas-fir or ponderosa pine, will not develop in a landscape and if the frequency of disturbance is too low, other types, such as aspen, will be lost from landscapes. If disturbances are too small, then shade intolerant types of old growth, such as alder or long-leaf pine, will not occur. Biodiversity planning and management will necessarily involve evaluating tradeoffs between different old-growth types.

Some landscapes are exceptionally diverse and any attempt to cover the range of habitats within one or more protected areas would imply excessive withdrawal of land and water resources from other human purposes, along with major impacts upon local societies and economies (Miller 1996). Many key habitat and ecosystem types, including patches of old-growth forest and wetland, are found on farms and forest holdings, often in private, community or corporate hands. These patches need to be managed as components of the overall landscape under arrangements that are equitable to land owners.

2.3 Ecological Functions

Hydrological flows, nutrient cycles, migratory patterns, and other ecological processes typically span greater scales than most managed areas (Miller 1996). Entire watersheds from mountain sources to shorelines require management strategies that address quantity and quality of stream flow. This involves control of pollution and contaminants, preservation of wetlands, erosion control, and mitigation of impacts from land uses including grazing regimes that often contribute excessive nutrient fluxes to soils, streams and inshore habitats.

A focus on process, function, and interaction provides a basis for making the difficult choices among species to which salvage efforts are to be allocated (Willson 1996). Protection strategies using a single species perspective can result in an array of complex management strategies likely to contain contradictory prescriptions (Pearson et al.1996). It may be more effective to focus our efforts on species that have strong interactions with several others, because changes in those species will have a greater impact on total community function than those interacting only weakly or with only one other (Willson 1996). Perhaps, the unit of preservation/conservation should be the interactions/processes themselves, rather than single species. By focusing more on ecological networks, sets of interacting populations in a variable environment, instead of single species, our chances of maintaining biodiversity are likely to improve (Willson 1996). Taking a broader scale perspective and implementing protection at the landscape level is a more robust protection strategy. An intact landscape will preserve the natural processes and patterns of heterogeneity that its ecosystems and species require (Pearson et al. 1996).

2.4 Ecological Dynamics

Change has profound implications for land management. Historically, the quest for stability and preventing change in areas where productivity was maintained by dynamic events has led to the declining quality and quantity of many of our most desired habitats (Szaro 1990; Reid, 1996). Much prior management of habitats has viewed systems as being immutable and that all that is necessary is to put a fence around an area and it can be saved forever. Yet, there is abundant evidence that this may seldom be so, both within particular mini-systems and on geographic scales (Willson 1996). Ecosystems by their very nature are in a constant state of flux. Environmental change is inevitable and in a sense the one constant under which ecosystems exist; it is unrealistic to embrace the status quo because it does not exist (Lugo and Brown 1996). Whether people cause the change or not most ecosystems are now influenced in some way by human activities or the artifacts of those activities. Change per se is not necessarily something to be avoided. Ultimately it may be the underlying motivating factor in management decisions. We may wish to alter vegetation structure or composition to emphasize rarer or endangered species or communities (Szaro 1990).

The maintenance of natural levels and patterns of biodiversity requires serious consideration of many kinds of ecological interactions and of variation in their intensities and outcomes (Willson 1996). A central strategy for addressing the process of change is to establish landscape configurations that provide the set of habitat options required for survival of all organisms (Miller 1996). In addition to protected areas that will predictably cover between 5 and 10% of a particular region, the major part of the larger landscape can be managed through forestry, agriculture, grazing, fisheries, and wildlife regimes that mimic to some extent the dynamic patch-work quilt of nature. Flexibility is key. The nature and extent of perturbations, such as wind, flood, or earthquake, are often unpredictable.

2.5 Human Settlements and Land Use

Removal of vegetation, atmospheric warming, increased pollution loads, and quite sharp and sudden shifts in the pattern of land use are of tremendous importance in conservation practice (Bridgewater et al. 1996). Human cultures have selected and domesticated plant and animal varieties that are found in agricultural, pasture, forest, urban environments, and others. Some traditional cultures continue land-use practices that maintain progenitor races of plants and animals of considerable value to humans world-wide. Some varieties and environments were established through human enterprise back several millennia. It is feared that some of these varieties will remain available only to the extent that traditional cultivation practices are continued (CLADES 1991). The territorial units of land and water needed to embrace the range of species and genetic traits, ecosystems, ecological functions, and the space within which to adapt to change extend well beyond scales that can reasonably be contained in protected areas.

The factor of scale, coupled with the recognition of the role of human settlements and land uses in the development and maintenance of varieties and landscapes of importance to human interests, argue for their inclusion in bioregional management programs (Miller 1996). Protected areas, within the concept of biosphere reserves, and elements of remnant vegetation connected by corridors or "greenways" can provide the reservoirs of biodiversity and avenues for adaptive responses to environmental change (Bridgewater et al. 1996). Management strategies should be developed to look at the components of landscape and at the impact of possible changes in a landscape ecological framework.

3. IMPLEMENTING ECOSYSTEM MANAGEMENT

3.1 Using Multiple Boundaries for Organizing Information

How do many diverse organizations and groups arrive at common ground in defining a standard boundary or border for conducting ecosystem management (Szaro and Sexton 1998)? The short answer is they don't. Debates over which particular stratification or approach is best and should therefore be the single information organizing device for ecosystem management are not useful or constructive. A more useful question is which set of ecological approaches, and their associated boundaries, provide information that best addresses the concerns related to a particular resource management situation. Individuals and organizations experienced in implementing the basic elements of ecosystem management have gone beyond a focus on selecting a single boundary for organizing information necessary to support the approach. The appropriate question is what kind of information organizing systems and related boundaries are needed to provide the information necessary to support ecosystem management analyses and subsequent decision processes.

Implementing ecosystem management requires learning to work with multiple factors, at multiple scales, using multiple boundaries and borders for organizing information. Traditional approaches often oversimplified information collection and analyses by relying on a limited set of classifications and information constructs. Experience has consistently shown that attempting to constrain analyses and assessments to one or a few organizing systems and related boundaries results in less than satisfactory information to support an ecological approach. Experience also continues to show the complimentary nature of a number of spatial ecological systems for developing useful information. Even among systems where lines at the national scale are not strongly related, coincidence is often found at the regional or sub-regional scales. In general, individual spatial approaches tend to be the least compatible at the project level scales where agency interests are site specific, purpose specific and highly data specific. However, at most higher scales goals, objectives and information needs tend to be more general, offering opportunities for collaboration and data collection partnerships. Where common spatial frameworks exist at the higher geographic scales, data from and about individual projects can be consolidated and examined at that more general level, providing information in a shared system that would not otherwise be available between organizations. The challenge is for various organizations to work together effectively, using multiple boundaries and available information to produce the best understanding of the human-environment interaction in developing sustainable management options.

3.2 Ecosystem Management Toolkit

Ecosystem management is a result of changing scientific, social, political, cultural and economic information and values (Sexton 1998). It evolved in response to the need for a better approach to managing natural resources, building upon the knowledge drawn from a wide range of scientific disciplines and the experience of resource managers. Ecosystem management consolidates a number of well established concepts and principles and combines them in a new framework for understanding and managing lands and resources. It is an approach to ecological stewardship that reflects a shift in the national paradigm of how society views and values landscapes, public lands, and environmental health.

Ecosystem management can be thought of as the minimum set of tools a land manager should have available in attempting to define sustainable alternatives for the interactions of people and the environment. It is a term that specifically refers to a process or set of activities for addressing resource management, not a prescribed outcome. As a pre-decisional process, ecosystem management amends and expands the resource management tool kit that field-level professionals rely on to understand and manage lands and resources in an ecological context. It is not a panacea for current natural resource management issues. The fact remains that natural resource systems have certain limits and capacities. There is a wide range of competing values and uses. There are more people with more demands than resources can sustainably provide for. Difficult choices have to be made. Humans are faced with difficult choices in determining how they will interact with the environment to provide for essential materials and services, and maintain a healthy environment. The approach does not necessarily make hard choices any easier but it does support making those choices in the most informed and professional manner possible.

The list of "tools" is in itself representative of the multi-faceted nature and complicated operational challenges associated with implementing an ecosystem management approach (Table 1). The activities are complex and highly related. Most are related to or based on scales. All rely on using the best available science. All are improved by collaboration with others. Activities are listed under a series of headings as a means to simplify explaining the type of ecosystem management tools being implemented in field situations. It is clearly recognized that most tools do not fit neatly into any one category and that there are many equally practical ways to define categories.

Table 1. Ecosystem Management Toolkit (Adapted from Sexton 1998).

Institutional "Tools"
Organizational norms and behaviors   • using highly participatory processes for public involvement throughout the approach.   • seeking and forming as many partnerships as possible.   • developing, seeking out, utilizing, and transferring the very best scientific information.
Operational "Tools"
Analyses and assessments   • conducting information assessments and analyses across administrative and political borders to develop necessary information at multiple scales.   • developing information about species, associated communities and their structure, composition and function and about ecological processes: carbon cycle, nutrient cycle, hydrologic cycle, biological diversity, succession, population dynamics.
Multiple scales and levels   • evaluating information at multiple scales, at a minimum at least one scale above and one scale below the project or issue being reviewed.   • evaluating structures, functions, patterns and spatial relationships over long time spans.
Multiple borders and boundaries   • using a wide range of assessment and analytical boundaries to organize and collect information necessary to address problems and issues related to sustainable solutions.   • developing common boundaries and approaches when appropriate to facilitate the collaborative accumulation of information and its subsequent sharing and use.
Inventory, ecological classification and information   • integrating inventories and information within and between scales.   • developing common ecological classification systems and cooperating with partners on developing common data standards and collection protocols.
Risk, uncertainty and complexity   • embracing complexity, in the environment, in human values, in the information needed to conduct ecosystem management.   • recognizing uncertainty as a normal part of systems and building it into models and options.
Monitoring and evaluation / adaptive management   • defining key features and their related threshold levels of change and adopting strategies that protect systems from reaching those levels (e.g., species extinction, soil degradation).   • defining and monitoring practical criteria,, indices and parameters to address sustainability, biodiversity and ecosystem function and process.   • using an adaptive management approach for resource management activities.

Human dimensions
• examining factors and relationships representing human influences and incorporating that information in analyses addressing resource management options
• defining key elements and relationships within social system functions and processes.

Disturbance factors, landscape patterns, historic range of variation
• attempting to work within and mimic natural processes and patterns.
• defining and characterizing major disturbance factors and their frequency, distribution, pattern, proportion, distribution across and within landscapes

Desired conditions / landscape and system goals
• describing the conditions, trends, patterns, structures, components and processes and functions desired to be maintained, or reached at some point in the future, for specific areas.
• describing the key features associated with measuring and assessing sustainability and its f variability over what time and space.

Decision support systems, geographic information systems
• using geographic information systems and related decision support systems to evaluate information over time and space.
• developing models and predictions that support resource management information needs.

4. CONCLUSIONS

An ecosystem approach requires that natural resource managers, scientists, and the public share a vision for the future of a world in which societal and economic decisions are consolidated with an increasingly comprehensive, integrated understanding of the environment. Partnerships among governments, agencies, industry, environmental groups and other interested parties can help integrate management operations into an ecosystem-wide approach, collaborate in monitoring efforts and assessments to give better information to decision makers, provide education and outreach to increase public understanding, and take a more proactive approach to understanding and maintaining biodiversity. In order to move toward this vision, the paradigm for managing natural resources and ecosystem health must shift from a fragmented to an integrated multi disciplinary approach, from a site-specific to an ecosystem-wide context, and from a reactive to a proactive mode.

The inadequacy of the traditional resource management paradigm, which primarily focused on site based management strategies, to deal with multiple scales and larger areas that encompass both public and private lands coupled with the growing concern over decreasing biodiversity and loss of ecosystems gave rise to the concept of ecosystem management. Ecosystem management remains an evolving force that must yet respond and adapt to numerous challenges. A formal process of adaptive management, a continuing process of action-based planning, monitoring, researching and adjusting with the objective of improving the implementation and achieving the desired goals and outcomes, is needed to maximize the benefits of any option for land and natural resource management. For practical purposes ecosystem management is generally synonymous with sustainable development, sustainable management, sustainable forestry and a number of other terms being used to identify an ecological approach to land and resource management. Ecosystem management is a goal-driven approach to restoring and sustaining healthy ecosystems and their functions and values while supporting communities and their economic base (Szaro et al., 1996). It is based on a collaboratively developed vision of desired future ecosystem conditions that integrates ecological, economic, and social factors affecting a management unit defined by multiple boundaries including ecological and political ones (Sexton et al.1998).

5. REFERENCES

Bridgewater, P., D.W. Walton and J.R. Busby (1996): Creating Policy on Landscape Diversity. *In* R.C. Szaro and D.W. Johnston (eds.) Biodiversity in Managed Landscapes. Oxford University Press, New York, pp. 711-726.

CLADES.(1991): Status and Trends in Grass-Roots Crop Genetic Conservation Efforts in Latin America. University of California, Berkeley, California.

Crow, T.R. (1989): Biological Diversity and Silvicultural Systems. *In* Proceedings of the National Silvicultural Workshop: Silvicultural Challenges and Opportunities in the 1990's. USDA Forest Service, Timber Management. Washington, D.C., pp. 180-185.

Lugo, A.E. and S. Brown (1996): Management of Land and Species Richness in the Tropics. *In* R.C. Szaro and D.W. Johnston (eds.) Biodiversity in Managed Landscapes. Oxford University Press, New York, pp. 280-295..

Miller, K.R. (1996): Conserving Biodiversity in Managed Landscapes. *In* R.C. Szaro and D.W. Johnston (eds.) Biodiversity in Managed Landscapes. Oxford University Press, New York, pp 425-441..

Pearson, S.M., M.G. Turner, R.H. Gardner and R.V. O'Neill (1996): An Organism-based Perspective of Habitat Fragmentation. *In* R.C. Szaro and D.W. Johnston (eds.) Biodiversity in Managed Landscapes. Oxford University Press, New York, pp 77-95.

Reid, W. V. (1992): The United States needs a national biodiversity policy. Issues and Ideas. World Resources Institute, Washington, D.C.

Reid, W.V. (1996): Beyond Protected Areas: Changing Perceptions of Ecological Management Objectives. *In* R.C. Szaro and D.W. Johnston (eds.) Biodiversity in Managed Landscapes. Oxford University Press, New York, pp.442-453.

Spies, T.A. and J.F. Franklin (1996): The Diversity and Maintenance of Old-growth Forests. Pages 296-314 In R.C. Szaro and D.W. Johnston (eds.) Biodiversity in Managed Landscapes. Oxford University Press, New York.

Sexton, W.T. (1998): Ecosystem Management: Expanding the Resource Management "Tool Kit". Landscape and Urban Planning, Vol. 40, No.1-30, *in press*.

Sexton, W.T., C.W. Dull, and R.C. Szaro (1998): Implementing ecosystem management: A framework for remotely sensed information at multiple scales.

Landscape and Urban Planning, Vol. 40, No.1-30, *in press*.

Sexton, W.T. and R.C. Szaro (1998): Implementing ecosystem management: Using multiple boundaries for organizing information. Landscape and Urban Planning, Vol. 40, No.1-30, *in press*.

Szaro, R. C. (1990): Management of dynamic ecosystems: Concluding remarks. *In* J. Sweeney (ed.), Management of Dynamic Ecosystems. North Cent. Sect., The Wildl. Soc., West Lafayette, Ind., pp 173-180.

Szaro, R.C. (1995): Biodiversity Maintenance. *In* A. Bisio and S.G. Boots (eds.) Encyclopedia of Energy Technology and the Environment, First Edition. Wiley-Interscience, John Wiley & Sons, Inc., New York, pp 423-433..

Szaro, R.C. (1996): Biodiversity in Managed Landscapes: Principles, Practice, and Policy. *In* R.C. Szaro and D.W. Johnston (eds.) Biodiversity in Managed Landscapes. Oxford University Press, New York, pp. 727-770.

Szaro, R.C., G.D. Lessard and W.T. Sexton (1996): Ecosystem Management: An Approach for Conserving Biodiversity. *In* F. DiCastri and T. Younès (eds.) Biodiversity: Science and Development - Towards a New Partnership. CAB International, Oxon and International Union of Biological Sciences, Paris, pp.369-384.

Szaro, R.C., W.T. Sexton and C.R. Malone (1998): The emergence of ecosystem management as a tool for meeting people's needs and sustaining ecosystems. Landscape and Urban Planning, Vol. 40, No.1-30, *in press*.

Underwood, A.J. (1998): Relationships between ecological research and environmental management. Landscape and Urban Planning, Vol. 40, No.1-30, *in press*..

Willson, M.F. (1996): Biodiversity and Ecological Processes. *In* R.C. Szaro and D.W. Johnston (eds.) Biodiversity in Managed Landscapes. Oxford University Press, New York, pp. 96-107.

Decomposition Processes of Litter Along a Latitudinal Gradient

Hiroshi TAKEDA

Laboratory of Forest Ecology, Graduate School of Agriculture
Kyoto University, Kyoto 606-8502, Japan

ABSTRACT   Primary production of plant is controlled by the relative availability of light (energy source) and nutrients. Decomposition processes are constrained by the relative availability of carbon (energy source) and nutrients. Thus, decomposition of litter depends upon the primary production of plants. Decomposition of litter is a sequential processes by which organic mater is mineralized for plant uptake. During the decomposition processes, chemical transformations are performed by microbial populations. Thus the processes is constrained by the availability of energy and nutrient resources for microbial populations in the decomposing litter. The climatical conditions influence on the microbial activities and then decomposition processes are different along a latitudinal gradient.
Keyword: carbon and nitrogen dynamics, decomposition process,litterfall, litter quality

1.INTRODUCTION

In terrestrial ecosystems, climatic condition is the most important determinant of the primary productivity and decomposition of organic matter. Primary productivity of plant is limited by light(solar energy) and soil nutrient resources. Decomposition of organic matter is a biochemical process carried out by a diverse decomposer including microbial and animal populations (Heal, Anderson, & Swift, 1997). Activities of decomposer are also limited by carbon (energy) and nutrient resources. So plant and decomposer are competitive in their use of nutrient resources. Plant is autotrophic in their energy use , while heterotrophic decomposer depend upon plant for their energy resource. Decomposers release nutrient from decomposing litter for the primary production of plants. So the photosynthesis and decomposition processes are both fundamental function for the maintenance of biodiversity of terrestrial ecosystems.

Decomposition of litter is a sequential process of resource utilization by microbial and animal populations under a given environmental conditions (Heal, Anderson, & Swift, 1997) and is limited by energy and nutrient resources. Soil and climate condition are both important determinant for the amount and quality of plant litter and cause also the variation in litter decomposition rates. Decomposition rates are higher in the tropical than in the temperate forest ecosystems. The climatical and soil conditions influence the decomposition rates through the decomposer activities and the composition of decomposer. Thus, the climatical conditions causes the variations in decomposition processes along a latitudinal gradient.

2. CHANGES IN PRIMARY PRODUCTION AND DECOMPOSITION RATES ALONG A LATITUDINAL GRADIENT.

2.1. Litterfall

In forest ecosystems,. literfall is an important flux of carbon (energy) and nutrients into the soil decomposer system where nutrients are mineralized for plant growth. The amount of litterfall represents the net primary production of forest ecosystems. Net primary production is determined by climatical conditions, i.e., temperature and precipitation. Bray and Gorham(1964) showed an inverse linear relationship between annual litter fall and latitude. Since the review of Bray and Gorham, there has been a tremendous increase in the number of papers on carbon and nutrient circulation in forest ecosystems. Vogt et al(1986) synthesized and analyzed available data on litterfall on a global scle. They showed that broad leaved litter production had a significant correlation with latitude($r=0.58$ with 84 case studies).

Table. 1 show the changes in litter fall in various forest ecosystems along a latitudinal gradients. The amounts of litterfall are about two times higher in tropical forests than in temperate forests (Takeda, 1996). So climatical conditions are the main determinants of primary production of plant. These data show that soil decomposer systems in tropical are provided with higher amounts of litter along a latitudinal gradient.

Along a latitudinal gradient, various plant-soil systems are developed. Climatical conditions influence on the species composition of plant communities. Plant species composition is related to the litter quality. Resource quality of plant materials is defined by the major life forms such as annuals, deciduous and evergreen perennial, woody plants. The woody plants are separated into coniferous and broad leaved trees. Plant species composition influence on the resource quality of litter (Torreta et al. 1998)

Table 1 Changes in A_0 layer and litterfall along a latitudinal gradient (original data from Vogt et al.1986)

Forest type	Mean temperature (℃)	Precipitation (mm)	Amount of A0 layer (Kg/ha)	Litterfall (kg/ha yr)
Tropical deciduous forests	23	2,147	8,789	9,438
Tropical evergreen forests	26	2,504	22,547	9,369
Sub-tropical evergreen forests	13	1,705	22,185	5,098
Warm temperate deciduous forests	14	1,391	11,480	4,236
Warm temperate broadleaf evergreen	14	1,409	19,148	6,484
Cold temperate broadleaf deciduous	5	875	32,207	3,854
Cold temperate needle deciduous	10	1,806	13,900	3,590
Cold temperate neeedleleaf evergreen	8	1,275	44,574	3,144
Boreal nedleleaf evergreen	2	694	44,693	204

2.2   Decomposition rates along a latitudinal gradients

Vogt et al (1986) synthesized and analyzed available data on decomposition rates on a global scale and concluded that there were no significant correlation between latitude and decay rates. Meentemeyer(1978) presented a model incorporated actual evapotranspiration, lignin concentration, and AET/lignin concentration. Vogt et al (1986) apply the model for their data sets and showed that substrate lignin was better correlated with litter decay than climate in subtropics and tropics . Climate was more strongly correlated with decay rates than lignin in warm temperate to subpolar regions.

The amount of litterfall show a clear correlation with latitude. In contrast to this clear correlation , the decomposition rates were rather variable along a latitudinal gradients.This compariosn suggests that decomposition rates are controlled not only by climatical condition, but also by the litter quality and decomposer compositions.

3. LITTER QUALITY

Litter quality reflect the life form , plant organs, and plant species and have been expressed by physical and chemical properties. The physical properties and chemical properties are often correlated each other.   Decomposition processes carried out by heterotrophic organisms, thus the availability of carbon is an important driving factor of decomposition processes. Since Waksmann (1924), carbon to nitrogen ratio is a general index of the quality of litter. This indes also represent the relative availability of carbon energy and nutrients for the decomposers (Yamashita et al, 1998).

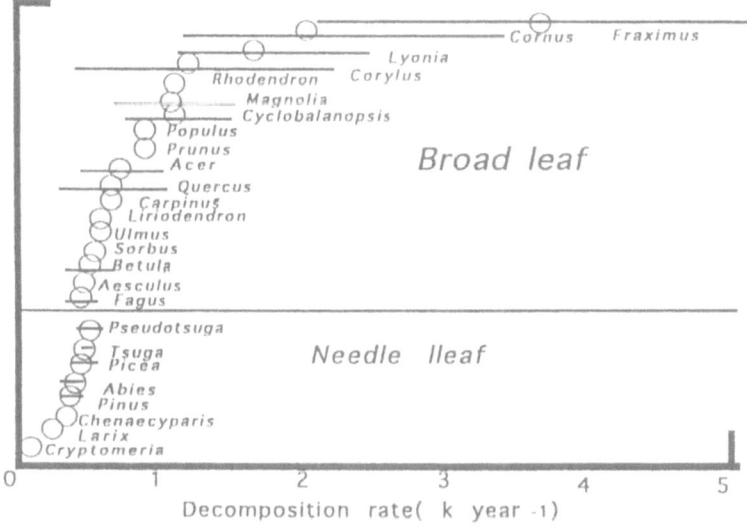

Fig.1 Decomposition rates of broad leaved and coniferous litter in a cool temperate forest (Takeda et al 1978). Horizontal bars indicate the standard errors.

200

Decomposition rates were significantly lower in the coniferous than in the broad
leaved litter(Fig. 1). Further, the variability of decomposition rates were greater in
the broad leaved than in the coniferous litter. These comparison suggested the
importance of litter quality on decomposition rates. in a same climatical zone. A
comparison of decomposition rates of leaf litter between the temperate and tropical
forest is shown in Table2. When the decomposition rates were compared between
the tropical and temperate broad leaved litters, climatical conditions are apparent
determinant for the decomposition rates. Decomposition rates was two times
higher in the tropical than in the temperate forest species. (Takeda, 1996)

Fig. 2. A comparison of decomposition rates of leaf liter between the temperate and
tropical forests. Decomposition rates are expressed by an exponential model:
$w=w_0exp(rt)$.

	Decomposition rates	Number of species
Tropical forests	1.85	51
Temperate forests	0.93	115

4. CARBON AND NITROGEN DYNAMICS

The decomposition processes have been studied in relation to the dynamics of
nitrogen, phosphorus and potassium (Tian and Takeda, 1998). Among these
nutrients, dynamics of nitrogen have been well studied together with carbon
dynamics (Staaf & Berg, 1982, Berg & Staaf, 1981, Berg, 1988, Tokuchi, et al
1998). During the decomposition processes, quality of litter is changed as a result
of carbon and nitrogen utilization by the heterotrophic microbial and animal
populations .
During the decomposition processes , nitrogen dynamics consist of leaching,
immobilization and mobilization phases. The decomposition phases from leaching
to mobilization can be explained by the relative availability of carbon and nitrogen(
nutrients. Fig.2 shows the dynamics of carbon and nitrogen in litter bags. The
decomposition processes are characterized by leaching and immobilization and
mobilization phases of nitrogen. These nutrients often limit microbial populations in
soil. A review of published data in decomposition processes shows that carbon is
used by the heterotrophic decomposer with the reservation of nitrogen during the
immobilization phase (Berg & Soderstrom, 1979).
So the decomposition progresses with the retention of nitrogen by microbial
immobilization until the critical value of carbon to nitrogen ratio(C/N ratio) is seen.
In this phase, carbon is non limiting factor for the decomposer activities. Nitrogen
may be a limiting factor in this stage From the critical value of C/N about 20,
carbon and nitrogen are both mineralized from the decomposing litter in the
temperate forest (Takeda, 1987). In contrast this, tropical litter show a
mineralization at a higher C/N ratio about 30-40(Takeda, 1996).

5. DECOMPOSITION AS A CONVERGENCE PROCESS

Litter quality reflect the life form and organs of plants. The litter quality are
variable between organ and species and litters are finally changed into humic

materials. Fig.3 show the changes in C/N ration in the initial to decomposed litter. C/N ratio show a variability reflecting plant species. C/N ratio converge to 20-30 and variability of decomposing litter C/N ratio is small between species (Fig. 3).

Decomposition processes in tropical forests are characterized by the rapid decomposition rates. In the decomposition studies in the Pasoh forest, Malaysia, I studied the decomposition rates of leaf litter of eight tree species including pioneer to climatic species. Initial C/N ratios ranged from 38 to 71 and are comparable with those of temperate tree species. But the nitrogen mineralization commenced at critical values about 23 to 48. The results show that the critical value of nitrogen mineralization is higher in tropical (C/N, 23 to 48) than in temperate forests (C/N about 20) and qualitative differences in decomposition processes between the temperate and tropical forests.

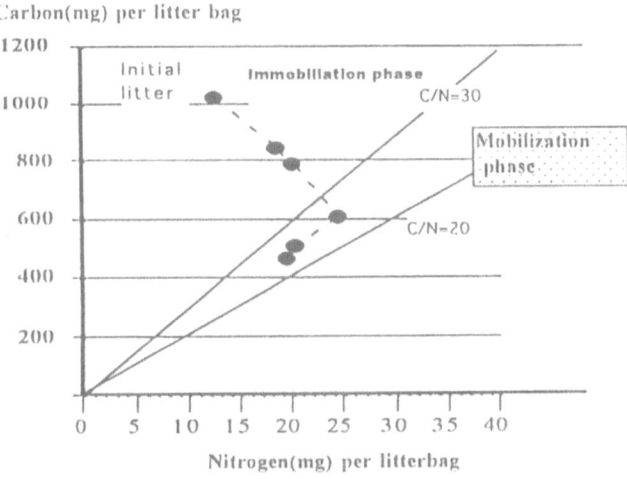

Fig. 2. Carbon and nitrogen dynamics during litter decomposition of *Betula grossa* over a 16 month. In the area above C/N=30, relative availability of carbon is higher than that of nitrogen. So, nitrogen is limiting factor for the decomposer activities. Below C/N=20, carbon is limiting factor for the decomposer activities. Mobilization of nitrogen occurred in the ranges of C/N=20 to 30.

6. DECOMPOSITION AND RESERVOIR OF SOIL ORGANIC MATTER

The maintenance of the carbon and nutrient sink is dynamically sustained by nutrient recycling systems of microbial decomposers in the soil. The forest soil studies have demonstrated the development of carbon and nutrient reservoirs in organic and soil layer in the temperate forest (Takeda, 1995c). Temperate forests develop two types of reservoir systems of carbon and nutrients. One type has been termed mull humus form where carbon and nitrogen are reserved in mineral soils with a C/ N ratio of 10-12, and another type, i.e. mor or moder humus with C/N ratio of 20-30.

The high critical value of C/N ratio in the decomposition processes in tropical

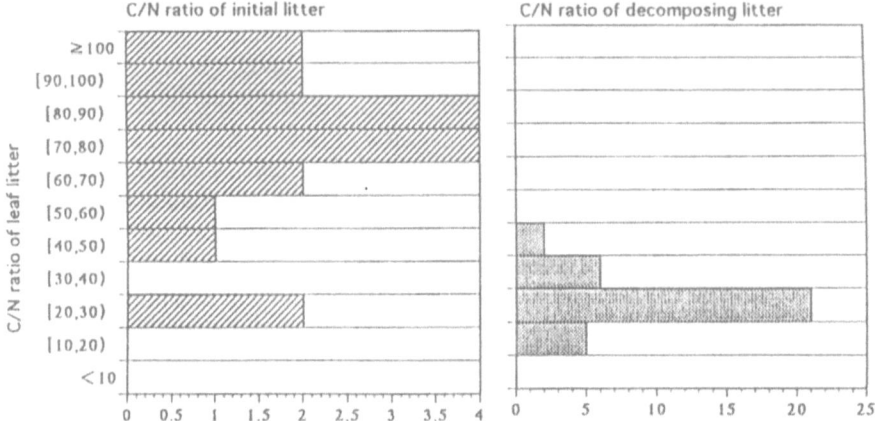

Fig3. Convergence processes of C/N ratio in decomposing litter. Initial litter qualities (C/N) are variable ,but C/N ratio converges to 20-30 of C/N ratio after 16 month decomposition.

forests suggests a poorly developed recycling systems in tropical soil systems. This result explains why the floor of tropical forests consists mainly of a litter layer. In contrast, temperate forests have thick layers of organic soil consisting of L, F and H layer and mineral soils.

Here we can classified three types of soil nutrient reservoir:

1.   Mineral soil reservoir (Mull type) where nutrient and carbon are stored in mineral soil (A) layer with poorly developped A_o layer. This type of soil is dominated in the fertile temperate forests.

2.   Soil organic layer reservoir (Moder type) where nutrient and carbon are reserved in A_0 layer with a poorly developped mineral soil (A) layer. This type is dominated in infertile coniferous forests in temperate regions and boreal forests.

3.   Plant mass reservoir (Non reserver type) where nutrient and carbon reserver is poorly developped. Plant biomass is main reservoir of carbon and nutrients. This type is dominated in the tropical forests.

7. NITROGEN DYNAMICS IN FOREST ECOSYSTEMS

In forest ecosystems, two types of soil reservoir are developed, i.e. mull and mor. This categories is not distinct, but a mull-mor continuum is developed in various forest soils (Green et al., 1993, Takahashi, 1998). Decomposition processes are limited by both nutrient and carbon. At the final mineralization stages, carbon and nutrients are both limiting factor for the microbial growth and the carbon to nitrogen ratio is decreased to C/N=20-30 in temperate forests.

In the soils with limitation of carbon energy sources, nitrogen mineralization is progresses from ammonification to nitrification. In the mull type soils, nitrification is dominated in the nitrogen economy. Mineralization rates depend upon the amount of nitrogen accumulated in soils. While the nitrification rates are dependent upon the relative availability of carbon and nitrogen. So, nitrification rates are related with C/N ratios (Fig. 3).

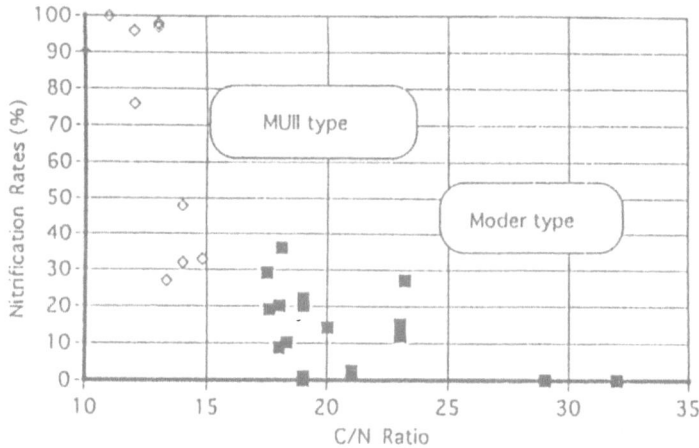

Fig.4 .Nitrification rates. Mull type soil with C/N =12-15, show high nitrification rates. In the moder or mor types with C/N =20-30 , ammonification is a dominant form of nitrogen mineralization. Samples of 36 forest stands in Japan.

8. PLANT-SOIL SYSTEMS ALONG A LATITUDINAL GRADIENT

Basing on the information of decomposition processes and soil organic matters, forest ecosystems are categorized into three types, i.e. 1.Plant mass reservoir type: poorly developed intra- N cycling, 2. Soil organic layer reservoir: Forest with tight intra- N cycling by microbial immobilization potentials. 3. Mineral soil reservoir: Forests with intra- N cycling by plant uptake (Fig. 5).Characteristics of plant soil systems in the mull and moder systems are given in Table.3.

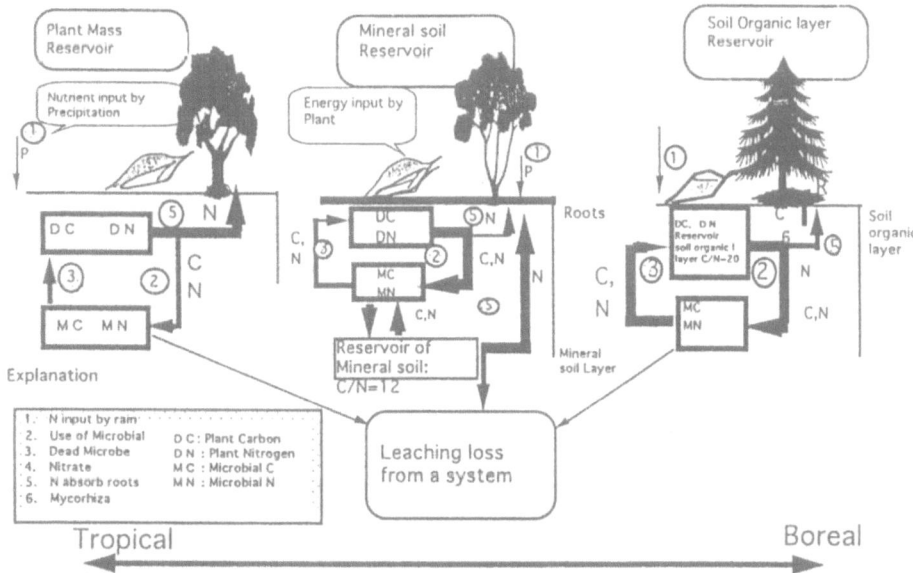

Fig. 5. Three types of carbon and nutrient reservoir developed in forest ecosystems

8.1 Plant mass reservoir type

At the commence of the primary succession, forests depend on their nutrient from precipitation and have poorly developed organic and mineral soils. Tropical rain forests are generally nutrient poor and depend upon the nutrient from precipitation. Decomposition studies of tropical forest show a rapid decomposition rates and a low potential of nutrient immobilization in the decomposing litter. Plant biomass is a reservoir of nutrients and carbon.

8.2 Soil organic layer reservoir

In this type forest, carbon and nutrient are reserved in soil organic matter. The soil is characterized by a thick development of soil organic layer (mor or moder) and shallow mineral soils. In this forest ecosystems, C/N ratio of soil organic layer is about 20-30. Nitrogen mineralization is mainly ammonification. Ammonium is positively charged and tightly bounded to the soil organic layers. So leaching loss of nitrogen is smaller Nitrogen is used by mycorrizae symbiosis. This system is characterized by a tight intra-system N cycle within a forest ecosystem and is common in the nutrient poor forests and coniferous forests.Litter quality of coniferous forest may influence on the development of humus layer with a high C/N about 20-30.

Table 3. Characteristics of Plant-soil systems in Mull and moder soils.

	Mull type soil	Mor or Moder type so
1 Nitrogen supply	Nitrogen is reserved in meneral soils with C/N ratio about 10-12. Nitrification is dominant in nitrogen economy Forageing efficiecy of nitrate is higher in plant with rooting systems than in microbial poulations	Nitrogen is reservd in the soil organic layers with C/N ration about 20-30. In this system, nitrogen is anmmonium form and is tightly bounded to organic matter.So leaching loss of nitrogen is small.
2 Nitrogen utilization	The rates of nitrogen uptake are higher by plant root than by the microbial populations. Nitrates is easily leached out without the plant uptake.	The efficiency of nitrogen intra-cycling dependent upon the nitrogen immobilization by microbial populations. Nitrogen foraging efficiecy may be higher in microbial populations than in plant. So.microbial populations such as fungi is superior competitor for nitrogen use.
3. Tree strategies	Rapidly growing trees with large size of modules, such as Pterocarya sp. Aesculus sp, early growing trees. Exploitive type of trees are adapted in this system.	Tree species adapted for poor soil conditions, and have a high nutrient using efficiency. Trees and microbial populations are competitive in nutrient use. Tree species with small module size, such as coniferous species are dominated with the symbiotic relation with mychorizae.

8.3 Mineral soil reservoir

Nitrification processes are often dominant nitrogen mineralization in the mineral soils with C/N =10-12(Hirobe et. Al. 1998). In soil systems with C/N =12, carbon is a limiting factor for the activities of microbial populations. In this system, nitrification is dominate in nitrogen dynamics. Nitrification processes are carried out by autotrophic microbial , such as *Nitrosomanas* and *Nitrobactor*. The energy of these microbes derived from the oxidation of NH_4 or NO_2. The dominance of chemoautotrophic bacteria in the C/N=12 reservoir systems, suggesting the limitation energy derived from organic carbon.

In this soil reservoir, nitrate is easily leached out from the ecosystems. without the effective uptake of nitrate by plant roots. Generally, plant growth is better in the fertile soil with nitrification than in the poor soils with ammonification. In this system, effective intra-cycling of nutrient may be maintained by the tight relationship between the nutrient release from decomposition and nutrient uptake by plants.

9. CONCLUSION

In forest ecosystems, nutrient cycling is functioning by energy derived from decomposing organic matter. Decomposition rates of organic matter are constrained by the relative availability of carbon and nutrient in the decomposing litter. The decomposition processes consists of leaching, immobilization and mobilization phase in which relative availability of nutrients and carbon energy is different states. Decomposition is a convergence processes from diverse plant organic matter to a homogenous humus materials and thus is a energy loosing process. While the nutrient is recycled within the decomposer system, carbon is non-recycle within the ecosystems. Availability of carbon energy sources characterized the three types of forest ecosystems. In mor or moder soils, soil organic matter is stored as organic debris with C/N ratio of about 20-25. While in the mull soils, majority of nutrient and carbon are stored in the mineral soils with C/N =10-12. The C/N=20 reservoir is predominant in the nutrient poor forests , coniferous and boreal forest. The C/N =10 soil is dominant in deciduous forest and coniferous forest in temperate to warm temperate forests. Tropical forest show a poor development of soil reservoir systems.

10. REFERENCE

Berg, B. (1988). Dynamics of nitrogen(15N) in decomposing Scot pine(Pinus sylvestris) needle litter. Long-term decomposition in a Scot pine forest. VI. Can. J. Bot., 66, 1539-1546.

Berg, B., & Soderstrom, B. (1979). Fungal biomass and nitrogen in decoposing socts pine needle litter. Soil. Biol. Biochem., 11, 339-341.

Berg, B., & Staaf, H. (1981). Leaching, accumulation and release of nitrogen in Decomposing Forest forest litter. In F. Clark & T. Rosswall (Eds.), Terrestrial nitrogen cycles. Processes, ecosystem strategies and management impacts. (pp. 163-178). Stockholm.

Bray, J. R., & Gorham, E. (1964). Litter production in forests of the world. Adv. Ecol. Res., 2, 101-158.

Heal, O. W., Anderson, J. M., & Swift, M. J. (1997). Plant litter quality and decomposition : An Histrical overview. In G. Cadish & K. E. Giller (Eds.), Driven by nature Plant litter quality and decomposition (pp.3-30). Cambridge: CAB international.

Green , R.N, R.L. Trowbridge, and K. Klinka(1993): Towards a taxonomic classification of humus forms. Forest Sience Monograph 29, 1-49.

Hirobe, M., N. Tokuchi, and G. Iwatsubo (1998).: Topographic differences of soil N transformation pattern along a forest solope. (In this volume)

Meentemeyer, V. (1978). Macroclimate and lignin control of litter decomposition. Ecology, 59, 465-472.

Staaf, H., & Berg, B. (1982). Accumulation and release of plant nutrients in decomposing Scots pine needle litter. Long-term decomposition in a Scots pine forest II. Can. J. bot., 60, 1561-1568.

Takahashi, M (1998): Size distribution and carbon-to-nitrogen ratios of size fractionated organic matter in the forest floor of coniferous and broadleaved stands. (In this volume)

Takeda, H., Ishida, Y., and T. Tsutsumi (1987). Decomposition of leaf litter in relation to litter quality and site conditions. Memooirs of the college of Agriculture, Kyoto University, 130, 17-38.

Takeda, H. (1996). Templates for the organization of soil animal communties in tropical forests. In I. M. Turner, C. H. Diong, S. L. Lim, & P. Ng K.L. (Eds.), Biodiversity and the dynamics of ecosystems (pp. 217-226). Kyoto: The international Network for DIVERSITAS in Western Pacific and Asia (DIWPA).

Tian ,Xing-jun, and H. Takeda(1998): Decomposition process of leaf litter in a coniferous forest. (In this volume).

Torreta, N. K.(1998): Changes in nitrogen and in the carbon fractions of the decomposing litters of bamboo. In this volume)

Tokuchi, N, M. Hirobe, and K. Koba (1998): Gross soil N transformations in a coniferous forest in Japan. In this volume

Vogt, K., Grier, C., & Vogt, D. (1986). Production, turnover and nutrient dynamics of above- and below-ground detritus of world forests. Adv Ecol Res, 15, 303-377.

WaksmanS.A. (1924): Influence of microorganisms upon the carbon-nitrogen ratio in soil. Journal of Agricultural Science 14, 555-562

Yamashita, T, H. Tobita, and H. Takeda. (1998): Nitrogen dynamics of decomposing japanese cedar and japanese cypress litter in plantation forests. In this volume.

Size Distribution and Carbon-to-Nitrogen Ratios of Size-Fractionated Organic Matter in the Forest Floor of Coniferous and Broadleaved Stands

Masamichi TAKAHASHI

Hokkaido Research Center, Forestry & Forest Product Research Institute
Sapporo 062-8516, Japan

ABSTRACT   The C/N ratios of the forest floor under coniferous stands usually converge to higher values than those under broadleaved stands. Because litter decomposition progresses with physical disintegration and chemical deterioration, if disintegration is delayed so that larger organic matter having a high C/N ratio accumulates, the C/N ratio of the organic layer is hypothesized to be high. To test this hypothesis, a comparison of the size distribution of forest floor organic matter was made between a coniferous stand group and a broadleaved stand group. The results showed that no clear difference exists in the size distribution between the two groups. In the H layers the coniferous group showed significantly higher C/N ratios than the broadleaved group even in the finer size of the organic matter.
Keywords: forest floor, litter decomposition, organic matter, physical fractionation

1. INTRODUCTION

Forest litter decomposition processes can be divided into physical disintegration and chemical deterioration (Babel 1975). In the fields physical disintegration occurs simultaneously with chemical deterioration. A close relationship exists between size and chemical composition. Finer size of organic matter usually shows more decomposed characteristics, for example, a lower C/N ratio, lower contents of labile carbon like wax and sugar, and higher contents of recalcitrant carbon like lignin and cellulose (Ohta and Kumada 1977ab; Kanazawa et al. 1977ab). Therefore size distribution of disintegrating organic matter would also be an important factor to determine the chemical characteristics of a forest floor.

Nutrient concentrations in fresh litter vary with species and soil conditions (Handley 1954; Kawada 1966; Morita 1972; Vitousek et al. 1994). Various nutrient concentrations in fresh litter converge to a certain level through decomposition because of the cumulative effect of nutrient release or immobilization by soil biota (Staaf and Berg 1980). The carbon-to-nitrogen ratio (C/N ratio), which has been used as an indicator of decomposition stages, decreases with decomposition and then finally converges. However, a difference exists in the C/N ratio convergence in the late stage of decomposition between leaf litter and needle litter, when the C/N ratio converges to about 20 for leaf litter compared with about 30 for needle litter (Takeda 1997; Takahashi 1997).

If the disintegration process is delayed and larger size of the organic matter that has a high C/N ratio is dominant, the C/N ratio of the layer is hypothesized to be high. To date, many studies have quantified the changes in chemical composition during decomposition using the litterbag method, but a few studies have characterized size distribution of the forest floor. Physical fractionation by size or by density is recognized as a useful method to show the decomposition processes of organic matter (Stevenson and Elliott 1989; Christensen 1992).

In this study I compared the changes in the size distribution of organic matter in the forest floor between needle litter and leaf litter. The relationship between size of organic matter, carbon and nitrogen contents, and the C/N ratio were also determined to test the hypothesis.

2. MATERIALS AND METHODS

Forest floor samples were collected from nine broadleaved stands for the leaf litter group and from nine coniferous stands for the needle litter group (Table 1). The forest floors were separated into L, F, and H layers. The mull type did not have an H layer. Gravel, coarse mineral soil aggregates, roots, coarse woody materials, such as branches, twigs, and barks, were eliminated by handpicking. The macro-nutrient contents of organic layers in some sites were as described in Takahashi (1997).

The fractionation method in this study was: air-dried F layers were fractionated

Table 1. Vegetation and humus types at sampling sites.

Location	Dominant Species	Humus type
	Broadleaved group	
Ibaraki 1	*Quercus acutissima, Quercus serrata*	Moder
Ibaraki 3	*Quercus serrata*	Mull
Fukushima 1	*Quercus crispula*	Moder
Chiba 1	*Quercus acuta, Castanopsis sieboldii,*	Moder
Chiba 3	*Quercus salicina*	Mull
Shizuoka	*Castanopsis sieboldii, Quercus acuta*	Moder
Niigata 1	*Fagus crenata*	Moder
Niigata 3	*Fagus crenata*	Moder
Fukushima 2	*Fagus crenata*	Moder
	Coniferous group	
Aomori 2	*Thujopsis dolabrata*	Mor
Nagano 1	*Sciadopitys verticillata*	Moder
Aomori 3	*Pinus densiflora*	Moder
Aomori 4	*Pinus densiflora*	Mull
Ibaraki 5	*Pinus thunbergii*	Mor
Tochigi 2	*Larix kaempferi*	Mull
Tochigi 3	*Chamaecyparis obtusa*	Moder
Tochigi 4	*Cryptomeria japonica*	Mull
Niigata 2	*Cryptomeria japonica*	Moder

from large organic matter to small sizes using a set of sieves. The opening sizes of the sieve screens were 6 mm, 4 mm, 2 mm, 1 mm, 0.5 mm, 0.25 mm, 0.1 mm, and 0.044 mm. The H layers were not dried because air-dry treatment often hardens the aggregates. Moist H layer (1 kg) was soaked in 5 L distilled water and was stirred. Coarse aggregates, which were often attached to roots, were crushed gently with the fingers. Roots were picked out by hand, but fine root fragments could not be eliminated completely. This aqueous suspension of the H layer was fractionated by the same set of sieves that was used for the F layers. The organic matter on the sieve was washed with distilled water at every sieving. The filtrate of washing was collected and added to the suspension before the next sieving. Accordingly, the volume of the suspension increased during the sieving, but it did not exceed 8 L. The filtrate suspension that passed through 0.044 mm opening screen was left to stand for 24 hours at 4 °C, and then the precipitate and the suspension were separated using a siphon. Distilled water was added to the precipitate to separate the suspension again. This procedure was repeated twice. The suspended organic matter was recovered by centrifuging at 10,000 g. Water-soluble organic matter in the supernatant was condensed to dryness by a vacuum rotary evaporator. Samples were dried at 70 °C in an oven and were weighed.

Dried samples were combined to five fractions for the F layers and to six fractions for the H layers. The fraction numbers and their sizes were: f1 >4 mm, f2 4 - 1 mm, f3 1 - 0.25 mm, f4 0.25 - 0.044 mm, f5 0.044mm - precipitation, f6 suspension, and f7 water-soluble. The F layers were comprised fractions f1 to f5. The f1 and f2 of the H layers were combined so that the H layers were comprized fractions f2 to f7. The fractions were ground to determine the total carbon and nitrogen contents using a C-N corder (Yanagimoto, MT-600). The ash contents were determined after ashing at 550 °C for 6 hours. The total carbon and nitrogen contents were expressed as ash and water-free (dried at 70 °C in an oven) bases. The ash contents were expressed as oven dry bases.

Statistical analyses were made by the T-test to compare each fraction of the leaf and needle litter groups. A comparison among fractions in each F and H layer was made for each litter group. If ANOVA was significant, Tukey's HSD method was used for a multiple comparison (SPSS 1997).

3. RESULTS

Because the size of fresh leaf litter was much larger than that of needle litter, the organic matter larger than 6 mm dominated in the F layers of the leaf litter group (Fig. 1). Organic matter smaller than 6 mm, however, showed similar size distribution patterns in both the litter groups, with two small peaks, 4 - 2 mm and 0.25 - 0.1 mm. Organic matter larger than 1 mm comprised 79% for the leaf litter group and 71% for the needle litter group in the F layers on average.

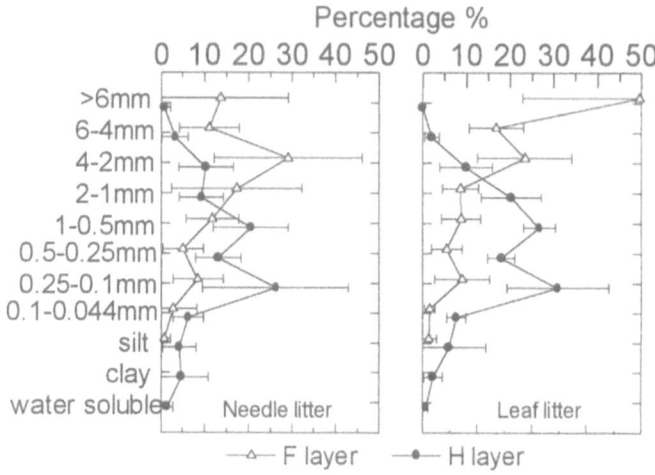

Fig. 1. The size distributions of F and H layers in coniferous stands group and broadleaved stands group. Error bars show SD of the mean.

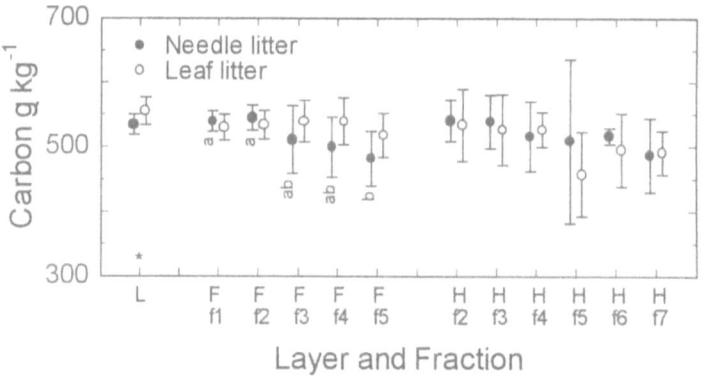

Fig. 2. The carbon contents in the fractionated organic matter. The same letter within each layer is not significantly different (P<0.05) by Turkey's HSD test. * is significantly different at P<0.05 by T-test between each fractions of the litter group.

Disintegration was rather advanced in the H layers for both litter groups. In the H layers, organic matter larger than 1 mm showed 26% for the leaf litter group and 24% for the needle litter group. The distribution patterns in the H layers, which had two peaks of 1 - 0.5 mm and 0.25 - 0.1 mm, were also similar for both groups.

No significant differences in carbon contents occurred between the litter groups for all fractions except for the L layers (Fig. 2). The carbon contents were very stable on ash and water free-bases among the layers and fractions. The average nitrogen contents of the leaf litter group was higher than those of the needle litter

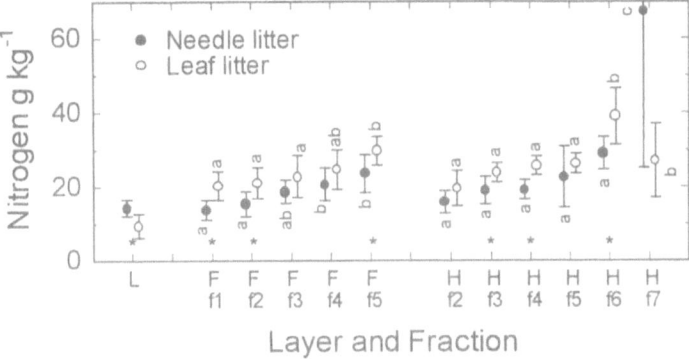

Fig. 3. The nitrogen contents in the fractionated organic matter. Legends are same as in Fig. 2.

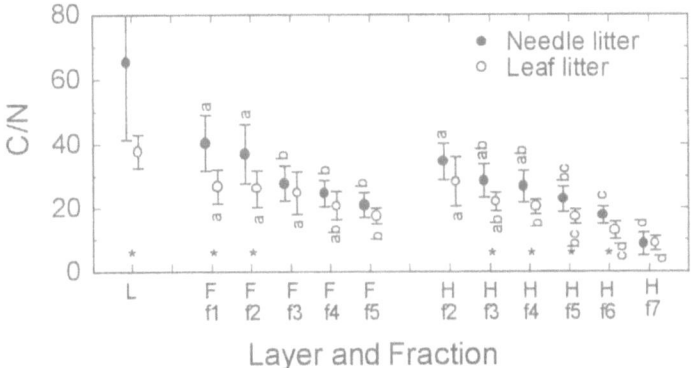

Fig. 4. The C/N ratio of the fractionated organic matter. Legends are same as in Fig. 3.

group for all of the layers and fractions, except for f7 in the H layer (Fig. 3). The nitrogen contents increased continuously with decreasing fraction size. In the H layer, f6 of the leaf litter group and f7 of both groups showed large increases in nitrogen contents compared with the larger size of fractions.

The C/N ratios decreased as the size of the fractions decreased (Fig. 4). The needle litter group showed higher C/N ratios than the leaf litter group for all the layers and fractions. In the L layers, the needle litter had much higher C/N ratios than the leaf litter. In the F layer, large fractions f1 and f2 showed significantly higher C/N ratios for the needle litter group, but the organic matter smaller than the f3 fractions show no significant differences at the 5% level. In the H layer most of the fractions from f3 to f6 showed significant differences in the C/N ratios. For fine fractions the differences in the C/N ratios between both litter groups were somewhat larger in the H layer than those in the F layer. For example, the differences in C/N ratios at fractions f3, f4, and f5 showed 3.0, 3.9, and 3.5 in the F layer, respectively, but 6.5, 6.3, and 5.4 in the H layer. The f7 fractions had very low C/N ratios, about 9 on average, and showed no significant difference between the litter groups.

4. DISCUSSION

The data set of this study included various needle and leaf litter, as well as humus forms like mull, moder, and mor, which indicates that decomposer communities vary among study sites (Babel 1975; Green et al. 1993). Despite such variable site conditions, specific patterns of size distribution for F layer and H layer existed. Vegetation and decomposer composition would not influence the physical disintegration process, although the disintegration rates may vary. In the F layer, the major f1 and f2 fractions of disintegrated organic matter showed significantly higher C/N ratios for the needle litter group than for the leaf litter group. In the H layer the needle litter group showed significantly higher C/N ratios than the broadleaved group for most size classes. These differences would be responsible for the higher C/N ratio of the needle litter group. In this context, the hypothesis that the difference in size distribution of needle litter results in a high C/N ratio of the H layer would be rejected.

If the progress of disintegration promotes chemical deterioration, the results of the F layer, which showed that the differences in C/N ratios between litter groups were negligible for organic matter finer than the f3 fraction, would be reasonable. However, the finding that these differences in C/N ratios increased for most fine fractions in the H layer is not likely to make sense. Even if the comparison made by the data set excluded mull types that do not have H layers, the statistics resulted in the same conclusion. The reason is not clear but roots may contribute higher C/N ratios of the needle litter group. Because of dense root mats in the H layers, input of organic matter through root turnover and exudate may influence the C/N ratios of the H layers. That mechanism would be affected by vegetation type.

The nitrogen contents increased and the C/N ratios decreased continuously as the size of organic matter became finer. Except for f6 and f7, changes in the C/N ratios during disintegration seem to be analogous to the whole litter decomposition process. Fine organic matter from agricultural soil and manure is considered to be of microbial origin, in particular bacterial cell walls, which become a significant substrate for mineralizing nitrogen (Cameron and Posner 1979; Aoyama 1991). The f6 of this study may also originate from microorganisms, but the C/N ratio of this fraction was still higher for the needle litter group than for the leaf litter group. Such fine and suspended organic matter has an important role for translocation of organic matter from organic layers to mineral horizons (Ohta et al. 1986). Further is needed to fully understand the nature and function of organic fine substances in the forest floor.

REFERENCES

Aoyama, M. (1991): Properties of fine and water-soluble fractions of several composts II. Organic forms of nitrogen, neutral sugars, and muramic acid in fractions, Soil Science and Plant Nutrition, Vol. 37, pp. 629 - 637.

Babel, U. (1975): Micromorphology of soil organic matter, In Soil components Vol.1 Organic components. eds. Gieseking, J. E., Springer-Verlag. pp. 369 - 473.

Berg, B. and H. Staaf (1980): Decomposition rate and chemical changes of Scots pine needle litter. II. Influence of chemical composition, Ecological Bulletin (Stockholm), Vol.32, pp.373 - 390.

Cameron, R. S. and A. M. Posner (1979): Mineralisable organic nitrogen in soil fractionated according to particle size, Journal of Soil Science, Vol.30, pp.565 - 577.

Christensen, B. T. (1992): Physical fractionation of soil organic matter in primary particle size and density separates, Advances in Soil Science, Vol.20, pp.1 - 90.

Green, R.N., R.L. Trowbridge and K. Klinka (1993): Toward a taxonomic classification of humus forms, Forest Science, Monograph, Vol.29, pp.1 - 49.

Handley, W. R. C. (1954): Mull and Mor formation in relation to forest soils, Forest commission bulletin, Vol.23, 115p.

Kanazawa, S., S. Wada, S. Takeshima and Y. Takai (1977a): The decomposition processes and existence forms of organic matter in subalpine forest soil. (part 1) Microscopic observation and carbon-nitrogen content of fractionated organic layer of Pwh soil type of Mt. Shigayama, Journal of the science of soil and manure, Japan, Vol.48, pp.181 - 186. (in Japanese with English summary)

Kanazawa, S., S. Takejima, H. Wada and Y. Takai (1977b): The decomposition processes and existence forms of organic matter in subalpine forest soil. (part 2) The proximate chemical composition of fractionated organic layer of Pwh soil type of Mt. Shigayama, Journal of the science of soil and manure, Japan, Vol.48, pp.187 - 192. (in Japanese with English summary)

Kawada, H. (1966): A study on change of nutrient composition of freshly fallen leaves in process of decomposition, Bulletin of the Government Forest Experiment station, Tokyo. Vol.194, pp.167 - 180. (in Japanese with English summary)

Morita, K. (1972): Mineral composition of the fresh litter of major tree species in Japan, Bulletin of the Government Forest Experiment station, Tokyo. Vol.243, pp.33 - 50. (in Japanese with English summary)

Ohta, S. and K. Kumada (1977a): Studies on the humus forms of forest soils. III. Elementary composition of fractionated horizons of forest soils, Soil Science and Plant Nutrition, Vol.23, pp.355 - 364.

Ohta, S. and K. Kumada (1977b): Studies on the humus forms of forest soils. IV. Humus composition of fractionated horizons of forest soils, Soil Science and Plant Nutrition, Vol.23, pp.503 - 511.

Ohta, S., A. Suzuki, and K. Kumada (1986): Experimental studies on the behavior of fine organic particles and water-soluble organic matter in mineral soil horizons, Soil Science and Plant Nutrition, Vol.32, pp.15 - 26.

SPSS (1997): SPSS Base 7.5 for Windows user's guide, SPSS Inc.

Stevenson, F. J. and E. T. Elliott (1989): Methodologies for assessing the quantity and quality of soil organic matter. *In* Dynamics of soil organic matter in tropical ecosystems, edited by D. C. Coleman, J. M. Oades, and G. Uehara, University of Hawaii Press, pp.173 - 199.

Takahashi, M. (1997): Comparison of nutrient concentrations in organic layers between broad-leaved and coniferous forest, Soil Science and Plant Nutrition, Vol.43, pp.541 - 550.

Takeda, H. (1997): Carbon and nutrient cycling in forest ecosystems, KASEAA, Vol.35 (1), pp.26 - 31. (in Japanese)

Vitousek, P. M., D. R. Turner, W.J. Parton and R.L. Sanford (1994): Litter decomposition on the Mauna-Loa environment matrix, Hawaii: Patterns, mechanisms and models, Ecology, Vol.75, pp.418 - 429.

Nitrogen Dynamics of Decomposing Japanese Cedar and Japanese Cypress Litter in Plantation Forests

Tamon YAMASHITA*[1], Hiroyuki TOBITA*[2] and Hiroshi TAKEDA*[2]

*1 University Forest, Faculty of Life & Environmental Sciences
   Shimane University, Matsue, Japan
*2 Laboratory of Forest Ecology, Graduate School of Agriculture
   Kyoto University, Kyoto 606-8502, Japan

ABSTRACT Decomposition processes of Japanese cedar (sugi) and Japanese cypress (hinoki) litter were studied using the litter bag method in the plantation forest over a 49-month period. Weight loss patterns were modelled with exponential curves. Decay constant k [yr^{-1}] was -0.306 for sugi and was -0.239 for hinoki. Sugi and hinoki lost 52% and 47% of their original mass after 49-month, respectively. While N concentration of both litters increased to about two holds of initial value, net immobilization of N was observed only in hinoki litter. Hinoki litter produced only NH_4^+ over a 49-month period. Sugi litter produced NH_4^+ at first. After that nitrification was occurred. The N_2ase activities were significant in both litter but were much higher in sugi litter than in hinoki.
Key words: litter decomposition, N mobilization, N immobilization, N_2 fixation, nitrification

1. INTRODUCTION

The forested area is about 25 million ha in Japan. This means the forest occupies two-thirds of Japanese land. Ten million ha of total forested area is the plantation forest mainly of Japanese cedar and Japanese cypress. The plantation ecosystems of these two coniferous species are very important in Japan because they are main source for the forest products.

Forest soil is characterized by the existence of organic layer on the surface mineral soil. There is large standing stock of organic matter in the forest floor and surface soil of the forested ecosystems. Organic matter in the forest floor of plantation forest is uniform in species but varied in decomposition stage. Plant growth in forest depends on the nutrient storage in and on the surface soil. Nutrient cycling in such soil subsystem have to be clarified to sustain the high forest productivity and to solve the environmental issue.

Decomposition of freshly fallen litter is the first step of nutrient cycling in the forested ecosystem. Since coniferous litter decomposes slowly in the temperate region (Takeda et al. 1987), long term experiments are required to determine the accurate decomposition processes. Macro-nutrients like N contained in degrading organic matter will be mobilized or immobilized by microbes depending upon the decomposition stages (Berg & Staaf 1981). Furthermore, it is known that some forest litter including Japanese cedar litter fix N_2 in atmosphere asymbiotically (Nioh 1980).

In this paper, we will show the mass loss pattern of decomposing coniferous litter in relation to litter quality at first. Then we will discuss the C and N dynamics during decomposition processes with the special reference to the asymbiotic N_2 fixation and mineralization potential over 49-month period.

2. STUDY SITE AND METHODS

2.1 Study Site

This study was carried out in a Japanese cedar (sugi; *Cryptomeria japonica* D. Don, Taxodiaceae) and a Japanese cypress (hinoki; *Chamaecyparis obtusa* Endl., Cuppressaceae) plantation. The plantations were established on a slope of Mt. Hieizan (135° 51' E, 35° 06' N), Shiga, central Japan. Stand age was about 40 yr at the beginning of this study. Mean altitude of the plantations is about 400 m ASL. The annual precipitation is around 2000 mm and the annual mean temperature is 11 °C. Soil type is brown forest soil derived from granites, and the humus type was a moder. A loose litter layer (L) covers several cm of humus horizon (F-H). Some parts of forest floor in hinoki plantation exposed bare soil with rare organic layer. Thin (5-10 cm) A horizon and thick (more than 50 cm) B horizon lay below organic layer.

2.2 Litter Bag Experiment

Needle litter of both tree species was collected using litter traps in each plantation from October, 1992, to January, 1993. The litter bags with an area of 10 cm x 10 cm were made of nylon net with a mesh size of 1 mm. We enclosed 3.00 g of the needle litter in each litter bag. Bags were set in February, 1993. We retrieved 10 or 14 bags every 3 months during the period from May 1993 to February 1995, and February 1996 and March 1997.

Litter bag samples were transported to the laboratory and cleaned off moss and other extraneous organic remnants. Cleaned samples were treated as following three ways; 1) extracted for analysis of mineral N content, 2) incubated in laboratory incubator at 30 °C for another 30 days in the dark to measure the N mineralization potential, 3) incubated in laboratory incubator at 30 °C for 24 hrs in the dark to estimate the N_2ase activity in the litter and then dried for weighing and chemical analyses.

2.3 Chemical Analysis

Three or four of the field only and the field and laboratory incubated samples were extracted with a volume of 100 ml of 2M-KCl per bag for an hour. The NH_4^+-N was analyzed by the indophenol blue method (Keeney & Nelson, 1982). The

NO_3^--N was reduced to NO_2^--N using zinc powder and determined by the modified Griess-Ilosvay method (Keeney & Nelson, 1982).

We used the acetylene reduction assay (Hardy et al. 1973) to estimate N_2ase activity in sugi and hinoki litter. We adopted the theoretical value for conversion ratio 3.0 (Hardy et al. 1968).

The air-dried samples were ground using a laboratory mill to pass 0.5 mm screen. The milled samples were used to analyze for total C and N on a CN-Coder (MT-600, Yanaco, Japan).

3. RESULTS AND DISCUSSION

3.1 Mass Loss and Organic Matter Dynamics

Fig.1 shows the mass loss patterns of decomposing sugi and hinoki litter. Sugi litter lost more weight than hinoki after 49 months. The decay constant k (yr^{-1}) acquired from a single exponential model (Olson 1963) was -0.306 (r^2=0.892) for sugi and -0.239 (r^2=0.656) for hinoki. Takeda et al. (1987) reviewed the decay constant k, and pointed out that k for coniferous litter ranged from -0.111 to -0.499 depending upon tree species and site condition. Tsutsumi et al. (1961) performed the fence method for sugi litter decomposition over a 30-month period and found that k was -0.111. Kawahara (1975) conducted the litter bag method to study the decomposition rate of hinoki for a 17-month period and reported that k was -0.346. The decay constant k varied with the advance of decomposition, -0.357 for the first year to -0.113 for whole period of a 5-year hinoki litter decomposition (Takeda 1995) and -0.467 for the first year to -0.276 for whole period of a 4-year pine litter decomposition (Hasegawa & Takeda 1996). The coefficient of determination (r^2) for a single exponential model decreased with the advance of decomposition over a 5-year period (Takeda 1995). This means decomposition experiment for short term tends to acquire the higher decay constant than for long term. And long term experiment causes lowering of the coefficient of determination (r^2).

Table 1 Decay constants [yr^{-1}] and initial pool sizes [% dry mass] calculated from a double exponential model. k_1: decay constant for labile pool, k_2: decay constant for recalcitrant pool, A_0: labile fraction, $100-A_0$: recalcitrant fraction.

species	k_1	k_2	A_0	$100-A_0$	r^2
sugi	-1.60	-0.110	34.0	66.0	0.9738
hinoki	-2.33	-0.058	30.8	69.2	0.9792

Table 1 summarizes the decay constants and the initial pool sizes calculated based on a double exponential model. This model provides closer fit than a single exponential model in this case and offers the information on the quality and amount of labile (A_0) and recalcitrant ($100-A_0$) fraction (Weider & Lang 1982). The recalcitrant fraction exceeded the labile fraction by twice in both litter. Sugi litter contains higher labile pool and lower recalcitrant pool than hinoki litter. Decay constant k_1 for sugi is smaller than hinoki, and k_2 is larger. According to this model, labile fraction of sugi is larger in pool size but its decomposability is lower than hinoki. On the other hand, recalcitrant pool of sugi is smaller in size and its decomposability is higher than hinoki.

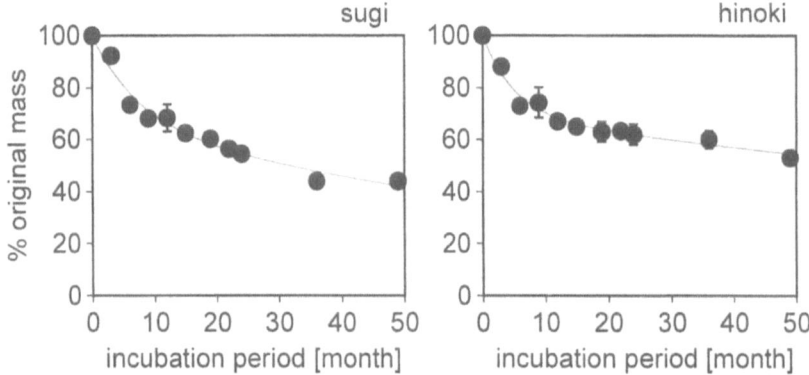

Fig.1 Mass loss processes of sugi and hinoki litter. Left is sugi litter and right is hinoki litter. Regression line is based on the double exponential model. Black circle is average value and bar shows one standard error.

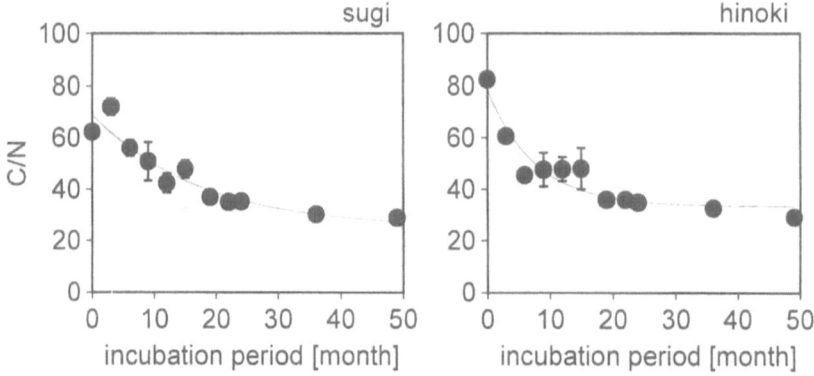

Fig.2. Changes in C/N ratio of sugi and hinoki litter. Left is sugi litter and right is hinoki litter. Regression line is based on the asymptotic model. Black circle is average value and bar shows one standard error.

3.2 Carbon and Nitrogen Dynamics

The changes in C concentration was gradual. The initial C concentration was 58 % dry mass in both litter and decreased to 48 % for sugi and 44 % for hinoki after 49 months. The concentration of total N increased from 0.9 % dry mass to 1.7 % in sugi and from 0.7 % to 1.5 % in hinoki during 49-month incubation. During first 3 months total N concentration decreased by leaching in sugi litter then it increased. Total N of hinoki litter increased consistently throughout the period. Fig. 2 shows the changes in the C/N ratio. The C/N ratio of sugi litter increased during first 3 months reflecting the decrease of N content at that time. The C/N ratio is converging to 25 for sugi and 33 for hinoki according to an asymptotic model. Net immobilization of N was observed in hinoki litter. Sugi litter revealed no significant net immobilization of N. Both litter showed ups and downs in the N content through first two years. After that the mobilization of N occurred, when the C/N ratio decreased to about 35.

3.3 Mineralization and Fixation of N

Mineralization potential of N of sugi and hinoki litter was shown in Fig.3. It was very low during the early stage of both litter. Especially, sugi litter produced almost no mineral N. Hinoki litter produced mineral N from earlier stage than sugi litter. Maximum rate of N mineralization was 500 to 600 μgN gDM^{-1} 30days^{-1} in both litter. From quantitative points of view there is no deference in mineralization potential of both litter. But qualitative deference was observed between sugi and hinoki litter. Hinoki litter produced only NH_4^+-N throughout the period. Sugi litter produced NH_4^+-N and oxidized some of NH_4^+-N to NO_3^--N especially in later stage. It is hard to retain NO_3^--N in litter material because of its high mobility. Production of NO_3^--N is to accelerate the N mobilization in sugi litter.

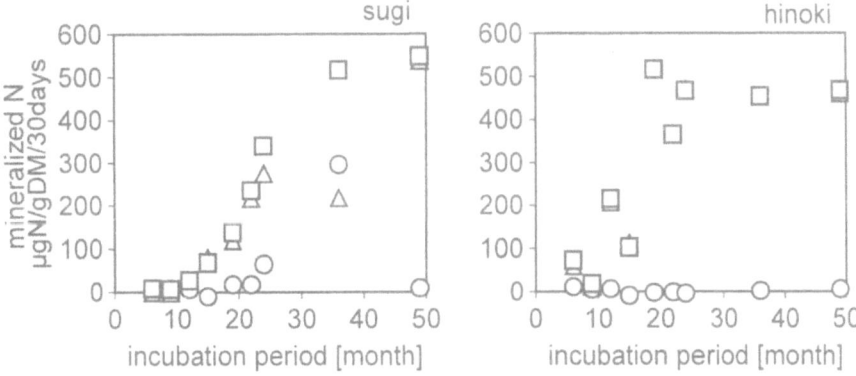

Fig.3. Mineralization potential of N of sugi and hinoki litter. Left is sugi litter and right is hinoki litter. Triangle: NH_4^+-N; circle: NO_3^--N; square: $(NH_4^+ + NO_3^-)$-N.

The changes in acetylene-reducing activities are shown in Fig.4. The activities appeared from 6-month sample in both litter. The activities of hinoki litter was low but significant through the period. Sugi litter revealed high activities till 12-month sample. No activities were observed in the sample of May 1994 because litter sample were dried up after long drought. The activities decreased gradually and stopped with the sample of February 1995, after two year incubation. Sugi litter with 60 to 40 of C/N ratio revealed high acetylene reduction rate. The rate slowed down as the C/N ratio decreased below 40. When the C/N ratio went down below

30, the acetylene reduction activity was hardly observed.

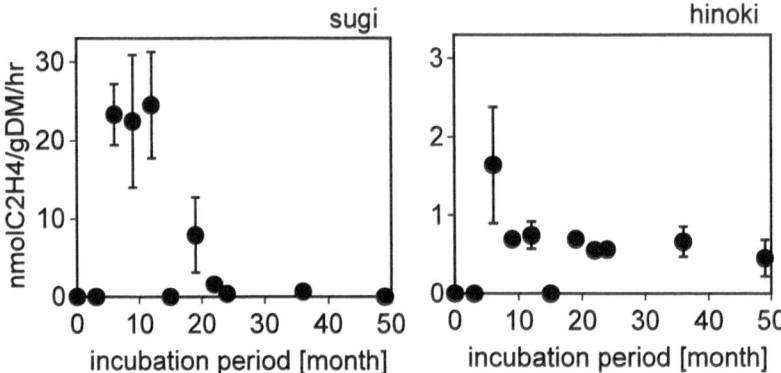

Fig.4. Changes in acetylene reduction activities of sugi and hinoki litter. Left is sugi litter and right is hinoki litter. Black circle is average value and bar shows one standard error.

The amount of nitrogen fixed to sugi and hinoki litter was estimated. Weight loss estimated by a double exponential curve was used for calculation of the total amount of nitrogen fixed by litter sample during the decomposition. The amount of nitrogen fixed by one gramme of sample was converted to the amount per litter bag by multiplying the estimated weight at that time. When one gramme of litter decomposed for two years sugi and hinoki litter fixed 0.46 mg N and 0.037 mg N, respectively.

4. CONCLUSION

Hinoki litter lost less weight than sugi after a 49-month incubation and revealed the consistent increase in total N concentration. In addition, hinoki litter produced only NH_4^+-N which is hard to leach out through the N mineralization processes. These factors contribute to cause the net immobilization of N.

Sugi litter fixed significant amounts of N_2 in atmosphere from 6-month to 22-month. This reflects the high activity of heterotrophic microbes in decomposing sugi litter during that period. Active heterotrophs retained mineralized N in litter and kept the N mineralization potential low during the early stage of decomposition. But they absorbed inadequate N to recover the initial leaching loss unlike the hinoki.

In later stage of the decomposition, the sugi and hinoki litter had high mineralization potential under the laboratory condition. In addition the sugi litter possessed the nitrification potential. Finally both of litter hardly released the mineralized N and remained at the high N level in view of concentration. This suggested that the microbes of later stage might utilize C from exogenous source for their activity whereas N_2 fixers might depend on endogenous C of the decomposing litter.

The dynamics of N in decomposing litter seems to be regulated by the carbon availability and the origin of carbon. It is important to clarify the C source on which the various decomposers depend. Hinoki litter showed very low ARA. This is also attributable to C source. Probably litter materials of hinoki produced less available C or inhibiting substances for N_2 fixers.

REFERENCES

Berg, B. & H. Staaf (1981) Leaching, accumulation and release of nitrogen in decomposing forest litter. Ecological Bulletin 33: 163-178.

Hardy, R.W.F., R.C. Burns, & R.D. Holsten (1973) Application of the acetylene assay for measurement of nitrogen fixation. Soil Biology and Biochemistry 5: 47-81.

Hardy, R.W.F., R.D. Holsten, E.K. Jackson, & R.C. Burns (1968) The acetylene-ethylene assay for N2 fixation: Laboratory and field evaluation. Plant Physiology 43: 1185-1207.

Hasegawa, M. & H. Takeda (1996) Changes in Collembola and Cryptostigmata communities during the decomposition of pine needles. Pedobiologia 41: 225-241.

Kawahara, T. (1975) Decomposition of litter in forest floor II. Effect of the mixture of two kinds of leaf-litter on their decomposition rates. Japanese Journal of Ecology 25: 71-76. In Japanese with English summary.

Keeney, D.R. & D.W. Nelson (1982). Nitrogen - inorganic forms. In Methods of Soil Analysis, Part 2, Chemical and microbiological properties. 2nd Ed. (A.L. Page, Ed), pp. 643-698. Agronomy, No. 9-2, ASA-SSSA, Madison, USA.

Nioh, I. (1980) Nitrogen fixation associated with the leaf litter of Japanese cedar (Cryptomeria japonica) of various decomposition stages. Soil Science and Plant Nutrition 26: 117-126.

Olson, J.S. (1963). Energy storage and the balance of the producers and decomposers in ecological systems. Ecology, 44, 327-332.

Takeda, H. (1995) A 5 year study of litter decomposition processes in a Chamaecyparis obtusa Endl. forest. Ecological Research 10: 95-104.

Takeda, H., Y. Ishida & T. Tsutsumi (1987) Decomposition of leaf litter in relation to litter quality and site condition. Memoirs of the College of Agriculture, Kyoto University 130: 17-38.

Tsutsumi, T., I. Okabayashi, & T. Shidei (1961) On the decomposition of forest litter (II). The Bulletin of Kyoto University Forests 33: 187-198. In Japanese with English summary.

Weider, R.K., & G.E. Lang (1982) A critique of the analytical methods used in examining decomposition data obtained from litter bags. Ecology 63: 1636-1642.

Decomposition Process of Leaf Litter in a Coniferous Forest

Xing-jun TIAN and Hiroshi TAKEDA

Laboratory of Forest Ecology, Graduate School of Agriculture
Kyoto University, Kyoto 606-8502, Japan

ABSTRACT   During the two-year decomposition period, mass loss of litter was about 41% and 56% for *Abies* needle and *Betula* leaf litter, respectively. Nitrogen was immobilized in both litters. Change pattern of Ca, P and Na was similar. The elements appeared higher in weight remaining than the organic matter at early stage, but lower in later stage. Mg and K as easily leachable elements lost lower to mass remaining of litter in whole decomposition period. *Abies* needle was decomposed by fungi first in L layer, followed by animals which led to the fragmentation of needles. The final remaining of litter deposited in H layer and may have been decomposed mainly by bacteria. *Abies* needle and *Betula* leaf litter disappeared in the forest floor within 6 cm and 4 cm, respectively.
Key words: coniferous forest, decomposition, morphology, nutrients, litter

1. INTRODUCTION

Forest ecosystem is considered as a unit of biological organization, just like the cell, the tissue, the organism, or the population, because of its functional integrity. The integrity of the ecosystem is maintained by the transfers of nutrients and energy. Functionally an ecosystem can be divided into two main subsystems namely plant subsystem and decomposition subsystem. Decomposition subsystem is very important in the recycling of nutrients and in providing energy for the soil organisms. Thus a great number of workers (e.g., Berg and Cartoon 1995, Blair et al. 1990, Melissa and Aber 1996, Swift et al. 1979, Tian et al. 1997) have recently contributed to the study of decomposition.

Decomposition is a process of releasing energy and nutrients. During the decomposition, litter changes in its physical, chemical and morphological properties. The changes can be modified with litter bag method and morphological approach. Decomposition pattern largely varies with forest ecosystems. In subalpine coniferous forest (or boreal forest) the decomposition rate is slow. The knowledge of decomposition mechanism is very useful for the forest management.

The aim of this study is to determine the decomposition process of litters in nutrient dynamics and morphological changes.

2. STUDY AREA

The study area is located on the north side (elevation 2000 m) of Ontake

Mountain 3067 m high), Gifu Prefecture, Japan (35 °48'N and 137° 28'E). Coniferous forests dominated in this zone. Mean annual rainfall is about 2000 mm. Mean temperature is 3-4 ℃.

Soil in the area was classified as a wet humus podzol (Pw(h)) (Editorial committee of Forest soil of Japan' 1983). It was acidic with pH about 4. The soil was about 1-2 cm of L layer, 2-3 cm of F layer and 5-6 cm of H layer in thickness .

Two species namely *Abies veitchii* and *Abies mariesii* dominated the site. Species of *Betula ermanii, B. corylifolia* and *Picea jezensis var. hodoensis* accompanied with the two dominated species. There was no shrub layer developed, bad developed herb layer and a 'feather' moss layer dominated by *Pogonatum japonicum* Sull. et Lesq. in the site.

The litters were composed of needles (66.14%), leaves (22.2%), branches (6.81), barks (0.92%) and the others (0.17%). The litter of needle and leaf accounted for 88.34% of the total litter fall at the site.

3. METHOD

The decomposition process of leaf litters was determined with litter bag method and morphological method. Three grams of the dried litter of *Betula* leaf (*Betula ermanii* and *B. corylifolia*) and 3g. of *Abies* needle (*Abies mariesii* and *Abies veitchii*) were each placed in 12 x12 cm litter bags with a mesh size of 1 mm. One hundred and ten sealed, labeled bags of each type were placed on the forest floor of the site in November of 1994. Ten litter bags of each were collected, dried (40 ℃) and weighed every month during growing seasons from June of 1995 to November of 1996. Total nitrogen and carbon of needle and leaf litter were measured by automatic gas chromatography. After an acid wet oxidation in HNO_3+HClO_4, the following analyses were performed: vanad-yellow method for P, atomic absorption method for K, Ca, Mg and Na (for details, see Tian et al. 1998).

Five soil samples for thin sections were collected with a cover can of 6 cm in diameter and 8 cm in height in July 1995. The soil samples were sliced by a sliding microtome with a rapid thin method (Takeda 1988). Five vertical thin sections and 80 horizontal sections of each 0.1 mm in thickness were prepared. The five vertical thin sections were used to observe the structure of a soil profile. The distribution pattern of *Abies* needles and their recognized decomposition products were determined with 80 horizontal thin sections (for details, see Tian et al. 1997).

4. RESULTS AND DISCUSSION

4.1 Decomposition rate

During two-year incubation, mass loss of *Betula* leaf and *Abies* needle was about 56% and 41%, respectively (Fig.1). The decay constant (Olson 1963) of *Abies* needle and *Betula* leaf in this study ranged from 0.24 and 0.49, respectively. It shows that decay rate for *Abies* needle is slower than that for *Betula* leaf. Mass loss

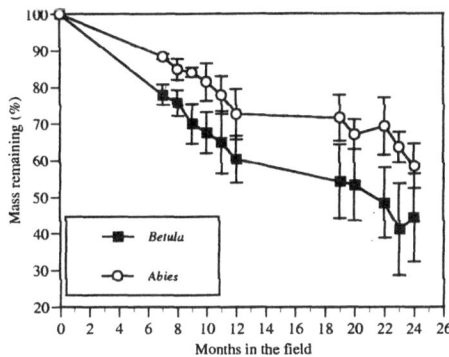

Fig.1 Changes in weight remaining of litters during two-year decomposition. Vertical bars represent standard error.

was 20.1% in *Betula* litter and 15.4% in *Abies* needle litter in the first winter. While in the second winter the *Abies* needle and *Betula* leaf litter lost only 9.4% and 1.6%, respectively. This indicates that the leaching process played an important role in the first winter, and the fresh litters had rich easily leachable materials. During the growing seasons, mass loss of litter was higher in the first year than in the second year, especially for *Betula* leaf which was richer in nutrients than *Abies* needle. It is obvious that the mass loss of *Betula* leaf was higher than that of *Abies* needle litter in the first year and second winter, but it became almost the same rate 12% and 11.3% respectively in the second growing season. So variations of decomposition in different resources mainly take place in the early decomposition stage.

4.2 Nutrient and carbon dynamics during decomposition of litters

Carbon concentration was about 50% initially in both litters and remained at that level during the decomposition period (Table 1). The absolute amount of carbon in two litters were significantly correlated with mass of litter (Fig.2).

Table 1. The initial concentrations of *Betula* leaf and *Abies* needle litter

Litter type	Concentration (%)						
	C	N	Ca	K	Mg	Na	P
Betula leaf	47.46	1.27	0.63	0.29	0.13	0.03	0.03
Abies needle	49.08	0.92	0.53	0.15	0.05	0.01	0.03

Mass remaining of nitrogen (N) was higher than that of litter in both litters, especially for *Abies* needle litter(Fig.2). This result suggests that the concentration of N increased during the decomposition period. Tian et al.(1998) reported that an increase in concentration in early decomposition stage was observed in almost all decomposition studies. The increase in nitrogen was attributed to: 1) the faster release rate of C and other elements (Lousier and Parkinson 1978; Tian et al.1997); 2) not easier leaching (Gosz et al. 1973) and 3) sometime immobilized by fungi (Swift et al. 1979; Takeda 1995; Hasegawa and Takeda 1995).

Changing patterns of mass remaining of Ca, P and Na for *Abies* needle litter and *Betula* leaf litter were similar (Fig.2). Generally, the mass remaining appeared higher in first winter and in early summer, then decreased lower to that of mass

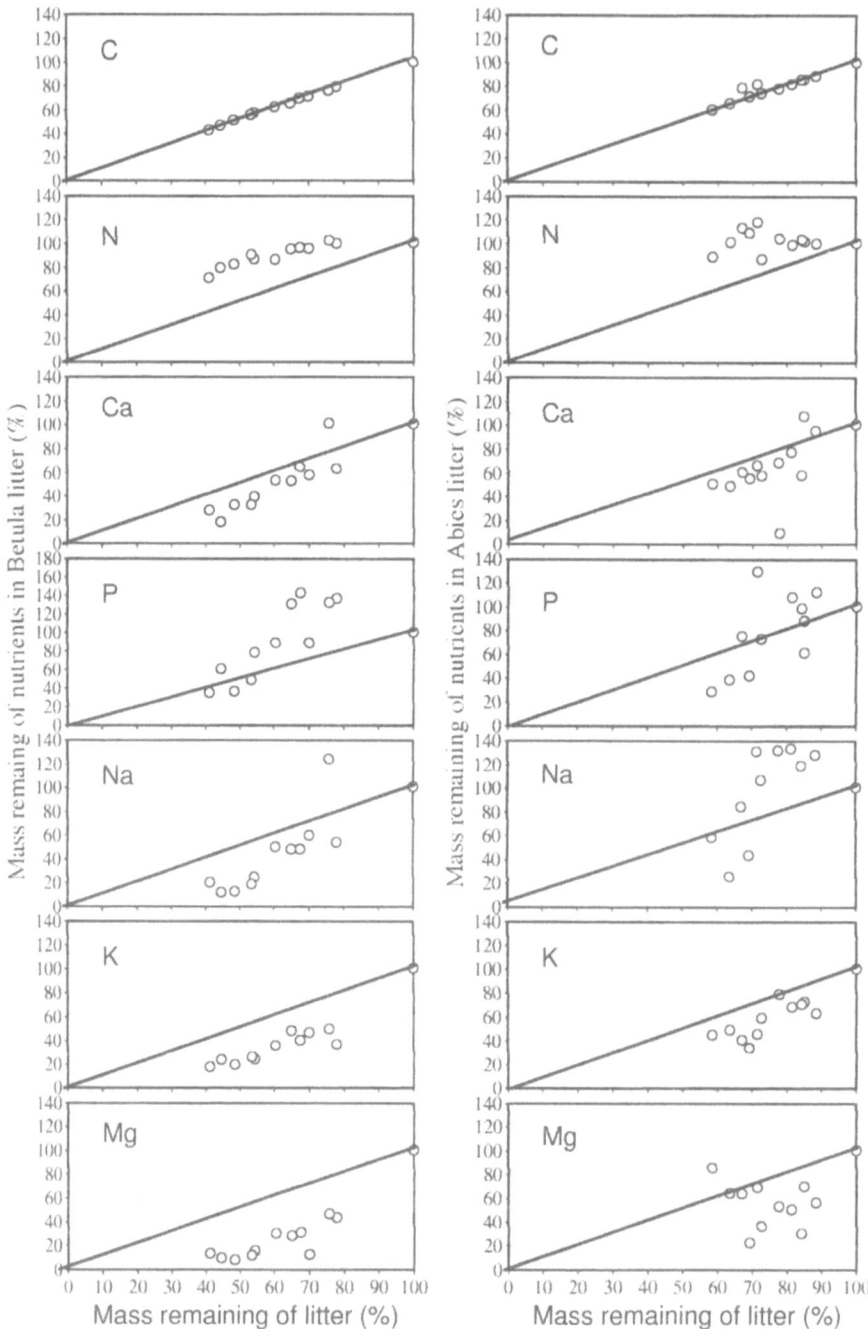

Fig. 2 Changes in mass remaining of carbon and nutrients compared with mass loss of litter

remaining in the latter stage of the study period. This may be due to that those three nutrients were structural elements (Blair 1988) or lower initial concentration (Table 1) without effect of leaching but sensitive to the decomposition by organisms (Tian et al. 1998).

Mass remaining of K and Mg for both *Abies* needle and *Betula* leaf showed a lower level throughout the decomposition process (Fig. 2). This is leading to suggest that two elements were easily leached or decomposed during decomposition period. Lousier and Parkinson (1978) suggested that K appeared to be completely water soluble element and to be readily leached during early decomposition stage. Much losses of K and Mg were also reported in many works (e.g. Gosz et al. 1973; Berg and Cortina 1995; Laskowski et al. 1995a).

4.3 Changes in morphology of leaf litters in the forest floor

The needles of *Abies*, just after falling, was first colonized by fungi. The fungi colonized in the surface and then penetrated into the mesophyll tissues (Fig 3a). In this stage, the needles changed into dark brown color in the L layer. As the decomposition advanced (the place of needle in the soil became deeper) the needle was decomposed gradually. Tian et al. (1997), based on the decomposition level of needle, divided the tissues into four types namely fresh, slightly decomposed, moderately decomposed and greatly decomposed types. They investigated the distribution of those four types and found that the relative decomposition rate of the tissues was in the order of mesophyll > vascular bundle > epidermis in *Abies* needles. The epidermis was decomposed first, but kept a great number of slightly decomposed and moderately decomposed type in the L and F layer. Its fragments (greatly decomposed) remained to H layer of soil (Fig 3c). Although decomposition of mesophyll was started later than epidermis was, it was rapidly disappeared due largely to the activities of animals (Fig. 3b). Compared with epidermis, the number of slightly decomposed and moderately decomposed mesophyll rapidly disappeared in the L and F layer, especially for the moderately type. The greatly decomposed type appeared in the L layer increased in upper F layer but less found in H layer, indicating that the rate of loss of mesophyll was faster than that of the epidermis.

Newly fallen leaves of *Betula* were yellow-brown in color but often showed black patches, indicating the colonization by fungi. Colonized fungi soon grew and penetrated into inner tissue, palisade parenchyma and spongy parenchyma, as well as vein (Fig. 3d). The epidermis layer was so thin that fungi could easily penetrate into inner tissues. When the condition was suitable for fungi growing, the leaves were decomposed completely by fungi and changed to hyphae or mycelium remains (Fig. 3e). In leaves, vascular tissues in midveins and petioles were separated from the leaf blades. These leaves showed reticulate type of venation through the disappearance of epidermis, mesophyll tissues. This skeletonization of leaf (Fig. 3f) was partially explained by the feeding of soil animals (Bal 1982). Leaf blades disappeared within the surface 2-3 cm, but the midveins and petioles of leaves remained in deeper layers. The disappearing order of tissues in *Betula* leaf litter was mesophyll $\geqq$ epidermis > vascular; Decomposition of leaf litter occurred mainly

228

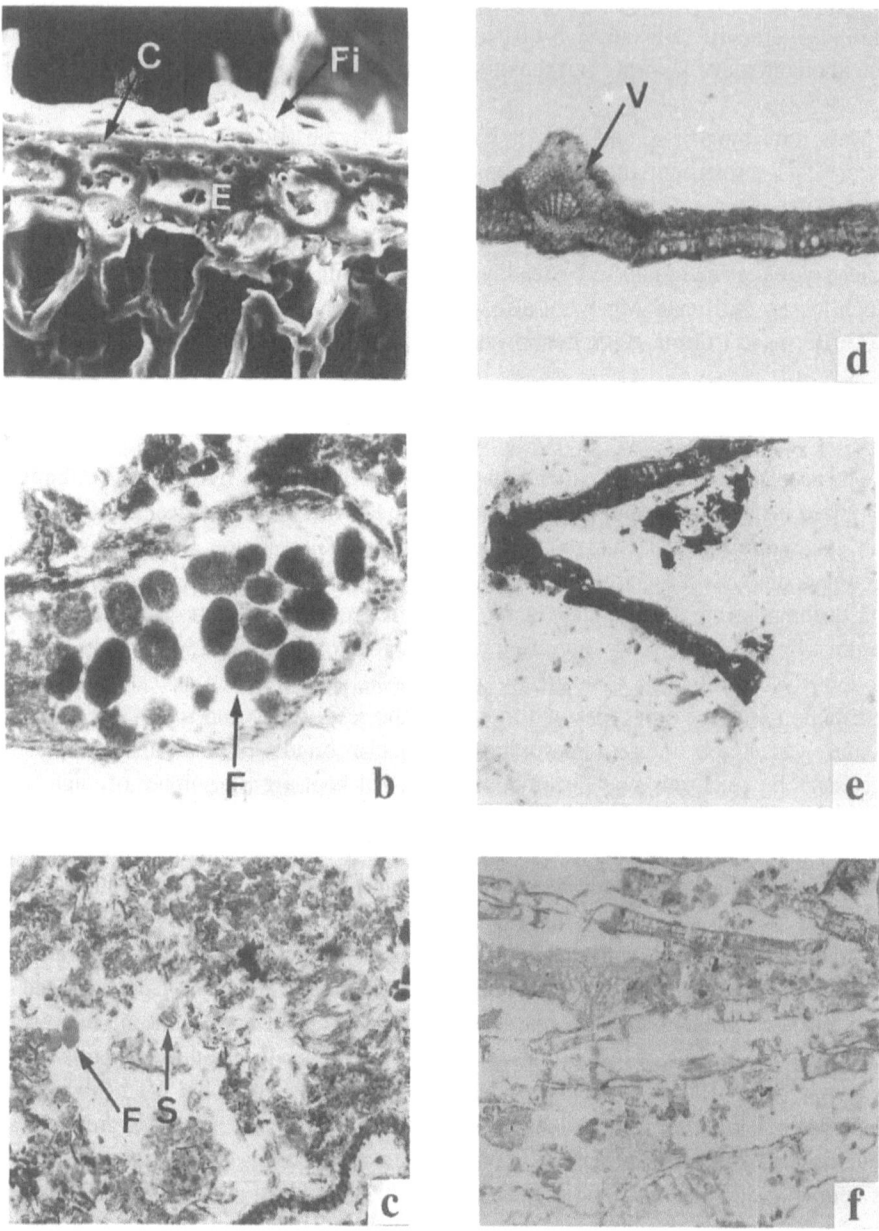

Fig.3 Morphological changes in *Abies* needle and *Betula* leaf during the decomposition. **a**: an electronic
microscope figure of needle showing the colonized fungi on the surface of needle and the epidermis(500
X); **b**: a vertical section of soil through a needle of *Abies* showing the needle was ingested by animal and
left faces in the needle (40 x in 2-2.5 cm layer depth); **c**: a horizontal section of soil showing the needle
decomposed strongly leading to the stomata left only in the soil (40 X, 5-5.5 cm layer of soil); **d**: The
Betula leaf deposited in surface layer showing the fungi colonized and penetrated into inner tissues. (40 X,
horizontal section of soil, 0-0.5 cm layer); **e**: a horizontal section of soil through *Betula* leaves
showing the leaves decomposed by fungi strongly (40 X, 0.5-1 cm layer); **f**: a vertical section of soil
through the leaves of *Betula* showing the leaves may be ingested by animal (40 X, 0.8 cm in depth).
Notes : C, cuticle layer; F, faeces; E, epidermis; Fi, fungi; V, vein; S, stomata.

within the L and F layers. *Abies* needles and *Betula* leaves completely disappeared within the 0 to 6 cm and 0 to 4 cm in depth, respectively.

4.4 Decomposition process of leaf litter in the forest

For the new fallen *Abies* needle, fungi first colonized the epidermis and grew on it through the growth phase and steady phase (Hasegawa and Takeda 1995). The cuticle layer was destroyed or separated from the epidermis(Fig.3a). As the cuticle was broken, fungi greatly penetrated into the inner tissue. In this stage the nitrogen were immobilized so that the concentration of nitrogen increased a lot. This phase was confirmed by the research of Hasegawa and Takeda (1995) in the pine forest and the research on *Abies alba* litter in soil (GourBiere 1986). Meanwhile, the animals also invaded into the inner tissue. The mesophyll was ingested gradually (Fig.3b). We observed that the mesophyll decomposed by fungi was deposited in the deep layer of the forest floor while that by animals was soon shrank. It indicates that the bulk of the mesophyll was ingested by animals and the decomposition by animals was faster than that by fungi alone. These processes took place in L and F layers. Due to the ingestion of animals and fungal decomposition some elements such as Ca and P were released and leached into soil leading to the mass remaining of those elements less than the mass of litter. Due to the inner tissue ingested and digested by animal, large number of needles was destroyed and disappeared gradually. The disappearance of needle took place mainly in F layer. The comminuted needles were converted into particles and faeces (Fig.3b). This led the resources to decrease in size but increase in surface area. Bacteria may colonized a lot in the detritus, contributing a function to the decomposition of litter in later stage or in deeper layer of forest floor. Decomposition of needle may be processed by a relative activity of fungi, animal and bacteria. Fungi played an important role in the early stage of decomposition; animals mainly contributed in the second stage; bacteria perhaps followed animals to play efficiency role in the last stage.

The leaf litter of *Betula* was colonized first then penetrated into inner tissue by fungi. Some times the litter was ingested by animals (Tian et al. 1997). The fungi may play an important role in decomposition of litter based on the morphological observation. Those process occurred in L and upper F layer. Due to the activity of fungi the mass remaining of N decreased slower than that of litter. Much of *Betula* leaf litter disappeared surface 4 cm depth of soil. The final remaining with the detritus of *Abies* needle left in deeper soil was decomposed by microorganisms.

5. REFERENCE

Bal, L.(1982): Zoological Ripening of Soils, Center for Agricultural Publishing and Documentation, Wageningen, 365p.

Berg, B.(1991): FDA-Active Fungal Mycelium and Lignin Concentrations in Some Needle and Leaf Litter Types, Scandinavian Journal of Forest Research, Vol.6, pp. 451-462.

Berg, B. and J. Cartoon (1995): Nutrient Dynamics in Some Decomposing Leaf and Needle Litter Types in a *Pinus Sylvestris* Forest, Canadian Journal of Forest

Research, Vol.10, pp.1-11.

Blair, J.M. (1988): Nutrient Release from Decomposing Foliar Litter of Three Tree Species with Special Reference to Calcium, Magnesium and Potassium Dynamics, Plant and Soil, Vol.110, pp.49-55.

Blair, J.M., R.W. Parmelee and H.M. Beare (1990): Decay Rates, Nitrogen Fluxes, and Decomposer Communities of Single- and Mixed-Species Foliar Litter, Ecology, Vol.75, No.1, pp.1976 -1985.

Editorial Committee of 'Forest soil of Japan' (1983): Forest Soil of Japan, Forestry Technological Society of Japan, Tokyo, 678p. (in Japanese)

Gosz,J.R., G.E. Likens and F.H. Dormann (1973.): Nutrient Release from Decomposing Leaf and Branch Litter in the Hubbard Brook Forest, New Hampshire, Ecological Monographs, Vol. 43, pp.173-191.

GourBiere, F. (1986): Microscopie de la Mycroflore des Aiquills de Sapin (*Abies alba*) Marasmius et Rosaceus, Canadian Journal of Botany, Vol. 65, pp. 131-136.

Hasegawa, M. and H. Takeda (1995): Changes in Feeding Attributes of Four Collembolan Populations During the Decomposition Process of Pine Needles, Pedobiologia, Vol.39, No.2, pp.155 169.

Loursier, J.D. and D. Parkinson (1978): Chemical Element Dynamics in Decomposing Leaf Litter, Canadian Journal of Botany,Vol.56, pp.2795-2812.

Laskowski, R., B. Berg, M. Johansson and C. McClaugherty (1995): Release Pattern for Potassium from Decomposing Forest Needle and Leaf Litter. Long-Term Decomposition in a Scots Pine Forest IX, Canadian Journal of Botany, Vol. 73, pp.2019-2027.

Melillo, J.M. and D. Aber (1996): Immobilization of a 15N-labeled Nitrate Addition by Decomposing Forest Litter, Oecologia, Vol.105, pp.141-150.

Olson, J.(1963): Energy Storage and Balance of Producers and Decomposers in Ecological Systems, Ecology, Vol.44, pp.322-331.

Swift, M., O. Heal and J. Anderson (1979): Decomposition in Terrestrial Ecosystems, Blackwell Scientific Publications, Oxford, London, 371p.

Takeda, H.(1988): A Rapid Method for Preparing Thin Sections of Soil Organic Layers, Geoderma, Vol.42, pp.159-164.

Takeda, H. (1995): A 5 Year Study of Litter Decomposition Processes in a *Chamaecyparis Obtusa* Endl. Forest, Ecological Research, Vol.10, pp. 95-104.

Tian, X., H. Takeda and T. Ando (1997): Application of a Rapid-Slicing Method for Observations on the Decomposing Litter in the Mor Humus Form in a Subalpine Coniferous Forest, Ecological Research, Vol.12, No.3, pp.289-300.

Tian, X., H. Takeda and T. Ando (1998): Dynamics of Carbon and Nutrients in *Abies* Needle and *Betula* Leaf Litters during the Two-Year Decomposition Period in a Subalpine Coniferous Forest, Applied Forest Science Kansai, (in press).

Changes in Nitrogen and Carbon Fractions of the Decomposing Litters of Bamboo

Nimfa K. TORRETA[*1], Hiroshi TAKEDA[*1] and Jun-Ichi AZUMA[*2]

[*1] Laboratory of Forest Ecology, Graduate School of Agriculture
Kyoto University, Kyoto 606-8502, Japan
[*2] Laboratory of Recycle System of Biomass, Faculty of Agriculture
Kyoto University, Kyoto, Japan

ABSTRACT: The changes in nitrogen and in the carbon fractions during the decomposition of the three bamboo litter parts were studied by litterbag method. All the litter components showed a significant inverse relationship between N concentration in residual materials and percent weight remaining. The content of all carbon fractions analyzed decreased through time. A strong influence of N in the early stages of decomposition and the take over of lignin and other acid insoluble sugars were observed after two years of decomposition.
keywords: Decomposition, Nitrogen, Carbon fractions, Bamboo

1. INTRODUCTION

Bamboos are plants of enormous importance and of historic economic value in Asia. Within the context of a steadily declining natural forest stock in Asia, bamboos are becoming increasingly scarce resources with many uses and excellent qualities seldom found in other plants. Continued and increased production of these plants are necessary to sustain a country's need for bamboos. One way of doing this is by investigating the dynamics of the decomposition of bamboo litters. The changes in nutrients and carbon fractions during decomposition influence nutrient dynamics and remarkably affect soil fertility and primary production of an ecosystem. The decomposition rate of these chemicals determines the rate at which nutients become available for renewed uptake by plants,thereby determining the ecosystems productivity.

Decomposition of litter is one of the most frequently studied processes in various ecosytems. However, the scope and type of ecosytem to which most related studies were based on, relied heavily on mass loss and dynamics of macronutrients of decomposing litters of various tree species found in the temperate and tropical forests. Only a few has been noted on nutrient dynamics of bamboo litters. Among these are those conducted by Tripathi S.K. and K.P. Singh (1992) in which nutrient immobilization and release patterns in various plant parts during decomposition were studied and another related study (Tripathi, S.K. and K.P. Singh, 1992) is on the abiotic and litter quality control during the decomposition of different plant parts in dry tropical bamboo savanna in India. A study on carbon stock and cycling in a bamboo stand has also been reported by Isagi (1994). Relatively, fewer studies have been conducted regarding changes in carbon fractions (i.e. lignin, soluble and insolublecomponents etc.) in decaying litter of various tree species and almost zero in bamboo species. However, results of those studies that have been done (e.g. Berg et al. 1982, 1984; Melillo et al. 1989) indicate that litter quality as defined by carbon fractions and nutrient content data are good predictors of decay rate and nutrient dynamics. Somehow, this paper focuses on this aspect. It is therefore the aim of this paper to present the results of the changes in nitrogen and in various carbon fractions and show how these components affected the rate of decomposition during the two-year decomposition period of the litters of bamboo.

2. EXPERIMENTAL SITE

This research study was conducted in the Simazu experimental forest at kansai Research Center , Forestry and Forest Products Research Institute at Momoyama , Fushimi, Kyoto, Japan from 1994 November to 1996 October. The experimental forest was believed to be planted to moso-chiku bamboo (7,000 culms/ha) since the Tokugawa period (1603-1867) with a total area of 7,045 m². The area has blocky, clay loam type of soil in the surface and sandy clay loam in the deeper layer of the soil. It is also characterized with gravel sand and mud sediments of unconsolidated diluvium soil. The geological formation is known as the Osaka formation. The mean annual precipitation from 1989 to 1993 was 1,547 mm. The rate of litterfall as determined by litter traps for leaf, leafsheath and twig was 4.38; 1.9 and 0.52 t ha $^{-1}$ respectively (Isagi 1994).

3. MATERIALS AND METHODS

3.1. Litterfal Collection and Sample Preparation

Freshly abscised bamboo litters were collected and ovendried at 40⁰C for several days and later segregated into leaf, leafsheath and twig. Litterbags measuring 10 cm x 10 cm with a mesh size of 3.0 mm were prepared. Each litterbag was filled separately with 3 g of leaves, 2 g of leafsheaths and 2 g of twigs and closed by sewing over the open ends. In 1994 November, 12 litterbags of leaf, 10 litterbags of leafsheath and 8 litterbags of twig were randomly placed in each of the 20 subplots (1m x 2m) of a 10m x 20 m study plot established in the bamboo stand and allowed to decompose under natural conditions.In the first year, collection of leaf and leafsheath litters was done monthly, while twig sampling was done quarterly. In the second year, quarterly sampling was done in all litter types. On each sampling occassion, 10 litterbags of each litter type were collected from 10 subplots.

3.2. Laboratory Analyses

After collection, litterbags were cleared of any attached plant materials , weighed and dried to a constant weight at 40°C . Mean values of mass loss were calculated for each sample set of 10 bags. Thereafter, the samples were pooled into a bulk sample for chemical analyses.

The composite samples were ground and analyzed for nitrogen and various carbon fractions . Total carbon and nitrogen content was determined by a CN analyzer. Content of alcohol benzene solubles was determined by the standard methods (P-8010) of the Japanese Industrial Standard (JIS). Lignin content was determined by the standard method (T222-os-74) of the Technical Association of the Pulp and Paper industry (TAPPI). The combination of the methods of Galambos (1967) and Blumenkrantz and Asboe-Hansen (1973) was followed for gluconic- uronic acid content analysis. Absorbances of reaction mixtures were read and recorded at 525nm with a single beam spectrophotometer.Neutral sugar analysis was based on the alditol acetate method originally described by Gunner et al. in 1961.The neutral monosaccharides converted into alditol acetates were analyzed by gas liquid chromatography. For phenolic acid determination, dried native sample was saponified with 2N NaOH to extract the phenolic acids. Derivatized phenolic acids were analyzed by gas liquid chromatography at 200°C using methyl margarate as an internal standard.

3.3. Calculation

The instantaneous annual decay rate (k) for each bamboo litter part was shown by the negative exponential decay model described by Olson (1963): $W_t / W_0 =$ exp $(-kt)$ where: W_0 is the initial wt of the organic matter and W_t is the weight of organic matter or nutrients remaining after time t.

4. RESULTS AND DISCUSSION

4.1. Mass Loss Through Time

A continuous decrease in mass loss through time was observed in all bamboo litter parts (Figure 1a). A negative correlation existed between time (months) and percent mass remaining. The correlation coefficient values were high and significant (i.e $r = 0.89$, P< 0.001 (leaf); $r = 0.93$, <0.001 (leafsheath); and $r = 0.94$, P< 0.001 (twig)).

After 2 year more than half of the leaf litter (63.7%) was lost, followed by the leafsheath (58.80%) and then by twig litter (51.05%). The instantaneous annual decay rate (k) was highest in the bamboo leaf (0.645), followed by the bamboo leafsheath (0.490) and bamboo twig (0.388).

Considering the percentage of original mass remaining of the three bamboo litter parts, the amount remaining apparently decreased for the first 3 months and started to vary considerably from the 4th month until the 8th month (Figure 1a). Beyond this period, a steady weight loss was observed throughout the decomposition period in all litter types.The continuous decrease in weight of litters for the first 3 months maybe attributed to the abiotic process of leaching , while in the succeeding months until the 8th month , the variable weight loss pattern may be due to the stimulatory effect of frequent rain showers which could have provided the favourable effect of optimum moisture content on the activity of the decomposers. It was observed that after frequent rains through days 225 (which marked the highest water content throughout the decomposition period) (Figure 1b), a significant loss in weight was observed for all bamboo litter parts. After this period, steady weight loss was noted. This is an indication that at this point , the nutrient controlled phase is almost over and the slowly degrading materials produced as by products of the decay process started to dominate.

4.2. Nitrogen Mobility

A progressive increase of N over the initial concentration was observed in all litter components (Figure 2a). A similar trend was noted by several authors (i.e. Gosz et al.1973; Lousier and Parkinson 1978; Schlesinger 1981) and they accounted this phenomenon as a result of microbial retention of nitrogen (maybe from atmospheric precipitation or imports of N in fungal hyphae and contaminating debris) which also acts to reduce the C/N ratio through time. A significantly inverse linear relationship between the percentage of original mass remaining and N concentration in the residual materials described this trend. On the other hand, the N content remaining in all litter components at different points in time of the sampling period either increased or decreased (Figure 2b). This pattern was apparently recognized in the early stages of decomposition especially in the leaf and leafsheath litters. A marked short term increase in N content was also noted from the 8th month until the 10th month in all litter components which coincided with the rainy period in the area. In most studies, increases in the amount of N have also been observed as a result of many factors (Berg and Staaf, 1980).This accumulation was followed by a slow N release in the later part of the decomposition process. This trend seemed to indicate that N level seemed to stimulate weight loss only initially (at least before 382 days) whereas in later stages, the lignin and lignin related level had a retarding effect that probably overtook the effect of N level as decomposition continued.

Narrow C:N ratio and increased N concentration observed in all litter components were results of decay in which N was probably immobilized by microorganism and C was respired and lost from the litter. This further suggested that at some point during the decomposition process, N was no longer limiting to microbial growth and activity.

4.3. Changes in Carbon Fraction Dynamics

The concentrations of alcohol benzene soluble, protein and lignin fractions increased with the level of decomposition in all substrates studied, while content of initial amount remaining decreased (Table 1). The increase in the concentration of alcohol benzene fraction may have been due to the incorporation of some materials from the environment into the litter components. Increase in the concentration of protein could be from the same source of N increase as mentioned earlier in this paper, since values for protein concentration were calculated from N values of this study. Increase in the relative amount of lignin fraction in all litter components maybe the result of low molecular materials converted into recalcitrant materials during the process of decomposition.

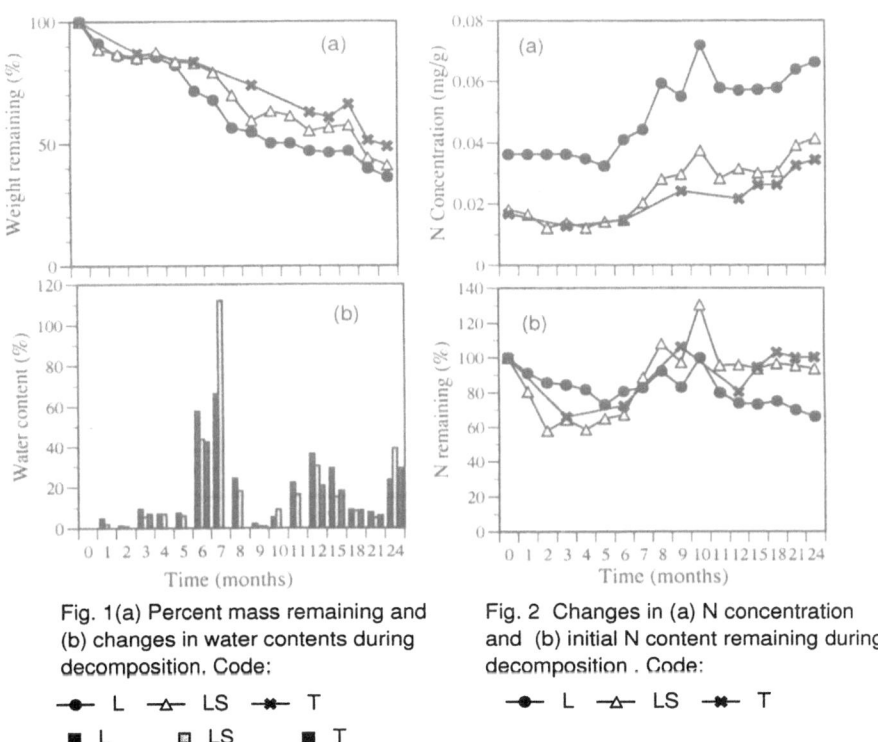

Fig. 1(a) Percent mass remaining and (b) changes in water contents during decomposition. Code:

—●— L   —△— LS   —✳— T
■ L   □ LS   ■ T

Fig. 2 Changes in (a) N concentration and (b) initial N content remaining during decomposition. Code:

—●— L   —△— LS   —✳— T

The general carbon fraction mobility in the three bamboo litter types was: phenolic acid>gluconic-uronic acid>alcohol benzene>protein>neutral sugar>lignin. As indicated in the mobility series, Phenolic acid (PA) was the fastest to decompose. This is rather unusual since PA are presumed to include tannins and polyphenols which retards decomposition (as cited by Schlesinger et al. 1981). However, studies of Suberkropp et al (1976) and Brunson (1977) lend support to this finding. In their study, they observed the rapid loss of PA and linked such phenomenon to leaching process. Suberkropp et al. (1976) suggested that this component is not actually lost from the litter but instead some of it may become complexed with protein and become isolated in the lignin fraction. Similarly, the loss of the gluconic-uronic acid fraction (GluUA) (acid sugar related to neutral sugar) being next to PA as indicated in the mobility series was assumed not to have been lost to leaching too. Some of these materials if not all, may have joined or added up to the neutral sugar fraction in the litters. With this, it can then be said that C fraction mobility of decomposing bamboo litter types still reflects the classical assumption that as decomposition progresses, easily decomposible litter fractions are depleted and structural components such as non-cellulose polysaccharides (hemicelluloses) and lignins are slowly decomposed (Berg et al. 1982).

4.4. Litter Quality Effects on Decomposition

The regression between the amounts of organic matter and C fractions and N remaining were all positive and mostly significant (Table 2). Correlation coefficient for N decreased as decomposition progressed , while coefficient for lignin showed a reverse trend. The other litter qualities (i.e. neutral sugar and gluconic uronic acid) likewise significantly affected the rate of decomposition. These results stressed the strong influence of N in the early stages of decomposition and the take over of lignin and acid insoluble sugars in the later stages of decomposition.

Table 1. Changes in the concentration (%) and in the initial amount remaining (g) in C fractions of bamboo litter parts during decomposition.

Days	ALbenzene	Neutral sugar	Lignin	Protein	GluUA	Phenolic acid
0	4.90 (0.15) a	52.36 (1.57)	27.94 (0.84)	7.41 (0.22)	1.91 (0.06)	0.05 (0.0014)
	4.10 (0.08) b	57.40 (1.15)	29.76 (0.60)	5.52 (0.11)	1.58 (0.03)	0.11 (0.0021)
	5.71 (0.11) c	51.22 (1.02)	30.17 (0.60)	5.25 (0.11)	2.00 (0.04)	0.10 (0.0021)
98	5.04 (0.13)	48.16 (1.22)	30.39 (0.77)	7.56 (0.19)	1.69 (0.04)	0.07 (0.0017)
	2.70 (0.05)	62.46 (1.06)	27.49 (0.47)	4.07 (0.07)	1.65 (0.03)	0.08 (0.0014)
	1.40 (0.02)	59.23 (1.03)	30.92 (0.54)	3.92 (0.07)	2.17 (0.04)	0.08 (0.0014)
262	7.19 (0.12)	31.38 (0.53)	38.98 (0.66)	12.38 (0.21)	1.47 (0.02)	0.05 (0.0009)
	4.51 (0.06)	46.25 (0.64)	34.26 (0.48)	9.38 (0.13)	1.87 (0.03)	0.08 (0.001)
382	7.50 (0.11)	29.57 (0.42)	39.01 (0.55)	11.42 (0.16)	1.39 (0.02)	0.05 (0.0007)
	4.05 (0.04)	42.95 (0.47)	36.47 (0.40)	9.40 (0.10)	1.59 (0.02)	0.06 (0.0006)
	6.84 (0.09)	49.09 (0.62)	32.49 (0.41)	6.55 (0.08)	1.58 (0.02)	0.09 (0.0011)
686	10.29 (0.11)	18.27 (0.20)	41.66 (0.45)	13.41 (0.15)	1.24 (0.01)	0.01 (0.0002)
	9.68 (0.08)	25.20 (0.21)	41.55 (0.34)	12.94 (0.11)	1.33 (0.01)	0.01 (0.0001)
	2.52 (0.02)	41.59 (0.41)	38.57 (0.38)	10.14 (0.10)	1.40 (0.01)	0.02 (0.0002)

Note: Values not enclosed in parenthesis are concentration values, while values in parenthesis are values for initial amount remaining . Code : leaf (a): leafsheath (b) ; twig (c)

Table 2. Coefficients of correlation (r) for regressions between the amount of organic matter remaining and the C fractions and N remaining of decomposing bamboo litter parts for the initial 98 days, 382 days and the entire period of decomposition (686 days).

C Frac./ N	98 Days	382 Days	686 Days
Alc. benzene sol.	0.6 a	0.42	0.01
	0.98 b *	0.91	0.53
		0.22 c	0.13
Neutral sugar	0.99**	0.99**	0.99***
	0.81	0.97**	0.98***
		0.93*	0.97***
Lignin	0.55	0.97***	0.98***
	0.97*	0.7**	0.86***
		0.99***	0.99***
Protein	0.99**	0.29*	0.56**
	0.95*	0.56*	0.55*
		0.16	0.31
Glu. uro. acid	0.99**	0.97***	0.98***
	0.89	0.93***	0.95***
		0.97**	0.97***
Phenolic acid	0.71	0.78**	0.86***
	0.53	0.69**	0.81***
		0.95**	0.96***
Nitrogen	1***	0.21	0.53*
	0.91	0.56*	0.52*
		0.07	0.4

Note : Leaf (a); Leafsheath (b); Twig (c)
*,p<0.05; **,p<0.01; ***, p<0.001

Acknowledgements: Special thanks are due to the members of the Forest Ecology laboratory for their constructive criticisms.

REFERENCES:

Berg, B. and H. Staaf (1980) : Decomposition Rate and Chemical Changes of Scots Pine Needle Litter.II. Influence of Chemical Composition. In : T.PERSSON (ed.), Structure and function of Northern Coniferous Forest - An Ecosystem Study. Ecological Bulletins 32, 373-390. Swedish Research Council, Stockholm.
Berg, B., Hannus,K.,POPOFF,T., and O. Theander (1982) : Changes in Organic Chemical Components of needle Litter During Decomposition. Long Term

238

Decomposition in a Scots Pine Forest. I.Can.J. Bot. 60:1310-1319.

Berg, B., G. Ekbohm, and C. A. McClaugherty (1984): Lignin and Holocellulose Relations During Long Term Decomposition of Some Forest Litters. Long-t e r m Decomposition in a Scots Pine Forest IV. Can.J. Bot. 62: 2540-2550.

Blumenkrantz,N., and G. Asboe-Hansen (1973) : Anal. Biochem. 54: 484-489.

Brinson, M. M. (1977) : Decomposition and Nutrient Exchange of Litter in an Alluvial Swamp Forest. Ecology 58 : 601-609.

Galambos, J.T. (1967) : Anal. Biochem. 19,119-132.

Gosz, J.R., G.E. Likens, and F.H. Bormann (1973) : Nutrient Release from Decomposing Leaf and Branch Litter in the Hubbard Brook Forest, New Hampshire. Ecol. Monogr. 47: 173 - 191.

Gunner, S.W., J. K. N. Jones and M. B. Perry (1961) : Analysis of sugar Mixtures by Gas Liquid Chromatography, Chemical. Ind. (London), 255 p.

Isagi, Y. (1994) : Carbon Stock and Cycling in a Bamboo *Phyllostachys bambusoides* stand. Ecological Research 9: 47 - 55.

Lousier, J. D. and D. Parkinson (1978) : Chemical Element Dynamics in Decomposing Leaf Litter. Can. J. Bot. 56 : 2795 - 2812.

Melillo, J.M. , J.D. Aber, A. E. Linkens, A.E. Ricca, A.B. Fry and K.J. Nadelhoffer (1989) : Carbon and Nitrogen Dynamics Along the Decay Continuum : Plant Litter to Soil organic Matter. In Ecology of Arable Land. Edited by M.Clarholm and L. Bergstrom. Kluwer Academic Publishers, Norwell, Ma.

Olson, J.S. (1963) : Energy Storage and Balance of Producers and Decomposers in Ecological Ecosystems. Ecology 44 : 322 - 331.

Schlesinger, W.H. and M. M. Hasey (1981) : Decomposition of Chaparral Shrub and Foliage: Losses of Organic and Inorganic Constituents from Deciduous and Evergreen Leaves . ecology 62 (3) : 762 -774.Suberkropp, K., G. L. Godshalk and M.J. Klug. 1976. Changes in the Chemical composition of Leaves during Processing in a Woodland Stream. Ecology 57: 720-727.

Tripathi, S.K. and K.P. Singh (1992): Nutrient Immobilization and Release Patterns During Plant Decomposition in a Dry tropical bamboo Savanna, India. Biol. Fertil. Soils. 14 : 191 - 199.

Tripathi, S.K. and K.P. Singh (1992) : Abiotic and Litter Quality Control During the Decomposition of Different Plant parts in Dry Tropical bamboo Savanna in India. Pedobiologia 36: 241 - 256.

Gross Soil N Transformations in a Coniferous Forest in Japan

Naoko TOKUCHI, Muneto HIROBE and Keisuke KOBA

Laboratory of Forest Ecology, Graduate School of Agriculture
Kyoto University, Kyoto 606-8502, Japan

ABSTRACT Topographic influence on gross soil N transformation was investigated along a slope on a conifer plantation forest. The half of minelarized N was immobilized and the half was left in $NH4^+$ pool at the upper part of the slope, while 93 % of minelarized N was used for immobilization or nitrification at the lower part of the slope. From the similarity of microbial biomass and microbial C/N, microbial flora was similar among the sites. However, the gross mineralization rate was four times faster below 15 m than above 30 m. It indicated that the substrate was decomposable below 15 m. At the upper part of the slope with Oa horizon, humified organic matter with high C/N would be resistible for microbes. It resulted in relatively slow N cycling.
Keywords: gross N transformation, nitrification, immobilization, C/N, topography.

1. INTRODUCTION

The topography influences nutrient cycling through the movement of water and solute. For instance, soil moisture changes systematically along toposequence. It directly affects N mineralization by regulating the activities of soil microorganisms and influencing the production of plant. In consequence, N cycling that is characterized by N pool and turnover rate, also changes along toposequence. Hirobe et al. (1998a) studied net N transformation along the short steep slope under Japanese red ceder (*Cryptomeria Japonica* D. Don) plantation forest in Japan. They showed that net N transformation changed along the slope. In their study net nitrification predominated in the footslope while there was little net nitrification in the ridge. However, net N transformation does not show N immobilization (Jackson et al. 1989; Hart et al. 1994). Davidson et al. (1992) compared soil N transformations between the young forest and the old forest. They showed the significant difference in net nitrification rates between sites, but gross nitrification rates were similar. It suggests the importance of NO3⁻ immobilization.

In this study we investigated topographical difference in gross soil N transformation along the slope using ¹⁵N dilution method (Hirobe et al., 1998a). The objectives are (1) to describe gross N transformation along the slope (2) to identify the important process in soil N transformation (3) to determine the controlling factors for soil N transformation.

2. STUDY SITE

The study was carried on Mt. Ryuoh, Shiga prefecture, Japan, the same site as Hirobe et al. (1998b) in this issue (35 ° 1' N, 136 ° 20 ' E, Tokuchi et al., 1993). The soil was Dystrochrepts. The overstory vegetation was a 45-year-old *Cryptomeria japonica* **D. Don** (Japanese red cedar) plantation that had reached canopy closure.Transect was constructed along the slope from 765 m to 841 m altitude. Nine soil sampling sites were also established in the transect at 15 m intervals. Each site was represented by the distance from the lowest site. Forest floor accumulation type was mull-like-moder in the footslope and mor in the ridge.

3. MATERIALS AND METHODS

Top 5 cm mineral soil were collected from nine sites in the transect on November, 1995. The soil samples were sieved through a 2 mm mesh to homogenize and to remove coarse fragments. The soil samples were dried and measured total C and N and 2 M KCl extractable organic C and N.

After sieving, fresh mineral soil from each site were incubated at 25 °C for 28 days to determine net N mineralization and net nitrification rates. The initial gravimetric moisture content was determined and then adjusted to 60 % of maximum water holding capacity. Gross mineralization and $NH4^+$ consumption rates were determined by $^{15}NH4^+$ dilution method for 1 day incubation (Tietema and Wessel, 1992). Before and after incubation, $NO3^-$ and $NH4^+$ concentrations were determined for net and gross rates (Keeney and Nelson, 1982). Atom % of ^{15}N enrichments were determined by a mass spectrometer (Tracer Mat, Finnigan Co., San Jose, USA). Microbial biomass (C and N) was determined by fumigation-extraction method (Jenkinson and Powlson, 1976).

Net mineralization and nitrification rates were the difference in pool size between pre- and post-incubation. Gross nitrification and $NO3^-$ immobilization rates were calculated after Nishio's equation (Nishio et al., 1985; Wessel and Tietema, 1992). The mean residence time of $NO3^-$ pool; MRT of $NO3^-$ was calculated the ratio of the $NO3^-$ pool size to gross $NO3^-$ immobilization plus gross nitrification (Schimel, 1988; Hart et al. 1994). MRT of $NH4^+$ was calculated the ratio of the $NH4^+$ pool size to gross $NH4^+$ immobilization plus gross nitrification.

4. RESULTS AND DISCUSSION

4.1 Two patterns of gross N transformation along the slope

Gross mineralization was significantly larger in the lower part below 15 m than the upper part above 30 m (Fig. 1a). Gross nitrification also showed the distinct difference between the lower part below 15 m and the upper part above 30 m. Net N transformation showed that the form of inorganic N was almost $NO3^-$ below 60 m whereas it was $NH4^+$ above 75 m (Fig. 1b, Hirobe et al., 1998a). The boundary was different in respect with gross N transformation because of incubation periods.

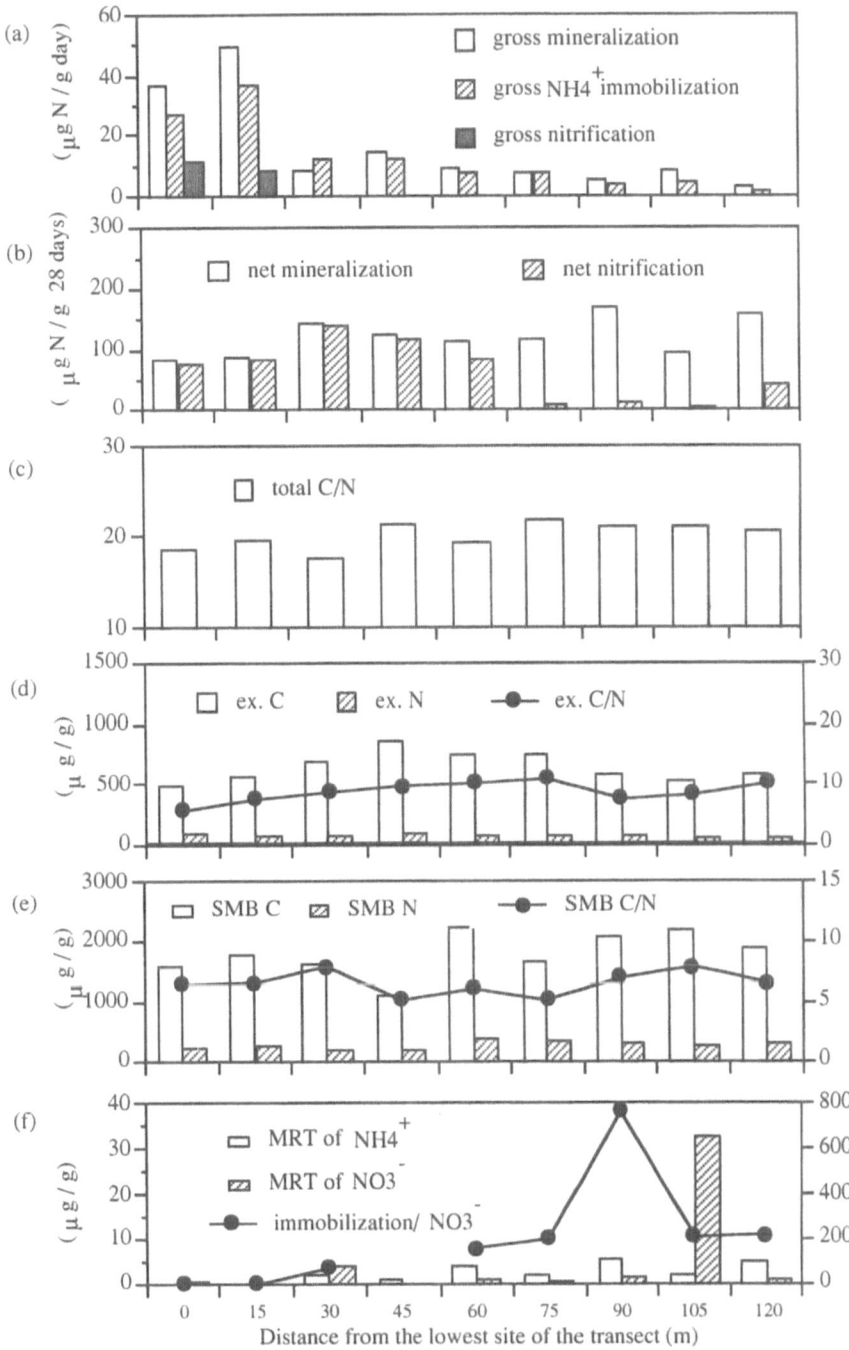

Fig. 1 Soil N transformation rate and the related properties along the transect. (a) gross rates, (b) net rates, (c) extractable C and N, (d) soil microbial biomass and (e) N transformation indecies.

Based on gross N transformation, soil N transformation is classified into two different patterns along the slope. One is found in the lower part below 15 m, the other is shown in the upper part above 30 m. Below 15 m gross N transformation was characterized by high gross mineralization and nitrification. While above 30 m it was characterized by low gross mineralization and nitrification. Two typical patterns of soil N transformation is illustrated in Fig. 2.

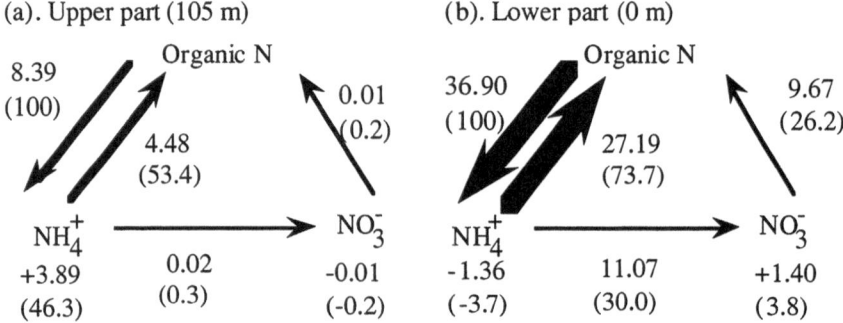

(a). Upper part (105 m)

(b). Lower part (0 m)

Fig. 2 N transformation pattern. Numbers without parentheses indicate gross N transformation rates and changes in pool size, units are μ g g^{-1} day^{-1}. Numbers with parentheses indicate the ratio to gross N mineralization.

4.2 Relative importance of each process in gross N transformation

In all sites, immobilization was dominant over nitrification as can be seen by the ratios between immobilization and nitrification as shown in other study (Fig. 1a, Tietema and Wessel, 1992). From the similar MRT of NO3$^-$ among the sites, turnover rate of NO3$^-$ pool was also high in the upper part above 30 m as in the lower part below 15 m despite the pool size. NO3$^-$ immobilization is important process on soil N transformation in the upper part above 30 m as reported by Davidson et al. (1992).

Above 30 m, gross mineralization and NH4$^+$ immobilization were the main processes of gross N transformation (Fig. 1a). However, it was shown that mineralized NH4$^+$ was left in soil during incubation and NH4$^+$ pool increased (Fig. 1a, 2). Schimel and Firestone (1989) also showed that about 20 % of NH4$^+$ was incorporated abiotically in forest floor. The mechanism is thought to involve reaction of NH4$^+$ with activated phenol or quinone rings (Stevenson, 1982). In the upper part of the slope, organic matter has higher C/N ratio than it in the lower part and humus accumulated mor type (Fig. 1c). It is considered litter contains more phenol and lignin in the upper part than in the lower part.

The ratio of immobilized NH4$^+$ to gross nitrification was lower in the lower part below 15 m than in the upper part above 30 m (Fig. 1f). It indicated two possibilities. One possibility is that nitrification is advantage over NH4$^+$ immobilization in the lower part below 15 m, the other is more NH4$^+$ remains available for nitrification. The close values between microbial biomass C/N and ex-C/N in the lower part below 15 m indicates that it falls in the C deficiency to N of the substrate (Fig. 1d, e). As nitrification was mainly autotrophic process in acidic forest soil (Tietema et al, 1992), the result was that autotrophic nitrifier completed over heterotrophs under the C deficient condition below 15 m.

4.3 Substrate influence on gross N transformation rate

There was no significant difference in microbial biomass and microbial C/N among sites (Fig. 1e). It suggests that soil microbial flora is similar among the sites. The relatively low microbial C/N ranged from 5.1 to 7.9 might mean more dominant of bacteria over fungi, because the C/N of bacteria and fungi are represented as 5 and 15, respectively (Killham, 1994).

There were no significant differences in extractable organic C and N concentrations; considered readily decomposable substrate, among sites (Hart et al., 1994, Fig. 1d). However, gross N mineralization rates was four times faster at 0 m and 15 m than the other sites (Fig. 1a). Under the similar microbial fauna, different gross N transformation pattern indicates the difference in the substrate availability for microbes among the sites despite of the amount of extractable organic matter. The substrate quality seems to be more decomposable for microorganisms in the lower below 15 m than the upper part.

5. REFERENCES

Davidson, E. A, Hart, S. C., and Firestone, M. K. (1992): Internal cycling of nitrate in soils of a mature coniferous forest. Ecology, Vol. 73, pp. 1148-1156.

Hart, S. C., Nason, G. E., Myrold, D. D. and Perry, D. A. (1994): Dynamics of gross nitrogen transformations in an old-growth forest: the carbon connection. Ecology, Vol. 75, pp. 880-891.

Hirobe, M., Tokuchi, N. and Iwatsubo, G. (1998a): Spatial variability of soil N transformations along a slope. (SSSA in reviewing).

Hirobe, M., Koba, K., Tokuchi, N. and Iwatsubo, G. (1998b): Topographic differences of soil N transformations along a slope. (in this issue).

Jackson, L. E., Schimel, J. P. and Firestone, M. K. (1989): Short-term partitioning of ammonium and nitrate between plants and microbes in an annual grassland. Soil Biology and Biochemistry, Vol. 21, pp. 409-415.

Jenkinson, D. S. and Powlson, D. S. (1976): The effects of biocidal treatments on metabolism. V A method for measuring soil biomass. Soil Biology and Biochemistry, Vol. 8, pp. 209-231.

Keeney, D. R. and Nelson, D. W. (1982): Nitrogen-inorganic forms. pp. 643-698. In Page, A. L. et al. (ed.) Methods of soil analysis. Part 2. American Society of Agronomy, Inc., Soil Sci. Soc. Amer. Inc. 1159 pp. Madison.

Killham, K. (1994): Soil ecology. Cambridge University Press. 242 p. Cambridge.

Nishio, T., Kanamori, T. and Fujimoto, T. (1985): Nitrogen transformations in an aerobic soil as determined by a $^{15}NH_4^+$ dilution technique. Soil Biology & Biochemistry, Vol. 17, pp. 149-154.

Schimel, D. S. (1988): Calculation of microbial growth efficiency from ^{15}N immobilization. Biogeochemistry, Vol. 6, pp. 239-342.

Schimel, D. S. and Firestone, M. K. (1989): Inorganic N incorporation by coniferous forest floor material. Soil Biology Biochemistry, Vol. 21, pp. 41-46.

Stevenson, F. J. (1982): Humus Chemistry. Wiley, New York.

Tietema, A. and Wessel, W. W. (1992): Gross nitrogen transformations in the organic layer of acid forest ecosystems subjected to increased atmospheric nitrogen input. Soil Biology Biochemistry, Vol. 24, pp. 943-950.

Tietema, A., De Boer, W., Riemer, L. and Verstraten, J. M. (1992): Nitrate production in nitrogen saturated acid forest soil: vertical distribution and characteristics. Soil Biology Biochemistry, Vol. 24, pp. 235-240.

Tokuchi, N., Takeda, H. and Iwatsubo, G. (1993): Vertical changes in soil solution chemistry in soil profiles under coniferous forest. Geoderma, Vol. 59, pp. 57-73.

Wessel, W. W. and Tietema, A. (1992): Calculation gross N transformation rates of ^{15}N pool dilution experiments with acid forest litter: Analytical and numerical approaches. Soil Biology Biochemistry, Vol. 24, pp. 931-942.

Topographic Differences in Soil N Transformation Patterns Along a Forest Slope

Muneto HIROBE[*1], Naoko TOKUCHI[*1] and Goro IWATSUBO[*2]

*1 Laboratory of Forest Ecology, Graduate School of Agriculture
   Kyoto University, Kyoto 606-8502, Japan
*2 Faculty of Agriculture, Kinki University, Nara, Japan

ABSTRACT The soil N transformation pattern depended on the topographic positions along a forest slope. Nitrate was the dominant inorganic N in the net mineralized N during incubation of the soil of the lower of the slope, and ammonium was the dominant in the soil of the upper of the slope. The middle of the slope was the boundary of the lower and upper soils. In the boundary, soil N transformation patterns seemed to depend on the local environmental conditions. Thus, the relationships between soil N transformation and soil properties were examined by correlation analysis. Net N mineralization rate highly correlated with total C, C/N ratio and pH, while net nitrification highly correlated with water content and 2M KCl extractable organic C and N.

Keywords: N mineralization, nitrification, spatial variability, slope.

1. INTRODUCTION

Nitrogen (N) cycle in forest ecosystems is dominated by the internal cycle between plant and soil. Plants take up N mainly in simple inorganic forms rather than organic N (Pastor et al. 1984). Organic N is the most abundant form of N in soil, while inorganic N concentration is generally low (Binkley and Vitousek 1989). Thus, N availability often limits plant growth in forest ecosystems (Vitousek et al. 1982). The major forms of inorganic N in soil are NH_4^+-N and NO_3^--N, and they are produced through N mineralization process. The difference in the forms of inorganic N will affect N cycling in forest ecosystems because of the relatively high availability of mobile NO_3^--N compared to NH_4^+-N (Binkley and Vitousek 1989). Therefore, it is important to investigate patterns and controlling factors of soil N transformations.

Soil N transformations show substantial variability at both global and small spatial scales (Vitousek et al. 1982, Robertson et al. 1988), and slope position is known as one of the topographic gradients affecting soil N transformations (Zak et al. 1991, Garten et al. 1994).

In this study, the spatial variability of soil N transformations due to a topographic gradient was investigated along a short steep slope in a plantation forest. Some soil properties were also studied to evaluate the factors affecting soil N transformation.

2. EXPERIMENTAL DESIGN

The study site was chosen on a south facing slope of a 45-year old *Cryptomeria japonica* D. Don. plantation. The slope length was 135 m, and ranging from 765 m to 851 m in elevation. A total of 10 sampling plots were established at 15 m intervals along a slope. A transect of 5 m wide and 135 m long was also placed along the slope starting from the bottom of the slope. The distance used in the succeeding section of this paper represents the distance from the bottom of the transect.

Triplicated soil samples were collected from 0-5 cm layer of mineral soil at the 10 sampling plots on 13 Oct. 1994. As mentioned later, soil N transformations changed drastically between 60 m and 75 m. Thus, to investigate the spatial variability of soil N transformations in these zones, additional soil samples were collected from 0-5 cm layer of mineral soil at every 3 m point from 45 to 90 m on the transect on 24 Aug. 1995. At each point, 6 soil samples were collected from every 1 m across the transect. Totally, ninety-six soil samples were collected on additional soil sampling.

Soil samples were sieved through 2 mm mesh and roots were removed by hand. Soil samples from 10 sampling plots were used for the measurement of soil N transformation. Each soil sample from 96 points was divided into two subsamples. One subsample was used for the measurement of soil N transformation, 2M KCl extractable organic C and N. The other subsample was air dried and was used for the measurement of total C and N, and pH.

Soil N transformation was measured by 28-d laboratory incubation at 25°C in the dark with moisture held constant at 60% of field capacity. Total C and total N of soil were measured using a CHN analyzer (Yanako, MT-3, Kyoto, Japan). Soil pH was measured by grass electrode in a 1:2.5 soil-water suspension. 2M KCl extractable organic C in the soil extracts was determined using a organic C analyzer (Shimadzu, TOC5000, Kyoto, Japan). 2M KCl extractable organic N was calculated as the difference between extractable total N and inorganic N in the soil extracts after determination of extractable total N in the soil extracts using a total N analyzer (Sumitomo Chemical Co., N-200, Osaka, Japan).

3. RESULTS AND DISCUSSION

3.1 Spatial variability of soil N transformations

The soil N transformation pattern depended on the topographic positions along a slope (Fig. 1, 2). Nitrate was the dominant inorganic N in the net mineralized N during incubation of the soil of the lower of the slope, and ammonium was the dominant in the soil of the upper of the slope (Fig. 1, 2). The middle of the slope was the boundary of the lower and the upper soils (Fig. 1, 2). The patterns of soil N transformations along the slope were similar in both sampling time (Fig. 1, 2). Similar patterns of N transformations on a slope have been reported by discontinuous sampling on different slope positions in a temperate secondary forest (Yoshida et al. 1980), and in a *Cryptomeria japonica* D. Don. and *Chamaecyparis obtusa* Endl. plantation forests in Japan (Takeda 1994, Toda and Haibara 1994), and also in a southern Appalachian oak forest (Garten 1993).

To evaluate the qualitative difference of soil N transformation pattern more clearly, percent nitrification was calculated by following equation (1).

percent nitrification = (nitrification / mineralization) * 100   (1)

Negative values of mineralization and nitrification were treated as zero in the calculation. Percent nitrification changed drastically in the middle of the slope, and the variations of percent nitrification were higher in the middle of the slope than those of the lower and the upper of the slope (Fig. 3). The variations of percent nitrification were also high at 78 m and 87 m where net nitrification were low (Fig. 3). From these results, it was suggested that the soil in the middle of the slope was not the intermediate characteristics between the lower and the upper soils, but the middle of the slope was the transition zone where the patches of the lower and the upper soils mixed depending on the local environmental conditions.

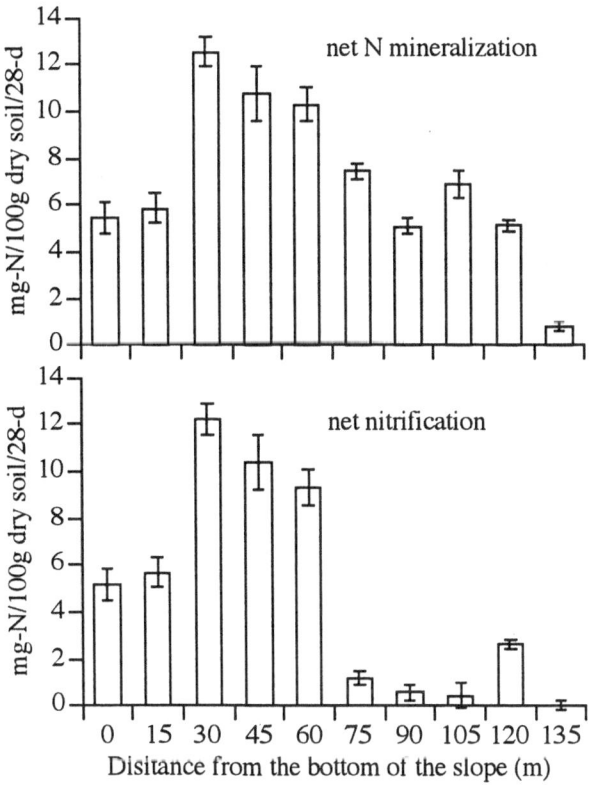

Fig. 1. Net N mineralization rate (above) and nitrification rate (below) of the surface soil at 10 plots along the slope (mean ± SE, n = 3).

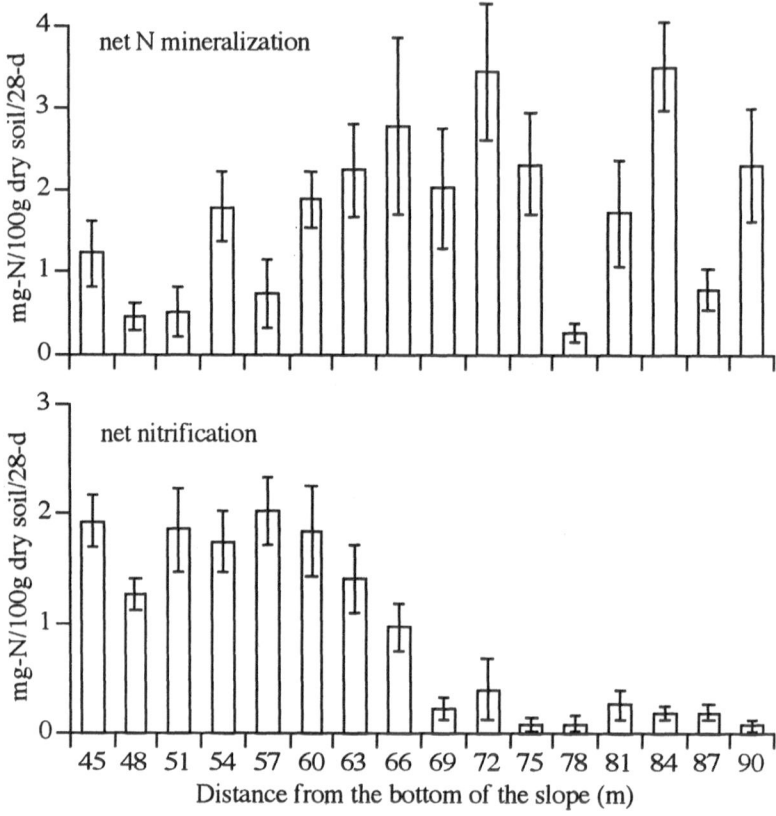

Fig. 2. Net N mineralization rate (above) and nitrification rate (below) of the surface soil at 96 points from 45 m to 90 m on the transect (mean ± SE, n = 6).

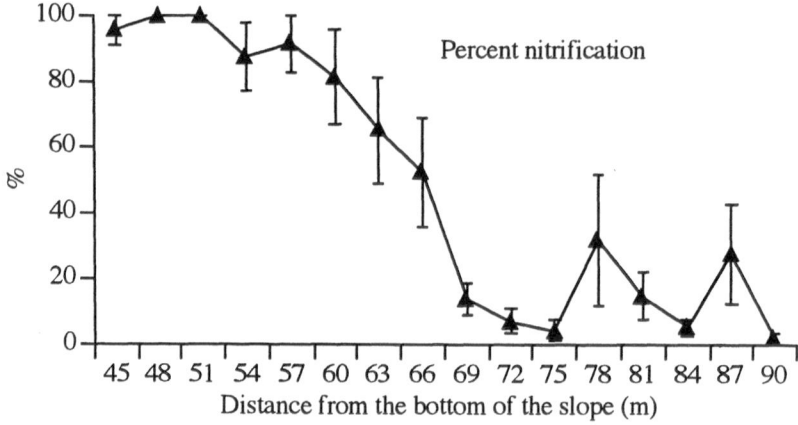

Fig. 3. Percent nitrification of the surface soil at 96 points from 45 m to 90 m on the transect (mean ± SE, n = 6).

3.2 Factors affecting soil N transformations

A large number of soil properties have been mentioned as factors on soil N transformations (Vitousek et al. 1982, White and Gosz 1987). The factors affecting soil N transformation can be divided into two groups; 1) environmental factors, and 2) resource factors (Killham 1994). In this study, following controlling factors were investigated about the soil samples collected from 96 points between 45 m and 90 m of the slope; 1) environmental factors such as water content and pH (Table 1), and 2) resource factors such as total C and N, C/N ratio, and 2M KCl extractable organic C and N (Table 1). 2M KCl extractable organic C and N are reported as the indices of the quantity of readily decomposable organic matter for microbial populations (Hart et al. 1994).

The relationships between soil N transformation and controlling factors were examined by a correlation analysis. Net N mineralization rate showed high positive correlation values with total C and C/N ratio ($r = 0.534$ and 0.655, respectively, $n = 96$, $P < 0.01$), and a high negative correlation value with pH ($r = -0.459$). Whereas, net nitrification showed high correlation values with water content and 2M KCl extractable organic C and N ($r = 0.511$, 0.518, and 0.517, respectively, $n = 96$, $P < 0.01$). These correlation analysis suggested that net N mineralization was strongly affected by the quantity and the quality of total organic matter, while net nitrification was strongly affected by water environment and the quantity of readily decomposable organic matter. Percent nitrification showed high correlation values with water content and C/N ratio ($r = 0.497$ and -0.481, respectively), and it was suggested that the qualitative difference of soil N transformation pattern along a slope was strongly affected both by the environmental factor and the resource factor.

ACKNOWLEDGEMENTS

We thank Drs. H. Takeda, H. Kawaguchi, and M.J. Mitchell for their valuable comments on the manuscript; Mr. T. Yamashita, Mr. N. Kasuya and all the members of the Laboratory of Forest Ecology, Kyoto University, for their help in the field sampling and the laboratory analysis; Dr. K. Nishimura for helpful advice on laboratory analysis; Dr. K. Yokota and Ms. Y. Kaneko for help in the laboratory analysis. This study was financially supported partly by a Grant from the Ministry of Education, Culture and Sports, Japan relating to JSPS Fellowship for Japanese Junior Scientists (No.2714).

REFERENCES

Binkley, D. and P.M. Vitousek (1989) : Soil nutrient availability, pp. 75-96. *In* Ed. R. Pearcy, H. Moony, J. Ehleringer and P. Rundel, Physiological Plant Ecology, Chapman & Hall, London, UK.

Table 1. Total C and N (%), C/N ratio, pH (H2O), water content (g g soil^{-1}), and 2M KCl extractable organic C and N (mg 100 g dry soil^{-1}) of surface soil sampled at 96 points from 45 m to 90 m on the transect. Distance show the distance from the bottom of the slope and values show mean (± SE, n = 6) at each point.

Distance	Total C	Total N	C/N ratio	pH (H2O)	Water Content	Ex.-O.-C	Ex.-O.-N
45	9.94 (0.45)	0.64 (0.02)	15.5 (0.29)	4.18 (0.04)	0.79 (0.05)	53.4 (2.19)	10.51 (0.42)
48	8.32 (0.45)	0.56 (0.02)	14.8 (0.27)	4.14 (0.05)	0.66 (0.03)	50.0 (2.16)	9.92 (0.49)
51	9.08 (0.69)	0.60 (0.04)	15.1 (0.28)	4.24 (0.05)	0.59 (0.04)	47.7 (2.84)	9.41 (0.39)
54	8.88 (0.61)	0.59 (0.04)	15.1 (0.26)	4.24 (0.05)	0.67 (0.08)	49.9 (1.88)	9.52 (0.39)
57	9.83 (0.46)	0.61 (0.02)	16.2 (0.42)	4.19 (0.03)	0.68 (0.03)	50.7 (0.91)	9.81 (0.94)
60	9.69 (0.59)	0.63 (0.03)	15.3 (0.26)	4.37 (0.03)	0.58 (0.04)	44.5 (1.61)	8.86 (0.38)
63	9.74 (0.57)	0.65 (0.03)	15.1 (0.32)	4.48 (0.05)	0.65 (0.02)	48.2 (2.47)	8.89 (0.37)
66	11.14 (0.84)	0.69 (0.05)	16.3 (0.47)	4.25 (0.06)	0.77 (0.03)	52.5 (1.80)	10.64 (0.54)
69	10.08 (0.60)	0.61 (0.04)	16.7 (0.40)	4.23 (0.07)	0.54 (0.04)	42.9 (1.02)	8.73 (0.35)
72	10.46 (0.56)	0.66 (0.03)	15.8 (0.37)	4.26 (0.04)	0.45 (0.02)	39.6 (1.12)	7.94 (0.37)
75	10.19 (0.84)	0.62 (0.05)	16.4 (0.36)	4.20 (0.04)	0.38 (0.03)	37.7 (2.65)	7.28 (0.30)
78	7.97 (0.32)	0.49 (0.02)	16.3 (0.21)	4.31 (0.03)	0.50 (0.03)	36.5 (1.29)	6.65 (0.27)
81	9.99 (0.45)	0.59 (0.04)	17.0 (0.27)	4.08 (0.09)	0.49 (0.05)	42.8 (2.53)	8.56 (0.45)
84	12.62 (1.38)	0.75 (0.08)	16.8 (0.18)	4.05 (0.08)	0.51 (0.06)	40.5 (3.96)	8.80 (0.89)
87	8.42 (0.63)	0.48 (0.04)	17.5 (0.30)	4.10 (0.07)	0.53 (0.03)	39.0 (2.91)	7.47 (0.45)
90	13.57 (0.75)	0.66 (0.03)	20.4 (0.50)	3.82 (0.05)	0.59 (0.07)	48.7 (4.05)	9.47 (0.39)

Garten Jr., C.T. (1993) : Variation in foliar 15N abundance and the availability of soil nitrogen on Walker Branch Watershed, Ecology, Vol. 74, pp. 2098-2113.

Garten Jr., C.T., M.A. Huston and C.A. Thoms (1994) : Topographic variation of soil dynamics at Walker Branch Watershed, Tennessee, Forest Science, Vol. 40, pp. 497-512.

Hart, S.C., G.E. Nason, D.D. Myrold and D.A. Perry (1994) : Dynamics of gross nitrogen transformations in an old-growth forest: the carbon connection, Ecology, Vol. 75, pp. 880-891.

Killham, K. (1994) : Soil Ecology, Cambridge University Press, Cambridge, UK. 242pp.

Pastor, J., J.D. Aber, C.A. McClaugherty and J.M. Melillo (1984) : Aboveground production and N and P cycling along a nitrogen mineralization gradient on Blackhawk Island, Wisconsin, Ecology, Vol. 65, pp. 256-268.

Robertson, G.P., M.A. Huston, F.C. Evans and J.M. Tiedje (1988) : Spatial variability in a successional plant community: Patterns of nitrogen availability, Ecology, Vol. 69, pp. 1517-1524.

Takeda, H. (1994) : Interactions between plant and decomposer populations in forest ecosystems - A mechanism of biodiversity maintenance -, Japanese Journal of Ecology, Vol. 44, pp. 211-222 (In Japanese).

Toda, H. and K. Haibara (1994) : Kinetics of mineralization of nitrogen in forest soils (I) Characteristics of soil nitrogen mineralization of different aged stands, slope positions, and soil depth, Journal of Japanese Forest Society, Vol. 76, pp. 144-151 (In Japanese with English summary).

Vitousek, P.M., J.R. Gosz, C.C. Grier, J.M. Melillo and W.A. Reiners (1982) : A comparative analysis of potential nitrification and nitrate mobility in forest ecosystems, Ecological Monograph, Vol. 52, pp. 155-177.

White, C.S., and J.R. Gosz (1987) : Factors controlling nitrogen mineralization and nitrification in forest ecosystems in New Mexico, Biology and Fertility of Soils, Vol. 5, pp. 195-202.

Yoshida, S., Y. Haruta and I. Nioh (1980) : Nitrogen dynamics in forest soils (II) Mineralization and nitrifying activity of nitrogen in the different soil types of brown forest under natural vegetation, Journal of Japanese Forest Society, Vol. 62, pp. 230-233 (In Japanese with English summary).

Zak, D.R., A. Hairston and D.F. Grigal (1991) : Topographic influence on nitrogen cycling within an Upland Pin Oak ecosystem, Forest Science, Vol. 37, pp. 45-53.

Dynamics of Soil Microbial Activities in Different Vegetation Types of the Seasonally Dry Tropics

Pitayakon LIMTONG*[1], Kazuhiro ISHIZUKA*[2],
Masamichi TAKAHASHI*[2], Vanlada SUNANTAPONGSUK*[1]
and Prasode TUMMAKATE*[1]

*1 Land Development Department, Ministry of Agriculture and Coorporatives,
   Bangkok, Thailand
*2 Forestry and Forest Products Research Institute, Tsukuba, Ibaraki
   305-8687, Japan

ABSTRACT   Seasonal changes in soil microbial populations and enzymatic activities were compared among a natural forest, a grassland, a young teak plantation, and an old teak plantation in western Thailand. Total and cellulolytic populations of bacteria and actinomycetes tended to increase in the rainy season. Total fungal populations tended to increase in the dry season, while cellulolytic fungal populations increased in the rainy season. For all plots cellulase activities were high from late dry season to the middle of the rainy season, but decreased from late rainy season to the early dry season. Seasonal changes in xylanase activities were inversely related to those of cellulase. Phosphatase activities varied with a relatively small range. Cellulase activities were low in grassland.
Keywords: soil microbial population, soil enzymes, mixed deciduous forest, teak plantation, grassland

1. INTRODUCTION

In tropical areas forests devastated by human activities is the main cause of declining forest areas. Large and intensive disturbances of forests create open canopy forests and grasslands (Brown 1993). On the other hand, trials of afforestation also have been done on the other types of lands (Dixon et al. 1994). Vegetation changes from forest to grassland or plantation cause differences in subterranean microbial and enzymological activities, which may alter nutrient cycling and soil organic dynamics (Srivastava and Singh 1991).

In this study we aimed to evaluate the changes in soil microbial populations and enzymatic activities in different vegetation types in the seasonally dry tropics. The land uses of natural forest, deforested grassland, young teak plantation, and old teak plantation were investigated in the same watershed area of western Thailand.

2. MATERIALS AND METHODS

2.1 Study site.

The study was done at the Mae Klong Watershed Research Station (Royal Forest Department), Thong Pha Phum, Kanchanaburi Province, in western Thailand (14°30'-14°45' N; 98°45'-99°00' E). The climate is tropical monsoon with a rainy (May – September) and a dry season (November – March). April and October usually comprise transition periods between them. The mean annual temperature is 26.8°C with the maximum in April (30.1°C) and the minimum in December (23.5°C). The annual rainfall is about 1,600 mm with a maximum in August (311 mm) and a minimum in January (0.1mm) (Suksawang and Seangkoovong 1991).

Four plots were established in a natural forest (NF), a grassland (GR) where the forest used to be devastated by shifting cultivation and now abandoned, a 2 year-old young teak plantation (TY, planted in 1992), and a 15 year-old teak plantation (TO, planted in 1977). The elevation of the plots ranges from 100 to 300m m.s.l. All the plots were located within 4 km of each other.

The vegetation of the natural forest is mixed deciduous forest (MDF type). The dominant species are *Shorea siamensis*, *Berrya ammonilla*, *Dillenia parviflora*, and *Xylia xylocarpa* var. *keruii* with the ground vegetation bamboo. The vegetation in the grassland, where we sampled, is mainly *Arundo donax* and *Imperata cylindrica*. The 15 year-old plantation has 70% teak (*Tectona grandis*) and 30% *Gmerina arborea* in terms of number of trees. Upland rice was cultivated with teaks for two years in the young plantation.

The soil in this area is classified into Ultisols, mainly derived from gneiss. Table 1 summarizes the chemical properties of the top 10cm of the soils (Limtong et al. 1995). The topsoils in the plots show a relatively similar chemical condition.

Table 1. Chemical characteristics of topsoils (0 – 10 cm)

Plot	pH	O.M.%	Total-N%	Exchangeable cations (cmol$_c$/kg)				CEC (cmol$_c$/kg)	Available P (mg/kg)
				K	Na	Mg	Ca		
NF	6.50	4.12	0.16	0.68	0.37	0.47	10.08	14.0	10.2
GR	5.75	4.13	0.21	0.71	0.36	0.45	5.56	12.3	10.7
TY	5.73	5.01	0.23	0.85	0.38	0.65	9.44	17.1	23.7
TO	5.97	4.23	0.19	0.57	0.33	0.75	9.04	16.9	8.1

2.2 Determination of soil biological activities

Soil samples of all four experimental plots were collected from the soil surface (0 - 15 cm in depth) for ten samples/plot in the first year (Dec. 1992 – Sep. 1993) and for five samples/plot in the second year (Dec. 1993 – Sep. 1994). Sampling was done every three months. Determination of the population of total number of bacteria, actinomycetes and fungi were done by a standard dilution plate count by the method of Murao et al. (1979), and the number of cellulolytic microorganisms were counted using cellulose powder as the carbon source in the media (Limtong et. al. 1990)

Soil samples in the first year were extracted by distilled water in the ratio of soil : water 1:10, and were centrifuged to separate the soil particles. Crude enzymes in the supernatant solution, as an estimation of the soil enzymatic activities, were phosphatase estimated by the method of Tabatabai and Bremner (1969), cellulase estimated by the method of Mandels and Sternberg (1976), and xylanase estimated by the method of Kitpreecharvanich et al. (1984).

All results are expressed as oven dry soil (105°C, 24hrs).

3. RESULT

Fig. 1 shows the water contents of soil samples. The samples on March 1993, usually in the dry season, were relatively moist because rain fell before sampling. Fig. 2 shows that the total population of bacteria, actinomycetes, and fungi fluctuated through the years we observed them. The populations of bacteria and actinomycetes fluctuated with similar patterns, but actinomycetes had wider ranges of population changes. The populations in the young teak plantation showed relatively narrow ranges. The fungal population was smaller than the bacterial and actinomycetes populations. No clear differences occurred in the microbial populations among different vegetation types. Although the seasonal changes in the microbial populations are not so apparent, the populations of actinomycetes and bacteria tended to increase from late dry season (March 1993), as typically found in the old teak plantation and the grassland plots. In contrast, fungal populations tended to increase in the dry season. The differences in these total and cellulolytic microbial populations among the plots were not so apparent.

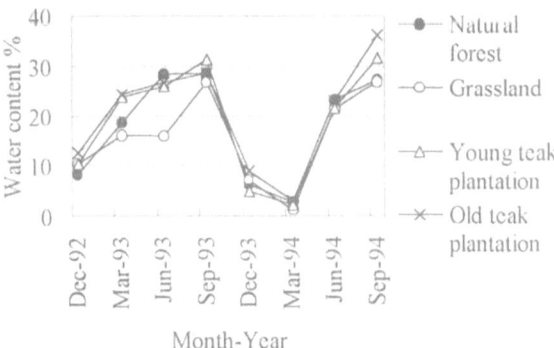

Fig. 1. Changes in water contents of surface soils in each plot.

The cellulolytic microbial populations were similar to the total populations for bacteria and actinomycetes (Fig. 3). However, the peaks of cellulolytic fungal population were somewhat different from those of the total population. For example, samples in December 1993 had large total fungal populations but had

256

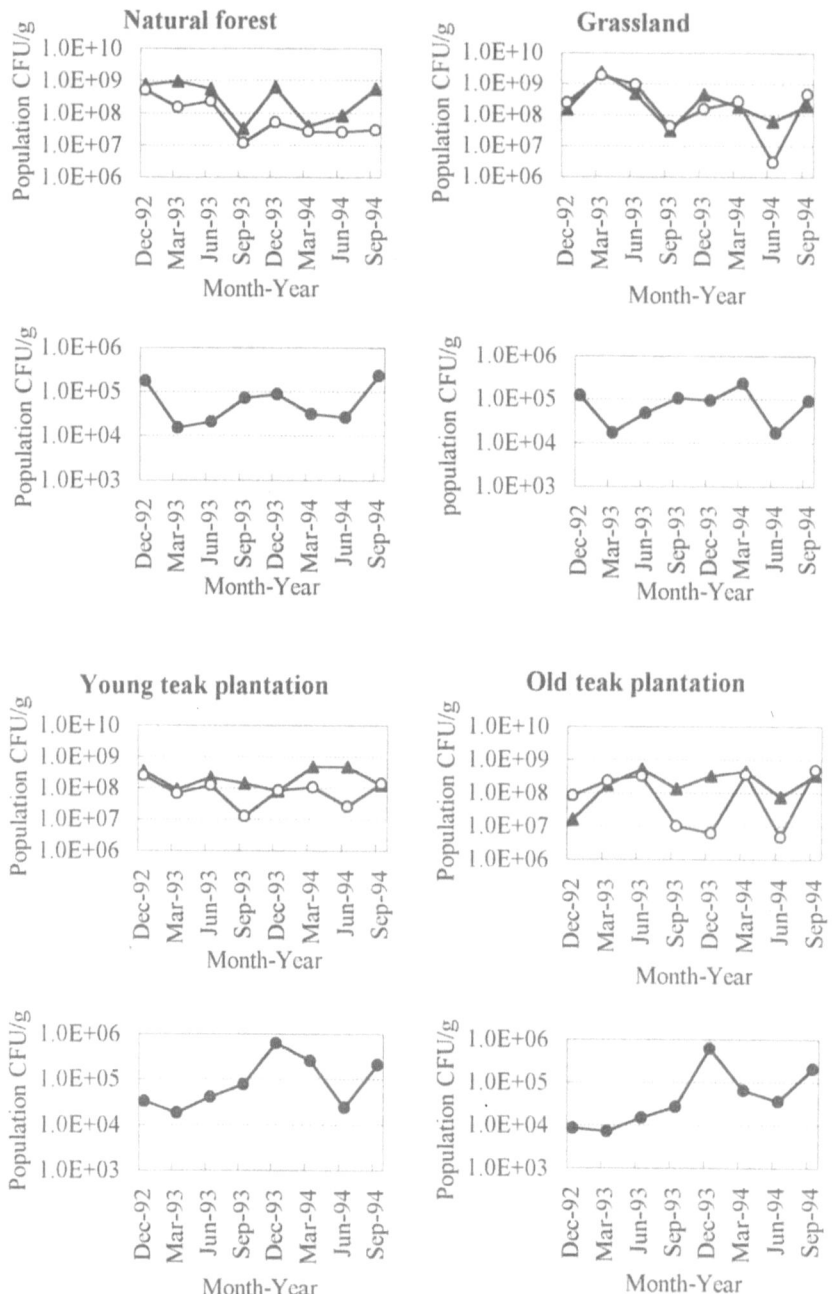

Fig. 2. The changes in total populations of bacteria, actinomycetes, and fungi in the surface soil of the natural forest. The triangles and circles in the upper figures are bacteria and actinomycetes, respectively. The lower figures show fungal populations.

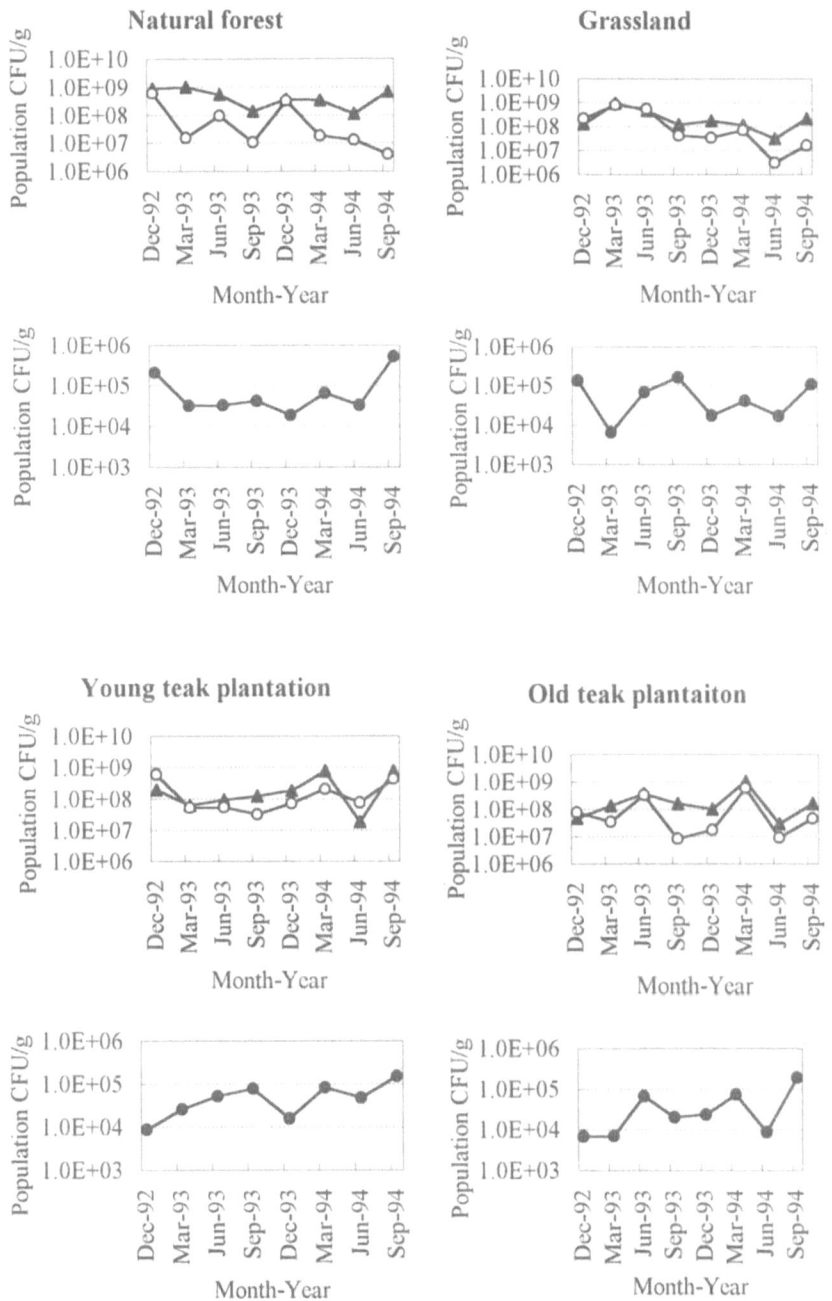

Fig. 3. The changes in celluloytic populations of bacteria, actinomycetes, and fungi in the surface soil of the natural forest. The triangles and circles in the upper figures are bacteria and actinomycetes, respectively. The lower figures show fungal populations.

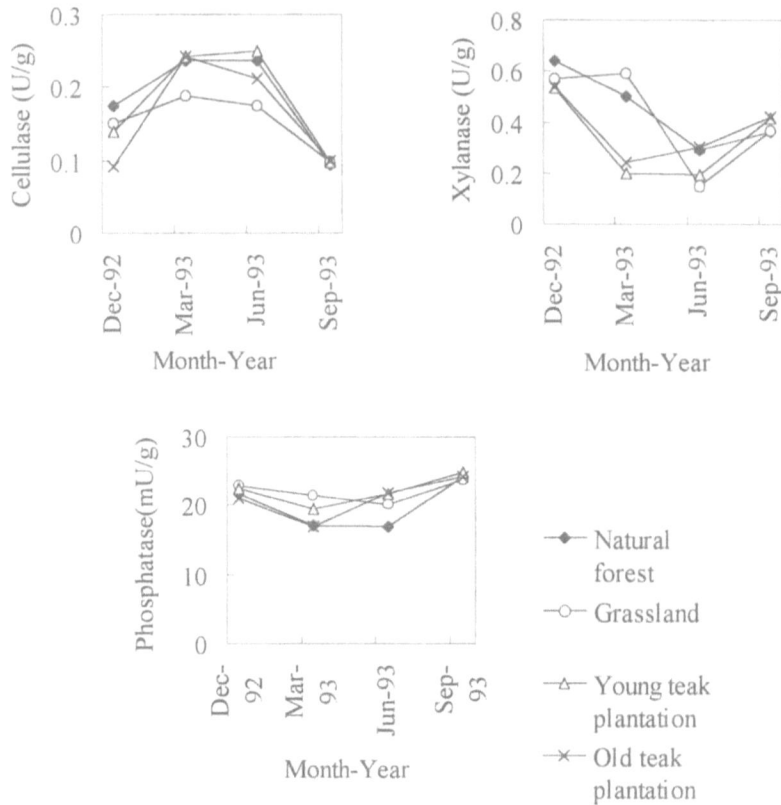

Fig. 4. Changes in enzymatic activities in each plot.

small cellulolytic fungal population for most plots. Cellulolytic fungal population tended to increase in the rainy season.

Enzymatic activities had similar trends among the plots (Fig. 4). Cellulase activities increased in March and they were still high in June. In the late stage of the rainy season (in September) the cellulase activities decreased. However, xylanase showed higher activities from the late stage of the rainy season to the early dry season, while in the middle of the rainy season, (June 1993), xylanase showed lowest activities. Phosphatase activities showed small fluctuation throughout year. When the plots were compared, cellulase in the grassland showed lowest activities in the rainy season.

4. DISCUSSION

The changes in vegetation may not have influenced the overall microbial populations pattern in the study area. The microbial populations fluctuated through the years, but the seasonal changes were not so clear. We made observation for two years, but the trends were sometimes different between the first and second years, such that samples in March 1992 were quite different from those in March 1993. One reason for this discrepancy may be the moisture conditions in the soil. Total populations of bacteria, actinomycetes, and cellulolytic fungi seem to increase in the rainy season, and the total fungal populations seem to increase in the dry season, although this is not clear. Because most fungi can sporulate for survival in unsuitable condition, increases in total fungal populations may result from fungal spore production in the dry season.

The vegetation changes in these study plots were relatively mild in comparison to a complete conversion to cultivated land. Duxbury and Tate (1981) showed several enzymes activities varied as much as 50-fold between noncultivated and cultivated soil with sugarcane or paragrass in the USA. In Indonesia, primary forests have the highest enzymatic activities, followed by secondary forests, and cultivated land showed very low activities (Salam et al. 1998). In this study only the grassland plot showed lower cellulase activities, even in the rainy season. Unless the land is used intensively as shifting or permanent cultivated land, enzymatic activities may not decrease, even though natural forest is converted to plantation forest.

The cellulase activity decreased in September, although cellulolytic microbial populations were still high in the rainy season. A small amount of litter on the soil before the dry season may result in lower cellulase activities. Cellulase is usually correlated with fungal activities (Hayano 1986, Rhee et al. 1987), but no apparent correlation was found between cellulolytic microbial populations and cellulase activities in our study sites.

REFERENCES

Brown, S. (1993): Tropical forests and the global carbon cycles: the need for sustainable land-use patterns, Agriculture, Ecosystems, and Environment, Vol. 46, pp.31-44.

Dixon, R. K., Brown, S., Houghton, R. A., Solomon, A. M., Trexler, A. M., Wisniewski, J. (1994): Carbon pools and flux of global forest ecosystem, Science, Vol.263, pp.185-190.

Duxbury, J. M. and R. L. Tate III (1981): The effect of soil depth and crop cover on enzymatic activities in Pahokee muck, Soil Science Society of American Journal, Vol. 45, pp.322-328.

Hayano, K. (1986): Cellulase complex in a tomato field soil: induction, localization and some properties, Soil Biology and Biochemistry, Vol.18, pp.215-219.

Kitpreecharvanich, V., M. Hayashi and S. Nagai (1994): Production of xylan degrading enzymes by thermophilic fungi, *Aspergillus fumigatus* and *Humicola lanuginosa*, Journal of Fermentation Technology, Vol 62, No 1, pp 63-69.

Limtong, P., S.Vangnai, V. Sunantapongsuk and S. Piriyaprin. (1990): Isolation and selection of thermophilic cellulolytic microorganisms for compost production in Thailand. The Kasetsart Journal. Vol. 24, No.1, pp.108-115.

Limtong, P., V. Sunantapongsuk, K. Ishizuka, P. Tummakate, A. Chalermpongse, K. Yantasath and S. Poonsawat (1995): The relationship of soil microorganisms and their activity with soil organic matter dynamics in different types of the tropical forest, *In* proceedings of the international workshop on " the changes of tropical forest ecosystems by El Niño and others", STA, NRCT and JISTEC, Kanchanaburi, Thailand, pp.151-169.

Mandels, M. and D. Sternberg. (1976): Recent advances in cellulase technology. Journal of Fermentation Technology, Vol. 54, No.4, pp.267-286.

Murao, S., J. Kanamoto and M. Arai. (1979): Isolation and identification of a cellulolytic enzyme producing microorganisms. Journal Fermentation Technology, Vol.57, No.3, pp.151-156.

Rhee, Y. H., Y. C. Hah and S. W. Hong (1987): Relative contributions of fungi and bacteria to soil carboxymethylcellulase activity, Soil Biology and Biochemistry, Vol. 19, pp.479-481.

Salam, A. K., A. Katayama, and M. Kimura (1998): Activities of some soil enzymes in different land use systems after deforestation in hilly areas of west Lampung, south Sumatra, Indonesia, Soil Science and Plant Nutrition, Vol. 44, pp.93-103.

Srivastava, S. C. and Singh, J. S. (1991): Microbial C, N and P in dry tropical forest soils: Effect of alternate land-uses and nutrient flux, Soil Biology and Biochemistry, Vol.23, pp.117-124.

Suksawang, S. and P. Seangkoovong (1991): Impact of selective logging system on water yield of a small tropical watershed in Thailand, Mae Klong Watershed Research Station, Research Note, Vol.20.

Tabatabai, M.A. and J.M. Bremner. (1969): Use of p-nitro-phenyl phosphate for assay of soil phosphatase activity. Soil Biology Biochemistry Vol. 1, pp.301-307.

The Role of Soil Microbial Biomass in Burned Japanese Red Pine Forest

Takahiro TATEISHI

Center for Ecological Research, Kyoto University, Kyoto 606-8502, Japan

ABSTRACT   This paper presents a review of the results of the estimation of microbial biomass in the soils of burned and unburned Japanese red pine forests, and discusses the effects of fire on soil microbial biomass and the role of microbial biomass in the early stages of the secondary succession after fires. Microbial biomass in the mineral soil layer at burned sites with different years after fires were similar to that at the unburned site.   This indicates fire did not affect the size of the microbial biomass in the mineral soil layer quantitatively. From the results that microbial biomass in burned soils contained a substantial amount of nutrients, it follows that soil microbial biomass can act as a nutrient pool available for plants which grow in the secondary succession after fires.
Keywords: Soil microbial biomass, forest fires, *Pinus densiflora*

1. INTRODUCTION

Forest fire is one of the actions disturbing forest ecosystems and affects biota along with the carbon and nutrient dynamics established in forest ecosystems. Many large forest fires have occurred in the *Pinus densiflora* forests along the coast of the Seto Inland Sea, western Japan.   In this area the changes of biota, and the water and nutrient salt cycling have been investigated after forest fires (Nakagoshi et al., 1987), however little information on microorganisms at burned forests was obtained (Horikoshi et al., 1986).   Microorganisms in soils play an important role in the decomposition of organic matter and mineralization of nutrients.   Microbial biomass has been defined as the living part of the soil organic matter which contains living microorganisms.   The roles of soil microorganisms described above might depend on the size of the soil microbial biomass.   Hence the estimation of microbial biomass in soils is the first step for further understanding of the role of microorganisms in decomposer systems.

The results of estimation of the soil microbial biomass carried out in burned Japanese red pine forests, the effects of fire on soil microbial biomass, and the role of microbial biomass in the early stages of the secondary succession after fires are presented in the following sections.

2. STUDY SITES

The study sites were six burned Japanese red pine forests according to the time in years after the fire and one unburned red pine forest (Table 1). The study sites were situated in the south of Hiroshima Prefecture, western Japan. This region has a warm-temperate monsoon climate. The average annual mean temperature and annual precipitation during the 10-year period from 1980 to 1989 in this area were 15.5℃ and 1,442 mm, respectively. The plant community in the forest vegetation along the coast of the Seto Inland Sea in Hiroshima Prefecture belongs to the Rhododendro-reticulati-Pinetum densiflorae H. Suzuki et Toyohara 1971 (Nakagoshi et al., 1987), and the pine forests before fires had the same association. In the unburned site at Ato, the dominant tree species is *Pinus densiflora* and the age of the forest about 30 years. In the burned sites, the dominant shrubs were *Quercus serrata* Murr. and *Rhododendron reticulatum* D. Don, while the dominant herbs were *Miscanthus sinensis* Andress. and *Dicranopteris dichotoma* Bernh. Further information about soil characteristics and vegetation at the burned sites have been described in previous papers (Tateishi et al., 1989b; Tateishi and Horikoshi, 1995).

Table 1   Description of study sites.

Site	Burning date	Texture	Thickness of soil layer (cm)			pH	Total C (%)	Total N (%)
			A_0	A	B			
Burned sites								
Hinourasan	March 90	S	1.7	1.6	3.0	4.9	1.1	0.05
Nagatani	Dec. 88	LS	0.2	0.3	0.7	4.4	0.7	0.03
Nigata	April 86	LS	11.0	2.0	3.0	4.9	2.7	0.12
Tennoh	Nov. 85	SL	0	0.5	1.0	4.4	4.6	0.21
Ato	May 84	SL	0	1.0	0.5	4.8	1.0	0.05
Nenoura	April 83	L	1.0	1.0	7.0	4.7	2.5	0.09
Unburned site								
Ato		L	5.0	1.0	5.5	4.3	2.0	0.10

3. ESTIMATION OF MICROBIAL BIOMASS IN THE SOILS OF BURNED AND UNBURNED JAPANESE RED PINE FORESTS

3.1 Method for estimating soil microbial biomass

Several methods for this estimation have been proposed. In these methods chloroform fumigation techniques were widely used in various soils with different land use. In this study we used the chloroform fumigation-incubation (CFI) method proposed by Jenkinson and Powlson (1976). This method could not always provide reliable estimates in acidic soils, soils recently treated with organic matter, and calcareous soils. All soils used in this study were acidic. In addition the soils immediately after fire were eutrophic and this condition was maintained for at least 6 months (Horikoshi et al., 1987; Horikoshi, 1989). These soils correspond to soils recently treated with organic matter. In order to obtain reliable estimates of the soil microbial biomass, we employed slightly modified CFI method. The details of the method and its validity were described in our previous papers (Tateishi et al., 1989a; Tateishi, 1995).

3.2 Microbial biomass in soils of burned and unburned pine forests and effects of fire on soil microbial biomass

Microbial biomass in the soil at burned sites according to the time in years after the fire and unburned site was determined with our modified CFI method. The microbial estimates were expressed as mg C/100 g dry soil, g C/m^2 in the upper 5 cm mineral soil, and percent of microbial-C to total C content in soils (Fig. 1. A, C, D).

The above-ground plant biomass was also shown in Fig. 1.B. The above-ground plant biomass was 0 g/m^2 immediately after fire, and then increased with months after fire and reached 381 g/m^2 53 months after fire, which was only one fifth that of unburned site.

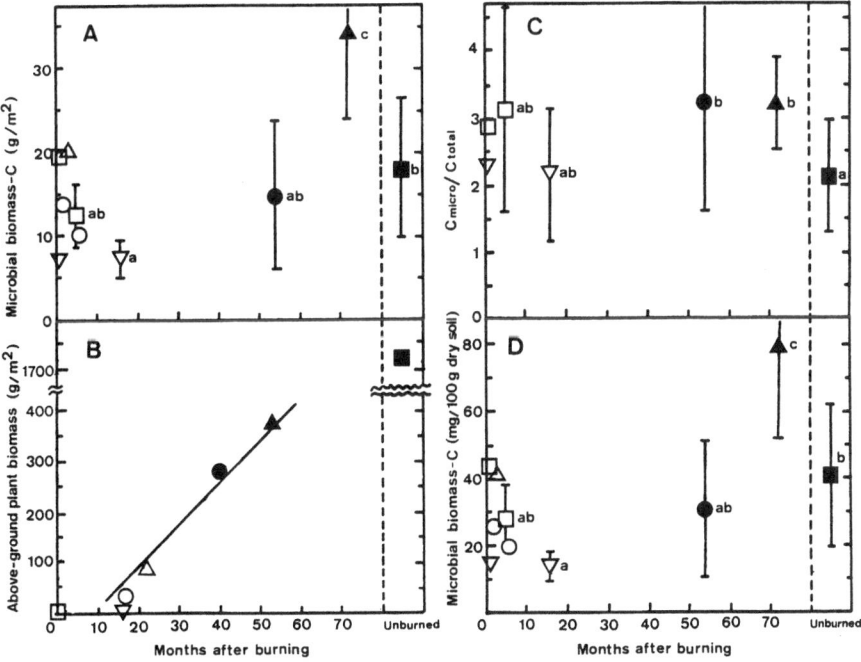

Fig. 1 Soil microbial biomass (A, D), above-ground plant biomass (B), and C_{micro}/C_{total} ratio (C) in the soils at the burned sites with different months after fire and at the unburned site. Same letters indicate the differences between values are not significant (P<0.05) by Dancun's new multiple range test. Vertical bars indicate standard deviation. Burned sites: □, Hinourasan; ▽, Nagatani; ○, Tennoh; △, Nigata; ●, Ato; ▲, Nenoura. Unburned site: ■, Ato.

Microbial biomass in the mineral soil layer at burned sites was not related to the time after fire and ranged from 14 to 79 mg/100 g dry soil, which corresponds to 7 to 34 g/m^2. Microbial biomass in the soils of burned sites was nearly the same or slightly higher than that in the unburned site, except at one burned site 16 months after fire. The phenomena observed at the study sites can be explained by the following three causes: (1) As soil temperature did not become higher during the fire, microbial death could be avoided; (2) Continuous supply of substrates for microorganisms had been retained in the soils of the burned sites; (3) The microorganisms in a Japanese red pine forest seem to be restricted by the availability of carbon, and the level of microbial biomass may not be greatly affected by fire.

The proportions of microbial biomass C to total C content in soils (C_{micro}/C_{total}) were recognized as a useful indicator of the metabolic status of a given microbial community (Coleman and Crossley Jr., 1996). The C_{micro}/C_{total} of the burned soils ranged from 2.2 to 3.3%. The ratios did not show significant differences among burned sites but were significantly higher than those of unburned site ($P<0.05$). This might be attributed to the difference in the substrate quality or the utilization efficiency of substrate by microbial communities between the burned and unburned soils.

From the above results it follows that fire did not affect the size of the soil microbial biomass at the pine forests quantitatively, but qualitatively.

4. THE ROLE OF SOIL MICROBIAL BIOMASS IN BURNED JAPANESE RED PINE FOREST

It is well recognized that microbial biomass in soils has a role of not only decomposition of organic matters but also serves as a source and a sink of nutrients available for plants. These roles of soil microbial biomass are very important in the burned sites following forest fires where above-ground plants were burned and the recycling system of nutrients ravaged.

The role of microbial biomass as a source of the nutrients was based on the results of decomposition of microbial cells after the death of microorganisms and the release of nutrients contained in them.

There are two types of nutrient release from microbial biomass. One is a temporal release of nutrients from microbial biomass when microorganisms in soils were killed by destructive actions such as heating or chemical treatments. A typical example of this is the close relationship between the decrease of microbial biomass and the increase of nitrogen mineralization in arable soils (Marumoto et al., 1982). In addition, Sakamoto et al. (1992) found that most decomposable nitrogen in soils was derived from microbial biomass nitrogen when soils were heated. This observation indicates that when microorganisms in soils were subject to destructive actions such as heating or chemical treatments a large amount of microorganisms died, microbial cells were decomposed, and the nitrogenous compounds they contained were released and mineralized in soils. Moreover, other nutrient elements might be released from microbial biomass in the same way.

The second type of nutrient release is long term, and it takes place through microbial turnover which occurs under steady state condition without soil disturbance. Turnover time and nutrient flux through soil microbial biomass have been estimated in arable and grassland soils (Jenkinson and Ladd, 1981; Okano et al., 1987). Jenkinson and Ladd (1981) determined nitrogen flux through biomass in arable soils to be 38 kg ha^{-1} yr^{-1}, which was calculated from turnover time of biomass nitrogen (2.5 yr). The nitrogen flux was of the same order as the annual uptake of nitrogen in crops.

On the other hand, the role of microbial biomass as a sink was demonstrated by estimation of the content of nutrient elements in microbial biomass. Baath and Soderstrom (1979) estimated nutrient content in fungal biomass in the soils of *Pinus sylvestris* forest. In their study the proportion of nitrogen content in fungal biomass to the total content in soils ranged from 2.2 to 19.6%. Anderson and Domsch (1980) also calculated the nutrient content in microbial biomass in arable soils from the concentrations of mineral elements in cultured microbial cells, and found that amounts of N, P, and K contained in microbial biomass were 108, 83, and 70 kg in the 0-12.5 cm mineral soil layer of arable soils, respectively. They concluded that soil microbial biomass represents a labile pool of nutrients which are available for plants.

Although the effects of fire or prescribed burning on soil microbial biomass were investigated in some burned forests (Fenn et al., 1993; Pietikainen and Fritze, 1993; Fritze et al., 1993; Hossain et al., 1995; Dumontet et al., 1996), the interaction between soil microorganisms and recovering plants in terms of nutrient pools and fluxes has not been investigated. In order to elucidate the contribution of soil microbial biomass to a nutrient pool for plants at burned and unburned sites, we compared the content of nutrient elements contained in soil microbial biomass and nutrient uptake by plants.

First, the content of microbial nutrients was calculated from microbial biomass-C estimates using the factors of conversion of microbial-C to biomass mineral content provided by Anderson and Domsch (1980). The conversion factors were modified to make them suitable for the burned soils in which the fungal to bacterial biomass ratio was different from those determined by Anderson and Domsch (1980). The conversion factors used in this study were: N:C=0.153, P:C=0.107, K:C=0.092. The contents of nutrient elements in soil microbial biomass at four burned sites and unburned site were shown in Table 2. Microbial biomass contained large amount of N, P and K in the mineral soil layer at both burned and unburned sites, while the total amount of microbial nutrients in the 0-5 cm mineral soil and FH horizon at the unburned sites were two- to ten-fold greater than that in the mineral soil layer at the burned sites.

Next, nutrients uptake by plants can be roughly calculated from the concentrations of nutrient elements in twigs and leaves grown in a year. We calculated the uptake of P and K by plants at the burned sites from the data on the concentrations of P and K in twigs and leaves of regenerating plants obtained by Nakane (personal communication) (Table 2). The uptake of P by plants was 3.2-5.2% of the content of P in soil microbial biomass at the burned sites, while the

Table 2   Quantities of nutrient elements in soil microbial biomass and nutrient uptake by plants at the burned and unburned Japanese red pine forests.

Site	Soil Depth (cm)	Months after fires	Microbial biomass C and nutrient contents in soils (kg/ha)				Above-ground plant biomass (kg/ha)[a]	Nutrient uptake by plant (kg/ha·yr)	
			C	N	P	K		P	K
Burned sites									
Hinourasan	0-5	1	191	29	20	18	0 (1)	0	0
		5	125	19	13	12			
Nagatani	0-5	1	76	12	8	7			
		16	72	11	8	7	31 (16)	0.40	4.4
Ato	0-5	54	148	23	16	14	2835 (40)	0.74	8.5
Nenoura	0-5	72	342	52	37	32	3812 (53)	1.16	12.7
Unburned site									
Ato	FH		535	82	57	49			
	0-5		181	28	19	17			
	Total		716	110	77	66	17200	6.5[b]	40.0[b]

a, Numbers in parentheses indicate months after burning when the above-ground plant biomass was investigated.

b, Referred to Kawahara (1971).

uptake of K ranged from 40 to 67 % of the K content in soil microbial biomass at the burned sites.   As we could not assess the uptake of nutrients in the unburned pine forest, we referred to the data in a 20-year-old *P. densiflora* forest obtained by Kawahara (1971).   The uptake of N, P and K were 70, 6.5 and 40 kg/ha, respectively.   These values correspond to 8.5-64 % of the total amount of nutrients contained in microbial biomass of the FH and the mineral soil layer.

From the above results we found that microbial biomass contained large amounts of nutrients in the upper 5 cm soil at the burned sites ranging from 1 to 72 months after the fire and the amount of P and K contained in microbial biomass was quite higher than the amount of nutrient uptake by regenerating plants at the burned sites. In some studies (McGill et al., 1986; Insam et al., 1991) significant correlation between crop yield and the size of soil microbial biomass was observed.   This implies that soil microbial biomass contributes to soil fertility and plant production. It is important for regenerating plants after fire that soil microbial biomass at the burned sites did not decrease by fire. Consequently soil microbial biomass, which contains a substantial amount of nutrients, can acts as a nutrient pool available for regenerating plants for at least 6 years after fires at the study sites.

5. ACKNOWLEDGMENTS

The author wishes to express sincere thanks to Dr. K. Nakane, Professor of

Hiroshima University and Mr. T. Shinoda for providing the data on the concentrations of nutrients in plant organs.

REFERENCES

Anderson, J.P.E. and K.H. Domsch (1980): Quantities of Plant Nutrients in the Microbial Biomass of Selected Soils, Soil Science, Vol. 130, pp. 211-216.

Baath, E. and B. Soderstrom (1979): Fungal Biomass and Fungal Immobilization of Plants Nutrients in Swedish Coniferous Forest Soils, Revue d'Ecologie et de Biologie du Sol, Vol. 16, pp. 477-489.

Coleman, D.C. and D.A. Crossley Jr. (1996): Fundamentals of Soil Ecology, Academic Press, San Diego, 205p.

Dumontet, S., H. Dinel, A. Scopa, A. Mazzatura and A. Saracino (1996): Post-Fire Soil Microbial Biomass and Nutrient Content of a Pine Forest Soil from a Dunal Mediterranean Environment, Soil Biology and Biochemistry, Vol. 28, pp. 1467-1475.

Fenn, M.E., M.A. Poth, P.H. Dunn and S.C. Barro (1993): Microbial N and Biomass, Respiration and N Mineralization in Soils beneath Two Chaparral Species along a Fire-Induced Age Gradient, Soil Biology and Biochemistry, Vol. 25, pp. 457-466.

Fritze, H., T. Pennanen and J. Pietikainen (1993): Recovery of Soil Microbial Biomass and Activity from Prescribed Burning, Canadian Journal of Forest Research, Vol. 23, pp. 1286-1290.

Horikoshi, T. (1989): Changes of Fungus Flora after Fires in Burned Japanese Red Pine Forests, Recent Advances in Microbial Ecology, Japan Scientific Societies Press, Tokyo, pp. 260-264.

Horikoshi, T, T. Tateishi and F Takahashi (1986): Changes of fungus flora after fires in *Pinus densiflora* forest, Transactions of Mycological Society of Japan, Vol. 27, pp. 283-295.

Horikoshi, T, T. Tateishi and F Takahashi (1987): Changes of Microflora after Fires in *Pinus densiflora* Forests, Researches Related to the UNESCO's Man and Biosphere Programme in Japan 1986-1987, pp. 47-49.

Hossain, A.K.M.A., R.J. Raison and P.K. Khanna (1995): Effects of Fertilizer Application and Fire Regime on Soil Microbial Biomass Carbon and Nitrogen, and Nitrogen Mineralization in an Australian Subalpine Eucalypt Forest, Biology and Fertility of Soils, Vol. 19, pp. 246-252.

Insam, H., C.C. Mitchell and J.F. Dormaar (1981): Relationship of Soil Microbial Biomass and Activity with Fertilization Practice and Crop Yield of Three Ultisols, Soil Biology and Biochemistry, Vol. 23, pp. 459-464.

Jenkinson, D.S. and J.N. Ladd (1981): Microbial Biomass in Soil: Measurement and Turnover, Soil Biochemistry, Vol. 5, Marcel Deckker, New York, pp. 415-471.

Jenkinson, D.S. and D.S. Powlson (1976): The Effects of Biocidal Treatments on Metabolism in Soil. V. A Method for Measuring Soil Biomass, Soil Biology and

Biochemistry, Vol. 8, pp. 209-213.

Kawahara, T. (1971): The Return of Nutrients with Litter Fall in the Forest Ecosystems (II) The Amount of Organic Matter and Nutrients, Japanese Journal of Forestry, Vol. 53, pp. 231-238, (in Japanese with English Summary).

Marumoto, T., J.P.E. Anderson and K.H. Domsch (1982): Mineralization of Nutrients from Soil Microbial Biomass, Soil Biology and Biochemistry, Vol. 14, pp. 469-475.

McGill, W.B., K.R. Cannon, J.A. Robertson and F.D. Cook (1986): Dynamics of Soil Microbial Biomass and Water-Soluble Organic C in Breton L after 50 Years of Cropping to Two Rotation, Canadian Journal of Soil Science, Vol. 66, pp. 1-19.

Nakagoshi, N., K. Nehira and F. Takahashi (1987): The Role of Fire on Pine Forests of Japan, The Role of Fire in Ecological System, SPB Academic Publishing, Hague, pp. 91-119.

Okano, S., M. Nishio and Y. Sawada (1987): Turnover Rate of Soil Microbial Nitrogen in the Root Mat Layer of Pasture, Soil Science and Plant Nutrition Vol. 33, pp. 373-386.

Pietikainen, J. and H. Fritze (1993): Microbial Biomass and Activity in the Humus Layer Following Burning: Short-Term Effects of Two Different Fires, Canadian Journal of Forest Research, Vol. 23, pp. 1275-1285.

Sakamoto, K., T. Yoshida and M. Satoh (1992): Comparison of Carbon and Nitrogen Mineralization between Fumigation and Heating Treatments, Soil Science and Plant Nutrition, Vol. 38, pp. 133-140.

Tateishi, T. (1995): Estimation of Microbial Biomass in the Soils of Burned and Unburned Japanese Red Pine Forests, Doctoral Thesis, Hiroshima University, 131p. (in Japanese).

Tateishi, T. and T. Horikoshi (1995): Microbial Biomass in the Soils of Burned and Unburned Japanese Red Pine Forests in the Setouchi District, Western Japan, Bulletin of Japanese Society of Microbial Ecology, Vol. 10, pp. 9-20.

Tateishi, T., T. Horikoshi, H. Tsubota and F. Takahashi (1989a): Application of the Chloroform Fumigation-Incubation Method to the Estimation of Soil Microbial Biomass in Burned and Unburned Japanese Red Pine Forests, FEMS Microbiology Ecology, Vol. 62, pp. 163-172.

Tateishi, T., T. Horikoshi, F. Takahashi and H. Tsubota (1989b): Microbial Biomass and Microflora in the Soils at Burned and Unburned Japanese Red Pine Forests, Bulletin of Japanese Society of Microbial Ecology, Vol. 4, pp. 77-87.

Allelopathic Interactions in Forestry Systems

Ravinder Kumar KOHLI

Department of Botany, Panjab University, Chandigarh 160 014, India

ABSTRACT A natural phenomenon of release of secondary metabolites by plants or micro-organisms in the environment, normally termed as allelopathy, assumes a significant role in different managed forestry systems. A large number of trees have been reported to exhibit this phenomenon. It influences the successional pattern of the understorey vegetation and the overall dynamics of the ecosystems. In agroforestry systems the allelopathy leads to loss in productivity and changes in the nutrient cycling. The phenomenon could be successfully exploited for the management of weeds and pests apart from understanding and overcoming the problem of soil sickness and conservation of nitrogen etc. It also provides a platform for the scientists of various disciplines to work together.
Keywords : Agroforestry, Allelopathy, *Eucalyptus, Leucaena, Populus,*

1. INTRODUCTION

Though, the systematic study of allelopathy is relatively new, yet statements pointing to this phenomenon are old. Theophrastus (372-285 BC) had observed the inhibition of pigweed by alfalfa plant due to chemical inhibition. Among trees, Black Walnut (*Juglans nigra*) is perhaps the first which in 1881 was reported to inhibit the adjoining plants (Stickney and Hoy 1881). Nevertheless, it was during the last two decades that the science of allelopathy exhibited by trees has gained interest of Scientists.

Allelopathy refers to both direct and indirect, beneficial and harmful effects of plants (including microbes) on other plants through the release of chemical compounds in the environment by the donor plant. If the donor and recipient plants happen to belong to same taxonomic group then another term Autotoxicity or Autoallelopathy is used. Though the significance of autotoxicity is not well understood, yet it is thought to enhance geographical distribution of plants, prevent seed decay and induce dormancy in the seeds (Friedman and Waller 1983).

The allelopathic chemicals which are responsible for bringing out this phenomenon are infact the *Secondary Metabolites.* Almost each morphological part i.e. leaf, stem, bark, root, inflorescence of the plant normally exhibits this phenomenon. However, the content varies. The mode of release may be through volatilization, root exudation, stem flow, microbial activity, litter decomposition and during ploughing of residues in the soil (Rice 1984; Putnam and Tang 1986).

2. ALLELOPATHIC RESEARCH IN FOREST ENVIRONMENT

The science of allelopathy has been studied more in man-made plantation systems than the natural forests. In areas with deficient natural forest resources, growing of predominantly fast growing trees in various plantation forestry programmes has been on the increase. Especially during the recent past, interest in this direction has gained momentum in understanding the interactions between trees and crops in agroforestry systems. Here, the woody perennials are intentionally incorporated with the agricultural crops in a variety of arrangements. The choice of tree species greatly influence the crop productivity and overall success of these systems. Allelopathy is known to play an inhibitory, regulatory or rarely stimulatory role in such environments, depending upon the input of allelochemicals. A large number of trees have been reported to exhibit this phenomena and the number of reports have increased tremendously especially during the last two decades (Rice 1984, 1995; Elakovich and Wooten 1995, 1996; Singh 1995). This phenomenal increase can be attributed to evidences of chemical inhibition, in agriculture as well as forestry, employing new techniques which are fast coming up.

Despite bottlenecks, the understanding of allelopathy in agroecosystems offers some challenging new aspects such as crop improvement, sustainable weed and pest management, control and management of plant pathogens and nitrogen conservation which can be applied to enhance productivity in a sustainable manner. There are a number of reports e.g. the plantations of *Juglans nigra* (Weidenheimer et al. 1989), *Leucaena leucocephala* (Chou and Kuo 1986; Suresh and Rai 1988), *Eucalyptus* (Kohli 1990; Singh et al. 1993), *Populus* (Kohli et al. 1996), *Pinus* spp. (Kil and Yim 1983; Melkania and Singh 1987) etc., indicating role of allelopathy in determining the structure and composition of understorey vegetation

3. ALLELOPATHY OF FAST GROWING TREES IN AGROFORESTRY ECOSYSTEMS

In India, fast growing exotic tree species like *Eucalyptus*, *Populus* and *Leucaena* have been promoted indiscriminately largely because of their early net returns. However, their indiscriminate promotion has evoked a mixed response from planters, developmentalists and environmentalists. In fact, in most of the agroforestry systems these are preferred over the indigenous ones. There are a number of reports which highlight the negative influence of trees on crops in agroecosystems (Table 1). The allelochemicals present in their foliage, litter, flowers, bark escape into the environment through various mechanisms and reduce crop productivity and overall sustainability of such ecosystems.

Table 1. Negative Tree-crop Interactions in Agroecosystems

Tree	Test Crops	Reference
Acacia arabica	*Triticum aestivum*	Sheikh and Haq (1978)
A. holosericea	*Pennisetum glaucum*	Lamers et al.(1993)
A. nilotica.	*Triticum aestivum.*	Sheikh (1988); Sharma (1992)
	Pennisetum glaucum	Lamers et al.(1993)
A. tortilis	*Pennisetum americanum*	Sharma et al.(1994)
Albizia saman.	*Oryza sativa*	Sae-Lee et al.(1992)
Azadirachta indica	*Pennisetum typhoides.*	Long and Persaud (1988)
		Brenner et al. (1993)
	Triticum aestivum	Puri and Bangarwa (1992)
Cassia siamea	*Sorghum spp.*	Jensen (1983)
Casuarina equisetifolia	*Vigna mungo,*	Srinivasan et al. (1990)
	V. radiata,	
	V. unguiculata,	
	Cajanus cajan,	
	Glycine max.	
Dalbergia sissoo.	*Triticum vulgare.*	Khan and Aslam (1974);
		Sheikh and Haq (1978);
		Puri and Bangarwa (1992)
Dipterocarpus obtusifolius	*Oryza sativa*	Vityakon et al. (1993)
Erythrina spp.	*Oryza sativa*	Salazar et al. (1993)
Eucalyptus camaldulensis	*Arachis hypogaea,*	Craig et al. (1988)
	Hibiscus cannabinus	
	Abelmoschus esculentus.	Igboanugo (1987)
	Vigna unguiculata,	Igboanugo (1988b)
	Zea mays,	
	Sorghum bicolor	
E. citriodora	*Capsicum annuum,*	Igboanugo (1988a,b)
	Vigna unguiculata,.	
	Zea mays,	
	Sorghum bicolor	
E. grandis	*Vigna unguiculata,.*	Igboanugo (1988b)
	Zea mays,	
	Sorghum bicolor	
E. tereticornis	*Triticum aestivum,*	Malik and Sharma (1990)
	Brassica spp.	
	Cicer arietinum,	Kohli et al. (1990);
	Lens esculentum,	Singh and Kohli (1992)
	Triticum aestivum,	
	Brassica oleracea,	

	Trifolium alexandrinum, Sorghum bicolor, Pennisetum typhoides	Rai et al. (1990)
Eucalyptus spp.	Vigna unguiculata,	Onyewotu (1985)
Holoptelea integrifolia	Cyamopsis tetragonoloba	Muthana and Arora (1977)
Inga edulis	Oryza sativa	Salazar et al. (1993)
Leucaena leucocephala	Ipomoea batatas	Karim et al. (1991)
	Oryza sativa.	Tomar and Srivastava (1986)
Populus deltoides	Saccharum officinarum	Sheikh and Haq (1986)
	Triticum aestivum	Ralhan et al. (1992); Singh et al. (1993)
Populus spp.	Solanum tuberosum	Anonymous (1984)
Prunus cerasoides.	Hordeum vulgare, Glycine max	Bhatt and Todaria (1990)
Ulmus spp.	Triticum aestivum	McMartin et al. (1974)
Zizyphus rotundifolia	Pennisetum americanum, Cyamopsis tetragonoloba	Sharma et al. (1994)

4. AN INDIAN PERSPECTIVE OF THE ALLELOPATHIC STUDIES OF SOME FAST GROWING EXOTICS

There are many reports on the allelopathic interactions among crops and trees. However, in this particular talk, emphasis is made on the three most commonly grown agroforestry trees in India i.e. *Eucalyptus, Populus* and *Leucaena.*

4.1 *Eucalyptus*

Eucalyptus in India has been grown on large scale because of its fast growth rate and 'claimed' quick monetary gains. However, because of its antisocial nature, especially under monoculture conditions, the popularity of the tree has suffered a great setback (Kohli 1987). Its monocultures have been reported to harbour little understorey vegetation (Bhaskar and Dasappa 1986, Singh et al. 1993). Most of these negative aspects of *Eucalyptus* on understorey vegetation and also the crop yield have been attributed to allelopathy which has been nicely demonstrated in this tree. Allelochemicals of the tree are both volatile as well as non volatile in nature (Kohli, 1990).

Among the volatiles, limonene, 1-8, cineole, citronellal, citronellol, α-pinene etc. are highly toxic (del Moral and Muller 1970; Al-Mousawi and Al-Naib 1975, 1976; Singh et al. 1991). Being heavier than air, the volatile terpenes upon release from leaves travel downwards and get adsorbed on the surface of soil, where they affect the adjoining vegetation. Baker (1966) demonstrated that the volatile oils of *E. globulus* inhibit the root growth of cucumber seedlings besides its own. Further, oils of *E. globulus* and *E. citriodora* and their dominant pure components, limonene and citronellol, respectively at the concentrations of 10,20,30 nl/cc vaporised in growth chambers/fumitoria reduced the germination, seed vigour and seedling growth of four crops - mung bean, oat, barley and lentil. The chlorophyll content, cellular respiration and water contents were also adversely affected when pots bearing these plants were placed in these chambers (Kohli 1990). In order to demonstrate the continuous addition of volatile terpenes of Eucalypts in soil underneath, seeds of mung bean were placed in Petri dishes having soil adsorbed with eucalypt oil and in Petri dishes having oil adsorbed

(a)

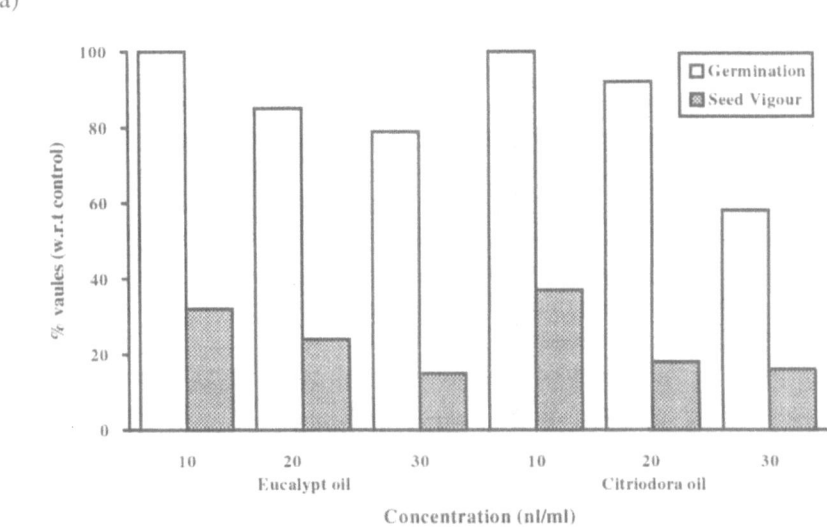

(b)

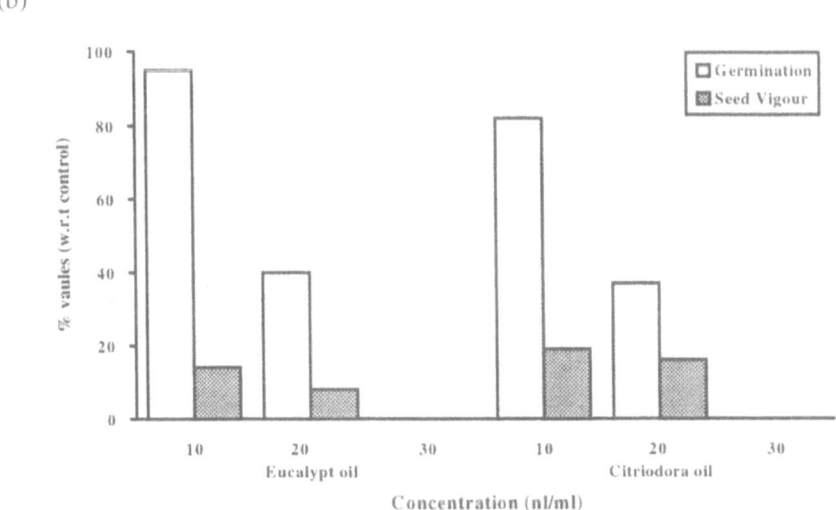

Fig. 1 Germination of mung bean in the soil adsorbed by eucalypt oil and citriodora oil without (a) or with (b) continuous column of oil vapours.

soil in continuous column of oils in the fumitoria. The germination and seedling length of mung bean was reduced under both the conditions and the effect was more in the Petri dishes having adsorbed soil and receiving continuous flush of oils (Fig. 1 a, b) (Singh et al. 1991).

Maximum amount of oils was extracted from the fresh leaves (424 µl/100g) followed by decomposing leaves (208 µl/100g) and the soil (62 µl/100g) compared to none in *Eucalyptus* free area (Singh 1991). The content of oil, however, changes with species, seasons and age etc.

The allelochemicals of eucalypts are continuously added to the soil system. Even if the soil is replaced with the one collected from the adjoining non-tree area, it gets enriched with these allelochemicals soon. In order to exclude the contribution of non-volatile allelochemicals from *Eucalyptus*, a number of summer and winter season crops were grown in (a) the soil under its plantations (first set) and (b) under the plantation but with the soil replaced with the soil from the eucalypt-free area by digging a pit and by layering it with polythene (second set). The crop performance in the first set of condition, was relatively poorer than the second set implying, thereby that in the second set of conditions, the non-volatile allelochemicals such as phenolics were excluded. In other words, *Eucalyptus* exhibits its allelopathic influence through volatile as well as non volatile allelochemicals released by it.

The soil under the canopy of *Eucalyptus* is also found to be rich in chemicals, the content of which varies with depth and distance from the tree. Upon analysis these were found to be 11 types of phenolic acids, 8 of which namely caffeic acid, cinnamic acid, *p*-coumaric acid, ferulic acid, gallic acid, gentisic acid, syringic acid and vanillic acid were identical with those found in leaves (Kohli, 1990). These chemicals were found to be toxic to germination, growth, development and yield of crops (Singh and Kohli, 1992).

The eucalypts are also extensively raised along the field boundaries as windbreaks, shelterbelts or simply scattered in the fields. Our study has shown that the shelterbelts of eucalypts are very harmful to the crops upto 12m from the tree line growing in the adjoining area (Table 2). It adversely affects the initial growth, biomass and crop yield (Kohli et al. 1990; Singh and Kohli 1992).

4.2 *Populus deltoides*

With the *Eucalyptus* surrounded with the controversies, *Populus deltoides* in the recent years has gained prominence in various agroforestry systems in India. However, negative allelopathic reports of this tree on the adjoining crops and the understorey vegetation have also started appearing. Understorey vegetation of *P. deltoides* due to allelopathy was found to be deficient in species diversity, species richness and evenness as well as biomass as compared to the indigenous plantations of *Albizzia lebbeck* and *Dalbergia sissoo* (Kohli et al. 1996). The leaf and soil leachates, and rain drops of the tree were found to reduce the germination and radicle growth of *Triticum aestivum* and *Cicer arietinum* (Melkania 1984) and *Oryza sativa* (Koul et al.1991). Its leaves as compared to the other parts contribute maximum amount of leachable allelochemicals into the environment (Kohli et al. 1997).

Table 2 Grain yield performance of wheat in the first 1x20m strip and the next 1x40m strip (i.e. from 1x20 to 1x60m) in the fields sheltered on the north by *P. deltoides* or *E. tereticornis* compared to unsheltered ones. Values in parenthesis represent yield per square meter.

Area	Grain Yield (in Kg) in the area		Total Grain	Loss in yield	
	1 x 20 m	1 x 40 m	Yield (Kg/ha)	Total	%
Unsheltered	8.27 ± 8.84	16.69±0.24	4165 ±131.2	-	-
	(0.4136)	(0.4172)			
P. deltoides	4.86 ±0.55	16.66±0.79	3819±133.01	346	8.31
	(0.2432)	(0.4165)			
E. tereticornis	4.11±0.95	16.08±0.87	3627±262.98	538	12.9
	(0.2055)	(0.4020)			

The seeds of different crops respond differently to leachable allelopathic chemicals from the leaves. Singh (1995) has shown a relationship of the ratio of the seed coat thickness and seed volume with the germinability in response to the treatment. Thus, the seeds with ratio exceeding 1.0833 show resistance to the allelochemicals from *Populus*. Lesser the ratio, more was the reduction in germination. This tree under shelterbelt or alley system adversely affect the crop yield. Under Agrisilvicultural practice, the introduction of *P. deltoides* result in the reduction of grain yield of wheat (Ralhan et al. 1992; Singh et al. 1993). *P. deltoides* reduces the biomass and grain yield significantly upto a distance of 12 m (Table 2). Shorter the distance more the loss (Singh et al. 1998). The comparative loss under similar set of conditions with *Dalbergia sissoo* a native tree instead was relatively less. Singh (1995) attributed the allelopathic influence of the tree as evidenced by the presence of phenolic acids namely gallic acid, chlorogenic acid, *p*-hydroxy benzoic acid, vanillic acid, syringic acid and *p*-coumaric acid as well as to the presence of salicin a phenol glucoside. Thus, indiscriminate plantation of *P. deltoides* also pose a question mark on their ever increasing popularity.

4.3 *Leucaena leucocephala*

Leucaena leucocephala is another fast growing, nitrogen fixing perennial tree which has been extensively promoted in India as well as in other tropical countries. In India, because of its fast growth rate, fodder and fuel value and above all green manure, it has become a favourable agroforestry tree. However, it is also reported to be environmentally harmful tree and thus cannot be recommended for sustainable ecosystems. Suresh and Rai (1987) reported allelopathic influence of the tree on crops such as sorghum, cowpea and sunflower through aqueous extracts of leaves and litter mulched in soil. Its plantations are reported to harbour very little understorey (Suresh and Rai 1988) but are characterized by the presence of its own seedlings Batish et al.1998 (personal communication) have shown that incorporation of both leaf and litter powder in the soil result in the reduction in growth of *Zea mays*. This was attributed to the presence of phenolic compounds in the decomposing litter, leaves as well as soil. Besides, mimosine - a non-protein amino acid present in its leaves is also known to be toxic in nature (Rizvi et al. 1990, Chaturvedi and Jha 1992). However, it does not seem to adversely affect the seedlings of its own type. That is why a plenty of Leucaena seedlings are found to be present under the plantations of *L. leucocephala*.

5.MANIPULATING ALLELOPATHY FOR SUSTAINABLE PRODUCTIVITY : SOME STRATEGIES

The allelopathic phenomenon as exhibited by a number of trees and crops can be successfully manipulated for enhancing the productivity of croplands in a sustainable manner. These include the identification of new plant based chemicals for the management of weeds and pests. Further, allelopathic companion crops can also be used to selectively suppress the weeds. Some environmentally safe weedicides/herbicides could be synthetically designed on the basis of natural chemicals so that the target of weed management is achieved without polluting the soil. Further, the allelopathic phenomenon can be successfully manipulated to conserve nitrogen in the nitrogen deficient areas.

5.1 Weed and Pest Management

The use of botanicals (chemicals of plant origin) is fast gaining preference over synthetic pesticides/weedicides, because the former are biodegradable, environmentally safe, cost effective, easy to be stored, require less space and have longer shelf-life. The science of allelopathy holds in store a great potential for weed and pest management. Phytotoxic leaf and litter leachates of some trees (*Abies balsamea, Picea mariana, Pinus divaricata, P. resinosa* and *Thuja occidentalis*) as reported by Jobidon (1986) could be utilized for controlling weeds and pests (Rice 1995). Based on the chemistry of allelochemicals, new herbicides can also be synthesized. The example of Cinmethylene whose chemical structure is based of cineole, a volatile terpene in the leaves of *Eucalyptus* and isolated from its plantation sites, *Artemisia* or *Salvia* spp. is well known to the scientists working in this area. Ailanthone - an allelochemical of *Ailanthus altissima* (Heaven's Tree) is currently being tested for use as a herbicide (Heisey 1996) A few more compounds of the tree origin possess the potential to suppress weeds. Cineole and Citronellol from *Eucalyptus* have recently been found to be promising based on their inhibitory effect on the germination of noxious weed *Parthenium hysterophorus* (Kohli et al.1998). Likewise, Rizvi et al.(1990) have reported the herbicidal properties of mimosine. Though Juglone - a natural product from *Juglans regia* as such has not been tried specifically for the purpose of weed management, yet based on its toxic effects it could be one of the novel compounds to be exploited for this purpose on commercial scale.

Neem (*Azadirachta indica*) - the wonder tree of India is probably one of the most promising tree species which could be successfully exploited for the sustainable management of pests and weeds. From times immemorial, this tree has been used as a source of medicine. Its leaves and fruits are being successfully used to keep away the insects and stored grain pests. Several compounds Nimbin, Nimbidin, Nimbinin, Protolimonoids, Limonoids, Tetranoterpenoids, Pentanoterpenoids, Hexanoterpenoids and Non triterpenoidal constituents from the tree possess a strong potential to suppress insects, pests and fungi etc. (Koul et al. 1990).

5.2 Control of Replant and Soil Sickness Problem

Understanding of the allelopathic phenomenon in agroforestry and forestry environment can largely overcome the problem of soil sickness and replant. The latter is very common in some trees. It results from the release of chemical substances which get accumulated in the soil and thus prevent their natural regeneration by suppressing either seed germination or interfering with the growth. Ward and McCormick (1982) reported poor regeneration of *Tsuga canadensis* due to allelopathy. The extracts of litter, foliage, roots, mycorrhizae and bark of the tree inhibited initial germination and killed six day old seedlings. This problem is also very common in the conifer forests and in orchards. Pellissier (1994) and Gallet (1994) reported this in case of *Picea abies* and attributed it to the presence of phenolic compounds particularly the p-hydroxy acetophenone. Forest litter also plays an important role in these ecosystems (Kuiters 1990). Autoinhibition due to the allelopathy has also been reported in *Abies balsamea* (Thibault et al. 1982), *Araucaria cunninghamia* (Bevege 1968), *Cunninghamia lanceolata* (Zhang, 1993), *Picea mariana* (Thomas 1974), and *Pinus radiata* (Chu-Chou 1978). In some cases phenolic compounds particularly *p*-hydroxy acetophenone have been implicated (Gallet, 1994).

The replant problem and soil sickness is very old in orchards particularly of *Prunus persica* (Proebsting and Gilmore 1941), *Pyrus malus* (Borner 1959) and Citrus (Burger 1981). By studying the soil chemistry, allelochemicals and role of micro-organism this problem has been largely understood and can, therefore, be controlled.

5.3 Nitrogen Conservation

Nitrogen is a vital nutrient which the plants require for their growth. Therefore, its conservation is very important particularly for the sites deficient of it. Of late, it is being realised that allelopathic phenomenon can be utilized for this purpose. Though the reports are very scanty, yet can be extremely useful for the future as model studies. The products of nitrogen which are known to suppress pests also inhibit nitrification and can thus conserve nitrogen (Koul et al. 1990). Some mulches are used to achieve this target. Jobidon (1986) reported that mulches of wheat, oat and barley enhance nitrogen level in black spruce leaves and thus reduced its loss.

Allelopathy, therefore, is not an accidental phenomenon but has been demonstrated in trees more convincingly. Further, due to longer life span of trees in nature, compared to other vegetation, the significance of this phenomenon gains more credit.

REFERENCES

Al-Mousawi, A.H. and F.A.G. Al-Naib (1975) : Allelopathic effects of *Eucalyptus microtheca* F Muell., Journal of University of Kuwait (Science), Vol. 2, pp. 59-66.

Al-Mousawi, A.H. and F.A.G. Al-Naib (1976) : Volatile growth inhibitors produced by *Eucalyptus microtheca*. Bulletin of Biological Research Centre, Vol. 7, pp.17-23.

Anonymous (1984) : Annual Research Report, Department of Forestry and Natural Resources, Punjab Agricultural University, Ludhiana, India.

Baker, H.G. (1966) : Volatile growth inhibitors from *Eucalyptus globulus*, Madroño, Vol. 18, pp. 207-210.

Bevege, D.I. (1968) : Inhibition of seedling hoop pine (*Araucaria cunninghamia* Ait.) on forest soils by phytotoxic substances from the root zones of Pinus, Araucaria and Flindersia, Plant and Soil, Vol. 29, No. 2, pp. 263-273.

Bhaskar, V and Dasappa (1986) : Ground flora in *Eucalyptus* plantation of different ages, In Eucalyptus in India - Past, Present and Future, eds. J.K. Sharma, C.S. Nair, S. Kedarnath and S. Konda, Kerala Forest Research Institute, Kerala, India, pp. 213-224.

Bhatt, B.P. and N.P. Todaria (1990) : Studies on the allelopathic effects of some Agroforestry tree crops of Garhwal Himalaya, Agroforestry Systems, Vol. 12, pp. 251-255.

Borner, H. (1955) : The apple replant problem I. The excretion of Phlorizin from apple root residues, Contributions of Boyce Thompson Institute, Vol. 20, pp. 39-56.

Brenner, A.J., R.J.V.D. Beldt and P.G. Jarvis (1993) : Tree-crop interface competition in a semi-arid sahelian windbreak, Proc. 4th International Symposium on Windbreaks and Agroforestry, Hedeselskabet, Viborg, Denmark, pp. 15-23.

Burger, W.P. (1981) : Allelopathy in Citrus orchards, Ph.D. Thesis University of Port Elizabeth, South Africa.

Chaturvedi, O.P. and A.N. Jha (1992) : Studies on allelopathic potential of an important agroforestry species, Forest Ecology and Management, Vol. 53, pp. 91-98.

Chou, C.H. and Y. Kuo (1986) : Allelopathic research of subtropical vegetation in Taiwan. III. Allelopathic exclusion of understorey by *Leucaena leucocephala,* Journal of Chemical Ecology, Vol. 12, pp. 1431-1448.

Chu-Chou, M. (1978) : Effects of root residues on growth of *Pinus radiata* seedlings and a mycorrhizal fungus, Annals of Applied Biology, Vol. 90, pp. 407-416.

Craig, I.A., Wasunan and M. Saenlao (1988) : Effect of paddy bund planted *Eucalyptus* trees on the performance of field crops. Paper presented at the Fifth Annual Farming Systems Conference, Kamphaengsaen, Thailand.

del Moral, R. and C.H. Muller (1970) : The allelopathic effects of *Eucalyptus camaldulensis,* American Midland Naturalist, Vol. 83, pp. 254-282.

Elakovich, S.D. and J.W. Wooten (1995) : Allelopathic woody plants. Part I. *Abies alba* through *Lyonia lucida,* Allelopathy Journal, Vol. 2, No. 2, pp. 117-146.

Elakovich, S.D. and J.W. Wooten (1996) : Allelopathic woody plants. Part II. *Mabea* through *Zelkova,* Allelopathy Journal, Vol. 3, No. 1, pp. 9-32.

Friedman J. and G.R. Waller (1983) : Caffeine hazards and their prevention in germinating seeds of coffee *Coffea arabica* L., Journal of Chemical Ecology, Vol. 9, pp. 1099-1106.

Gallet, C. (1994) : Allelopathic potential in bilberry-spruce forests: influence of phenolic compounds on spruce seedlings. Journal of Chemical Ecology, Vol. 20, pp. 1009-1024.

Heisey, R.M. (1996) : Identification of an allelopathic compound from *Ailanthus altissima* Simaroubaceae and characterization of its herbicidal activity, American Journal of Botany, Vol. 83, pp. 192-200.

Igboanugo, A.B.I. (1987) : Effect of some eucalypts on growth and yield of *Amaranthus caudatus* and *Abelmoschus esculentus,* Agriculture Ecosystem and Environment, Vol. 18, pp. 243-249.

Igboanugo, A.B.I. (1988a) : Morphology and yield of chilli (*Capsicum annuum*) in relation to distance from lemon-scented eucalyptus (*Eucalyptus citriodora*) stands, Indian Journal of Agricultural Sciences, Vol. 58, No. 4, pp. 317-319.

Igboanugo, A.B.I. 1988b. Effect of some eucalyptus on yields of *Vigna unguiculata* L. Walp., *Zea mays* L. and *Sorghum bicolor* L., Agriculture Ecosystem and Environment, Vol. 24, pp. 453-458.

Jensen, A.M. (1983) : Shelterbelt effects in Tropical and Temperate Zones, International Development Research Centre Manuscript Reports, IDRC-MR80-e, Ottawa, 61p.

Jobidon, R. (1986) : Allelopathic potential of coniferous species to old-field weeds in Eastern Quebec Canada, Forest Science, Vol. 32, pp. 112-118.

Karim, A.B., P.S. Savill, and E.R. Rhodes (1991) : The effect of young *Leucaena leucocephala* (Lam.) de Wit hedges on the growth and yield of maize, sweet potato and cowpea in an agroforestry system in Sierra Leone, Agroforestry Systems, Vol. 16, pp. 203-211.

Khan, G.S. and R.M Aslam (1974) : Extent of damage of wheat (*Triticum vulgare*) by Shisham (*Dalbergia sissoo*), Proc. Pakistan Forestry Conference, Pakistan Forest Institute, Peshawar, pp. 37-40.

Kil, B.S. and Y.J. Yim (1983) : Allelopathic effects of *Pinus densiflora* on undergrowth of red pine forest, Journal of Chemical Ecology, Vol. 9, pp. 1135-1151.

Kohli, R.K. (1987) : Eucalyptus - an antisocial tree for Social Forestry, In Social Forestry for Rural Development, eds. P.K. Khosla and R.K. Kohli, ISTS Publications, Solan, India, pp. 235-241.

Kohli, R.K. (1990) :Allelopathic Potential of *Eucalyptus*. MAB-DOEn Project Report, India, 199p.

Kohli, R.K., D.R. Batish and H.P. Singh (1998) : Eucalypt Oils for the Control of Parthenium *Parthenium hysterophorus* L., Crop Protection, In Press.

Kohli, R.K., Singh and R.C. Verma (1990) : Influence of eucalypt shelterbelt on winter season agroecosystems. Agriculture Ecosystems and Environment, Vol. 33, pp. 23-31 .

Kohli, R.K., H.P. Singh, and D.R. Batish (1997) : Phytotoxic potential of *Populus deltoides* Bartr. ex Marsh. I. Comparative contribution of different parts, Indian Journal of Forestry, Vol. 20, No. 3, pp. 300-304.

Kohli, R.K., H.P. Singh and D. Rani (1996) : Status of floor vegetation under some monoculture and mixculture plantations, Journal of Forest Research, Vol. 1, pp. 205-209.

Koul, O., M.B. Isman and C.M. Ketkar (1990) : Properties and uses of neem *Azadirachta indica*, Canadian Journal of Botany, Vol. 68, pp. 1-11.

Koul, V.K., A. Raina, Y.P. Khanna, M.L. Tickoo and H. Singh (1991) : Evaluation of allelopathic influence of certain farm grown tree species on rice *Oryza sativa* L cv PC-19, Indian Journal of Forestry, Vol. 14, pp. 54-57.

Kuiters, A.T. (1990) : Role of phenolic substances from decomposing forest litter in plant-soil interactions, Acta Botanica Neerlandica, Vol. 39, pp. 329-348.

Lamers, J.P.A., K. Michels, B.E. Allison and R.J. Vandenbeldt (1993) : Agronomic and socioeconomic aspects of windbreaks in Southwest-Niger. Proc. 4th International Symposium on Windbreaks and Agroforestry, Hedeselskabet, Viborg, Denmark. pp. 28-30.

Long, S.P. and N. Persaud (1988) : Influence of neem (*Azadirachta indica*) windbreaks on millet yield, microclimate and water use in Niger, West Africa. In Dryland Agriculture - a Global Perspective, eds. P.W. Unger, T.V. Sneed, W.R. Jordon, and R. Jensen, Texas Agricultural Experimental Station, Texas, pp. 313-314.

Malik, R.S. and S.K. Sharma (1990) : Moisture extraction and crop yield as a function of distance from a row of *Eucalyptus tereticornis*, Agroforestry Systems, Vol. 12, pp. 187-195.

McMartin, W., A.B. Frank and R.H. Heintz (1974) : Economics of shelterbelt influence on wheat yields in North Dakota, Journal of Soil and Water Conservation, Vol 29, pp. 87-91.

Melkania, N.P. (1984) : Influence of leaf leachates of certain woody species on agricultural crops, Indian Journal of Ecology, Vol. 11, pp. 82-86.

Melkania, N.P. and J.S. Singh (1987) : Allelopathy in Himalayan forest species, Proc. IX International Symposium Tropical Ecology and International Conference on Rehabilitation of disturbed ecosystems : A Global Issue, Banaras Hindu University, Varanasi, India.

Muthana, K.D. and G.D. Arora (1977) : Studies on the establishment and growth of and economic tree species, *viz. Holopetalia itegrifolia* in integrated land use pattern in arid environment, Annual Report, CAZRI, Jodhpur, India.

Onyewotu, L.O.Z. (1985) : Shelterbelt effects on the yield of agricultural crops: a case study of a semi-arid environment in Northern Nigeria, Centre de Recherches pour le Développement International, Rapport Manuscrit, IDRC-MR117e f, Ottawa.

Pellissier, F. (1994) : Effect of phenolic compounds in humus on the natural regeneration of spruce, Phytochemistry, Vol 36, No. 4, pp. 865-867.

Proebsting, E.L. and A.E. Gilmore (1941) : The relation of peach root toxicity to the re-establishing of peach orchards, Proc. American Society of Horticultural Science, Vol. 38, pp. 21-26.

Puri, S. and K.S. Bangarwa (1992) : Effect of trees on the yield of irrigated wheat crop in semi-arid regions. Agroforestry Systems, Vol. 20, pp. 229-241.

Putnam, A.R. and C.S. Tang (1986) : Allelopathy : State of the Science. In The Science of Allelopathy, eds. A.R. Putnam and C.S. Tang, John Wiley and Sons, New York, pp. 1-19.

Rai, R.S.V., C. Swaminathan and C. Surendran (1990) : Studies on intercropping with coppice shoots of *Eucalyptus tereticornis* Sm., Journal of Tropical Forest Science, Vol 3, No. 2, pp. 97 100.

Ralhan, P.K., A. Singh and R.S. Dhanda (1992) : Performance of wheat as intercrop under poplar *Populus deltoides* Bartr. plantations in Punjab, India, Agroforestry Systems, Vol. 19, pp. 217-222.

Rice, E.L. (1984) : Allelopathy, Academic Press, New York, 422p.

Rice, E.L. (1995) : Biological Control of Weeds and Plant Diseases - Advances in Applied Allelopathy, University of Oklahoma Press, Norman, USA, 439p.

Rizvi, S.J.H., R.C. Sinha and V. Rizvi (1990) : Implication of mimosine allelopathy in agroforestry, Proc XIX World Congress on Forestry, Vol. 2, pp. 22-27.

Sae-lee, S., P. Vityakon and B. Prachaiyo (1992). Effects of trees on paddy bund on soil fertility and rice growth in Northeast Thailand, Agroforestry Systems, Vol. 18, pp. 213-223.

Salazar, A., L.T. Szott and C.A. Palm (1993) : Crop tree interactions in alley cropping systems on alluvial soils of the Upper Amazon Basin, Agroforestry Systems, Vol. 22, pp. 67-82.

Sharma, B.M., S.S. Rathore and J.P. Gupta (1994) : Compatibility studies on *Acacia tortilis* and *Zizyphus rotundifolia* with field crops under arid conditions, Indian Forester, Vol. 120, No. 5, pp. 423-429.

Sharma, K.K. (1992) : Wheat cultivation in association with *Acacia nilotica* (L.) Willd. ex Del. Field bund plantation - a case study, Agroforestry Systems, Vol. 17, pp. 43-51.

Sheikh, M.I. (1988) : Planting and establishment of windbreaks in arid areas, Agriculture Ecosystem and Environment, Vol. 22/23, pp. 405-423.

Sheikh, M.I. and R. Haq (1978) : Effect of shade of *Acacia arabica* and *Dalbergia sissoo* on the yield of wheat, Pakistan Journal of Forestry, Vol. 29, pp. 183-185.

Sheikh, M.I. and R. Haq (1986) : Effect of size, placement and composition of windbreaks for optimum production of annual crops and woods, Final Technical Report (December 1979-November 1986), Pakistan Forest Institute, Peshawar, 125p.

Singh, H.P. (1995) : Phytotoxic effects of *Populus* in natural and agro-ecosystems, Ph.D. thesis Panjab University, Chandigarh, India.

Singh, A., R.S. Dhanda, P.K. Ralhan (1993) : Performance of wheat varieties under poplar *Populus deltoides* Bartr. plantations in Punjab, India, Agroforestry Systems, Vol. 22, pp. 83-86.

Singh, D. (1991) : Phytotoxic properties of *Eucalyptus* with special reference to the role of its volatile components, Ph.D. Thesis Panjab University, Chandigarh, India.

Singh, D. and R.K. Kohli (1992) : Impact of Eucalyptus tereticornis Sm shelterbelts on crops, Agroforestry Systems, Vol. 20, pp. 253-266.

Singh, D., R.K. Kohli and N. Jerath (1993) : Impact of Eucalyptus and other plantations on phytodiversity in India, Proc. International Conference on Forest Vegetation Management - Ecology, Practice and Policy, Auburn University, Auburn, Alabama, USA, pp. 152-159.

Singh, D., R.K. Kohli and D.B. Saxena (1991) : Effect of eucalyptus oil on germination and growth of *Phaseolus aureus* Roxb., Plant and Soil, Vol. 137, pp. 223-227.

Singh, G., N.T. Singh and I.P. Abrol (1994) : Agroforestry Techniques for the rehabilitation of degraded salt-affected lands in India, Land Degradation and Rehabilitation, Vol. 5, pp. 223-242.

Singh, H.P., R.K. Kohli and D.R. Batish (1998) : Shelter effects of Poplar *Populus deltoides* on the growth and yield of wheat in North India. Agroforestry Systems, In Press.

Smith, I.K. and C.J. Fowden (1966) : A study on mimosine toxicity in plants, Journal of Experimental Botany, Vol. 17, pp. 750-761

Srinivasan, K., M. Ramasan and R. Shantha (1990) : Tolerance of pulse crops to allelochemicals of tree species, Indian Journal of Pulses Research, Vol 3, pp. 40-44.

Stickney, J.S. and P.R. Hoy (1881) : Toxic action of black walnut, Transactions of Wisconsin State Horticultural Society, Vol. 11, pp. 166-167.

Suresh, K.K. and R.S.V. Rai (1987) : Studies on the allelopathic effects of some agroforestry tree crops, International Tree Crops Journal, Vol. 4, pp. 109-115

Suresh, K.K. and R.S.V. Rai (1988) : Allelopathic exclusion of understorey by a few multi-purpose trees, International Tree Crops Journal, Vol. 5, pp. 143-151

Theophrastus (ca 300 B.C.) : Enquiry into Plants and Minor Works on Odour and Weather Signs, Vol. 2 Translation to English by A. Hort. W. Heinemann, London, 1916.

Thibault, J.R., J.A. Fortin and W.A. Smirnoff (1982) : *In vitro* allelopathic inhibition of nitrification by balsam poplar (*Populus balsamifera*) and balsam fir (*Abies balsamea*), American Journal of Botany, Vol 28, pp. 478-485.

Thomas, A.S., Jr. (1974) : The effect of aqueous extracts of blue spruce leaves on seed germination and seedling growth of several plant species. (Abstr.) Phytopathology, Vol. 64, No. 5, pp. 587.

Tomar, G.S. and S.K. Shrivastava (1986) : Preliminary studies of rice cultivation in association with trees, In Agroforestry Systems : A new challenge, eds. P.K. Khosla, S. Puri and D.K. Khurana, Indian Society of Tree Scientists, New Delhi. pp. 207-212.

Vityakon, P., S. Sae-lee and S. Seripong (1993) : Effects of tree leaf litter and shading on growth and yield of paddy rice in Northeast Thailand, Kasetsart Jounal (Natural Science), Vol. 27, pp. 219-222.

Ward, H.A. and L.H. McCormick (1982) : Eastern hemlock allelopathy, Forest Science, Vol. 28, No. 4, pp. 681-686.

Weidenhamer, J.D., C.H. David and T.R. John (1989) : Density dependent phytotoxicity distinguishing resource competitions and allelopathic interference in plants, Journal of Applied Ecology, Vol. 26, pp. 613-624.

Zhang, Q. (1993) : Potential role of allelopathy in the soil and the decomposing root of chinese fir replanted woodland, Plant and Soil, Vol. 151, pp. 205-210.

Comparative Vegetation Analysis under Multipurpose Plantations

Ravinder Kumar KOHLI

Department of Botany, Panjab University, Chandigarh 160 014, India

ABSTRACT A study was conducted in and around Chandigarh, India to analyse the vegetation under the monoculture plantations of exotic (*Eucalyptus tereticornis, E. citriodora, Populus deltoides* and *Leucaena leucocephala*) and indigenous (*Albizia lebbeck, Dalbergia sissoo* and *Acacia nilotica*) tree species. Exotic plantations were observed to harbour less number of plant types compared to indigenous ones. In addition, but their density and biomass were also drastically reduced. Likewise, the indices of diversity, evenness and richness were also comparatively less under exotic plantations indicating the homogeneity and thus instability of the vegetation under them. Further, the soil collected beneath these exotic plantations was found to be rich in phytotoxic allelochemicals as indicated by their bioefficacy against *Phaseolus aureus* - a test plant.
Keywords : allelochemicals, ecological indices, exotics, indigenous, monocultures.

1. INTRODUCTION

Understorey vegetation plays an important role in the nutrient cycling and overall dynamics of both natural forests and man made plantations. The latter are being raised as monocultures in order to meet the needs of the people and to reduce pressures on the fast depleting natural forests. For plantations, selection of tree component is of vital importance for environmental sustainability. Nevertheless, under various afforestation programmes, this factor remains generally ignored. The selection is based on the consideration of fast growth rate, economic gains and resource demand, rather than ecological aspects. In order to fill the gap between the supply and demand of forest resource. the emphasis on plantations of exotic trees has increased tremendously especially during the last two decades.

Among the exotics, *Eucalyptus tereticornis, E. citriodora, Populus deltoides* and *Leucaena leucocephala* are being promoted at a large scale. They are being raised in blocks as monoculture and energy plantations along field bunds as shelterbelts/windbreaks and also in the wasteland areas. Because of their fast growth rate coupled with early economic returns and incentives from private entrepreneurs these are being preferred over the indigenous ones like *Albizia lebbeck, Dalbergia sissoo* and *Acacia nilotica*. These exotics are being raised without keeping in mind their ecological and environmental suitability and sustainability in terms of soil health, native flora and fauna, hydrological balance, vegetation dynamics etc. Generally, the plantations of these exotics are observed to have very sparse understorey vegetation, probably due to the allelopathic interference of these trees besides the competition. Keeping this in mind a study was performed to assess the status of ground vegetation under the plantations of these exotics in terms of plant diversity, density, richness in comparison to the plantations of indigenous trees. Besides the quantification and the bioefficacy of allelochemicals, if any, from the soil collected from these plantations also formed the objective of this study.

2. MATERIAL AND METHODS

2.1 Study Site

The study was conducted in and around Chandigarh (30° 55' N; 76° 54' E; 333 m MSL) where the plantations of *Eucalyptus tereticornis, E. citriodora, Populus deltoides* and *Leucaena leucocephala* (all exotics), *Albizia lebbeck, Acacia nilotica* and *Dalbergia sissoo* (all indigenous) were selected. The trees were nearly 8-10 years old, except in case of *D. sissoo* which were nearly 15 years old.

2.2 Vegetation Analysis

For this, ten quadrats each of 1x1 m were laid at random in each of the plantations. The plants falling in each quadrat were counted, harvested and identified. The biomass and density (number/m^2) of the plants under each plantation was determined. Besides, the vegetation was analysed using various ecological indices such as Shannon-Weiner index of diversity, Richness index, Evenness index and index of dominance as per the details given in Kohli et al. (1996).

2.3. Extraction of allelochemicals from soil

Soil was collected from a depth of 0-10 cm under each plantation, air dried and sieved. Allelochemicals were extracted from it using sodium salt of EDTA following the method of Kaminsky and Muller (1977). These were quantified and their bioefficacy was determined using mung bean - *Phaseolus aureus* as a test plant.

2.4 Bioefficacy Studies

A solution (concentration 0.25%) of the chemicals extracted from the different plantation sites was prepared in pure water (Conductivity <0.05 µS). Hundred seeds of *P. aureus* were imbibed in 25 ml of respective solutions in a 6″ diameter Petri plate for 8 h. These were then arranged on a Whatman # 1 filter paper circle underlined with a thin absorbent cotton wad moistened in the respective solutions. The treatment with pure water in a similar way served as control. The entire set-up was placed in a seed germinator maintained at controlled conditions of temperature (25 ± 2 °C), humidity (75 ± 3 %), and photoperiod (16h light and 8h dark). After a week the number of seeds that germinated, their seedling length and biomass were determined.

3. RESULTS

3.1 Vegetation Analysis

The exotic plantations supported very sparse vegetation as compared to the indigenous ones. Not only the plant types, but density and biomass were also reduced (Table 1). Compared to nearly 28 or 29 plant types under plantations of indigenous trees the exotics except *L. leucocephala* had 14 to 16 plant species only. *L. leucocephala* supported only 6 type of species. Likewise, the number of plants and biomass per unit area were also drastically reduced under exotics compared to indigenous plantations (Table 1). Maximum number of plants per unit area were found under the plantations of *D. sissoo* whereas minimum was observed in *L. leucocephala* plantations. However, the biomass was assessed to be minimum under *Eucalyptus* plantations (Table 1).

Table 1. Comparison of the number of plant types, total density and biomass of the vegetation under different plantations.

Plantation	Number of Plant Types	Density (no./ m^2)	Biomass (g/ m^2)
Exotic			
E. tereticornis	16 ± 2.5	175 ± 12.3	78.00 ± 5.6
E. citriodora	14 ± 1.8	154 ± 10.4	74.35 ± 6.7
P. deltoides	17 ± 3.2	180 ± 14.8	80.45 ± 4.8
L. leucocephala	6 ± 0.8	45 ± 7.6	150.00 ± 4.2
Indigenous			
A. lebbeck	28 ± 4.5	278 ± 19.8	170.35 ± 6.4
D. sissoo	29 ± 3.8	418 ± 22.5	295.09 ± 9.6
A. nilotica	29 ± 3.5	310 ± 18.9	235.10 ± 8.7

The reduced density and biomass under exotic plantations is further supported by the reduced values of the indices of species Richness and Evenness and the Shannon-Weiner index of Diversity under these, compared to indigenous plantations (Table 2). However, index of dominance was more under these plantations compared to indigenous ones. It was found to be maximum under *L. leucocephala* plantations indicating that the vegetation was dominated by a very few species as is also evident from the number of plant types.

Table 2. Comparison of the Various Ecological Indices under different Exotic and Indigenous Plantations. Different superscripts in a column represent significant difference at $p < 0.05$.

Plantations	Index of Evenness	Index of Richness	Shannon-Weiner Index	Index of Dominance
Exotic				
E. tereticornis	0.684^e	1.99^b	1.98^c	0.238^c
E. citriodora	0.672^f	1.86^c	1.85^d	0.245^b
P. deltoides	0.695^d	2.03^b	2.01^c	0.233^c
L. leucocephala	0.723^a	1.40^d	1.25^e	0.608^a
Indigenous				
A. lebbeck	0.706^c	3.55^a	2.60^a	0.143^e
D. sissoo	0.712^b	3.65^a	2.58^a	0.148$^{d\,e}$
A. nilotica	0.704^c	3.58^a	2.54^b	0.151^d

3.2 Content of Allelochemicals in Soil

The soil collected from exotic plantations was found to be rich in organic chemicals when compared to indigenous ones, where only negligible amount was present. Among the exotics maximum amount of the chemicals was present in the soil from *L. leucocephala* followed by *E. tereticornis*, *E. citriodora* and *P. deltoides*. By and large, the content of the chemicals in the exotics was more by 4-5 times than in indigenous ones (Fig. 1).

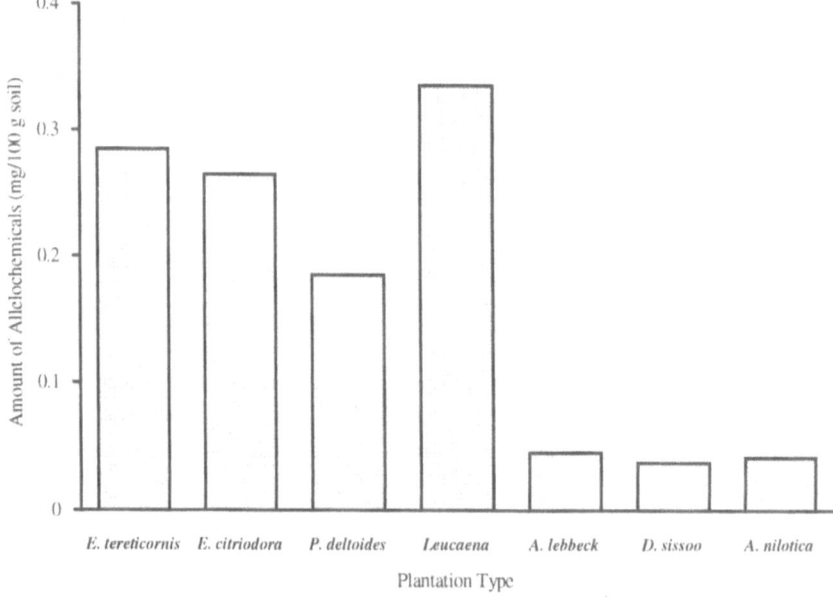

Fig. 1. Amount of allelochemicals extracted from soil of different plantations under study.

3.3 Bioefficacy of the Soil Chemicals

The Soil chemicals extracted from the floor of the exotic plantations were found to drastically reduce the germination of the *P. aureus*. Not only the germination, but even the seedling length of the tested seed after germination were also reduced (Table 3).

Among the exotics, the soil chemicals from the *L. leucocephala* were found to be most inhibitory followed by *E. tereticornis*. However, the soil chemical from the *P. deltoides* soil were comparatively less toxic (Table 3).

Table 3. Effect of 0.25 % organic chemicals extracted from soil of different exotic plantations on *P. aureus*. Values are in % with respect to control; In control germination 100 ± 0, seedling length 14.58 ± 1.63 cm.

Parameter	E. tereticornis	E. citriodora	L. leucocephala	P. deltoides
Germination	44.3 ±2.5	46.5 ± 3.3	37.7 ± 3.2	58.5 ± 3.4
Seedling length	35.8 ± 2.4	37.3 ±2.9	24.3 ±0.8	54.8 ± 5.2

4. DISCUSSION

The density, diversity and biomass of the understorey vegetation was drastically reduced under monocultures of exotic trees compared to indigenous ones. Several reports are available in the literature which indicate lesser vegetation under the exotic plantations e.g. *Eucalyptus tereticornis* (Singh et al. 1993), *E. globulus* (del Moral and Muller 1969), *E. camaldulensis* (del Moral and Muller 1970), *E. baxteri* (del Moral et al 1978), *Acacia tortilis* (Mehta and Sen 1994) and *Delonix regia* (Chou and Lee 1992). Further, the indices of richness, evenness and diversity were also comparatively less under exotic plantations, compared to indigenous ones. Contrarily, the index of dominance was higher under these exotic plantations, indicating thereby, the instability of the community under them. Since higher the value of index of dominance, more homogenous the community will be and thus less stable. On the other hand lower value of index of dominance in the indigenous plantations indicate heterogeneity and thus more stability of the community.

In order to observe whether this reduced vegetation under exotics was due to allelopathy, the floor soil was collected from these plantations. It was observed that soil collected from exotic plantations was rich in allelochemicals. These allelochemicals were found to inhibit the germination and growth of *Phaseolus aureus*. These may be released in to the atmosphere upon leachation, volatilization or decomposition of the fallen plant parts (Rice 1984). This does not seem surprising in the light of several reports indicating the release of allelochemical from these exotics into their surrounding environment and affecting the vegetation.

Eucalypts are also known to release volatile terpenes like *cineole, limonene* and *citronellal* which travel downwards, adhere to the soil particles and damage the vegetation (Kohli 1990, del Moral and Muller 1969, 1970, Singh et al 1991). Likewise, *Leucaena* plantations are known to release a toxic substance *mimosine* apart from a number of phenolic acids through its leaves and pods and cause considerable toxicity (Chou and Kuo 1986, Chaturvedi and Jha 1992). Similarly, allelopathy of *P. deltoides* has also been convincingly demonstrated through its various allelochemicals such as phenolic acids and *salicin* - a phenol glucoside (Singh 1995).

Thus, from the present study it is clear that monocultures of exotic plantations such as *E. tereticornis, E. citriodora, L. leucocephala, P. deltoides* harbour reduced understorey vegetation compared to the indigenous ones like *A. lebbeck, A. nilotica* and *D. sissoo*.

REFERENCES

Chaturvedi, O.P. and A.N. Jha (1992) : Studies on allelopathic potential of an important agroforestry species, Forest Ecology and Management, Vol. 53, pp. 91-98.

Chou, C.H. and L.L. Lee (1992) : Allelopathic substances and interactions of *Delonix regia* (Boj.) Raf., Journal of Chemical Ecology, Vol. 18, pp. 2285-2303.

Chou, C.H. and Y. Kuo (1986) : Allelopathic research of subtropical vegetation in Taiwan. III. Allelopathic exclusion of understorey by *Leucaena leucocephala,* Journal of Chemical Ecology, Vol. 12, pp. 1431-1448.

del Moral, R. and C.H. Muller (1969) : Fog drip : A mechanism of toxin transport from *Eucalyptus globulus,* Bulletin of the Torrey Botanical Club, Vol. 96, pp. 467-475.

del Moral, R. and C.H. Muller (1970) : The allelopathic effects of *Eucalyptus camaldulensis,* American Midland Naturalist, Vol. 83, pp. 254-282.

del Moral, R., R.J. Willis and D.H. Ashton (1978) : Suppression of coastal heath vegetation by *Eucalyptus baxteri,* Australian Journal of Botany, Vol. 26, pp. 203-219.

Kaminsky, R. and W.H. Muller (1977) : The extraction of the soil phytotoxins using a neutral EDTA solution, Soil Science, Vol. 124, No. 4, pp. 205-209.

Kohli, R.K. (1990) : Allelopathic Potential of Eucalyptus, Project Report, Department of Environment, Government of India, 199p.

Kohli, R.K., H.P. Singh and D. Rani (1996) : Status of Floor Vegetation Under Some Monoculture and Mixculture Plantations in North India, Journal of Forest Research, Vol. 1, No. 4, pp. 205-209.

Mehta, M. and D.N. Sen (1994) : Vegetation analysis under some tree stands of semi-arid zones, Proceedings of the National Academy of Sciences, India, Vol. 64 B, No. II, pp. 195-198.

Rice, E.L. (1984) : Allelopathy, Academic Press, New York, 422p.

Singh, D., R.K. Kohli and N. Jerath (1993) : Impact of Eucalyptus and other plantations on phytodiversity in India, Proc. International Conference on Forest Vegetation Management - Ecology, Practice and Policy, Auburn University, Auburn, Alabama, USA, pp. 152-159.

Singh, D., R.K. Kohli and D.B. Saxena (1991) : Effect of eucalyptus oil on germination and growth of *Phaseolus aureus* Roxb., Plant and Soil, Vol. 137, pp. 223-227.

Singh, H.P. (1995) : Phytotoxic effects of *Populus* in natural and agro-ecosystems, Ph.D. thesis Panjab University, Chandigarh, India.

Allelopathic Effects of Exotic Tree Species on Microorganisms and Plants in Galicia (Spain)

Manuel J. REIGOSA[*1]; Xosé C. SOUTO[*2]; Luis GONZÁLEZ[*1] and Juan C. BOLAÑO[*1]

*1 Lab. Ecofisioloxía Vexetal. Dep. Bioloxía Vexetal e Ciencia do Solo.
  Facultade de Ciencias de Vigo. Universidade de Vigo, Spain
*2 Producción Vexetal. Dep. Enxeñería dos Recursos Naturais e do Medio
  Ambiente. Escola Universitaria de Enxeñería Técnica en Industrias Forestais de
  Pontevedra, Universidade de Vigo, Spain

ABSTRACT During one year, the evolution of eleven groups of micro-organisms has been monitored in the soil in a plantation of oakwood (*Quercus robur*, the autochthonous tree species) and three alien trees (*Acacia melanoxylon*, *Eucalyptus globulus* and *Pinus radiata*). The main differences could be attributed to the different water regimes and soil pH, but also to some extent to the allelochemicals released. Also, the effect of the litter decomposing in the soils on germination and seedling growth was studied. The results are compatible with Rabotnov's hypothesis, that is, that allelopathy is mainly possible between plants that have not coevolved or in special conditions where phytotoxins are accumulated. The results are discussed regarding to this hypothesis.
Keywords: Allelopathy, *Acacia melanoxylon*, *Pinus radiata*, *Eucalyptus globulus*, *Quercus robur*

1. MATERIALS AND METHODS

The studied plots were an autochthonous oakwood (*Quercus robur*), an eucalyptus crop (*Eucalyptus globulus*), an acacia crop (*Acacia melanoxylon*) and a pine crop (*Pinus radiata*). These last three are non-indigenous species in the NW of the Iberian Peninsula.

1.1 Soil microorganisms

Microbial quantifications were made using the Most Probable Number (MPN) method (Alexander, 1982) and counting colonies growing on agar media. They were made monthly throughout a year. Nine soil samples were taken in each stand at three randomly chosen locations. Litter was removed from the surface and soil cores were taken (up to 10 cm depth). They were sieved (2 mm) and homogenised. Aliquots were saved for determining pH, soil moisture and microbial populations densities. Twenty grams of each soil were shaken in 180 ml of water for 25 minutes (10^{-1} dilution) and consecutive 10-fold serial dilutions were obtained (until 10^{-8} dilution). Four replicates were prepared for each soil. The following microbial groups were determined: dinitrogen fixers (Clark, 1965), proteolitic microorganisms, ammonifiers, starch hydrolyzers (Pochon and Tardieux, 1962), denitrifying bacteria (Focht and Joseph, 1973), *Nitrosomonas*, *Nitrobacter* (Alexander and Clark, 1965), bacteria, fungi (Parkinson *et al.*, 1971), cellulose hydrolyzers (Eggins and Pugh, 1962) and algae (Wilson, 1937).

1.2 Soil bioassays and phenolics in soil

The effects of soil solutions on seed germination and seedling growth and the phenolic composition of the soils from the four stands were determined monthly from February to July. This period coincides with germination and early growth of most of the understorey species in this part of the Iberian peninsula. Germination and growth of three species were measured: *Lactuca sativa* var. Great Lakes, *Dactylis glomerata* and *Trifolium repens* (the two former as understory species in old forests). Soil samples were taken as for microbial analysis and transferred in Petri dishes (approximately 0.5 cm high), with Whatman 3MM paper and 50 seeds of each one of the receptor species. Five replicates were used per soil and species. Plates were placed in a stove at 28 °C under constant humidity. After 60 h (*L. sativa*), 72 h (*T. repens*) or 120 h (*D. glomerata*) dishes were transferred to a cold chamber at -20°C to stop radicle growth. Germination percentage and emergent radicle length were then determined.

Soil samples for phenolics determination were taken as for the previous experiments, and pH was measured. The pH of the extractant was adjusted to the pH of each soil, because this parameter is essential in the quantity and chemical stage of phenolics in the soil (Kuiters and Denneman, 1987; Appel, 1993). Extraction and identification procedures were performed as previously described by Souto (1997).

2. RESULTS

Statistical analyses were made to determine the influence of different parameters on microbial growth. First of all ANCOVA (analysis of covariance) was used to know the importance of pH and soil moisture. Since both parameters significantly affected microbial growth rate, multiple regression analyses were performed for each microbial group each month, taking pH and soil moisture as independent variables. A new variable was then created, in which effects produced by pH or soil moisture were removed. ONEWAY ANOVA analysis and LSD test were then applied with the new variable to determine differences in microbial growth due exclusively to stand characteristics, once pH and soil moisture effects were eliminated.

Results of this experience showed important differences among soils in relation to the amount of microorganisms. Soil from *Acacia* stand had the lower densities in many groups of microbes, followed by soil from *Eucalyptus*. Soils from oakwood and *Pinus* stand had almost always the highest densities. Statistically significant differences among soils are showed in table 1 for nitrogen cycle microorganisms. It can be seen that most important differences appeared between oakwood and *Acacia* and *Eucalyptus* stands, mainly for *Nitrosomonas* and proteolitic microorganisms.

Table 1. Microbial densities in four soils, expressed as log (n+1). **Q, E, P** and **A** indicate stand from *Quercus, Eucalyptus, Pinus* and *Acacia* respectively. (P.M.= proteolitic microorganisms; D.F.= dinitrogen fixers; Ns.= *Nitrosomonas*; Am.= ammonifiers). Significant differences showed when letters are different (p<0.05), once effects produced by pH and soil moisture were removed

	Q	E	P	A		Q	E	P	A
P.M.					**Ns.**				
Sept.	6,30 a	5,31 b	5,76 ab	5,99 ab	Feb.	1,49 a	1,52 a	1,59 a	0,51 b
Oct.	5,69 ab	5,33 b	5,47 ab	5,84 a	Apr.	1,41 a	1,03 b	1,13 b	1,19 a
Nov.	5,98 a	5,84 a	5,49 ab	5,25 b	May	1,59 a	1,21 ab	1,52 a	1,04 b
Jan.	6,41 a	5,80 b	6,04 ab	6,09 ab	June	1,24 a	0,41 c	1,37 a	0,76 b
Feb.	6,30 a	5,71 b	5,67 b	5,85 b					
May	5,77 ab	6,00 a	5,45 b	6,17 a	**Am.**				
Jul.	6,36 a	6,10 ab	5,63 b	6,18 ab	Sept.	6,74 a	5,84 b	6,13 ab	6,12 ab
Aug.	6,28 a	5,92 bc	5,65 b	6,17 ac	Jan.	6,18 ab	6,29 a	5,95 ab	5,71 b
D.F.					Mar.	5,77 b	5,97 ab	6,27 a	5,45 c
May	3,31 b	3,39 b	3,33 b	3,56 a	Aug.	6,04 a	5,54 b	5,46 b	5,77 ab
Aug.	3,03 ab	3,16 ab	2,84 b	3,26 a					

We also found differences among soils when tested general microbial groups and carbon cycle microorganisms. Thus, lower densities of fungi, cellulose hydrolyzers, starch hydrolyzers and algae were found in the Acacia stand when compared with the other stands. These differences were again due to characteristics of the stand, since effects produced by pH and soil moisture were removed. Results of these groups are showed in table 2.

Table 2. Microbial densities in four soils, expressed as log (n+1). **Q, E, P** and **A** indicate stand from *Quercus, Eucalyptus, Pinus* and *Acacia* respectively. (Bac.= general bacteria; C.H.= cellullose hydrolyzers; S.H.= starch hydrolyzers). Significant differences showed when letters are different (p<0.05), once effects produced by pH and soil moisture were removed.

	Q	E	P	A		Q	E	P	A
Bac.					**C.H.**				
Oct.	6,68 a	6,49 b	6,48 b	6,42 b	Sept.	8,28 a	8,10 c	8,24 ab	8,19 b
Dic.	6,70 a	6,73 a	6,33 b	6,55 a	Oct.	6,63 a	6,84 b	6,80 ab	6,72 ab
Feb.	6,68 a	6,51 b	6,47 b	6,68 a	Dic.	7,29 a	6,81 a	6,53 b	6,34 a
Jul.	6,73 a	6,71 a	6,08 b	6,14 b	Feb.	6,63 a	6,55 a	6,66 a	6,33 b
Fungi					May	6,93 a	6,26 ab	6,28 a	6,41 b
Oct.	5,88 c	6,03 b	6,18 a	5,89 c	Aug.	6,90 a	6,43 b	6,31 ac	6,53 c
Feb.	5,99 b	5,84 c	6,19 a	5,71 d					
May	6,39 a	6,05 b	6,16 a	6,05 b	**S.H.**				
Jul.	6,55 a	6,34 b	6,20 c	5,98 d	Sept.	5,78 a	4,93 b	5,35 ab	5,27 ab
Aug.	6,15 a	5,97 b	6,11 a	6,14 a	Mar.	5,67 a	4,83 b	5,27 ab	5,25 ab

Algae					Jul.	5,39 a	4,36 a	4,25 a	5,46 b
May	4,59 a	3,82 ab	4,21 a	3,98 b	Aug.	4,31 a	4,05 b	5,06 a	4,39 b
Jul.	4,95 a	3,63 a	3,87 a	3,88 b					
Aug.	3,72 a	3,38 b	3,95 a	3,43 b					

2.1 Soil bioassays and phenolics in soil

Results are summarised in tables 3 and 4, as a percentage with respect to oakwood soil. There were no statistically significant differences between oakwood soil treatment and water control, so the first was taken as a control (because it is the soil in the climax ecosystem). In general, growth was stronger affected than germination.

The most sensitive species according to germination was *T. repens*. Soil from *Acacia* stand was the most toxic. The most critical months seem to be spring ones, from April to June, probably due to the presence of toxic substances released from flower residues on the stand ground. This effect can also be observed on growth, in which soil from *Acacia* stand still remained the most toxic.

L. sativa germination was affected essentially by soils from *Eucalyptus* and *Acacia* stands. The most important inhibitions are found, however, in growth. Almost all months presented significant inhibitions, the most notable being those produced by *Acacia* soil. This soil was toxic since February to June. Something similar happened with *Eucalyptus* and *Pinus* soils, but with less intensity and not so long.

With respect to *D. glomerata*, it can be seen that phytotoxic effects are less important. The most notable result is the inhibitory capacity of the soil from *Acacia* stand on both germination and growth.

Table 3. Effects of soil solution on germination of three species. Significant differences (p< 0.05) in each column with respect to the control are represented as asterisks.

	Germination					
	February	**March**	**April**	**May**	**June**	**July**
L. sativa						
E. stand	98.7 ± 1.3	89.7 ± 9.5	79.4 ± 6.2*	92.3 ± 4.3	90.7 ± 2.1*	97.8 ± 2.9
P. stand	88.7 ± 1.9*	98.9 ± 5.7	90.3 ± 1.9	96.5 ± 3.3	97.8 ± 1.2	102.6 ± 2.3
A. stand	79.1 ± 2.3*	96.4 ± 4.2	79.4 ± 7.2*	87.5 ± 2.2	98.3 ± 1.4	98.2 ± 2.6
D. glom.						
E. stand	72.0 ± 9.0*	—	84.3 ± 2.4	—	109.1 ± 13	117.4 ± 16.5
P. stand	63.0 ± 4.2*	—	80.0 ± 10*	—	72.7 ± 13.1	108.7 ± 17.2
A. stand	82.5 ± 5.3	—	67.0 ± 8.4*	—	60.6 ± 11*	121.7 ± 5.9
T. repens						
E. stand	105.3 ±2.9	103.1 ± 3.3	81.1 ± 5.2*	86.8 ± 6.6	85.4 ± 4.8*	101.5 ± 3.2
P. stand	104.3 ± 2.0	98.7 ± 6.1	98.4 ± 3.9	75.1 ± 6.5*	80.2 ± 4.1*	93.9 ± 3.3
A. stand	95.1 ± 2.4	94.5 ± 5.3	83.2 ± 4.7*	75.1 ± 2.8*	76.5 ± 2.9*	97.9 ± 4.6

Chromatographic analysis revealed very small quantities of phenolics in any of the soils. Results are showed in table 5. Oakwood soil presented only two phenolics, and in very small amounts. More phenolics were detected in *Eucalyptus* stand, where quantities were the highest of all soils, but concentrated in June and July. Despite this result, phenolics amounts were very small. The soil from *Acacia* stand also showed an important number of phenolics, but in such low concentrations that most of them appeared as traces, and similar results were observed for *Pinus* stand.

Table 4. Effects of soil solution on growth of three species. Significant differences ($p < 0.05$) in each column with respect to the control are represented as asterisks.

	February	March	April	May	June	July
Growth						
L. sativa						
E. stand	72.3 ± 3.9*	84.1 ± 3.3*	86.5 ± 2.9*	98.0 ± 3.6	103.6 ± 2.8	94.1 ± 1.8
P. stand	70.0 ± 4.2*	89.3 ± 6.1*	88.4 ± 2.5*	82.4 ± 2.7*	98.0 ± 2.3	107.5 ± 4.9
A. stand	76.7 ± 6.0*	72.0 ± 2.9*	68.4 ± 4.5*	83.9 ± 4.5*	81.3 ± 2.3*	100.1 ± 3.9
D. glom.						
E. stand	100.7 ± 9.3	—	84.7 ± 1.4	—	93.9 ± 10.1	94.9 ± 13.7
P. stand	114.5 ± 9.6	—	91.1 ± 7.2	—	87.0 ± 8.9	78.5 ± 9.2
A. stand	99.3 ± 8.1	—	52.7 ± 5.9*	—	54.1 ± 12.6*	93.9 ± 13.6
T. repens						
E. stand	90.0 ± 3.8*	93.2 ± 2.9	85.0 ± 2.4*	95.2 ± 2.9	97.4 ± 2.8	96.8 ± 2.9
P. stand	97.9 ± 1.3	105.6 ± 5.2	90.3 ± 1.8*	92.2 ± 4.2	97.8 ± 3.1	87.1 ± 3.4*
A. stand	93.0 ± 4.4	92.6 ± 3.7	83.9 ± 3.4*	84.7 ± 5.1*	78.9 ± 1.6*	88.3 ± 3.8*

Table 5. Phenolics in soils of oakwood, *Eucalyptus*, *Pinus* and *Acacia* stands (tr. indicates traces; n.d. indicates not detected; ?= lost data). Data in µg/g dry soil.

	Phenolics	Febr.	March	April	May	June	July
Oakw.	*p*-hydroxybenzoic acid	n.d.	n.d.	n.d.	n.d.	n.d.	0.0233
	Vanillic acid	tr.	tr.	n.d.	n.d.	n.d.	tr.
Eucal.	*p*-hydroxybenzoic acid	0.0167	n.d.	n.d.	n.d.	0.0655	0.0251
Stand	Vanillic acid	n.d.	n.d.	n.d.	n.d.	0.0639	tr.
	Gallic acid	n.d.	n.d.	n.d.	n.d.	0.0662	0.0154
	Ferulic acid	n.d.	n.d.	n.d.	n.d.	tr.	n.d.
	Vanillin	n.d.	n.d.	n.d.	n.d.	0.0302	tr.
Pinus	*p*-hydroxybenzoic acid	tr.	0.0078	n.d.	tr.	0.0161	0.0181
Stand	Vanillic acid	0.0363	tr.	n.d.	tr.	tr.	0.0389
	Chlorogenic acid	n.d.	n.d.	n.d.	n.d.	n.d.	tr.
	Protocatechuic acid	n.d.	n.d.	n.d.	n.d.	n.d.	tr.
Acacia	*p*-hydroxybenzoic acid	tr.	0.0455	tr.	n.d.	0.0204	?
Stand	Vanillic acid	tr.	tr.	tr.	n.d.	tr.	?
	p-coumaric acid	n.d.	tr.	n.d.	n.d.	tr.	?
	Gallic acid	tr.	tr.	n.d.	n.d.	0.0129	?
	Ferulic acid	n.d.	tr.	n.d.	n.d.	tr.	?
	p-hydroxybenzaldehide	tr.	n.d.	n.d.	n.d.	n.d.	?
	Vanillin	tr.	tr.	n.d.	n.d.	tr.	?

3. DISCUSSION

In this work we have studied eleven microbial groups throughout a year, related with nitrogen and carbon cycles, in four different soils. Results indicate the existence of important differences among the four soils with respect to some microbial groups. It was also demonstrated the great importance of edaphoclimatic parameters on microbial development. Soil pH was the main one, responsible for differences in many cases.

Soil pH and allelopathy are closely related. First, because pH depends on chemical substances that reach the soil, so minimal differences in pH could be due to the slower rate of chemical compounds released to the environment. Besides, the activity of some substances is very sensitive to pH (Kuiters, 1990; Appel, 1993). Several differences in microbial populations that are explained by pH could finally be due to different activities of chemical compounds.

Lower number of some microbial populations were found in the *Acacia* stand when compared with the other stands. The *Eucalyptus* stand also had low densities in many cases, although in a soft way than the *Acacia* stand. Finally, the *Acacia* stand appears clearly different from the other stands in number of microorganisms, so indicating the existence of chemical interference phenomena among the forest trees and edaphic flora. This result has been already shown in decomposition bioassays made by our group (Souto et al., 1994), with the same negative effects produced by *Acacia melanoxylon*.

Several authors have established allelopathic effects of phenolics on nitrifying bacteria and other microbial groups (Rice and Pancholy, 1972; Lodhi and Killingbeck, 1980; Olson and Reiners, 1983) and on many plant species (Li et al., 1993; Lehman et al., 1994; Blum, 1995; Inderjit and Dakshini, 1995). Sometimes it was thought that they could act through synergistic or additive effects. In this work we found very small amounts of phenolics in soils of any of the stands during the months in which inhibitions of microbial growth were detected. Thus, it seems unlikely that inhibitions are due to this kind of compounds. Other substances are probably responsible for the allelopathic effects.

Using soil bioassays we found that inhibitory effects were more important in spring, probably due to allelopathic compounds released by flowers. However, we also found inhibitions in other months, when flowers debris do not exist on soils, so they can be produced by decomposition of leaves and barks (Souto, 1997). Soils from *Acacia* and *Eucalyptus* stands were quite toxic, mainly in spring. In this season the germination and early growth of typical understory species occur, so allelopathy is specially favoured because they did not coevolve with alochthonous species (Rabotnov, 1974). However, as mentioned above for microorganisms, it seems unlikely phenolics are responsible for inhibitions found in the soil assays, because of the very small amounts in soils of the four stands.

REFERENCES

Alexander, M. (1982) : Most-probable number method for microbial populations, American Society of Agronomy, Madison (Wisconsin), 2nd Edition, pp. 815-820.

Alexander, M. and Clark, F.E. (1965) : Nitrifying bacteria, American Society of Agronomy, Madison (Wisconsin), pp 1477-1483.

Appel, H. M. (1993) Phenolics in ecological interactions: the importance of oxidation, Journal of Chemical Ecology, Vol. 19, No. 7, pp. 1521-1551.

Blum, U. (1995) : The value of model plant-microbe-soil systems for understanding processes associated with allelopathic interaction. One example, ACS Symposium Series 582, Washington, pp. 127-131.

Clark, F.E. (1965) : Agar-plate method for total microbial count, American Society of Agronomy, Madison (Wisconsin), pp 1460-1466.

Eggins, H.O.W. and Pugh, G.J.F. (1962) : Isolation of cellulose decomposing fungi from the soil, Nature, 193, pp. 94-95.

Focht, D.D. and Joseph, H. (1973) : An improved method for the enumeration of denitrifying bacteria, Soil Sci. Soc. Amer. Proc., 37, pp. 698-699.

Inderjit and Dakshini, K. M. M. (1995) On laboratory bioassays in allelopathy, Botanical Review, 61, pp. 28-44.

Kuiters, A. T. (1990) Role of phenolic substances from decomposing forest litter in plant-soil interactions, Acta Bot.Neerl., Vol. 39, No. 4, pp. 329-348.

Kuiters, A. T. and Denneman, A. J. (1987) Water-soluble phenolic substances in soils under several coniferous and decidous tree species. Soil Biology and Biochemistry, Vol. 19, No. 6, pp. 765-769.

Lehman, M. E., Blum, U., and Gerig, T. M. (1994) Simultaneous effects of ferulic and p-coumaric acids on cucumber leaf expansion in split-root experiments, Journal of Chemical Ecology, Vol. 20, No. 7, pp. 1773-1782.

Li, H. H., Inoue, M., Nishimura, H., Mizutani, J., and Tsuzuki, E. (1993) Interactions of trans-cinnamic acid, its related phenolic allelochemicals, and abscisic acid in seedling growth and seed germination of lettuce, Journal of Chemical Ecology, Vol. 19, No. 8, pp. 1775-1787.

Lodhi, M.A.K. and Killingbeck, K.T. (1980) : Allelopathic inhibition of nitrification and nitrifying bacteria in a ponderosa pine (Pinus ponderosa Dougl.) community, American Journal of Botany, 67, pp. 1423-1429.

Olson, R.K. and Reiners, W.A. (1983) : Nitrification in subalpine balsam fir soils: tests for inhibitory factors, Soil Biology and Biochemistry, 15, pp. 413-418.

Parkinson, D., Gray, T.R.G. and Williams, S.T. (1971) : Methods for studying the ecology of soil microorganisms, IBP Handbook 19, Blackwell Sci. Publ., Oxford.

Pochon, J. and Tardieux, P. (1962) : Techniques d'analyse en microbiologie du sol, La Tourelle, St. Mandé.

Rabotnov, T.A. (1974) : On the allelopathy in the phytocenoses, Izo Akad Nauk SSSR Ser. Biol., 6, pp. 811-820.

Rice, E.L. and Pancholy, S.K. (1972) : Inhibition of nitrification by climax ecosystems, American Journal of Botany, 59, pp. 1033-1040.

Souto, X.C. (1997) : Fenómenos alelopáticos de especies arbóreas sobre plantas del sotobosque. Un estudio comparativo, PhD. Thesis, University of Vigo.

Souto, X.C., González, L. and Reigosa, M.J. (1994) : Comparative analysis of allelopathic effects produced by four forestry species during decomposition process in their soils in Galicia (NW Spain), Journal of Chemical Ecology, 11, pp. 3005-3015.

Wilson, J.K. (1937) : Pure culture of algae from soil, Soil Sci. Soc. Amer. Proc., 1, pp. 211-212.

Soil Microorganisms and Phenolics : Their Implication in Spruce Natural Regeneration Failure.

X. Carlos SOUTO [*1] , Geneviève CHIAPUSIO [*2] and
François PELLISSIER [*2,3]

*1 University of Vigo, EUET Forestal, 36005 Pontevedra, Spain
*2 University of Savoie, LDEA, 73376 Bourget-du-Lac Cedex, France
*3 Corresponding author.

ABSTRACT Among other factors, allelopathy due to phenolics is a cause of natural regeneration deficiency in subalpine spruce (*Picea abies* L. Karst.) forests. Turnover of these allelochemicals in soil with and without microorganisms is reported herein. Microflora not only appeared to be able to metabolize some aromatic compounds naturally occurring in soil (*p*-hydroxy-benzoic and vanillic acids) but also being able to synthesize other molecules (ferulic and *p*-coumaric acids), probably by means of lignin degradation. Thus, results presented in this study suggest to pay attention to soil microorganisms in order to decrease the allelopathic potential and to improve natural regeneration.

Keywords : allelopathy, microorganism, phenolics, *Picea*, regeneration

1. INTRODUCTION

Allelopathy, the positive or negative effect of one plant upon another through the production of chemical inhibitors released into the environnement (Rice, 1984), is probably a common cause of forest natural regeneration failure (Fisher, 1987). In the Northern French Alps, subalpine spruce (*Picea abies* L. Karst.) forests are characterized by poor natural regeneration (André et al. 1987). Previous studies reported allelopathic inhibition of spruce germination and seedlings growth, due to soil phenolics (Pellissier 1993 and Pellissier 1994). On the other hand, it has been shown that some of these allelochemicals may be metabolized by some spruce mycorrhizal fungi (Boufalis and Pellissier, 1994). Further, Blum and Shafer (1988) shown that soil microbial populations such as fungi, bacteria and actinomycetes are able to metabolized some phenolics. In the same way, Souto (1997) shown that microbial populations were probably not too much affected by phenolics. Those preliminary observations led us to consider a direct allelopathy against spruce germination and seedlings growth and, an indirect allelopathy, mediated by soil microorganisms which could metabolize understory plants inactive phenolics into biologically active compounds. Thus, the objectives of this study were : 1) to identify and quantify phenolics in subalpine spruce forest soil and ; 2) to test the effect of soil microorganisms on these phenolics turnover.

2. MATERIAL AND METHOD

2.1. General Procedure (Figure 1)

Soil (3 randomly collected samples mixed and passed through 5 mm grill) was separated in three parts : the first one (A), non-sterilized was grouped into two equal parts - the first for phenolic quantification and the second for microbial inoculation - ; the second (B) and the third (C) were sterilised (at 120 C for 25 mm), supplied with phenolics as identified from the first part and with or without microorganisms (inoculated from the first half), respectively.

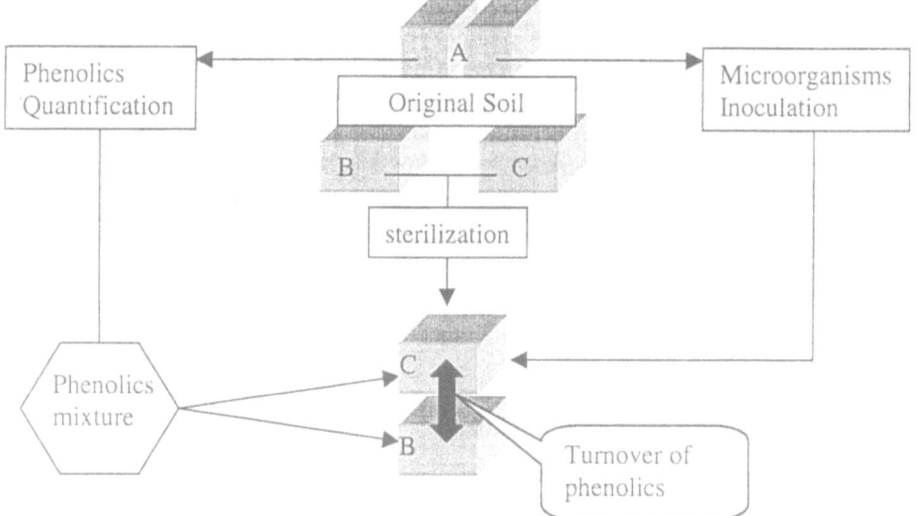

Fig. 1: Method used to study effects of microorganisms on phenolics pool in spruce forest soil.

2.2. Phenolics Quantification

Three replicates of 20 g of soil (fresh weight) were smoothly shacked in 80 ml demineralized water for 18 h (dark, 15 °C). After filtration, phenolics were extracted from acidified solution (by 2 N HCl) with ethyl ether and redissolved in ethanol before H.P.L.C. analysis. A short column of Novapak 18 C was placed immediately before a 300 mm x 3.9 mm ID column, filled with µBondapak 18 C. Linear gradient elution was carried out at a flow rate of 1.5 ml mn^{-1}. Solvent A was 0.5 % acetic acid in distilled water and solvent B, acetonitrile with 0.5 % acetic acid. A gradient from 0 % to 20 % B over 45 mn, followed by 15 mn reequilibration with A was used. Identification of phenolics was performed by comparison of retention times and wavelength detection. The amounts were determined by comparison with reference curves of mixtures of standard compounds (Gallet and Pellissier, 1997).

2.3. Bioassay

After sterilization (see above), 4 replicates of soil (20 g) were incubated with 2 ml of 2.10^{-1} dilution of microorganisms solution from original soil and 4 replicates received 2 ml sterile-demineralized water as treatment without microorganisms. Both of these replicates received 2 ml of phenolics mixture (Sigma Chemical Co.). After 3 days incubation, phenolics mixture was added once more. At the end of the incubation (6 days), phenolics have been quantified as described above (2.2). Phenolics were quantified just after sterilization, after 3 days incubation and at the end of the experiment. Because bioassay results didn't give homogeneous standard deviations, effects of microorganisms on phenolics were analysed using the non-parametric Mann Whitney U test (P<0.05).

3. RESULTS

Original soil respectively contained $4.7 \ 10^{-5}$, $4.8 \ 10^{-5}$, $1.1 \ 10^{-3}$ and $0.9 \ 10^{-3}$ μMol g^{-1} dry weight soil of p-OH-benzoic acid, p-OH-benzaldehyde, vanillic acid and vanillin (61.39% soil moisture). Because soil sampling has been done during winter (under 1.5 m snow) and the objective of this study was to describe effects of microbial populations on phenolics, mixture used into bioassay was more concentrated, according method proposed by Blum and Shaffer (1988). Thus, phenolics mixture was prepared in order to add respectively 0.6, 0.42, 7.4 and 4.2 μMol g^{-1} dry weight soil of p-OH-benzoic acid, p-OH-benzaldehyde, vanillic acid and vanillin.

Figure 2 presents turnover of phenolics with and without microorganisms, expressed as percent of input (addition of mixture) and the difference between presence and absence of microbial population in the third set of columns. Microorganisms were responsible for a significant decrease of p-OH-benzoic (-14%) and vanillic (-22%) acids, compared to the data without microorganisms.

Further, with as without microorganisms, protocatechuic acid appeared after the 6 days incubation (respectively 274.3 and 278.7 μM). Amounts of this neoformed compound are not significantly different with and without microorganisms, suggesting that its apparition was due to physico-chemical process. Two other phenolics were also neoformed but only in presence of microorganisms (Table 1).

Table 1 : *De novo* phenolics identified after 6 days incubation, with and without microorganisms.

	Neoformed phenolic acids (μM)		
	Protocatechuic	Ferulic	p-coumaric
With microorganisms	274.3 ± 2.4	6.7 ± 0.8	4.7 ± 0.7
Without microorganisms	278.7 ± 2.4	no	no

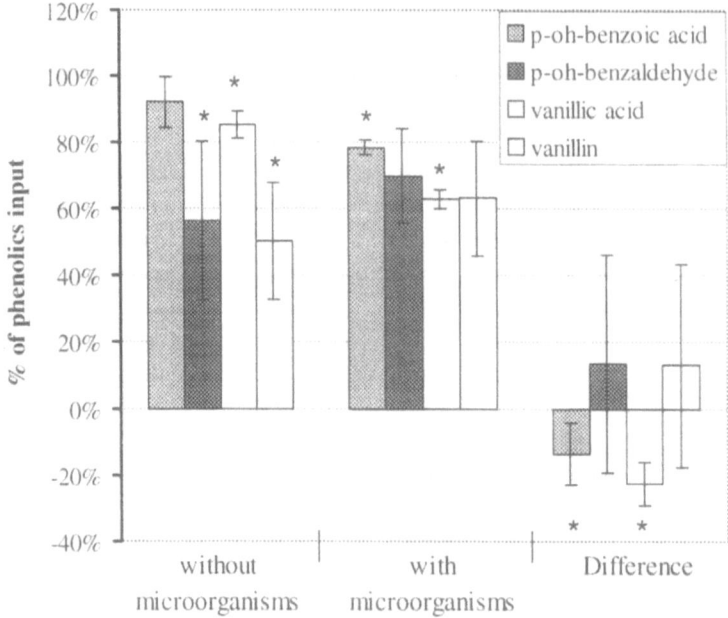

<u>Fig. 2:</u> Phenolics turnover in absence and presence of microorganisms. The first two sets of columns give phenolics contents of soil after 6 days incubation, expressed as % of phenolics input (sum of days 0 and +3 additions). The third set of columns is the difference between the second and the first one. * indicates a significant difference according to Mann Whitney U test for $p<0.05$.

4. DISCUSSION

4.1. Turnover of phenolics in spruce forest soil.

Numerous bacteria and fungi have been shown to catabolize simple substituted derivatives of phenol (Barz and Weltring, 1981). This potential for the disintegration of exogenous carbon compounds (originated from autotrophic plants) assigns to the soil microflora an important biological function as part of the carbon cycle in forest ecosystems.

Implication of microorganisms in this process is confirmed by results presented in this study. Sutherland et al. (1983) described the catabolic pathway of p-coumaric and ferulic acids by *Streptomyces setonii* as the following sequences : i) p-coumaric $\rightarrow$ p-hydroxy-benzaldehyde $\rightarrow$ p-hydroxy-benzoic acid $\rightarrow$ protocatechuic acid and, ii) ferulic $\rightarrow$ vanillin $\rightarrow$ vanillic acid $\rightarrow$ protocatechuic acid. Such pathway also physico-chemically occurs in aerobic conditions.

According to observed differences in phenolics pool with and without microorganisms, two facts suggested an identical sequence : i) the significant decrease of *p*-hydroxy-benzoic and vanillic acids due to microflora and, ii) the formation of protocatechuic acid. Further, two other phenolics have been synthesized only in presence of microbial populations : ferulic and *p*-coumaric acids. These two compounds are well-known to be products of lignin degradation by microorganisms (Turner and Rice, 1975). Thus, there is no doubt that soil microflora in spruce forest is implicated in turnover of aromatic compounds, both as decomposers than as producers. By such a way, they can be considered as mediators of allelochemicals production.

4.2. Implication for spruce natural regeneration.

Natural regeneration failure is one of the characteristics of subalpine spruce forests in the Alps. This problem was reported nearly a century ago (Schaeffer, 1911) and since then numerous factors have been proposed to be responsible for it viz., competition for nutrients and water (Duchaufour, 1953), deficiency of energy (Brossier, 1977) and seed parasitism (Roques and Trosset, 1986). In spite of the relevance of such research, the problem remained unchanged. However, Fisher (1987) showed that allelopathy in many instances played a major role in the natural regeneration failure. According to this report, therefore, the allelopathic hypothesis was investigated in subalpine spruce forests (André et al., 1987). By means of field and laboratory experiments, it has been shown that phenolics produced by understory species such as *Vaccinium myrtillus* L. and *Athyrium filix-femina* L. Roth are able to inhibit spruce seeds germination (Pellissier, 1993), seedlings growth (Pellissier, 1994) and growth and respiration of some of *P. abies* mycorrhizal fungi (Boufalis et al., 1994; Boufalis and Pellissier, 1994). Relevant observation has been done in Newfoundland spruce forest with *Picea mariana* as target species and *Kalmia angustifolia* as allelochemicals producer (Mallık, 1992). Nevertheless, both of these studies mainly focused on direct allelopathy (producer plant → phenolics → target plant).

On the other hand, soil represents a tank of phenolics wherein biologically active concentrations can vary throughout the year according to various factors such as microbial activity, sorption mechanisms on colloids, humidity... (Kuiters, 1987). By this way, soil should be considered as the "Black Box" of the forest ecosystem (Reigosa et al., 1996) and microflora activity on phenolics pool (and other potentially allelochemicals) should not be underestimated. Results presented herein showed that a better understanding of allelopathic interactions in spruce forests pass through a better knowledge of microbial populations activity. Research is in progress in order to identify and quantify main groups of microorganisms and to precize their role in the phenolics turnover, notably to identify the more detoxifying groups. One of the applied aspects of such investigation is to control the more interesting microbial groups by means of nitrogen and carbon regulation, in order to decrease the allelopathic potential of soil and subsequently to improve natural regeneration.

ACKNOWLEDGMENTS

This work was supported by the Rhône-Alpes Region which provided funds for experiments and post-doctoral fellowship to Dr. C.X. Souto, by means of "Emergence" programme.
Thanks to Dr. C. Gallet for her technical assistance in H.P.L.C. analysis.

REFERENCES

André, J., P. Gensac, F. Pellissier and L. Trosset (1987) : Régénération des peuplements d'épicéa en altitude : recherches préliminaires sur le rôle de l'allélopathie et de la mycorhization dans les premiers stades du développement, Revue d'Ecologie et de Biologie du Sol, Vol. 24, No. 3, pp. 301-310.

Barz, W. and Weltring K.-M. (1981) : Biodegradation of aromatic extractives of wood, in : The biochemistry of plants. A comprehensive treatise, P.K. Stampf and E.E. Conn Eds., Academic Press, New-York, pp. 608-666.

Blum, U. and S.R. Shafer (1988) : Microbial populations and phenolic acids in soil, Journal of Chemical Ecology, Vol. 20, No. 6, pp. 793-800.

Boufalis, A. and F. Pellissier (1994) : Allelopathic effects of phenolic mixtures on the respiration of two spruce mycorrhizal fungi, Journal of Chemical Ecology, Vol. 20, No. 9, pp. 2283-2289.

Boufalis, A., F. Pellissier and L. Trosset (1994) : Responses of mycorrhizal fungi to allelopathy : *Cenococcum geophilum* and *Laccaria laccata* growth with phenolic acids, Acta Botanica Gallica, Vol. 141, No. 4, pp. 547-550.

Brossier, J. (1977) : Evolution des idées en matière forestière et conséquences sur la gestion des forêts alpines, Revue Forestière Française, Vol. 29, pp. 153-161.

Duchaufour, P. (1953) : Régénération de l'épicéa et pédologie, Revue Forestière Française, Vol. 4, pp. 257-268.

Fisher, R.F. (1987) : Allelopathy : a potential cause of forest regeneration failure, in : Allelochemicals : role in agriculture and forestry, G.R. Waller Ed., Washington : American Chemical Society Symposium Series, pp. 176-184.

Gallet, C. and F. Pellissier (1997) : Extracellular persistence of phenolic compounds in natural solutions of a coniferous forest, Journal of Chemical Ecology, Vol. 23, No. 10, pp. 2401-2412.

Kuiters, L. (1987). Phenolic acids and plant growth in forest ecosystems, Ph.D. Thesis, Amsterdam : Free University.

Mallik, A.U. (1992) : Possible role of allelopathy in growth inhibition of softwood seedlings in Newfoundland, in : Allelopathy. Basic and applied aspects, S.J.H. Rizvi and V. Rizvi Eds., Chapman & Hall, London, pp.321-340.

Pellissier, F. (1993) : Allelopathic inhibition of spruce germination. Acta Oecologica, Vol. 14, No. 2, pp. 211-218.

Pellissier, F. (1994) : Effects of phenolic compounds in humus on the natural regeneration of spruce, Phytochemistry, Vol. 36, No. 4, pp. 865-867.

Reigosa, M.J., L. González, X.C. Souto and A.M. Sánchez (1996) : Measuring allelopathic effects of trees on ecophysiological parameters in the understory. Methodological considerations from a plant physiologist's point of view, Proc. of the First World Congress on Allelopathy, Cadiz, in press.

Rice, E.L. (1984) : Allelopathy, Academic Press, Orlando, 422p.

Roques, A. and L. Trosset (1986) : Analyse de la distribution altitudinale des insectes ravageurs des cônes d'épicéa (Picea abies (L.) Karst.) dans les Alpes françaises du Nord, in : Proc. of the Second Conference on the cone and seed insects, Besançon, pp. 83-90.

Schaeffer, A. (1911) : Régénération de l'épicéa dans les forêts des Hautes Régions, Bulletin Trimestriel de la Société Forestière de Franche-Comté et Belfort, Vol. 11, pp. 292-300.

Souto, X.C. (1997) : Fenómenos alelopáticos de especies arbóreas sobre plantas del sotobosque. Un estudio comparativo, Ph. D. Thesis, University of Vigo, Spain, 237p.

Sutherland, J.B., D.L. Crawford and A.L. Pometto III (1983) : Metabolism of cinnamic, p-coumaric and ferulic acids by Streptomyces setonii, Canadian Journal of Microbiology, Vol. 29, pp. 1253-1257.

Turner, J.A. and E.L. Rice (1975) : Microbial decomposition of ferulic acid in soil, Journal of Chemical Ecology, Vol. 1, No. 1, pp. 41-58.

Allelopathy and Competition in Coniferous Forests

A. U. MALLIK

Department of Biology, Lakehead University
Thunder Bay, Ontario, Canada P7B 5E1

ABSTRACT The paper deals with the growth inhibitions and interference potential of ericaceous shrubs in the coniferous forests, the possible biological and chemical mechanisms involved and suggests the control measures to combat the problems of growth inhibition. Competition, allelopathy, resource-toxin hypothesis and the involvement of ericoid mycorrhizae have been suggested to explain growth inhibition in conifers. Control of ericaceous shrubs through integrated management practices involving use of herbicides, mulching for destroying vegetative buds depending on the site and the extent of shrub spread apart from the use of mycorrhizal fungi and/or repeated fertilizer application are recommended.
Keywords: allelopathy, conifers, ericaceous shrubs, growth inhibitions, ericoid mycorrhizae

1. INTRODUCTION

Inadequate natural regeneration and growth check of planted conifers are observed in many temperate forests with dense ericaceous understory (Mallik 1995). This paper summarizes the characteristics of ericaceous induced conifer growth inhibitions, describes the possible biological and chemical mechanisms involved in the growth inhibition process and explores the control measures of the ericaceous shrubs in order to enhance conifer regeneration. *Vaccinium myrtillus* can induce germination inhibition of conifers such as Norway spruce (*Picea abies*), *Kalmia angustifolia* do not affect germination but inhibits primary root growth of black spruce (*Picea mariana*), red pine (*Pinus resinosa*), salal (*Gaulthera shallon*) induces growth-check in conifers such as Sitka spruce (*Picea sitchensis*), western hemlock (*Tsuga heterophylla*), western red cedar (*Thuja plicata*) and amabilis fir (*Abies amabilis*)) and heather (*Calluna vulgaris*)-induced growth-check in Sitka spruce has been known for a long time in Europe.

2. MECHANISMS OF CONIFER GROWTH INHIBITION

In a majority of the cases, the ericaceous plants that induce growth inhibition in conifers possess very efficient vegetative regeneration strategies; they are quite resilient to disturbances such as forest fire and logging and their rate of recovery after disturbance is fast (Mallik 1993, 1994; Bunnell 1990). Their rapid vegetative growth results in high litter biomass on the forest floor and the litter of ericaceous plants contain a wide range of secondary metabolites, some of which are phyto- and fungi-toxic (Jalal and Read 1983 a, b). The litter is characterized by high N capital but low nutrient availability. Ericaceous plants grow well in relatively nutrient poor soils and are equipped with ericoid mycorrhizae that are adapted to nutrient deficient sites. The following mechanisms are suggested to explain the ericaceous-induced growth inhibition in conifers:

2.1 Competition

Most understory ericaceous plants grow vigorously by dense sprouting after removal of forest canopy and can attain almost complete ground cover in a matter of four to six years. This rapid aboveground growth is accompanied by the proliferation of an extensive network of roots and rhizomes, making them highly competitive in the initial phase of forest succession (Mallik 1993, 1994). Vegetative sprouting and seed regeneration in many ericaceous plants are stimulated by cutting and fire (Mallik and Gimingham 1985; Mallik 1991). It has been suggested that competition for soil nutrients, particularly N is the main cause for reduced growth of western hemlock, Sitka spruce and western red cedar in presence of salal (Weetman et al. 1990). Messier (1992) suggested that competition for light is a major reason for salal-induced check of Sitka spruce and western red cedar. By having morphological and physiological plasticity in relation to available light some ericaceous plants such as *Kalmia*, *Vaccinium* spp. and salal equipped with sun and shade leaves are able to change the above and below ground biomass allocation pattern and become a better competitor (Bunnell 1990; Smith 1991).

2.2 Allelopathy

Allelopathic inhibition of seed germination and primary root growth of germinants has been suggested as the main reason for failure of natural regeneration of Norway spruce in the presence of *Vaccinium myrtillus* (Pellissier 1994), Sitka spruce in the presence of *Empetrum hermaforditum* (Zackrisson and Nilsson 1992), black spruce in the presence of *Kalmia* (Mallik 1987) and red pine in the presence of *Kalmia* (Mallik and Roberts 1994). Growth-check of planted seedlings of these and other tree species is also thought to be due to allelopathic interaction between the understory shrubs and the conifers. A number of phenolic compounds such as *o*-hydroxyphenylacetic, *p*-hydroxybenzoic, vanillic, *p*-coumaric, ferulic, syringic and *m*-coumaric acids, are found in the leaves and humus of *Kalmia* and salal and have been implicated in growth inhibition of black spruce (Zhu and Mallik 1994), and western red cedar and western hemlock (de Montigny 1992). Germination and root growth inhibition of Norway spruce in the presence of fresh leaves and humus of *V. myrtillus* are attributed to high concentrations of caffeic acid in them (Pellissier 1993, 1994; Gallet and Lebreton 1994). High concentrations of tannins in leaves and inflorescence of salal are thought to exert growth inhibit the growth of western hemlock, western red cedar and amabilis fir (Preston et al. 1998).

2.3 Resource-toxin Hypothesis

Most sites with ericaceous dominance have deficiency in available N. Inderjit and Mallik (1997) demonstrated that water leachates of *Kalmia* can bring about significant changes in nutrient concentrations of humus and mineral soil solutions. They found significantly higher pH, lower concentrations of total phenols and available N and higher concentrations of Fe, Zn, K, Ca, Mg and Mn in soil amended with *Kalmia* leaves compared to the unamended control soil. These results support the views of Damman (1971) that a long term occupancy of *Kalmia* in a site may bring about permanent change in soil nutrient characteristics which is unfavorable for conifer regeneration.

2.4 Ericoid Mycorrhizae

Ericaceous plants are in symbiotic association with a variety of mycorrhizae that are specifically adapted to nutrient-poor habitats (Read 1983, 1991). Ericaceous plants produce large quantities of polyphenols that can bind soil organic N as calcitrant protein-phenol complexes (Tackechi and Tanaka 1987). Ericoid mycorrhizae are able to utilize protein N that is complexed with tannic acid by means of enzymatic degradation whereas ectomycorrhizal fungi associated with of conifers could not obtain N from the same source (Bending and Read 1996). Under field conditions, some or all of these mechanisms can work simultaneously and/or sequentially to induce the growth inhibitory effects. Moreover, the site characteristics and climatic conditions play important roles in the growth inhibition process. Dynamics of conifer-ericaceous interactions where competition, allelopathy, nutrient sequestration and differential nutrient uptake all play a part can be presented in a conceptual diagram (Fig. 1).

3. VEGETATION CONTROL

Traditional vegetation management strategies such as herbicides, scarification, cutting and burning are of limited success in controlling the ericaceous plants. Among the herbicides tested, Garlon and Bialaphos provide some control when applied in late summer. Mulching can control some ericaceous plants by destroying their vegetative buds but the high cost of treatment and the difficulty of operating a mulching machine in stony soil makes this option impractical. Repeated application of fertilizer-N (600-1344 kg per ha) was found to be effective in reducing *Kalmia* and salal growth and releasing conifer growth. Black spruce seedlings pre-inoculated with mycorrhizal fungi such as *Paxillus involutus*, *Laccria laccata* and E-strain, were able to overcome *Kalmia*-induced growth inhibition in greenhouse experiments.

312

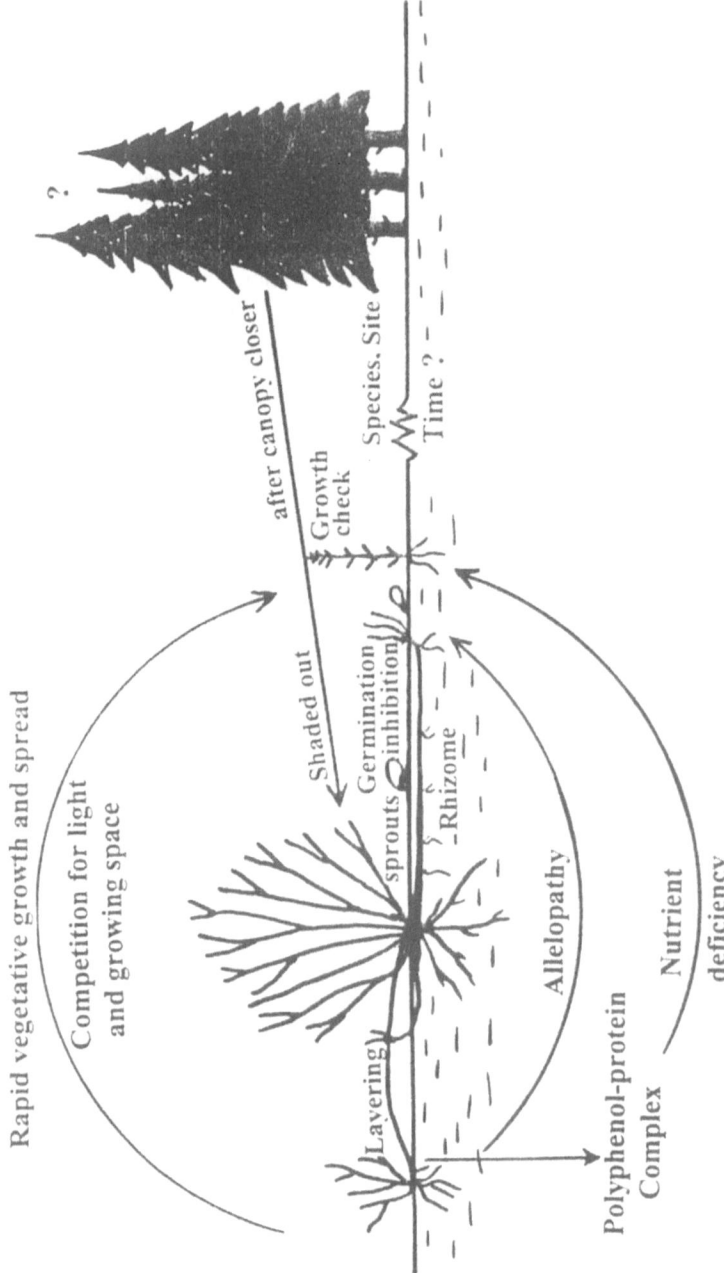

Figure 1. Conifer growth inhibition resulting from the combined effects allelopathy, competition, phenol-bound nutrient sequestration and ericoid mycorrhizae of the ericaceous plants.

4. CONCLUSIONS

(1) Certain ericaceous understory plants grow rapidly by vegetative regeneration after forest-clearing and inhibit the growth of conifers; (2) the type and extent of conifer growth inhibition are dependent on site factors, autecological characteristics of the species involved and the disturbance regimes; (3) multiple ecological processes such as competition, allelopathy, allelopathy-nutrient interaction, the assart flush, the abilities of ericoid, ecto- and endomycorrhihzae to gain access to N from protein-phenol/polyphenol complexes as well as the regeneration response of ericaceous plants to disturbance are factors involved in the ericaceous-induced conifer growth inhibition; (4) application of N-fertilizer can release conifers from growth-check, and repeated application of high rates of N can reduce the cover and abundance of certain ericaceous plants; (5) the commonly used herbicides such as Vision and Velpar are not effective in controlling the ericaceous plants but herbicides such as Garlon and Bialaphos have potential in controlling some ericaceous plants; (6) black spruce seedlings pre-inoculated with certain mycorrhizal fungi can overcome *Kalmia*-induced growth-check; (7) it is necessary to establish appropriate field trials to validate results of controlled experiments under greenhouse conditions.

REFERENCES

Bending, G.D. and J.R. Read (1996) : Nitrogen mobilization from protein-polyphenol complex by ericoid and ectomycorrhizal fungi, Soil Biology and Biochemistry, Vol. 28, pp. 1603-1612.

Bunnell, F.L. (1990) : Reproduction of salal (*Gaultheria shallon*) under forest canopy, Canadian Journal of Forest Research, Vol. 20, pp. 91-100.

Damman, A.W.H. (1971) : Effects of vegetation changes on the fertility of a Newfoundland forest site, Ecological Monograph, Vol. 41, pp. 253-270.

de Montigny, L. (1992) : An investigation into the factors contributing to the growth check of conifer regeneration on northern Vancouver Island, Ph. D. Thesis, Forest Sciences Department, University of British Columbia, Vancouver, Canada.

Gallet, C. and P. Lebreton (1994) : Evolution of phenolic patterns and associated litter and humus of a mountain forest ecosystem, Soil Biology and Biochemistry, Vol. 27, pp. 157-165.

Inderjit and A.U. Mallik (1996) : The nature of interference potential of *Kalmia angustifolia*, Canadian Journal of Forest Research, Vol. 26, pp. 1899-1904.

Jalal, M.A.F. and D.J. Read (1983a) : The organic acid composition of *Calluna* heathland soil with special reference to phyto- and fungitoxicity. I. Isolation and identification of organic acids, Plant and Soil, Vol. 70, pp. 257-272.

Jalal, M.A.F. and D.J. Read (1983b) : The organic acid composition of *Calluna* heathland soil with special reference to phyto- and fungitoxicity. II. Monthly quantitative determination of the organic acid content of *Calluna* and spruce dominated soils, Plant and Soil, Vol. 70, pp. 273-286.

Mallik, A.U. (1987) : Allelopathic potential of *Kalmia angustifolia* to black spruce, Forest Ecology and Management, Vol. 20, pp. 43-51.

Mallik, A.U. (1991) : Cutting, burning and mulching to control *Kalmia* : results of a greenhouse experiment, Canadian Journal of Forest Research, Vol. 67, No. 5, pp. 1309-1316.

Mallik, A.U. (1993) : Ecology of a forest weed of Newfoundland : Vegetative regeneration strategies of *Kalmia angustifolia*, Canadian Journal of Botany, Vol. 71, pp. 161-166.

Mallik, A.U. (1994) : Autecological response of *Kalmia angustifolia* to forest types and disturbance regimes, Forest Ecology and Management, Vol. 65, pp. 231-249.

Mallik, A.U. (1995) : Conversion of temperate forests into heaths : role of ecosystem disturbance and ericaceous plants, Environmental Management, Vol. 19, pp. 675-684.

Mallik, A.U. and C.H. Gimingham (1985) : Ecological effects of heather burning II. Effects on seed germination and vegetative regeneration, Journal of Ecology, Vol. 73, pp. 633-644.

Mallik, A.U. and B.A. Roberts (1994) : Natural regeneration of red pine on burned and unburned sites in Newfoundland, Journal of Vegetation Science, Vol. 5, pp. 179-186.

Messier, C. (1993) : Effects of neutral shade and growing media on growth and, biomass allocation and competitive ability of *Gaultheria shallon*, Canadian Journal of Botany, Vol. 70, pp. 2271-2276.

Pellissier, F. (1993) : Allelopathic inhibition of spruce germination, Acta Oecologica, Vol. 14, pp. 211-218.

Pellissier, F. (1994) : Effect of phenolic compounds in humus on the natural regeneration of spruce, Phytochemistry, Vol. 36, pp. 865-867.

Preston, C.M., J.A. Trofymow, B.G. Sayer and J. Niu (1998) : 13C CPMASS NMR investigation of the proximate analysis fractions used to assess litter quality in decomposition studies, Canadian Journal of Botany (In press).

Read, D.J. (1983) : The biology of mycorrhiza in Ericales, Canadian Journal of Botany, Vol. 61, pp. 985-1004.

Read, D.J. (1991) : Mycorrhizas in ecosystems, Experientia, Vol. 47, pp. 31-36.

Smith, N.J. (1991) : Sun and shade leaves: clues to how salal (*Gaultheria shallon*) responds to over-story stand density, Canadian Journal of Forest Research, Vol. 21, pp. 300-305.

Tackechi, M. and Y. Tanaka (1987) : Binding of 1,2,3,4,6-pentagalloylglucase to proteins, lipids, nucleic acids and sugars, Phytochemistry, Vol. 26, pp. 95-97.

Weetman, G.F., R. Fournier, J. Baker, E. Schnorbus-Panozzo, and A. Germain (1989) : Foliar analysis and response of fertilized chlorotic Sitka spruce plantations on salal dominated cedar-hemlock cutovers on Vancouver island, Canadian Journal of Forest Research, Vol. 12, pp. 1501-1511.

Zackrisson, O. and M.C. Nilsson (1992) : Allelopathic effects by *Empetrum hermaphorditum* on seed germination of two boreal tree species, Canadian Journal of Forest Research, Vol. 22, pp. 1310-1319.

Zhu, H. and A.U. Mallik.(1994) : Interactions between *Kalmia* and black spruce: isolation and identification of allelopathic compounds, Journal of Chemical Ecology, Vol. 20, No. 2, pp. 407-421.

Role of Allelopathy in Regulating the Understorey Vegetation of *Casuarina Equisetifolia*

Daizy R. BATISH and Harminder Pal SINGH

Department of Botany, Panjab University, Chandigarh 160 014, India

ABSTRACT A study conducted under the 14 year old plantations of *Casuarina equisetifolia* reveals the reduction of understorey vegetation in comparison to the adjoining open areas. The number of plants, species type and the biomass were greatly reduced under the plantations. The leachates from fresh leaves and litter and the understorey soil were found to be rich in phenolics and exhibited phytotoxic effects against the *Bidens pinnata* and *Parthenium hysterophorus* which were obscured from the plantations but were present in the adjoining area. Both the germination as well seedling growth of these two plants was significantly reduced in response to the different leachates. Thus, allelopathy was observed to play a significant role in regulating the understorey vegetation dynamics in *C. equisetifolia* stands.
Key words : allelopathy, phenolics, litter, *Bidens pinnata*, *Parthenium hysterophorus*.

1.INTRODUCTION

Casuarina equisetifolia (family Casuarinaceae) is a fast growing, evergreen, nitrogen fixing multipurpose tree which has recently gained prominence in India. It is being grown in blocks as energy plantations, along rivers as riparian plantations and along the field boundaries as shelterbelts/windbreaks. It is also being grown in the undeveloped wasteland areas in order to reclaim the soil. However, the plantations are observed to support little vegetation, perhaps due to allelopathy. Several studies are available in literature which indicate reduced floor vegetation under the monoculture plantations (del Moral and Muller 1970; Al-Mousawi and Al-Naib 1975, 1976; Melkania and Singh 1987; Suresh and Rai 1987). In majority of these cases, besides competition for nutrients. light and space, allelopathy has been implicated as an additional factor. The *C. equisetifolia* plantations are characterized by the absence of a number of plant species under its canopy which are otherwise present around it. Further, the floor of these plantations is characterized by a very conspicuous layer of litter composed mainly of needle like leaves, small twigs and cone like fruits. This may release allelochemicals in the soil and hinder the germination and growth of understorey vegetation.

The present study was, therefore, undertaken to estimate the vegetation biomass, amount of leaf litter and allelopathic influence of intact leaves, litter (both fresh and decomposed) and its soil medium on the vegetation which fails to invade these plantations.

2. MATERIAL AND METHODS

2.1 Vegetational Analysis

For the present study a block plantations of *C. equisetifolia* was selected in Chandigarh (30° 42′ N, 76° 54′ E, 330 m msl), India. The trees were nearly 14 years old with intra row tree to tree distance of 4 m and inter row distance of 5 m. The vegetation analysis was done using a 1 m quadrat. Twenty quadrats were laid randomly and all plants falling in them were counted, sampled and identified. Type of species, their density and biomass were also determined. Besides, the amount of freshly fallen raw litter as well as the finely degraded litter per unit area was also quantified.

2.2 Preparation of Leachates

Fresh green leaves, decomposed and fresh litter and the soil were collected from the *C. equisetifolia* sites. Their aqueous leachates were prepared by soaking 20g of each material in 100 ml of pure water (conductivity < 0.05 μS, obtained through Millipore RO - Milli Q Water Purification System) for 24h at 25°C. These leachates were filtered through a muslin cloth followed by Whatman # 1 filter paper and were kept at 4°C until used. For each set of repetitive experiment fresh leachates were prepared.

2.3 Determination of Phenolic content

The content of total phenolics was estimated in the freshly prepared aqueous leachates of green leaves, fresh and decomposed litter and soil using Folin-ciocalteu reagent as per the method of Swain and Hillis (1959). These were expressed in terms of dry weight of the respective plant part or soil used.

2.4 Germination Studies

For germination bioassay, seeds of *Bidens pinnata* and *Parthenium hysterophorus* were collected locally from the wildly growing stands. The germination studies were carried out in 6″ diameter Petri dishes lined with Whatman # 1 filter paper and underlined with a thin absorbent cotton wad. These were moistened with 10 ml of each of the leachate solutions. 25 seeds of each were imbibed for 10 h in the respective solutions and then equidistantly placed on the Petri dishes. Three replications were maintained for each treatment. Likewise, treatment with pure water instead of leachates served as control. All the Petri dishes were kept in a seed germinator maintained at nearly 25°C temperature, 75% humidity. After a period of 10 days, per cent germination, seedling length and dry weight of seedlings were measured.

3. RESULTS

Under *C. equisetifolia* the number of species recorded in a square meter quadrat were found to be very less in comparison tree free control site. Likewise, the density of plants was also less (Table 1). The biomass of floor vegetation was determined to be nearly 55 % of that in the open area control. The amount of fresh and decomposed litter which formed a matrix on the floor was measured to be 296 ± 33.2 g/m^2 and 688 ± 60.2 g/m^2, respectively. The vegetation was also seen to be either devoid of or with very low density of *Bidens pinnata* and *Partheniun hysterophorus* which were otherwise present abundantly in the nearby open area and also along the border of the plantations. Thus, these were selected as bioassay plants.

Table 1. Effect of aqueous leachates on the per cent germination of *B. pinnata* and *P. hysterophorus* and the amount of phenolics present in them.

Parameter	*C. equisetifolia* Site	Control Site
No. of Species (per m^2)	12 ±2	39 ±3
Density (plants/m^2)	56 ± 6	115 ± 8
Biomass (g/m^2)	212 ±33.9	385 ±23.4
Amount of fresh litter (g/m^2)	296 ±33.2	-
Amount of Decomposed litter (g/m^2)	688 ±60.2	-

The germination of both *B. pinnata* and *P. hysterophorus* was markedly decreased in the aqueous leachates of leaves, litter and the soil. Maximum effect on germination was seen in case of intact leaves and least in the soil leachates. Further, *B. pinnata* plants were more sensitive than *P. hysterophorus* (Fig.1a). Not only the germination, but the seedling length was also appreciably reduced in both the plants (Fig. 1b). Fresh litter leachates reduced the seedling length of both *B. pinnata* and *P. hysterophorus* to the maximum extent. It was reduced to nearly 10% of the control in both *B. pinnata* and *P. hysterophorus.*

320

(a)

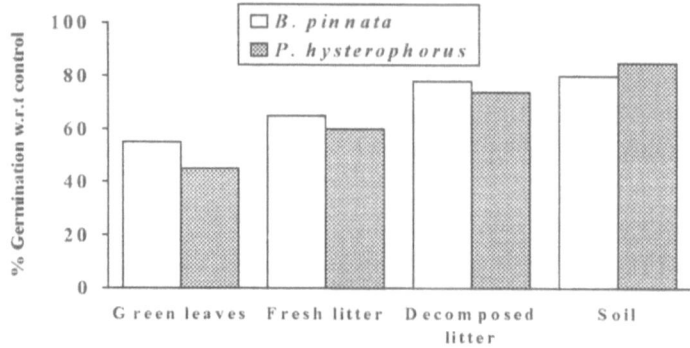

(b)

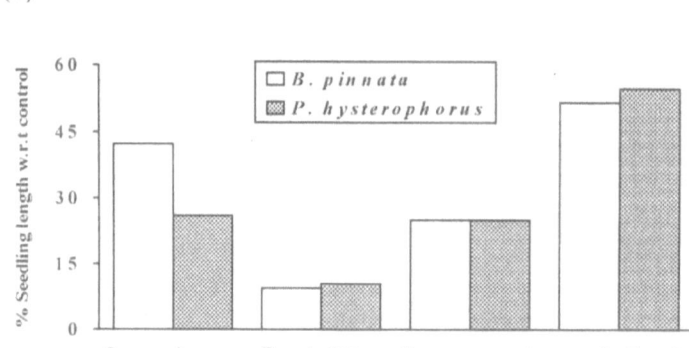

Fig. 1. Effect of soil, leaf and litter leachates of *C. equisetifolia* on the germination (a) and seedling growth (b) of *B. pinnata* and *P. hysterophorus*.

The leachates prepared from the green leaves, fresh and decomposed litter and soil were found to contain appreciable amount of phenolics. It was found to be maximum in the green leaves followed by fresh litter, decomposed litter and soil (Table 2).

Table 2. Content of phenolics in leachates of parts under study.

Type of Aqueous Leachates	Amount of Phenolics (mg/100g dry wt.)
Green Leaves	36.67 ±3.05
Fresh Litter	15.50 ±1.23
Decomposed Litter	10.22 ±0.67
Soil	6.46 ±0.47

4. DISCUSSION

It is evident from the present study that number of species, density and biomass of vegetation is reduced under the monoculture plantations of *C. equisetifolia* as compared to the open area showing thereby that the status of vegetation is not heterogenous and thus stable. The plants of *B. pinnata* and *P. hysterophorus* were present in very low densities or were completely absent. These were, therefore, chosen as sensitive species. There are several reports in literature which show that monoculture plantation support reduced or little understorey vegetation for example *Cedrus deodara* (Melkania and Singh 1987), *Pinus densiflora* (Kil and Yim 1983), *Cassia siamea* (Goel and Sareen 1986), *Delonix regia* (Chou and Lee 1992) and *Populus deltoides* (Kohli et al. 1996). In most of the cases the plausible reason for the restrained understorey vegetation has been given to be allelopathy (Suresh and Rai 1988, Kohli et al. 1996).

The plantations of *C. equisetifolia* are characterized by the presence of a thick matrix of litter composed mainly of the needle like leaves, small twigs and cone like fruits. The litter was composed of two regions, the upper containing raw and undecomposed litter and the lower layer of decomposed litter which consists of fine pieces. The present results indicate that they fall in sufficiently higher amount. Kuiters (1989) reported that that litter of the forest floors releases allelochemicals through decomposition and leachation. It was, therefore, thought that that it might be playing a significant role in imparting allelopathic phenomenon to the tree.

The leachates prepared from leaves, fresh litter, decomposed litter as well as soil beneath litter exhibited phytotoxicity against *B. pinnata* and *P. hysterophorus*. Not only germination but seedling length was found to be less as compared to the control. Thus, leachation seems to be one mechanism of release of allelochemicals from the tree to the environment. In coniferous forests, the extracts prepared from the fresh leaves and litter significantly reduce the germination of native grasse species (Jobidon 1986).

Decomposition of the litter may also contribute sufficiently towards the release of allelochemicals (Rice 1984). Litter of several trees have been shown to be toxic to native vegetation (Molina et al. 1991, Inderjit and Mallik 1996, Singh 1996). Phenolics have been found to be the major components of the allelochemicals particularly in the coniferous forests where replant problem is very prevalent (Gallet 1994, Pellissier 1994). In the present study also, phenolics were found to be the major component of the leachates which were found to be maximum in the leaves followed by litter and soil.

Thus, from the present study it could be concluded that the allelopathy plays a significant role in regulating the understorey vegetation in the plantations of *C. equisetifolia*.

322

ACKNOWLEDGEMENTS

H.P. Singh is thankful to Council of Scientific and Industrial Research (CSIR), New Delhi, India for financial assistance in the form of Post doctoral Research Associateship.

REFERENCES

Al-Mousawi, A.H. and F.A.G. Al-Naib (1975) : Allelopathic effects of *Eucalyptus microtheca* F Muell., Journal of University of Kuwait (Science), Vol. 2, pp. 59-66.

Al-Mousawi, A.H. and F.A.G. Al-Naib (1976) : Volatile growth inhibitors produced by *Eucalyptus microtheca*. Bulletin of Biological Research Centre, Vol. 7, pp.17-23.

Chou, C.H. and L.L. Lee (1992) : Allelopathic substances and interactions of *Delonix regia* (Boj.) Raf., Journal of Chemical Ecology, Vol 18, pp. 2285-2303.

del Moral, R. and C.H. Muller (1970) : The allelopathic effects of *Eucalyptus camaldulensis*, American Midland Naturalist, Vol. 83, pp. 254-282.

Gallet, C. (1994) : Allelopathic potential in bilberry-spruce forests: influence of phenolic compounds on spruce seedlings. Journal of Chemical Ecology, Vol. 20, pp. 1009-1024.

Goel, U. and T.S. Sareen (1986) : Allelopathic effect of trees on the understorey vegetation, Acta Botanica Indica, Vol 145, pp. 162-166.

Inderjit and A.U. Mallik (1996) : The nature of interference potential of *Kalmia angustifolia*, Canadian Journal of Forest Research, Vol 26, No. 11. pp. 1899-1904.

Jobidon, R. (1986) : Allelopathic potential of coniferous species to old-field weeds in Eastern Quebec Canada, Forest Science, Vol. 32, pp. 112-118.

Kil, B.S. and Y.J. Yim (1983) : Allelopathic effects of Pinus densiflora on undergrowth of red pine forest, Journal of Chemical Ecology, Vol. 9, pp. 1135-1151.

Kohli, R.K., H.P. Singh and D. Rani (1996) : Status of floor vegetation under some monoculture and mixculture plantations, Journal of Forest Research, Vol. 1, pp. 205-209.

Kuiters, A.T. (1989) : Effect of phenolic acids on germination and early growth of herbaceous woodland plants, Journal of Chemical Ecology, Vol 15, No. 4, pp. 467-479.

Melkania, N.P. and J.S. Singh (1987) : Allelopathy in Himalayan forest species. In : Proc. IX International Symposium Tropical Ecology and International Conference on Rehabilitation of Disturbed Ecosystems : A Global Issue, Banaras Hindu University, Varanasi, India.

Molina, A., M.J. Reigosa and A. Carballeira (1991) : Release of allelochemical agents through fall, and top soil in plantations of *Eucalyptus globulus* Labill. in Spain, Journal of Chemical Ecology, Vol 17, pp. 147-160.

Pellissier, F. (1994) : Effect of phenolic compounds in humus on the natural regeneration of spruce, Phytochemistry, Vol. 36, No. 4, pp. 865-867.

Rice, E.L. (1984) : Allelopathy, Academic Press, New York, 422p.

Singh, H.P. (1995) : Phytotoxic Effects of *Populus* in Natural and Agro-ecosystems, Ph. D. thesis Panjab University, Chandigarh, India.

Suresh, K.K. and R.S.V. Rai (1987) : Studies on the allelopathic effects of some agroforestry tree crops, International Tree Crops Journal, Vol. 4, pp. 109-115

Suresh, K.K. and R.S.V. Rai (1988) : Allelopathic exclusion of understorey by a few multi-purpose trees, International Tree Crops Journal, Vol. 5, pp. 143-151.

Swain, T. and W.E. Hillis (1959) : The phenolic constituents of *Prunus domestica* I. - the quantitative analysis of Phenolic constituents, Journal of Science Food and Agriculture, Vol. 10, pp. 63-68.

Long-term Response of Radiata Pine to Phosphate Fertiliser on a Strongly P-Fixing Soil

Peter HOPMANS and David W. FLINN

Center for Forest Tree Technology, Department of Natural Resources and Environment, Heidelberg, VIC 3084, Australia

ABSTRACT Growth of Radiata pine on a strongly P-fixing soil in south-eastern Australia was severely limited by P deficiency. Phosphate fertiliser was applied at rates of 0, 60, 90, 120, 180 and 240 kg P/ha as split applications at age 4 and 7 years to determine the response in growth and changes in tree nutrient status. Application of at least 120 kg/ha was required to increase the rate of growth from 15 to 31 m^3/ha/yr and to improve total volume by approximately 370 m^3/ha at age 22 years. Growth response to fertiliser over the 18 year period was strongly correlated with levels of P in foliage 4 years after treatment. At 22 years, trees were again deficient in P indicating the need for another application of fertiliser to maintain growth over a 35 year rotation.
Keywords: Radiata pine, nutrition, fertiliser, growth, forest soil.

1. INTRODUCTION

In the 1970's, plantations of Radiata pine (*Pinus radiata* D. Don) were established mainly on land cleared of native eucalypt forest often growing on highly weathered soils. Natural availability of phosphorus (P) for these soil types is generally low, and P deficiency is common in young pine plantations on these sites. Several studies have shown the need for P fertilisers at planting to ensure satisfactory establishment and early growth (Flinn *et al.*, 1979a & b; Flinn *et al.*, 1982; Flinn and Aeberli, 1982). While spot applications of P fertilisers are an effective means of increasing the uptake of P by young trees, it was recognised that additional treatments may be required to maintain growth, particularly on soils with a high P adsorption capacity. This hypothesis was tested in a trial established in 4 year-old Radiata Pine on a strongly P-fixing soil at Narbethong in Victoria, Australia.

2. METHODS

A general description of the native vegetation, climate, soils and geology at the Narbethong trial is given by Flinn *et al.* (1982). The soil is a well-structured, reddish-brown clay loam grading to a heavy red clay at depth (0.6 - 0.8 m). Availability of P in the surface soil (0 - 0.1 m) was low and P adsorption capacity was high as shown by pH (5.5), available P (Bray-P, 1.4 mg/kg), P adsorption maximum (3430 mg/kg) and the high levels of free iron oxides (39 g/kg) and aluminium oxides (13 g/kg).

The trial was planted in 1973 at a nominal spacing of 2.4 m x 2.4 m. Trees were fertilised with superphosphate at a rate of 16 g P/tree soon after planting. After 4 years, seven treatments were applied to 216 m^2 plots arranged in a randomised block with three replications. Each block comprised one control plot (spot application of 16 g P/tree at planting only) and three pairs of plots each with one of the following treatments: superphosphate broadcast at 700, 1050 and 1400 kg/ha equivalent to elemental P at 60, 90, and 120 kg/ha. These treatments were repeated three years later to half the plots at the same initial rates. Therefore the total amounts of superphosphate broadcast were 0, 700, 1050, 1400, 1400 (2 x 700), 2100 (2 x 1050), and 2800 (2 x 1400) equivalent to 0, 60, 90, 120, 180 and 240 kg/ha of elemental P. Heights and diameters (DOB at 10 cm above ground) were measured at 4 and 7 years and these results were reported by Flinn *et al.* (1982). Thereafter, diameters (over bark at approximately 1.3 m height) were measured in subplots of around 125 m^2 at 9, 12, 16 and 22 years. In addition, bark thickness and height were measured at 22 years. Foliage samples were collected at prior to treatment and at 6, 8, 10, 12, 14 and 22 years for chemical analysis of essential plant nutrients.

3. RESULTS AND DISCUSSION

3.1. Growth Response to Fertiliser

Basal area growth in response to applications of P fertiliser showed that a single spot application of superphosphate at planting (25 kg P/ha) was insufficient to maintain good long-term growth (Fig. 1). Single applications of P at rates of 60 and 90 kg/ha improved growth substantially but these treatments did not achieve full site potential. Dual applications of superphosphate at relatively high rates were required on this strongly P-fixing soil to attain near maximum growth. Final measurements were taken at age 22 years and results for height, basal area, and under-bark volume are summarised in Table 1.

Total volume increased substantially in response to fertiliser (Table 1, Fig. 2). Single applications of 60 and 90 kg P/ha increased volume by 244 m^3/ha and 279 m^3/ha, respectively. This is equivalent to a 75% and 86% improvement in growth compared with the control treatment consisting of a spot application (25 kg P/ha) at planting. Volume responses at higher rates of P fertiliser were indistinguishable statistically and the average improvement for these treatments was 367 m^3/ha or 113% compared with the control.

Table 1. Height, basal area, and under-bark volume of 22 year-old Radiata Pine at Narbethong treated with a spot application of P at 25 kg/ha at planting followed by broadcast superphosphate at rates of 0, 60, 90, and 120 kg P/ha at age 4 and 7 years.

Rate (kg/ha)	Height (m)	Basal Area (m^2/ha)	Volume (m^3/ha)
nil	26.2	43	325
60	30.5	67	569
120 (2x60)	28.9	81	697
90	28.7	73	604
180 (2x90)	29.2	79	670
120	31.5	71	671
240 (2x120)	30.3	80	731
F level[a]	NS	***	***
PLSD[b]	-	12	114

[a] Statistical level of significance: NS, not significant; ***, $P \cdot 0.001$.

[b] PLSD, Fisher's Protected Least Significant Difference.

These results indicate that follow-up fertiliser at a rate of at least 120 kg P/ha was required to maintain maximum growth on this strongly P-fixing soil. This input of fertiliser at age 4 years has increased the rate of growth two-fold from 15 m^3/ha/yr to 31 m^3/ha/yr and it is expected that this rate can be maintained by further applications of P on the basis of foliage diagnostic testing.

Volume responses to P fertiliser at Narbethong compare quite favourably with results from long-term field trials at other locations in Victoria and New South Wales. Application of P at similar rates to 9 year-old Radiata pine on podzolised sands at Rennick increased volume at age 19 years by 127 and 146 m^3/ha respectively (Hopmans et al., 1993). Likewise, broadcast application of P at around 100 kg/ha to 9 and 18 year-old pine on yellow duplex soils at Scarsdale increased stand volume by 100 m^3/ha over a period of 11 to 13 years (Flinn et al., 1979b). Long-term volume responses of similar magnitude were also reported for Radiata pine at the Belanglo State Forest in NSW (Turner, 1982) where merchantable volumes increased more than 3 fold from 144 m^3/ha to 400 m^3/ha over a 26 year period following broadcast applications of P at rates of 75 and 100 kg/ha at age 4 years.

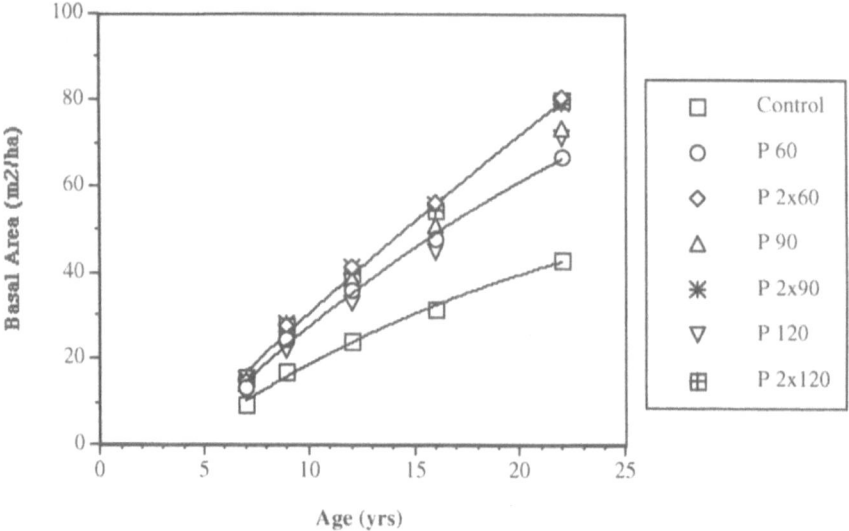

Fig. 1. Basal area growth of Radiata pine at Narbethong treated with a spot application of P at planting at a rate of 25 kg/ha followed by broadcast superphosphate at rates of 0, 60, 90, and 120 kg P/ha at age 4 and 7 years.

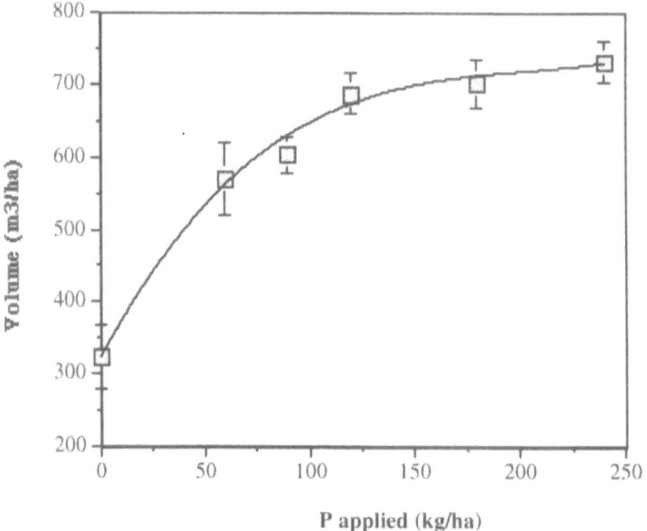

Fig. 2. Total under-bark volume of 22 year-old Radiata pine treated with various rates of P as superphosphate at age 4 and 7 years at Narbethong. Bars show standard errors of treatment means (n = 3).

3.2. Nutrition

Levels of nutrients in foliage at Narbethong were generally satisfactory indicating that apart from P, there were no other nutrient deficiencies at this site. Application of P fertiliser not only raised levels of P in foliage, but also increased uptake of Ca and Mn. A slight decrease in K levels was also evident. At 22 years, trees had again declined to P deficiency, but in addition levels of N were also marginal indicating the need for N as well as P fertiliser to attain maximum growth. The change in foliar P levels in response to fertiliser treatment is shown in Fig. 3. Pre-treatment assessments showed severe P deficiency across all plots indicating that a spot application of P at 25 kg/ha (16 g P/tree) at planting was inadequate to maintain foliar P above deficiency level to age 4 years.

Broadcast applications of superphosphate raised foliar concentrations of P, but levels barely reached concentrations considered necessary for satisfactory growth of Radiata pine (Raupach, 1975; Will, 1978), even at the higher rates of fertiliser. This is a reflection of the strong adsorption of fertiliser P by this soil. A steady decline to P deficiency was observed for the lowest rate of P fertiliser (60 kg/ha). In contrast, dual applications did maintain foliar concentrations of P above deficiency level to age 22 years (Fig. 3). However, foliar levels of P are approaching deficiency even at the high rate of 240 kg P/ha indicating the need for further treatment to maintain satisfactory growth until the end of the rotation. Results indicate that a single application of 120 kg P/ha at age 4 years can be expected to maintain foliage concentrations above the deficiency level for approximately 15 years. Clearly, further treatment is required to maintain a satisfactory P status for the remainder of the rotation. In contrast, response of Radiata pine at the Belanglo State Forest in NSW (Turner, 1982) to applications of fertiliser P at rates of 75 and 100 kg/ha at age 4 years continued for the duration of the rotation (30 years). This is probably a reflection of the differences in P adsorption capacity of the soils at Narbethong and Belanglo.

3.3. Tree Nutrition and Response to Fertiliser

Growth at Narbethong was clearly limited by low availability of soil P as demonstrated by the very low levels of P in foliage prior to treatment at 4 years. Response to applied P varied significantly between the lower rates of 60 and 90 kg/ha and the higher rates (Table 1). Levels of P ranged from deficient to satisfactory while levels of other nutrients were generally satisfactory. It is therefore assumed that P was the only nutrient limiting growth early in the rotation. However at the final assessment, foliage analysis showed that levels of N were either marginal or deficient demonstrating the need for N as well as P fertiliser later in the rotation.

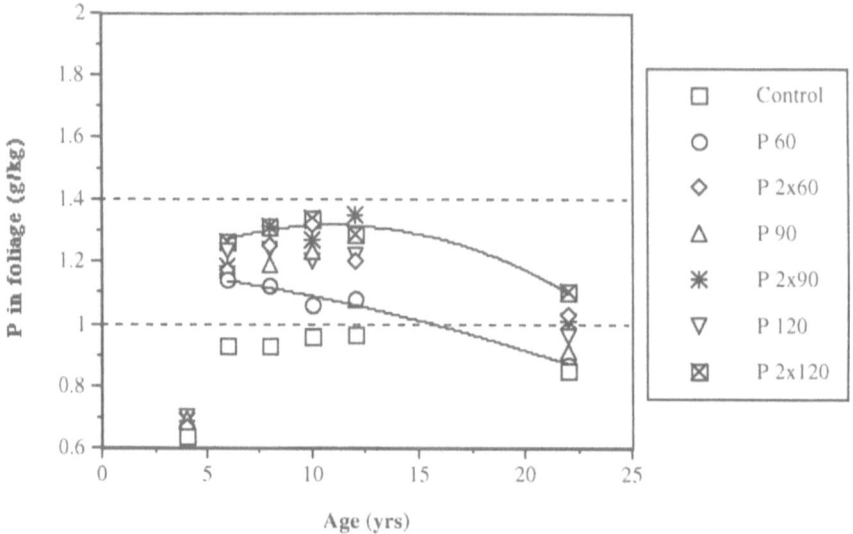

Fig. 3. Concentrations of P in foliage of Radiata pine at Narbethong treated with various rates of P as superphosphate at age 4 and 7 years. Lines (- - -) indicate deficient (1.0 g/kg) and satisfactory (1.4 g/kg) levels of P.

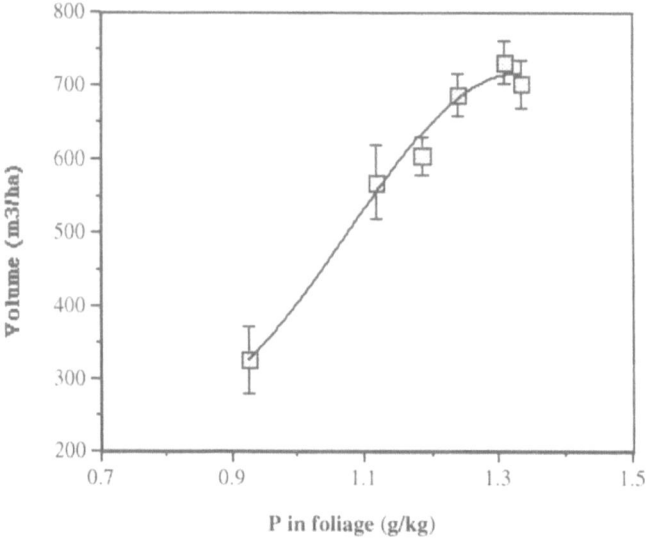

Fig. 4. Total under-bark volume of 22 year-old Radiata pine in response to treatment with superphosphate at age 4 and 7 years and concentrations of P in foliage at age 8 years at Narbethong. Bars show standard errors of treatment means (n = 3).

Regressions of pre- and post-treatment P levels in foliage and total volume at age 22 years were conducted to evaluate the use of foliage diagnostic testing for the prediction of volume response to fertiliser at Narbethong. This showed that the very low and deficient levels of P in foliage prior to treatment were a weak predictor of long-term growth responses to superphosphate applied at different rates. In contrast, levels of P in foliage 4 years after treatment, which reflect the uptake of fertiliser P, were strongly correlated with volume at 22 years (Fig. 4). These results indicate that foliage analysis can be used post-treatment to predict long-term volume responses to fertiliser. Treatment needs to raise P levels in foliage to at least 1.3 g/kg to maintain growth at near maximum levels (Fig. 4). The experimentally determined threshold level for P of 1.3 g/kg is consistent with the concentration of 1.4 g/kg considered to be necessary for satisfactory growth of Radiata pine in Australia (Raupach, 1975) and New Zealand (Will, 1978).

4. CONCLUSIONS

Growth responses and changes in tree nutrition of Radiata pine following broadcast applications of superphosphate in a field trial on a strongly P-fixing soil showed that:

- Spot applications of 16 g P/tree (25 kg/ha) applied at planting were inadequate to maintain foliar P levels above deficiency to age 4 years.

- Broadcast application of superphosphate at rates of at least 120 kg P/ha at 4 years increased total volume at 22 years from 325 m^3/ha to 690 m^3/ha. This represents an improvement in MAI from 15 m^3/ha/yr to 31 m^3/ha/yr.

- Volume response to fertiliser was strongly correlated with levels of P in foliage 4 years after fertiliser application. This showed that foliar levels need to be raised to at least 1.3 g/kg to maximise growth. Based on the duration of response it is expected that two broadcast applications will be needed for plantations on strongly P-fixing soils in order to maintain the P status of Radiata Pine at a satisfactory level over a rotation of 35 years.

5. ACKNOWLEDGMENTS

The authors gratefully acknowledge the contributions from past and present staff of the Department of Natural Resources and Environment to the maintenance, field assessments, laboratory analysis, and data processing conducted as part of this study. The authors also thank the Victorian Plantations Corporation for the financial support and collaboration during the latter stage of the project.

6. REFERENCES

Flinn, D.W., Hopmans, P., Moller, I., and Tregonning, K. (1979a): Response of Radiata pine to fertilisers containing N and P applied at planting, Aust. For., Vol. 42, pp. 125-131.

Flinn, D.W., and Aeberli, B.C. (1982): Establishment techniques for Radiata pine on poorly drained soils deficient in phosphorus, Australian Forestry, Vol. 45, pp. 164-173.

Flinn, D.W., James, J.M., and Hopmans, P. (1982): Aspects of phosphorus cycling in Radiata pine on a strongly phosphorus-adsorbing soil. Australian Forest Research, Vol. 12, pp. 19-35.

Flinn, D.W., Moller, I., and Hopmans, P. (1979b): Sustained growth responses to superphosphate applied to established stands of *Pinus radiata*, New Zealand Journal of Forestry Science, Vol. 9, pp. 201-211.

Hopmans, P., Tomkins, I.B., and Geary, P. (1993): Growth and nutrition of Radiata pine on podzolised sands in response to phosphate fertiliser, Research Report No 355, Department of Natural Resources and Environment, Victoria, 25p.

Raupach, M. (1975): Soil and fertiliser requirements for forests of *Pinus radiata*, Advances in Agronomy, Vol. 19, pp. 307-353.

Turner, J. (1982): Long-term superphosphate trial in regeneration of *Pinus radiata* at Belanglo State Forest, NSW, Australian Forest Research, Vol. 12, pp. 1-9.

Will, G.M. (1978): Nutrient deficiencies in *Pinus radiata* in New Zealand, New Zealand Journal of Forestry Science, Vol. 8, pp. 4-14.

Monitoring Population Viability in Declining Tree Species Using Indicators of Genetic Diversity and Reproductive Success

Alexander MOSSELER[*1] and O. P. RAJORA[*2]

*1 Canadian Forest Service, Atlantic Forestry Centre, P. O. Box 4000
Fredericton NB, Canada E3B 5P7
*2 University of Alberta, Department of Renewable Resources, Edmonton AB
Canada T6G 2H1 and BioGenetica Inc., P. O. Box 60201
University of Alberta
Postal Outlet, Edmonton, Alberta, T6A 2S5 Canada

ABSTRACT Biodiversity monitoring involves recurrent observations and recording changes and trends based on established biological benchmarks. The development of biological indicators at each level of the biodiversity hierarchy (landscape, community, species/genetic) is required to monitor and assess human impacts on natural populations and habitats. Genetic diversity is the ultimate source of biological diversity, providing the raw material for adaptation, evolution and survival of individuals and species - and becomes especially important to species survival in changing environments. Some indicators for monitoring population viability (which are defined here in the narrow sense as genetic diversity and reproductive success) in small, isolated populations of white pine, *Pinus strobus* L., red pine, *Pinus resinosa* Ait., and white spruce, *Picea glauca* (Moench) Voss, from the Island of Newfoundland (NF), Canada, are discussed as examples for monitoring the effects of small population size and increasing fragmentation on genetic diversity and reproductive behavior. Results with these populations suggest that reproductive success may provide better indicators for monitoring the effects of recent population declines, particularly in the early phases of population decline, than many of the standard genetic indicators.
Keywords: population viability, indicators, genetic diversity, reproductive success

INTRODUCTION

The sustainable use of natural resources and the conservation of biological diversity have gained increasing significance and urgency as global-scale environmental issues. These two issues, together with the equitable sharing of genetic resources, are the central focus of the Convention on Biodiversity (CBD).The sustainable use of natural resources has also been identified as a national research priority in Canada by the Natural Sciences and Engineering Research Council of Canada, and by both provincial and federal land management agencies and natural resources departments (e.g., Canadian Forest Service). Monitoring the sustainable use of natural resources and the conservation of biodiversity requires that meaningful, cost-effective measures or indicators be developed that can be used to monitor society s performance in sustainable development. International progress towards the elaboration and implementation of a set of criteria and indicators (C&I) for the conservation of biological diversity has been identified as a priority in the work program being developed by the signatories (Conference of the Parties) to the CBD. Several international initiatives (e.g. Montreal Process, Helsinki Process, Tarrapoto, etc.) are underway to support the development of C&I for monitoring biodiversity conservation at the landscape, ecosystem, and species/genetic levels of biodiversity. The work program supported by both the CBD and the Intergovernmental Panel on Forests (IPF and its successor agencies) are promoting these international and regional initiatives to achieve a common understanding of the concepts used in formulating C&I for sustainable forest management and the development of a set of national level indicators. These many initiatives have clearly shownthat indicators for monitoring biological diversity need to be improved.

Forest biodiversity monitoring is aimed at understanding and managing the impacts of human activities, particularly forest management practices on forest health and biodiversity. At the species level, the effects of forestry practices revolve largely around declining breeding population sizes, declining population densities, increasing fragmentation of populations on a landscape scale, and disruption of biological processes and structures (e.g., connectivity, propagule dispersal) affecting population viability (Fig. 1). Biodiversity monitoring involves recurrent observations and recording changes and trends based on established biological benchmarks for indicators at each level or scale of biodiversity. Diversity at the species and population levels can largely be ensured by maintaining genetic diversity and reproductive success.

For managing terrestrial biodiversity, trees often have special significance as keystone or flagship species because they define forested communities or habitats and represent a disproportionate amount of the biological resources in terrestrial ecosystems. By virtue of their size and dominance, trees form the habitat for many other associated organisms and may thus represent useful indicator species for monitoring habitat conditions. Monitoring begins with the accumulation of benchmark data on key indicators (Fig. 1) for selected species. Our emphasis is currently on tree species that are at risk: either rare, vulnerable, threatened, endangered or otherwise of special commercial or ecological value. From Figure 1, the following objectives can be defined for population management: (i) avoidance of inbreeding and genetic drift, (ii) maintenance of interpopulational dispersal and gene flow, (iii) maintenance of genetic diversity and structure, (iv) maintenance of reproductive capacity and normal patterns of mating behavior, and (v) ecological restoration of species and habitats. At present, there are few, if any, comprehensive programs designed to monitor the status of forest-dependant species at risk. This weakens the ability of land management agencies to set conservation objectives at the species, population and genetic levels of biodiversity.

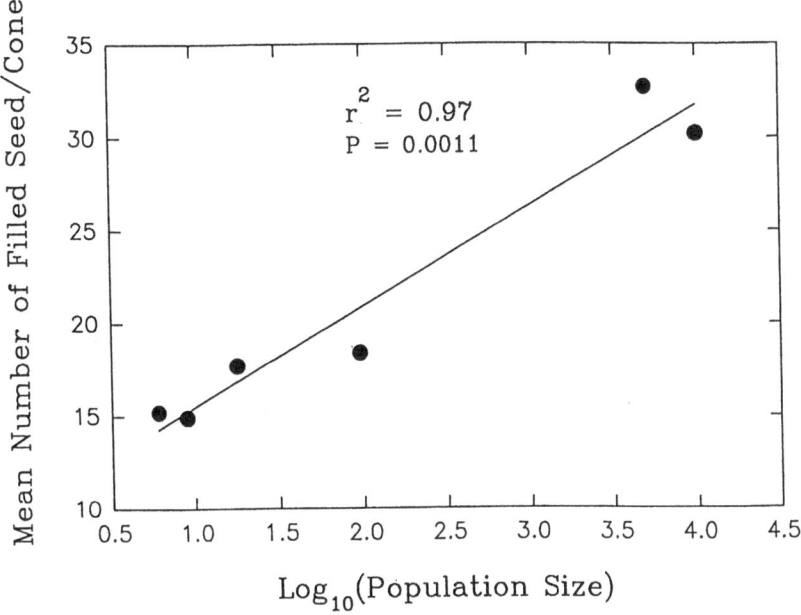

Fig. 1 Indicators for measuring and monitoring population viability

INDICATORS OF POPULATION VIABILITY

The biological criteria that maintain viability (e.g., genetic diversity and reproductive success) are affected by the biological processes listed in Figure 1. Genetic diversity and reproductive success are closely linked biologically and can be usefully linked in the monitoring and assessment processes. While reproductive success can reflect either abiotic features of the environment or the demographic and/or genetic consequences of small population size, particularly in rapidly declining species, reproductive success will ultimately determine the genetic status of a species. Therefore, Figure 1 recognizes reproductive success as a central component of population viability assessment. Furthermore, reproductive behavior provides a suite of useful, meaningful, and cost-effective indicators for monitoring population health and status.

Useful indicators are usually quantitative and unambiguous. Practicality and simplicity are also major considerations in choosing indicators that are cost-effective, technically feasible, and scientifically valid. The population biological processes that maintain population viability are relatively well understood in theory. The theoretical foundations of population genetics are very well developed; perhaps better developed than in any other area of population ecology (Wright 1978), and if the population biology of a species is reasonably well understood then the genetic parameters of populations of that species are generally quite predictable. Thus, genetic indicators may be useful in determining the health and status of populations and the biological processes that maintain their viability. However, a major challenge for indicator development is understanding how the distributional (population dispersion, fragmentation, connectivity, density, etc.) and demographic (population size and structure) changes imposed by human activities on the landscape affect these biological processes and at what level these landscape changes begin to have adverse effects. These relationships need to be quantified for several major groups of species with common life history features (e.g., conifers, tropical hardwoods, etc.). The concept of a minimum viable population size may provide a useful framework for the development and use of indicators for monitoring population viability (Gilpin and Soule 1986).

The spatial patterns of genetic organization and the transmission of genes from one generation and/or location to another is affected by the reproductive success of individuals and/or populations, which in turn is affected by changes in the numbers and distributions of organisms within and among populations. Thus, population declines resulting from recent human activities may affect the processes (e.g., gene flow, dispersal, inbreeding, genetic drift) that determine population viability. The genetic and reproductive consequences of changes in these processes can be measured and serve as indicators of change in population status, allowing us to monitor these changes relative to established benchmarks for a species or a species group/guild. Several genetic and reproductive indicators for these processes are available. Some of these indicators represent direct measures; whereas other indicators may provide useful indirect or surrogate measures. For instance, measures of species distribution and abundance may provide a good indication of levels of genetic diversity, patterns of population genetic structure, and reproductive behavior. These indicators may be operationally useful in managing populations, for developing policies related to conservation and the sustainable use of resources, and for the restoration of species and habitats. Since adult densities, mating systems, and seed and pollen dispersal distances affect the distribution and genetic structure of a species, conservation strategies must be developed with regard to these attributes (Hamrick and Nason 1996). For instance, the distribution of genetic diversity in small populations of selfing species may require more populations to preserve genetic diversity than large, dense populations of cross-pollinating species.

Population fragmentation can reduce total genetic diversity if the resulting fragments experience the combined effects of the genetic drift and inbreeding that can result either from small size or reduced gene flow through disruptions in movement of pollinators and seed dispersers. Under this scenario, genetic diversity may be lost as populations experience increased probabilities of extirpation due to small population size. If the fragmented populations experience no increase in the rates of extirpation, the combined effects of genetic drift and mild inbreeding may be neutral or even increase total species genetic diversity as population structure increases (e.g., the proportion of genetic diversity attributable to the among-population component may increase as the within-population component of diversity decreases). Until more is known about the direction of the impacts of fragmentation on particular species with respect to changes in population genetic structure, inbreeding, genetic drift, and increased probabilities of local extirpation, it will be difficult to make generalizations on the effects of population fragmentation on genetic diversity (Hamrick and Nason 1996).

The physical structure of populations of a species across the landscape can be described by several basic attributes: habitat specificity, average population sizes, population density, and fragmentation. Some of these attributes have also been used to define the rarity of a species (Rabinowitz et al. 1981). In practical terms, direct measurements of genetic diversity and structure will be used primarily for developing the recovery or restoration strategies for species at risk. For operational management and monitoring purposes it may be enough to classify species at risk according to their distributional characteristics (e.g., population sizes and density) in a landscape. >From this information some reasonably good inferences may be made regarding the genetic structure and reproductive behavior of a species and its susceptibility to genetic erosion in a landscape dominated by human activities.

The distribution of genetic variation within and among populations is closely tied to the physical distribution and structure of populations, the dispersal capabilities of species, and the resulting degree and pattern of dispersal and gene flow among populations. Although pollen and seed dispersal distances decrease rapidly from a point source (a leptokurtic distribution approximated by a negative exponential function; see Wright 1953; Wang et al. 1960; Silen 1962; Lanner 1966; Levin and Kerster 1974, Muller 1977; Burczyk et al. 1996), in the extensive, continuous populations of many boreal trees, gene flow through pollen dispersal may be much less dependant on the nearest neighbours (Yazdani et al. 1989; Adams and Birkes 1991). Long-distance dispersal from the pollen cloud created by the large, extensive populations typical of many boreal zone species may be considerable (Burczyk et al. 1996). However, as population sizes and numbers decline beyond certain thresholds, self-fertilization and fertilization by close relatives is expected to result in increased inbreeding. Species or populations experiencing rapid population decline may not be able to adjust to the effects of inbreeding depression; whereas the mild inbreeding experienced under conditions of gradual decline (e.g., at the margins of the geographic range of a species) may not adversely affect reproductive or genetic status and may in fact have some beneficial effects. However, rapid inbreeding can depress seed yield and quality and reduce both reproductive success and genetic diversity. The adverse effects of inbreeding depression on seed yields has been observed in small, isolated populations of white spruce in Newfoundland (Fig. 2).

MONITORING GENETIC DIVERSITY AND REPRODUCTIVE SUCCESS IN SMALL ISOLATED TREE POPULATIONS

Genetic diversity is the ultimate source of biological diversity, providing the raw material for adaptation, evolution and survival of individuals and species; and can be described using biochemical and molecular markers by parameters such as the proportion of polymorphic loci, the number of alleles per polymorphic locus, and the expected heterozygosity at Hardy-Weinberg equilibrium (Hamrick and Godt 1990). How genetic diversity is structured within and among populations can reveal important information on the evolutionary forces or processes that determine levels and patterns of genetic diversity, and the vulnerability to erosion of genetic diversity. Furthermore, the organization of genetic diversity within a species is of central importance to the development of conservation strategies and restoration activities (Falk and Holsinger 1991). The total genetic variance maintained by a species is a function of the partitioning of the total variance among and within the populations and subpopulations of a species (Wright 1978; Chesser et al. 1996). Therefore, some understanding of how the total genetic diversity is structured is important for conservation of these genetic resources. Inherent or existing levels of genetic diversity become especially important for adaptation in rapidly changing environments. By monitoring the accumulation of inbreeding within individuals or the presence of genetic drift using biochemical marker techniques, losses of genetic variance within populations and the impacts of human activities can be assessed.

Under normal stand conditions, reproductive success is affected among other things by variation in cone production, floral phenology, variations in fecundity, seed quality (e.g., germination capacity, speed, etc.), soil and seedbed conditions, interspecific competition, and climate (Yazdani et al. 1985; Burczyk et al. 1996). However, in declining populations near the margins of the geographic range or those that have been highly fragmented by human activities, the physical structure of populations may take on increasing importance. For instance, poor pollination conditions (see Yazdani et al. 1989) affected by small population size, low within population density, and population fragmentation may increase the ratio of self- versus cross-pollination leading to increased inbreeding and inbreeding depression in reproductive fitness.

Inbreeding depression is commonly observed in seed derived from self-

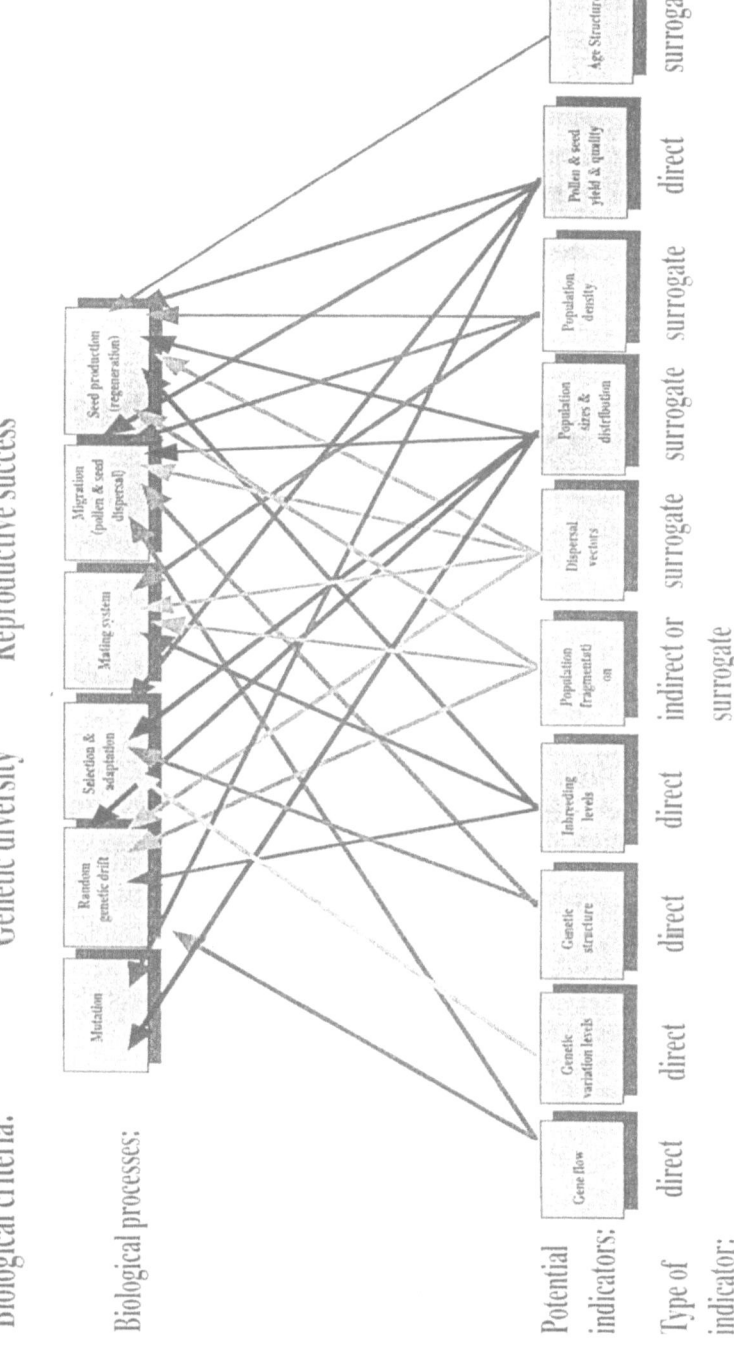

Fig 2. Correlation between population size and average seed yield per cone from six Newfoundland populations of white spruce.

fertilization or matings between close relatives, particularly in tree species that normally rely on outcrossing or cross-fertilization (Sorensen 1969; Bramlett and Popham 1971; Franklin 1970). Filled seed yields per cone and seed quality in six populations of varying size from central Newfoundland showed a strong positive correlation with population size and isolation with the smallest and most isolated populations showing significant reductions in seed yields (Fig. 2). Seedlings from the three smallest populations in this study also showed higher proportions of chlorophyll-deficient progeny following seed germination and reduced seedling height growth performance at age two - typical observations associated with the effects of inbreeding depression. These results can be related to the effects of inbreeding depression in white spruce (Mergen *et al.* 1965; Coles and Fowler 1976; Fowler and Park 1983).

The geographically disjunct populations of white pine, *Pinus strobus* L., on the island of Newfoundland, provide a useful case for monitoring the effects of small population size and population fragmentation on genetic diversity and reproductive behavior following a dramatic population decline. This decline resulted from liquidation harvesting for sawlogs at the turn of the century (1890-1920) followed by the introduction in the 1930s of a devastating fungal disease caused by the white pine blister rust, *Cronartium ribicola*. For Newfoundland, the decline of white pine represents a dramatic change on the landscape and the loss of a distinct forest community that provided both wildlife habitat and industrial wood supply. Biochemical genetic markers were used to establish some benchmark data on genetic diversity and reproductive behavior (i) to compare the disjunct NF populations with ancestral mainland populations near the center of the species geographic range, (ii) to demonstrate the use of indicators for monitoring the effects of small population size and fragmentation on genetic and reproductive status, and (iii) to provide the foundation for developing a conservation strategy directed towards ecological restoration (Rajora and Mosseler 1998).

Seeds collected from six small natural white pine populations from NF and three populations from the center of the geographic range in central Canada were used to assess embryo development, seed yields, genetic diversity, and mating system parameters. Isozyme analyses were used to determine population genetic structure and diversity based on 20 gene loci coding for 12 enzymes and mating system parameters based on eight polymorphic loci (Rajora and Mosseler 1998). The embryos in mature white pine seeds from Newfoundland were poorly developed at maturity in comparison with populations at the center of the geographic range, suggesting either inbreeding or climatic effects associated with the northern margins of the geographical range. Estimates of genetic diversity indicated that prior to population collapse the disjunct NF populations were reasonably similar to those from populations at the geographic centre of the range, with the latter showing somewhat higher latent genetic potential. All populations showed higher heterozygosities than expected under the Hardy-Weinberg equilibrium and only slightly higher inbreeding levels in the NF populations. An hierarchical analysis of population differentiation indicated comparable levels of population differentiation at the regional and provincial levels. Genetic distances among populations within and among regions were small and did not correlate well with geographic distances. Contrary to our expectations, and despite at least 8,000 years of post-glacial geographic isolation, NF white pine populations are as genetically diverse as populations from the center of the geographic range, and do not differ in genetic diversity levels reported for other parts of the range. Interpopulation gene flow was high (3.9 to 5.2 migrants per generation). Nevertheless, alleles unique to individual populations were observed. Estimates of heterozygosity and mating system parameters in the progeny (seed) population indicated only a slight increase in inbreeding in some of the smaller, isolated NF populations. If this is an emerging trend and population decline persists over subsequent generations, then decreases in genetic diversity may be expected.

In species such as red pine which are not affected by inbreeding depression, the adverse effects of reduced pollination on reproductive success (e.g., seed set) have been observed in small isolated populations in Newfoundland where variations in stand density are positively correlated with seed yields per cone indicating that the size of the local pollen cloud can have significant effects on seed production apart from any inbreeding depression effects. Red pine lacks genetic diversity (Mosseler *et al.* 1992; DeVerno and Mosseler 1997), and carries none of the genetic load of deleterious genes associated with inbreeding depression following self-fertilization (Fowler 1965). Thus, in declining populations we observe both the effects of reduced pollination per se and the effects of inbreeding depression on reproductive capacity. The generalized patterns for correlations between seed production and stand sizes and stem densities within stands are reviewed in Mosseler (1998).

SUMMARY

While a significant deterioration in reproductive capacity may be evident in declining populations as reduced seed yield and seed quality, adverse effects on such adaptive traits may not be reflected in biochemical estimates of genetic diversity and inbreeding levels, suggesting that direct measures of reproductive success may provide more useful indicators for monitoring the effects of recent population declines, particularly in the early phases of population decline. In conjunction with genetic analyses of adaptive trait variation (e.g., long-term common garden studies), estimates of genetic structure and diversity based on biochemical and/or molecular marker techniquess may provide useful indicators for the longer term monitoring of population decline. Future research aimed at developing indicators for species monitoring should attempt to relate direct measures of genetic and reproductive status to indirect or surrogate measures based on the demographic and distributional characteristics of species. Such surrogate measures may provide useful, cost-effective, timely and more operationally feasible indicators for monitoring species status where resources and expertise are limited.

REFERENCES

Adams, W.T. and D.S.Birkes (1991): Estimating mating patterns in forest tree populations.pp 157-172 In: Fineschi, S., M.E. Malvolti, F. Cannata and H.H. Hatemer (eds.)Biochemical markers in the population genetics of forest trees. SPB Academic Publishing, The Hague, Netherlands.

Bramlett, D.L. and T.W. Popham (1971): Model relating unsound seed and embryonic lethal alleles in self-pollinated pines. Silvae Genetica. 20: 192-193.

Burczyk, J., A.T. Adams, and J.Y. Shimizu (1996): Mating patterns and pollen dispersal in a natural knobcone pine (*Pinus attenuata* Lemmon.) stand. Heredity 77: 251-260.

Chesser, R.K., O.E. Rhodes, Jr., and M.H. Smith (1996). Gene conservation. pp. 237-252. *In* Population dynamics in ecological space and time, O.E. Rhodes, Jr. R.K. Chesser, and M.H. Smith (eds.). University of Chicago Press, Chicago. 388p.

Coles, J.F. and D.P. Fowler (1976): Inbreeding in neighbouring trees in two white spruce populations. Silvae Genet. 25: 29-34.

DeVerno, L.L. and A. Mosseler (1997): Genetic variation in red pine (*Pinus resinosa*) revealed by RAPD and RAPD-RFLP analysis. Can. J. For. Res. 27:1316-1320.

Falk, D.A. and K.F. Holsinger (eds.). (1991): Genetics and conservation of rare plants. Oxford University Press, New York.

Fowler, D.P. (1965): Effects of inbreeding in red pine, *Pinus resinosa* Ait. Silvae Genetica13:170-177.

Fowler, D.P and Y.S. Park (1983): Population studies of white spruce. I. Effects of self-pollination. Can. J. For. Res. 13: 1133-1138.

Franklin, E.C. (1970): Survey of mutant forms and inbreeding depression in species of the family *Pinaceae*. Southeast For. Exp. Stn., U.S.D.A. For. Serv., Res Paper SE-61, 21 p.

Gilpin, M.E. and M.E. Soule (1986): Minimum viable populations: processes of species extinction. pp. 19-34. *In* M.E. Soule (ed.) Conservation biology: the science of scarcity and diversity. Sinauer Associates, Sunderland MA.

Hamrick, J.L. and M.J.W. Godt (1990): Allozyme diversity in plant species. pp. 43-63.*In* A.D.H. Brown, M.T. Clegg, A.L. Kahler, and B.S. Weir (eds.), Plant population genetics, breeding and genetic resources. Sinauer Assoc. Inc, Sunderland, Massachusetts.

Hamrick, J.L. and J.D. Nason (1996): Consequences of dispersal in plants. pp. 203-236. *In* Population dynamics in ecological space and time, O.E. Rhodes, Jr. R.K. Chesser, and M.H. Smith (eds.),. University of Chicago Press, Chicago. 388p.

Lanner, R.M. (1966): Needed: a new approach to the study of pollen dispersion. Silvae Genetica 15:50-52.

Levin, D.A. and H.W. Kerster (1974): Gene flow in seed plants. Evolutionary Biology7:139-220.

Mergen, F., J. Burley, and G.M. Furnival (1965): Embryo and seedling development in *Picea glauca* (Moench Voss. after self-, cross-, and wind pollination. Silvae Genet. 14: 188-194.

Mosseler, A. (1998): Minimum viable population size and the conservation of forest genetic resources. pp.191-205. *In* Sunil Puri (ed.) Tree improvement: applied research and technology transfer. Science Publishers Inc., Enfield, NH.

Mosseler, A., K.N. Egger and G.A. Hughes (1992): Low levels of genetic diversity in red pine confirmed by random amplified polymorphic DNA amrkers. Can. J. For. Res. 22:1332-1337.

Muller, G. (1977): Cross-fertilization in a conifer stand inferred from enzyme gene markers in seeds. Silvae Genetica 26: 223-226.

Rabinowitz, D., D.S. Cairns, and T. Dillon (1981): Seven forms of rarity. pp. 205-217. *In* H. Synge (ed.), The biological aspects of rare plant conservation, John Wiley and Sons, New York.

Rajora, O.P. and A. Mosseler (1998): Genetic diversity and population structure of the disjunct Newfoundland and central Ontario populations of eastern white pine (*Pinus strobus* L.). Can. J. Bot. 76: (In press).

Silen, R.R. (1962): Pollen dispersal considerations for Douglas-fir. J. Forestry 60: 790-795.

Sorensen, F. (1969): Embryonic genetic load in coastal Douglas-fir, Pseudotsuga menziesii var. menziesii. Am. Nat. 103: 389-398.

Wang, C.-W., T.O. Perry, and A.G. Johnson (1960): Pollen dispersion of slash pine (*Pinus elliotii* Engelm.) with special reference to seed orchard management. Silvae Genetica 4: 78-86.

Wright, J.W. (1953): Pollen dispersion studies: some practical applications. J. Forestry51: 114-118.

Wright, S. (1978): Evolution and the genetics of populations, vol. 4 of Variability within and among natural populations. University of Chicago Press, Chicago.

Yazdani, R., O. Muona, D. Rudin, and A.E. Szmidt (1985): Genetic structure of a *Pinus sylvestris* L. Seed-tre stand and naturally regenerated understory. Forest Science31: 430-436.

Yazdani, R., D. Lindgren, and S. Stewart (1989): Gene dispersion within a population of *Pinus sylvestris*. Scand. J. For. Res. 4:295-306.

Using GIS to Review Wildlife Habitat Information in the Kii Peninsula, Japan

Tsuyoshi YOSHIDA*[1], Manabu ISHIZAKI*[2] and
Kazuhiro TANAKA*[1]

*1 Forest Planning Laboratory, Faculty of Agriculture,
   Kyoto Prefectural University, Kyoto, Japan
*2 Department of Forest Resources, Faculty of Bioresources,
   Mie University, Mie, Japan

ABSTRACT: Over the last decade, various type of models, collectively referral to as "habitat models" have been developed to address different wildlife management issues of species-habitat relationships.   Although there are demands in developing various "habitat models", significant efforts tare not yet described in Japanese wildlife management.   It is very difficult to propose a good "habitat model", because lack of data sets for establishment of a habitat model is mortal in Japanese wildlife ecology.   However, many species are facing to extinction in mountainous forests environment of Japan.   This paper describes "present status" of wildlife management efforts and wildlife ecological data, then seeks an effective method in habitat modeling to maintain biodiversity of the study area, the Kii Peninsula, Japan.

Keywords: habitat model, the Kii Peninsula, GIS (Geographic Information System), AHP (Analytical hierarchy process)

1. INTRODUCTION

The use of quantitative models in various aspects of wildlife management (inventory, impact assessment, mitigation, population dynamics, and others) has increased rapidly in many countries, especially in the United States and Canada. Over the last decade, various type of models, collectively referral to as "habitat models" have been developed to address different wildlife management issues such as nesting, food, and many aspects of species-habitat relationships.   Several modeling techniques have been developed for habitat suitability analysis including the US Fish and Wildlife Services habitat suitability models (Fish and Wildlife Service 1981).   These models can provide necessary predictions in natural resources planning and decision making processes (Schamberger and

345

O'Neil 1986).

Although there are strong demands in developing various "habitat models" in Japan, significant efforts have not been described in Japanese wildlife management. Due to various reasons, it is very difficult to propose a good "habitat model" in Japan. For instance, lack of adequate data set in wildlife ecology is mortal. There are no effective method and tool to establish a model in large scale of wildlife ecology and management in Japan, even if many wildlife species, especially mountainous environments are identified as threaded or endangered. To obtain the protection afforded biodiversity pictures on the wide range of forest environments in Japan, it is necessary to establish a prototype of wildlife habitat model. Thus, the objectives of this paper are (1) to explain "present status" of wildlife management efforts and wildlife ecological data in the Kii Peninsula, Japan, and (2) to seek an effective method in habitat modeling to maintain biodiversity of the study area.

2. STUDY AREA: THE KII PENINSULA

The Kii Peninsula is located in the Kinnki Area, the central part of Japan. It is close to several major cities of Japan such as Osaka, Kyoto, and Nagoya (Fig. 1). The Kii Peninsula is said to be including three prefectures of Wakayama, Nara and Mie.

The Pacific Ocean

Fig 1: The Kii Peninsula in Kinki area of Japan

The Kii Peninsula has unique features of topography and elevation. The Oodai-Gahara Mountains are the highest peaks of the area with only from 1500 to 1600m above sea level, but the mountains are steep. As a result, these areas have unique forest environments. The Peninsula is also known for its rainfall. The area's annual precipitation can be over 4,500 mm. September 14, 1923, the Oodai-Gahara Mountains had 1,011mm for daily precipitation. If it compares with annual precipitation of 1,360 in Osaka in 1923, it makes clear that the area is the most heavily rained area in Japan .

2-1. FORESTS AND FORESTRY OF THE KII PENINSULA

The Kii Peninsula has been the place for the center of Japanese forest industries. Table 1 and 2 describe forest areas of the Kii Peninsula. In recent years, many forest owners, researchers, governmental agencies, and NGO believe in shifting from traditional "utilization forestry" to "multiple use of forests", because most Japanese domestic timbers are no longer profitable because of the deterioration in price caused by the relatively cheaper imported timber (Tanaka 1997). As a result, well-managed timber stands are rarely seen in many of Japanese mountainous area. However, just like Table 1 and 2 show that most parts of the Peninsula is still forested.

Table 1: The area of three prefectures in the Kii Peninsula (km^2)

Prefectures	Forested Areas	%	Total Area
Wakayama	3,640	77.0	4,725
Nara	2,840	76.9	3,692
Mie	3,770	65.2	5,778
Total of the Kii Peninsula	11,250	79.2	14,195

* Including all types of forests such as bamboo

Table 2: Forest area of the Kii Peninsula (km^2)

Prefectures	National Forests		Privately Owned Forests	
	Man-made forests	Natural Forests	Man-made forests	Natural Forests
Total	570	300	6,000	3,540
Nara	140	60	1,670	990
Mie	240	120	2,210	1,250
Wakayama	190	120	2,120	1,300

● Not including bamboo and shrubs

The Kii Peninsula is known for the oldest tradition of timber management in Japan. The region still produces high quality of Sugi (Japanese red cedar) and Hinoki (Japanese cypress). In addition to the timber quality, the peninsula contains wide diversity of wildlife. However, they are causing negative impacts on forest management. For example, Japanese serows feed on immature stands of Sugi and Hinoki, and the damage could cost millions of Yen for a landowner every year. Bear barks trees at many locations. Japanese monkeys cause significant damages to mushroom farmers.

2-2. WILDLIFE OF THE KII PENINSULA

Effective management of wildlife largely depends on understanding their habitat's needs (Clark et al 1993). In Japan, wildlife management is one of "developing studies". Not many species are known for their population, habitat, or even status of the threat. Extending nearly 3,800 kilometers from northeast to southwest, Japan has a generally mild temperate climate with a rich variety of local habitats. This expansiveness has resulted in regional variations in wildlife ecology. With 45 percent of mammal species in the nation are only seen in Japan. However, 68 Mammalians are listed in JRDB (Japanese Red Data Book of Threatened Animals). For birds, 132 species are listed in JRDB. In addition, according to the International Union for Conservation of Nature and Natural Resources (IUCN), 29 mammals, 33 birds, 8 reptiles, 10 amphibians, 7 fishes, and 45 invertebrates are listed in the "Red Data Book". These numbers are usually less than its of a typical tropical nation because of differences in diversity, but Japan has much higher rate of threat wildlife comparing to other temperate nations of the world. Table 3 lists several examples of wildlife of the Kii Peninsula and their present status. These are only several examples of species. Many must agree hundreds of species in the Kii Peninsula are facing to extinction. Yet, any significant efforts to protect the diversity of the Peninsula are not developed. For instance, there are 30 species of Asiatic land salamander (the most primitive living salamanders) in the world; however, ten of these are either endangered or rare in mountain streams of Japan. In the Kii Peninsula three are facing to extinction because of drastic changes on their mountainous forest habitats in last few decades.

2-3. EXSISTING WILDLIFE DATA

In many western nations, wildlife data are shared among governmental agencies, NGO, academic facilities and others. In addition, there are digitized geo-spatial images of many natural resources over extended area. Unfortunately, the network among these organizations in Japan is still poorly developed. Almost any wildlife data on population, habitat, or any other significant ecological data are not digitized for effective tools in modeling, like GIS (Geographic Information System).

Based on the reality of "Japanese wildlife information", it has to start to from zero to work on habitat modeling. If any scientists or biologists develop a habitat model in wildlife ecology over extended area of Japan like the Kii

Peninsula, they have to be begin to trace all paper printed data into the computer in order to establish adequate wildlife habitat model.

Traditional approaches to indicate wildlife distribution maps are (1) dot maps, (2) grid-base maps, and (3) range maps. Most of Japanese maps of wildlife distribution are either grid or dot types. Table 4 shows that a few examples of wildlife distribution map in the Kii Peninsula.

Table 3: Some major species of the Kii Peninsula and their present status

Common name	Species	Status in JRDB
Japanese black bear	Selenarctos thibetanus japonicus	E
Japanese serow	Capricornis crispus	Lp
Japanese flying squirrel	Pteromys momonga	R
Japanese dormouse	Glirulus japonicus	R
Japanese golden eagle	Aqulia chrysaetos japonica	E
Japanese falcon	Falco peregrius japonesis	E
Japanese harrier	Ciurus aeruginosus	R
Mandarin duck	Aix galericulata	K
Loggerhead (Marine turtle)	Caretta caretta	E
Asiatic salamanders	Hynobidae spp	E/R
Japanese giant salamander	Megalobatrachus japonicus	R

E = Endangered, V = Vulnerable, R = Rare,

K = Insufficiently known, Lp = Local population

Table 4: Examples of wildlife distribution maps in the Kii Peninsula

Examples of organizations	Example of Wildlife distribution maps
Environmental Agency of Japan	Grid data on major wildlife habitat
Mie, Nara, and Wakayama Prefecture, Department of Education	Population, habitat, and range of antelopes in the Kii Peninsula by grid maps
Mie Prefecture, Department of Environment	Population and habitat of major waterfowl by dot maps

3. GIS (Geographic Information System) BASED WILDLIFE MANAGEMENT

Use of multivariate statistics to assess wildlife habitats has increased because of multidimensional space and nature of wildlife habitat (Caspen et al 1986). Necessary assumptions can be very difficult to satisfy, but complex multivariate calculations on a landscape level are now possible due to the analytical and spatial capabilities of GIS (Clark et. al 1993). In the North America, recent advances in GIS technology and the increasing availability of digital database have provided the necessary tools to develop, test, and apply multivariate habitat models to large scales (Manen and Pelton 1997). Although many wildlife-habitat relationships on mammals, birds, herpes, and even fish are described with GIS in the North America and other countries, only few studies have been reported in wildlife-habitat relations with GIS in Japan. Without GIS or some other effective tools, it is very difficult to obtain an adequate model for wildlife management in a large scale. Even if the studies describe in the wide range of wildlife food, habitat, population, or any other aspects of ecology, many are not based on sustainable management, or insufficient for an effective natural resource management.

4. THE NEED FOR ALTERNATIVE APPROACH IN HABITAT MODELING

Habitat modeling for endangered species has received considerable attention over the last few years, because the model allows to forest managers to seek alternative approach to maintain biodiversity of the modeled area.

The initial study that formed the building for GIS analysis of endangered wildlife habitat in Japan includes Murakami and Hirata (1997). The study focuses on a habitat suitability model for the nesting selection of red-crowned crane (*Grus japonesis*). The study used buffering capabilities of GIS to seek potential nesting sites for cranes in the Eastern Hokkaido, based on the following conditions (Murakami and Hirata 1997);

(1) Distance from the nearest river network must be less than 265 m;
(2) Distance from the nearest road network and the nearest building must be larger than 104 m and 364 m respectively;
(3) Distance from the existing nests must be larger than 2640 m; and
(4) Nests must be located within the wetland area.

Murakami and Hirata (1997) developed a model based on a simple set of rules using available data of geography. And as Herr and Queen (1993) and Murakami and Hirata (1997) pointed out that a perfect nesting suitability model is difficult to develop; however, it could limit error and increase efficiency by addressing adequate techniques in habitat modeling, like Herr and Queen (1993) used a series of chi-square analysis.

Work of the methodological developments dealing with the quantitative analysis of habitat suitability model has their roots or rigorous statistical methods, most notably regression analysis. The use of multivariate statistical methods in wildlife-habitat model has some technical merit. For the most, these models allow different factors affecting habitats to be treated as variable inputs to habitat analysis. In this context, impacts of different factors can be determined and predicted. The strengths of well-studied statistical techniques can be put to bear on the analysis of various factors and their impacts to the habitat. Nevertheless, it is impossible to develop statistical analysis on habitat model in Japan because of deficiency of data set in wildlife management. On the other hands, we simply must provide a model for protection of wildlife species within limited time because of threaded status of species in the Kii Peninsula. Thus, we need to use alternative approach to develop decent habitat model with insufficient data set.

5. WHY USE ANALYTICAL HIERARCHY PROCESS (AHP) IN THE MODELING

Habitat modeling has inherent characteristics that lend itself well to AHP analysis. First and foremost, it is multi-factorial; that is, it involves examination of a number of factors that must be assessed relative to their impacts on the habitat. Second, the AHP can exhibits flexibility in dealing with both the quantitative and qualitative factors in a multicriteria evaluation problem (Banai-Kashani 1989). That is, some factors are inherently qualitative and defy any attempt to devise explicit quantitative measures. Others, while they may be amenable to quantification, they are too difficult or expensive to generate. Furthermore, some quantifiable factors may in fact be too loosely defined and hence could only be generated with high degree of error data inconsistency. Such data may actually be naturally vague and could only be captured using proxy variables on other surrogate measures, especially in Japan. Third, the relationships between species and its habitat may not be well understood. This lack of understanding in Japan presents a major difficulty in developing an adequate formulation and explicit specification of the relationship between the species and its habitat.

6 AHP IN HABITAT MODEL

Zahedi (1986) and Saaty (1987) provide excellent and brief overviews of AHP and its theoretical underpinnings.

Step 1 of AHP, involves the construction of a decision problem into a hierarchy of interrelated decisions   (Saaty 1977, and Mendoza and Sprouse 1989). At the top of the hierarchy is the goal of the analysis (e.g. selecting the best or most suitable option). The elements at the lower level hierarchies include the attributes such as objectives - perhaps even more refined attributes follows at the next lower level   - until the last level which typically contain the options or alternatives.

Step 2 involves the comparison of the attributes or elements in one level relative to their contribution or significance to the elements of the next higher level.   This step constitutes much of the evaluation (quantitative) or assessment (quantitative) of the decision problems and its hierarchy.   However unlike other quantitative decision-making tools, the evaluation and assessment process in Step 2 are easily within the grasp of the decision maker (DM) and the information required of the DM are transparent and are not difficult to provide.   A hierarchical structure of AHP provides a powerful method of partitioning the elements of a complex system into different sets know as the level of hierarchy (Saaty 1987).   The basic unit of the procedure involves setting up a weight consisting of judgments based on expert knowledge of relative importance between and among elements.   For the following example, the top of hierarchy is habitat suitability index (HSI) of Japanese dormouse in the Kii Peninsula, that is final HSI is a comparison among each variables on Level 2 with respect to their contribution to habitat suitability of dormouse (Table 5).

Table 5: An example of AHP in habitat model for Japanese dormouse

Level 1 (the Goal)	Level 2		Level 3	
Variables	Variables	Weights	Variables	Weights
HIS	Overstory richness	0.2	High	0.7
			Intermediate	0.2
			Low	0.1
	Understory richness	0.2	High	0.8
			Intermediate	0.1
			Low	0.1
	Distance to major roads	0.1	>1000m	0.5
			500~1000m	0.3
			<500m	0.2
	Potential food	0.4	Insects	0.6
			Fruits	0.2
			Pollen	0.1
			Others	0.1
	Ave. Winter temperature	0.1	$<-10°$ C	0.6
			$>-10°$ C	0.4

* Weights on each variable should be decided based on the expert judgements

7.   CONCLUSION

The Kii Peninsula is located at the center of Japan.   The region has a long history and tremendous cultural values. In addition, the Peninsula maintains a great environment for its forests and wildlife. Because the place is close to major central cities such as Osaka, Kyoto, and Nagoya, it has distinguished recreational values. Although the region attracts a great interest from the public, deterioration of timber price, deforestation caused by recreational expansions, and radical decreases in young population at the mountainous areas are effecting its forests and management regime. As a result, many forests stands are no longer properly managed, and it causes significant losses in biodiversity.

GIS is a computer-assisted system for the acquisition storage, analysis and display of geographic data. These capabilities of GIS paved the way for its widespread use in habitat analysis (Pereira and Itami, 1991; Chang, et al, 1992; and Chang, et al. 1993). Nevertheless, as this paper indicates examples in the Kii Peninsula, much wildlife habitat information are not yet consolidated for a spatial analysis of GIS. Because a lot of wildlife species are threatened in the Kii Peninsula at this moment, we are simply not able to waste a time to producing an "adequate data" for typical GIS wildlife–habitat relationship model such as statistical analysis. Consequently, we would conclude that combining GIS with AHP offers some advantages in wildlife habitat model in Japan. For the most, the spatial features and functionality of GIS allows habitat analysis to be location on site specific. The ability of GIS to perform simple spatial analysis such as map overlays, classification, and reclassification, generation of factors, distance and context operators (e.g. buffering, filtering, neighborhood analysis) make it a convenient platform for implementing AHP, particularly by expert judgment-based parameters and its given weights in the hierarchy.

REFERENCES

Banai-Kashani, R. (1989). A new method for site suitability analysis: The analytic hierarchy process. *Environmental Management*. Vol, 13, pp. 685-693

Caspen, D. E., J. W. Fenwick, D. B. Inkley, and A. C. Boyton (1986). Multivariate models of songbird habitat in New England Forests. pp. 171-175 in J. Verner, M. L. Morrison, and C. J. Ralph, eds. *Wildlife 2000: modeling habitat relationships of terrestrial vertebrates*. Univ.Wisconsin Press, Madison.

Chang, K., Yeo, J. J., and Verbyla, D. L., and Li, Z. (1992). Interfacing GIS with wildlife habitat analysis: a case study of Sitka black-tailed deer in southeast Alaska. *GIS/LIS 92 Proceedings*. pp.95-104

354

Chang, K., Verbyla, D. L. and J. J.Yeo. (1993). Deer habitat analysis at two spatial scales. *GIS/LIS 93 Proceedings.* pp.109-117

Clark, J. D., J. E. Dunn, and K. G. Smith (1993). Multivariate model of female black bear habitat use for a geographic information system. *J. Wildl. Manage.* Vol, 57, pp. 519-526.

Herr, A. M. and L. P. Queen. (1993). Crane habitat evaluation using GIS and remote sensing. *Photogrammetric Engineering and Remote Sensing.* Vol, 59, pp. 1531-1538.

Manen, F. T. and Pelton, M. R. (1997). A GIS model to predict black bear habitat use. *J. For.* Vol, 95, pp. 6-12.

Mendoza, G.A. and W. Sprouse. (1989). forest planning and decision making under fuzzy environments: An overview and analysis. *Forest Science.* Vol 35, No. 2, pp. 481-502.

Murakami, H. and Hirata, K.(1997). Potential nesting site analysis of red-crowned cranes in Kushiro wetland using GIS. *Theory and Applications of GIS.* Vol. 5, No.2 pp9-64

Saaty, T.L. (1977). A scaling method for priorities in hierarchical structures. *Journal of Mathematical Psychology,* Vol. 15, No. 3. 234-281.

Saaty, T.L. (1987). The Analytic Hierarchy Process - What it is and how it is used. *Mathematical Modelling.* Vol, 9, pp. 161-176.

Schamberger, M. L. and L. J. O'Neil. (1986.) *Concepts and constrains of habitat-model testing.* pp 5-10 in J. Verner, L. Morrison, and C. J. Ralph, eds. Univ. Wisconsin Press, Madison.

Tanaka, K. (1997). Long rotation forestry supported by the RVFP theory. *Proc. of IUFRO.* Symp. on Sustainable Management of Small Scale Forestry in Kyoto, 1997. pp 222-228.

Pereira, J. M. C. and R. M. Itami (1991). GIS-based habitat modeling using logstic multiple regression: a study of the Mt. Graham red squirrel. *Photogrammetric Engineering and Remote Sensing.* Vol, 57, pp. 1475-1486

U.S. Fish and Wildlife Services. (1981). Standards for the development of suitability index models. *Ecological Manual Service Manual 103.* U.S. Department of Interior, Fish and Wildlife Service, Northern Prairie Wildlife Research Center, Jamestown, ND. 203p.

Zahedi, F. (1986). The Analytic Hierarchy Process - A Survey of the Method and Its Applications". *Interfaces* Vol, 16, pp. 96-108.

Analysis of Species Hyper-diversity in the Tropical Rain Forests of Indonesia : the Problem of Non-observance

Keith RENNOLLS*[1,*3] and Yves LAUMONIER*[2,*3]

*1 University of Greenwich, London SE18 6PF, UK
*2 CIRAD-Forêt, Baillarguet, PO Box 5035, 34032 Montpellier, Cedex, France
*3 FIMP, INTAG, Manggala Wanabakti, PO Box 7612, Jakarta 10076, Indonesia

ABSTRACT The main interest in the assessment of forest species diversity for conservation purposes is in the rare species.   The main problem in the tropical rain forests is that most of the species are rare. Assessment of species diversity in the tropical rain forests is therefore often concerned with estimating that which is not observed in recorded samples.   Statistical methodology is therefore required to try to estimate the truncated tail of the species frequency distribution, or to estimate the asymptote of species/diversity-area curves. A Horvitz-Thompson estimator of the number of unobserved ("virtual") species in each species intensity class is proposed.   The approach allows a definition of an extended definition of diversity, (or generalised Renyi entropy). The paper presents a case study from data collected in Jambi, Sumatra, and the "extended diversity measure" is used on the species data.
Keywords: species diversity, virtual species, extended species diversity.

1. INTRODUCTION: NON-OBSERVANCE OF SPECIES

For a long time there has been much interest in the speciation process in tropical rain forests, and how it depends on environmental factors, Ashton(1969,1976), Lieberman et.al(1996), Tilman(1994). Species diversity has been discussed and defined in several ways, Hurlbert(1971), Hill(1973), Peet(1975), Pielou(1975). Methods of definition and assessment of biodiversity are important but difficult areas of much current activity, Magurran (1988), Boyle and Boontawee(1995), Vanclay(1996).   One of the main problems in the definition and assessment of diversity is that it is scale dependent, and it is not clear that estimates of local diversity from small plots tell us much about the diversity at a larger scale.

Suppose that species identification of observed trees is perfect.   When a ground sample plot is used to collect data on the species of tree in a tropical rain forest, it is found that some species have many trees in the plot, but many species have few trees in the plot. For a tropical rain forest a relatively high proportion of the species observed are represented by a few trees; that is, many of the observed species are relatively rare.

An observed abundant species was almost certain to be observed, so we can reason that it will be relatively unlikely that there are equally abundant similar species which were not observed. However, a relatively rare species which is observed, had quite a high chance of not being observed, and it is therefore quite likely that there are other similarly relatively rare species which were not observed. These species which were not observed, but which we may argue most probably exist, we term "virtual species".

From a sample area of forest trees we therefore have information on observed species, and inferences about unobserved virtual species. Calculation of a measure of tree species diversity which takes account of both observed species and virtual species provides us with an extended definition of species diversity. In this paper we formulate such an extended definition of species diversity

2. AN ESTIMATE OF TOTAL NUMBER OF SPECIES; VIRTUAL SPECIES

Suppose there are n trees observed in a plot, and that a particular species, s say, has r trees observed. Making the simplifying assumption either that the trees are randomly distributed, or that the n selected are a random sample of the population of trees, then we may estimate the proportional representation of species s in the population as r/n . Hence the probability of species s *not* being observed is

$$P_0(s) = (1-(r/n))^n \qquad (1)$$

Hence the probability of species s being observed can be estimated as

$$\pi_s = (1-P_0(s)) \qquad (2)$$

If we suppose the sample is a sample of species, and the response variable $y_s = 1$ when a distinct species is observed, ($y_s = 0$ otherwise), then the Horvitz-Thompson estimator of the total number of species is given by

$$\hat{S} = \sum_s \frac{y_s}{\pi_s} = \sum_s \frac{1}{(1 - P_0(s))} \qquad (3)$$

(Cochran(1977)). For each observed species we are estimating that there are $\{ \frac{1}{(1 - P_0(s))} - 1 \}$ unobserved "virtual" species. It should be noted that this estimate of the number of species in the population takes no account of the species that are so rare in the population their chance of being observed is much less than $1/n$. Adjustment of the total species number estimator for this effect needs to make use of truncation models, or equivalently area-species curves, both of which are discussed elsewhere, Rennolls and Laumonier (1998).

3. AN EXTENDED ENTROPY MEASURE OF SPECIES DIVERSITY

3.1 Renyi's Generalised Entropy: α-entropy

An α-entropy **index of diversity** may be defined on data collected **from a single plot**, by

$$H_\alpha = \frac{\log(\sum \hat{p}_i^{\,\alpha})}{(1-\alpha)} \qquad : \alpha \geq 0 \qquad (4)$$

where $\hat{p}_j$ is the observed proportion of trees in the jth of s observed species. H_α has the properties that: the total species count on the plot, s = exp(H_0), where H_0 is the maximal value of H_α ; the Shannon-Weaver index $H' (\equiv H_1) = -\sum_{i=1}^{s} \hat{p}_i \log \hat{p}_i$; and Simpson's index of concentration, λ = $\sum_{i=1}^{s} \hat{p}_i^2 = \exp(-H_2)$. H_α/H_0 is a measure of even-ness.

3.2 An Extended α-entropy

For each observed and identified species i, we may estimate that there are v_i species in the region, where

$$v_i = \frac{1}{(1 - P_0(i))} \tag{5}$$

The usual sample-based estimates of species diversity are known to be biased (maximum likelihood) estimates of the region species diversity, the bias arising partly from the non-observance of all the species in the sample. We may therefore consider generalisations of standard entropy measures, making use of estimates of virtual species in the region. An alternative form of the α-entropy is therefore suggested:

$$\tilde{H}_\alpha = \frac{\ln(\sum (\tilde{p}_i)^\alpha)}{1 - \alpha} \quad where \quad \tilde{p}_i = \frac{v_i}{\sum v_i} \tag{6}$$

which includes generalisations of the Shannon-Weaver entropy, and Simpson's concentration index.

4. THE PROBLEM OF NON-IDENTIFICATION

It seems that all the previous theoretical work available and mentioned in papers concerned with the definition and estimation of tree species diversity in forests, and methods using species-area curves make the assumption that trees are uniquely and correctly identified. Of course, identification of all observed trees to unique classes would be sufficient for analytical purposes. The actual species names are labels, which while they may be of particular interest to floristic experts, possibly for inter-regional comparisons, do not add any information to the analysis of diversity in a region.

In practice, floristic surveys can have relatively high proportions of trees classified as "unidentified". This issue of "unidentifiability" has a major consequence for the validity of the estimators used to estimate tree species diversity. The presence of non-identifiability will introduce a negative bias to the estimation of Renyi generalized entropy measure of diversity, H_α , and in particular, the total number of species in a region. These comments also apply to the extended generalised entropy defined in section 3. Rare and unknown species are very likely to be included in the "unidentified group". Hence the increased observance of such rare or unknown species will be masked by their being subsumed under the "unidentified" group. This will have the effect of eliminating the upward asymptotic trend of the species-area curve, and will therefore undermine the estimation of total number of species, by extrapolative and truncation methods.

A simple model of the non-identification has been developed in Rennolls(1998) and it has been used to modify the equation (1), and hence equations (4) and (5). The theory is not presented here. The actual effect on the data from the sites described in the next section are practically negligible.

5. CASE-STUDY

Use has been made of (4) and (6) for data collected in Batang Ule, a research site in Sumatra.

5.1 Batang Ule

The data used in the analysis was from the research site based at Jambi which has been measured by Dr.Yves Laumonier, his students and others. The site is 3ha in area, with dimensions of 300mx100m, with the longer dimension being oriented along the line of maximal topographic gradient. 1885 trees of diameter greater than or equal to 10cms were observed to fall into 504 species, including an "unidentified" species class. For the purposes of the analysis reported in this paper this class is treated just as another species label. Each of the other species observed had less than 20 trees observed over the 3ha plot.

The numbers of species with only one, two or three tree observations were 221, 94, and 50 respectively, (accounting for 30% of the observed trees). If we define a species to be "relatively rare" if its observed rate of occurrence is less than of equal to 1 tree/ha., then this site has 68% of its species which are relatively rare, a percentage which highlights the importance of rare species in the analysis of the diversity of tropical rain forests, and justifies a label of "hyper-diversity".

For the species diversity analysis reported in this paper, the site has been divided into 30 100m x 10m strip-plots which are oriented across the line of maximal topographic gradient. Plot 1 is the first strip at the bottom of the valley; Plots 1-8 are essentially in the valley; Plots 9-14 are on the slope from the valley to the plateau/ridge; Plots 15-16 are top of slope ridge plots. Plots 17-26 cover the "plateau" region of the site, which consists of a web of ridges with local valleys and slopes between them; Plots 27-28, are again almost entirely ridge plots.

5.2 Virtual Species Estimates

In Batang Ule, where the non-identification rate was very low at 1%, the virtual number of species for every observed species with just one tree observed was 1.59. The corresponding figures for the two- and three- observed-tree species were 1.16 and 1.05 respectively. The observed number of species was 504 (=s), and the estimate of S, from (3), was 641. This is unexpectedly large since it had previously been considered that the species carrying capacity would not be much greater than the observed species count of 504 on the 3ha site. The reason for the relatively large inflation in the estimate is that the species identification was very vigorous in this study, resulting in a very long tail of observed and identified rare species. The estimate of the total number of species, (dia.>10cm.), using extrapolated species-area curves is 676, (96% C.I. (631,720)), (Rennolls and Laumonier(1998a)), consistent with the virtual-species estimator. The virtual species estimatator will be affected by the possible inappropriateness of the random spatial distribution model for the rare species. Certainly the more frequently occurring species do not occur randomly. However, the data are so sparse for the rare species that the hypothesis of Complete Spatial Randomness (CSR) for the rare species is untestable on a single species basis. The hypothesis may be testable on a pool of rare species. H_1 is 5.4 from the sampled data, but the value of $\tilde{H}_1$ is 6.2.

5.3 Graphical Analysis of α-entropy

The α-entropy graphs for the 30 plots are shown in Figure 1, Orloci (1991). Note that the value for α = 0 corresponds to log(S), where S is the total number of

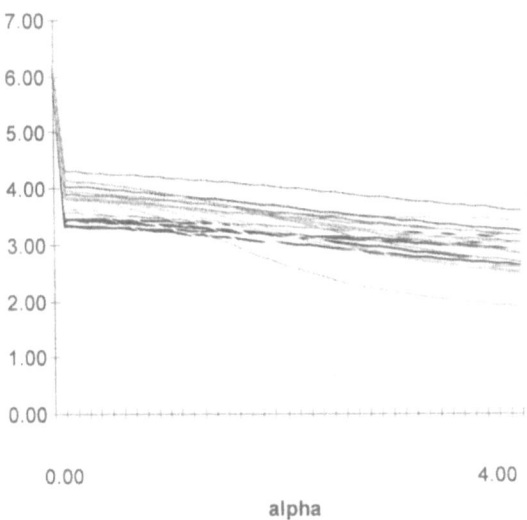

Figure 1. Batang Ule: α-entropy graphs for 30 plots.

species observed in all of the 30 plots. The 0.1 α-entropy value indicates log(s_i) where s_i is the number of species observed in plot i. Whilst the majority of the plots retain their relative rank over the α-range, there are five plots, 14, 15, 16, 17, and 18, which exhibit anomalously low diversity measures with increasing α. These contiguous plots range from the first which is at the top of the slope, through 15 and 16 which are the plots on the ridge above the slope to the valley, to 17 and 18 which are the first two plots on the "ridgey plateau". Whilst they would not be distinguishable from other plots in terms of number of species, or the Shannon Weaver measure of diversity, they do have substantially reduced high-order diversity measures. For example, they would be distinguished from the other plots in terms of Simpson's Diversity measure, corresponding to $\alpha = 2$, but use of H_3 would provide even better discrimination. In terms of the effect of land-facet on species diversity it might be suggested that it could be important to have a class corresponding to "a ridge above a slope to a valley", though "well-drained ridge" might be adequate.

5.4 Graphical analysis of extended α-entropy

The extended α-entropy graphs for the 30 Batang Ule plots are shown in Figure 2. Although the α-axis has been extended from Figure 1, very high order extended entropies do not appear to form any clearer groupings than in Figure 2.

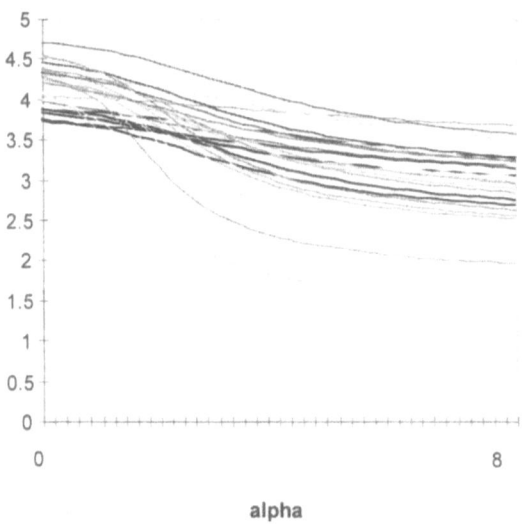

alpha

Figure 2. Batang Ule: extended α-entropy graphs.

5.5 Ordination of Extended α-Entropy Graphs

If the extended α-entropy value at $\alpha=0$ is plotted against the difference in value for $\alpha=0$ and 8, we obtain Figure 3, a simple ordination. Ordination using Factor Analysis and Multidimensional scaling have been reported elsewhere, Rennolls(1997), Rennolls and Laumonier(1997), Rennolls and Laumonier(1998b). By following the plot number in sequence from 1 through to 30 in Figure 11, we find that the ordination clearly distinguishes the valley plots form other plots, and to a lesser extent the slope-plots from the ridge-plots. However the discrimination between the differing land-facet classes, in terms of species diversity, is not as clear as that obtained using the ordinary α-entropy values. This might be due to the fact that extended α-entropy gives more weight to the rare species.

$H_0 - H_8$

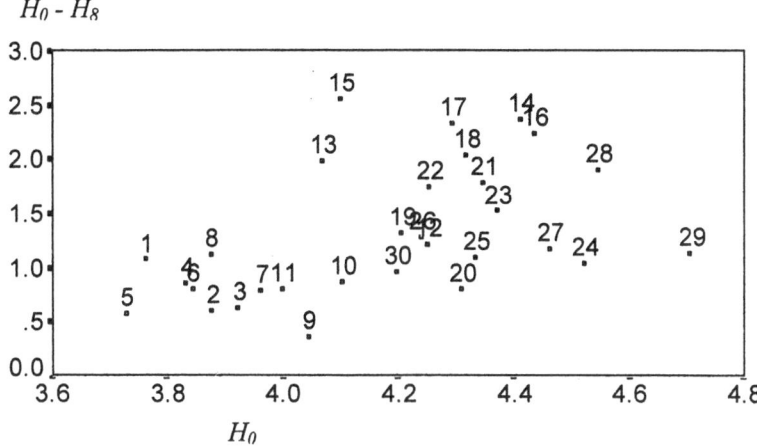

H_0

Figure 3.   Scatter plot of the H_0 (species richness) and ($H_0 - H_8$) values.

Use of extended α-entropy analysis does not seem to improve the ordination of the diversity structure in Batang Ule, and it might well be that the standard α-entropies are more suited to diversity structure analysis in general, but this conclusion will have to await the results of the analysis of other data sets.

6. REFERENCES

Ashton,P.S. (1969) Speciation among tropical trees: some deductions in the light of recent evidence.   Biological Journal of the Linnean Society. 1: 155-196.

Ashton,P.S. (1976) Mixed Dipterocarp Forest and its variation with habitat in Malayan lowlands: a re-evaluation of Pasoh. Malayan Forester 39: 56-72.

Boyle, T.J.B. and Boontawee, B. (editors) (1995)   Measuring and Monitoring Biodiversity in Tropical and Temperate Forests, CIFOR, Bogor.

Hill, M.O.   (1973)   Diversity and evenness: a unifying notation and its consequences. Ecology, 54(2), 427-431.

Hurlbert, S.H.   (1971)   The nonconcept of species diversity: a critique and alternative parameters. Ecology, 52(4), 577-586.

Lieberman,D., M.Bieberman, R.Peralta, G.S.Hartshorn (1996)   Tropical forest structure and composition on a large-scale altitudinal gradient in Costa Rica. Journal of Ecology, 84: 137-152.

Laumonier, Y. (1997). The vegetation and physiography of Sumatra. Geobotany 22. Kluwer Academic Publishers. Dordrecht. 222 pp.

Magurran, A.E. (1988) Ecological Diversity and Its Measurement. Chapman and Hall.

Nohr, H., and A.F.Jorgensen (1997) Mapping of biological diversity in Sahel by means of satellite image analysis and ornithological surveys. Biodiversity and Conservation, 6, 545-566.

Orloci, L. (1991) Entropy and Information. SPB Academic Publishing.

Peet, R.K. (1975) Relative Diversity Indices. Ecology, 56: 496-498.

Pielou,E.C. (1975). Ecological Diversity. John Wiley.

Rennolls, K. (1997). Diversity issues in the Forest Inventory and Monitoring Project. FIMP-INTAG Tech. Ser. Report n°7. Jakarta. 157 pp.

Rennolls, K. and Laumonier, Y. (1997) Revealing the Structure of Tree-Species-Diversity in the Tropical Rain Forest Ecosytem. Bulletin of the International Statistical Institute, Proceedings of the 51st Session, Instambul, Contributed Papers, Vol.1, pp25-26.

Rennolls, K. and Laumonier, Y. (1998a) Species-Area and Species -Diameter Curves for Three Forest Sites in Sumatra. Submitted to Journal of tropical Forest Science.

Rennolls, K. and Laumonier, Y. (1998b) Species Diversity Structure Analysis at Two Sites in the Tropical Rain Forest. Submitted to the Journal of Tropical Ecology.

Tilman, D. (1994) Competition and Biodiversity in Spatially Structured Habitats. Ecology, 75(1), 2-16.

Trichon, V. (1996). Hétérogénéité spatiale des structures en forêt naturelle de basse altitude à Sumatra. Thèse Doctorat Université Toulouse III. 260 pp.

Vanclay, J.K. (1996) Towards more rigorous Assessment of Biodiversity. Presented at UIFRO 4.11, Monte Verde, September 1996. In press.

Mixed Stands Between Description and Modelling

Alain FRANC

Engref, Département MAI, Paris, France
Laboratoire ESE, Orsay, France

Keywords : Mixed multicohort stands, Algorithmic Complexity, Modelling, Description.

1. THE COMPLEXITY OF A MIXED FOREST

It might appear at first glance that deciding whether a given number is a random number or not is very far from scientific questions relevant for a better understanding or description of mixed forest stands. In this section, a link will be woven between algorithmic information theory, which deals with the definition of random numbers, and modelling of mixed stands.

1.1 WHAT IS A RANDOM NUMBER ?

In the mid 60's, three scientists named Solomonov, Kolmogorov and Chaitin simultaneously and independently proposed a definition of a random number. Let a number be represented as a sequence of 0' and 1'. What is a random sequence ? Surely, the sequence

$$a = 01\ 01\ 01\ 01\ 01\ 01\ 01\ ...$$

is not a random sequence, whereas the sequence

$$a = 01\ 10\ 00\ 11\ 01\ 11\ 10\ ...$$

has been generated by a random number generator (more correctly, it is a pseudo-random number, as it will be seen within a few paragraphs).

The idea behind the definition of a random number is the following. Let us have a number, say $x = 01010101010101$ as a binary sequence and a programme whose output is the number x. For example, x is the output of the programme 'print the sequence 01 seven times', written in a proper language. Now, imagine the set of all programmes whose output is precisely x, and compute the length of each programme in the set. There is at least one programme whose length is minimal (they may be several ones). The ratio between the length of this minimal programme and the length of the binary sequence is called the algorithmic complexity of the sequence. It can be shown that, when the sequence is very long, it does not depend on the computing language.

If the algorithmic complexity of a binary sequence is close to zero, there exist a short programme to generate it as the output of the programme. An example is the programme 'Print one billion times the sequence 01', whose output has a low algorithmic complexity. A random number is a number represented by a sequence whose algorithmic complexity is equal to one. The length of the shorter programme which can generate the sequence is equal to the length of the sequence itself. It is then equivalent to the programme 'print x', whose length is equal to the length of the sequence *x*, at least approximately if this sequence is very long. The shortest way to communicate a random number is to describe it digit by digit.

1.2 FROM NUMBERS TO DATA SETS

Let us imagine an object which is known by a sequence of numbers. It can be a tree, and the sequence of numbers be its height, basal area, age, etc. ... . A model of this data set is a small programme, much smaller than the length of the data set itself, which reconstructs the data set.

There is a slight change with algorithmic information theory, where data set is recovered exactly as an output of the programme. In modelling, approximations are allowed. A model is a short programme which can reconstruct a good approximation of the sequence of numbers. More often, the tolerance is tuned by the modeller according to a given objective. This point of vue on modelling is close to the etymological sense of the word 'model', which means 'imitation'. A model is an imitation of a data set by an output of a small programme. This idea has been proposed in the 60' by Solomonov, and is presented in details in Li & Vitanyi (1997). They claim that 'Science may be regarded as the art of data compression' (Li & Vitanyi, p. 585).

The ratio of compressibility of a sequence of numbers can vary in the set]0,1] of real numbers. If the complexity is close to 0, the programme is a reliable representation of the data set, and the object can be modelled. If the complexity is close to one, there exist no short programme which will reconstruct the data set as an output. It means that the best exact knowledge of the data set is a mere description of it, item by item. There is no way to compress the description into a model.

1.3 MODELLING AND NATURAL HISTORY

Let us imagine a forest stand, known by a sequence of variables (species, age, size, shape, etc. ...) attached to each tree, including its position. This is a description of a stand. If the data set is compressible, there exist a model which allows a more compact description of the stand. This compression will usually be built from redundancies between variables, such as allometric relationships, growth curves, etc. ... .

A point of vue from Natural History will emphasise the diversity of individual situations between trees, and remain close to description. A point of vue from mathematical modelling will emphasise similarities between different situations and transform them into a compression algorithm which will be called a model. Both points of vue are often presented as conflicting vues on the same object as, say, a piece of forest in the Amazonian Basin. If we follow the algorithmic information theory, the knowledge of the complexity of an object should lead rather towards Natural History and description if the complexity is high, or rather towards mathematical modelling if the complexity is low. The choice should then depend on the object, and not on the observer.

But, as there is a 'but ...', a theorem has been demonstrated which states that, in general, the complexity of an object known as a sequence of numbers cannot be computed (Delahaye, 1994). Scientists are faced with a very uncomfortable situation : the complexity of a sequence does exist, but is not computable. More precisely, it has been demonstrated that any algorithm which has as an input a sequence and as output its complexity will deliver an answer in finite time for a finite number of sequences only. It is then impossible in front of data only to decide whether natural history or modelling is a better point of vue.

1.4 ON COMPLEXITY AND DIVERSITY

Description is sometimes a relevant way of having knowledge on a mixed forest, and modelling cannot be the only scientific study of it, when it is complex such as in the Tropics. But complexity is not associated solely with the Tropics. Complexity is the rule as well in Temperate forest if the entire ecosystem is taken into account, namely flora, fauna, including insects, soil fauna, etc. ... . It is the case as well in Temperate forest, for the canopy layer only, if the entire life cycle of a stand is taken into account, and not only a few years of growth stage. Life cycle incorporates pollination, fructification, fruit dispersal, germination, etc. ..... . An entire cycle of a very simple mixed Atlantic forest, composed of beeches, oaks, maples, ashes and cherry trees cannot be modelled entirely.

This should reinforce the position of Natural History as a way of building relevant descriptions of complex heterogeneous forests, but the reality is more subtle, and much remains to be understood.

On one hand, algorithmic complexity is associated with randomness, and it cannot be stated that living objects (and mixed forests) are random associations of features. The diversity of life gives us an example of structures which resist compression with models, but which are anything but randomness. This very association between incompressibility and structure which escapes the algorithmic definition of complexity has been recognised as a weakness of algorithmic information theory when applied to life (Nicolis & Progogine, 1992), and is a weakness when applied to forest too. There is an inner logical contradiction is studying biological diversity with indexes, trying to summarise the whole diversity of a community into a single index, whatever it is. If such a compression is possible, then the diversity is low, by definition. It is proposed here that the diversity of a structured community is the incompressible part of its description. But we are left without a mathematical definition of a structure, or of diversity. Proposing a mathematical definition of a complex but structured data set still is an open question.

On the other hand, the diversity of a mixed forest as approached by individual descriptions such as in Natural History, points out the role of individuals, and is poorly linked with an approach of a stand as a community of trees. Each tree is an open system, connected to other compartments within the ecosystem. A forest is not a collection of trees, as inventories point it out, but is a network of interacting trees. These interactione, recognised as central in modelling mixed forests, are poorly taken into account in inventories, and in description. Knowing the size, the age and the species of a tree, for example, gives valuable information on the size and the species of neighbours, if those variable are not allocated at random. The interaction through competition for resources means that size distribution in small plots can be modelled, namely compressed into a given number of plot types, and actually is. This ability to be compressed is the reflection of interaction. Or, put it the other way, interaction between individual within a set can be defined as the possibility to compress the simultaneous description of these individuals.

1.5 FOUR EXAMPLES

Let us imagine a forester who observes a forest and wishes to convey its picture to a distant colleague, over a costly transmission channel. His/her colleague may be on the star Sirius, and the forester has been allowed to transmit the picture but at the lowest possible cost. A description of a forest will be said reliable if the colleague on Sirius, who cannot see the forest, can draw a reliable picture from the description only. The cost if the transmission varies greatly with the stand type, as the following example show.

Let us imagine a fir plantation in Europe. A sentence like 'a plantation of silver fir, at density 300 stems per hectare, at age 80, each fir being approximately 25 m tall, and 35 cm wide at beast height' conveys a fairly good picture of this forest to the other end of the transmission channel.

Let us imagine a seminatural forest, like a mixture between oak and beech. The same sentence : ' a seminatural oak-beech forest, even-aged, of age 100 years, at 150 stems per hectare, with trees being approximately 35 m tall and 60 cm wide at breast height whatever the species' conveys only a partial picture of the forest. There may exist different types of mixture : species can be mixed stem by stem, randomly, or as monospecific aggregates. It is them necessary to precise the type of mixture, with quantitative features such as the size of aggregates, in order to convey a reliable picture if the forest.

Let us imagine an Appalachian mixed forest, with about 20/30 species per hectare. It is no longer possible to convey a reliable picture of the forest with a few words only : the species, their mixtures, the number of stems per species, the vertical and horizontal structure have to be specified, as such a mixed forest will probably be a multicohort stand.

Let us imagine now a Guyanese forest. It is no longer possible to compact the description in few sentences : every tree is different from any other tree in the stand, and the description of the stand has to be made tree by tree for it to be reliable. It is no longer possible to compact the information.

2. THREE LEVELS OF DESCRIPTION AND MODELLING

Looking more closely at previous examples suggests that producing a compact reliable picture of a forest relies on the possibility to aggregate several individual trees into one stand type which can be described as such.

2.1 DESCRIPTION BY MEAN VALUES

The compactness of the description of a homogeneous stand comes from the fact that each tree is very close to a virtual tree which has the mean values for each variable : mean height, mean basal area, mean age, etc. ... . Describing the mean tree provides as well a good description of the stand, as the programme to draw the picture of the stand can be written *(i)* draw the mean tree from the transmitted data set *(ii)* repeat the same description for each tree. Going from the tree level to the stand level adds only a few lines in the programme, and does not really increases its length, whatever the size of the forest. Variations between the virtual mean tree and any actual tree are described as fluctuations, namely as the realisation of a random variable. It is usually a Gaussian variable, of mean 0 and standard deviation σ. If the trees are labelled $i = \{1, \cdots, n\}$ in the stand, we have

$$h_i = \overline{h} + N(0, \sigma)$$

Looking more carefully at this formula, it can been noticed that the information in the data set representing the stand has been divided into two parts :

- the mean value, whose complexity is very low, as it has the same value for each tree
- the random fluctuation around the mean value, whose complexity is maximum.

This procedure of separating the information in an object into two parts, one accounting for the regularities which is highly compressible and considered as useful and the second as incompressible remaining random information is very general and has been proposed by Kolmogorov. It is known as 'Kolmogorov minimal sufficient statistics' (Li & Vitanyi, p. 114). The random component can usually be ignored to transmit a reliable picture of the forest, and this leads to an efficient way to compact the description of the forest.

Models at the stand level have been developed for over one century in forestry, including the so-called yield tables which are a link between modelling and management tool. It is often possible to follow the evolution of the mean variable while ignoring fluctuations around it. The fluctuations do not influence the evolution of the mean. A typical stand growth model has the form

$$\begin{cases} \dfrac{dh}{dt} = f_h(h, g) \\ \dfrac{dg}{dt} = f_g(h, g) \end{cases}$$

where h is the height and g is the basal area. There is no diversity nor in the values of the variables which describe a stand at the tree level nor in the individual trajectories within a stand : all tree share the same evolution. Such a compactness of possible evolution of a system simplified as a unique time evolution of a set of mean variable is common in ecology, where population diversity is often compacted into one component, as size or biomass, whose evolution is described by a set of coupled differential equations, as in Lotka-Volterra models. Most of those models have been produces in the so called 'Golden Age of Theoretical Ecology' (Scudo & Ziegler, 1978).

2.2 DESCRIPTION BY A DISTRIBUTION FUNCTION

Let us imagine a temperate mixed stand composed of several tree species per hectare (say twenty to forty) , which can be found in many warm temperate climate, such as Southern China, Japan, Chile, Southern Appalachian, Southern shore of Caspian Sea, etc. ... . Some of these stand are composed of patches of single cohort stands. A cohort is a group of trees regenerating after a single disturbance, and single cohort stands are considered as less complicated than multicohort stands, and are more commonly managed by foresters (Oliver & Larson, 1996, p. 147). Most of these single cohort stands can be faithfully described by histograms of relevant variable, such as species composition, diameter distribution, age distribution, etc. ... . Many different stand described at the individual level can fit into the same more compact description with distribution function : any permutation of variables between trees will change the individual description, but keep the distribution description unchanged. The diversity within a stand can be divided into two parts :

- the distribution function
- the allocation of given values to given trees

When the allocation can be ignored, a set of distribution functions for each relevant variable and their interaction is sufficient to convey a reliable picture of a given forest. Models dealing with distribution functions have been imported in ecology from human demography, where they have developed in the 40's. A seminal role is often attributed to a paper written by Leslie (Leslie, 1945), and a category of models on distribution function is called Leslie model. The importation into forestry is often associated with the name of Usher (Usher, 1966). The distribution function of a given variable is a vector, and the evolution of this vector is often formalised by a matrix model. The family of models dealing with the evolution of distribution functions are often referred to as matrix population models (Caswell, 1989).

2.3 INDIVIDUAL BASED DESCRIPTION AND MODELLING

Many of recent models in forestry belong to the so called 'distance dependent tree model' (DDTM). It means that the data set on which the model works consists in individual variable for each tree, as well as the location of each tree. The state of a forest as given by a DDTM is a mere description. The compaction occurs at two stages :

- for interaction between trees
- for the description of trajectories

Basically, a DDTM relies on the assumption that the neighbourhood of a tree drives its growth. It is a model in algorithmic information theory sense as it is assumed that local interactions belong to the same class whatever the tree. More precisely, individual trees are described, and not compressed, whereas interaction rules are compressed into a set of 'a priori rules', often as reaction to competition indexes.

Those models are mixed between description and algorithmic modelling. They have 'exploded' in many fields of ecology since the time where computing became affordable, efficient, and powerful. A recent survey is DeAngelis & Gross (1992).

3. FROM INDIVIDUAL TO POPULATION

Trying to understand how the precise description of each tree and it interaction with the environment can lead to a better understanding of the evolution of a stand as a whole is a very difficult question, still open for most of real cases. It is by now very easy to implement an individual based model in a computer, and run simulation, with very realistic features. But simulation does not drive progress into understanding. For example, stand models are known to be robust and useful for even-aged, monospecific and closed stands. But there is no explanation on this miraculous fact that they work. To understand why they work, namely which are the conditions which are at the basis of the possibility to summarise a stand by mean values of variable, has utmost utility. It can produce all conditions, less restrictive then initial experimental conditions where this result has been established, for which such a simplification is reliable.

An example of such an understanding is given below from the solid state physics.

3.1 THE ISING MODEL IN SOLID STATE PHYSICS

Let us have a crystal, made of regularly spaced atoms which can be considered as magnet. The magnet can have two states : either pointing upwards, or pointing downwards. This is an idealisation of a more complex reality, but will be a useful example for our purpose. A physicist will called it a set of atoms, or a system. An ecologist would call it a population of magnets, and a forester would call it a stand. Of course, those words are irrelevant out of their disciplinary context. They are given here to stress the links between those field. The link is a progress toward a better understanding on how a collective behaviour at the population level can emerge from interactions at the individual level.

Let us call s_i the state of an individual magnet labelled as $i \in \{1,n\}$. We will write $s_i = +1$ if the magnet is pointing upwards, and $s_i = -1$ if the magnet is pointing downwards. The mean magnetisation of the solid is $\bar{s} = \frac{1}{n}\sum_i s_i$. We will call x the fraction of nodes i pointing upwards. We have $\bar{s} = \frac{1}{n}(nx \times (+1) + n(1-x) \times (-1)) = 2x - 1$. The solid can be described at three levels :

- an individual level by $\omega = (s_1, \cdots, s_n)$
- a distribution function level by $(x, 1-x)$
- a global level by $\bar{s} = \frac{1}{n}\sum_i s_i$.

As $\bar{s} = 2x - 1$, the distribution function level and the global level are the same (this is due to the fact that the individual variable s can take two values, +1 and -1, only).

Let us suppose that the solid is known by its global level x only. There are C_n^{nx} individual states which fit to this distribution function. The logarithm of the number of individual states which boils down to the same distribution function is the entropy of the system. Using Stirling formula and dividing by the size n of the system, we have $H = -x \log x - (1-x)\log(1-x)$ which is the same formula as the Shannon diversity index for the population of magnets. The entropy, or the diversity index, is a measure of the number of individual states which can be compacted into the same distribution function.

The Ising model is a stochastic evolution model for magnetisation, where the probability for a cell i to be in state +1 or -1, which is noted $p(s_i = +1)$ and $p(s_i = -1)$ is a function of the state of neighbouring cells : $p(s_i = 1) = f(s_{i-1}, s_{i+1})$ if the lattice is a one dimensional grid. The same can be done in any dimension. If the neighbourhood of a given cell is simplified as being in the average state $\bar{s}$, we can write $p(s_i = 1) = f(\bar{s}, \bar{s})$. As we have $\bar{s} = 2x - 1$ and $p(s_i = 1) = x$, this leads to an implicit function in x, the solution of which is the mean magnetisation at equilibrium. This model has been proposed in 1925 by Ising to compute mean magnetisation from simple rules in a crystal lattice, and was further deeply investigated in two and three space dimensions.

It has been shown that this approximation is a very precise one provided the system is not close to a transition phase, which corresponds to a bifurcation for the solution of the implicit function. In the vicinity of a bifurcation point, the approximation $s_i \cong \bar{s}$ is no longer valid, but outside this vicinity, it is very precise. Now, it is known that bifurcation points in solution of an implicit functions usually occupy a very thin portion of the phase space of a system (in mathematical words, the measure of the set of bifurcation points is equal to zero).

The consequence is very optimistic : in most cases, if the equilibrium of the system has not settled close to a bifurcation point of the solution of the implicit function, it is possible to assume that each cell is in interaction with cells whose state is the mean state over the whole system.

3.2 THE MEAN FIELD APPROXIMATION

The approximation in Ising model is rather common in modelling spatially distributed systems and the technique can be extended far beyond Ising model (Manneville & Al., 1986). It is called mean field approximation, and is based on the assumption that each neighbourhood of an element in a system can be approximated by the mean neighbourhood, computed as if each element in the neighbourhood was a mean element. In most of the cases, it leads to an implicit function. Indeed, let us suppose that the system is described by a variable x_i in cell i. The evolution of a given cell is driven by the state of neighbouring cells. Previous computation can be copied item by item, and we are left with an implicit function for the global variable $\bar{x} = \dfrac{1}{n}\sum_i x_i$. It is relevant provided the solution is not close to a bifurcation point.

The good surprise comes from the genericity of the mean field approximation, which is well known in modelling physical systems. It is proposed here that the genericity of the mean field approximation is a clue for understanding the genericity of global stand models. It should be possible to extend global stand models far beyond the limited domain of regular stands (even-aged, monospecific and closed), provided a mean neighbourhood still is meaningful.

3.3 MODELLING AS SUMMARISING

Mean field approximation is one way among many to summarise data, and opens some perspectives for understanding why modelling works for such a complicated system as a mixed forests. Let us imagine a complicated object, which is described with many variables, say n with n being big (many thousands is not unrealistic). Mean field approximation asserts that it is possible to build a far simpler object from the data describing the initial and complicated one, with following properties : *(i)* this simple object is a summary of the initial one *(ii)* it is possible to follow its history without bothering with all the details of the initial one. The role of summary is played by the mean value of a relevant variable, such as mean height, mean basal area, mean age, etc. ... . In distribution models, where the stand is described by a set of distribution functions, there are several individual based description which can be summarised by the same distribution function, especially when the diversity of the stand, in the Shannon sense, is high. The role of the summary is played by the distribution function.

The existence of a summary with good previous property is not trite. Let us imagine a set of mixed multicohort forests which share at a given time the same summary, such as diameter distribution function. Each of these forests evolves according to its own biological logic, and can be observed again several decades afterwards. Each forests has its own history and trajectory. The distribution function is a good summary if all forests share the same distribution function at final time, whatever their detailed history at the individual level inbetween. The details can be ignored, and it is possible to focus on the summary only. This assumption is very strong, and is true for mean values in mean field theory. When this assumption is true or not is an open question in forest ecology.

4. MODELLING WITHIN DIVERSITY

Let us suppose a mixed, multicohort forest which is known by a huge set of data : species, age, position, size, shape, for each tree. The variability between trees makes it difficult to reconstruct the whole set as an output of a small programme. Modelling as compressing data might seem hopeless, as it is in contradiction with the very notion of diversity, which is basic to biological sciences. The importance of diversity in biological sciences for centuries has been wonderfully emphasised by Mayr (Mayr, 1982). Mean field theory shows that a common way may exist between modelling and emphasising diversity. The real difficulty is to find the relevant summary for a given mixed stand. This question is not solved yet. Mean field approximation gives a solution for objects which are simple in algorithmic sense : the mean value is a good summary. It cannot be extended without precaution to more complex objects. This is a field for research : how to observe a mixed multicohort forest, which are the relevant variables to summarise its history, or, in brief, how to look at it ?

References

Caswell H (1989) : Matrix population models. Construction, analysis and interpretation. *Sinauer Associate, MA.*
DeAngelis D. L. & Gross L. (Ed.) (1992) : Individual-based models and approaches in ecology. Populations, Communities and Ecosystems. *Chapmann & Hall, New York,* 524 pp.
Delahaye P. (1994) : Information, Complexité et hasard. *Hermès, Paris,* 234 pp.
Leslie P. H. : - 1945 - On the use of matrices in certain population mathematics. *Biometrika,* 33, 183-212.
Li M. & Vitanyi P. (1997) : An introduction to Kolmogorov Complexity and its Applications. *Springer Verlag, Berlin,* 637 pp.
Mannevile P., Boccara N., Vichniac G. Y. & Bidaux R. (Ed.) (1989) : Cellular automata and modelling of complex physical systems. *Springer Verlag, Berlin,* 318 pp.
Mayr E. (1982) : The growth of biological thought. Diversity, Evolution and inheritance. *Harvard University Press.* Edition française : Histoire de la

biologie. Diversité, Evolution et Hérédité. 1989, *Livre de Poche, Paris,* 1205 pp.

Nicolis G. & Prigogine I. (1992) : A la rencontre du complexe. *Presses Universitaires de France, ¨Paris,* 382 pp.

Oliver C. D. & Larson B. C. (1996) : Forest stand dynamics. *John Wiley & Sons,* 520 pp.

Real L. A. & Brown J. H. (1991) : Foundations of Ecology : classic papers with commentaries. *Chicago University Press,* 904 pp.

Scudo F. M. & Ziegler J. R. (1978) : The golden age of theoretical ecology : 1923 - 1940. *Springer Verlag, Lecture Notes in Biomathematics,* Berlin, 490 pp.

Usher M. B. (1966) : A matrix approach to the management of renewable resources, with special reference to the selection forests. *J. Appl. Ecol.,* 8, 355-367.

Quantifying Biodiversity
The Effect of Sampling Method and Intensity on Diversity Indices

Dieter R. PELZ and Paul LUEBBERS

Abteilung Forstliche Biometrie, University of Freiburg, Germany

INTRODUCTION

Biodiversity has become a rather important measure for the evaluation of ecosystems. Frequently, diversity is considered a major indicator for the well-being of an ecosystem. The measurement and quantification of biodiversity is not easy. For comparisons of biodiversity of different ecosystems or for monitoring over time appropriate estimators have to be defined.

Diversity consists of two components, the variety (i.e. species richness) and the relative abundance of species. In many studies only the species richness is considered, i.e. the number of species present. However, the relative abundance should also be considered in defining biodiversity.

In the literature a number of indices are suggested as measure of species diversity. The underlying assumptions of these indices are quite diverse and vary considerably for a given population. In this study frequently used indices were reviewed and the effects of sampling methods and sampling intensities on the indices were examined.

Five different indices were included in the study:

(1) **Shannon and Weaver:** This index is most frequently used. It calculates the relative proportions of species and ranges in general from 1 to 4.5:

$$H_s = -\sum_{i=1}^{S} p_i \ln p_i$$

(2) **Brillouin:** This index is used in cases where random sampling cannot be guaranteed, it has the form:

$$HB = \frac{2,302385}{N} (\log_{10} N! - \sum \log_{10} n_i!)$$

(3) **Simpson:** This index is a dominance measure and less a measure of species richness:

$$D = \sum \frac{(n_i(n_i - 1))}{(N(N-1))}$$

(4) **McIntosh:** This index is not a dominance index, but diversity and evenness can be calculated with it. It is defined as:

$$U = \sqrt{(\sum n_i^2)}$$

SIMULATION STUDY

In some cases, diversity will be measured for the entire population, however in most studies some sampling methods will be used. In this study the effect of sampling method and sampling intensity on the indices was ivestigated. The sampling methods most often used in forestry were simulated: plot sampling with a radius of 12.61 m (500 m^2), tree distance methods with 4, 5, 6, 7 and 8 trees (AB4 to AB8) and point sampling with a basal area factor of 1, 2, and 4 (WZ1 to WZ4).

With these sampling methods the selection probabilities of trees included in the sample differ. Fir fixed area sampling, the selection probabilities are constant, for point sampling the probability of selection is proportional to the basal area of a tree, and for tree distance methods the probabilities are inversely proportional to size. In addition, the size of sampling units were varied as well, from very small units (D4 - the distance to the 4 th nearest neighbor) to larger units (D8 - the distance to the 8 th nearest neighbor) and from point sampling with basal area factor 4 to basal area factor 1.

Basis for the simulations was a complete enumeration of a 4 ha stand close to Freiburg. A total of 2708 trees were measured, with 14 species. All trees were measured for x and y coordinates and dbh, the species was also recorded. These data constituted the basis for computer simulations. For this population the location of the sampling units was randomly generated. For fixed area sampling, all trees which were located closer than 12.61 m to the plot center were recorded for species and dbh. For tree distance methods the distance to the 4th (5th, 6th, 7th, 8th) constituted the plot boundary, it is in effect a point- tree distance measure (the sample with the distance to the 6th tree is widely used in inventory). Finally, the Bitterlich point sampling method was simulated for three basal area factors.

The 10 variations in sampling methods were simulated for 4 different sample sizes, with 20, 30, 50 and 100 for the 4 indices described above, for each sample 20 iterations or replications were used. In the following the results are discussed for a sample size of 20 and 100.

RESULTS AND DISCUSSION

In Figure 1 results for a sample size of n=20 are presented for the Shannon index. The results presented are based on 20 replications in order to avoid any random effect. All sampling methods underestimate the true value from the population, with point sampling with basal area factor 2 and 4 yielding the lowest results. The most accurate estimates are those with the largest sampling units, the smaller the sampling units, the lower the estimate for the Shannon Index. This indicates that with small sample sizes only those sampling methods using larger sampling units yield accurate results.

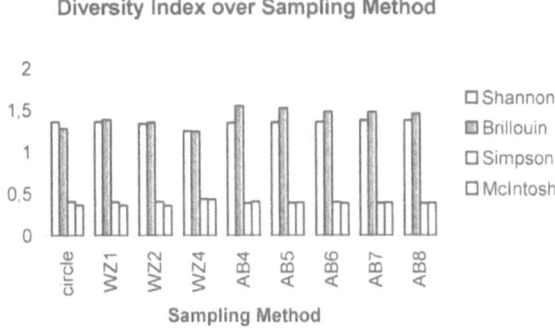

Figure 1: Shannon Index over Sampling Method (20 iterations with n=20)

In Figure 2 the results for a sample size of n = 100 are presented. For large sample sizes (it approaches full enumeration) all sampling methods yield good results, the estimate of the value of the Shannon Index is accurate for fixed area sampling and point sampling. There is a small deviation for the tree distance methods, they overestimate the true value.
Based on this study, it can be concluded that fixed area sampling and point sampling yield accurate results, but that point sampling with a high basal area factor tends give results with high variation. It should be noted that point sampling with a basal area factor 4 gives highly variable results depending on the sample size.

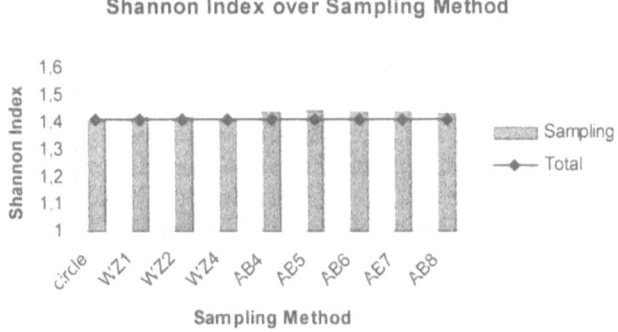

Figure 2: Shannon Index over Sampling Method (20 iterations with n=100)
For comparing the four indices and the effects of different sampling methods on the indices simulations were performed. In order to reduce the variation among the

results, the same randomly chosen points were used as basis for the calculations, i.e. for a given location and sampling method, the four indices were calculated.

In Figure 3 the values of the diversity indices Shannon, Brillouin, Simpson and McIntosh are given for sample sizes of n = 20. The absolute values of the indices are quite different: Shannon 1.415, Brillouin 1.403, Simpson 0.399, and McIntosh 0.275. The variation among the sampling methods are quite apparent. The tree distance methods tend to overestimate the Shannon and the Brillouin Index considerably, an effect that is less apparent for larger sample sizes.

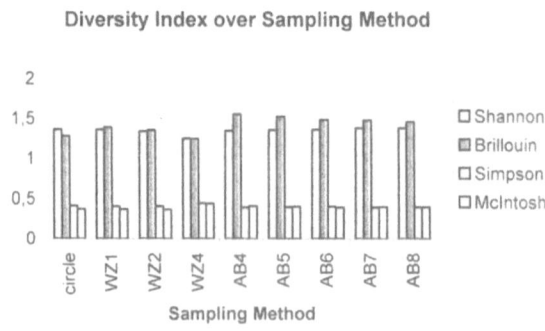

Figure 3 Diversity Index over sampling method for n=20

In Figure 4 the results are presented for a sample size of n = 100. The numerical values of the Shannon and Brillouin indices are quite similar, this is also true for the Simpson and McIntosh value, but between these groups is a gap. This could be expected, it illustrates the fact that the numerical values of different indices really cannot be compared, as they express different aspects of biodiversity.

For the large sample size the differences in indices for alternative sampling methods are diminished. It can be seen that for all sampling methods the results are quite stable, as the sample size approaches the total population.

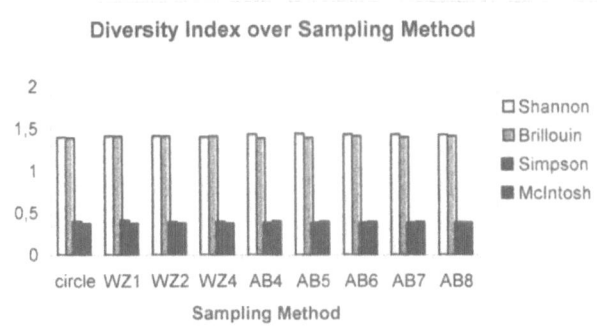

Figure 4 Diversity Index over sampling method for n=100

The sampling methods showed a significant effect on the results, especially for small sample sizes, as the sampling elements are of different size and the selection probabilities of the trees are different. In a fixed area plot, the selection probabilities for the trees are constant, this method shows the least effect for the different indices. For tree distance methods, the selection probabilities are inversely proportional to tree size. This will favor smaller trees over larger, i.e. the corresponding indices can be biased. For point sampling, the selection probabilities are proportional to the basal area of each tree, i.e. in the selection process larger trees will more likely be included.

In addition to the selection probabilities, the size of the elements have to be considered. In fully stocked stands, point sampling and tree distance methods (with i = 4-8) will result in much smaller sampling units, fewer trees are included in each element. This will be an important fact in comparing indices. There is a definite interaction between sampling method and diversity index, especially for small sample sizes.

The effect of sampling methods on the indices studied are apparent for small sample sizes which are commonly used for forest inventories. With increasing sample sizes the effects become less apparent, the results approximate the true value for the entire population for sufficiently large samples. For practical applications it has to be noted that, with the commonly used sampling intensities, the diversity indices become rather variable and are influenced by the sampling method used.

REFERENCES

BOYLE, T.J.B. AND B. BOONTAWEE eds. (1995): Measuring and Monitoring Biodiversity in Tropical and Temperate Forests. CIFOR.
CLARK P.J. AND EVANS F.C. (1954): Distance to nearest neighbour as a measure of spatial relationship in populations. Ecology 35:445-453

HEYWOOD, V.H. ed. (1995): Global Biodiversity Assessment. Cambridge University Press.

MAGURRAN, A.E. (1991) : Ecological Diversity and its Measurement. Chapman and Hall London

McINTOSH R,.P. (1967): An index of diversity and the relation of certain concepts to diversity. Ecology 48:392-404

PIELOU E.C. (1975): Ecological Diversity. Wiley and Sons, New York 1975

SHANNON C.E. AND WEAVER, W (1949): The mathematical theory of communications, Univ. Of Illinois Press, Urbana, 1949

SIMPSON. E.H. (1949): Measurement of diversity. Nature 163:688

Photosynthetic Light Environment of Tropical Lowland Forest and Growth Response of *Shorea Leprosula*

Muhamad AWANG*[1], Ahmad Makmom ABDULLAH*[1] and
Akio FURUKAWA*[2]

*1 Department of Environmental Sciences, Universiti Putra Malaysia
  Selangor Darul Ehsan, Malaysia
*2 Faculty of Science, Nara Women's University
  Kitauoya-Nishimachi, Nara, Japan

ABSTRACT Photon flux densities (PFD) of different microsites within a tropical lowland forest have been measured and compared spatially and temporally. The aims of the study were to determine the forest light environments at different microsites and the growth responses of its hardwood species which were grown under these different light regimes. Results indicated that the daily total PFD in the big gap, medium gap, small gap, smallest gap and understorey sites during the rainy season of November 1991 to January 1992 were 51.5, 17.7, 8.1, 8 and 4.5 % respectively, of the daily total PFD in the biggest gap. Results of this study also demonstrated that the frequency histograms of PFD were negatively skewed at the big open area, were bimodal within the medium gap and small gaps sites, and positively skewed under the dense canopy. Relative growth rate of height and leaf area ratios of *S. leprosula* seedlings grown under different light conditions showed that the seedlings in a big sized gap grew substantially more than those raised in smaller gaps and exhibited higher leaf area ratios and photosynthetic rates. This study also indicated that the photosynthetic capacity of the seedlings grown in the big gap were higher than seedlings in smaller gaps.
Keywords: Photon flux density, photosynthetic light response curves, tropical forest species.

1. INTRODUCTION

Photosynthetic photon flux density (PFD) in the waveband 400-700 nm is crucial in the growth, survival, and regeneration of the tropical forest species (Chazdon and Fetcher, 1984, Pearcy, 1988; Chazdon, 1984 and Chazdon *et al.*, 1986). The ecological and evolutionary successes of the forest species are largely due to their ability to capitalize on pattern of variation in light which is a major limiting resource in most forest types (Chazdon, 1988). In the tropical forest with a multilayered canopy, the light received at the forest floor is controlled by the interception of the layers. The variations in PFD distribution over forest floor and forest clearing or gaps were determined by gap size, shape and orientation (Chazdon, 1988; Chazdon and Fetcher, 1984; Pearcy, 1988). Bazzaz (1984) revealed that the larger gaps received more light than do small gaps and elongated gaps received more light when they are oriented east-west than when oriented north-south. On the other hand, the PFD being intercepted by the canopy can be utilised for the net dry matter production (Baker and Thomas, 1992). However, the ability of a plant canopy to intercept radiation or the conversion efficiency of the plant is influenced by several important factors such as canopy architecture, leaf structure and density of the light harvesting system within the leaves (Baker and Thomas, 1992; Nobel and Long, 1985).

This study was undertaken to compare diurnal and seasonal patterns of PFD in a range of rain forest habitats within a single lowland evergreen forest in Pasoh, Malaysia. This study determined the light environment in the lowland tropical forest and light intercepted and measured the utilisation of light intercepted by the canopy through growth and photosynthesis measurements. The above ground dry weight of the selected species were measured and compared between different light conditions.

2. STUDY AREA

The study was conducted at Pasoh Forest Reserve, which is located at 2°5'N and 102°18'W (Manokaran, 1990), with an altitude of 300-600 m in Negeri Sembilan of Malaysia. The mean annual rainfall was about 2000 mm with two peaks during November and December and April to May which is usually considered as wet season or rainy season and dry season in between. The monthly average temperature was 27°C.

3. MATERIALS AND METHODS

3.1 Equipment and Sampling

Measurements of PFD at different microsites were made using quantum sensors connected to a battery powered data-logger (Kona System Model Kadecup U2). The PFD measurements were monitored at 10-s intervals and were automatically computed 5 minute averages of PFD. This five minute averages were used because this time interval is shorter than the time that is sufficient to indicate sunflecks events while keeping the quantity of data which must be analyzed to manageable levels. Measurements of PFD were made in the dry season of August-September-October 1991 and in the rainy season of November-December 1991-January 1992.

3.2 Sites

Twenty five uniform sized seedlings of *S. leprosula* were transplanted into the four microsites with gap sizes of 25 m^2 (smallest), 40 m^2 (small), 100 m^2

(medium), 400 m^2 (big). These sites were chosen representing the actual range of light environments in Pasoh Forest Reserve.

3.3 Data Analysis

Daily averages were calculated from 8:00 hr to 18:00 hrs to reduce the effect of high frequency of low PFD in the early morning and late afternoon. The daily total and daily mean of PFD were determined for each site during the rainy and dry season. Frequency distribution of 5-min averages were calculated for each of the sampling days, using intervals that were considered relevant for measurable photosynthetic parameters. Leaf photosynthesis was measured using a portable leaf gas exchange system (Awang *et al.*, 1994). Subsequently, photosynthetic light response curves of the species were established using the hyperbolic equation (Thornley and Johnson, 1990).

4. RESULTS AND DISCUSSION

Photon flux density in the six sites during the rainy season of 1991 are summarised in Table 1 and in dry season of 1992 in Table 2. Daily mean PFD ranged from more than 25 μmol m^{-2} s^{-1} in the heavily shaded understorey to 1221 μmol m^{-2} s^{-1} in the biggest gap. The maximum mean daily total PFD recorded in the dry season and the rainy season of the biggest gap were about 50.4 mol m^{-2} day^{-1} and 44.3 mol m^{-2} day^{-1}, respectively. The daily total PFD in the big gap was about 51.6% and 24.9% during the rainy season and dry season, respectively, of the PFD in the biggest gap. The daily total PFD in the medium, small, smallest gaps and understorey were 17.8, 8.1, 8.0 and 5.5 % during the rainy season and 2.6, 3.9, 1.1 and 0.8 % during the dry season, respectively, of the PFD in the biggest gap. The daily total PFD obtained in the big gap site between the seasons differed by two times in magnitude during the dry season compared to rainy season. The daily total PFD in the medium gap is about 2 times greater than in the small gap and the smallest gap.

The frequency distributions of 5-min average PFD readings illustrated the extreme differences in the light environment of the sites as shown in Figures 1 and 2. In the rainy season of 1991, frequency distribution of PFD showed that more than 80% of PFD readings were above 500 μmol m^{-2} s^{-1} at the biggest gap, while most of the PFD readings were below 50 μmol m^{-2} s^{-1} at the understorey site. This suggested that there was higher diffuse radiation and the 5-min readings above 50 μmol m^{-2} s^{-1} were generally due to sunflecks.

During the dry season of 1992, about 64.5% of the 5-min average PFD readings at the understorey were below 10 μmol m^{-2} s^{-1}. In the 400-m^2 gap, the PFD concentrated between 100 μmol m^2 s^1 and 500 μmol m^2 s^1. The frequency histograms of the daily average PFD in big gap showed unimodal distribution with a peak at the higher PFD class of more than 500 μmol m^{-2} s^{-1}. The frequency distributions for the medium, small, and smallest gaps showed monomodal distribution with high frequency distribution at the lower PFD class.

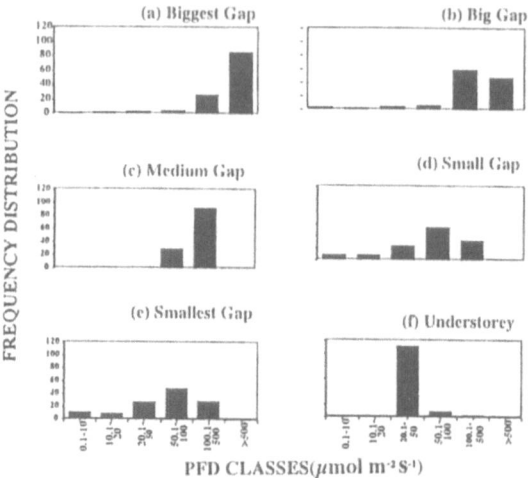

Fig. 1 Frequency distributions of PFD under different light conditions in the rainy season of November-December 1991 - January 1992.

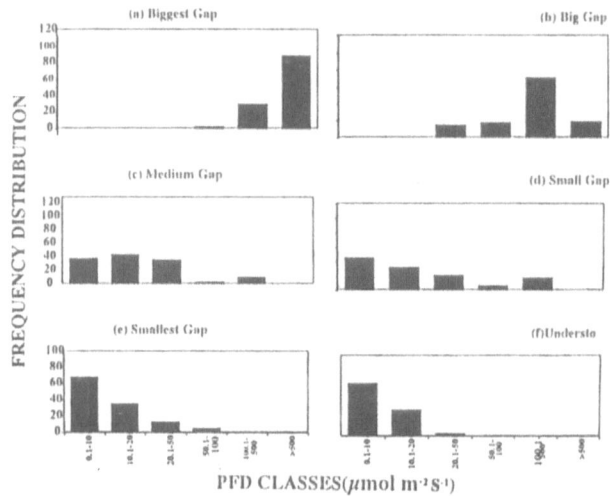

Fig. 2 Frequency distributions of PFD under different light conditions in the dry season of August-September-October 1992.

Generally, the frequency histograms showed that they were negatively skewed for the biggest gap site and bimodal within the medium size gap and small gap sites and positively skewed under the dense canopy. The variability in the light environment between the six sites indicated that the biggest gap site showed the highest variation compared to the other sites. Seasonal variability in PFD was clearly shown by the biggest gap site with less cloud cover during the dry season resulting in the higher direct solar radiation shown by higher frequency distribution in higher PFD class.

Table 1. Daily total PFD, daily mean PFD and standard deviation of PFD under different light conditions during the rainy season of November-December 1991-January 1992 at Pasoh Forest Reserve.

PFD	Microsite	Min.	Average	Max.
Daily total PFD (mol m^{-2} day^{-1})	Big Gap	5.8	15.9	25.2
Mean PFD (μmol m^{-2} s^{-1})		160.7	437.7	692.8
Standard deviation		100.1	326.7	611.0
Daily total PFD (mol m^{-2} day^{-1})	Medium	3.1	5.5	8.0
Mean PFD (μmol m^{-2} s^{-1})	Gap	84.8	151.3	218.9
Standard deviation		30.8	92.5	223.5
Daily total PFD (mol m^{-2} day^{-1})	Small	1.7	2.5	3.1
Mean PFD (μmol m^{-2} s^{-1})	Gap	46.1	68.3	86.2
Standard deviation		26.0	40.0	50.1
Daily total PFD (mol m^{-2} day^{-1})	Smallest	1.3	2.5	3.4
Mean PFD (μmol m^{-2} s^{-1})	Gap	36.5	67.9	92.9
Standard deviation		21.1	47.9	114.1
Daily total PFD (mol m^{-2} day^{-1})	Under	0.9	1.3	1.9
Mean PFD (μmol m^{-2} s^{-1})	storey	25.8	38.3	52.7
Standard deviation		2.0	22.7	94.8
Daily total PFD (mol m^{-2} day^{-1})	Biggest	14.5	30.8	44.3
Mean PFD (μmol m^{-2} s^{-1})	Gap	398.2	849.0	1221.0
Standard deviation		222.8	597.2	864.5

Table 2. Daily total PFD, daily mean PFD and standard deviation of PFD under different light conditions during the dry season of August-September-October 1992 at Pasoh Forest Reserve.

PFD	Microsite	Min	Average	Max
Daily total PFD (mol m^{-2} day^{-1})	Big Gap	6.6	10.2	12.7
Mean PFD (μmol m^{-2} s^{-1})		183.1	280.8	350.8
Standard deviation		123.9	226.3	280.7
Daily total PFD (mol m^{-2} day^{-1})	Medium	0.5	1.1	1.6
Mean PFD (μmol m^{-2} s^{-1})	Gap	14.4	29.5	44.2
Standard deviation		10.0	51.7	85.5
Daily total PFD (mol m^{-2} day^{-1})	Small	0.5	1.6	2.5
Mean PFD (μmol m^{-2} s^{-1})	Gap	14.2	44.2	70.2
Standard deviation		15.8	85.3	139.5
Daily total PFD (mol m^{-2} day^{-1})	Smallest	0.1	0.5	1.8
Mean PFD (μmol m^{-2} s^{-1})	Gap	4.0	12.8	48.5
Standard deviation		2.7	18.4	99.5
Daily total PFD (mol m^{-2} day^{-1})	Under	0.2	0.3	0.5
Mean PFD (μmol m^{-2} s^{-1})	storey	5.8	8.9	13.1
Standard deviation		2.0.	6.8	21.1
Daily total PFD (mol m^{-2} day^{-1})	Biggest	24.2	40.9	50.4
Mean PFD (μmol m^{-2} s^{-1})	Gap	665.8	1127.4	1389.4
Standard deviation		598.6	774.0	941.9

384

Figures 3 and 4 describe the height and above ground dry weight increment of *S. leprosula* seedlings grown under different light conditions. This study revealed that the seedlings grown in the biggest gap, which experienced high midday peaks and a heterogeneous distribution of daily high intensities grew substantially more than those raised in smaller gaps with lower midday peaks. Subsequently, seedlings of *S. leprosula* grown in the big gap had higher rates of leaf production, and were taller than seedlings grown in the smaller gap sites. In addition, this study also indicated that the biggest gap exhibited the highest rates in terms of height increment, mean leaf dry weight per plant and mean assimilatory area per plant.

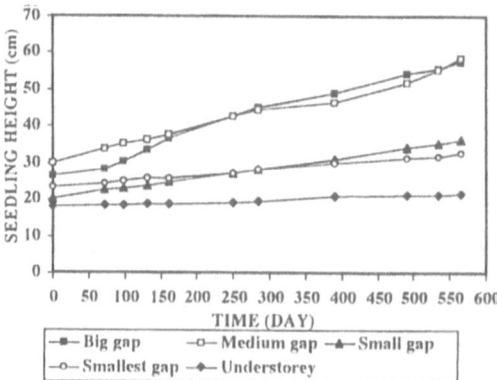

Fig. 3 Height increment of *S. leprosula* seedlings grown under different light conditions.

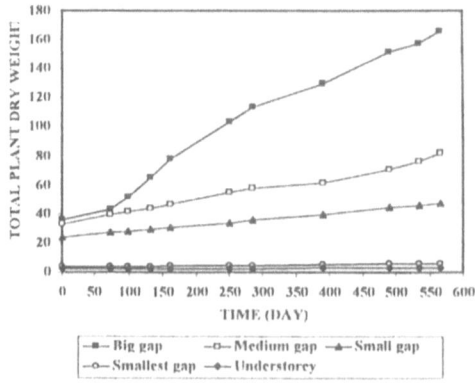

Fig. 4 Biomass increment of *S. leprosula* seedlings grown under different light conditions.

Figure 5 shows the diurnal courses of photosynthetic rates of *S. leprosula* seedlings grown under different light conditions. The figure illustrates the high variations in diurnal courses of photosynthetic rates. The maximum and minimum values of photosynthetic rates in the big gap were 7.5 mmol m^{-2} s^{-1} and 0.5 mmol m^{-2} s^{-1}. Figure 6 shows the photosynthetic light response of *S. leprosula* seedlings grown under different light conditions. The big gap which received daily total PFD ranging from 9.6 to 12.7 mol m^{-2} day^{-1} throughout the year showed the highest response with a maximum photosynthetic capacity (P_{max}) more than 8.4 mmol m^{-2} s^{-1}. The maximum photosynthetic rates of S. $leprosula$ seedlings grown in small and smallest gaps were about 4.1 and 3.2 mmol m^{-2} s^{-1}, respectively. In addition, this study also found that there was a linear relationship at low daily mean PFD between the daily carbon gain and dry matter production.

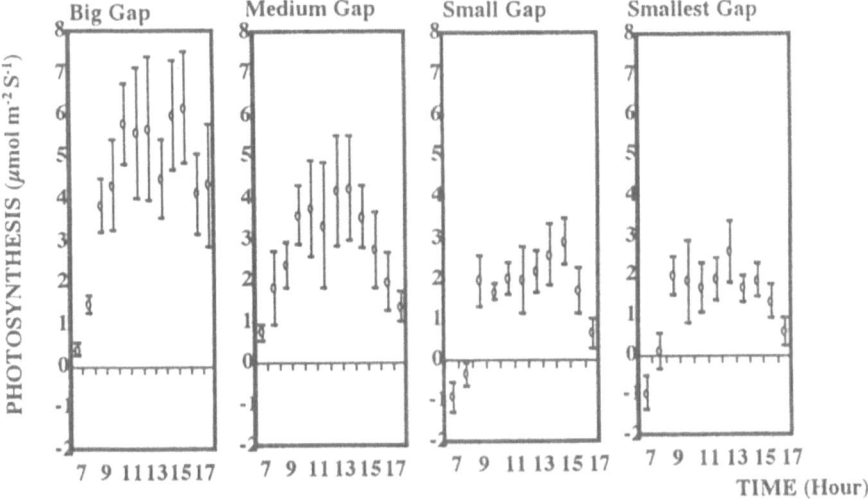

Fig. 5 Diurnal courses of photosynthesis of S. $leprosula$ seedlings grown under different light conditions at each point represented by an average of 6 replicates.

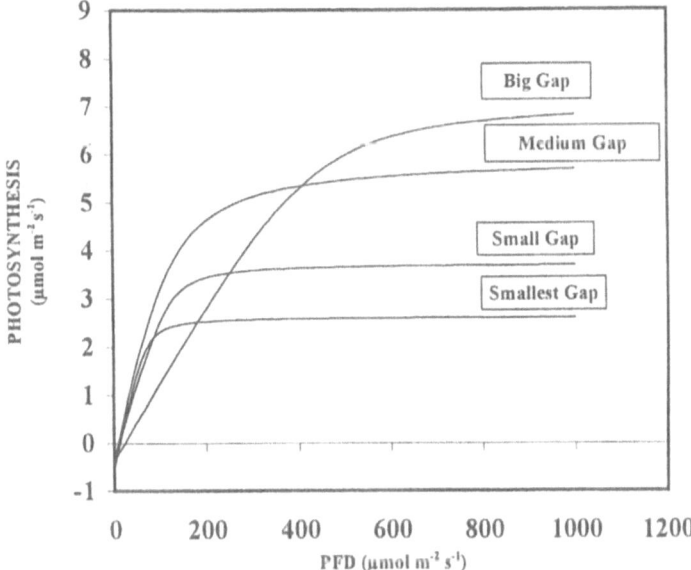

Fig. 6 Photosynthesis light response of S. $leprosula$ seedlings grown under different light conditions. Big gap (400m^2), medium gap (100m^2), small gap (40m^2), and smallest gap (25m^2).

Finally, it is concluded that the penetration of direct beam of solar radiation through holes in the canopy and gaps resulted in sunflecks which are common features of light environment under canopy of the tropical forest. It is also concluded here that the photosynthetic capacity may differ greatly under different constellations of external variables even if it is assumed that the photosynthetic pathways are identical.

REFERENCE

Awang, M.B., A.M. Abdullah, A. Furukawa, K. Ogawa and A. Hagihara (1994). In-situ CO_2 gas Exchange in Leaves and Reproductive of *Durio zibethinus* Murray. Transactions of the Malaysian Society of Plant Physiology. Vol.5.

Baker, N.R. and Thomas, H. (1992). *"Crop Photosynthesis: Spatial And Temporal Determinants"*. Elservier Science Publishers B.V., Netherlands. pp 453.

Bazzaz, F.A. (1983). "Characteristics of Population in Relation to Disturbance in Natural and Man Modified Ecosystems". In: Mooney, H.A. and Godron, M. (eds.). *Disturbance and Ecosystem-Components of Response*. Springer Verlag, Berlin. pp 259-75.

Bazzaz, F.A. (1984). "Dynamics of Wet Tropical Forests and Their Species Strategies". In: Medina, E., Mooney, H.A. and Vasquez-Yanes, C. (eds.). *Ecology of Plants of the Wet Tropics*. Dr. Junk, The Hague. Pp. 233-43.

Chazdon, R. and Fetcher, N. (1984). Photosynthetic Light Environments in a Lowland Tropical Rain Forest in Costa Rica. *Journal of Ecology* (1984), 72, 553-564.

Chazdon, R. (1988). "Sunflecks and Their Importance to Forest Understorey Plants". In: *Advances in Ecological Research* Vol. 18. Academic Press Inc. (London) Limited.

Manokaran, N. (1990). "Stand structure of Pasoh Forest Reserve, a Lowland Rain Forest in Peninsular Malaysia". *J. Tropical Forest Sciences* 3(1): 14-24. Nobel, P.S. and Long, S.P. (1985). *"Techniques in Bioproductivity and Photosynthesis"*. 2nd Ed. (Coomb, J., Hall, D.O., Long, S.P., and Scurlocks, J.M.O. eds.). Pergamon Press, Oxford. pp. 41-49.

Pearcy, R.W. (1988). "Photosynthetic Utilization of Light Flecks by Understorey Plants". *Aust. J. Plant Physiol.*, 1988: 223-38.

Thornley, J.H.M. and Johnson, I. A. (1990). *"Plant and Crop Modelling: A Mathematical Approach to Plant and Crop Physiology"*. Oxford University Press New York.

Growth Performance of Malaysian Tropical Trees under Different Light Regimes

Toshihiro YAMADA*[1], Taketo YOKOTA*[2], Akio FURUKAWA*[2],
Makmon ABUDULLA*[3], Samusddin JOHAN*[3] and
Muhamad AWANG*[3]

*1 Environmental Biology Division, The National Institute for Environmental
   Studies, Ibaraki 305, Japan
*2 Faculty of Science, Nara Women's University, Nara 630, Japan
*3 Department of Environmental Sciences, Universiti Putra Malaysia
   Selangor Darul Ehsan, Malaysia

Abstract   Growth and survival rates of indigenous Malaysian tree species were monitored between December 1996 and December 1997 following their establishment in an experimental farm in Universiti Putra Malaysia. The species studied consisted of 21 tree species ranging from large emergent trees to shrubs and from pioneer to climax trees and also one liana species. The mean relative growth rate of height of all species studied was 0.0012 cm cm^{-1} day^{-1} and the mortality rate was 26.1%. *Macaranga hypoleuca* showed the highest relative growth rate and *Caesalpinia sappan* exhibit the highest survival rate. Pioneer species favored a high light intensity and showed high growth and mortality rates. On the contrary, climax species had low growth and mortality rates. A partial shading condition was favored by some climax species. These results indicate that optimum light conditions varied between the species, suggesting that the selection of species to plant with respect to the light condition of a focal planting site needs to be considered for the successful end of a plantation program.
Keywords: climax species, mortality, pioneer species, relative growth rate, Malaysia.

1. INTRODUCTION

Traditionally, tropical rain forest resources have been divided into two main groups, timber resources and non-wood or minor forest products. Most financial appraisals of tropical rain forests have exclusively focused on the timber resources. Consequently, there has been a strong market for destructive logging and widespread forest clearing, and forest plantations have been solely aimed for the supply of timber. Although the area and the rate of deforestation in the tropical area are not well known, nor are there quantitative measurements of the effect of deforestation on habitat degradation, the other multiple functions serving forests such as the storage of genetic and medicinal resources and environmental roles preventing soil erosion and excessive water run-off have been neglected. A forest

387

plantation which consists of both timber and nontimber species and serves the multiple functions will be necessary to establish.

The choice of the species planted inevitably depends on the amount of information about species concerning the basic features of the woods such as growth performance, physiognomy, and uses of timber in literature. By virtue of prior research efforts, information about some timber species in the Malaysian tropics has so far been clarified (Appanah and Weinland, 1993). However, a majority of nontimber species has yet to be studied. Due to the fact that planting nontimber species is essential for the establishment of the forest serving multiple functions, the information about nontimber species is also needed to be clarified.

One of the factors that strongly affects growth of plants is the light condition. Many scientists found an optimum growth under partial shading in some tropical tree species (Itoh 1995, Ashton and de Zoya 1989, Barnard 1954, Watson 1935). Assuming that plants have their own species specific optimum light condition, selecting a suitable species to plant in respect to the light conditions of the focal planting site is ideal, and thus the information about a species specific optimum light condition for growth is of paramount importance.

The objectives of this study are to determine the growth and mortality rates of 22 Malaysian indigenous tree species which included both timber and nontimber species in plantation environments and the species specific growth responses to different light intensities. These will offer a basis for the evaluation of species potential in plantation environment and selection of the indigenous species suitable for forest plantation in the Malaysian tropics.

2. METHODS

2.1 Study Site and Species

The study was performed in Universiti Putra Malaysia (UPM), ca 27 km south of Kuala Lumpur. An experimental farm of 0.9 ha was established in the UPM campus and subsequently 26 plots of ca. 5 m X 5 m were set in the experimental farm. The seedlings studied were planted to the plots in December 1996. Each plot was designed to have several species and each species in a plot consisted of 10 seedlings. The number of the species as well as species mixture in a plot differed among the plots.

Growth performance of 2406 seedlings of 22 Malaysian indigenous species which consisted of one liana and 22 trees ranging from huge emergent trees to shrubs and from pioneer to climax trees (Table 1) were studied. Mature seeds of climax species were collected from Pasoh Forest Reserve, Negeri Sembilan, Malaysia in August 1996, while seedlings of pioneer species were collected from Puchong Forest Reserve, Selangor, Malaysia. They were maintained and established under the shade of a black nylon mesh cloth prior the transplanting to the experimental farm.

The light conditions of the plots differed among them in accordance with the

Table 1. Description of species studied.

Species	Code	maximum Dbh (cm)	Climax or pioneer species	Types of the mean total weight to RLI relationships
Caesalpinia sappan L.	Ca	n.a.[a]	Liana	no correlation
Dipterocarpus cornutus Dyer	Dc	100	Climax	optimum curve
Dipterocarpus crinitus Dyer	Dcri	100	Climax	n.a.[a]
Dipterocarpus sublamelatus Foxw.	Ds	100	Climax	no correlation
Xanthophyllum amoenum Chodat	Xa	50	Climax	no correlation
Elateriospermum tapos Bl.	Et	50	Climax	no correlation
Elaeocarpus nitidus Jack	En	20	Pioneer	n.a.[a]
Endsperum malaccense M. A.	End	30	Pioneer	optimum curve
Shorea lepidota (Korth.) Bl.	Sl	100	Climax	positive correlation
Shorea macroptera Dyer	Smc	100	Climax	optimum curve
Shorea maxima (King) Sym.	Smx	100	Climax	n.a.[a]
Shorea multiflora (Burck) Sym.	Smu	100	Climax	no correlation
Shorea paucifolia King	Spa	100	Climax	optimum curve
Macaranga gigantia (Rchb. f. & Zoll.) M. A.	Mg	30	Pioneer	positive correlation
Macaranga hypoleuca (Rchb. f. & Zoll.) M. A.	Mh	30	Pioneer	positive correlation
Macaranga lowii King ex Hk. f.	Ml	15	Climax	optimum curve
Macaranga trioba (Bl.) M. A.	Mt	30	Pioneer	n.a.[a]
Neobaranopsis hemii (King) Ashton	Nh	100	Climax	no correlation
Palaquium maingayi (Clarke) K. & G.	DK	60	Climax	n.a.[a]
Sapium baccatum Roxb.	Sap	50	Pioneer	positive correlation
Scapium macroppodum (Miq.) Beumee ex Heyne	Scp	70	Climax	positive correlation
Vatica bella V. Sl.	Vm	50	Climax	no correlation

a: data not available.

degree to the crown coverage by the naturally invaded pioneer trees of ca. 15 m in height. Photon flux density of four corners and a center of each plot at a height of 30 cm from the ground were measured with quantum sensors in April 1997. The photon flux density under full daylight was also measured simultaneously in an entirely open site adjacent to the plots. The relative light intensity (RLI, %) of the plots were calculated from these measurements.

2.2 Analyses of Growth Rates of Individual Seedlings

The seedling heights were measured at bimonthly intervals until December, 1997 and subsequently the relative growth rates of height (RGRH) were determined. As Blackman (1919) pointed out, organisms often exhibit an exponential growth during their initial growth periods. Therefore, a simple equation described below was used to estimate the relative growth rate of height (RGRH) of individual seedlings,

$$H = H_o e^{rt}, \qquad \text{Equation (1)}$$

where H and t are height and time, respectively, and r and H_0 are coefficients of the equation and are designated as the relative growth rate of height (RGRH) and initial height at planting, respectively. The r and H_0 were determined by a linear regression between t and logarithmic transformed H for all individual seedlings except for those died, injured, and severely infected by insects during the measuring period.

3. RESULTS AND DISCUSSION

3.1 Growth and Survival Rates

The mean initial height of all seedlings at the planting stage was 8.96 cm and the mean final height was 30.21 cm. The fastest growing individual seedling was *Macaranga hypoleuca* located at the brightest plot and attained a height of 210 cm for one year.

Although the RGRH was not determined for 896 seedlings, Equation (1) was successfully applied to determine the relationships between H and t. A degree of the fitting was measured by the coefficient of determination (r^2) between observed and calculated heights, and the mean value of r^2 was 0.84. The average RGRH of all seedlings was 0.00118 cm cm^{-1} day^{-1}.

In the analysis of pooled data obtained from all plots regardless of their light regimes, *M. hypoleuca* appeared to be the fastest growing species and followed by *Sapium baccatum* and *M. triloba*. This clearly means that pioneer species grow faster than climax species. Among the climax species, the fastest growing species was *Shorea maxima,* and followed by *M. lowii.* Most of dipterocarp species could not grow fast. *Vatica bella, Dipterocarpus sublamellatus,* and *Neobaranopsis hemii* of dipterocarp trees were among the slowest growing species.

During the course of the study, 843 of 2406 seedlings planted have died and thus the survival rate of all seedlings was 63.9%. *Endospermum malaccense* exhibited

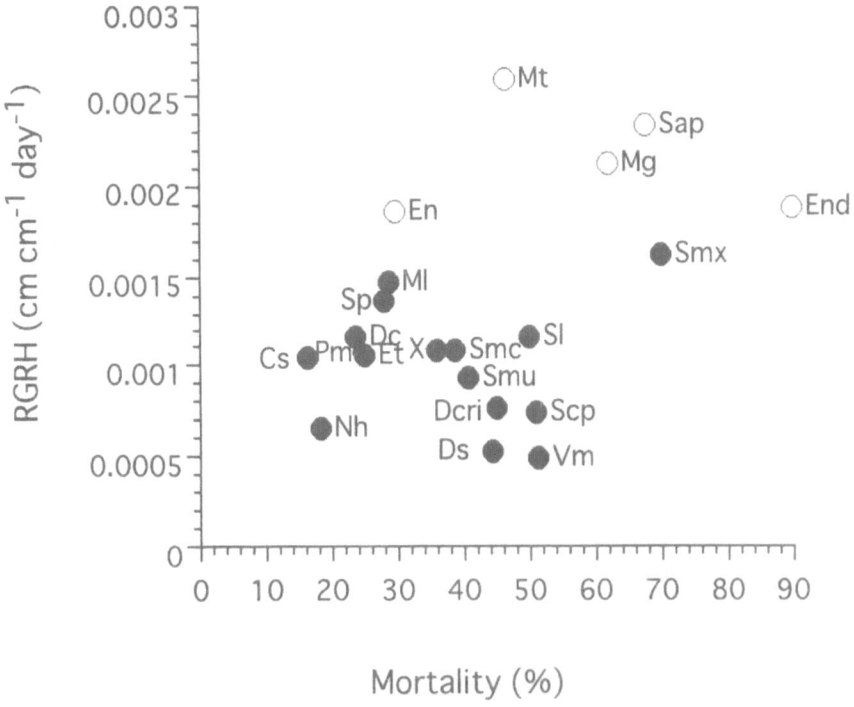

Fig. 1. Relationship between the mean values of RGRH and moralities of species examined. Open and closed circles represent pioneer and climax species, respectively. Characters show species name of the nearest symbol.

the least survival rate and *Caesalpinia sappan* had the highest values of survival rate.

Figure 1 shows the relationship between the mortality and RGRH of the species studied. *M. hypoleuca* was excluded from this figure as the number of sample size was too small (n = 7). For the plantation purpose, a fast growing species with a low mortality rate is ideal, however, the present study suggests that there was no indication of species that meet those criteria.

3.2 Species Specific Relationship between Growth and Light Conditions

The value of RLI in the plots ranged between 67% and 5%. The relationships between the mean total weight of seedlings in a plot and the RLI in the plot were analyzed for 17 species which were planted in more than 3 plots of different light regimes. The total weight of a seedling was estimated from its height by using an allometric equation derived by Lee et al. (1997). The results were summarized in Table 1. Seven, five, and five species showed no correlation (Fig. 2A), a positive correlation (Fig. 2B), and a optimum growth curve which has a peak of growth at a

392

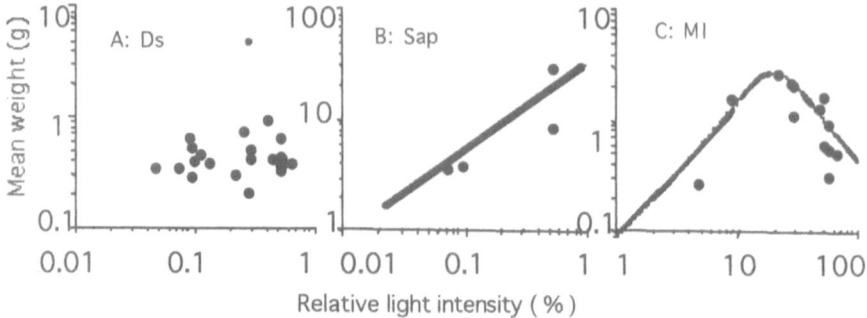

Fig. 2. Relationship between the relative light intensity (RLI) in a plot and the mean seedling dry weight in the plot of *Dipterocarpus sublamellatus* (A), *Sapium baccatum* (B), and *Macaranga lowii* (C). The straight line in B shows the linear regression on a double logarithmic scale. The curve in C shows the estimated optimum curve proposed by Hozumi et al. (1960).

moderate light condition (Fig. 2C), respectively, indicating that the growth of the first group is independent from light conditions, although there still remains some possibilities that their optimum light conditions are situated out of the range of light conditions which we examined, whereas that of the other two groups is strongly affected by light regimes. Therefore, the growth response to different light regimes can be concluded to vary between species. The results also demonstrate that the growth of the third group is inhibited by extremely low and high light intensities. Namely, a partial shading condition is favored by their growth. In short, their growth can be dramatically increased by planting under their optimum light conditions. Therefore, the species used for forest planting should be chosen in accordance with the light condition of a focal planting site.

This study clarified the growth performance and relationship between growth and light condition of 22 species in thousands of tropical trees. The information about many other tropical trees still remain unknown, and further clarification is needed.

ACKNOWLEDGMENTS

We thank Mr. N. Osada of UPM for his generous support to our research. We are grateful to Prof. T. Yamakura of Osaka City University for his logistic supports and Miss Zulina Zakaria of UPM for her review of the draft. This work is part of the Malaysia-Japan joint research project between FRIM (Forest Research Institute of Malaysia), UPM (Universiti Putra Malaysia), and NIES (National Institute for Environmental Studies), and is financially supported by the grant of Grovel Environmental Research Program (grant No. E-4) from Environment Agency, Japan.

REFERENCES

Appanah, S. and Weinland, G. (1993): Planting Quality Timber Trees in Peninsular Malaysia. Forest Research Institute Malaysia, Kepong.

Ashton, P. M. and Zoya, N. D. (1989): Performance of *Shorea trapezefolia* seedlings growing in different light regimes. Journal of Tropical Forest Science 1, pp. 356-364.

Barnard, R. C. (1954): A Manual of Malaysian Silvicalture for Inland Lowland Forests. Research Pamphlet No. 14. Forest Research Institute Malaysia, Kepong.

Hozumi, K., Shinozaki, K., and Kira, T. (1960): Concentration of mineral nutrients as an optimum factor for plant growth. I. Analysis of the optimum curve of growth in leaf vegetables under different levels of nitrogen supply. Physiology and Ecology 9, pp. 57-69 (in Japanese with English summary).

Itoh, A. (1995): Regeneration Processes and Coexistence Mechanisms of Two Bornean Emergent Dopterocarp Species. Thesis for Ph.D., Kyoto University.

Lee, H. S., Itoh, A., Kanzaki, M., and Yamakura, T. (1997): Height growth of Engkabang Jantong, *Shorea macrophylla* (De Vr.) Ashton, in a plantation forest in Sarawak. Tropics 7, pp. 65-78.

Watson, J. G. (1935): Plantation experiments at Kepong. Malaysian Forester 4, pp. 110-119.

Hydrological Effects of Afforestation and Pasture Improvement in Montane Grasslands, South Island, New Zealand

Barry FAHEY, Rick JACKSON and Lindsay ROWE

Landcare Research, Lincoln 8152, New Zealand

ABSTRACT A paired catchment study was established at Glendhu Forest in 1979 in the southern South Island of New Zealand to assess the hydrological impacts of afforesting tussock grassland. After a 3-year calibration period (1980–1982) one catchment was planted in pines over 67% of its area and the other was left in tussock. By 1989 the difference in annual water yield from the planted catchment was 130 mm, and from 1991–1996 it averaged 270 mm (31% of total runoff from the control). Differences in low flows showed a similar trend, and suggest that in dry periods, afforestation of tussock grasslands can reduce water yields by 0.11 mm/day. Two examples of the application of these data to the resolution of local and regional resource management issues are discussed. They demonstrate the importance of long-term catchment monitoring programmes.
Key words: hydrology, forests, grasslands, afforestation

1. INTRODUCTION

The hydrological effects of forestry have traditionally been evaluated by catchment experiments in which runoff and rainfall are monitored in selected catchments for specific periods around a planned alteration in land use. In 1979, a paired catchment study was established at Glendhu in the headwaters of the Waipori River in upland east Otago, on New Zealand's South Island. It was designed to investigate the stream flow behaviour and water balance of lightly grazed tussock grasslands (O'Loughlin et al. 1984; Pearce, et al. 1984), and to assess the hydrological impacts of afforestation (Fahey and Watson 1991; Fahey and Jackson 1997). The present paper summarises the long-term effects of converting tussock grassland to pine plantation on annual water yield and low flows at Glendhu. It then demonstrates how this information can assist in the resolution of local and regional water resource management issues in larger catchments through the application of water balance models.

2. FIELD AREA

The Glendhu experimental study is located in an area of mid-altitude tussock grassland in the upper Waipori River catchment (lat. 45° 50' S) 70 km west of the city of Dunedin (Fig. 1). It comprises two catchments, one left in tussock grassland (GH1) as a control (218 ha) and the other (GH2) planted in *P. radiata* (310 ha). The catchments are both north-facing with steep-to-rolling terrain ranging in elevation from 460 to 670 m. Bedrock is quartzo-feldspathic schist with some colluvium. Soils on the interfluves and slopes are silt loams (Dystrochrepts) (Hewitt 1982). In the valley bottoms they tend to be poorly drained Aquepts and Histosols. The dominant indigenous species is snow tussock (*Chionochloa rigida*) which is now associated with an extensive ground cover of introduced grasses. Red tussock (*C. rubra*) is found in the wetter valley bottoms. Approximately 10 000 ha of exotic forest plantation (*P. radiata* and *Pseudotsuga menziesii*) have been planted in the Waipori catchment. The nearest official climatological station is at Lake Mahinerangi, 20 km to the east at 400 m elevation. It has a mean annual temperature of 8.6 °C (January mean, 12.7 °C; July mean, 3.6 °C) and an average annual rainfall of 960 mm (New Zealand Meteorological Service 1983). The Glendhu mean annual rainfall (1980–1997) is 1364 mm. It occurs in many small events of long duration and low intensity (Campbell 1987).

3. METHODS AND INSTRUMENTATION

In 1982 catchment GH2 (310 ha) was planted into tussock along rip lines over 67% of its area with *P. radiata* at 1250 stems/ha. In 1989, 34 ha of trees in the lower reaches of the catchment were pruned and thinned to 270 stems/ha. Catchment GH1 (218 ha) was left in tussock as the control but has been lightly grazed at a stock density of one sheep/ha.

Runoff at Glendhu is measured by broad-crested v-shaped concrete weirs. Annual runoff is measured to an accuracy of ±5% (*c.*±40 mm) (Pearce et al. 1984). A network of manual gauges and two tipping bucket gauges are used to calculate catchment area rainfall. A climate station is located mid-way down the boundary between the two catchments. It records daily maximum and minimum temperatures, relative humidity, solar radiation, and wind data.

A daily water balance model has been used to evaluate the effects of land use on streamflow and ground water recharge elsewhere in New Zealand (Jackson & Rowe 1997). It is designed for use where detailed weather data are not available and takes the approach commonly used in models of crop water use to calculate evaporation (Doorenbos & Pruitt 1974). It calculates the energy-limited transpiration rate from a reference value e.g., Penman or Priestley-Taylor evaporation, multiplied by a "crop constant" (k). The reference evaporation rate may be calculated from available weather data, or more simply, may be assumed to be equal to published long-term average values (e.g., New Zealand Meteorological Service 1986). We used a value of k for *P.radiata* forest based on micro-meteorological and catchment studies elsewhere in New Zealand. For improved pasture it was assumed to be 1, and for tussock grassland 0.6, based on the work of Campbell (1987, 1989). The model requires daily gross rainfall totals, and (if available), daily runoff. Interception loss is assumed to be a constant fraction of the gross rainfall.

Fig. 1   Map showing location of Glendhu experimental catchments and the headwaters of the Waipori River and Deep Stream.

4. RESULTS

4.1 Changes in Water Yield after Afforestation of Tussock Grassland

There was little difference in measured water yield from the two catchments from 1980 to 1988 (Fig. 2). In 1989, however, 7 years after planting, annual runoff from GH2 had fallen by 130 mm. This trend has continued and for the period 1991–1997 the annual reduction in water yield has averaged 270 mm, or 31% less than GH1.

4.2 Changes in Minimum Low Flows after Afforestation of Tussock Grassland

The lowest average flow for 7 consecutive days has been proposed as a useful parameter to establish the impact of land-use change (Riggs 1972). At Glendhu the minimum 7–day low flows in the control catchment are in the range 0.5–1.0 mm/day in nearly all years. The impact of afforestation is shown by plotting the difference between the lowest average 7–day flow for the two catchments in each year (Fig. 3). As noted with annual water yields, afforestation begins to have a sustained influence on low flows from 1987 onwards. Between 1994 and 1997 the annual average reduction in the lowest average 7–day flow for the planted catchment was 0.11 mm/day, which is about 15% of the mean of the lowest 7–day low flow for the same period at GH1.

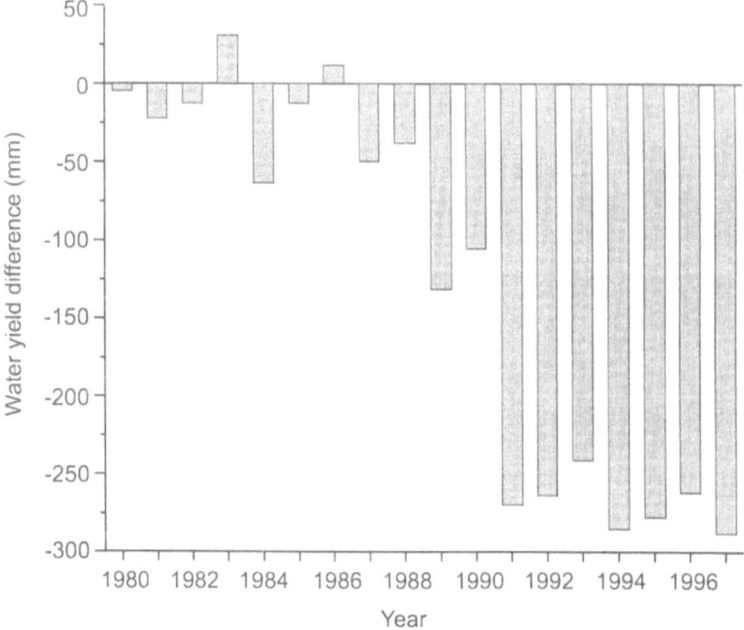

Fig. 2 Differences in annual water yield between the tussock catchment (GH1) and the planted catchment (GH2) at Glendhu, 1980–1997.

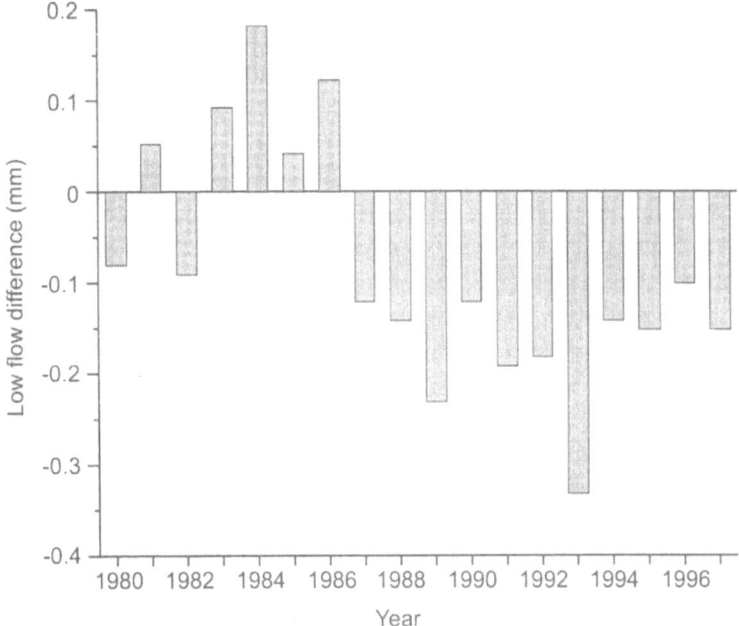

Fig. 3 Differences in minimum annual 7–day average low flow between the tussock catchment (GH1) and the planted catchment (GH2) at Glendhu, 1980–1997.

5. DISCUSSION

5.1 Application of Experimental Catchment Data to Resource Management Issues

5.1.1 Conversion of tussock grassland to pasture.

Long-term catchment studies of the type at Glendhu provide important background data against which to assess the effects of land-use change, and with which to formulate and test predictive models. For example, the city of Dunedin obtains up to 50% of its daily water requirements via pipelines from the Deep Stream catchment 10 km north of Glendhu (Fig. 1). Much of the original tussock cover remains, especially in the upper reaches, but large tracts have been modified by burning, grazing, over-sowing, and top-dressing, leading in some cases to the complete conversion from tussock grassland to pasture. The long-term impact of these changes on water yield is crucial to the sustainability of these catchments as water sources for the city. The hydrological effects of tussock depletion and conversion to pasture have been assessed by considering only evaporation differences between the two vegetation types, with reference to the tussock catchment at Glendhu.

Several studies in New Zealand have demonstrated that evaporation from pasture, when it is well-supplied with water, can be predicted by the Penman or Priestley-Taylor methods (e.g., Clothier, et al. 1982; McAneney & Judd 1983). The crop coefficient, k, for pasture is usually close to 1.0 (cf., Doorenbos & Pruitt 1974). The Priestley-Taylor method was used to calculate the monthly evaporation for pasture over the period 1990–1993 using data from the Glendhu climatic station (Fig. 4). Annual total evaporation from pasture was calculated to be 590 mm.

Total evaporation from tussock grassland has been estimated previously from the catchment water balance and from a lysimeter study. Annual evaporation averaged 470 mm/year for the calibration period at Glendhu (1980–1982) (Pearce et al. 1984), and 620 mm in the 1-year lysimeter study (Campbell 1987). Here we have calculated monthly evaporation for the control (tussock) catchment at Glendhu during the period 1980–1993 as the residual of the water balance by assuming that E = P - R, where E = wet and dry canopy evaporation, P = precipitation, and R = runoff. Mean monthly evaporation for tussock is shown in Figure 4. The mean annual total was 520 mm. It is clear from Figure 4 that, although there is little difference between the annual total evaporation for pasture and tussock there are important seasonal differences. Pasture extracts more water from the soil in summer than tussock thus delaying the recharge of water in the soil and regolith. Low-flow periods therefore tend to be longer and reach lower discharges from pasture compared with tussock areas.

5.1.2 Conversion of tussock grassland to plantation forestry.

A second example relating to the importance of retaining long-term catchment monitoring programmes comes from the direct application of the Glendhu results to the question of how much water is being lost to hydro electric power generation in the upper Waipori area because of afforestation. The total land area above the dam on Lake Mahinerangi that supplies water to the hydro electric generating plants in the middle reaches of the Waipori river is just under 30 000 ha, 5000 of which is in plantation forestry. The daily water balance model described earlier was used to estimate the likely reduction in water yield associated with current afforestation above the Mahinerangi dam.

The predictive ability of the model was first checked using rainfall and runoff data from Glendhu. The model predicted annual water yields for the tussock catchment (GH1) to within ±10% (Fig. 5). The input parameters for interception and transpiration were then altered in the model to simulate the presence of a mature forest cover (an interception factor of 0.30 and a crop factor of 0.6), and the model re-run for GH2 (the planted catchment) for the period 1991–1996 (Fig. 6). The average annual difference in flow between the two catchments as a result of afforestation for the 6–year period calculated using the model was 230 mm. The recorded average annual difference in flow between the two catchments for the same period was 270 mm. Thus the model can reasonably represent the hydrological changes accompanying afforestation at Glendhu. Moreover, since the physical character and general climate of the Glendhu catchments are broadly representative of the upper Waipori area, the model should be able to predict any changes arising from forestry at the larger catchment scale as well.

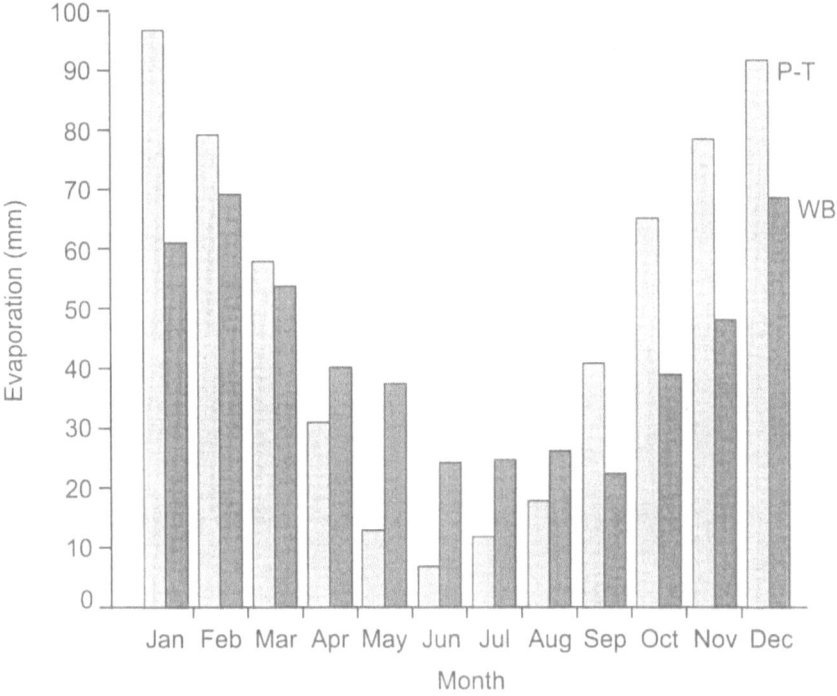

Fig. 4 Monthly evaporation estimates for the tussock catchment (GH1) at Glendhu derived from the monthly water balance (WB) for the period 1980–1993 (assuming E = P - R) compared with the Priestley-Taylor estimates (P-T) of monthly evaporation.

As a first approximation of the impact of the current forest estate on water yield from the upper Waipori catchment, the model was run using the 17-year Glendhu rainfall record (1980–1996). It was run for the first time with input parameters for tussock grassland, and for the second with input parameters for pasture. Finally, it was run assuming that all of the catchment was planted in a mature *P. radiata* forest of one age class with an interception factor of 0.30 and a crop factor of 0.6.

The current water yield for the upper Waipori catchment was then calculated by weighting the model predictions according to the proportion of the contributing areas of the three main cover types in the catchment excluding Lake Mahinerangi (tussock grassland 49%, pasture and depleted tussock grassland 34%, and plantation forestry, 17%). Based on the Glendhu rainfall record, the mean inflow to Lake Mahinerangi for the 17–year period of record was estimated at 580 000 m³/day. If the forest was not present, an extra 29 000 m³/day (5%) would be available for runoff if the land had originally been in pasture, and an extra 46 000 m³/day (8%) if the land had originally been in tussock grassland.

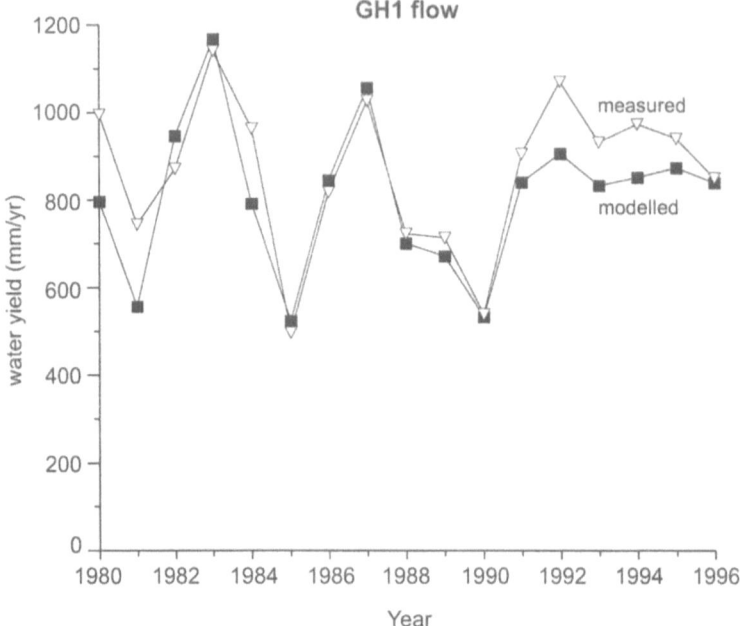

Fig. 5 Measured and modelled annual water yield for the tussock catchment (GH1) at Glendhu, 1980–1996.

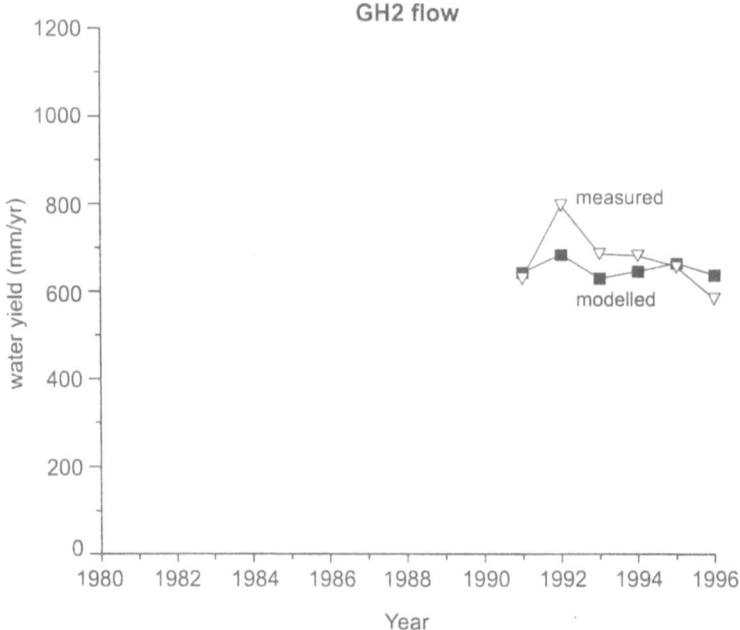

Fig. 6 Measured and modelled annual water yield for the planted catchment (GH2) at Glendhu, 1991–1996.

6. CONCLUSIONS

The need for the continued monitoring of rainfall and runoff from experimental catchments like those at Glendhu has often been questioned by funding agencies. However, as demonstrated here the data produced are crucial in assisting with the resolution of resource management issues. Experimental catchments also serve as a focus for process studies of the type undertaken by Campbell (1987, 1989) on evaporation and transpiration by tussock grassland. Combining the results of these studies with high quality catchment-based hydrological data fosters the development and testing of water balance models which in turn can help resource managers establish defensible guidelines for land-use change.

ACKNOWLEDGEMENTS

We thank the staff of Rayonier New Zealand Ltd, Invercargill for permission to use Glendhu Forest for our experimental catchment work. We also are indebted to Alex Watson for data collection and analysis, and to Brian Newman for servicing the field sites. This work was supported in part by grants from the Foundation for Science and Technology, New Zealand.

REFERENCES

Campbell, D.I. (1987): Evaporation, energy, and water balance studies of narrow-leaved snow tussock in Otago, New Zealand. Unpublished PhD thesis, Department of Geography, University of Otago. 173 p.

Campbell, D.I. (1989): Energy balance and transpiration from tussock grassland in New Zealand. Boundary Layer Meteorology, Vol 46, pp. 133–152.

Clothier, B.E., J.P. Kerr, J.S. Talbot and D.R. Scotter (1982): Measured and estimated evapotranspiration from well-watered crops. New Zealand Journal of Agricultural Research, Vol. 25, pp. 301–307.

Doorenbos, J. and W.O. Pruitt (1974): Guidelines for predicting crop water requirements. FAO Irrigation and Drainage Paper No. 24, FAO, Rome.

Fahey, B.D. and R.J. Jackson (1997): Hydrological impacts of converting native forests and grasslands to pine plantations, South Island, New Zealand. Agricultural and Forest Meteorology, Vol. 84, pp. 69–82.

Fahey, B.D. and A.J. Watson (1991): Hydrological impacts of converting tussock grassland to pine plantation, Otago, New Zealand. Journal of Hydrology (NZ), Vol. 30, pp.1–15.

Hewitt, A.E. (1982): Soils of Waipori farm settlement, east Otago, New Zealand. New Zealand Soil Survey Report No. 65. DSIR, Lower Hutt, New Zealand, 44p.

Jackson, R.J. and L.K. Rowe (1997): Effects of rainfall variability and land use on streamflow and groundwater discharge in a region with summer water deficits, Canterbury, New Zealand. International Association of Hydrological Sciences, Fifth Scientific Assembly, Poster Proceedings, Rabat, pp. 53–56.

McAneney, K.J. and M.J. Judd (1983): Pasture production and water use measurements in the Central Waikato. New Zealand Journal of Agricultural Research, Vol. 26, pp 7–13.

New Zealand Meteorological Service (1983): Summary of climatological observations to 1980. New Zealand Meteorological Service Miscellaneous Publication 177, 172p.

New Zealand Meteorological Service (1986): Summaries of water balance data for New Zealand stations. New Zealand Meteorological Service Miscellaneous Publication 189, 102p.

O'Loughlin, C.L., L.K. Rowe and A.J. Pearce (1984): Hydrology of mid-altitude tussock grasslands, upper Waipori catchment, Otago. I - Erosion, sediment yields, and water quality. Journal of Hydrology (NZ), Vol. 23, pp. 45–59.

Pearce, A.J., L.K. Rowe and C.L. O'Loughlin (1984): Hydrology of mid-altitude tussock grasslands, upper Waipori catchment, II: water balance, flow duration, and storm runoff. Journal of Hydrology (N.Z), Vol. 23, pp.60–72.

Riggs, H.C., (1972): Low flow investigations. In: Techniques of water-resources investigations of the United States Geological Survey - Water Supply Paper 1542-A, 18 p.

Water Balance Modelling on Small Forested Catchments

Pavel KOVAR

Land and Water Engineering Department, Forestry Faculty
Czech University of Agriculture, Prague, Czech Republic

ABSTRACT   The role of catchment management is very important in rainfall-runoff processes, particularly on small torrential catchments with extremely fluctuating discharges and sediment transport. Before taking measures in torrent channels one should first look at a catchment with special emphasis on land use, proper forestry and agricultural management and on area soil erosion control to reduce surface runoff by infiltration. To achieve better quantification of the significant components of the water balance, the WBCM (Water Balance Conceptual Model) was applied. This model can simulate a water balance quite well, when runoff data is available. In this study, a small catchment MALY POTOK, in the Czech Republic, was analyzed. Unfortunately, the outlet of Maly Potok is ungauged (small catchments are usually not measured). Three basic model parameters were therefore adjusted according to those determined on the Rokytenka catchment, which includes the Maly potok catchment area and which is gauged. The WBCM model was further implemented in a simulation of hypothetical scenarios representing changes in grassland, with the aim of quantifying the extent of the influence on surface runoff and on subsurface recharge. This simulation provides evidence that a 10% change in arable land area can significantly affect subsurface water recharge by infiltration of (plus or minus) 17% to 25%. This influence was computed in an „average" year. In a „dry" year it is still more significant. The trends in changes of surface runoff are even more considerable.

Keywords: Torrent control, water balance, scenario simulation.

1. CATCHMENT AND METHOD

In this paper, the data of a small torrential catchment at Maly Potok, East Bohemia, Czech Republic was implemented. The main catchment characteristics are as follows:

Catchment area: 0.59 km^2 Average area slope: 0.07
Length of main river: 0.700 km   Average altitude: 560 m a. s. l.
River slope: 0.06   Average annual rainfall: 924 mm
Land use: Forested area: 36% - spruce (Picea Abies)
            Pastures, meadows: 34%
            Arable land :   28%
            Miscellaneous: 2%

The Maly Potok catchment (6th order) is a left side tributary of Horsky Potok, then Rokytenka, Divoka Orlice and Labe (Elbe) river. The average discharge is 10 $1.s^{-1}$, annual values of rainfall and runoff are 924 mm and 500 mm, respectively. It is a natural infiltration area which provides ground water resources for drinking purposes (Orlicke Hory protected area). The catchment is geologically a Palaeozoic age with granit and gneiss. The soils are shallow, with a high content of gravel, with a significant representation of brown-podzolic types. The curve number (CN) varies between 71 to 78 (average CN = 75).

The WBCM model (version 4), which was implemented with the aim of quantifying the water balance on the Maly Potok catchment is a lumped model with probability parameter distribution over the area. It is based on the integrated storage approach. Each storage element represents the natural storages of interception, soil surface, root (or active) zone, whole unsaturated zone and ground water zone (if it is not very deep). The model with a daily step considers the storage of individual zones and treats their daily values, including input and output rates in line with physical regularities as reflected by the system of recursive finite difference and algebraic equations balancing the following processes (Kovar, 1981, Kovar, Vesely, 1996):

- Potential evapotranspiration
- Interception and throughfall
- Surface runoff recharge
- Active soil moisture zone dynamics
- Soil moisture content and actual evapotranspiration
- Ground water dynamics, base flow, total flow

The individual parameters of the WBCM-4 have the following physical meaning:

AREA   catchment area (km^2)

FC   parameter characterizing „average" value of a field capacity of active zone (-)

POR   parameter characterizing „average" value of soil porosity of active zone (-)

DROT   depth of active zone (mm)

WIC   upper limit of interception capacity (mm)

SMAX   parameter representing maximum capacity of the unsaturated zone (mm)

ALPHA   parameter expressing non-linear filling process of the unsaturated zone(-)

CN   runoff curve number (-)

P1,P2, P7   parameters affecting unsaturated zone dynamics (filling and exhausting processes) (-)

GWM   parameter expressing capacity of the active part of ground water zone (mm)

BK   linear transformation parameter of base flow process (days)

For daily values of potential evapotranspiration computation we used the modified Monteith-Penman method, as well as the Priestley-Taylor method, or alternatively the Hamon method. Selection of one of these depends on input data availability. The model unit that computes actual interception and throughfall is based on simulation of irregular distribution of local interception capacities around their mean value, WIC. These capacities vary between zero (bare soil) and a multiple of WIC. To avoid an abrupt threshold concept, a linear distribution around the WIC-value was accepted.

For quantifying surface runoff recharge, the US Soil Conservation Service (SCS) method based on the Curve Number (CN) assessment was used. The standard procedure for an initial CN value is accepted, and the daily storages of active zone, SS are computed by the SCS procedure. The recharge of the active (root) and then whole unsaturated zones depends greatly on the previous soil moisture content and is controlled by the FC parameter. The evaluation procedure is based on the assumption that the distribution of local FC-values around their average is non-linear (parameters P1,P2, P7, ALPHA). Only where FC has not yet been reached can the recharge replenish these zones (one by one) up to the FC-limiting value. The one-dimensional Richard's equation is used in the finite difference form. Simultaneously, exhaustion from this zone by evapotranspiration is computed. To simulate this procedure, an approach was applied which estimates the proportion between actual and potential evapotranspiration according to the soil moisture content and to particular physical properties of the soil.

The saturated zone is filled by groundwater recharge and depleted by base flow. It is simulated only within the framework of short-term groundwater participation in water balance. The Williams and La Seur method (Williams, La Seur, 1976) was applied here. In cases where possible control of a model efficiency can be achieved either through runoff or ground water table fluctuations, automatic optimization is applied. The parameters SMAX, GWM and BK were optimized by minimizing the sum of least squared differences between the computed and observed 10-day runoff depths on the Rokytenka gauged catchment, to which the Maly Potok belongs.

Four main groups of data are required for the model:

I. Hydrometeorological data: Daily values of rainfall as well as the data necessary for a potential evapotranspiration calculation

II. Hydrological data: Observed daily (monthly) runoff data if parameter optimization is required. Alternatively a daily record of ground water levels in a characteristic cross-section can be used.

III. Hydrological assessment of the „range values" of porosity, field capacity of both active and unsaturated zones, their depths, soil classification.

IV. Land use data: Forestry and agricultural data on land use, types of forest and crops, cropping pattern, watershed management.

Then the water balance equation controls the volumes of the main components of water balance:

$$SRAIN = AE + STF + (\Delta WP + \Delta WZ) = AE + STF + \Delta W$$

where SRAIN rainfall depth (mm)
  STF   total runoff depth (mm)
  AE   actual evapotranspiration (mm)
  ΔWP change in soil moisture content (mm)
  ΔWZ change in ground water storage (mm)
  ΔW change in subsurface storage (mm)

2. RESULTS

For the Maly potok catchment the following daily data were used:
- Daily rainfall data from the Rokytnice n. O. station
- Daily data on free water evaporation from the Usti n. O. station

- Hydrological and soil parameters from maps and other sources:
  CN <71, 79>, mean CN=75   POR= 0.45, FC= 0.35, DROT= 1000.0 mm
  P1= 0.1, P2= 0.2, P7= 0.7   WIC= 1.5 mm, ALPHA= 1.0,
  Optimized parameter values (Rokytenka):
  SMAX= 445.0 mm, GMAX= 2000.0 mm, BK= 1.7 days
- Initial soil moisture contents at the beginning of growing season (01/05) were assessed according to antecedent precipitation (March, April).

The model was implemented with data from 1981 to 1990 just to simulate the growing seasons from May 1 to October 31. To gain a broad spectrum of simulation, Tab. 1 shows the characteristic years of 1981 (wet year), 1983 (dry year) and 1986 (normal year) with the main water balance components. The same values of parameters for all 10 years were applied and the negligible errors in the balances provide evidence of acceptable simulation.

Tab. 1

GROWING SEASON WATER BALANCE IN TYPICAL YEARS MALY POTOK			
Balance component	WET 1981 [mm]	DRY 1983 [mm]	NORMAL 1986 [mm]
Rainfall depth SRAIN	669.4	360.1	515.7
Total runoff STF	106.7	57.5	57.1
Surface runoff SOF	19.1	20.4	14.8
Potential evapotranspiration PE	330.6	454.9	409.3
Actual evapotranspiration AE	274.7	339.2	331.1
Change in soil moisture content ΔWP	141.5	-83.2	42.3
Change in g. w. storage ΔWZ	140.6	45.6	85.0
Change in subsurface storage ΔW	282.1	-37.6	127.3
Error in water balance (mm)	5.93	0.94	0.20
Ditto (%)	0.89	0.26	0.04

The main important components of water balance are undoubtedly those expressing subsurface storage changes ΔW:

wet year 1981:   ΔW = 282.1 mm
dry year 1983:   ΔW = -37.6 mm
normal year 1986: ΔW = 127.3 mm

The surface runoff was not exceptionally high, even in the wet years. For illustration, the graph (Fig.1) is enclosed to show the course of water balance in the extreme years of 1981 and 1983.

The final step was to use the model in the implementation of hypothetical scenarios simulating changes in land use.

Scenario A represents the existing land use, while **scenario B** simulates the reduction of arable land by reducing its area by 10 percent, replacing it by pastures. This change can be roughly expressed by changes in parameters as follows.

Scenario	Parameters: CN (-)	WIC (mm)	DROT (mm)
A	75.0	1.5	1000.0
B	73.0	1.7	1050.0

Table 2 shows the influence of land use and management on water balance for the normal year 1986.

Tab. 2

WATER BALANCE CHANGES USING SCENARIOS A, B 1986, MALY POTOK			
Balance component	Scenario		Changes
	A (existing)	B (changed land use)	
	mm	mm	%
Rainfall depth SRAIN	515.7	515.7	
Total runoff STF	57.1	39.5	-30.8
Surface runoff SOF	14.8	10.3	-30.4
Actual evapotranspiration AE	331.1	326.2	- 1.0
Change in soil moisture content ΔWP	42.3	49.7	+17.5
Change in storage ΔWZ	85.0	100.4	+18.1
Change in subsurface storage ΔW	127.3	150.1	+17.9
Error in water balance (mm)	0.2	0.1	
Error in water balance (%)	0.04	0.02	

3. DISCUSSION

Formation and propagation of surface runoff and of subsurface recharge are mostly influenced by the following factors:
- Intensity and duration of rainfall
- Physiographic catchment parameters
- Land use and management
- The length of the river, its shape, slope, hydraulic radius and roughness.

From these four groups of factors, rainfall parameters always remain constant (namely storm rainfalls that can hardly be considered by a water balance model because of their short duration). Concerning a change in channel route and its slope reduction, a flood wave could be consequently transformed a little. However, increased hydraulic resistance limits the discharge capacity of the channel cross-section, which reduces the transformation effect. In conclusion, the most significant effect in reducing surface runoff in favour of infiltration is **land use and watershed management.**

4. CONCLUSIONS

The simulations provide evidence that the reduction of arable land area by 10% can significantly affect the subsurface water recharge (by infiltration) by 17% to 25%. This influence was computed in an „average" year (normal year 1986). In a „dry" year (1983) it was still more significant. Afforestation of former arable land on steep areas could obviously have a similar effect. These changes in land use in the Czech Republic are now very common, as previous agricultural intensification often led to over-production and excessive land exploitation.

410

REFERENCES

Kovar, P. (1981): Konzept-Modell fuer die Wasserhaushaltbilanz kleiner Einzugsgebiete. Zeitschrift fuer Kulturtechnik und Flurbereinigung N.22/1981, P. Parey, Hamburg, BRD (pp 341-352).
Kovar, P., Vesely, R. (1996): Use of water balance models in master urban plans. Proceedings of International Conference ENVIRO NITRA, Slovakia, (pp 106-112)
Williams, J. R., La Seur, W. R. (1976): Water yield model using SCS curve numbers. Journal of Hydraulic Division, HY9, (pp 1241-1253).

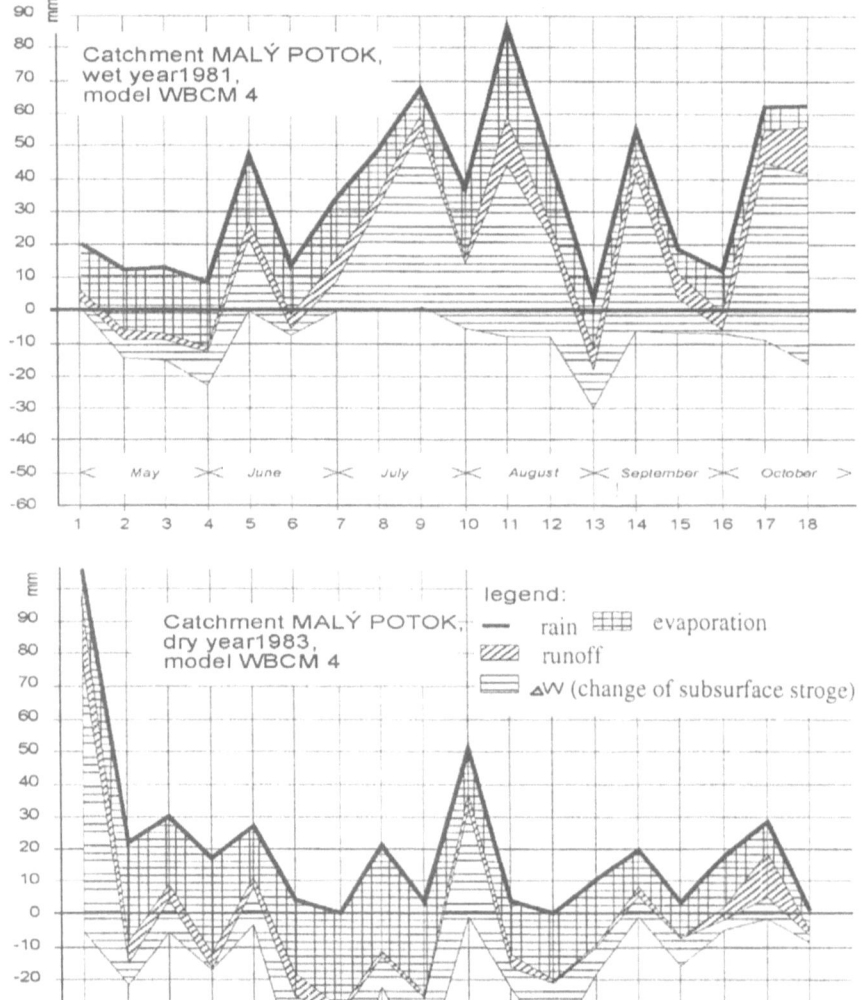

Fig.1: Main components of water balance of the Maly Potok catchment

Evaluation of Forest Canopy Shape from Standpoint of Thermal Exchange above Forest

Koji TAMAI*[1], Shigeaki HATTORI*[2] and Yoshiaki GOTO*[1]

*1 Kansai Research Center, Forestry and Forest Products Research Institute
  Kyoto, Japan
*2 Faculty of Agriculture, Nagoya University, Nagoya, Japan

ABSTRACT   Shape of forest canopy surface was evaluated in a deciduous forest from a viewpoint of thermal exchange.   The analysis of canopy shape with Morishita's dispersion index showed that projected size of dominant convex and concave shapes on the forest canopy surface were around 5.0m square in a foliate season and 5.0-7.5m square in a defoliate season.   On the other hand,   each vertical size of the convex and concave shapes was estimated to be larger in a foliate season than in a defoliate season. Thus, this study concludes that the canopy shape was smoother in a defoliate season than in a foliate season in this case.
Keywords: Morishita's dispersion index, convex shape, concave shape,
          PCM method, Deciduous forest

1.   INTRODUCTION

Exchange efficiency of heat and mass between a forest community and the atmosphere depends on leaf quantity and forest canopy structure.   Watanabe *et al.* (1990) clarified the relationship between a bulk coefficient of momentum and dimensionless leaf area density using a model. Evaluation of the exchange efficiency has been carried out in various forest communities using aerodynamic resistance values derived from the wind profile.   However, the relationship between the   aerodynamic resistance and forest structure has been considered mainly with respect to the tree height.   The exchange efficiency depends on the vertical wind exchange through the canopy surface of which has many uneven shapes.   For example, canopy of a large tree forms swelling shape and space among the canopies resemble a hollow.   These uneven shapes may give larger effects on the vertical wind exchange and the exchange efficiency than the tree height itself.   Nonetheless there has been no examples investigating the relationship between forest canopy shapes and exchange efficiency.   As a first step to evaluate this relationship,   characteristics of forest canopy shape were investigated in foliate and defoliate seasons in a deciduous broad-leaved secondary forest.

There are no examples of measurement and evaluation of forest canopy shape from the thermal exchange efficiency standpoint.   In this study, the forest canopy shape was measured by the PCM method (Sumida, 1993 and 1995) and its characteristics were analyzed using the Morishita's index of dispersion, I_δ (Morishita, 1959). I_δ was originally derived as an index to evaluate the concentration degree of living thing or plant and the colony size where they are concentrated. Although there are some similar indices, Morishita's I_δ has several important characteristics: first, it is not affected by the size of analysis objects appearing. Second, it is not affected by blockage area and number of analysis objects. By replacing the number of appearing plants with forest canopy elevation, that is the sum of canopy height and forest floor elevation, we concluded it possible to evaluate the level of unevenness in the forest canopy shape vertically and horizontally, using Morishita's I_δ. The shapes with resemblance to swellings and hollows on the canopy surface were named convex and concave shapes, respectively in this study.

2. MORISHITA'S DISPERSION INDEX, I_δ

The value of I_δ is calculated based on the number of an appearing plant in block elements with Eq. (1)

$$I_\delta = q \Sigma \{n_i(n_i-1)\}/N(N-1) \tag{1}$$

where q is the number of blocks, n_i is the number of individuals of an appearing plant in a block element i, and N is the total number individuals appearing in the entire block.

Based on the calculated I_δ in various size of block element, we can analyze the distribution characteristics of a plant. A relationship between the size of block element and I_δ often shows a triangular-shaped curve. In this case, as the maximum of I_δ is getting larger, the concentration degree of living thing or plant increases. The size of a block element having the maximum I_δ indicates the colony size where the appearing plant concentrates. In order to use this method for investigation of forest canopy shapes, we replace n_i with H_h and H_l that indicate non-dimensional canopy elevations.

$$H_h = (H_i - H_{min})/H. \tag{2}$$
$$H_l = (H_{max} - H_i)/H. \tag{3}$$

where H_i indicates the canopy elevation of an individual block element, and H_{max}, H_{min} are the maximum and minimum values of H_i, respectively, and H. is a standard height for non-dimensionalization.

H_h and H_l represent the concave and convex shapes, respectively. Thus, when investigating convex and concave shaped curves, H_h and H_l are used, respectively. When comparing calculated I_δ results for various sized block elements, the larger maximum value of I_δ means that the convex shape higher, or conversely, the concave shape is deeper. H_h and H_l values for one big sized block element is calculated as the respective sum (not average) of H_h and H_l

values for small sized elements composing it.   This summing process was used because the value of N has to be constant through the calculation.

3.   SITE DESCRIPTION and METHOD

3.1   Outline of Forest Basin

Our measurements were conducted in a forested study basin (the area of 1.6 ha) named the Kitadani   Basin in the Yamashiro Experiment Site, located in Yamashiro-cho, Soraku-gun, Kyoto, Japan (NL 34° 47', EL 135° 51') (Fig. 1). Deciduous broad-leaves like *Quercus serrata* and *Lyonia neziki* dominate as tall trees and shrubs, but evergreen trees like *Ilex pedunculosa* and *Eurya japonica* also coexist mainly as shrubs.   The basal area at breast height is 13.3 m^2/ha for deciduous broadleaf species, and 6.3 m^2/ha for evergreens (Table1). The soil layer is generally dry, shallow and immature and underlain with the thick weathered

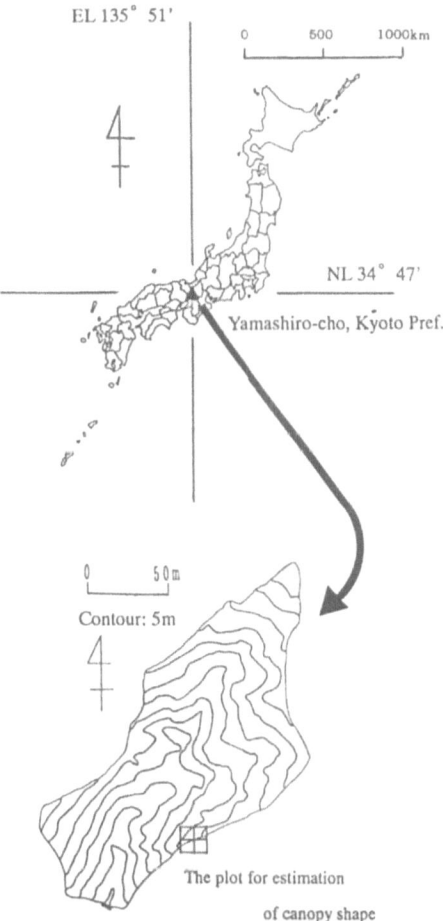

Fig. 1 Location of Exp. Site and plot
for measurement of canopy shape.

granite.   The soil texture is sand.

The annual average temperature is 15.8°C, the annunal average relative humidity is 74.6%, and the annual precipitation is 1627.0 mm.

In the study basin, deciduous broad-leaved trees occupy about 66% of the total (Table1),   and the forest canopy structure is remarkably different in the foliate and defoliate seasons.   Deciduous broad-leaved trees open their leaves from April to May, and drop their leaves from October to November.   Sky view factors through the forest canopy were estimated from photographs of the canopy taken by a fisheye type lens from the forest floor.   The sky view factors were about 15% in the foliate season and about 50% in the defoliate season. In addition, the leaf area indices were estimated from a plant canopy analyzer (LI-COR Inc., LAI-2000) as 4.42 in the foliate season and as 2.70 in the defoliate season.

Table 1   Dominant species and their basal
area in Kitadani Basin.

Basal area (m²/ha)

Total   19.60

Ever green species Sub total 6.29	
Ilex pendunculosa Mig.	2.95
Pinus densiflora Sieb. et Zucc.	1.22
Eurya japonica	0.73
others	1.39

Deciduous species Sub total 13.31	
Quercus serrata Thunb. ex Murry	4.48
Lyonia japonica elliptica	1.93
(Wall.) Drude var.	
(Sieb. et Zucc.) Hand.-Mazz.	
Alnus sieboldiana Matsumura	1.44
Clethra barvinervis Sieb. et Zucc.	1.25
Robinia pseudoacacia L.	0.83
others	1.39

3.2 PCM method

The shape of the forest canopy surface was measured using the PCM method. First, we set up a 27.5m × 27.5m square plot so that the four sides faced east, west, south and north.   The plot was divided into grids with 22 lines in total, located in 2.5m intervals.   Then, measurements were made at 144 intersections points of the east-west line and the south-north line using the PCM method (Sumida, 1993 and 1995).   In the PCM method, one measures the height of the upper and lower boundaries of a leaf group with a measuring pole and determines the layer structure.   The upper boundary of a highest leaf group was defined as the canopy surface and measured carefully, surveying the location of measuring pole from a tower platform above the forest canopy set at the center of the plot. However, for a point not covered by the forest canopy of evergreen trees in the defoliate season, we considered the forest floor as the forest canopy surface. In this case, although there are no leaf groups in the upper part, branches and trunks of deciduous trees do exist above the forest canopy. The observation periods were September in 1994 for the foliate season, and February in 1995 for the defoliate season.

3.3 Analysis Method of I_δ

By measuring the forest canopy height with the PCM method and referring to the forest floor elevation, 12 x 12 grid data of canopy elevation were obtained with the intervals of 2.5m. Because one canopy elevation datum represents 2.5m square. Then the smallest block, which corresponds to I_δ of 2.5m square, results in 144 total blocks. Different sized block areas were yielded five stages: block areas x block totals of 5.0m square x 36, 7.5m square x 16, 10.0m square x 9, and 15.0m square x 4. Canopy elevation at each block element was indicated by an elevation difference from the weir site. H_{max} and H_{min} in the foliate season were 45m and 29m, and in the defoliate season, 42m and 24m, respectively. H. was 1m in this study.

4. RESULT and DISCUSSION

4.1 Comparison of Forest Canopy Shape in Foliate and Defoliate Seasons

Figs. 2(a), (b) and (c) show contour maps of the forest floor and the canopy elevation, respectively. According to the forest floor elevation contour map (Fig. 2a), there is a ridge from the northeast to the west. The southern and northwest sides are slopes.

In the foliate season, there are large convex shapes in the northeastern part and the western part of the ridge (Fig. 2b). They correspond to the canopy of *Quercus serrata* , which is a deciduous tree. On the slopes of the southern and northwestern sides, some uneven shapes are found. When the defoliate season comes, the large peak due to *Quercus serrata* canopy disappears and some lower peaks appear on the northeastern side slope (Fig. 2c). They correspond to the canopy of evergreen trees, such as *Ilex pedunculosa*.

4.2 Evaluation of Forest Canopy Shape Characteristics

Figs. 3(a) and (b) show results of I_δ analysis. There are similar trends in both of the convex and concave shapes. First, the both maximum values of I_δ in the foliate season are larger than those in the defoliate season. This indicates that the vertical sizes of dominant convex and concave shapes are larger in the foliate

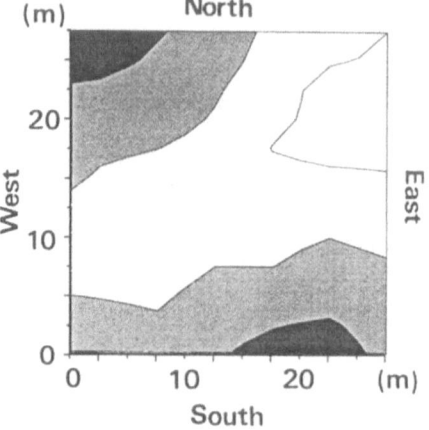

(a) Forest floor elevation map

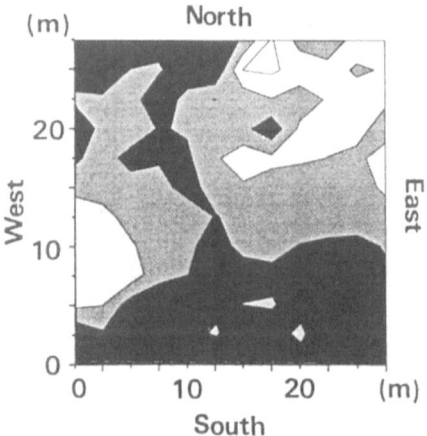

(b) Contour map of canopy elevation
in foliate season

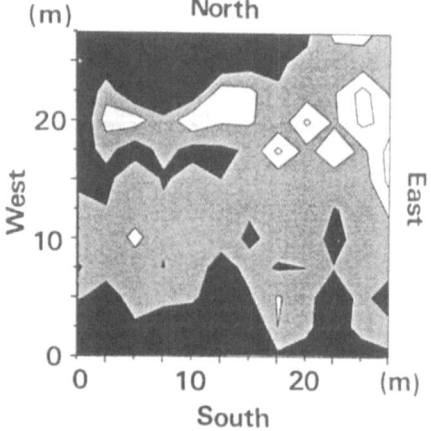

(c) Contour map of canopy elevation
in defoliate season

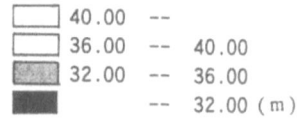

Fig. 2   Contour maps of forest floor and canopy elevation
measured from the elevation of the weir site

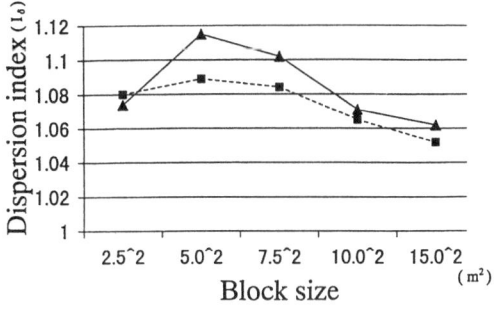

(a) Convex shape

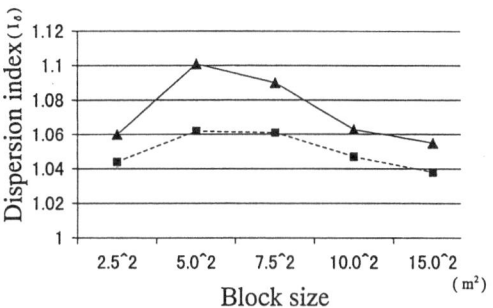

(b) Concave shape

Fig. 3 Size of the dominant convex and concave

shapes in the forest canopy

triangle: foliate season   square: defoliate season

season. The values of I_δ for convex and concave shapes reach maximums remarkably when the block area is 5.0m square in the foliate season. This shows that the projection sizes of dominant convex and concave shapes are about 5.0m square in the foliate season. Regarding the defoliate season however, the maximum values of I_δ for the both shapes corresponding to a 5.0m square block decrease to almost the same values corresponding to a 7.5m square block. This means that the projection sizes of dominant convex and concave shapes are around 5.0-7.5m square in the defoliate season. Thus, the projection sizes of the dominant convex and concave shapes are slightly larger in the defoliate season than in the foliate season. This may be caused by a fact that an influence of the forest floor elevation on I_δ is relatively larger in the defoliate season since the values of H_i are composed by elevations of low height evergreen trees and forest floor. As a result, the horizontal and vertical sizes of dominant convex and concave shapes tend to be larger and smaller in the defoliate season than in the foliate season. It can be concluded that the forest canopy shape is a fairly smoother in the defoliate season than in the foliate season in this forest.

Vertical wind and thermal exchanges seem to be controlled not only by the quantity of leaves and tree heights, but also by canopy shapes. Finding on different shapes between foliate and defoliate seasons obtained from this study may provide information on the effects on the wind and thermal exchanges. More detail observations are necessary for assessing the effects.

REFERENCES

Morishita, M. (1959): Measuring of the dispersion of individual and analysis of the distributional patterns, Mem. Fac. Sci., Kyushu Univ., Ser. E (Biol), 2, 215-235.

Sumida, A. (1993): Growth of tree species in a broadleaved secondary forest as related to the light environments of crowns. J. Jpn. For. Soc., 75, 278-286.

Sumida, A. (1995): Three-dimensional structure of a mixed broadleaved forest in Japan, Vegetatio, 119, 67-80.

Watanabe, T. and Kondo, J (1990): The influence of canopy structure and density upon the mixing length within and above vegetation, J. Meteor. Soc. Jpn., 68, 227-235.

Estimating Rates of Nutrient Recovery Following Timber Harvesting in a Second Growth Forest of Peninsular Malaysia

ZULKIFLI Yusop, BAHARUDDIN Kasran and ABDUL RAHIM Nik

Forest Research Institute Malaysia, Kepong 52109, Kuala Lumpur, Malaysia

ABSTRACT Rates of catchment nutrient recovery of a second growth forest following logging were estimated using a mass balance approach. The study site is located in Selangor, Peninsular Malaysia. The rainfall samples seemed to be enriched by solutes from local sources. This results in an over-estimation of the atmospheric nutrient inputs. However, lower rates of inputs were obtained when median concentrations were used for calculating the inputs, instead of the mean. Inputs from weathering were estimated from the catchment input-output budgets. Assuming 40 m^3/ha of timbers are to be removed, the time spans needed to compensate losses of nutrient in biomass and enhanced leaching would be 20, 35 and 30 years for Ca, Mg and K, respectively. The recovery time depends largely on the chemistry of the local rainfall. When nutrient inputs derived from three other forested stations were included in the analysis, the recovery spans range from 25 to 90 years for Ca, 20 to 35 years for Mg and 10 to 55 years for K.
Keywords: Nutrient Inputs, Nutrient Loss, Input-Output Budgets, Recovery Span.

1. INTRODUCTION

Peninsular Malaysia will eventually rely on her second growth forests for the supply of timbers. One of the pre-requisites to sustainably managed this ecosystem is the preservation of soil fertility, especially when a short rotation forestry is practiced (Nykvist, 1994). At an ecosystem level, the maintenance of site fertility could be achieved by optimising the harvesting rate to a level at which nutrient losses in the removed biomass and via enhanced leaching will not impair forest growth. In other words, the combined inputs of nutrients in rainfall and weathering must at least match the losses. If this is not satisfied, site fertility would be gradually depleted in the subsequent cutting cycles, leading eventually to poor regeneration and low productivity. An attempt was made to estimate nutrient recovery rates following selective logging by considering the catchment nutrient input and ouput budgets and losses in biomass.

2. EXPERIMENTAL SITE

Our experimental site is located in the Compartment 41, Bukit Tarek Forest Reserve in Kerling, Selangor, Malaysia at approximately 3°31' N latitude and 101°35" E longitude (Figure 1). The area is underlain by metamorphic sedimentary rocks from Arenaceous Series which were deposited together with the

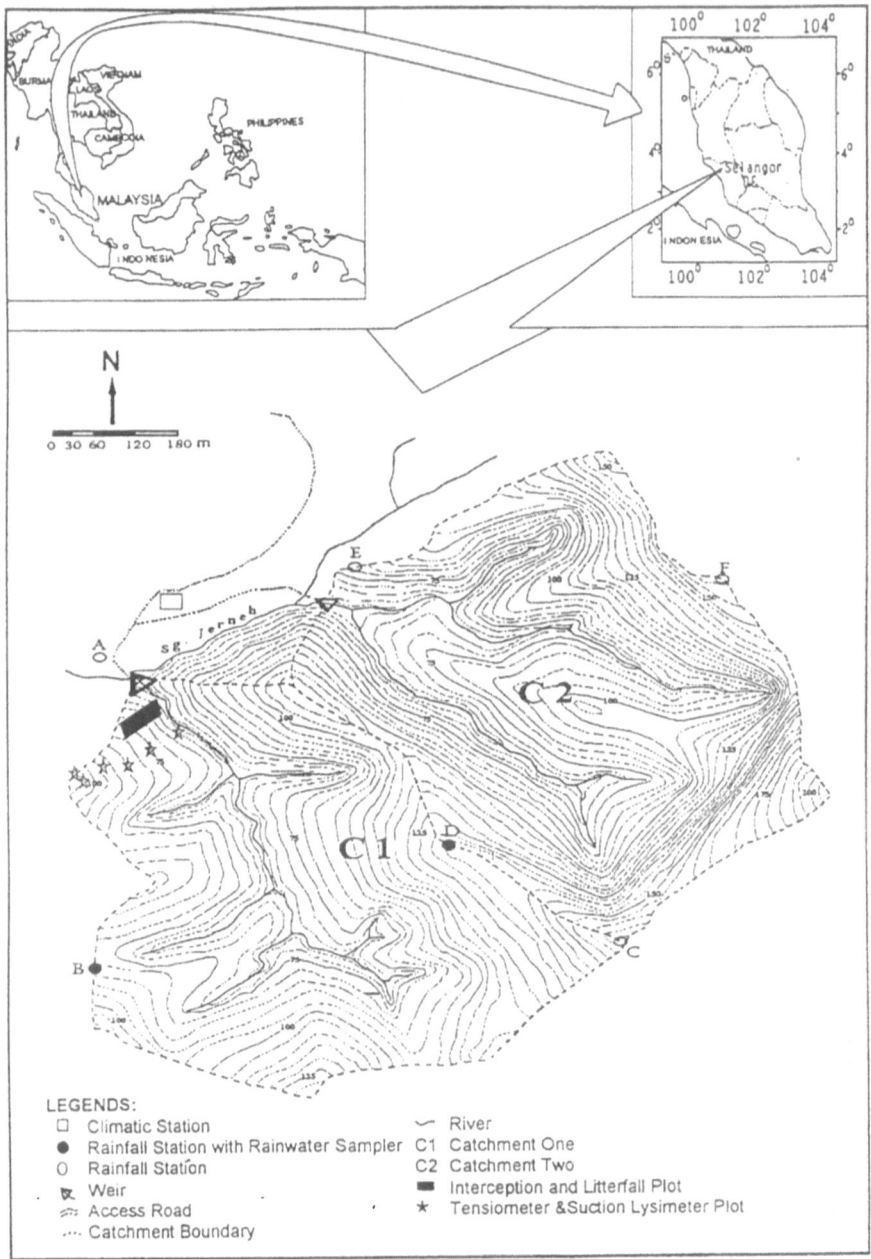

Figure 1: The location and layout of the Bukit Tarek
Experimental Watershed

argillaceous sediment during the Triassic Age (Roe, 1951). The site receives an annual rainfall average of 2763 mm with two peaks in the months of April/May and October/November (Zulkifli, 1996). Two soil series were found namely, Kuala Brang (*Orthoxic Tropudult*) covering 90% of the area, and Bungor (*Typic Paleudult*), mainly at lower elevations. The forest stocking is relatively poor and resembling to the Kelat-Kedondong forest type (Wyatt-Smith, 1963). The first forest logging operation was carried out 30 years prior to the investigation started.

3. METHODS

The quantity and the quality/chemistry of rainwater and streamflow at BTEW were examined from July 1991 to June 1994. Rainwater was sampled by mounting stainless steel funnels at 1 meter above the ground and routed by silicon and polypropylene tubings into a 4.5 l polyethylene bottle. This sampling technique collects "bulk precipitation" sample which include both, dry fall-out and wet deposition (Likens et al. 1977). Streamwaters were grab sampled on a weekly and storm basis, and complemented by an intensive storm event sampling using automatic water samplers (SIGMA SL800). A rainfall intensity sensor was installed to activate the sampler during storms. Samples were refrigerated prior to analysis to suppress biological activities and stabilise its chemical compositions. After being filtered through a 0.45 μm membrane filter, 50 ml of the filtrate from each sample was treated with 1 ml concentrated HNO_3 and kept in a refrigerator for chemical analyses within not more than six months. Ca and Mg were determined using an atomic absorption spectrophotometer whereas K by a flame photometer.

4. RESULTS AND DISCUSSION

4.1 Input-Output Budgets

The atmospheric inputs of Ca, Mg and K were calculated on a monthly basis, by multiplying the monthly mean concentration with the corresponding rainfall volume. The monthly inputs were then summed up to a yearly total. Similarly, the outputs were calculated as the products of the monthly mean nutrient concentration and the streamflow volume. The inputs and outputs of these nutrients and their net gains or losses are given in Table 1 together with results from other tropical forest sites. Confining results to Malaysian sites only (location 12 to 16), the inputs at BTEW were generally on the higher side. Much larger variation was observed when results from other regions were considered ranging from 0.3 to 37.8 kg/ha/yr for Ca, 0.1 to 8.4 kg/ha/yr for Mg and 0.5 to 14.4 kg/ha/yr for K. Among factors that determine rate of nutrient deposition are the rainfall regimes, distance from the coast, seasonal variation (dry or wet), surrounding land uses and the occurrence of cyclone (Bruijnzeel, 1991; Waterloo, 1994).

The major sources of K, Ca and Mg mostly relate to geochemical weathering. As

Table 1: Input-output budgets of water (mm) and major nutrients (kg/ha/yr) in selected lowland (sub) tropical forests

Location	Ann. Pre. (mm)	Annual Runoff (mm)	Calcium			Magnesium			Potassium		
			I	L	I - L	I	L	I - L	I	L	I - L
Spodosol/Psamments											
1. San Carlos, Venezuela	3565	1860	5.2	2.8	+2.4	0.7	0.6	+0.1	6.9	3.5	+3.4
Oxisol/Ultisol											
2. Caura River, Venezuela	3850	2425	1.3	15.5	-14.2	0.3	6.0	-5.7	1.0	14.6	-13.6
3. San Carlos, Venezuela	3565	2100	5.2	5.6	-0.4	0.7	1.5	-0.8	6.9	7.0	-0.1
4. Tai Lam Chung, Hong Kong	1900	1235	7	18	-11.0	6.5	6.0	+0.5	7.0	18.0	-11.0
5. Jari, Brazil	2300	1225	15.8	16.8	-1.0	3.3	8.1	-4.8	10.2	12.7	-2.5
6. Adiopodoume, Ivory Coast	2130	1000	37.8	46.6	-8.8	8.4	30.4	-22.0	6.3	69.1	-62.8
7. Banco (valley site), Ivory Coast	1800	630	30	43	-13	7	4.3	+2.7	5.5	1.3	+4.2
8. Tonka, Surinam	2145	515	16.2	7.6	+8.6	2.4	4.4	-2.0	14.4	2.4	+12.0
9. Ducke Reserve, Brazil	2475	450	0.3	0.9	-0.6	0.2	0.5	-0.3			
10. Ibidem	2075	400	tr	tr		tr	Tr		2.1	0.4	+1.7
11. Tulasewa, Fiji α	1932	167	<2.3	0.2	+2.1	<1.1	1.5	<-0.4	<6.1	0.4	+5.7
Malaysian Sites											
12. Ulu Gombak, Malaysia	2500	750	14.0	2.1	+11.9	3.2	1.5	+1.7	12.5	11.2	+1.3
13. Bt. Berembun, Malaysia η	2000	575	3.6	14.8	-11.2	0.6	9.2	-8.6	6.8	20.5	-13.7
14. Mendolong, Sabah (W3)	3215	1962	1.5	19.7	-18.2	0.3	17.3	-17.0	1.7	12.4	-10.7
15. Ibidem (W6)	3490	1950	1.7	3.1	-1.6	0.4	3.2	-2.6	1.8	3.2	-1.5
16. Danum Valley, Sabah β	2800	1400	0.3	1.2	-0.9	0.1	1.1	-1.0	0.5	0.8	-.2
17. This study (Catchment 1)	2750	1227	8.4	4.7	+3.7	3.3	4.4	-1.1	6.7	7.1	-0.4

Notes:
See Bruijnzeel (1991) for details of locations 1 to 10; 11, Waterloo (1994); 12, Kenworthy (1971); 13, Bruijnzeel (1991) & Zulkifli (1990); 14 & 15, Grip *et al.* (1994); 16, Burghouts (1993)

α - pre-cyclone event; β - limited rainfall data, baseflow concentration only; η - using rainfall chemistry of a nearby site of Pasoh.

such, these cations are expected to show net losses in their budget analysis, especially when a catchment has reached a state of dynamic equilibrium. This was indeed observed for Mg and K. Calcium, however, shows a net gain. This may suggest vegetation response to the relatively low Ca contents in soil compared to K and Mg (Zulkifli, 1996), thus, demands a longer time before a sufficient quantity can be generated from internal cycling. Furthermore, Ca tends to be assimilated in a much larger quantity for biomass production compared to Mg and K (Nykvist, 1994). Accumulation of Ca was also observed at Ulu Gombak (Table 1; site 12) with the net gain almost three times higher than the present site. In contrast, net losses of Ca, Mg and K were observed in forest catchments recently (<10 years) affected by selective logging and forest fire in Sipitang, Sabah (site 14 and 15). Such results are rather exceptional for a newly disturbed ecosystem since it is expected to be still aggrading nutrients for new biomass.

Discrepancies with regard to budget analysis of major cations may arise from variation in the weathering influence. For example, Grip et al. (1994) found that large differences in the net budgets of K, Ca and Mg between two nearby catchments in Sabah. They ascribed this to the contrasting characteristic in the bedrock nutrient contents and weathering rates over a small areal extent. Other difficulty in calculating nutrient budget pertains to the reliability of the atmospheric inputs. Even with strict quality control with regard to sampling, storage and chemical analysis, the atmospheric input may still deviate from the long term trend especially when massive changes in the land-use pattern coincide with sampling period. The apparently high inputs of nutrient at the present site could partly responsible for the positive budget of Ca and only marginal net losses of K. Bruinzeel (1991) cautioned the difficulties of comparing budget analyses from tropical forests without sufficient knowledge on the site characteristics.

4.2 Nutrient Recovery after Forest Harvesting

Assessment of nutrient recovery at ecosystem unit can be achieved by solving a mass balance equation considering the inputs and outputs of nutrients, and changes in the ecosystem storages. For non-gas elements the general equation can take the following form:

$$P_x + W_x = B_x + Q_{xs} + Q_{xp} + \Delta S_{xv} + \Delta S_{xl} + \Delta S_{xs}$$

where P_x, and W_x are the amounts of nutrient x supplied by atmospheric deposition and mineral weathering, respectively; B_x denotes the amount removed in biomass; the exports as solute, Q_{xs} and the loss in undissolved material, Q_{xp}. Changes in the amount of nutrients stored in the biomass, litter layer and soil reserve are denoted by ΔS_{xv}, ΔS_{xl}, and ΔS_{xs} respectively. At BTEW, internal cycles through litterfall and throughfall account for more than 90% of the total nutrient transfer to the forest floor (Zulkifli, 1996; Zulkifli et al. 1997). Therefore, it is sensible to assume that the present ecosystem has reached or is nearly reaching a climax in terms of nutrient capital so that ΔS_{xv}, ΔS_{xl}, and ΔS_{xs} are negligible and can be dropped from the equation.

Therefore, it is sensible to assume that the present ecosystem has reached or is nearly reaching a climax in terms of nutrient capital so that ΔS_{xv}, ΔS_{xl}, and ΔS_{xs} are negligible and can be dropped from the equation.

The atmospheric inputs at BTEW, calculated from the monthly mean concentrations of Ca, Mg and K at BTEW amounted to 8.6, 3.3 and 6.7 kg/ha/yr, respectively (Table 1). For a climax ecosystem, nutrient releases from weathering can be approximated from the input-output budgets of the respective element (Likens et al. 1977; Clayton, 1979; Creasey et al. 1986). This gave estimates of weathering inputs of 1.1 and 0.4 kg/ha/yr for Mg and K, respectively (Table 1). Since the exports of Ca were smaller than the input (net gain budget), the source of Ca from weathering can be regarded as too small or equal to zero.

Nutrient concentrations in logs, consisting of bark and wood, were obtained from nine lowland rainforest sites in the tropics, mostly on *oxisol* and *ultisol* soils (Nykvist, 1994, Table 2). The values vary greatly, ranging from 0.8 to 11.6 mg/g

Table 2: Concentrations of Plant Nutrient (mg/g) in Logs

Location	Ca	Mg	K	References
1. Ivory Coast, Banco	2.6	1.1	1.3	Benhard-Reversat (1975)
2. Ivory Coast, Yapo	4.9	0.5	1.0	Benhard-Reversat (1975)
3. Venezuela, San Carlos	0.8	0.2	0.8	Uhl and Jordan (1984)
4. (Terra firme forest)				
5. Venezuela, Western Llanos (Banco forest)	7.4	0.6	3.4	Hase and Fölster (1982)
6. Venezuela, San Carlos (Experimental site)	0.8	0.2	0.7	Jordan (1989)
7. Venezuela, San Carlos (Control site)	1.0	0.3	1.0	Jordan (1989)
8. Malaysia, Sabah	2.4	0.5	1.2	Sim and Nykvist (1991)
9. Papua New Guinea (Floodplain)	11.6	0.8	3.6	Lamb (1990)
10. Papua New Guinea (Hill forest)	11.6	0.7	3.6	Lamb (1990)

Source: After (Nykvist, 1994)

for Ca, 0.2 to 1.1 mg/g for Mg and 0.7 to 3.6 mg/g for K. The corresponding means were 4.79, 0.54 and 1.84 mg/g for Ca, Mg and K, respectively. These were used to estimate nutrient removal during forest harvest. Under the presently practiced Selective Management System (SMS), the projected stocking of a second growth forest when it is ready for another harvest ranges between 30 and 40 m^3/ha (Thang, 1991). Assuming that selective logging at BTEW extracts 40 m^3/ha (24 ton/ha) of timber, the losses of Ca, Mg and K in biomass would be 115, 13 and 44 kg/ha, respectively.

Rates of enhanced leaching due to logging were reported at two sites in study. Here, extracting 33% of the stocking resulted in 1.8-fold increase in Ca, 1.6-fold for Mg and 2.5-fold for K in the first year. In view of differences in the lithology and forest composition, it is certainly crude to apply these values

Table 3:Calculated Atmospheric Inputs (kg/ha/yr) and Concentrations (mg/l, in parentheses) of Ca, Mg and K for Forested Stations in Malaysia

Sites	Ca	Mg	K	Source of Rain chemistry Data
1. Pasoh, Negeri Sembilan	5.1 (0.18)	0.8 (0.03)	8.2 (0.29)	Manokaran (1980)
2. Sipitang, Sabah	1.4 (0.05)	0.3 (0.01)	1.4 (0.05)	Malmer (1993)
3. Danum Valley, Sabah	2.5 (0.09)	0.8 (0.03)	4.5 (0.16)	Burghouts (1993)
4. BTEW (mean values)[#]	8.4 (0.37)	3.3 (0.10)	6.7 (0.30)	This study
5. BTEW[$] (median values)	6.5 (0.23)	0.3 (0.01)	2.8 (0.10)	This study

Notes: # - Inputs were calculated from the monthly mean concentrations.
$ - Calculated from the median concentration

directly. But, it is equally naive to simply ignore the leaching losses. To account for losses in the undissolved fraction which were not quantified in the Berembun study (Zulkifli, 1990), higher increment ratios were applied in the present analysis. Selective logging at BTEW was therefore assumed to cause 2-fold increases in the exports of Ca and Mg and 3-fold for K. These would result in excesses leaching in the first year after logging by 4.1, 4.5 and 16.2 kg/ha for Ca, Mg and K, respectively. Supposing an annual reduction factor of 0.3, the total losses due to enhanced leaching calculated over 10 years would be 12.5 kg/ha for Ca, 12.7 kg/ha for Mg and 47.9 kg/ha for K.

Table 4: Time Span (yr) for the Atmospheric and Weathering Inputs to Compensate Losses of Nutrient in Biomass and Enhanced Leaching

Source of Rain Chemistry Data		Ca	Mg	K
1. Pasoh, Negeri Sembilan	a	25	30	10
	b		20	10
2. Sipitang, Sabah	a	90	90	65
	b		35	50
3. Danum Valley, Sabah	a	50	30	20
	b		20	20
4. BTEW (mean values)	a	15	10	15
	b		10	15
5. BTEW (median values)	a	20	90	35
	b		35	30

Notes: (a) - atmospheric inputs only; (b) - atmospheric and weathering inputs. Losses of Ca, Mg and K due to extraction of 40m³/ha of timber were estimated at 115, 13 and 44 kg/ha, respectively. The corresponding losses in enhanced leaching over 10 years were 12.5, 12.7 and 47.9 kg/ha. Values have been rounded to the nearest 5.

When the atmospheric inputs were computed from the monthly mean concentrations of rainfall solute, it would take only 10 to 15 years to compensate losses of Ca, Mg and K (Table 4, no. 4). Arguably, this might be too short a time for the ecosystem to attain a recovery stage and again points to the possible enrichment of rainfall samples. Further check on the distribution plot of the nutrient concentration data showed that the curves were generally skewed to the right, suggesting that the mean values may not be the best measure of central tendency. Alternatively, the median concentration were used to estimate the long term nutrient inputs. Lower rates of nutrient input were obtained; 0.28, 2.81 and 6.47 kg/ha/yr for Ca, Mg and K, respectively (Table 3). The lower inputs resulted in longer recovery spans (Figure 2). The time spans needed for the atmospheric inputs alone to compensate losses of Ca, Mg and K would be 20, 90 and 33 years, respectively. Much faster recovery was obtained for Mg when the weathering inputs were considered, to 33 years whereas span for K reduced to 28 years.

In view of uncertainties with regard to the rates of nutrient inputs at BTEW, the recovery spans were also estimated using inputs calculated from rainfall chemistry data at Pasoh in Negeri Sembilan (Manokaran, 1980), and two sites in Sabah; Sipitang (Malmer and Grip, 1994), and Danum (Burghouts, 1993). By putting aside the inputs obtained from the mean concentrations at BTEW (Table 4, no. 4) differences in the recovery spans for Ca were 25 to 90 years; Mg - 20 to 35 years and K - 10 to 50 years. While the range is reasonably small for Mg, large uncertainties still exist for Ca and K. This accentuates the need to have region specific data based on site fertility class and factors controlling the source and transport of rainfall solutes.

5. CONCLUSIONS

Rates of catchment nutrient recovery following logging must be assessed in relation to cutting rotation. The Selective Management System (SMS) currently practiced in Peninsular Malaysia prescribes a cutting rotation of 30 years for the richer forests and 55 years for the poorer ones. On this basis, the present analysis generally suggests that catchment nutrient recovery might be attained if the area is to be logged over 55 years rotation. However, if the shorter cutting cycle of 30 years is to be adopted for such forests, site fertility especially the Ca level would be depleted and may eventually affects forest productivity.

The above analysis, however must be regarded as a preliminary in view of uncertainties, especially in determining reasonable estimates of long term nutrient inputs. It is suspected that the atmospheric nutrient inputs at BTEW were overestimated due to enrichment of rainfall samples by dust and ash from land use and burning activities in the nearby sites. Unfortunately, the computation of rainfall inputs, combining rainfall chemistry from other forested sites in Malaysia still resulted in considerably large differences in the recovery spans especially for Ca. This underscores the need to establish a monitoring station remote enough

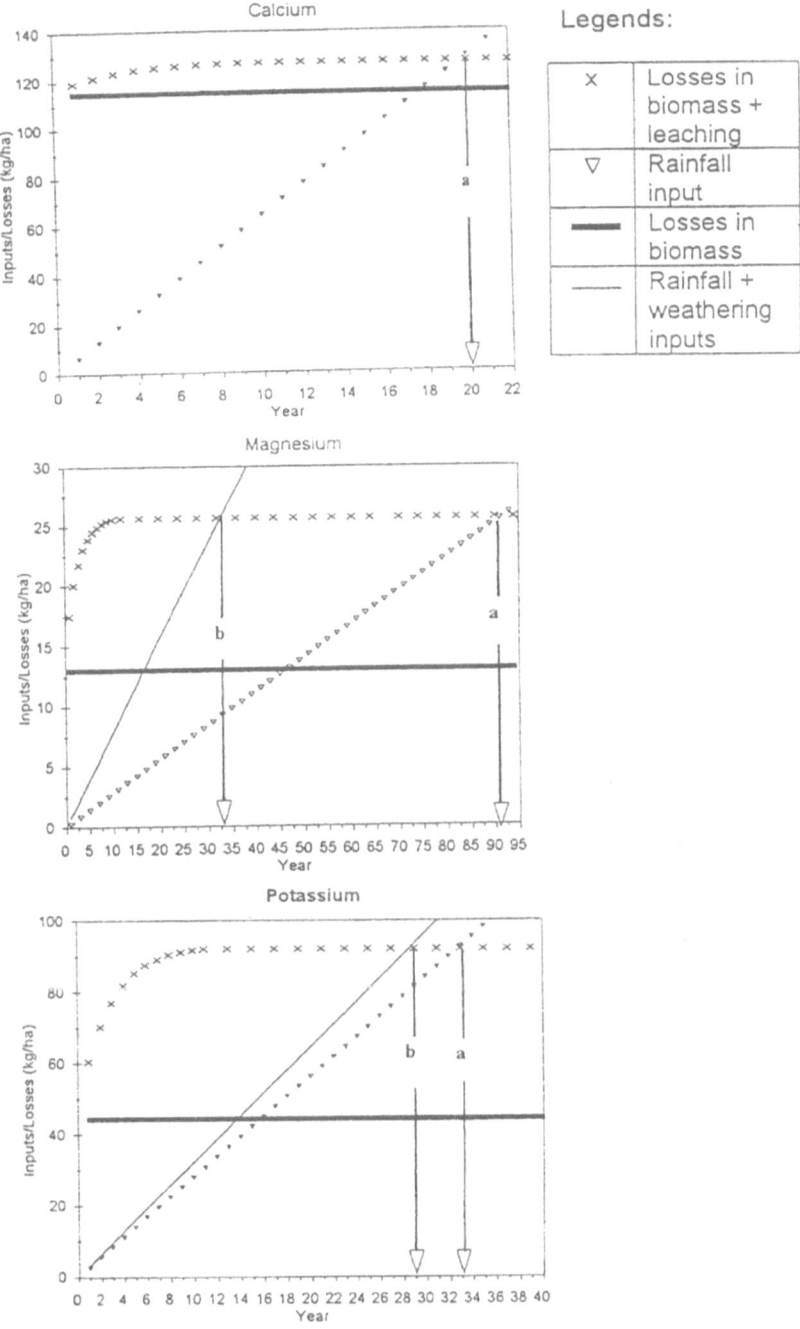

Figure 2: Losses of Ca, Mg and K in biomass and enhanced leaching and time (yr) required by "a" rainfall and "b" rainfall and weathering inputs to replenish the losses

uncertainties, especially in determining reasonable estimates of long term nutrient inputs. It is suspected that the atmospheric nutrient inputs at BTEW were overestimated due to enrichment of rainfall samples by dust and ash from land use and burning activities in the nearby sites. Unfortunately, the computation of rainfall inputs, combining rainfall chemistry from other forested sites in Malaysia still resulted in considerably large differences in the recovery spans especially for Ca. This underscores the need to establish a monitoring station remote enough from human influences for modeling catchment recovery. In addition, the total nutrient reserve in biomass and soil need to be incorporated in the analysis. Preferably, the data set must be region specific selected on the basis of soil fertility class and factors controlling the source and transport of rainfall solutes.

REFERENCES

Bernhard-Reversat, F.(1975): Recherches sur L'ecosysteme de la Foret Subequatoriale de Base Cote-d'lvoire. VI. Les Cycles des Macroelements, La Terre et la Vie, Vol. 29, pp. 229-254.

Bruijnzeel, L.A. (1991): Nutrient Input-Output Budgets of Tropical Forest Ecosystem : a Review, Journal of Tropical Ecology, Vol. 5, pp. 1-24.

Burghouts, Th.B.A. (1993): Spatial Variability of Nutrient Cycling in Bornean Rain Forest, Ph.D. Thesis, Vrije Universiteit, Amsterdam, 156p.

Clayton, J.L. (1979): Nutrient Supply to Soil by Rock Weathering, In: A.L. Leaf (ed.) Proceeding of the Symposium on Impact of Intensive Harvesting on Forest Nutrient Cycling, New York State University, New York, pp. 75-96.

Creasey, J., A.C. Edwards, J.M. Reid, D.A. Macleod and M.S. Cresser (1986): The Use of Catchment Studies for Assessing Chemical Weathering Rates in Two Contrasting Upland Areas in northeast Scotland, In: S.M. Colman & D.P. Dethier (eds.) Rate of Chemical Weathering of Rocks and Minerals, Academic Press, Orlando, pp. 467-502.

Grip, H., A. Malmer and F.K. Wong (1994): Converting Tropical Rainforest to Forest Plantation in Sabah, Malaysia, I. Dynamic and Net Losses of Nutrients in Control Catchment Streams, Hydrological Processes, Vol. 8, pp. 179-194

Hase, H and P.J. Fölster (1982): Bioelement Iventory of a Tropical (Semi-) Evergreen Seasonal Forest on Eutrophic Alluvial Soils, Western Llanos, Venezuela, Oecologica Plantarum, Vol. 3, pp. 331-346.

Jordan, C.F. (1989): An Amazonian Rain Forest. The Structure and Function of a Nutrient Stressed Ecosystem and the Impacts of Slash and Burn Agriculture, Man and Biosphere Series No. 2, UNESCO, Paris.

Kenworthy, J.B. (1971): Water and Nutrient Cycling in a Tropical Rain Forest, In: J.R. Flenly (ed.) Proceeding of the First Aberdeen-Hull Symposium on Malesian Ecology, University of Hull, Miscellaneous Series No. 11, pp. 49-65.

Lamb, D. (1990): Exploiting the Tropical Rain Forest: An Account of Pulpwood Logging in Papua New Guinea. Man and Biosphere Series 3, UNESCO,

Agricultural Science, Umeå, Sweden, 181p.

Malmer, A. and H. Grip (1994): Converting Tropical Rainforest in Sabah, Malaysia, II. Effects on Nutrient Dynamics and Net Losses in Streamwater, Hydrological Processes, Vol. 8, pp. 195-209.

Manokaran, N. (1980): The Nutrient Contents of Precipitation, Throughfall and Stemflow in a Lowland Tropical Rainforest in Peninsular Malaysia, The Malaysian Forester, Vol. 43, pp. 266-289.

Nykvist, N. (1994): Removal of Plant Nutrient in Logs when Clear-Felling Tropical Rainforests – A Literature Review, In: Wan Razali W.M., Shamsuddin I., S. Appanah and M. Farid A. R. (eds.) Proceeding of the Symposium on Harvesting and Silviculture for Sustainable Forestry in the Tropics, Forest Research Institute Malaysia, 5-9 Oct 1992. pp. 192-196.

Roe F.W. (1951): The Geology and Mineral Resources of the Fraser's Hill Area, Selangor, Perak and Pahang, Federation of Malaya with an Account of Mineral Resources, Caxton Press ltd, Kuala Lumpur.

Sim, B.L. and N. Nykvist (1991): Impact of Forest Harvesting and Replanting, Journal of Tropical Forest Science, Vol. 3, pp.251-284.

Thang, H.C. (1991): Forest Management Practices in Malaysia, Paper Presented at ASEAN Seminar on Land Use Decisions and Policies: Will Tropical Forests Survive their Impacts??, ASEAN Institute of Forest Management, Penang, 28-30 October.

Uhl, C. and C.F. Jordan (1984): Succession and Nutrient Dynamics following Forest Cutting and Burning in Amazonia, Ecology, Vol. 65, pp. 1476-1490.

Waterloo, M.J. (1994): Water and Nutrient Dynamics of *Pinus Caribaea* Plantation Forest on Former Grassland Soils in southwest Viti Levu, Fiji. Ph.D. Thesis, Vrije Universiteit, Amsterdam, 478p.

Wyatt-Smith J. (1963): Manual of Malaysian Silviculture for Inland Forest, Malayan Forest Record. No. 23.

Zulkifli Y. (1990): Effects of Logging on Streamwater Quality and Input-Output Budgets in Small Watersheds in Peninsular Malaysia, M.Sc. Thesis, Universiti Pertanian Malaysia

Zulkifli Y. (1996): Nutrient Cycling in Secondary Rain Forest Catchments of Peninsular Malaysia, PhD. Thesis, University of Manchester.

Zulkifli Y. and Baharuddin K. (1997): Nutrient Fluxes via Litterfall and Canopy Leaching in a Secondary Tropical Rainforest Site, Paper Presented at 2nd International Conference on Environmental Chemistry and Geochemistry in the Tropics, Kuala Lumpur, 7-11 April.

Hydrological Variations of Discharge, Soil Loss and Recession Coefficient in Three Small Forested Catchments

Kyongha KIM and Yongho JEONG

Forestry Research Institute, Seoul, Republic of Korea

ABSTRACT   This study aims to understand hydrological variation of discharge, soil loss and recession coefficient in three small forested catchments using the long-term hydro-data from 1983 to 1992. The study catchments include the natural matured deciduous, artificial planted coniferous and erosion-control worked mixed forest. The amount of discharge and soil loss varied with the rainfall and forest type. Especially it was concerned on the variation of the recession coefficient closely related to storage   capacity in three forested watersheds to clarify the effects of forest development on water resource augmentation. In erosion-control worked mixed forest, the recession coefficients for interflow and groundwater have gradually reduced since erosion control work finished. This can be interpreted to result from the increment of the storage capacity caused by improvement of the soil physical properties in catchment after erosion control work. In artificial planted coniferous forest, the recession coefficients related to surface runoff and interflow slowly decreased as the forest grew. However all recession coefficients in natural matured deciduous forest did not show any evident trends because of not change of the forest stand structure.

Key Words : Recession coefficient, Lapsed variation, Forest type.

1. INTRODUCTION

In Korea, the amount of rainfall shows high seasonal and annual variation. Most of rainfall precipitates intensely in July to August as typical monsoon type. The discharge in the wet season runs fast off to ocean. On the other hands, water resource usually is short of supply in the dry season. Also the mountainous area is very steep and soil depth is shallow. Disadvantageous situation for water resource management results in low percentage of water utility. The proportion of the water usage is no more than 24 percent of total amount of water resources, 126.7 billion tons year^{-1}.

More than 65 percent of the land in Korea is forest covered, and forested drainage basins serve as water supplies. Generally it is well known that forest can modulate flows, moderate hydrologic extreme such as floods and droughts. Many devastated areas in Korea have been rehabilitated for soil and water conservation

431

since 1970's. Some coniferous species such as *Pinus koraiensis, Larix leptolepis* were planted to establish artificial forest. Nowadays, forest in the head water catchment needs to be managed positively for a storm flow control and augmentation of low flow. The change of a storage volume in a forested catchment after reforestation and erosion control work have to be clarified for the water conservation technique through the forest management.

In this study, the recession curve analyses uses to find how forest change have influence on the storage capacity of a forested catchment. The recession curve tells in a general way about the natural storage feeding the stream. Accordingly, it contains valuable information concerning storage properties and aquifer characteristics(Tallaksen, 1995). If the recession curve is plotted on semi-logarithmic paper, the result is usually not a straight line but a curve with gradually decreasing slope(Linsley et al., 1982). By plotting the logarithm values, the three types of the recession coefficient : surface runoff, interflow, and groundwater can be determined by finding an inflection point. This work aims to understand the influence of forest change on the variation of the recession coefficient with the lapse of time.

2. MATERIAL AND METHOD

2.1 Study area

The experimental catchments locate at Kyonggi-do near Seoul metropolitan. Table 1 shows the condition of the topography and vegetation in the three experimental sites.

Table 1. The conditions of topography and vegetation in experimental catchments.

Forest type	Area (ha)	Elevation (m)	Parent material	Tree Height (m)	DBH (cm)	Remark
Mixed	5.2	130~210	Granite	6.0	11	Recovered in '74
Coniferous	13.6	160~290	Gneiss	4.0	12	Planted in '76
Deciduous	22.0	280~470	Gneiss	16.0	30	Natural forest

The natural deciduous forest is mature about 60-year old and covered predominantly with *Quercus serrata* and *Carpinus laxiflora*. The coniferous forest which consists of *Pinus koraiensis* and *Abies holophylla* was planted at a stocking rate of 3,000 stems ha^{-1} in 1976. The mixed forest has been devastated since an erosion control work was established in 1974. As shown in Fig. 1(a)-(c) deciduous and coniferous forest catchments shape laterally wide while the mixed forest is longitudinally long.

(a)   (b)

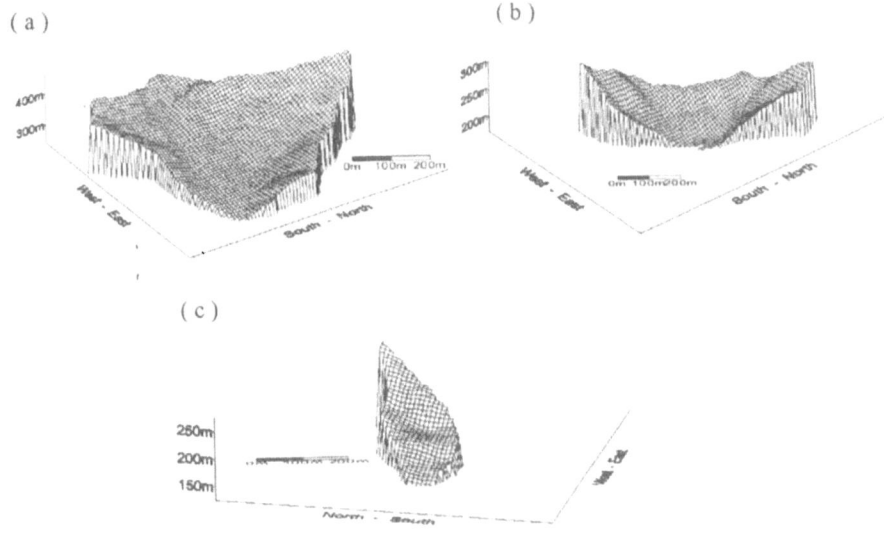

(c)

Fig. 1. Topographic views of deciduous(a), coniferous(b) and mixed(c) forest catchment.

2.2 Instrumentation

Rainfall was recorded continuously by tipping bucket recorder at every 0.5 mm tip^{-1} on the chart and snowfall also is measured as an equivalent depth. Water level has recorded by long-term recorder since 1979 at 90° sharp crested V-notch weir.

2.3 Data analysis

Of the rainfall over 50 mm, hyeto-hydrographs were plotted in order to select an individual event showed clear recession curve during the period of 1983 to 1992. The recession curves selected are plotted on a semi-logarithmic paper to analyse 3 components of the hydrograph on the basis of the inflection point. The recession curve has traditionally been separated into the linear components of surface runoff, interflow and groundwater(Barnes, 1939).

$$Q_t = Q_0 e^{-\alpha t} \qquad\qquad (1)$$

where Q_t is the discharge t time units after Q_0, e is the napierian base, and α is recession coefficient.

Here, α_1 and α_2 represented the slope of first and second inflection point on the recession curve will be defined as the recession coefficient related to surface runoff and interflow. α_3, the slope after second inflection point is the recession coefficient related to groundwater. The components represent different flow paths in the catchments, each characterized by different residence time, storage volumes and drainage functions.

3. RESULTS AND DISCUSSION

3.1 Variability of the discharge and soil loss

Annual discharge in three catchments responded in accordance with the variation of rainfall (Fig. 2). The amount of discharge varied depending on the forest type. The highest annual discharge was recorded in the mixed forest (71.5 %) while the lowest was in the coniferous forest (51.0 %), with deciduous forest (65.6 %) between the two forests. Main factors caused to show difference of discharge are thought to an interception loss and transpiration. Annual interception loss of the mixed forest is 18% and 29%, 32% in deciduous and coniferous forest respectively.

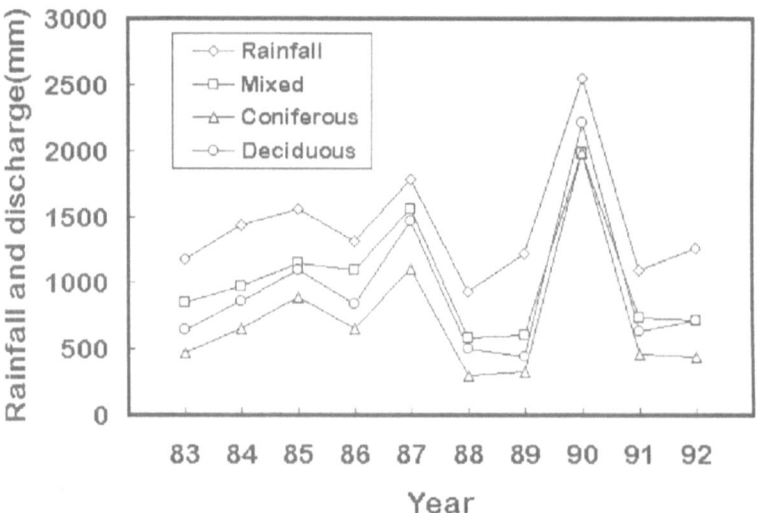

Fig. 2. Variations of annual rainfall and discharge in the study catchments for 10 years

The amount of monthly rainfall shows big difference in Fig. 3. Almost rainfall bursts in July to August. In the mixed forest, monthly discharge showed large discrepancy between the dry and rainy season. The forest stand structure of the mixed forest is still so poor that the amount of interception loss and transpiration is less than others. Also the soil physical properties are worse than others. On the other hands, the variation of monthly runoff in the deciduous forest was less than that in the mixed forest. The coniferous forest resulted in least runoff ratio of three forest types for the whole year. The amount of soil loss from the catchment

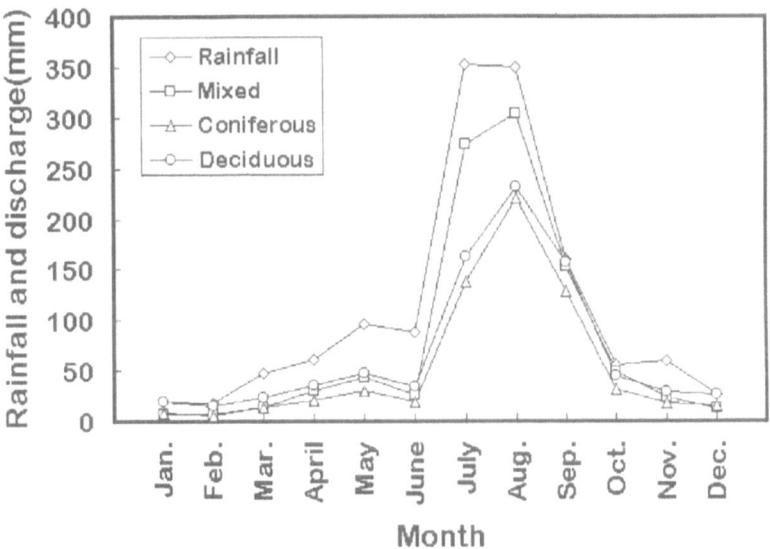

Fig. 3. Variations of monthly average rainfall and discharge in the study catchments for 10 years.

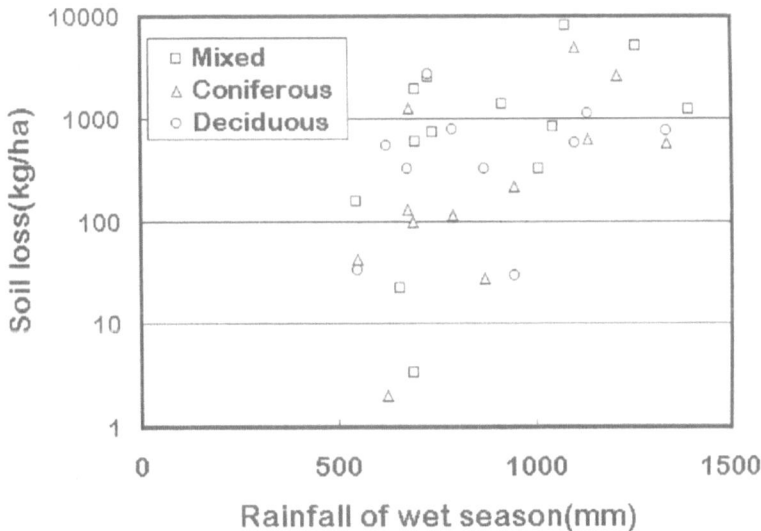

Fig. 4. Relationship between rainfall of wet season and soil loss in the study catchments.

tended to increase in proportion to the amount of rainfall (Fig. 4). Variation of soil losses showed wide range because of influences of the amount of soil loss in just previous year. The threshold annual rainfall occurred soil loss estimated about 500 mm. The mixed forest is more vulnerable to soil loss compared to others.

436

3.2 Lapsed variation of the recession coefficient

Physically based variation in the recession rate is cased by difference in climate during the time of recession, but is also determined by the conditions prevailing prior to the start of the recession (Tallaksen, 1995). Several workers have recognized a seasonal variation in the recession behavior which follows the seasonal change in evapotranspiration (Ando et al., 1986; Ambroise, 1988; Brandesten, 1988). All kinds of the recession coefficients in this study showed very large the coefficient of variability ranged from 20 % to 50 % during the study period.

Fig. 5 shows the variation of the recession coefficient of surface runoff(α_1) for 10 years. α_1 tended to gradually decrease in the coniferous forest whilst didn't show any tendency in others. That may be caused by the change of forest structure in the coniferous forest after planting. The amount of the initial loss by a interception and transpiration has increased greatly since 1976 as trees grow. Whilst the forest structure in others has rarely changed since 1983.

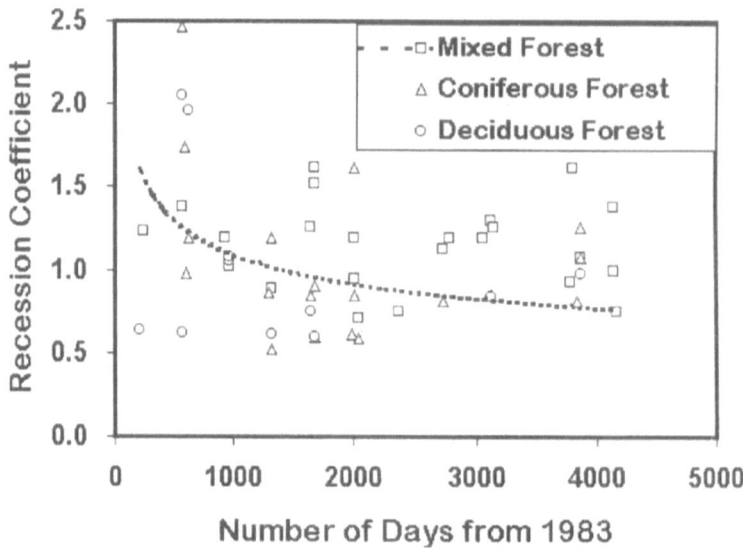

Fig. 5. Variation of the recession coefficient of surface runoff in the study catchments for 10 years.

The recession coefficient of interflow(α_2) reduced in the coniferous and mixed forests as time passed (Fig. 6). This can be interpreted by an increase of soil storage capacity after the planting and erosion control work. As the amount of evapotranspiration augments, the storage opportunity of rainfall in the soil increases. Increment of the storage capacity results in delaying the releasing time of interflow from the soil.

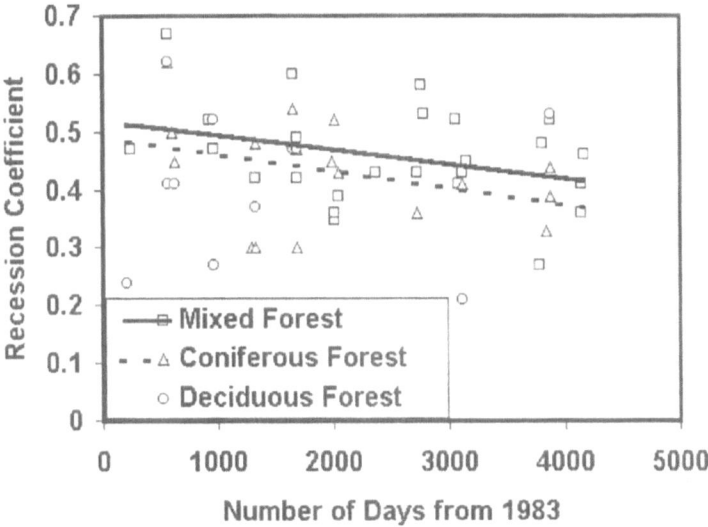

Fig. 6. Variation of the recession coefficient of interflow in the study catchments for 10 years.

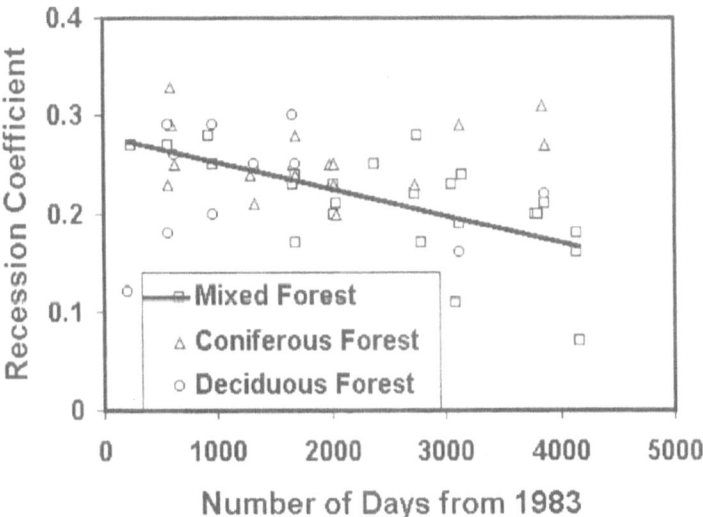

Fig. 7. Variation of the recession coefficient of groundwater in the study catchments for 10 years.

In case of the recession coefficient of groundwater(α_3), only mixed forest showed to reduce gradually during 10 years (Fig. 7). The mixed forest has been devastated land until erosion control work finished in 1974. After work, the soil layer formed rapidly and the soil physical properties improved.

438

4. CONCLUSION

The variation of hydrological properties included the discharge, soil loss and recession coefficient showed very wide range depending on the amount of rainfall during the study period. The ratio of monthly discharge fluctuated more in the mixed forest recovered by erosion control work than in the matured deciduous forest. The lapsed variations of the recession coefficient were different depending on the forest conditions. In case of the coniferous forest, a reforestation resulted in the reduction of the recession coefficient for surface runoff and interflow. In the mixed forest recovered by erosion control work, the recession coefficients for interflow and groundwater tended to decrease with the lapse of time.

REFERENCES

Ando.Y., Y. Takahasi, T. Ito and K. Ito. 1986. Regionalization of parameters by basin geology for use in a groundwater runoff recession equation. IAHS Publ., 156: 151-159.

Barnes, B.S. 1939. The structure of discharge-recession curves. Trans. Am. Geophys. Union. 20:721-725.

Brandesten, C.O. 1988. Seasonal variation in streamflow recession in the mire complex Komisse, southern central Sweden. In: Hydrol. Of Wetlands and Man's Influence in It, Proc. Int. Symp., June 1988. Publ. Academy of Finland, Helsinki, pp. 84-91.

Linsley, R.K., M.A. Kohler and J. L. Paulhus. 1982. Hydrology for engineers. McGraw-Hill. pp. 206-207.

Tallaksen, L.M. 1995. A review of baseflow recession analysis. J. Hydrol. 165:349-370.

Evaluating the Effectiveness of Forest Crop to Mitigate Erosion Using a Sediment Delivery Distributed Model

Vito FERRO[*1], Paolo PORTO[*1], Giovanni CALLEGARI[*2]
Francesco IOVINO[*2], Vittoria MENDICINO[*2] and
Antonella VELTRI[*2]

*1 Dipartimento di Ingegneria e Tecnologie Agro-Forestali, Sezione Idraulica,
   Facoltà di Agraria, Università di Palermo, Viale delle Scienze, 90128
   Palermo, Italy
*2 Istituto di Ecologia e Idrologia Forestale, National Research Council (C. N. R.),
   Via Cavour, 87030 Roges di Rende(CS), Italy

ABSTRACT In this paper sediment yield data, measured from 1978 to 1997 in a small experimental Calabrian basin reafforested with Eucalyptus trees (*Eucalyptus occidentalis* Engl.), and RUSLE (Revised Universal Soil Loss Equation) coupled with a sediment delivery distributed model are used to evaluate the antierosive effects of this forest cover. At first, the soil loss measurements carried out in two experimental plots, located in the basin, are used to evaluate the crop and management factor C of RUSLE for Eucalyptus coppice. The reliability of the selected C factor value is verified by comparing, at an event scale, the measured and the calculated sediment yield values at the basin outlet. Then, a Monte Carlo technique is used for evaluating the effects of the *knowledge uncertainty* and the *stochastic variability* of the model parameters on calculated sediment yield. Finally, at basin scale, the spatial variability of the crop cover is considered in order to estimate the sediment yield corresponding to scenarios having different percentages of the crop cover.
Keywords: soil erosion, sediment delivery, crop factor, forest antierosive effectiveness

1. INTRODUCTION

Accelerated soil erosion is a serious problem to consider for the development of a sustainable agriculture. Long-term erosion monitoring allows to observe different scenarios due to changes in land use. At this aim, the Food and Agriculture Organization of the United States sponsored a project to assess the effects of deforestation on the water regime in the region of the Kafue River Basin in Zambia (Mumeka 1986). In particular, hydrological observations were carried out in two small basins under both their natural conditions (woodland) and agricultural use with accompanying deforestation. For the two basins, after the deforestation treatment a significant increase of the runoff volume, due to the change in land use, was recognized. The absence of a dense forest cover also determined a decrease of the concentration time and peak discharges nearly doubled. In other words, forest cover affecting soil hydraulic properties (water retention and hydraulic conductivity) and modifying the hydrological hillslope processes is able to reduce erosion and to favour sediment delivery processes (Oyarzum and Peña 1995).

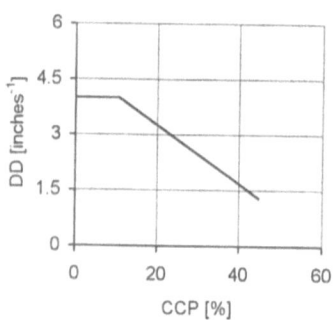

Fig. 1 Relationship between drainage density and crop cover percentage

The effectiveness of the forest cover as a control for erosion is due to the direct effect in the reduction of soil erodibility and rainfall detachment. In fact, under forest cover, topsoil accumulates a high organic matter content which enhances the stability of soil structure while crop residue and biological activity of animal organisms improve the infiltration capacity. The forest cover intercepts rainfall and collects water on its foliage reducing the kinetic energy of the rainfall. Some of the intercepted rainfall reaches the soil surface as stemflow while the drops falling from the canopy have an energy less than that of rainfall impacting on the surface of a no-covered area. The brushwood and litter provide a further soil protection able to dissipate residue kinetic energy. Physically based models for erosion processes, like EUROSEM (Morgan 1994) and LISEM (De Roo et al. 1994), deal in turn with interception during rainfall, the rainfall reaching the ground surface as direct throughfall and

leaf drainage and the volume of stemflow. In USLE (Universal Soil Loss Equation) derived models the influence of forest cover on soil erosion processes is represented by the crop management factor C which is estimated using information on ground cover, canopy, plant height and other factors such as the presence of a litter (Dissmeyer and Foster 1981). Recently the Agricultural Research Service of the United States Department of Agriculture and the Soil Conservation Service are developing a physically-based technology named WEPP (Nearing

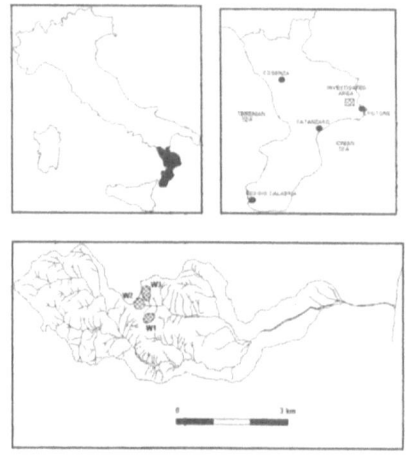

Fig. 2 Experimental Calabrian basins

et al. 1989) which distinguishes hydrological and erosion phenomena happening on interrill areas and on rills. The model simulates the processes of detachment, transport and deposition of soil particles on interrill areas by a single equation in which a ground cover adjustment factor and a canopy cover adjustment factor are introduced.

The vegetation cover also affects both the rill pattern developing on the hillslopes and depth and velocity of concentrated flows. The experimental runs of Rogers (1989), carried out on laboratory plots having slopes equal to 10 and 20% and crop cover percentages ranging from 0 to 45%, showed that the drainage density DD of the rill pattern decreases with increasing crop cover percentage CCP. In other words, for increasing crop cover percentage values, the sediment transport efficiency of the rill pattern decreases because of increasing the probability that the eroded particles stop before arriving at the nearest rill (Ferro 1997). Fig. 1, obtained from reanalysing the experimental data of Rogers (1989), clearly shows that a *minimum* CCP value (equal to 10%) limiting the effectiveness of vegetation to reduce erosion exists. For studying the relevant soil erosion phenomena in the mountainous Calabrian region (South Italy) (Aronica and Ferro 1997), in 1978 the "Soil Conservation Project" of the Italian National Research Council (CNR) equipped three small basins (W1, W2 and W3 in Fig. 2) to monitor the effects of afforestation on hydrological response and sediment yield (Cinnirella et al. 1998). In this paper, the experimental data measured from 1978 to 1997 in W2 basin, reafforested with Eucalyptus trees, and soil loss measurements carried out in two experimental plots located in the basin will be used. The plot measurements will allow to evaluate the crop and management factor C. Then the available sediment yield measurements at the basin outlet and the Revised Universal Soil Loss Equation (RUSLE) coupled with the sediment delivery distributed model will be used to evaluate the effectiveness of Eucalyptus coppice on erosion mitigation. Finally, Monte Carlo simulations will be carried out to evaluate the effects of the *knowledge uncertainty* and the *stochastic variability* of the model parameters on calculated sediment yield.

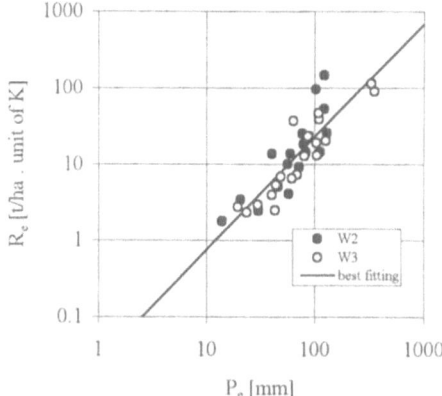

Fig. 3 Relationship between rainfall erosivity factor and rainfall depth at event scale

2. EXPERIMENTAL BASIN AND MEASUREMENT TECHNIQUE

The studied area is located in the ephemeral basin of Crepacuore stream which drains to the Ionian sea (Fig. 2). The basin W2 was planted in 1968 with

Eucalyptus occidentalis Engl. and was coppiced twice in 1978 and 1990 (coppicing cycle equal to 12 years). The forest cover is discontinuous and 20% of the basin area is bare.

The basin outlet is monitored by a H-flume weir and measurement of flow depth is carried out at the end of a rectangular channel by a mechanical recording water level gauge. The sampling device is constituted by a Coshocton Wheel collecting a sample (~1/200) of the flow volume. Each collected water sample is led into appropriately sized tanks. At the end of each event the collected suspension is well mixed and suspension samples at different height, of given volume (one litre), are drawn out. The suspended solid content [g] of each sample is processed by oven-drying at 105 °C. The ratio between the mean value of the suspended solid content [g] and the sample volume (one litre) is assumed as mean suspension concentration $C_{s,m}$ [g/l]. The sediment yield of each event is calculated by the product of the mean concentration $C_{s,m}$ and the measured runoff volume.

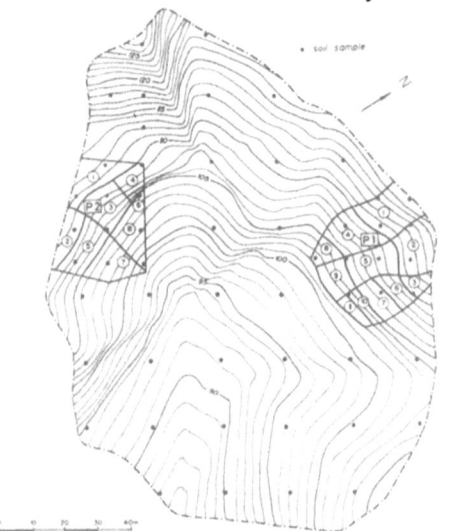

Fig. 4 W2 experimental basin divided into morphological units and soil sampling sites

Rainfall is measured by a recording raingauge and a rainfall erosivity factor R_e (Wischmeier and Smith 1965) is calculated for each rainfall event e. Knowledge of the kinetic energy of the rainfall event and the maximum 30 min rainfall intensity within the event are necessary to calculate R_e. Simplified methods for evaluating R_e by readily obtainable climatic variables, like the total rainfall depth P_e of the event, have been already developed (Cooley 1980). The following power relationship (Bagarello and D'Asaro 1994) is often used to estimate R_e if only a non-recording raingauge is available in a studied area:

$$R_e = a P_e^n \qquad (1)$$

in which a and n are empirical constants. Fig. 3 shows a good agreement between eq. (1) with a = 0.0249 and n = 1.4789 and the pairs (P_e, R_e) measured in two recording raingauges installed in W2 and W3 basins (Fig. 2). In the following analysis, eq. (1) allows to use the rainfall events for which only the total rainfall depth P_e is available. The soil erodibility factor K is calculated by using data of 55 soil samples uniformly distributed over the basin area (Fig. 4). Fig. 5 shows that K factor is lognormally (LN2) distributed with mean and standard deviation equal to 0.1872 and 0.1035, respectively. Two large plots, P1 and P2, (Fig. 4) were equipped within W2 basin to carry out sediment yield measurements.

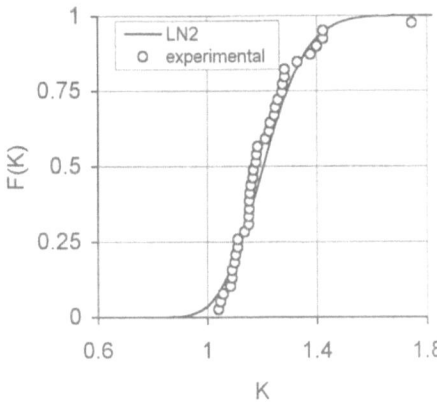

Fig. 5 Comparison between empirical cumulative frequency distribution function of soil erodibility factor and lognormal law

Each plot, having an area of 1000 m², is hydraulically isolated and is equipped by tanks with suitable sizes to store the runoff and the sediment yield of each event.

P1 plot has a mean slope of 48% and is covered with a *grass cover* which is classified, according to Wischmeier and Smith (1978), as a G type with no appreciable canopy and with a ground cover GC ranging from 0 to 10%. P2 plot has a mean slope of 35% and is covered with a *forest cover* of *Eucalyptus occidentalis* Engl.. The analysis is developed for 39 events measured at P1 plot, 23 events measured at P2 plot and 55 events measured at W2 basin outlet.

3. THE SEDIMENT DELIVERY DISTRIBUTED (SEDD) MODEL

Predicting the sediment yield at basin scale, i.e. the quantity of sediment which is transferred, in a given time interval, from eroding source through the *hillslopes* and the *channel network* to the basin outlet, can be carried out coupling a soil erosion model with a mathematical operator expressing the sediment transport efficiency of the hillslopes and the channel network (Walling 1983).

Sediment transport and deposition in the basin hillslopes is a physical process distinct from transport and deposition within the channel network. Therefore hillslope sediment delivery processes and channel ones have to be modelled separately. At a *mean annual temporal* scale the sediment delivery channel component is neglected because Playfair's law of stream morphology (Boyce 1975) establishes that << *over a long time a stream must essentially transport all sediment delivered to it* >>. For a small basin, having an ephemeral channel network and with no well-developed floodplains, the channel sediment delivery component can be neglected even at an *event* scale.

Richards (1993) suggested that sediments are produced from different sources distributed throughout the basin and that sediment delivery processes have to be modelled employing a spatially distributed approach which consideres individual fields. To apply a spatially distributed strategy at the basin scale requires the choice of both a soil erosion model and a spatial disaggregation criterion for the sediment delivery processes.

The existing difficulties of physically based modelling (numerous input parameters, differences between the scale of measurement of the input parameters and the scale of basin discretization, uncertainties of the selected model equations, etc.) increased the attractiveness of a parametric soil erosion model, like RUSLE (Renard et al. 1994).

For modelling the spatial disaggregation of the sediment delivery, the basin has to be discretized into *morphological units* (Bagarello et al. 1993) i.e. areas of clearly defined aspect, length and steepness (Fig. 6).

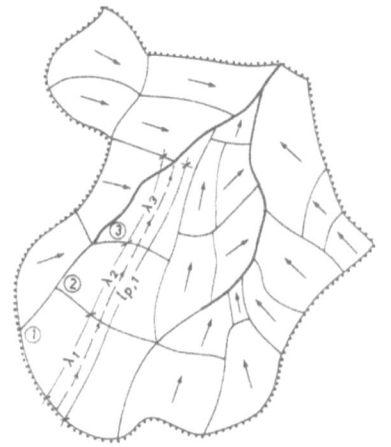

For a basin discretized into morphological units and neglecting the channel sediment delivery component, Ferro and Minacapilli (1995) suggested to take into account the within-basin variability of the sediment delivery processes by calculating the sediment delivery ratio SDR_i of each morphological unit, i, into which the basin is divided. According to the authors, SDR_i, measuring the probability that the eroded particles arrive from the considered i morphological unit to the nearest stream reach, has the following expression:

Fig. 6 Scheme for calculating the travel time of each morphological unit

$$SDR_i = \exp\left(-\beta\, t_{p,i}\right) = \exp\left(-\beta\, \frac{l_{p,i}}{\sqrt{s_{p,i}}}\right) = \exp\left[-\beta\left(\sum_{j=1}^{N_p} \frac{\lambda_{i,j}}{\sqrt{s_{i,j}}}\right)\right] \qquad (2)$$

in which $t_{p,i}$ is the travel time of each morphological unit, $l_{p,i}$ and $s_{p,i}$ are the length and the slope of the hydraulic path from the i-th morphological unit to the nearest stream reach, β is a coefficient which assumes, for a given basin, a different value β_e for each event, N_p is the number of morphological units localized along the hydraulic path, $\lambda_{i,j}$ and $s_{i,j}$ are length and slope of each morphological unit i localized along the hydraulic path j (Fig. 6).

The sediment production Y_i [t], of each morphological unit i into which the basin is divided, is simply calculated by the following equation:

$$Y_i = SDR_i\, A_i\, S_{u,i} \qquad (3)$$

in which A_i is the soil loss [t ha^{-1}] from the i-th morphological unit, which has to be estimated by the selected erosion model, and $S_{u,i}$ is the area [ha] of the morphological unit. A_i is estimated by the following equation:

$$A_i = R_e\, K_i\, L_i S_i\, C_i\, P_i \qquad (4)$$

in which K_i is the soil erodibility factor [t h/kg m^2], C_i is the cover and management factor, P_i is the support practice factor and $L_i S_i$ is the topographic factor calculated by RUSLE (Renard et al. 1994) (McCool et al. 1989) and Moore and Burch (1986) model. For RUSLE model L_iS_i is calculated by the following relationships:

$$L_iS_i = \left(\frac{\lambda_i}{22.13}\right)^{m_i} (10.8\sin\alpha_i + 0.03) \qquad\qquad \text{if } \tan\alpha_i < 0.09 \qquad (5a)$$

$$L_iS_i = \left(\frac{\lambda_i}{22.13}\right)^{m_i} (16.8\sin\alpha_i - 0.5) \qquad\qquad \text{if } \tan\alpha_i \geq 0.09 \qquad (5b)$$

in which λ_i is the slope length of the i-th morphological area, α_i is the slope angle,

$m_i = f_i/(1 + f_i)$ and $f_i = \sin\alpha_i / 0.0896\left(3\sin^{0.8}\alpha_i + 0.56\right)$.

For Moore and Burch model the following equation is used:

$$L_iS_i = \left(\frac{A_{s,i}}{22.1}\right)^{0.6} \left(\frac{\sin\alpha_i}{0.0896}\right)^{1.3} \qquad\qquad\qquad\qquad (6)$$

in which $A_{s,i}$ is the ratio between the area of the i-th morphological unit and the width measured along the contour line.

4. EVALUATING FOREST CROP ANTIEROSIVE EFFECTIVENESS

For applying the sediment delivery distributed model, the value C_i of the crop factor of each morphological unit and the value of β coefficient for each event have to be fixed. Since the two experimental plots P1 and P2 are large and are morphologically complex, the sediment delivery processes for each plot have to be modelled. Table 1 lists the total rainfall depth P_e [mm], the runoff D_e [mm], the rainfall erosivity factor R_e [t m/ha h] and the sediment yield Y_e [t], for each event e measured at P1 and P2 plots.

For P1 plot, the crop factor C_b is assumed variable with the ground cover percentage GC [%] (Table 2) according to the suggestions of Wischmeier and Smith (1978). The P1 plot is divided into 11 morphological units and for a fixed C_b value, corresponding to a given GC value, and for each event e, the β_e coefficient is estimated by the following equation:

$$Y_e = R_e C_b \sum_{i=1}^{N_u} K_i L_i S_i \exp\left(-\beta_e t_{p,i}\right) S_{u,i} \qquad\qquad\qquad (7)$$

in which N_u (equal to 11) is the number of morphological units into which P1 plot is divided. The median value β_m of the 39 β_e values is calculated for a fixed C_b value and for a selected expression of the topographic factor (eqs. (5) and eq. (6)) and is listed in Table 3.

The P2 plot is divided into 9 morphological units and for a fixed β_m value the median value C_m of the 23 crop factor values is calculated, corresponding to a GC value of the bare plot and a selected expression of the topographic factor, and is listed in Table 4.

Table 1 Experimental data measured at P1 and P2 plots

	P1					P1			
EVENT	P_e [mm]	D_e [mm]	R_e [t/ha per K unit]	Y_e [t]	EVENT	P_e [mm]	D_e [mm]	R_e [t/ha per K unit]	Y_e [t]
09/22/92	47.0	9.5	7.4	0.063	10/03/96		161.1	78.2	18.688
10/2-4/92	45.8	4.7	7.1	0.289	10/08/96	103.0	69.2	94.1	5.986
12/25/92-1/5/93	353.8	95.2	146.4	0.294	10/14/96	123.2	99.6	145.3	1.992
02/19/93	29.6	1.5	3.7	0.031	12/10/96	20.2	0.5	1.7	0.006
02/23/93	30.4	1.5	3.9	0.018	12/17/96	33.8	1.1	5.2	0.014
03/01/93	29.4	1.5	3.7	0.121	01/11/97	56.6	2.5	9.9	0.034
03/07/93	5.6	0.6	0.3	0.013					
3/25-27/93	46.0	1.5	7.2	0.035			P2		
05/07/93	40.0	4.0	8.3	0.258	EVENT	P_e [mm]	D_e [mm]	R_e [t/ha per K unit]	Y_e [t]
10/22/93	20.2	1.5	9.4	0.047	11/19-24/93	25.6	3.4	3.0	0.009
10/26/93	12.2	1.0	3.7	0.026	01/17/94	35.6	8.6	4.9	0.026
11/19-24/93	25.6	0.6	3.0	0.008	01/28/94	58.4	11.2	10.2	0.005
11/25-30/93	167.6	79.8	48.5	0.294	02/15/94	31.8	15.8	4.2	0.042
12/1-4/93	15.8	1.3	1.5	0.020	02/18/94	8.4	2.6	0.6	0.004
01/09/94	15.8	2.5	6.4	0.089	10/26/93	12.2	0.1	3.7	0.000
01/17/94	35.6	0.8	4.9	0.014	01/09/94	15.8	0.1	6.4	0.002
01/20/94	59.8	12.4	13.6	0.448	01/20/94	59.8	6.0	12.6	0.033
01/28/94	58.4	12.4	10.2	0.037	02/07/94	44.0	7.3	10.0	0.034
02/07/94	44.0	9.2	10.0	1.158	08/14/95	70.4	0.5	27.9	0.020
02/15/94	31.8	9.1	4.2	0.126	08/20/95	15.8	0.1	14.1	0.007
02/18/94	8.4	1.4	0.6	0.013	11/25/95	79.0	1.6	18.2	0.017
02/23/94	12.4	2.5	1.0	0.044	12/08/95	109.4	1.7	45.9	0.036
04/07/94	18.4	6.3	1.8	0.021	12/13/95	151.8	11.7	12.5	0.056
08/14/95	70.4	2.9	27.9	0.130	01/06/96	45.0	4.1	7.8	0.010
08/20/95	15.8	1.1	14.1	0.038	02/07/96	91.4	9.8	10.0	0.028
11/13/95	15.2	0.8	2.5	0.018	3/8-11/96	120.2	14.3	25.1	0.097
11/25/95	79.0	9.5	18.2	0.771	03/15/96	38.0	9.5	9.0	0.032
12/08/95	109.4	12.1	45.9	1.390	10/03/96		18.1	78.2	0.295
12/13/95	151.8	134.5	12.5	16.678	10/08/96	103.0	6.3	94.1	0.091
01/06/96	45.0	4.0	7.8	0.136	10/14/96	123.2	17.4	145.3	0.339
02/07/96	91.4	58.5	10.0	1.673	12/17/96	33.8	0.2	5.2	0.002
3/8-11/96	120.2	98.5	25.1	3.231	01/11/97	56.6	2.4	9.9	0.037
03/15/96	38.0	20.5	9.0	0.431					

Table 2 Crop factor values for bare plot

GC [%]	C_b
0	0.4500
2.5	0.4188
5	0.3875
7.5	0.3563
10	0.3250

The analysis at a basin scale is developed using a crop factor C_b corresponding to a ground cover percentage of 5 % ($C_b = 0.3875$) for the bare areas. A crop factor C_v of 0.1808 for RUSLE (eqs. (5)) and C_v of 0.1616 for Moore and Burch expression (eq. (6)) of the topographic factor are used for the forest cover. The aims of the basin analysis are both to verify the reliability of the determined crop factor C_v for the Eucalyptus coppice and to test the predictive capability of the model at an event scale. At first, an estimate of the median value β_m is carried out by using the first 10 events in the recording period. Fig. 7 shows, for each selected expression of the topographic factor (eqs. (5) and eq.

(6)), a good agreement between measured (Y_e) and calculated ($Y_{c,e}$) sediment yields by the following equation:

$$Y_{c,e} = R_e C_b \sum_{i=1}^{N_b} K_i L_i S_i \exp\left(-\beta_e t_{p,i}\right) S_{u,i} + R_e C_v \sum_{i=1}^{N_v} K_i L_i S_i \exp\left(-\beta_e t_{p,i}\right) S_{u,i} \quad (8)$$

Tab. 3 β_m values for P1 plot corresponding to the selected ground cover and topographic factor relationship

GC	β_m	
[%]	eqs. (5)	eq. (6)
0	0.1203	0.1626
2.5	0.1154	0.1564
5	0.1102	0.1499
7.5	0.1049	0.1430
10	0.0992	0.1357

Tab. 4 C_m values for P2 plot corresponding to the selected ground cover of P1 plot and topographic factor relationship

GC	C_m	
[%]	eqs. (5)	eq. (6)
0	0.2348	0.2205
2.5	0.2067	0.1896
5	0.1808	0.1616
7.5	0.1570	0.1365
10	0.1352	0.1140

in which N_b is the number of bare morphological units and N_v is the number of forested morphological units. For comparing model results $Y_{c,e}$ with measured value Y_e, the following index R^2, proposed by Nash and Sutcliffe (1970), is used:

$$R^2 = 1 - \frac{\sum_{e=1}^{N_e} \left(Y_e - Y_{c,e}\right)^2}{\sum_{e=1}^{N_e} \left(Y_e - \mu(Y_e)\right)^2} \quad (9)$$

in which N_e is the number of events (55) and $\mu(Y_e)$ is the mean of the measured values.

The R^2 index shows the *efficiency* of a model, the value of one indicates a perfect model, that of zero indicates the model results are not better than the mean, and a value ranging from zero to one indicates that using model prediction is better than using the mean. Table 5 shows that the selected estimate criterion of β_e ($\beta_e = \beta_m$) demonstrates a model efficiency high and greater than 0.45.

For improving β_e estimate, for each selected expression of the topographic factor (eqs. (5) and eq. (6)), the β_e values of the 55 events are calculated by eq. (8)

Tab. 5 R^2 index values for the selected β_e estimate criterion and topographic factor relationship

Topographic factors	β_e estimate	R^2
eqs. (5)	β_m	0.4553
eq. (6)	β_m	0.4553
eqs. (5)	function of D_e/P_e	0.5995
eq. (6)	function of D_e/P_e	0.6020

with $\beta_m = \beta_e$. Fig. 8 shows, as an example for RUSLE model, that β_e can be reliably estimated by the runoff coefficient D_e/P_e. The comparison between Fig. 9 and Fig. 7 shows that estimating β_e by the runoff coefficient gives a better

448

agreement between measured and calculated sediment yields. This result is confirmed by the R^2 index values listed in Table 5.

The randomness of hydrologic variables in addition with the circumstance that actual values are not known with certainty, has suggested to include the uncertainty analysis (Haan 1977) (Hession et al. 1996) in this modelling activity. According to MacIntosh et al. (1994) the major types of uncertainty are *knowledge uncertainty* and *stochastic variability*.

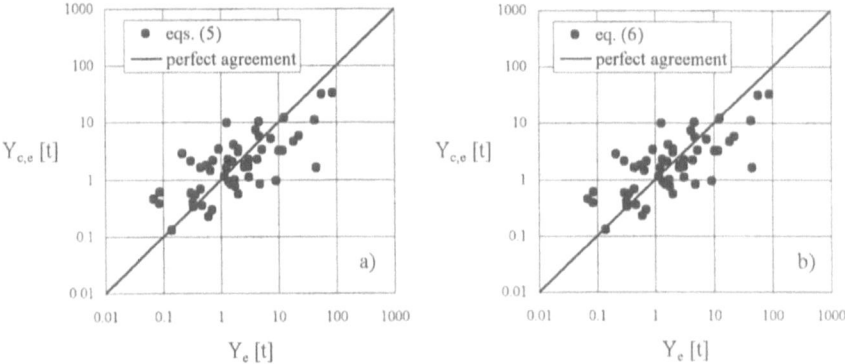

Fig. 7 Comparison between measured and calculated sediment yields by eq. (8) with $\beta_e = \beta_m$

Knowledge uncertainty is due to incomplete understanding of studied phenomena, inadequate measurement of system properties and input data availability. Stochastic variability is due to random variability of the studied natural environment and can be divided into temporal and spatial variability.

Monte Carlo simulation is the most robust method for propagating uncertainty through models and allows numerically operating a complex system with random components.

Table 6 Parameters of theoretical laws for the characteristic variables at event scale

VARIABLE	THEORETICAL LAW	PARAMETERS		LS MODEL
R_e	LN2	$\mu = 2.0558$	$\sigma = 1.1851$	both
K	LN2	$\mu = 0.1872$	$\sigma = 0.1035$	both
C_b	uniform	$C_{b,min} = 0.325$	$C_{b,max} = 0.45$	both
C_v	uniform	$C_{v,min} = 0.1352$	$C_{v,max} = 0.2348$	eqs. (5)
C_v	uniform	$C_{v,min} = 0.114$	$C_{v,max} = 0.2205$	eq. (6)
β_e	Gauss	$\mu = 0.0329$	$\sigma = 0.0228$	eqs. (5)
β_e	Gauss	$\mu = 0.0497$	$\sigma = 0.0261$	eq. (6)
D_e/P_e	Beta	$\alpha = 0.962$	$\varepsilon = 1.7096$	both
		$A = 0.032$	$B = 0.525$	

Repeated simulations are performed with the model using randomly selected input parameter values. For each simulation, the input parameter values are chosen using their known, predetermined, probability distribution.

The simulation process is repeated for a number of iterations sufficient to individuate an estimate of the probability distribution of the output variable.

In order to perform the Monte Carlo simulations, a probability distribution is preliminarly searched for each uncertain parameter. The values of the rainfall erosivity factor R_e calculated for each event are found to be lognormally distributed with mean μ and standard deviation σ listed in Table 6.

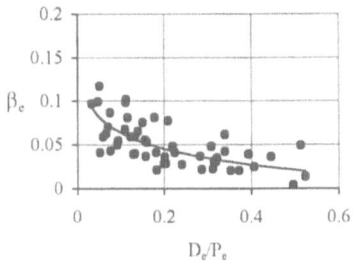

Fig. 8 Relationship between β_e and runoff coefficient

The crop factor for the bare, C_b, and that for the forested areas, C_v, are treated as having only knowledge uncertainty representing the range of possible values available from literature or determined by the measurements at P2 plot. The knowledge uncertainty for the examined crop cover, due to the difficulty to establish an appropriate value for the application in our model, is considered by a uniform distribution which describes crop factor values ranging from a minimum to a maximum value listed in Table 6. The β_e coefficient is normally distributed with mean and standard deviation listed in Table 6. The topographic factor is treated as a constant deterministic value of each morphological unit, under the assumption that the length and the slope of the units are controlled.

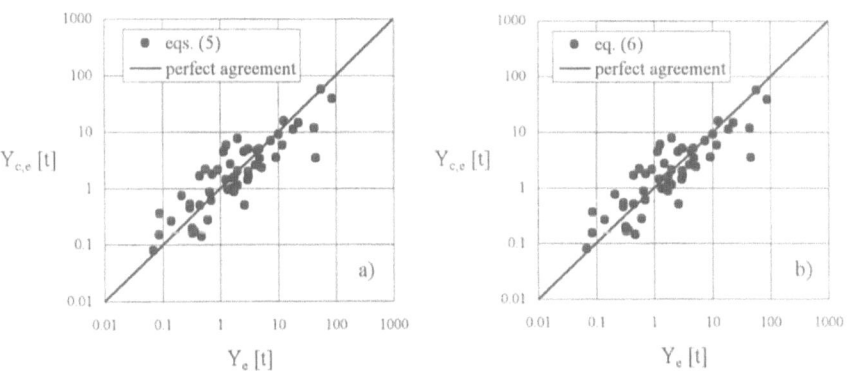

Fig. 9 Comparison between measured and calculated sediment yields by eq. (8) with β_e estimated by runoff coefficient

Each Monte Carlo simulation is carried out drawning at random a value from each of the distributions of rainfall erosivity, soil erodibility, crop factors and β_e coefficient. These values are then used with the constant topographic factor of each morphological unit as input to the model (eq. (8)), whose output represented one iteration of the simulated scenario. The resampling is repeated 15000 times, resulting in 15000 estimates of the output basin sediment yield $Y_{c,e}$. The 15000 values of $Y_{c,e}$ are used to calculate the theoretical probability distribution $P(Y_{c,e})$ which has to be compared with the empirical frequency distribution $F(Y_e)$ of the measured values Y_e of the basin sediment yield. Fig. 10a shows, for the case of topographic factors calculated by eqs. (5), a good agreement between the two distributions. Fig. 10b shows that the mean value μ and the standard deviation σ estimate of the sediment yield can be considered invariable for a number of iterations greater than 3000.

The resampling is also carried out estimating β_e by runoff coefficient D_e/P_e (Fig. 8) which is distributed according to a beta distribution whose parameters are listed in Table 6. Fig. 11 shows that the agreement between the two distributions, for β_e estimated by runoff coefficient, remains good and that little departures from the coincidence between $P(Y_{c,e})$ and $F(Y_e)$ happen only for probability values greater than 0.8.

Finally, the effectiveness of the forest cover to mitigate erosion is tested calculating the basin sediment yield (eq. (8)), with R_e equal to the median value of the rainfall erosivity factor, corresponding to an increasing number of bare morphological units. In other words, starting from a completely forested basin, having a percentage of bare area S_b equal to zero ($N_b = 0$), different scenarios

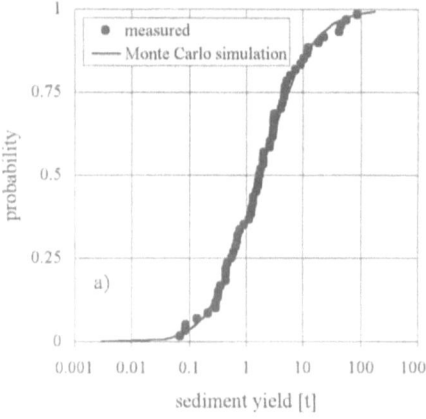 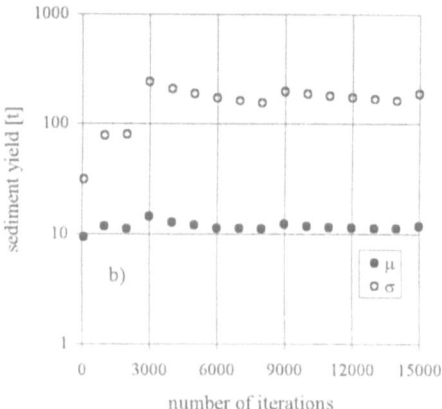

Fig. 10a Comparison between measured sediment yield frequency distribution and that generated by Monte Carlo technique with $\beta_e = \beta_m$

Fig. 10b Mean and standard deviation values of generated sediment yield sequences

corresponding to an increasing number of bare morphological units are simulated. As reference value, Y_{ref}, the sediment yield of the completely forested basin is used. The simulations reproduce a possible real situation in which a completely forested basin is subjected to deforestation. For simulating the deforestation activity, two different and opposite criteria are selected: a) the bare areas are obtained *deforestating* covered morphological units starting from the morphological unit having the *minimum* value of the topographic factor and going on for *increasing* L_iS_i values; b) starting from the morphological unit having the *maximum* value of L_iS_i and going on for *decreasing* L_iS_i values.

The selected criteria establish two *extreme conditions* represented by the two curves plotted in Fig. 12. The two curves define the area of the plane $(S_b, Y_c\text{-}Y_{ref}/Y_{ref})$ in which fall all the relationships $Y_c-Y_{ref}/Y_{ref} = \phi\ (S_b)$, in which ϕ is the functional symbol, representative of different forest management activities which will be possible to carry out in W2 basin.

Fig. 12 clearly shows that the criterion "increasing L_iS_i value" allows to drastically limit the increasing of sediment yield respect to the reference situation. In particular, for $S_b = 40\ \%$ and the "increasing L_iS_i" management criterion,

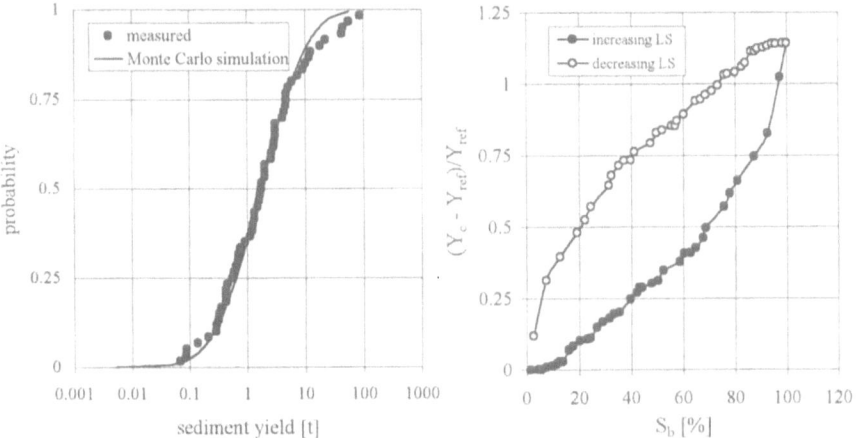

Fig. 11 Comparison between measured sediment yield frequency distribution and that generated by Monte Carlo technique with β_e estimated by runoff coefficient

Fig. 12 Comparison between Y_c-Y_{ref}/Y_{ref} parameter and percentage of bare area

an increasing of 25% of the calculated basin sediment yield Y_c is observed. While the criterion "decreasing L_iS_i" is characterized by an increasing of 75% of the calculated sediment yield.

Fig. 12 also shows that, for the "increasing L_iS_i" criterion, S_b values less than 15% produce negligible effects on the sediment yield respect to the reference situation (totally forested basin).

The optimal deforestation management criterion determined by the previous analysis, which is to bare morphological units starting from that having the minimum topographic factor and going on for increasing L_iS_i values, is also tested for the real crop cover of W2 basin which is characterized by a S_b value approximately of 25% corresponding to an area located in the upper part of the basin.

Using as Y_{ref} the basin sediment yield corresponding to the hypothesis of totally forested basin, different scenarios corresponding to increasing S_b values up to 25% are simulated.

Fig. 13, confirming the results of the previous analysis, shows that using the "increasing L_iS_i" deforestation criterion, when S_b values are less than 20%, the increments of the basin sediment yield is less than 6%.

5. CONCLUSIONS

In many parts of the world soil erosion is the most important problem of land management. In the past, forest has often been destroyed to permit agricultural land use. Nowadays, in many countries, the contemporary society requires that forests provide not only wood but also non-productive environmental functions like water and soil conservation. Forest ecosystems modifying the hydrological hillslope processes (rainfall interception, infiltration, runoff) are able to reduce erosion and to support sediment delivery processes. The available sediment yield measurements, carried out in a small Calabrian experimental basin, covered with *Eucalyptus occidentalis* forest, and RUSLE coupled with a sediment delivery distributed model are used to evaluate the effects of Eucalyptus coppice on sediment yield.

For applying the sediment delivery distributed (SEDD) model, the W2 experimental basin is divided into morphological units and the topographic factors of each unit are calculated by eqs. (5) and eq. (6). Then, the soil loss measurements carried out in two experimental plots, located in the basin, are used to estimate the crop and management factor of Eucalyptus coppice.

The analysis at basin scale is developed both to verify the reliability of the selected β_e coefficient, appearing in the SEDD model, and crop factor values. The analysis showed that the selected crop factor for Eucalyptus coppice and the median value β_m, estimated by using the first 10 events in the recording period, allow to obtain a good agreement between calculated and measured basin sediment yield at an event scale. The agreement between measured and calculated sediment yield values is improved by estimating β_e by the event runoff coefficient. In order to represent the model limitations and uncertainties, the effect of parameter uncertainty on calculated sediment yield is also studied. Monte Carlo simulations are performed using known, predetermined, probability distributions of the input parameters. The resampling of the model factors produced 15000 estimates of the output basin sediment yield which are used to calculate the theoretical probability distribution which has to be compared with the empirical frequency distribution of the measured sediment yield. The uncertainty analysis, developed at an event scale, showed that SEDD model has a good predictive ability also when the β_e coefficient is estimated by runoff coefficient. Finally, starting from a completely forested basin, different scenarios

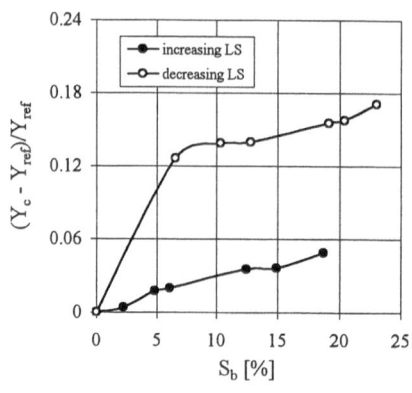

Fig. 13 Comparison between Y_c-Y_{ref} / Y_{ref} parameter and percentage of the real crop cover of W2 basin

corresponding to an increasing percentage of bare area are simulated. The analysis shows that a deforestation action carried out starting from the morphological unit having the minimum value of the topographic factor and going on for increasing L_iS_i values allows to drastically limit the increasing of the basin sediment yield respect to the reference situation (totally forested basin).

ACKNOWLEDGEMENTS

The research was supported by grants from Ministero Università e Ricerca Scientifica e Tecnologica (MURST), Governo Italiano, and from Consiglio Nazionale delle Ricerche (CNR).

REFERENCES

Aronica G. and Ferro V. (1997): Rainfall erosivity over Calabrian region. Journal of Hydrological Sciences, Vol. 42, n. 1, pp. 35-48.

Bagarello V. and D'Asaro F. (1994): Estimating single storm erosion index. Trans. of ASAE, Vol. 37, pp. 785-791.

Bagarello V., Baiamonte G., Ferro V. and Giordano, G. (1993): Evaluating the topographic factors for watershed soil erosion studies. Proc. Workshop on "Soil erosion in semi-arid mediterranean areas", Taormina, Ed. Morgan R.P.C., pp. 3-17.

Boyce R.C. (1975): Sediment routing with sediment delivery ratios. In: "Present and Prospective Technology for Predicting Sediment Yield and Sources", U.S. Dept. Agric. Publ. ARS-S-40, pp. 61-65.

Cinnirella S., Iovino F., Porto P. and Ferro V. (1998): Antierosive effectiveness of Eucalyptus coppices through the cover management factor estimate. Hydrological Processes, Vol. 12, n. 4, pp. 635-649.

Cooley K.R. (1980): Erosivity values for individual design storm. Journal of the Irrigation and Drainage Division (ASCE), Vol. 106, IR3, pp. 136-145.

De Roo A.P.J., Wesseling C.G., Cremers H.D.T. and Offermans R.S.E. (1994): LISEM: a new physically based hydrological and soil erosion model in a GIS-environment, theory and implementation. Proc. of Canberra Symp., IAHS Publ. n. 224, pp. 439-448.

Dissmeyer G.E. and Foster G.R. (1981): Estimating the cover-management factor (C) in the Universal Soil Loss Equation for forest conditions. Journal of Soil and Water Conservation, Vol. 36, n. 4, pp. 235-240.

Ferro V. (1997): Further remarks on a distributed approach to sediment delivery. Journal of Hydrological Sciences, Vol. 42, n. 5, pp. 633-647.

Ferro V. and Minacapilli M. (1995): Sediment delivery processes at basin scale. Journal of Hydrological Sciences, Vol. 40, n. 6, pp. 703-717.

Haan C.T. (1977): Statistical mathods in hydrology. Ames: Iowa State University Press.

454

Hession W.C., Storm D.E. and Haan C.T. (1996): Two-phase uncertainty analysis: an example using the Universal Soil Loss Equation. Trans. of ASAE, Vol. 39, n. 4, pp. 1309-1319.

McCool D.K., Foster G.R., Mutchler C.K. and Meyer L.D. (1989): Revised slope length factor for the Universal Soil Loss Equation. Trans. of ASAE, Vol. 32, n. 5, pp. 1571-1576.

MacIntosh D.L., Suter II G.W. and Hoffman F.O. (1994): Uses of probabilistic exposure models in ecological risk assessments of contaminated sites. Risk analysis, Vol. 14, n. 4, pp. 405-419.

Moore I.D. and Burch F.J. (1986): Physical basis of the lenght-slope factor in the Universal Soil Loss Equation. Soil Science Society of American Journal, Vol. 50, pp. 1294-1298.

Morgan R.P.C. (1994): The European soil erosion model: an update on its structure and research base. In "Conserving Soil Resources - European Perspectives", Ed. R.J. Rickson, CAB International Wallingford, Oxfordshire, pp. 286-299.

Mumeka A. (1986): Effect of deforestation and subsistence agriculture on runoff of the Kafue River headwaters, Zambia. Journal of Hydrological Sciences, Vol. 31, n. 4, pp. 543-554.

Nash J.E. and Sutcliffe J.E. (1970): River flow forecasting through conceptual models. Part 1. A discussion of principles. Journal of Hydrology, Vol. 10, pp. 282-290.

Nearing M.A., Foster G.R., Lane L.J. and Finkner S.C. (1989): A process-based soil erosion model for USDA-Water Erosion Prediction Project Technology. Trans. of ASAE, Vol. 32, n. 5, pp. 1587-1593.

Oyarzum C.E. and Peña L. (1995): Soil erosion and overland flow in forested areas with pine plantations at coastal mountain range, Central Chile. Hydrological Processes, Vol. 9, pp. 111-118.

Renard K.G., Foster G.R., Yoder D.C. and McCool D.K. (1994): Rusle revisited: status, questions, answers, and the future. Journal of Soil and Water Conservation, Vol. 49, pp. 213-220.

Richards K. (1993): Sediment delivery and drainage network. In: "Channel Network Hydrology", Eds. Beven K. and Kirkby M.J., John Wiley, New York, pp. 221-254.

Rogers R.D. (1989): Influence of sparse vegetation cover on erosion and rill patterns: an experimental study. Dissertation for the Degree of Master of Science, Colorado State University, Fort Collins, Colorado.

Walling D.E. (1983): The sediment delivery problem. Journal of Hydrology, Vol. 65, pp. 209-237.

Wischmeier W. H. and Smith D.D. (1978): Predicting rainfall erosion losses. A guide to conservation planning. U.S.D.A., Agr. Handbook, 537.

A Concept for Runoff Processes on a Steep Forested Hillslope

Makoto TANI

Forestry and Forest Products Research Institute, Ibaraki 305-8687, Japan

ABSTRACT   Runoff generation processes on a steep forested hillslope are discussed based on findings from recent observational studies and analyses of the processes by a conceptual model. The observations demonstrate an important role of rapid lateral saturated flow and an insignificant function of bypass flow in the unsaturated soil layer. The model represented by a surface soil layer with preferential flow pathways at the bottom can produce two types of discharge under an application of Darcy's law contributing to stormflow and baseflow. The results suggest that the soil matrix within surface soil layer may control the distribution of rain water into stormflow and baseflow and that the preferential flow pathways may support a rapid transportation of water discharged from the soil layer downslope.
Keywords: hillslope, runoff process, soil matrix, macropores, Darcy's law

1. INTRODUCTION

Understanding detailed runoff processes on a hillslope has been a subject requesting various kinds of observational and modelling studies. One of the mysterious issues on the processes is to produce storm runoff although all the rain water infiltrates into soil and no overland flow occurs. Many observational studies challenging this problem has emphasized role of macropores (eg. Mosley, 1979), and have tried to explain the reason why the most part of stormflow is occupied by pre-event old water although water is quickly discharged through preferential flow pathways composed by macropores (McDonnell, 1990). Such an important role of the preferential flow pathways must indicate that a momentum equation like Darcy's law is hardly applied to the water movement in the soil. This may have strong influences on a modelling strategy for runoff processes on a hillslope. This brief paper attempts to explain a concept for runoff generation processes on a steep forested hillslope based upon some observational findings and a hillslope model considering a role of the preferential flow pathways.

2. OBSERVATIONAL FINDINGS

Some examples for observational studies will be shown first to understand basic characteristics on runoff processes on steep forested hillslopes.

Most part (.90%) of rainfall during a storm event with wet antecedent conditions is discharged as storm runoff in some catchments such as Maimai 8 in New Zealand (Mosley, 1979) and Tatsunokuchi-yama in Japan (Tani and Abe, 1987). The large stormflow occurrences demonstrate that the area involved in the contribution to stormflow is not limited within small zones near streams or with a poor permeability of soil surface but that almost all the catchment area must contribute to the stormflow production. To find the macropores and fast lateral flows within them from observations in Maimai 8 (Mosley, 1979) could account for large productions of stormflow without any occurrence of overland flow. McDonnell (1990) emphasized the contribution of fast bypass flow through cracks to vertical movement during a storm event as well as the contribution of lateral pipe flow. He pointed out the importance of displacement process of new event water by old water retained within soil matrix to account for a large occupation of old water in runoff discharge through macropores.

Tani (1997) has studied on runoff responses to storm rainfall in the Tatsunokuchi-yama Experimental Forest, Japan. In two catchments there, almost all the rainfall contributed to stormflow in the wettest conditions when enough large cumulative rainfall was given. Observations on a steep hillslope with a thin soil layer within one of the catchments showed the following findings: 1) tensiometric responses near bedrock demonstrated a quick vertical propagation of rainfall pulse after the wet soil condition was once yielded. 2) A large hydrograph of stormflow discharged from the study slope in the wettest condition was sufficiently simulated by a kinematic wave runoff model applied to the lateral flow that was developed on the bedrock by soil water supply due to the quick vertical propagation. 3) In the previous dry conditions, rain water was mostly retained within soil matrix making a wetting front and only a little amount of stormflow was discharged from the slope. Tani (1997) agreed that such a small hydrograph response was produced by a bypass flow in macropores. However, he emphasized that the quick vertical propagation of rainfall pulse within soil matrix in a wet condition was a key controller for storm runoff generation. Once the wetting front reaches the bedrock, the specific moisture capacity within the wet transition zone becomes a small value, which ensures a large celerity of vertical propagation in the Richards equation even though the vertical movement is described as the Darcian flow. This observation strongly suggests that major stormflows are produced by a system of fast lateral saturated flow within macropores receiving vertical quick propagations of rain water within unsaturated soil matrix.

A recent study conducted in a steep hillslope, the CB1 catchment, in Oregon Coast Range, USA, demonstrated clear differences in flow mechanisms between unsaturated and saturated zones (Anderson et al., 1998). Labeled water with deuterium was given to the soil surface during the catchment-scale sprinkler experiments with a constant intensity to trace water flow within unsaturated vadose zone. In addition, bromide injection into water at the bedrock/colluvium interface at 1.9 m depth was conducted to trace water flow in saturated materials. The water labeled by deuterium moved slowly through the vadose zone as plug flow the velocity of which was controlled by rainfall rate and water content. In the plug flow, old soil water was pushed out at the base of the soil column by incoming new rain water. Any preferential flow of new water was not observed during either low-intensity irrigation or a natural storm. On the other hand, the bromide peak velocity of 0.6 cm/s calculated from its arrival at a measuring weir 19 m downslope from the injection point suggested that a rapid saturated flow occurred through shallow fractured bedrock.

The rapid propagation of rain pulse in a wet unsaturated condition stated by Tani (1997) is a typical property of plug flow detected by the tracer experiment by Anderson et al. (1998). The role of bypass flow in vertical water movement was not emphasized by Tani (1997) or denied by Anderson et al. (1998) though it was important for the conception of McDonnell (1990). On the other hand, all the three observational studies agree with a quick downslope transportation of water out of the saturated zone. Such findings from these observational studies should be utilized to model runoff processes on a steep hillslope.

3. CONCEPTUAL MODEL

3.1 Modelling strategy

Observational studies have suggested important roles of preferential flow in runoff processes on a steep hillslope. However, the hydraulic characteristics of preferential flow is still unknown even though those of a single natural pipe can be described based on an experiment of soil samples with these pipes (Kitahara, 1993). This uncertainty strongly depends on quite heterogeneous spatial distributions of the preferential flow pathways. It should be noted that only water movement within soil matrix can be formulated by Darcy's law though various kinds of water movements are involved in runoff processes on hillslope. Thus, one of the strategies for modelling runoff processes may lie in an efficient utilization of Darcy's law in consideration of the heterogeneous characteristics of preferential flow. The observational studies by Tani (1997) and Anderson et al. (1998) mentioned above showed that a preferential flow did not play an important role in the vertical water movement in an unsaturated zone while the downslope water movement within saturated zone was controlled by a fast preferential flow. This suggests that Darcy's law may be applicable to the unsaturated zone though its application to the downslope saturated flow is not rational.

3.2 Model description

A conceptual model proposed by Tani and Abe (1996) illustrated in Fig. 1 is based on a difference in water movement between unsaturated and saturated conditions derived from the observational findings mentioned above. The outline is described in this section.

Let us consider that a soil layer underlain with impermeable bedrock on a hillslope includes abundant macropores in it. Rain water falling onto the soil surface must infiltrate through the soil matrix regardless of the macropores because it can not enter them under the negative pressure condition in an unsaturated zone. On the other hand, water can easily exfiltrate to macropores in a saturated zone generated on the impermeable bedrock due to water supply from the upper unsaturated zone. If preferential flow pathways consisting of macropores efficiently drains water from the saturated zone downslope, rising of the water table of the saturated zone is strongly controlled by the drainage. In consideration of this control, we assume that the following bottom boundary condition of the soil layer is given to our model: when the matric potential ψ at the bottom is negative, no water is drained from the soil layer to the preferential flow pathways, and when ψ is zero, water is discharged to the pathways so that ψ does not become positive. As a result, the soil layer will be always unsaturated except for the bottom boundary and water will move only within soil matrix in the layer. Therefore, Darcy's law can be applied to the water movement there even though macropores are included in the unsaturated zone. Introduce additional assumptions that the soil layer has a uniform thickness, the continuity equation of the flow can be expressed by a two-dimensional form and the physical properties of soil matrix in the layer is homogeneous.

The model described above yields two types of discharge through a calculation process: one is Q_L, the discharge from the downslope end, and another is Q_P, the discharge from the bottom of the layer into the preferential flow pathways. The detailed equations were listed in the previous paper (Tani and Abe, 1996).

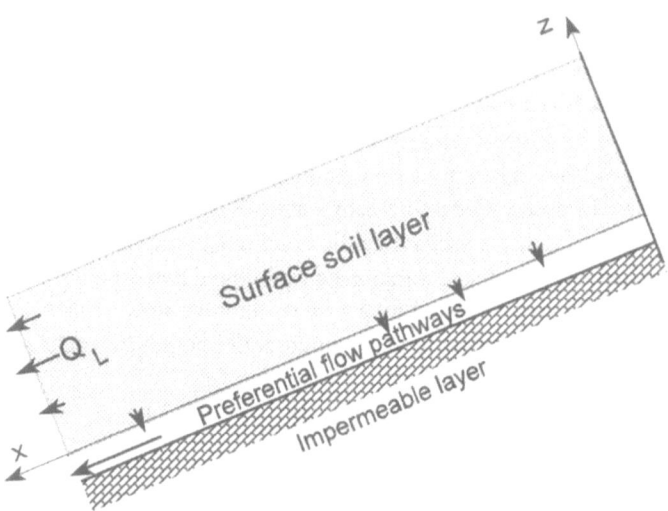

Fig. 1 Conceptual model for a surface soil layer with a preferential flow pathways at the bottom

3.3 Calculation results

Recession stage from the wettest condition was calculated through the model by employing equations of soil physical properties used by Tani (1982). Parameter values used in the calculation were defined based on investigations in a study catchment in Japan, Minamitani (MN) catchment in the Tatsunokuchi-yama Experimental Forest (Tani, 1997).

Recession hydrographs of runoff from the mean slope of the study catchment calculated by the hillslope model are shown in Fig. 2. The discharge is expressed with values in mm/hr in a unit catchment area. An early stage of the recession and the long recession are displayed in the left and right figures, respectively. First, Q_V, discharge from the bottom of the soil layer, begins its recession at a large initial quantity but decreases very quickly in several hours. Q_L, discharge from the downslope end, is kept almost constant for several days after Q_V disappears. A longer-term recession is produced only by Q_L.

The recession properties of runoff obtained from the hillslope model were compared with those observed in MN catchment. In this comparison, a recession hydrograph was approximated by an exponential function, and a value of half life, t_h, was employed for easier understanding of a recession property. Two periods of runoff recession stages observed in MN, in September 1976 and from November to December 1976, were used for the comparison. Relationships of half life to the runoff discharge for both the calculation and the observation are shown in Fig. 3, where the relationship for Q_L is plotted only after the duration of constant discharge rates. Values of t_h for Q_V are very small compared to those for the observed runoff throughout the runoff range, whereas Q_L has generally larger t_h values than the observed runoff though they have similar values in the lowest runoff range.

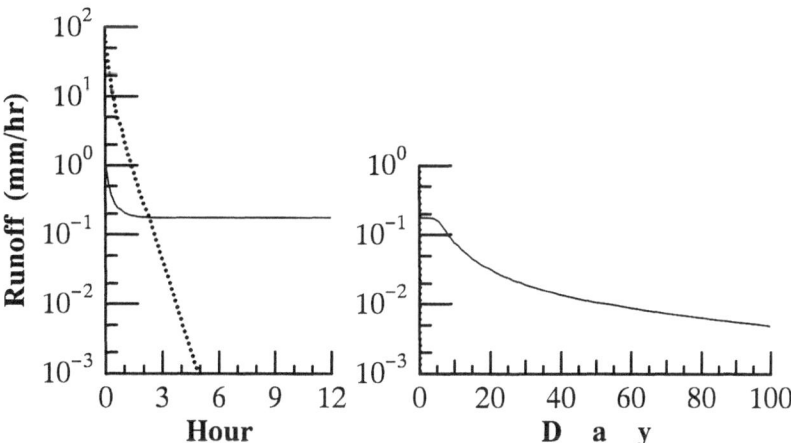

Fig. 2 Recession hydrographs calculated by the conceptual model.
Left: in the early stage, right: in the long duration.   — Q_L, $\cdots Q_V$

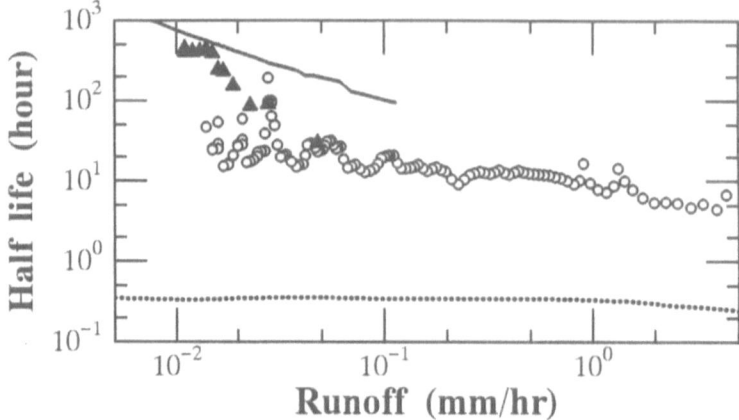

Fig. 3 Relationships of half life to runoff rate in the recession periods.
— Q_L(calculated), ⋯Q_V(calculated).
○ observed for September in 1976.
▲ observed from November to December in 1976.

The comparison of the model calculation with the observational results in the period of low runoff rates suggests that the downslope flow produced from the unsaturated soil layer can account for the baseflow recession considering that evapotranspiration neglected in the calculation may make the runoff recession steep (Suzuki, 1984). On the other hand, the comparison in the period of high runoff rates shows that the discharge produced from the vertical gradient of matric potential has a very short time lag compared to that of the observed storm runoff. Therefore, the observed runoff needs some more additional time lag, which may be attributed to a lateral preferential flow bringing the drained water from the soil layer downslope. Although it is difficult to describe the hydraulic properties of the preferential flow system, the velocity of 0.76 cm/s for a single natural pipe measured by Kitahara (1994) and the bromide peak velocity of 0.6 cm/s conducted in the CB1 catchment mentioned earlier (Anderson et al., 1998) may support understanding the characteristics of the lateral preferential flow. Because flow with the velocity of 0.76 cm/s can bring water 100 m downslope only within 4 hours, it is rational that coupling the vertical discharge with the lateral flow account for the time lag of the observed runoff at high rates. Thus, one can conclude that the soil layer produces two types of discharge by an application of Darcy's law to the soil matrix in it: one is the discharge from the bottom of the layer contributing to stormflow through the lateral preferential flow, and another is the discharge from the downslope end contributing directly to the baseflow.

4. DEVELOPING PREFERENTIAL FLOW PATHWAYS

Observational findings on hillslopes and comparisons of model calculations with observed hydrographs mentioned earlier suggested an important role of lateral saturated flow through macropores in runoff generation processes on a steep forested hillslope. However, it is not easy to see through such a preferential flow pathway system under the ground although an experiment of tracer injections by Anderson et al. (1998) provided an evidence of fast saturated flow. Surveying the system at a hillslope scale must be one of the key subjects on hillslope hydrology.

Another important subject is to elucidate the creation of preferential flow pathways responding to the development of surface soil layer from a geomorphological point of view. Interesting findings related to this viewpoint have been obtained by Shimokawa (1984) from his investigations of revegetation and topsoil reformation processes on landslide scars in the south part of Japan. He estimated recurrence interval of landslide from the reformation of topsoil after the previous landslide occurrence. For example, this was estimated at about 200 years in a study area composed by deeply weathered granodiorite.

One possible process for creating preferential flow pathways under the ground may be considered here corresponding to the cycle of landslide occurrences, quite a short term from a geomorphological point of view. Many rills on the surface of bedrock are soon generated as pathways for downslope flow after a landslide occurrence. As fresh soil is produced by the weathering process and the surface soil layer covers the bedrock, the erosion may continue within the pathways like pipes due to the concentration of rain water into the pathways, whereas soil beside the pathways is made stable due to the growth of roots of invaded plants. This contrast can insure the development of pathways where rain water concentrates. Thus, this consideration suggests that downslope flow is 'preferential' throughout a geomorphological life-cycle of the surface soil layer on hillslope. Such a rough sketch for the development of preferential flow pathways may request further strategies for coupling hydrological studies on runoff processes with those on geomorphological processes of surface soil development.

5. CONCLUSION

Understanding runoff generation processes on a steep forested hillslope has been discussed based on findings from recent observational studies and analyses of the processes by a conceptual model. This model consisting of a surface soil layer with preferential flow pathways at the bottom showed that the discharge from the bottom of the layer contributes to stormflow through the lateral preferential flow and that the discharge from the downslope end contributes directly to the baseflow. Seeing through the preferential flow pathways and understanding their development reflecting a geomorphological background are strongly required in the forthcoming studies.

REFERENCES

Anderson, P.A., W.E. Dietrich, D.R. Montgomery, R. Torres, M. Conrad and K. Loague (1998): Subsurface flow paths in a steep, unchanneled catchment, Water Resour. Res., Vol. 33, pp. 2637-2653.

Kitahara, H. (1993): Characteristics of pipe flow in forested slopes, Exchange processes at the Land Surface for a Range of Space and Time Scales (Proc. Yokohama Symp.), IAHS Publ. No.212, Wallingford, pp. 235-242.

Kitahara, H. (1994): A study on the characteristics of soil pipes influencing water movement in forested slopes, Bulletin of the Forestry and Forest Products Research Institute, Vol. 367, pp. 63-115.

McDonnell, J.J. (1990): A Rationale for old Water discharge through macropores in a steep, humid catchment, Water Resour. Res. Vol, 26, pp. 2821-2832.

Mosley, M.P. (1979): Streamflow generation in a forested watershed, New Zealand, Water Resour. Res., Vol. 15, pp. 795-806.

Shimokawa, E. (1984): A natural recovery process of vegetation on landslide scars and landslide periodicity in forested drainage basins, Symposium on Effects of Forest Land Use on erosion and Slope Stability, University of Hawaii, Honolulu, pp.99-107.

Suzuki M. (1984): The properties of a base-flow recession on small mountainous watersheds (2) Influence of evapotranspiration on recession hydrographs, J. Japanese Forestry Society, Vol. 66, pp.211-218. (in Japanese with English summary)

Tani, M. (1982): The properties of a water-table rise produced by a one dimensional, vertical, unsaturated flow, J. Japanese Forestry Society, Vol. 64, pp. 409-418. (in Japanese with English summary)

Tani, M. (1997): Runoff generation processes estimated from hydrological observations on a steep forested hillslope with a thin soil layer, J. Hydrol., Vol. 200, pp.84-109.

Tani, M. and T. Abe (1996): Roles of matrix flow in runoff recession produced from a hillslope with preferential pathways, Proc. International Conference & Environment research: Towards the 21st Century, Vol. 1, Kyoto, pp. 85-92.

Flow Pathways on Steep Forested Hillslopes: the Tracer, Tensiometer and Trough Approach

Jeffrey MCDONNELL*[1], Dean BRAMMER*[1], Carol KENDALL*[2], Niclas HJERDT*[1], Lindsay ROWE*[3], Mike STEWART*[4] and Ross WOODS*[5]

*1 SUNY-ESF, 1 Forestry Drive, Syracuse, NY 13210, USA
*2 U.S. Geological Survey, 345 Middlefield Rd., Menlo Park, CA 94025, USA
*3 LandCare Research New Zealand, Christchurch, New Zealand
*4 Institute for Geological and Nuclear Sciences, Lower Hutt, New Zealand
*5 National Institute for Water and Atmospheric Research, Christchurch, New Zealand

ABSTRACT The physical controls on the downslope movement of water and solutes on steep hillslopes are still poorly understood. We conducted combined tracer, tensiometer and trough experiments at the Maimai Catchment, New Zealand to determine subsurface flow pathways and the relative importance of surface vs. bedrock topography in water and tracer flux. The bedrock topographic surface exerted the largest control on spatial patterns of subsurface flow and Br tracer movement. The paper presents a conceptualization of flow at the Maimai hillslope that accounts for preferential flow of tracer and displacement of isotopically old water during rainfall.
Keywords: Hillslope hydrology, tracers, tensiometers, troughs, subsurface flow.

1. INTRODUCTION

The subsurface movement of water and solutes within catchments is not well understood. Subsurface flow is not visually apparent and point measurements are difficult to integrate up to the hillslope and catchment scale. Flow paths are therefore difficult to delineate and subsurface flow response often varies greatly within and between catchments (Bonell 1993). Field and laboratory experiments using hydrometric and environmental tracer studies indicate that no one single mechanism adequately describes flow in small catchments (Buttle 1994). Tracer experiments generally reveal a predominance of old water (Sklash et al. 1986); tensiometer experiments generally show evidence of preferential flow to depth (McIntosh et al. 1998); trough studies have yielded highly equivocal results regarding the link between flow and upslope contributing area (Woods and Rowe 1996). Recent process studies have shown strong correlations between soil wetness and surface topography (e.g. Nyberg 1996). Many modeling studies also have assumed that hillslope flow directions can be related to hydraulic gradients derived from surface topography (e.g. Quinn et al. 1991). Freer et al. (1997) note that this assumption leads to believable maps of predictions of soil saturation but that the link between topography and subsurface flow at the hillslope scale has not been fully assessed.

Currently, there is a need for detailed studies of hillslope-scale subsurface flow processes to assess the assumption used in most process modeling studies, that hillslope flow is controlled by hillslope topography. It is at this scale where the details of point-scale features like macropore flow may be scaled up to larger landscape elements. One approach is to combine integrated hillslope measurements (using hillslope-scale trough systems) with more traditional point measurements (tensiometers), and tracer approaches (^{18}O and Br). In this study, we test the fundamental hypothesis that surface topography controls flow and tracer movement at the hillslope scale.

2. STUDY SITE

A subsurface collection system along the base of a hillslope hollow on the left bank of the stream draining the M8 catchment was constructed in 1992 and reported on by Woods and Rowe (1996). A vertical face 60 m long and 1.5 m high was cut across the toe of the hillslope (Figure 1). Thirty subsurface flow collection troughs were installed end-to-end, across the base of the excavated face at the soil-bedrock interface. The troughs were sealed to the cut face and covered. Flow collected in the troughs was routed to tipping bucket flowmeters and recorded continuously using a Campbell Scientific CR10 data logger.

Hillslope surface topography was mapped by Landcare Research and NIWA (National Institute of Water and Atmospheric Research) of New Zealand in two separate surveys. The first survey was conducted in 1992 for the work of Woods and Rowe (1996). It consisted of 756 distinct survey points and was used to estimate subcatchment areas draining to each 1.7 m trough. Seventy soil depth points were also obtained across the hillslope by inserting an aluminum rod into the soil profile until refusal. A second survey of 37 additional soil depth points was conducted in August 1995 for this study. Additionally, 99 soil depth readings were obtained in the study grid.

Soil depth is 0.15 - 0.30 m on the hillslope ridges, but increases to 0.60 - 2.0 m along the mid-slopes and in the central hollow. Lower slope soils within the hollows display mottling and gleying indicative of frequent saturation. Previous studies (McKie 1978; McDonnell 1990) show that soil catenary sequences increase in thickness downslope and in the hollows; these soils are less well drained and have generally higher porosity and lower hydraulic conductivity. As a result, the soils are strongly weathered and leached yellow brown earths, with low natural fertility. The thin nature of the soils promotes the lateral development of root networks and channels at the soil-bedrock interface, similar to that reported by Tani (1997) and Peters et al. (1995). Soil profiles reveal extensive macropores and preferential flow pathways at vertical pit faces along the trough system that form along cracks, holes and root channels (Mosley 1979, 1982; McDonnell 1990). Lateral root channel networks are evident in the numerous tree throws that exist on the hillslope.

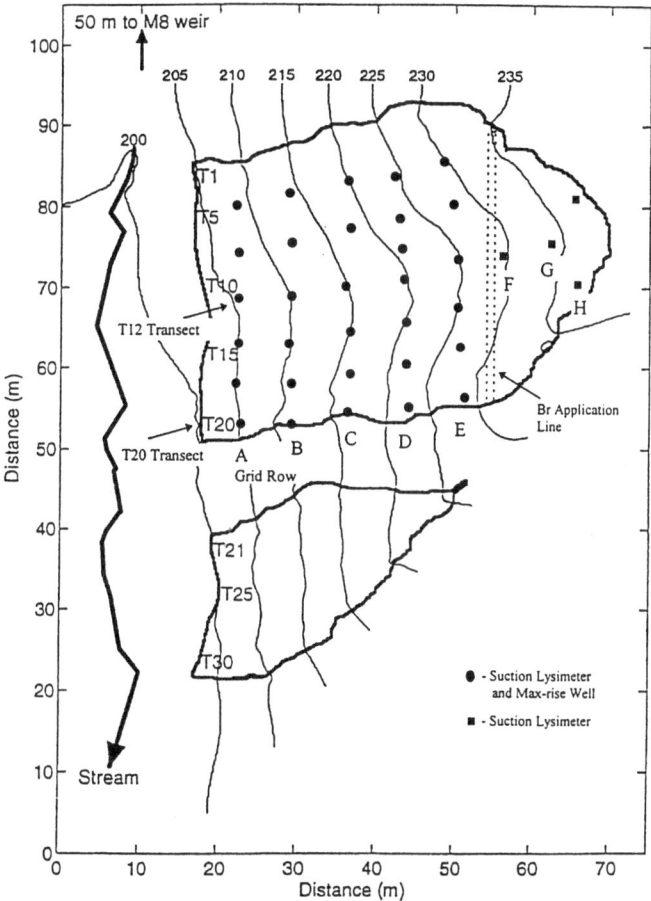

Figure 1. Map of the Maimai hillslope showing Br injection line and 5 m grid of wells and suction lysimeters (modified after Woods and Rowe, 1996).

3. METHODS

Rainfall, soil water, groundwater and trough water were sampled for ^{18}O. In addition to the use of the natural tracer, a line source of 3 kg LiBr was applied at 30 m upslope from the trough face (Figure 1). Application concentration of the solution was 150 mg l^{-1}. Thick vegetation cover and a variable depth humus layer did not allow sprayer or furrow tracer application. A borehole method was used to inject the tracer solution into the mineral soil layer, at a depth of approximately 0.10 - 0.15 m, following methods outlined in Mikovari et al. (1995). A slug of 333 ml of tracer solution was applied in each hole at 0.50 m along the surveyed application lines and then re-covered with soil. Br samples were collected from the suction lysimeters (soil water), maximum-rise wells (groundwater), troughs (throughflow) and stream (runoff). Br was analyzed in the field using an ion selective probe.

A 5 m grid of maximum-rise wells was installed upslope of the troughs to measure the variation of maximum event water table response across the hillslope (Figure 1). The wells were made of 19 mm diameter PVC pipe. The lower 100 mm was slotted and screened with vinyl mesh. A 17 mm diameter inner graduated tube and cork dust were used to record maximum water table heights. The wells were completed to the soil-bedrock interface and water table depths recorded daily. Suction lysimeters were co-located with the maximum-rise wells and installed in mineral soil layer at the nodes at 0.20 - 0.60 m depth. Five additional lysimeters were installed along a transect at 1 m depth, up the central hollow of the hillslope (Figure 1). Recording pressure-transducer tensiometers were located at 5 and 10 m upslope of the troughs at two topographic positions: gully and ridge. Rainfall amount, duration and intensity were recorded using a tipping bucket raingauge connected to a Campbell Scientific CR10 data logger. Sequential rainfall samples were collected at 5 mm intervals using a rainfall sampler described by McDonnell et al. (1991).

4. RESULTS

We report data from three separate sampling periods: December 1993 - February 1994; January - February 1995; March - May 1995. This last period was when the Br tracer addition was made. Seven separate storm events were recorded and sampled during this 45 day period -- rainfall characteristics, runoff amounts and tracer volumes are summarized in Table 1.

4.1 Tracer Data

The downslope transport of Br was tracked as a function of time and hillslope grid position for the seven rain events. Br recovery at the troughs varied by rainfall intensity/duration and time since tracer application. The first event accounted for the largest recovery of Br (30%). The next two events were characterized by low rainfall amounts and intensities (Table 1); tracer recovery was only 2.5%. Subsequent events (#4-#7) were more representative of local precipitation. After 26% and 21% recovery respectively in events #4 and #5, the last two measured events resulted in decreasing amounts of tracer recovery (6 - 8%).

Figure 2 shows the estimated amount of Br tracer that was applied directly to each trough subcatchment. This was calculated based on how much Br was deposited in each soil plug and what subcatchment area was computed for the trough, applying the Freer et al. (1997) multiple flow path algorithm to surface topography. Figure 2 also shows the calculated total recovery of Br across the face for each trough subcatchment. Recovery for the entire hillslope was extremely high (82%), reflecting the rapid transport mechanisms at the site. The amounts recovered in the various troughs did not correspond with the application amounts for individual subcatchments. T6, T9 and T18, which received 700 g of tracer input did not result in any tracer recovery (Figure 2). Troughs with subcatchments which did not receive large or any input of tracer (T10, T11, T13 and T14) were responsible for large percentages of Br recovery (7 - 35%). There was no significant correlation between percentage of Br recovered and subcatchment areas and volumes ($r^2 = 0.44$ and 0.50 respectively).

Table 1. Summary of the intensive study period (March - May 1995) summarizing rainfall, antecedent wetness and Br recovery data.

Event	Date	Duration (hrs)	Peak Intensity (mm/hr)	Total Amount (mm)	API₇	% Peak Subsurface Saturation	% Br Recovery
1	25-26 Mar	18.3	6.6	41.5	10.8	80	30.1
2	1-2 Apr	28.2	5.0	30.2	6.4	30	1.1
3	7-8 Apr	16	5.2	24.4	5.1	25	1.4
4	9-12 Apr	74.3	9.8	83	4.1	50	26.1
5	19-20 Apr	22.2	10.0	62	0.1	60	20.9
6	24-27 Apr	63	10.4	74	9.1	65	8.5
7	5-6 May	37.8	10.8	61	0	60	6.7
	Total Rainfall for Entire Study Period (45 days): 376.1						

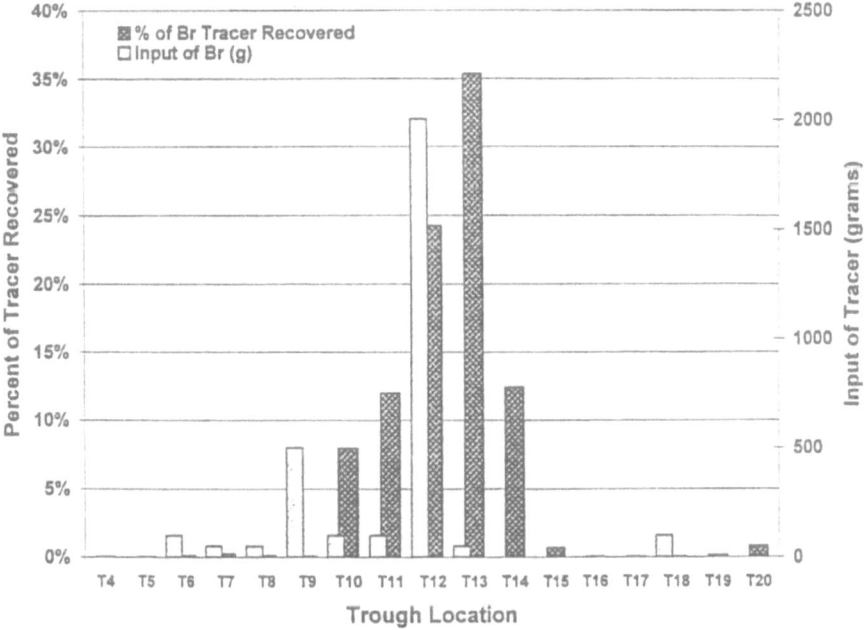

Figure 2. Input and recovery of the Br tracer by subcatchment.

Despite the high tracer recovery, there was little evidence of new water in high flow troughs during the 7 events. Figure 3 shows a plot of δ ^{18}O deflections in rainfall, hillslope piezometer water and trough water. The hydrograph separation shows negligible new water for the trough that carried the largest flow, even though 30% of the Br was recovered during this single event.

4.2 Tensiometer Data

Hillslope Ridge Positions: Soil depths at hillslope ridge positions were approximately 45 cm – similar to most areas upslope of low-response troughs. Available mixing volume for the new water input was small. Pre-storm matric potentials in areas upslope of low-response trough sites were on the order of -20 H_2O cm (Figure 4a). Given a soil porosity of 50% and conversion of matric potential to volumetric water content (based on soil water characteristic curves in McDonnell, 1990), yields roughly 200 mm of stored pre-event water. Assuming complete mixing, new water from the 40 mm rain event should account for roughly 25% of the trough flow (similar to new water amounts for low flow troughs). In fact, mixing was not

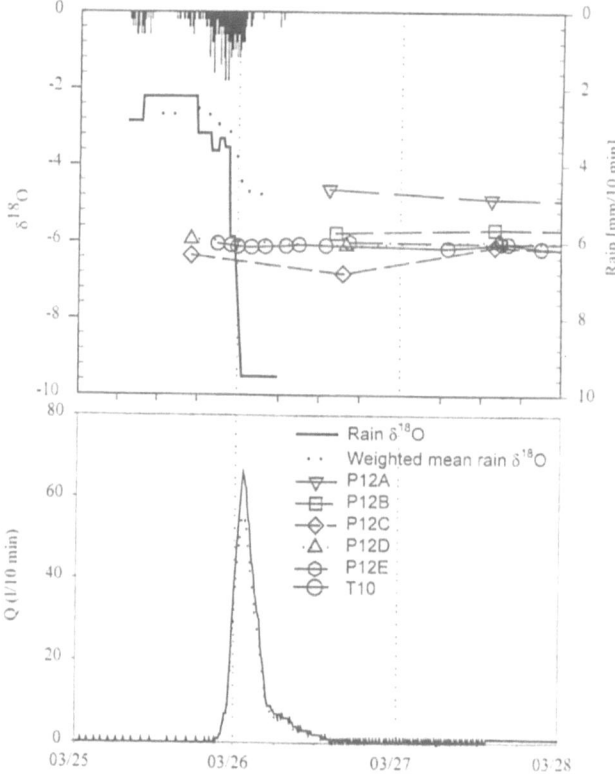

Figure 3. δ ^{18}O values from rainfall, maximum rise wells and trough 10 and a two component hydrograph separation for trough flow. The faint dashed line below the hydrograph denotes the split between old and new water.

complete, as shown in Figure 4a – tensiometer data indicate a disequilibrium in matric potentials and strongly downward flow during times 1800 to 2300 hr. Bypass flow to depth is evident between 2000-2200 hr, based on the close coincidence of paired tensiometer movement and uniformity of measured potentials. Tensiometers at 44.6 cm and 26.7 cm below the ground show a sensitive response to applied rainfall during the 18th/19th event and during events in the following days. Up to 20 cm of water table was set-up during the event – positive pore pressures lasted until 1800 hr on the 20th, coincident with the cessation of flow at T6.

Hillslope Gully Positions: Soil water responses in areas upslope from the high flow troughs were quite different to those data described above. Soils were much thicker – on average 150 cm. Given this greater soil depth and similar pre-storm matric potential conditions, one would expect much lower amounts of new water. Following the same logic as applied to the low-response trough sites, one would expect less than 10% new water, as seen in Figure 3. Tensiometer data in Figure 4b suggest that 25 cm of water table was set-up during the event.

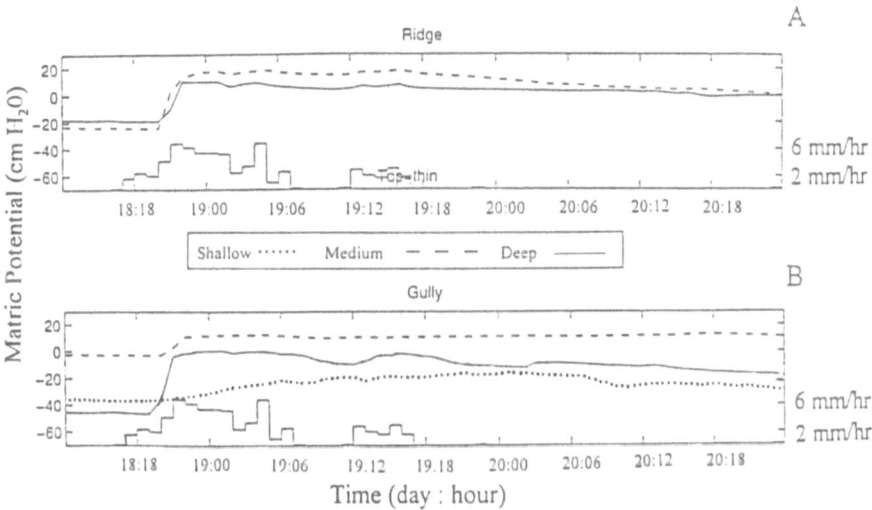

Figure 4. Tensiometer output for an event during January 1994 showing matric potential response at a ridge site (A) and gully site (B) on the experimental hillslope.

4.3 Trough Data

Variability of flow across the trough face was related to the extent of hillslope subsurface saturation. Troughs T11, 12 and 13 were very responsive to rainfall input (Figure 5). Immediate and large outflow off the hillslope occurred in and around the hollow upon reaching a saturation threshold, normally < 1 hr after the onset of rainfall. Events #2 and #3 illustrate the trough face response to low rainfall input (Figure 6). These rainfall events resulted in a maximum flow response of only 40 l 10 min[-1] and subsurface saturation of 25 - 30% of the hillslope (Table 1). Flow recorded from T12 and 13 was only 5-10% of flow levels seen in other events.

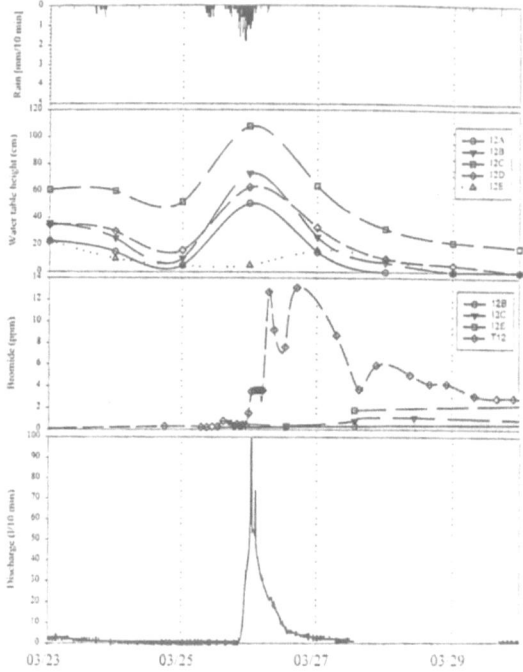

Figure 5. Br concentration in maximum rise wells and trough discharge for event 1 (41.5 mm).

Bromide samples taken from wells and suction lysimeters on the hillslope display a significant dampening and time lag of maximum concentrations compared to the trough outflow concentrations (Figure 5). The dampening effect and time lag is more pronounced downslope, towards the trough face, suggesting a more focused transport of bromide on the lower sections of the hillslope. The Br signal is more muted downslope (Figure 5) as Br is dispersed into the matrix and intersecting macropores. Figure 6 also shows a dilution effect in the downslope direction as more rain is added to the hillslope.

4.4 Topographic Data

Surface- and bedrock-predicted drainage is shown in Figure 7. The areas were calculated by Freer et al. (1997) who used the multiple flow algorithm of Quinn et al. (1991). The greater amounts of accumulated area for the bedrock surface in T12 and T13 agree with the larger Br recovery percentages at these locations (See Figure 2). Total bromide recovery (82% of the line source input) for the complete study period was more highly correlated with the bedrock index distribution.

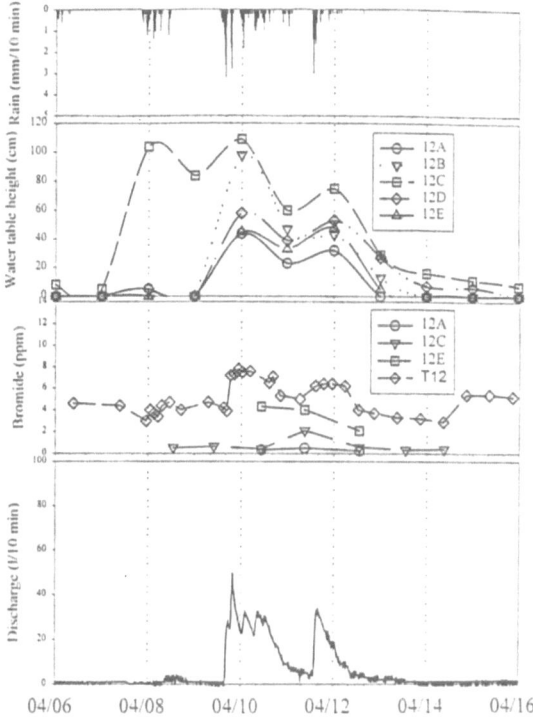

Figure 6 Br concentration in maximum rise wells and trough discharge for events 3 (24.4 mm) and 4 (83 mm) (B).

5. DISCUSSION AND CONCLUSIONS

Recent work by Woods and Rowe (1996) at the Maimai hillslope showed that small trenches (1-3 m wide) do not capture or represent hillslope-scale flow. Previous studies at the site, including Mosley (1979) and McDonnell (1990) were naive in their attempts to compute catchment-wide estimates of subsurface flow from single small trench sites. Woods and Rowe (1996) clearly demonstrated that flow may vary considerably across an otherwise planar looking hillslope section. They also observed (Woods and Rowe, 1996, p. 68) that there is "some positive correlation between flow and area but trough 12, 9 and 21 do not dominate flow as might be expected." This observation may be due to the fact that the bedrock surface controls flow direction during rainfall events on the hillslope, as described in recent studies by McDonnell et al. (1996) and Freer et al. (1997). McDonnell (1990) showed that water perches at the soil bedrock interface during most rainfall events at Maimai and that this saturated flow controls the rapid hillslope runoff response. Thus the process of interest in understanding subsurface flow is saturation from below; a wetting up from the bedrock surface into the soil profile (McDonnell 1997). Differences in flowpaths not immediately related or predicable by standard topographic surveys or topographically-based modeling approaches may result from the discrepancy between the bedrock surface and the soil surface topography.

Data from this study show that water tables on the Maimai hillslope were extremely short-lived. Therefore between events, the bedrock surface may not be an important control on lateral unsaturated flow. Under these conditions, the surface

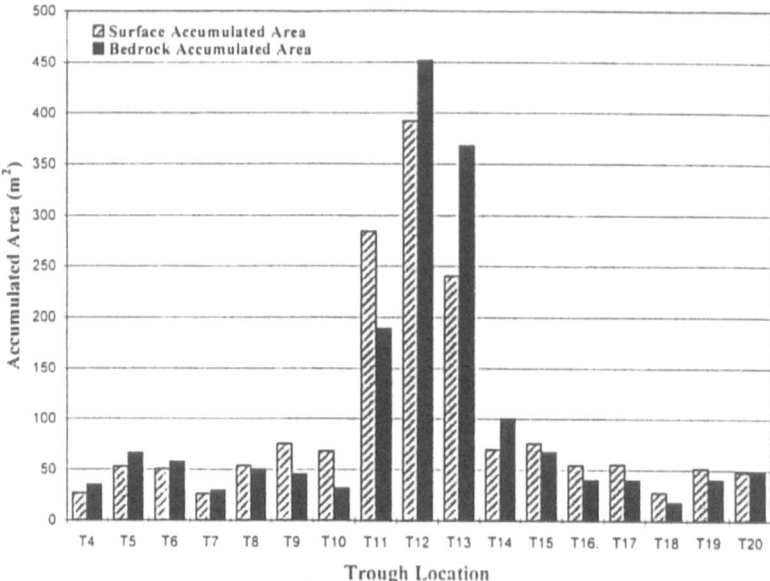

Figure 7. Surface vs bedrock upslope accumulated areas (from Freer et al., 1997). flow patterns (Table 1).

topography may be the best surrogate for flow direction since gravity and matric potential together control the form of the equipotential net and resulting unsaturated flow direction. A possible explanation for the variability observed in Woods and Rowe (1996) may be the switching over between bottom-up bedrock induced flow (during periods of transient water table development during events) and back to topographically-controlled unsaturated flow between events where a top-down drainage process resumes.

Total Br tracer recovery during the study period was extremely high: over 82% (Brammer and McDonnell 1996). Thus, sorption of tracer to the highly weathered soils at Maimai was very weak, and Br was quickly remobilized on each rainfall event. Br appeared in the trough outflow only 6 hr after the application 35 m upslope, which suggests non- Darcian flow transport on the hillslope (given measured K_{sat} of 5-300 mm hr^{-1} in McDonnell 1990). We hypothesize the development of an extensive network of preferential flow paths that allowed Br to be flushed out of the matrix and transported rapidly down the hillslope towards the trough face via saturated flow through macropores. We further speculate that near the application line, at the upper part of the hillslope, many macropores intersect the Br tracer plume. During rainfall events, water in these pores will quickly display an increase in Br concentration. Relatively concentrated Br solution is then transported downslope by gravity. The efficiency of this transport is governed by the extent of the interconnected macropore structure, i.e. extent of subsurface saturation, and hydraulic head. During high intensity and/or long duration rainfall events, most pores are connected and transport large quantities of tracer to the trough face (Figure 5). For smaller events, fewer pores are interconnected, and only portions of the hillslope contribute significant water to the troughs (Figure 6). During such events, smaller quantities of Br appeared at the trough face -- this Br probably originated from the lower portions of the hillslope, and not from the plume source upslope.

Our data suggest that the Br breakthrough and isotopic composition of hillslope waters is a result of differential matrix storage with depth, due to increased soil bulk density, flattened soil water characteristic curve and decreased matrix hydraulic conductivity with depth. Thus our conceptualization of hillslope flow is that only a very small amount of bypass flow to depth is required to fill storage in the deepest soil layers and drive saturated lateral preferential flow at eluviated zones at the soil-bedrock interface. The key to the conceptualization are the multiple constraints placed upon the model; that all of the hillslope flow is old water and the upper soil layers have the ability to store most rainfall due to their increased pdf curve-inferred storage capability. Our model may also help to identify common responses seen in other steep humid hillslope with thin soils (Peters et al. 1995; Tani 1997).

REFERENCES

Beven, K. J., M.J. Kirkby, N. Schoffield, and A. Tagg (1984) : Testing a Physically-Based Flood Forecasting Model TOPMODEL for Three UK Catchments, J. Hydrol., Vol. 69, pp.119-143.

Bonell, M. (1993) : Progress in the Understanding of Runoff Generation Dynamics in Forests. J. Hydrol., Vol. 150, pp. 217-275.

Brammer, D. D. and J.J. McDonnell (1996) : An Evolving Perceptual Model of Hillslope Flow at the Maimai Catchment. In M. G. Anderson and T. Burt (Editors) Advances in Hillslope Processes. John Wiley, New York, pp.35-60.

Buttle, J. (1994) : Isotope Hydrograph Separations and Rapid Delivery of Pre-event Water From Drainage Basins. Prog. In Phys. Geog., Vol. 18, pp.16-41.

Freer, J., J.J. McDonnell, K. Beven, D. Brammer, D. Burns, R. Hooper and C. Kendall (1997) : Topographic Controls on Subsurface Storm Flow at the Hillslope Scale for Two Hydrologically Distinct Small Catchments. Hydrol. Proc, Vol 11, pp.1347-1352.

McDonnell, J .J. (1990) : A Rationale for Old Water Discharge Through Macropores in a Steep, Humid Catchment. Water Resour. Res., Vol. 26, pp.2821-2832.

McDonnell, J. J., M. Stewart and I.F. Owens (1991) : Effects of Catchment-Scale Subsurface Watershed Mixing on Stream Isotopic Response. Water Resour. Res., Vol. 26, pp.3065-3073.

McDonnell, J.J., J. Freer, R. Hooper, C. Kendall, D. Burns, K. Beven and N. Peters (1996) : New Method Developed for Studying Flow on Hillslopes, EOS Trans. Amer. Geophys. Union, Vol. 77, pp.465.

McDonnell, J.J. (1997) : Comment on "The Changing Spatial Variability of Subsurface Flow Across a Hillside" by R. Woods and L. Rowe, J. Hydrol. (NZ), Vol. 36, pp.103-106.

McKie, D. A. (1978) : A Study of Soil Variability Within the Blackball Hill Soils, Reefton, New Zealand. M.S. Thesis, University of Canterbury, Christchurch, N.Z., 180 pp.

McIntosh, J., J.J. McDonnell and N. Peters (1998) : The Effects of Preferential Flow on Soil Water and Conservative Solute Transport in large Intact Field Cores. Hydrol. Proc., in press.

Mikovari, A., C. Peter and Ch. Leibundgut (1995) : Investigation of Preferential Flow

Using Tracer Techniques. IAHS Publ., Vol. 229, pp.87-97.

Mosley, M. P. (1979) : Streamflow Generation in a Forested Watershed, New Zealand. Water Resour. Res., Vol. 15, pp.795-806.

Mosley, M. P. (1982) : Subsurface Flow Velocities Through Selected Forest Soils, South Island, New Zealand. J. Hydrol., Vol. 55, pp.65-92.

Nyberg, L. (1996) : Spatial Variability of Soil Water Content in the Covered Catchment at Gardsjon, Sweden. Hydrol. Proc., Vol. 10, pp.89-103.

Peters, D. L., J. Buttle, C. Taylor and B. LaZerte (1995) : Runoff Production in a Forested, Shallow Soil Canadian Shield Basin. Water Resour. Res., Vol. 31, pp.1291-1304.

Quinn, P., K. Beven, P. Chevallier and O. Planchon (1991) : The Prediction of Hillslope Flow Paths for Distributed Hydrological Modeling Using Digital Terrain Models. Hydrol. Process., Vol. 5, pp.59-79.

Sklash, M. G., M. Stewart and A. Pearce (1986). Storm Runoff Generation in Humid Headwater Catchments, 2, A Case Study of Hillslope and Low-order Stream Response. Water Resour. Res., Vol. 22, pp.1273-1282.

Tani, M. (1997). Runoff Generation Processes Estimated From Hydrological Observations on a Steep Forested Hillslope With a Thin Soil Layer. J. Hydrol., Vol. 200, pp.84-109

Tsuboyama, Y., R. Sidle, S. Noguchi and I. Hosoda (1994) : Flow and Solute Transport Through the Soil Matrix and Macropores of a Hillslope Segment. Water Resour. Res., Vol., 30, pp.879-890.

Woods, R. and L. Rowe (1996) : The Changing Spatial Variability of Subsurface Flow Across a Hillside. J. Hydrol. (N.Z.), Vol. 35, pp.51-86.

Intrastorm Fluctuations of Piezometric Head and
Soil Temperature within a Steep Forested Hollow

Yoshio TSUBOYAMA[*1], Shoji NOGUCHI[*1], Toshio SHIMIZU[*1],
Roy C. SIDLE[*2] and Ikuhiro HOSODA[*3]

[*1] Forestry and Forest Products Research Institute, Kukizaki, Ibaraki, Japan
[*2] University of British Columbia, Vancouver, Canada
[*3] Tohoku Research Center, FFPRI, Morioka, Iwate, Japan

ABSTRACT   Coupled measurements of storm runoff, piezometric response, and thermal fluctuations were conducted at a steep forested hollow (0.25 ha) in Hitachi Ohta, Japan.   Short-term variations in soil temperature during an intensive storm indicate that convergent subsurface flow caused rapid piezometric rise in the head of the hollow.   Soil temperature changed rapidly even during saturated conditions, indicating that aggregate contributions of intensive rainfall and convergent flow cause considerable mixing between rainwater and groundwater.
Keywords: zero-order basin, stormflow, subsurface flow, piezometric head, soil temperature.

1. INTRODUCTION

Headwater areas in mountainous terrain are usually characterized by geomorphic hollows located in the middle to upper portions of the hillslopes. These hollows are known to be sites of active hillslope processes (e.g., Sidle et al. 1985) as well as source areas of runoff (e.g., Ohta 1994) and solute transport (e.g., Luxmoore at al. 1990).   Their role in catchment hydrology, however, remains less understood compared to relatively planar hillslopes where detailed hydrometric and geochemical investigations have been conducted (e.g., Brammer and McDonnell 1996).

A recent study at Hitachi Ohta, Japan (Sidle et al. 1995) showed that a steep hollow with shallow soils contributed stormflow of similar magnitude as a linear hillslope (on unit area basis) once a threshold of saturation was reached. Such non-linear response may have been overlooked by "hillslope hydrologists" but is still important in understanding catchment hydrology.   The objectives of this study are to further elucidate internal processes which may control or affect such hydrological response within the hollow.   Specifically, results from coupled measurements of storm runoff, piezometric response, and thermal fluctuations in the hollow are presented and discussed.

2. METHODS AND MATERIALS

2.1. Site Description

The study was conducted at the Hitachi Ohta Experimental Watershed in Japan. Details on this site are given in Tsuboyama et al. (1994a) and Sidle et al. (1998). The hollow studied in this watershed is deeply incised with steep sideslopes (mean side slope gradient: 34°) and covered with mixed stand of *Cryptomeria japonica* and *Chamaecyparis obtusa* planted around 1920. As a result of knocking-pole penetration tests conducted at 70 points within the hollow, soils appeared generally shallower along the longitudinal axis of the hollow and deeper along the topographic divides (Tsuboyama et al. 1994b). Soil depth to bedrock ranged from 0.44 to 4.15 m with an arithmetic mean and standard deviation (SD) of 1.42 and 0.45 m, respectively.

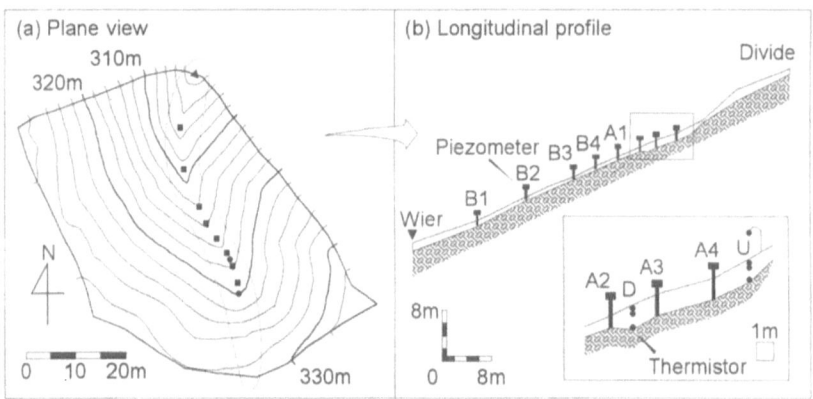

Fig. 1 Topography and instrumentation of the hollow at Hitachi Ohta.

2.2. Instrumentation and Field Methods

A 60° V-notch gauging weir was installed at the outlet of the hollow to monitor discharge. Eight float-type recording piezometers were installed along the longitudinal axis of the zero-order basin (B1-B4 and A1-A4; Fig. 1). All piezometer tubes (43 mm inner diameter) were perforated in the lower 10 cm and depth to the bottom of piezometer ranged from 0.5 to 1.2 m (Tsuboyama et al. 1994b). Soil temperature was monitored at the two profiles in the upper portion of the hollow (D and U; Fig. 1). At each profile, thermistors were installed at depths of 0.2 and 0.5 m as well as at the soil-bedrock interface (1.4 and 1.3 m deep for D and U, respectively). Air temperature was measured at the upper profile (U), 1.5 m above the ground (but under the forest canopy) using the same type of a thermistor. All data described above were collected every 30 minutes.

Air and soil temperatures at 1.5 m above and 0.2 m below the ground, respectively, were measured at an open meteorological station located 250 m north of the hollow. Because the station is located on a hilltop with no upslope contributing area, effects of lateral flow on soil temperature should be minimal. Precipitation was measured at the same site using a tipping bucket and storage rain gauges. All data at the station were collected every 10 minutes.

3. RESULTS AND DISCUSSION

3.1. Thermal Fluctuations During Fair Weather

There was a long dry period at the site with little precipitation (<1 mm) during two weeks from late August to mid-September in 1994. Fig. 2 shows thermal fluctuations during the last four days of this dry period. Neither outflow from the hollow nor the groundwater in piezometers was observed during these four days.

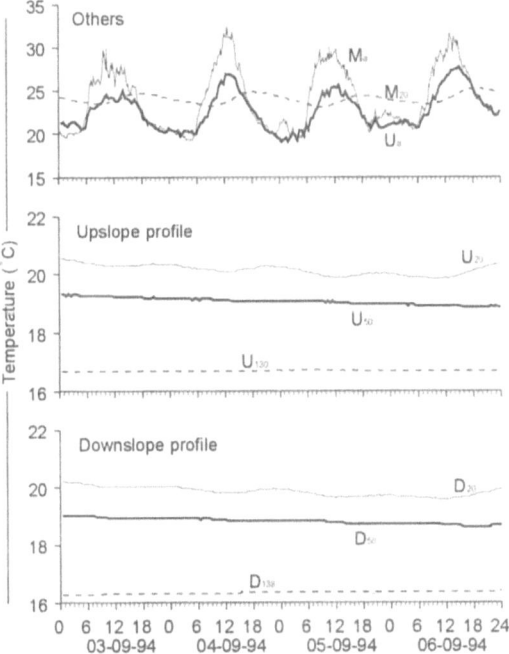

Fig. 2 Thermal fluctuations during 3-6 September 1994. Symbols "M", "U", and "D" refer the locations of measurements, i.e., "M". the meteorological station, "U" and "D": the upper and lower profiles in the hollow (Fig. 1). The subscript "a" denotes air temperature measured 1.5 m above the ground. Other subscribed numbers denote depths (unit: cm) below the ground.

Air temperatures both at the meteorological station (M_a) and below the canopy (U_a) exhibited a daily, nearly sinusoidal cycle (Fig. 2). Soil temperature at the meteorological station (M_{20}) also exhibited a daily cycle while the amplitudes were considerably damped and fluctuations were delayed compared to M_a and U_a. As to soil temperatures within the hollow, no significant diurnal fluctuations were noted at either depth.

Clearly thermal fluctuations near the soil surface (shallower than 0.2 m) within the hollow would not propagate to depths of 0.5 m or deeper for this short time period (~hours) when effects of subsurface flow on soil temperature were assumed to be minimal. In other words, if soil temperature in these deeper positions fluctuated during this short time period, the changes should have resulted from heat convection caused by water flux. This is a key assumption to the subsequent analyses.

3.2. Frontal Storms with Dry Antecedent Conditions

Fig. 3 shows hydrologic and thermal response of the hollow to a series of rainstorms attributed to the stationary front. There was only 9 mm of precipitation during 10 days prior to these storms. Neither discharge from the hollow nor groundwater in piezometers was observed before these storms. Precipitation during these storms totaled 110 mm.

During the initial rainstorm (2100h on 12th to 0500h on 13th), only the lower piezometers (B1 to A1) responded. Piezometer B2 was an exception but some operational problems were found. Among the piezometers which responded, the peak piezometric rise was the highest at B1 and the lowest at A1 (Fig. 3). This pattern suggests that a wedge of groundwater accumulated from the base of the hollow. During the subsequent rainstorms (from 1400h on 13th), the upper piezometers became more responsive to individual storms, while the lower piezometers gradually became to show flatter peaks as they approached saturation (Fig. 3).

The discharge responded rapidly to the first peak of the rainfall (2200h-2400h on 12th), while stormflow increased in terms of both volumes and peak rates during subsequent rainstorms (Fig. 3). Sidle et al. (1995) reported that this hollow became more responsive during a period of increasing wetness over one month. A similar pattern of response occurred during the shorter time period (Fig. 3). Rapid increases in stormflow appear to coincide with response of the middle to upslope piezometers, suggesting that shallow groundwater developing along the trough of the hollow (although not necessarily saturated) facilitated the hydrologic linkage between upper and lower portions of the hollow.

During these storms soil temperatures fluctuated only gradually (Fig. 3), possibly because the rainwater temperature (inferred by air temperatures) was within the range of vertical variations of temperatures within the profile. In such a case, thermal information could not be used as a signature for hydrological tracing (Sklash and Farvolden 1982).

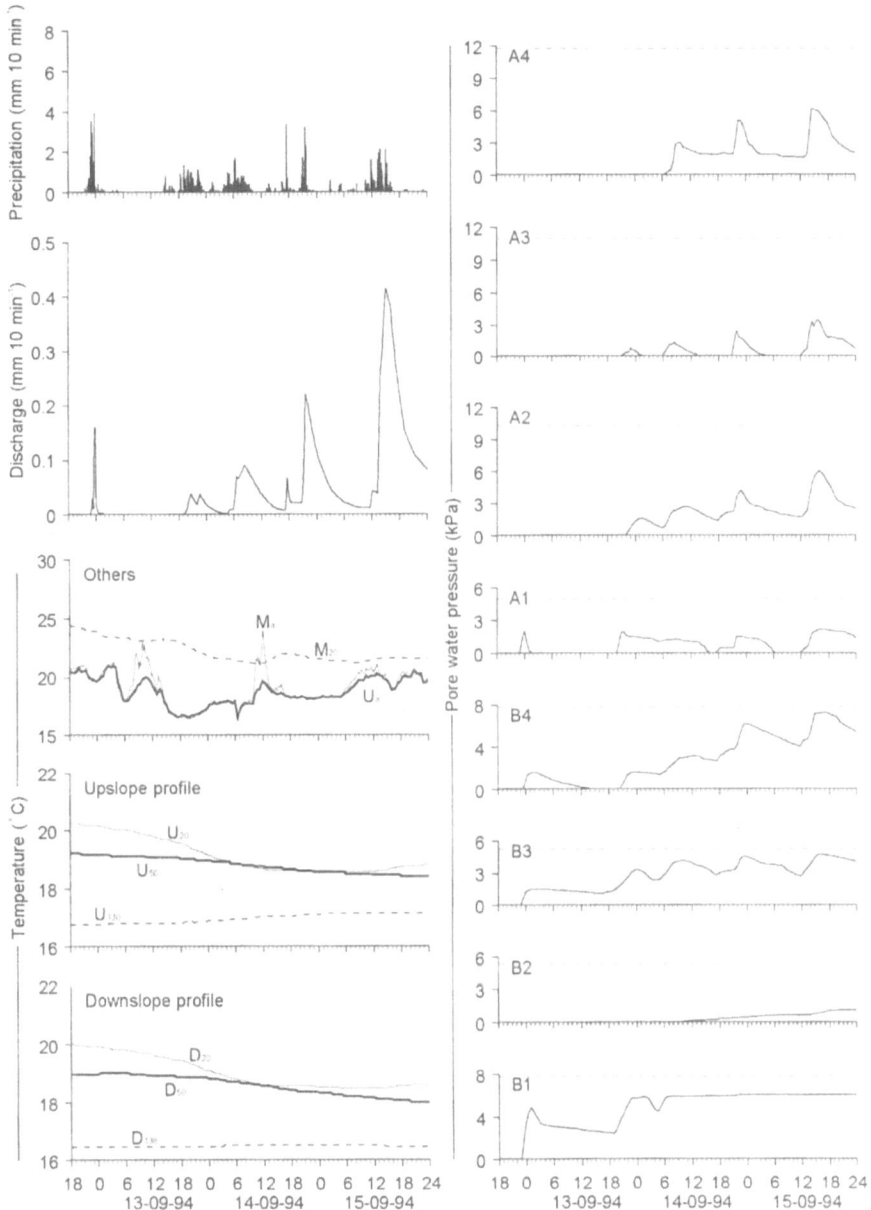

Fig. 3   Hydrologic and thermal response of the hollow to a series of the frontal storms.   Symbols A4 to B1 represent the locations of piezometers shown in Fig. 1.   Broken lines in the graphs of piezometric response reveal pore-water pressure when the water table inside the piezometer is at the same level as the soil surface. Symbols and subscripts related to temperatures are used similarly as in Fig. 2

480

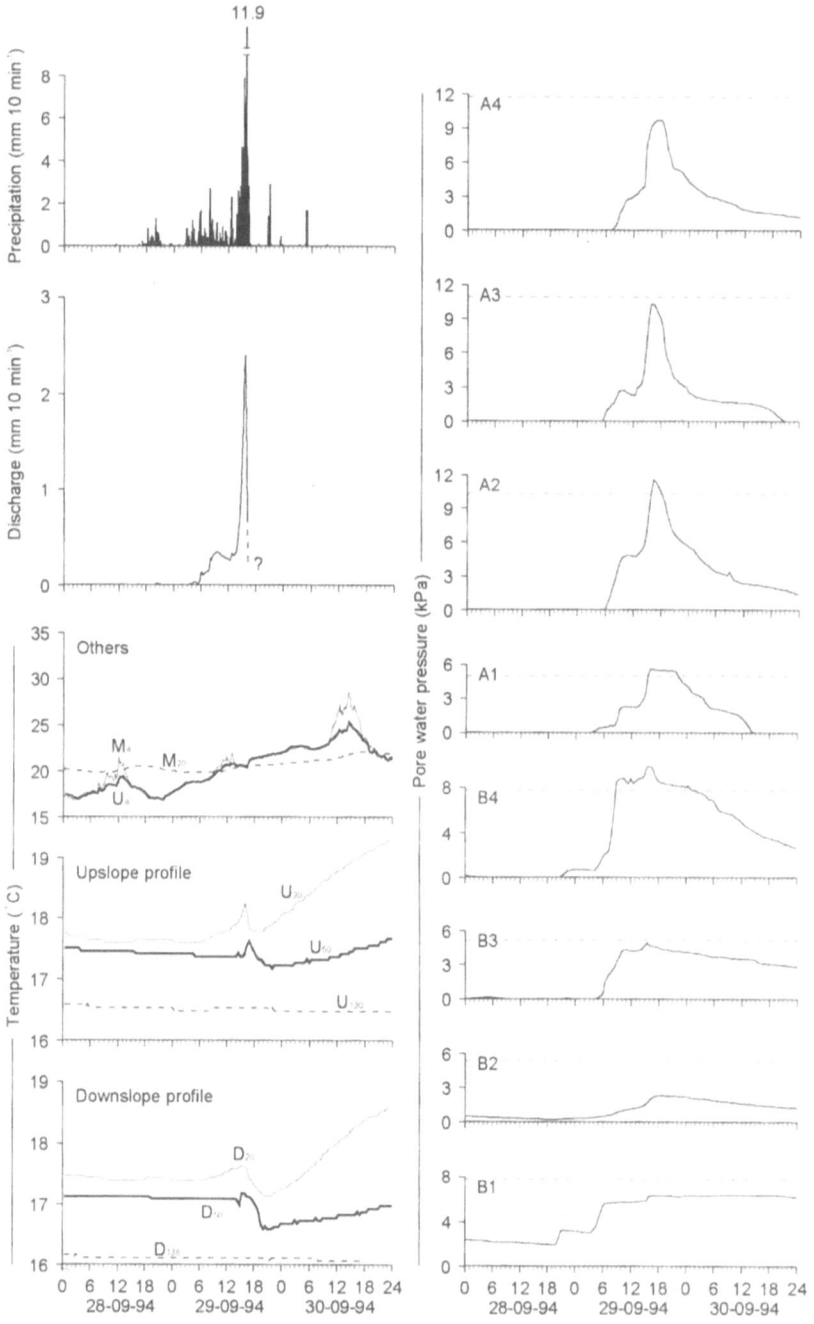

Fig. 4   Hydrologic and thermal response of the hollow to typhoon storm T9426.
Symbols and subscripts are used similarly as in Fig. 3.

3.3. A Typhoon Storm with Wet Antecedent Conditions

Fig. 4 shows hydrologic and thermal response of the hollow to typhoon storm T9426. Antecedent 10- and 5-day precipitation for this storm was 65 and 40 mm, respectively. The storm deposited 132 mm of rainfall during 38 h, while more than half (69 mm) fell during 3 h from 1400h on 29 September. Maximum 1-h intensity was 40 mm h^{-1}, which is one of the highest intensities ever recorded at this site. Due to excessive sedimentation at the gauging weir, the later portion of the stormflow data was unrecorded (Fig. 4).

In contrast to the previous example (Fig. 3), the timing of piezometric response was less affected by the slope position (Fig. 4). Except for the lower two piezometers (B1 and B2), where the water table persisted from the previous storms, most of other piezometers simultaneously exhibited an initial rapid rise during 3 h from 0600h on September 29th. As rainfall intensified (1400h-1700h on 29th), pore-water pressures in the upper three piezometers increased rapidly, while pore-water pressures in the lower piezometers exhibited flatter peak.

During this storm soil temperature at U_{20} increased until the rainfall peak (~1600h) and then began to decline quickly while temperature at the soil-bedrock interface (U_{130} and D_{138}) was nearly constant (Fig. 4). The abrupt change from an increase to a decrease was also observed at U_{50}, D_{20}, and D_{50} (Fig. 4).

Three important considerations related to temperature were noted: (1) the vertical patterns of temperature were consistent before and during the storm, i.e., higher in the air and lower in the deeper part of the soil profile; (2) the soil temperature at the meteorological station continuously increased during the storm, suggesting that rainwater temperature did not changed enough to cause the thermal fluctuations in the subsurface of the hollow; and (3) the horizontal distribution of soil temperature at the surface might be equalized through the extended pre-wetting period. Based on these considerations, the thermal fluctuations appear to result from percolation of "warm" rainwater followed by an invasion of "cold" groundwater converging from upslope.

4. CONCLUSION

Although the applicability of temperature as a signature for hydrological tracing depends on prestorm and intrastorm thermal conditions, data from our study clearly indicate that in some cases convergent flow rather than percolating rainwater would contribute to rapid build-up of pore water pressure in the head of the hollow. Also short-term fluctuations of soil temperature even during saturated conditions (e.g., D_{50} from 1400h to 1800h on 29 September; Fig. 4) suggest that considerable mixing occurred between rainwater and shallow groundwater. This soil/topography-controlled mixing which occurs in the head of the hollow would imply alternation of flow paths of presumed hydrograph components such as rainwater, soilwater, and groundwater before they discharge to the stream, thus questioning unique relationship between presumed flow components and actual flow pathways.

ACKNOWLEDGEMENTS   We thank Motohisa Fujieda, Mayumi Hosoda, and Shigeki Murakami for their help in conducting the field work.

REFERENCES

Brammer, D.D., and J.J. McDonnell (1996): An evolving perceptual model of hillslope flow at the Maimai catchment, In: M. G. Anderson and S. M. Brooks (eds.), Advances in Hillslope Processes, John Wiley & Sons, New York, Vol. 1, pp. 35-60.

Luxmoore, R.J., P.M. Jardine, G.V. Wilson, J.R. Jones, and L.W. Zelazny (1990) Physical and chemical controls of preferred path flow through a forested hillslope, Geoderma, Vol. 46, pp. 139-154.

Ohta, T. (1994): "Zero order basin" and forest hydrological researches - an outline of forest hydrology in Japan -, In: Proc. of the Inter. Symp. on Forest Hydrology, 1994, University of Tokyo,Tokyo, pp. 1-10.

Sidle, R.C., A.J. Pearce, and C.L. O`Loughlin (1985) Hillslope Stability and Land Use, Water Resour. Mono. 11, Am. Geophys. Union, Wash., D.C., 140p.

Sidle, R.C., Y. Tsuboyama, S. Noguchi, I. Hosoda, M. Fujieda, and T. Shimizu (1995): Seasonal hydrologic response at various spatial scales in a small forested catchment, Hitachi Ohta, Japan, J. Hydrol., 168: 227-250.

Sidle, R.C., Tsuboyama, Y., S. Noguchi, I. Hosoda, M. Fujieda, and T. Shimizu (1998): Progress Towards Understanding Stormflow Generation in Headwater Catchments, In: Environmental Forest Science, Proc. of IUFRO Div. 8 Conf., 19-23 October 1998, Kyoto, Japan, Kluwer Academic Publisher (this issue).

Sklash, M.G., Farvolden, R.N. (1982): The use of environmental isotopes in the study of high-runoff episodes in streams, In: E.C. Perry Jr. and C.W. Montgomery (eds.), Isotope studies of hydrologic processes, Northern Illinois University Press, DeKalb, Illinois, pp. 65-74.

Tsuboyama, Y., R.C. Sidle, S. Noguchi, and I. Hosoda (1994a) Flow and solute transport through the soil matrix and macropores of a hillslope segment, Water Resour. Res., Vol. 30, pp. 879-890.

Tsuboyama, Y., I. Hosoda, S. Noguchi and R.C. Sidle (1994b): Piezometric response in a zero-order basin, Hitachi Ohta, Japan, In: Proc. Inter. Symp. on Forest Hydrology, 1994, Univ. of Tokyo, Tokyo, pp. 218-225.

Progress Towards Understanding Stormflow Generation in Headwater Catchments

Roy C. SIDLE*[1], Yoshio TSUBOYAMA*[2], Shoji NOGUCHI*[2],
Ikuhiro HOSODA*[3], Motohisa FUJIEDA*[2] and Toshio SHIMIZU*[2]

*1 Departments of Forest Resources Management and Geography, University of
   British Columbia, Vancouver, British Columbia, V6T 1Z4, Canada
*2 Forestry and Forest Products Research Institute, Tsukuba Norin Kenkyu,
   Ibaraki 305, Japan
*3 Tohoku Research Center, Forestry and Forest Products Research Institute,
   Morioka 020-01, Japan

ABSTRACT   An emerging concept of stormflow generation in steep, forested
headwaters is presented that incorporates dynamic hydrologic linkages among and
within distributed geomorphic units. The hydrogeomorphic concept is based on
field studies at Hitachi Ohta watershed in Japan. For drier antecedent conditions,
stormflow is generated primarily as saturation overland flow from narrow riparian
zones and as direct channel interception. As moisture increases, subsurface flow
from the soil matrix augments stormflow. Further increases in wetness induce
incipient runoff from hollows with shallow soils as well as macropore drainage.
During very wet conditions, self-organized preferential flow networks develop
that efficiently drain hillslopes; hollows also contribute significantly to stormflow.
Keywords: hillslope hydrology, streamflow generation, hydrogeomorphic concept,
macropores, zero-order basin.

1. INTRODUCTION

Headwater catchments have largely been ignored in managing temperate and
tropical forests. Independent studies have noted the significance of headwaters as
sites of sediment production and streamflow generation (e.g.; Dietrich and Dunne
1978; Pearce et al. 1986; Sidle et al. 1995). Additionally, these sites are of
ecological interest because they represent sources and sinks for nutrients (Feger et
al. 1990; Wilson et al. 1991). During the past few decades, different paradigms
have emerged to attempt to explain stormflow and streamflow generation. Even
today, considerable debate continues regarding the importance of various hydrologic
pathways and source areas in headwater catchments as well as the interaction of
these pathways and source areas with respect to peak runoff, solute transport, surface
erosion, and mass wasting.

The variable source area concept advanced the understanding of streamflow generation in vegetated catchments (Tsukamoto 1963; Hewlett and Hibbert 1967; Kirkby and Chorley 1967). This conceptual model invokes a dynamic riparian source area that shrinks and expands in response to rainfall and fluctuating water tables. Although the model recognizes subsurface flow from adjacent hillslopes, specific hydrologic mechanisms operating at different spatial scales are not specified. Several studies in moderate to gently sloping basins have cited saturation overland flow and return flow within broad, relatively flat riparian areas as the dominant stormflow generation mechanisms (e.g., Dunne and Black 1970; Eshleman et al. 1993), whereas Sklash and Farvolden (1979) attributed stormflow generation to a groundwater 'ridging' effect. Studies in steeper basins revealed several dominant stormflow generation mechanisms, including the subsurface translatory flow (Hewlett and Hibbert 1967), capillary fringe response (Gillham 1984), pneumatic effect (Yasuhara and Marui 1994), and preferential flow associated with macropores and soil pipes (Mosley 1979; Tsukamoto and Ohta 1988; McDonnell, 1990; Tsuboyama et al. 1994a). All of these investigations in steep forested terrain can safely ignore Hortonian overland flow because of the high infiltration capacity of forest soils. Thus, lateral subsurface runoff is at least partly promoted by the existence of a hydrologic impeding layer (e.g., bedrock, till) below the soil profile (Harr 1977; Megahan 1983).

While the importance of subsurface flow in steep forested catchments is generally acknowledged, the specific pathways are the topic of much debate. Large discharges from soil macropores and pipes during natural and simulated storms have been measured or inferred at several sites (Whipkey 1965; Mosley 1979; Tsukomoto and Ohta 1988; Kitahara and Nakai 1991; Turton et al. 1992). Studies with nonreactive chemical tracers have shown that macropore systems increase in importance (Chen and Wagenet 1992) and may expand during wetter conditions by interacting with surrounding mesopores (Luxmoore and Ferrand 1993; Tsuboyama et al. 1994a). Isotope and other natural tracer studies (Sklash and Farvolden 1979; Pearce et al. 1986; McDonnell et al. 1990; Wilson et al. 1991) have enhanced our understanding of the relative proportion of 'old' and 'new' water contributing to streamflow and, in some cases, have questioned the importance of preferential flow pathways. However, pathway specification based on end member mixing analysis invoked in these tracer studies is fraught with difficulties if intercompartmental mixing within the regolith occurs (Luxmoore and Ferrand 1993; Sidle et al. 1995; Noguchi et al. 1997; submitted; Tsuboyama et al. 1998).

The significance of understanding flow pathways in headwater catchments has very practical implications for forest management. By knowing the active hydrologic zones in such basins, it is possible to better plan the temporal and spatial aspects of various forest management activities. Of particular interest are activities that could cause compaction in hydrologically sensitive zones where flow paths may be diverted. Information could also guide the use of chemicals within watersheds as well as predict the impacts of acid deposition. Furthermore, better and more meaningful estimates of important parameters used in determining peak flow production and downstream routing may evolve from an improved understanding of stormflow pathways. Finally, knowledge of flow pathways is critical to addressing the spatial and temporal scaling issues inherent in cumulative watershed effects analysis.

The research reported in this paper is a summary of recent findings from the Hitachi Ohta Experimental Watershed in Japan. We develop insights into dynamic hydrologic response at various spatial scales ranging from a hillslope segment to a 2.48 ha forested watershed. Emphasis in our discussion focuses on flow response and pathways during a sequence of storm events with increasing antecedent wetness.

2. METHODS AND MATERIALS

2.1 Study Area

An ongoing series of investigations at Hitachi Ohta Experimental Watershed in Japan provide the basis for the development of a hydrogeomorphic model for deeply incised headwater basins. Stormflow generation studies focused on a series of nested experimental units within a 2.48 ha first-order drainage.

The study site is located on the east side of the main island of Honshu at a latitude of 36°34'N and a longitude of 140°35'E (Fig. 1). Prior to the 20th century the watershed was covered with a natural hardwood forest. Following clearcutting of this natural forest, the watershed was replanted with Sugi (*Cryptomeria japonica*) and Hinoki (*Chamaecyparis obtusa*) around 1920. Hardwood and various understory species coexist in gaps within this current conifer stand.

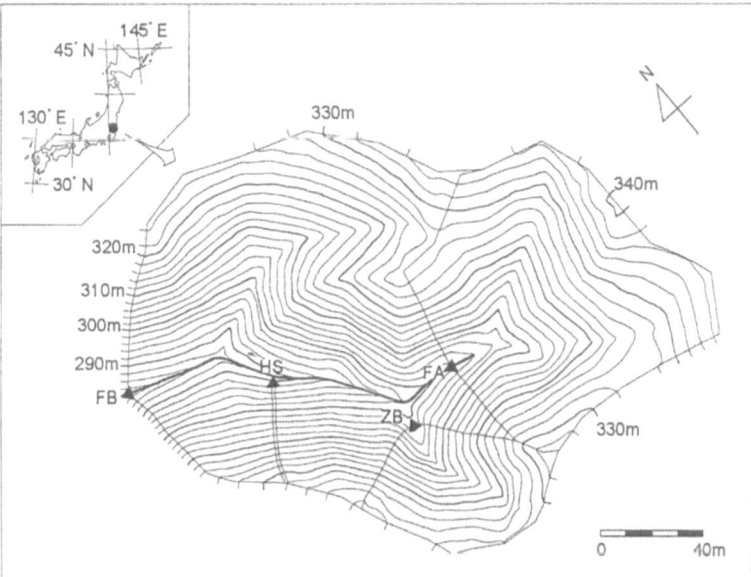

Fig. 1 Topographic map of the 2.48 ha catchment at Hitachi Ohta Experimental Watershed, Japan. Nested sub-catchments and the hillslope segment are specified.

Basin elevation ranges from 283 to 341 m. Average channel gradient is 8.7° and sideslope gradients ranged from 8.5 to 50.6° (mean gradient 32.4°). Because of the deeply incised channels, riparian corridors throughout the basin are very narrow (1.0-3.6 m wide; average width 2.1 m). Such conditions are quite typical in temperate and tropical headwater basins. Average annual precipitation is 1457 mm with two major rainfall seasons: (1) the Baiu season from early June to mid-July and (2) the typhoon season from September to late October. Summers are typically hot and humid. Winter snowfall is sporadic and a persistent snowpack does not develop.

Soils, derived from volcanic ash, are well aggregated and overlain by a thin organic horizon. Surface lithology is metamorphic, primarily schist and amphibolite. Total soil depth to bedrock within the watershed ranges from about 0.3 to 4.7 m, with mean depth of 1.6 m (Sidle and Tsuboyama, 1992; Tsuboyama et al., 1994b). Soil macropores with diameters ≥ 2 mm were described in nine soil pits excavated within the watershed. Average density of macropores is 25.7 m^{-2} (Noguchi et al. 1997). Macropore diameters range from 2 to 40 mm with a mean of 12.2 mm. Subsurface erosion, decayed and live root channels, and interactions between subsurface erosion and root channels accounted for about 92% of the described macropores.

The high infiltration capacity of the soils, shallow depth to bedrock, abundant macropores, and steep slope gradients promote lateral subsurface flow in hillslopes at the exclusion of overland flow. Although the narrow riparian zone is generally at or very near saturation, it has a limited capacity to generate saturated overland flow during storms compared to wider corridors in more gently sloping basins (Dunne and Black 1970; Eshleman et al. 1993; Fujieda et al. 1997).

2.2 Field Methodology

Stormflows during the 1992 typhoon season were continuously monitored in a series of nested hydrologic units consisting of: a 2.48 ha first-order drainage (FB); an incipient 0.84 ha first-order drainage (FA); a 0.25 ha zero-order basin (ZB); and various soil components of a 0.0045 ha hillslope segment (HS) (Fig. 1). While other temporal runoff data were available for these nested catchments, the 1992 season represented a rather unique situation because very little rainfall occurred between the end of the Baiu season and the slightly late onset of the typhoon season. Thus, the temporal sequence of storms during the 1992 typhoon season represent a progressive increase in antecedent wetness. Precipitation was measured by recording and storage rain gauges located in the northern portion of the greater watershed.

Storm discharge was monitored at V-notch weirs installed at the outlets of FB, FA, and ZB (Fig. 1). Subsurface discharge from a hillslope segment (HS) was monitored at a pit excavated near the perennial channel (but about 0.5 m outside the riparian zone) (Fig. 1). Flow from the pit was segregated into five components: the organic and A horizons combined (C1); the mineral soil matrix (C3, excluding visible macropores); and three separate groups of macropores (C2, C4, and C5). The hillslope above the pit was 49 m long with an average gradient of 39°. The estimated contributing area above the pit based on topographic surveys, pit width, and slope length is 45 m². It is recognized that the monitored hillslope segment does not represent flow conditions for all linear hillslope segments in the drainage. In fact, a nearby soil pit experienced considerably less (but unmonitored) subsurface flow during rainfall events. Thus, water yields obtained at the soil pit must be extrapolated with care. Details of the continuous monitoring system for subsurface drainage are presented by Tsuboyama et al. (1994a) and Sidle et al. (1995). Storm discharge from all watershed components was calculated on a unit contributing area basis and expressed as mm of runoff to facilitate direct comparisons.

A series of conservative tracer tests was conducted in the lower 2 m (slope distance) of the same hillslope segment (HS) to elucidate flow pathways and the response of macropores during controlled hydrologic inputs and antecedent conditions. Chloride tracer was applied continuously until outflow concentrations from the pit face approached steady-state. The resulting data collected for these controlled step-change tracer experiments were analyzed using miscible displacement techniques (e.g., Rose and Passioura 1971). Experimental details of these tracer experiments have been described by Tsuboyama et al. (1994a) and Sidle et al. (1994).

Later investigations within the experimental slope section (HS) employed a combination of staining agents to further elucidate preferential flow pathways (Noguchi et al. 1997; submitted). Approximately one pore volume of a dilute white paint solution was applied at the line irrigation source (located 2 m slope distance above the pit face). Following dilute paint application, the 2 m segment of the hillslope was carefully dissected in 10 cm intervals to ascertain routing of preferential flow in the soil and weathered regolith (Noguchi et al., submitted). Additionally, individual macropores > 2 mm in diameter were "traced" upslope during excavation (starting at the original pit face) by spraying colored powdered chalk into the macropore cavities (Noguchi et al. 1997). The combination of these two staining techniques allows for a an interesting comparison of actual preferential flow paths (white paint staining) with macropore location and distribution (colored chalk).

Eight float-type recording piezometers were installed at the soil-bedrock interface along the longitudinal axis of ZB (Sidle and Tsuboyama 1992; Tsuboyama et al. 1994b). This zero-order basin is one of six such hollows in FB. Soils in ZB are typically 0.7 m shallower than soils in neighboring basin FA. Basin FA is primarily comprised of two zero-order basins that converge 19 m upstream of the weir. Three other zero-order basins exist in FB. No hydrologic measurements were conducted within these basins. By definition, all of these zero-order basins or hollows are unchanneled (Tsukamoto 1963; Dietrich and Dunne 1978). Within ZB, a series of coupled measurements of stormflow, shallow groundwater response, and soil temperature fluctuations were conducted during the 1994 typhoon season (Tsuboyama et al. 1998). Soil temperature was monitored at several depths and at several positions within ZB using thermistor probes. Temperature data were collected at 30 min intervals and calibrated with results from laboratory tests (Tsuboyama et al. 1998).

3. RESULTS AND DISCUSSION

3.1 General Hydrologic Response

Rainfall and runoff data for seven successive storms during the 1992 typhoon season are summarized in Table 1. Storms were included if total precipitation exceeded 10 mm and maximum 30-min rainfall intensity exceeded 4 mm h^{-1}. Antecedent wetness as indexed by both 10- and 30-day antecedent rainfall (API_{10} and API_{30}) ranged from 0.5 to 109.6 mm and 11 to 187.1 mm, respectively. Maximum 30-min rainfall intensity ranged from 4.6 to 23.6 mm h^{-1}.

Table 1 Rainfall characteristics and runoff or outflow values (in mm) for each component of Forest Basin B, Hitachi Ohta, Japan (adapted from Sidle et al. 1995)

Storm Date	Storm Length (min)	Total Precip. (mm)	API_{10} (mm)	API_{30} (mm)	Max 30 min int. (mm h^{-1})	Runoff or Outflow (mm)			
						FB	FA	ZB	HS
19 Sep.	930	11.7	0.5	11.0	4.6	0.117	0.003	0.000	0.004
26 Sep.	1390	19.5	12.2	21.6	6.8	0.316	0.016	0.000	0.199
29 Sep.	90	12.2	19.6	41.1	16.8	0.348	0.036	0.032	0.156
30 Sep.	570	23.1	34.0	55.6	8.4	1.113	0.260	1.003	2.893
8 Oct.	900	61.4	57.0	88.8	18.6	7.068	4.579	16.018	13.974
9 Oct.	110	15.6	106.0	150.2	23.6	1.779	0.937	4.144	4.000
20 Oct.	1970	48.0	49.1	187.1	10.4	9.383	7.533	15.487	25.075

Water yields at FB during the seven storms increased progressively from only 1% at the onset of the typhoon season (driest conditions) to almost 20% during the final storm (Fig. 2a). These influences reflect the cumulative effects of antecedent moisture on stormflow production. The highest water yield observed at FB is lower than the yield from a similar sized storm in a larger catchment in New Zealand (McDonnell et al. 1990).

The contribution of the narrow riparian corridor to stormflow, including direct channel interception, was assessed throughout the storm season assuming all rainfall on this corridor would be completely converted to streamflow. At the beginning of the typhoon season, the riparian corridor contributed all of the measured stormflow from FB (Fig. 2b). As the season progressed, contributions from other geomorphic components in FB increased water yields. Thus, the riparian contribution to stormflow declined rapidly. After the fourth storm, riparian contributions were < 10% of the runoff hydrograph from FB (Fig. 2b). By the end of the typhoon season, the riparian corridor contributed only 5.4% of the stormflow response. These results show that narrow riparian corridors in such dissected mountain topography cannot deliver sufficient saturated overland flow to account for wet season stormflow response. Subsurface transport from adjacent hillslopes together with discharge from zero-order basins is clearly necessary to explain observed stormflow from FB (Table 1).

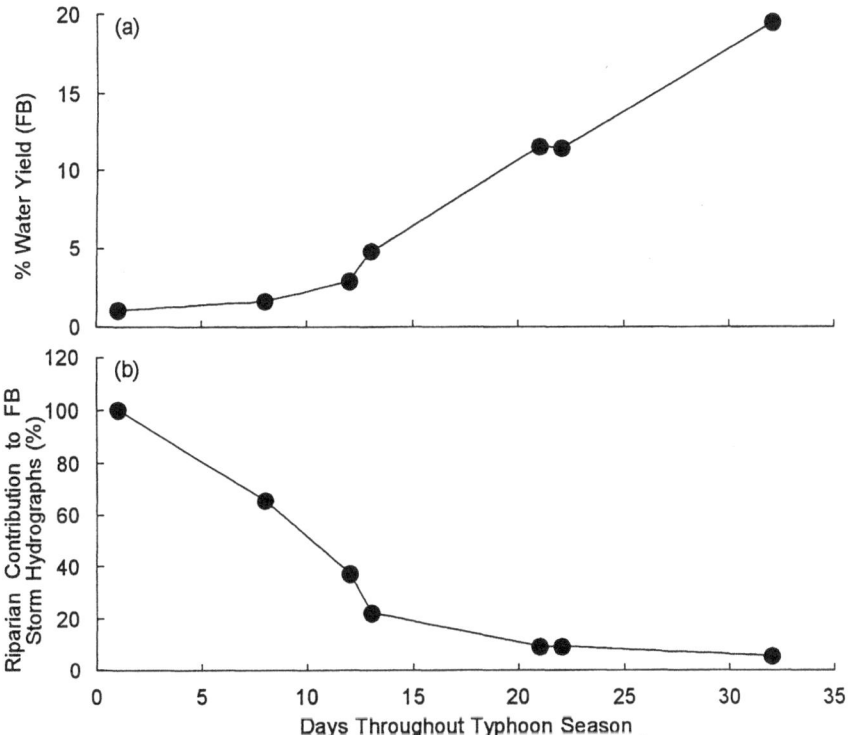

Fig. 2 Temporal changes in water yield from basin FB as well as in the riparian zone contribution to outflow from FB from the onset to the end of the 1992 typhoon season.

3.2 Hillslope Response

Subsurface stormflow within hillslopes was directly related to antecedent wetness. During the driest antecedent conditions (storm on 19 September), subsurface drainage from the linear slope segment comprised only 3.4% of storm runoff from FB (based on equivalent areas; Table 1). This contribution progressively increased, and by the fourth typhoon storm (30 September) onwards, the hillslope segment produced at least twice the stormflow on an equivalent area basis as the entire FB. During the last typhoon storm, the hillslope segment generated 3.3 times more stormflow compared to an equivalent area of FB.

Pathways of water transport in hillslopes reflect a complex interaction among the following watershed attributes: antecedent wetness, macroporosity, bedrock characteristics, topography, soil organic matter, perched water tables, matrix permeability, and soil depth. No overland flow on relatively planar hillslopes was observed even during the largest and most intense storm events.

During the first two typhoon storms (September 19th and 26th), essentially all of the subsurface flow from the hillslope segment emanated from the soil matrix. Estimated hillslope travel lengths for matrix flow (outside of the riparian zone) ranged from 1.8 to 3.6 m based on the product of average pore velocity (derived from tracer tests, Tsuboyama et al. 1994a) and lag time from the mass centroid of the rainfall hyetograph to the mass centroid of the subsurface flow hydrograph (Sidle et al. 1995). Such small travel distances would be consistent with recharge to the riparian zone described in the variable source area concept (Tsukamoto 1963; Hewlett and Hibbert 1967; Kirkby and Chorley 1967).

More complex subsurface flow conditions developed starting in the third and fourth typhoon storms (which occurred back to back on September 29th and 30th). During the high intensity, short duration event on September 29th, almost all subsurface flow emanated from the soil matrix. However, some minor response was observed in several monitored macropore groups. During the longer, but less intense September 30th storm, several macropore groups (including the porous organic-rich horizon) contributed almost 2% of the total subsurface runoff. These macropore contributions were primarily during the recession limb of the hydrograph from FB. Although the estimated travel lengths for matrix flow during these two events was similar to lengths calculated for the earlier storms, the onset of preferential flow adds a new dimension of conceptual complexity which is difficult to resolve within the context of the variable source area model. Measured dye velocities at our hillslope site (Noguchi et al. submitted) indicate that maximum preferential flow velocity may be as much as 20 times higher than average pore velocities estimated for the soil matrix in tracer tests (Tsuboyama et al. 1994a). Thus, resulting preferential flow systems could have the potential to extend further upslope.

The next series of typhoon storms (8-9 October) produced proportionally greater macropore flow. Rainfall during the October 8th event was less than three times that of the September 30th event, but macropore flow was typically at least an order of magnitude higher (Fig. 3). Corresponding increases in matrix flow were only 3-fold. During the peak of the October 8th storm, macropore group C2 (dominated by flow from one macropore) contributed up to 21% of the slope drainage with < 2.4% coming from other macropores (including the organic-rich layer). Smaller, but significant, macropore flows were measured during the high intensity, short duration storm on October 9th (Fig. 3). For these two storms macropore flow contributed to the later portion of the rising limb, but mainly to the peak and falling limb of the hydrograph at FB (Fig. 3). Once again, matrix stormflow appeared to be confined to the lower few meters of slope by the riparian zone, but isolated preferential flow networks could extend further upslope.

The nature of the increased macropore response that was observed during these wetter site conditions warranted further examination. A series of tracer tests and staining tests were conducted in the lower 2 m of the hillslope segment to further elucidate preferential flow pathways. Tracer tests showed that for wet conditions, applied chloride was diluted in certain macropores due to connections with upslope drainage (Tsuboyama et al. 1994a; Sidle et al. 1994). Additionally,

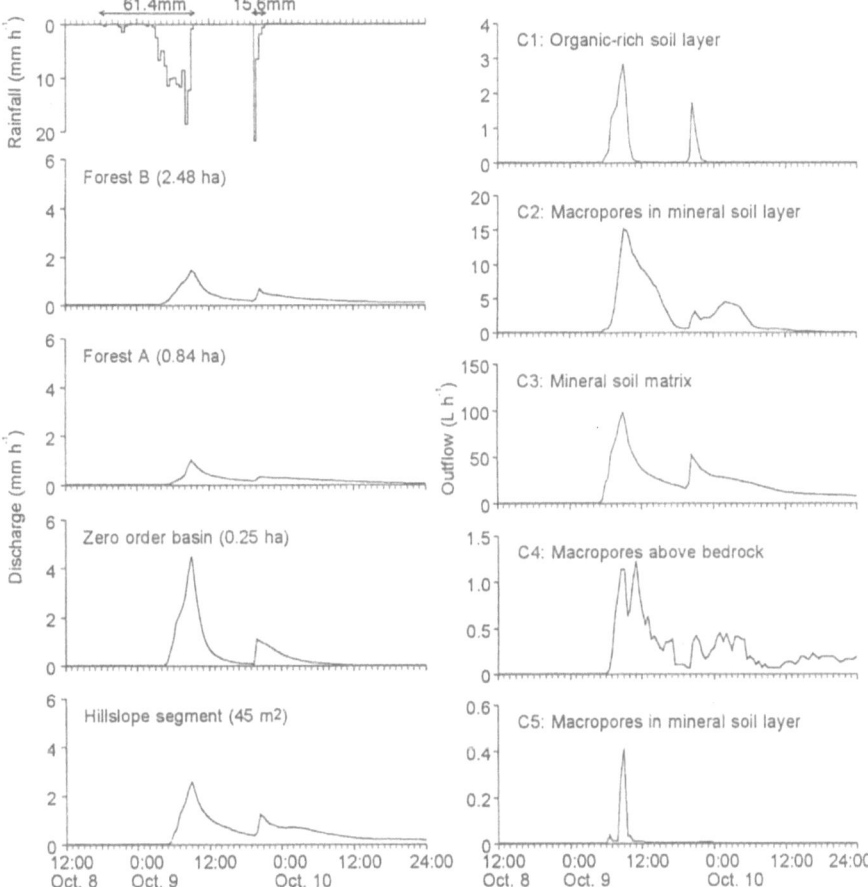

Fig. 3 Stormflow hydrographs for the four nested drainage areas in Hitachi Ohta together with discharge components of subsurface flow, 8-9 October 1992 storms (adapted from Sidle et al. 1995).

the effective pore volumes calculated for the flow-averaged breakthrough data from the entire profile were much less than total pore water volumes (measured *in situ* with tensiometers). Such point tensiometric measurements cannot capture the complex nature of the preferential flow paths. These results from the conservative tracer tests imply both an expansion of individual macropores via interaction with the surrounding soil matrix and mesopores (Wilson et al. 1991; Luxmoore and Ferrand 1993) as well as a lateral extension and branching of macropore networks.

Extensive soil pit investigations in the Hitachi Ohta watershed together with dye staining experiments in the same hillslope segment revealed that most macropores were < 0.5 m in length and originated from subsurface erosion, live or decayed root channels, and interactions between subsurface erosion and root channels (Noguchi et al. 1997; submitted). Certain macropores interacted with the surrounding soil matrix (possibly mesopores) to enlarge these preferential flow paths during wet conditions. Although few macropores were directly connected to one another, staining tests revealed connection of macropores via zones of buried organic matter, fractures in bedrock, and intersection with a perched water table provided the connective linkages for these complex preferential flow systems. An example of this complex network of interconnection is the most active macropore in C2. Extensive interaction with this macropore and the surrounding soil was evident. Additionally, even though this macropore emerged from the upper left hand side of the excavated pit face, further upslope excavations revealed that it was linked to fractures in the rough bedrock surface between 16 to 30 cm upslope. Other staining evidence indicated that these fractures were being fed by water flowing along low 'troughs' in the rough bedrock surface as well as root and subsurface erosion channels.

Several previous studies that inferred macropore flow response in forested slopes based on tracer partitioning of 'old' and 'new' water are worth noting. Wilson et al. (1991) used electrical conductivity measurements to separate 'old' and 'new' water. They attributed high concentrations of 'new' water early in the runoff hydrograph to macropore flow and later increases in 'old' water to rapid declines in preferential flow contributions. On the other hand, McDonnell (1990) hypothesized that isotopically 'old' water would be rapidly displaced from interconnected macropore systems once a perched water table accreted and intersected these interconnected channels. According to this theory, macropores at greater depths are most active and continuity of the macropores is essential. Macropore flow was not directly measured in either study nor were macropore systems mapped immediately following the tracer investigations.

In our study, macropore flow is clearly dependent on antecedent moisture conditions. Earlier in the storm season (i.e., drier antecedent conditions) macropore flow emerges largely during the recession limb of the runoff hydrograph. As antecedent wetness increases, macropore flow initiates late in the rising limb of the hydrograph and continues throughout the recession limb. These somewhat delayed and non-linear temporal responses do not follow Wilson et al's (1991) hypothesis of an early emergence of 'new' water, however, their inferences related to interactions between macropores and surrounding mesopores appears relevant. Our findings also differ with the concept of a uniformly rising perched water table that intersects a connected network of macropores (McDonnell 1990). Instead we offer an explanation of a complex network of connectivity that allows rather short macropore segments to spatially link with other preferential flow pathways. These systems appear to behave in a self-organizing manner that is facilitated at various junctions or nodes by increasing antecedent wetness. Factors that promote the connectivity of such systems appear to include shallow soil depth, buried organic matter, high soil permeability, and a hydrologic discontinuity in the subsurface. These nodes of interconnection may form a backbone of switching mechanisms from which macropore networks may or may not expand. Such mechanisms could explain the rather erratic spatial behavior in subsurface flow response noted at some sites (McDonnell 1990; Woods and Rowe 1996).

The final storm of the typhoon season produced the largest macropore response. During the second peak of this storm, macropore group C2 contributed in excess of 25% of the total subsurface flow compared to <1% for all other macropore groups (Fig. 4). Although peak response from other macropore groups coincided rather closely with the peak runoff from FB, the contribution from C2 was delayed to the recession limb of FB. This delay may correspond to the time required for nodes in this preferential flow system to condition and connect. For this storm it is estimated that matrix flow travel lengths could extend as far as 9 m upslope of the riparian zone, whereas preferential flow paths could easily extend much further upslope.

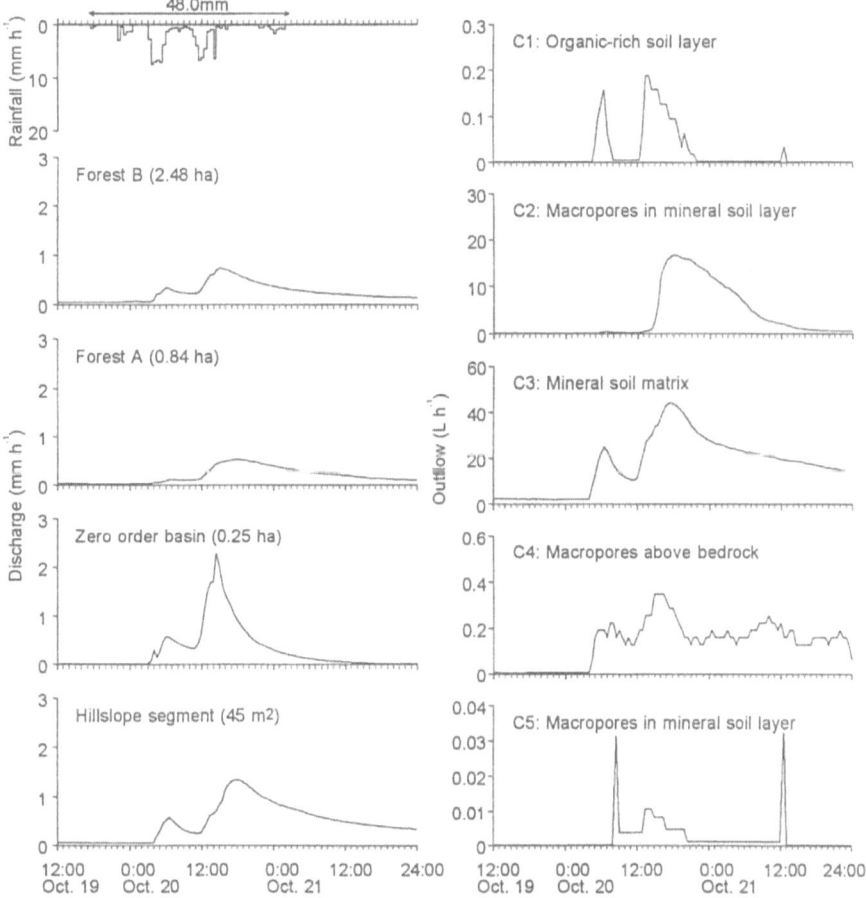

Fig. 4 Stormflow hydrographs for the four nested drainage areas in Hitachi Ohta together with macropore contributions from C2, 20 October 1992 storm (adapted from Sidle et al. 1995).

3.3 Zero-order Basin Response

Zero-order basins represent important geomorphic and hydrologic components of headwater systems. These hollows are known to be source areas for streamflow as well as sites of episodic loading and unloading as the result of mass wasting processes (e.g., Tsukamoto 1963; Dietrich and Dunne 1978; Tsukamoto and Ohta 1988; Sidle and Tsuboyama 1992; Tsuboyama et al. 1994b). However, the connections between zero-order basin hydrology and downstream response are largely unknown.

Neither ZB (shallower soils) nor the two hollows included in FA (slightly deeper soils) produced any runoff during the first two storms of the 1992 typhoon season. Although a small increase above baseflow conditions was noted in FA, it could be attributed to channel influences (interception and saturated overland flow) in the lower 12% of FA that was located downslope of the two hollows. The next two successive storms (September 29th and 30th) generated a transition from shallow groundwater storage to downstream water supply in ZB. During the short-duration September 29th event, a hydrologic threshold was reached facilitating an incipient discharge from ZB (Fig. 5). At this time, the groundwater profile in the hollow was wedge-shaped. Once this threshold was reached, runoff from ZB increased markedly to levels similar to those from FB (on a unit contributing area basis) (see September 30th storm response in Fig. 5 and Table 1). Most all runoff from FA during the September 29th and 30th storms could be attributed to direct contributions to the perennial channel; i.e., no runoff from zero-order basins (Table 1). This lack of response is related to the deeper soils in FA compared to ZB.

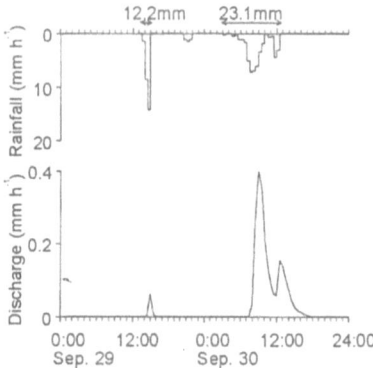

Fig. 5   Stormflow hydrograph for the zero-order basin (ZB) in Hitachi Ohta for the two storms during 29-30 September 1992 (adapted from Sidle et al. 1995)

During the last three storms of the 1992 typhoon season, runoff from ZB was both rapid and abundant (Figs. 3 and 4). For these three storms, water yields from

ZB ranged from 26.1% to 32.3% compared with yields of 11.4% to 19.5% from FB. Groundwater profiles in ZB were approximately parallel to the slope during this wetter and more hydrologically responsive period of the typhoon season. Peak flows from ZB were synchronized almost exactly with those from FB, indicating that these responsive zero-order basins contribute significantly to peak runoff at the catchment scale (Figs. 3 and 4). This wetter period also triggered a runoff response from FA (deeper soils). For the October 8th and 9th storms water yields were in the range of 6.0 to 7.5%, compared to yields typically < 1% for earlier typhoon storms. During the final typhoon storm, water yield from FA (15.6%) was similar to the yield from FB (19.5%), indicating that the deeper soils in the hollows of FA were recharged above their threshold response level. However, even for this wettest period, runoff from FA was less than half of that measured from ZB on an equivalent area basis. The degree of water storage and release from zero-order basins appears to be highly dependent upon soil depth. These hollows, which are sometimes remote from the main channel, provide important runoff sources when they become connected to the stream during wet antecedent conditions.

3.4 Conceptual Hydrologic Linkages

To develop an improved understanding of headwater stormflow generation it is necessary to specify appropriate spatial and temporal linkages among various hydrologic components. Such linkages within a hillslope system have already been discussed in section 3.2 for hillslope-scale response. Specifically, a conceptual model of self-organizing preferential flow pathways was presented based on physical evidence from dye staining, conservative tracer, and hydrometric tests at our field site. Such preferential flow mechanisms likely occur within zero-order basins and appear partly responsible for the rapid piezometric responses observed along the axes of these hollows (Sidle and Tsuboyama 1992; Tsuboyama et al. 1994b) as well as the rapid runoff produced once threshold groundwater conditions are met.

A better understanding of the hydrologic linkage between zero-order basins and streams is needed to specify the timing of runoff contributions from these often remote geomorphic features. Detailed synoptic measurements of stormflow, shallow groundwater response, and soil temperature fluctuations were collected during selected storms in ZB to elucidate this linkage (Tsuboyama et al. 1998). Results of this research presented in these proceedings indicate a change in the dominant flow process from infiltration to throughflow corresponding to thermal fluctuations at the 0.2 m soil depth (Tsuboyama et al. 1998). These short-term responses together with evidence presented from macropore staining and tracer investigations (Tsuboyama et al. 1994a; Sidle et al. 1994; Noguchi et al. 1997; submitted), suggest that hydrologic mixing occurs within the soil profile. Such hydrologic complexity brings into question the assumption of temporal invariance of compartmental water signatures, commonly used in end member mixing analysis studies (e.g., Pearce et al. 1986; McDonnell 1990; Eshleman et al. 1993).

The abrupt transition from recharge to discharge measured in ZB and supported by these detailed synoptic studies, elucidates the temporal hydrologic contributions of these remote basins to catchment runoff.

Saturated overland flow from the narrow riparian zones in such first-order watersheds together with direct channel interception of rain water can only supply a very limited amount of the stormflow from FB during wet conditions. While these contributions represent the majority of the storm runoff during the first two small typhoon storms, their more or less constant supply of water was proportionally diminished as catchment water yields increased later in the season. As riparian contributions decreased throughout the typhoon season, upper watershed linkages (i.e., hillslope to channel and zero-order basin to channel) developed and increased in relative magnitude.

4. SUMMARY

Results from hydrometric, conservative tracer, dye staining, and subsurface thermal response tests are summarized for the Hitachi Ohta Experimental Watershed. Based on these investigations, an improved understanding of stormflow generation in deeply incised headwater catchments is emerging. During drier antecedent conditions, most of the storm runoff appears to originate as saturated overland flow from the narrow riparian zone or as direct channel interception. As moisture increases, the soil matrix in the lower few meters of the hillslope augments up to 40% of this storm runoff. With further increasing wetness, zero-order basins with relatively shallow soils begin to contribute to storm runoff corresponding to the development of 'threshold' groundwater levels in these hollows. Additionally, preferential flow systems begin to develop in hillslopes (and also probably in hollows), facilitating subsurface transport over greater slope lengths during storms. During the wettest conditions, runoff contributions increase from zero-order basins, including basins with slightly deeper soils. Preferential flow systems begin to self-organize into complex networks involving discrete macropores, lithic contacts, bedrock fractures, buried organic matter, and subsurface topography. From this better understanding of the temporal and spatial attributes of hydrologic pathway, a hydrogeomorphic concept for stormflow generation is emerging.

REFERENCES

Chen, C. and R.J. Wagenet (1992): Simulation of water and chemicals in macropore soils. Part 1. Representation of the equivalent macropore influence and its effect on soilwater flow, J. Hydrol., 130: 105-126.

Dietrich, W.E. and T. Dunne (1978): Sediment budget for a small catchment in mountainous terrain, Z. Geomorphol. Suppl. Bd., 29: 191-206.

Dunne, T. and R.D. Black (1970): Partial area contributions to storm runoff in a small New England watershed, Water Resour. Res., 6: 1296-1311.

Eshleman, K.N., J.S. Pollard and A.K. O'Brien (1993): Determination of contributing areas for saturation overland flow from chemical hydrograph separations, Water Resour. Res., 29: 3577-3587.

Feger, K.H., G. Brahmer and H.W. Zöttl (1990): Element budgets of two contrasting catchments in the Black Forest (Federal Republic of Germany), J. Hydrol., 116: 85-99.

Fujieda, M., T. Kudoh, V. de Cicco and J.L. de Calvarcho (1997): Hydrologic processes at two subtropical forest catchments: the Serra do Mar, Sãn Paulo, Brazil, J. Hydrol., 196: 26-46.

Gillham, R. W. (1984): The effect of capillary fringe on water-table response, J. Hydrol., 67: 307-324.

Harr, R. D. (1977): Water flux in soil and subsoil on a steep forested slope, J. Hydrol., 33: 37-58.

Hewlett, J.D. and A.R. Hibbert (1967): Factors affecting the response of small watersheds to precipitation in humid areas, In: W.E. Sopper and H.W. Lull (Editors), Proc. of the Internat. Symp. on Forest Hydrology, Pergamon, New York, pp. 275-290.

Kirkby, M.J. and R.J. Chorley (1967): Throughflow, overland flow and erosion, Bull. Internat. Assoc. Sci. Hydrol., 12: 5-21.

Kitahara, H. and Y. Nakai (1992): Relationship of pipe flow to streamflow on a first order watershed, J. Jpn. For. Soc., 74: 318-323 (in Japanese).

Luxmoore, R.J., and L.A. Ferrand (1993): Towards a pore-scale analysis of preferential flow and chemical transport, In: D. Russo and G. Dagan (Editors), Water Flow and Solute Transport in Soils, Springer-Verlag, Berlin, pp. 45-60.

McDonnell, J.J. (1990): A rationale for old water discharge through macropores in a steep, humid catchment, Water Resour. Res., 11: 2821-2832.

Megahan, W.F. (1983): Hydrologic effects of clearcutting and wildlife on steep granitic slopes in Idaho, Water Resour. Res., 19: 811-819.

Mosely, M.P. (1979): Streamflow generation in a forested catchment, Water Resour. Res., 15: 795-806.

Noguchi, S., Y. Tsuboyama, R.C. Sidle and I. Hosoda, Morphological characteristics of macropores and distribution of preferential flow pathways in forested slope segment, Soil Sci. Soc. Am. J. (submitted, 1998).

Noguchi, S., Y. Tsuboyama, R.C. Sidle and I. Hosoda (1997): Spatially distributed morphological characteristics of macropores in forest soils of Hitachi Ohta Experimental Watershed, Japan, J. For. Res. 2: 207-215

Pearce, A.J., M.K. Stewart and M.G. Sklash (1986): Storm runoff generation in humid headwater catchments 1. Where does the water come from?, Water Resour. Res., 22: 1263-1272.

Rose, D. A. and J. B. Passioura (1971): The analysis of experiments on hydrodynamic dispersion, Soil Sci., 111: 252-257.

Sidle, R.C. and Y. Tsuboyama (1992): A comparison of piezometric response in unchanneled hillslope hollows: coastal Alaska and Japan, J. Jpn. Soc. Hydrol. & Water Resour., 5: 3-11.

Sidle, R.C., Y. Tsuboyama, S. Noguchi and I. Hosoda (1994): Subsurface flow through the soil matrix and macropores: results of tracer tests at Hitachi Ohta, Japan, In: Proc. Inter. Symp. on Forest Hydrology, 1994, Univ. of Tokyo, Tokyo, pp. 225-232.

Sidle, R. C., Y. Tsuboyama, S. Noguchi, I. Hosoda, M. Fujieda and T. Shimizu (1995): Seasonal hydrologic response at various spatial scales in a small forested catchment, Hitachi Ohta, Japan, J. Hydrol., 168: 227-250.

Sklash, M.G. and R.N. Farvolden (1979): The role of groundwater in storm runoff, J. Hydrol., 43: 45-65.

Tsuboyama, Y., R.C. Sidle, S. Noguchi and I. Hosoda (1994a): Flow and solute transport through the soil matrix and macropores of a hillslope segment, Water Resour. Res., 30: 879-890.

Tsuboyama, Y., I. Hosoda, S. Noguchi and R.C. Sidle (1994b): Piezometric response in a zero-order basin, Hitachi Ohta, Japan, In: Proc. Inter. Symp. on Forest Hydrology, 1994, Univ. of Tokyo, Tokyo, pp. 218-225.

Tsuboyama, Y., S. Noguchi, T. Shimizu, R.C. Sidle and I. Hosoda (1998): Intrastorm fluctuations of piezometric head and soil temperature within a steep forested hollow, In: Environmental Forest Science, Proc. of IUFRO Div. 8 Conf., 19-23 October 1998, Kyoto, Japan, Kluwer Academic Publisher (this issue).

Tsukamoto, Y. (1973): Study on the growth of stream channel, J. Jpn. Soc. Erosion Control Engr., 87, 4-13, (in Japanese).

Tsukamoto, Y. and T. Ohta (1988): Runoff processes on a steep forested slope, J. Hydrol., 102: 165-178.

Turton, D.J., C.T. Haan and E.L. Miller (1992): Subsurface flow responses of a small forested catchment in the Ouachita Mountains, Hydrol. Processes, 6: 111-125.

Whipkey, R.Z. (1965): Sub-surface stormflow from forested slopes, Bull. Internat. Assoc. Sci. Hydrol., 10: 74-85.

Wilson, G.V., P.M. Jardine, R.J. Luxmoore, L.W. Zelazny, D.A. Lietzke and D.E. Tod (1991): Hydrogeochemical processes controlling subsurface transport from an upper subcatchment of Walker Branch watershed during storm events. 1. Hydrologic transport processes, J. Hydrol., 123: 297-316.

Woods, R.A. and L.Rowe (1996): The changing spatial variability of subsurface flow across a hillside. J. Hydrol. (NZ) 35: 51-86.

Yasuhara, M. and A. Marui (1994): Groundwater discharge from a clayey hillslope, In: Proc. Inter. Symp. on Forest Hydrology, 1994, Univ. of Tokyo, Tokyo, pp. 241-248.

Mechanisms of Landslide Triggered Debris Flows

Kyoji SASSA

Disaster Prevention Research Institute, Kyoto University, Kyoto 611-0011, Japan

ABSTRACT   Debris flows are often caused by heavy rainfalls. Warning of debris flows is issued by rainfall pattern, rainfall intensity and cumulative rainfall(or duration) in the countries such as Japan, USA et al.   However, some debris flows are not regulated by rainfall. Two big debris flow disasters occurred in Japan. One (Gamahara debris flow) in Nagano Prefecture in 1996 killed 14 persons, one (Harihara debris flow) in Kagoshima Prefecture killed 21 persons in 1997. Both took place without any rainfall intensity at the time of occurrence, and especially the Gamahara debris flow occurred in the day of almost no precipitation in a dry season, and cumulative rainfall was not high. The field investigation and dynamic loading undrained ring shear tests for these two debris flows presented two mechanisms of landslide triggered debris flows which do not need full saturation of debris for the high speed motion of debris with a very low shear resistance.
Keywords: landslides, debris flows, undrained loading, liquefaction, ring-shear test.

1. INTRODUCTION

Two catastrophic events of debris flows with 14 and 21 fatalities gave a shock to the Japanese society. Two events were investigated by special budgets by the Japanese Ministry of Education, Science, Culture and Sports, and also by the Science and Technology Agency, Japan. The Disaster Prevention Research Institute, Kyoto University has developed a new dynamic loading undrained ring shear apparatus to mechanically simulate the stress and shear displacement of shear zone under dynamic condition. The 1995 Kobe earthquake caused a very disastrous landslide that killed 34 persons living at the toe of a gentle slope. The earthquake attacked in very dry season after the historically most dry year in the Kobe area. The ground surface was quite dry. But still the landslide showed a high mobility (9.6° as the average apparent friction angle during motion) and destroyed 11 houses. The research using the dynamic loading ring shear apparatus developed by Sassa et al (Sassa 1995) revealed that the sliding surface liquefaction phenomenon paid key role of this landslide. The phenomenon is explained as shearing can cause grain crushing in sandy materials under a certain pressure in the shear zone, and it results in volume shrink and causes local liquefaction due to excess high pore water pressure when the shear zone is saturated (Sassa 1996, Sassa et al. 1996). Based on this research, a special budget (nearly one million US dollars) to develop a much improved type of dynamic-loading ring shear apparatus was accepted and built in 1996. This paper presents the first application of this apparatus to study the mechanism why debris flows occurred in dry season or without enough water to saturate the landslide mass.

2. APPARATUS AND SAMPLES

Figure 1 shows a front view of the apparatus before the setting in the laboratory. It is too tall (5.1 m) for sample setting, so the lower half was installed in the pit of the laboratory. It has a mechanical part (backside) and an electronic part (front side) in this photo. The mechanical part has two servo-control motors for shearing, two servo-control oil pistons for normal stress loading and adjusting the gap between the upper shear box and the lower shear box. The electronic part has two computers for stress/speed control and data recording, and also one computer for monitoring of system functions and automatic safeguard and alarm system, and one computer for manual of testing and

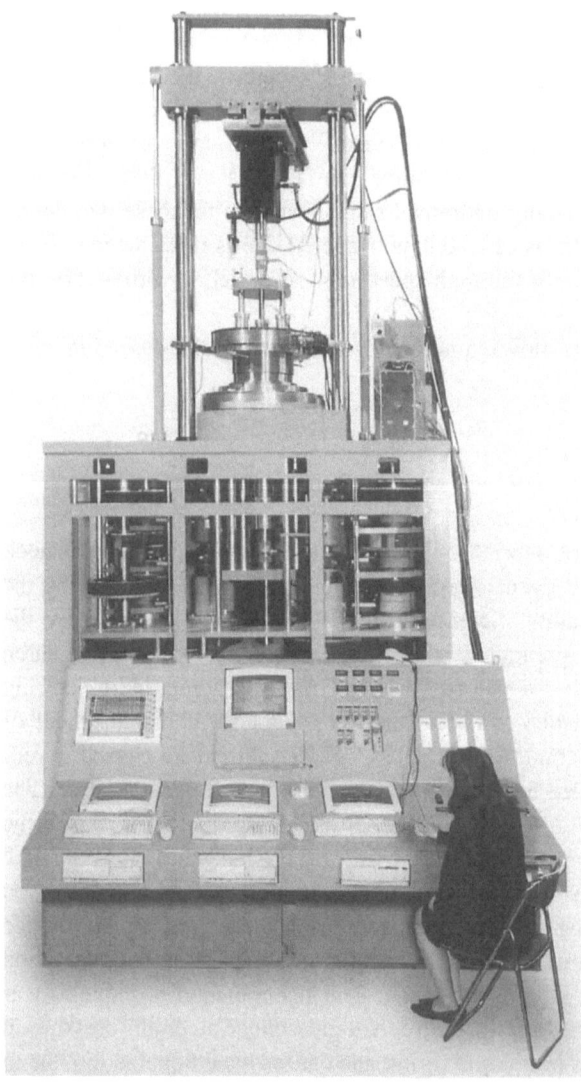

Figure 1 New dynamic loading ring shear apparatus.

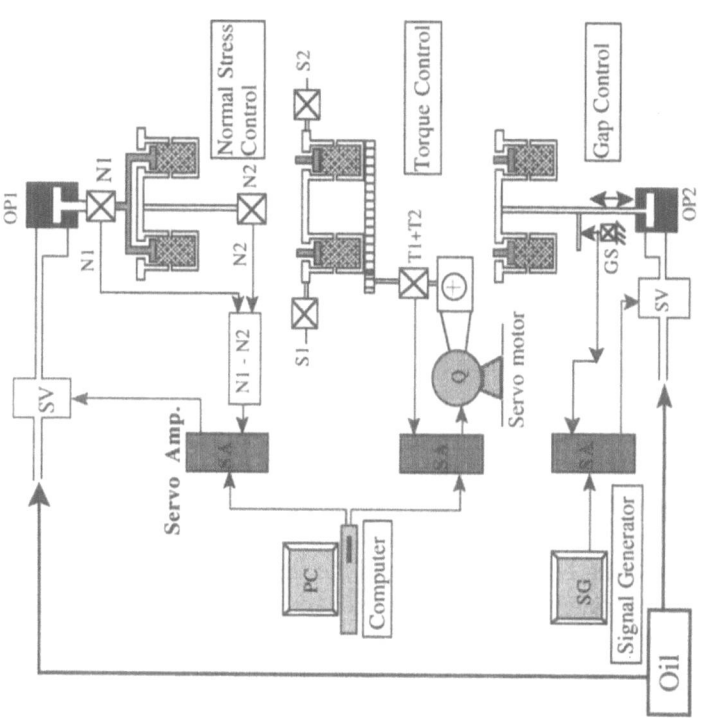

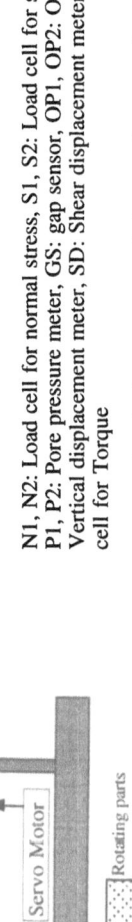

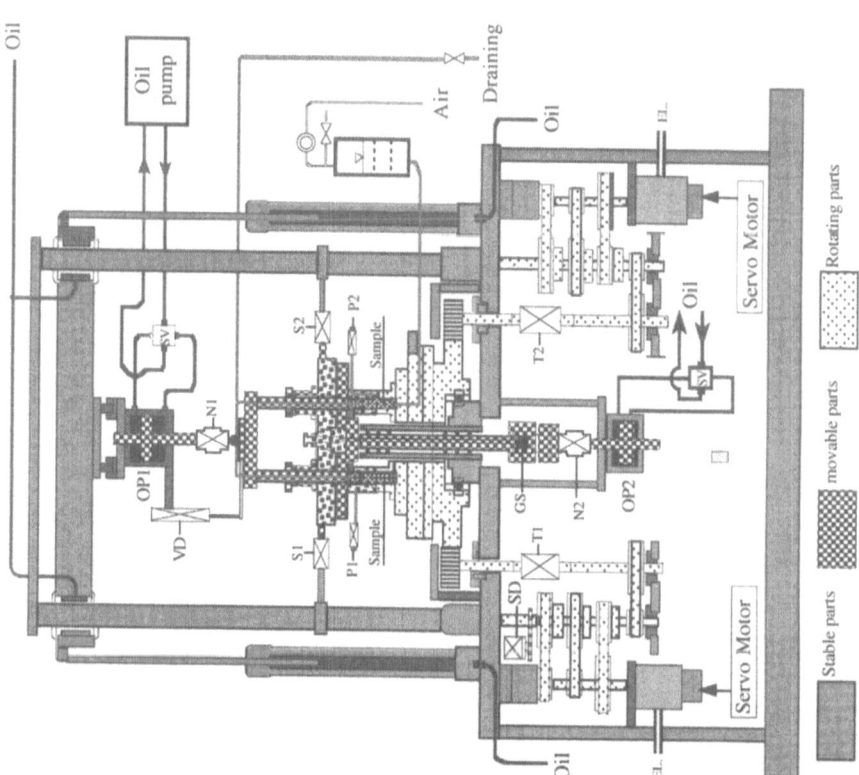

N1, N2: Load cell for normal stress, S1, S2: Load cell for shear resistance, P1, P2: Pore pressure meter, GS: gap sensor, OP1, OP2: Oil piston, VD: Vertical displacement meter, SD: Shear displacement meter, T1, T2: Load cell for Torque

Figure 2 Mechanical structure and electronic control system.

analysis.

Figure 2 illustrates the structure of system. OP1 is the oil piston for normal stress loading. OP2 is the oil piston to adjust location of the upper cap plate on which the upper shear box is fixed. N1 and N2 are load cells to monitor normal stress. The side friction to the upper shear box from the sample is included in N2 value. Therefore, $(N1-N2)$ value represents the real normal force acting on the shear surface of the sample. The value of $(N1-N2)$ are used as a feed back signal to servo-control amp. And normal stress is automatically controlled as the pre-decided control signal. The size of shear box is 250 mm inside diameter, 350 mm outside diameter. The maximum speed of shearing is 224 cm/sec at the first gear and 33 cm/sec at the second gear.

Shear stress is given by a servomotor (there are two motors, but usually one motor is used). Torque is monitored by torque meters T1 and T2 that are used to supply a feed back signal for shear stress control. The torque is transferred through the shear surface and tends to rotate the upper shear box and the cap plate. But the cap plate is retained to the fixed pillars through two load cells for shear resistance (S1 and S2). Shear resistance acting on the shear surface is monitored by S1 and S2. Shear stress calculated from monitored by T1 and T2 and by S1 are the same except the period of acceleration and deceleration.

Shear stress and normal stress are controlled as the pre-decided time-dependent value given by the computer. The lower shear box is rotated after shear failure, while the upper shear box is retained. However, any sample and pore water should not leak from the gap between two shear boxes. The detailed section of half shear box is illustrated in Figure 3. Rubber edge is pasted to the lower shear box and pressed to the upper shear box. To seal pore water and sample, the contact pressure of rubber edge is automatically adjusted by OP2

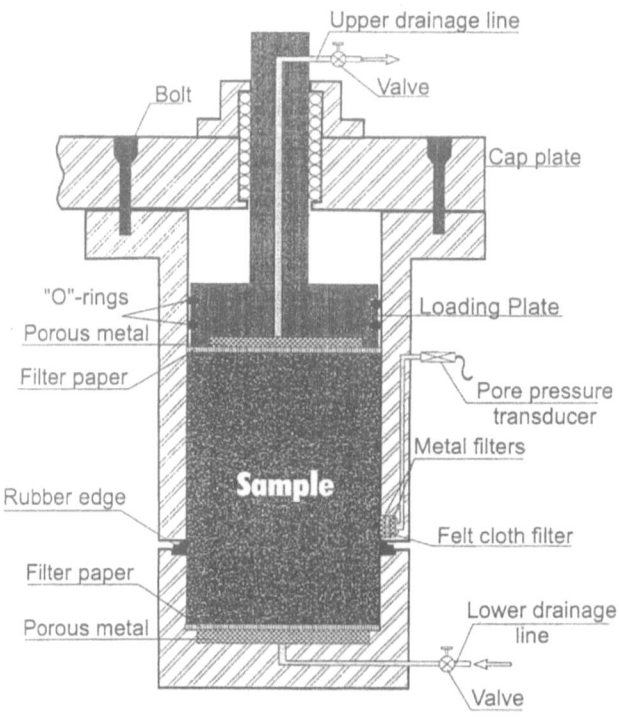

Figure 3 Schematic illustration of shear box.

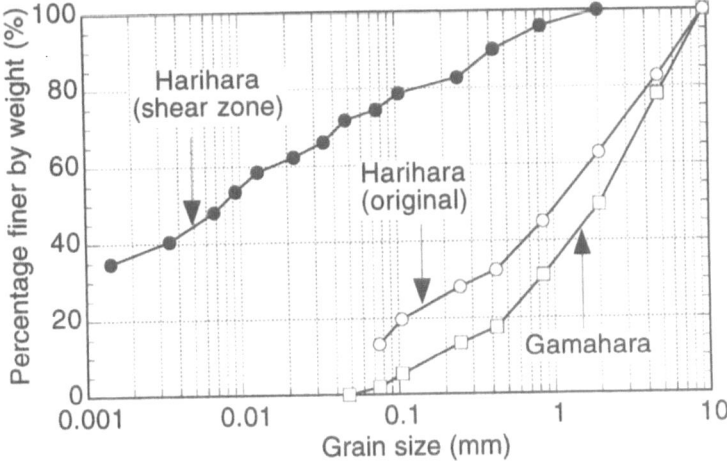

Figure 4 Grain size distribution of samples taken from Gamahara, Harihara and the shear zone of the Harihara after test.
A: Debris (Jurassic sedimentary rocks) taken from the source area of the Gamahara landslide triggered debris flow, B: Andesitic debris taken from the source area of the Harihara landslide triggered debris flow, C: Sample of the shear zone after ring shear test of B.

through the feed back signal from the gap sensor GS in Figure 2, which monitors the position of the cap plate and the upper shear box by 1/1000 mm precision. The signal generator provides control signal necessary for prevention of leakage. The constant value of rubber edge friction is subtracted from the monitored shear resistance. Teflon (polytetrafluorethylene) is sprayed to the surface of rubber edge to reduce friction before each test.

Figure 4 shows the grain size distribution of samples used in this paper. The soils were sieved to remove grains larger than 9.52 mm in the field. It is necessary to test in the apparatus. Gravel and boulders are included in debris, but they do not contact each other. A shear surface will be formed through the matrix of finer grains to avoid such large stones. Therefore, grains less than about 1cm will be enough to obtain the reasonable value of shear resistance.

3. TYPE OF GAMAHARA DEBRIS FLOW

3.1 Outline of the Gamahara Landslide-Triggered-Debris Flow

A landslide triggered debris flow occurred in the Gamahara torrent, Otari village, Nagano Prefecture, Japan on 6 December 1996. It killed 14 persons and injured 8 persons working for the construction of Sabo dams to prevent debris flows. This debris flow gave shock to the Japanese Government and Society, because it occurred and killed 14 persons without major rainfalls and earthquakes. It was the period usually very safe for debris flows, and the day of debris flow was almost no rainfall (1 mm/day) and rainfall in a day before the debris flow was 49 mm (Marui et al. 1997, Sassa et al. 1997). Occurrences of most debris flows

504

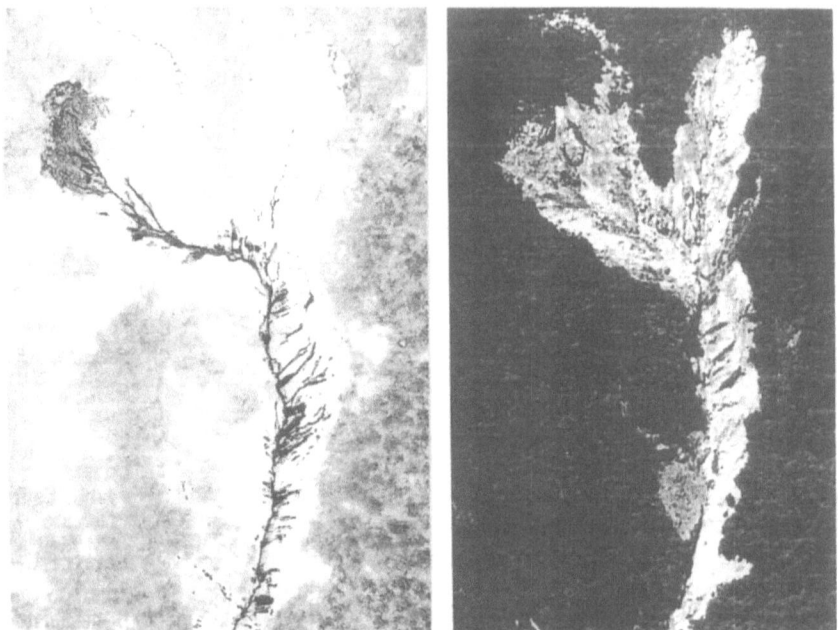

Figure 5 Air-photos before and after the debris-flow disaster.
Left photo was taken on 7 December 1996(courtesy of Pasco Company).
Right photo was taken on 19 July 1996 (courtesy of Nakanihon Kouku
Company).

during heavy rainfalls are usually regulated by rainfall intensity and cumulative
rainfalls. Takahashi proposed the initiation mechanism of debris flows by the
appearance of overland flow on the torrent deposit(Takahashi 1978). It is a
concept that rainfall (water discharge) regulates the occurrence of debris flow.
And warning of debris flows is issued based on monitoring of rainfalls in Japan
(as written below), USA (Wieczorek et al. 1990). Time prediction of debris flow
guided by the Ministry of Construction in Japan is based on both parameters of
the rainfall intensity and the effective cumulative precipitation (The Japan Society
of Erosion Control Engineering, 1992).   The effective cumulative precipitation
Re is calculated as below (Takei, 1993).

$$Re = (1-\alpha) R_1 + (1-\alpha)^2 \; R_2 + \text{-------} + (1-\alpha)^n Rn \qquad (1)$$

Here, α : a constant expressing the decrease of effect.
        R: hourly precipitation
        n of suffix of R: n hour before the hour in consideration.
The value of α is different in areas. Each prefectural government is requested
to decide the value appropriate for the region. However, this is not easy for
practical cases. Then, for the example of the Gamahara debris flow in Nagano
Prefecture in 1996, a simplified criterion was used for precaution for debris flow

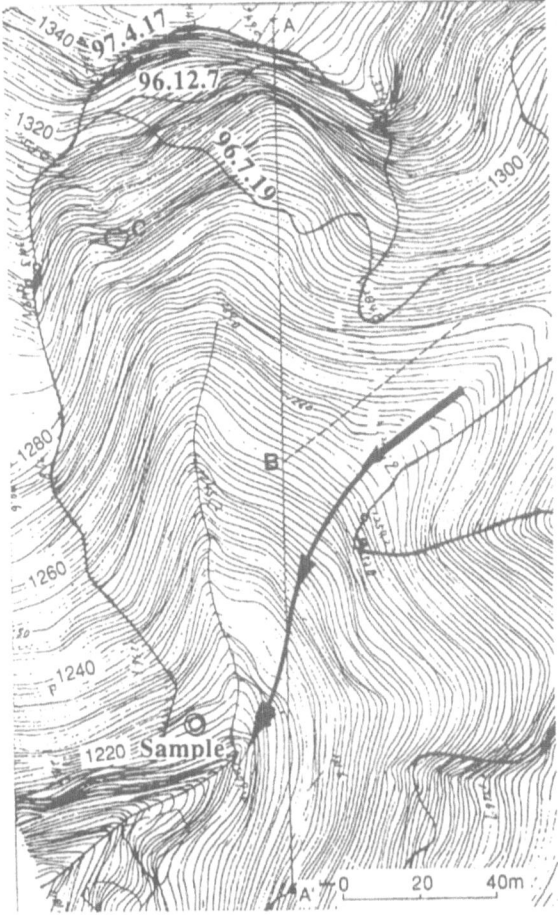

Figure 6 Plan of the initial landslide and the Gamahara torrent (made from the air photo of 1997.4.17).

A- A': survey line by non-mirror total station.

→ : Main stream of the Gamahara torrent.

◎: Sampling point.

based on previous records in this area. The criterion to stop construction works to avoid debris flow disasters was decided as 50 mm in the cumulative precipitation for 6 hours, or 15 mm per hour (Nikkei Construction 1997). This number seems to be safe referring the recent debris flow disasters in this area on 11 July 1995. The total rainfall at this disaster was 376 mm for one day, and the maximum rainfall intensity was 48 mm/hour. Many debris flows were caused in the northern part of Nagano Prefecture including this torrent and many parts of railways and roads and bridges were destroyed by debris flows.

Figure 5 shows two air photos before/after the debris flow. The left photo taken one day after the debris flow showed the trace of debris flow. A landslide is located in the top of the debris flow trace; it is apparently the cause of this debris flow. The right photo shows the same area in July 1996. Two previous landslides were observed. Comparing two photos, the landslide that triggered the 1996.12.6 debris flow is a retrogressive failure at the head scarp of the previous landslide (45 m wide and 65 m long).

The head scarp is very steep and the rainfall is limited, so liquefaction of this initial landslide was unlikely to take place. Overland stream water on the torrent bed so that may cause debris flow is not expected because of low precipitation and no trace of natural landslide dam and its failure. So this debris flow is estimated to

506

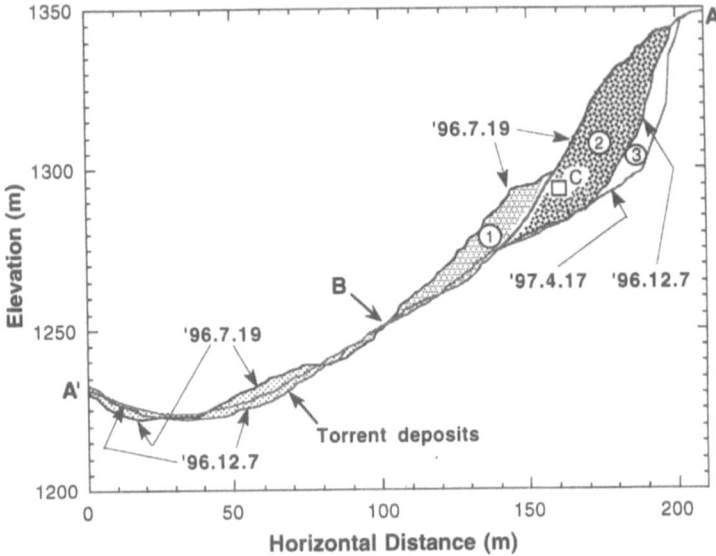

Figure 7 A-A' sections made from air-photos taken on 1996.7.19, 1996.12.7 and 1997.4.17.

have been triggered by the failed mass from the landslide.

3.2 Field Investigation

We investigated the initial landslide seen in Figure 5 and surveyed using no-mirror total station and took samples on 4 May 1997. Figure 6 shows the plan of initial landslide and the Gamahara torrent. The counter lines were made based on the air photo of 1997.4.17. Three borders of landslides were drawn based on the air photos of 1996.7.19 (before the 1996.12.6 debris flow), 1996.12.7 (one day after the debris flow) and 1997.4.17 (after another retrogressive slide). The Gamahara torrent stream enters into this figure from the right side and goes out to the left side along three thick arrows. After the landslide, a branch stream from the landslide is formed and drawn as a curved line in Figure 6. A-A' line was the survey line of non-mirror total station on 4 May 1997. The section is shown in Figure 7 that is drawn from three air photos with reference of 1997.5.4 direct survey. The profile of 1996.7.19 shows the landslide(with the sliding surface expressed by a dashed line) on the 11 July 1995 heavy rainstorm and probably its left debris ①. The mass of ② is estimated to be the landslide mass at the 1996.12.6 debris flow. The mass of ③ will be a retrogressive slide early April in 1997. The part lower than B in Figure 6 and Figure 7 is the torrent bed and deposits of Gamahara. Figure 7 indicates that about 6 m torrent deposits between two ground surface lines of ˋ96.7.19 and ˋ96.12.7 were transported together with the landslide mass. The landslide mass of ② slid because of steepness of slope, it was probably neither saturated nor liquefied. However, the torrent deposits downstream from the point B were saturated by the stream flow. Therefore, this

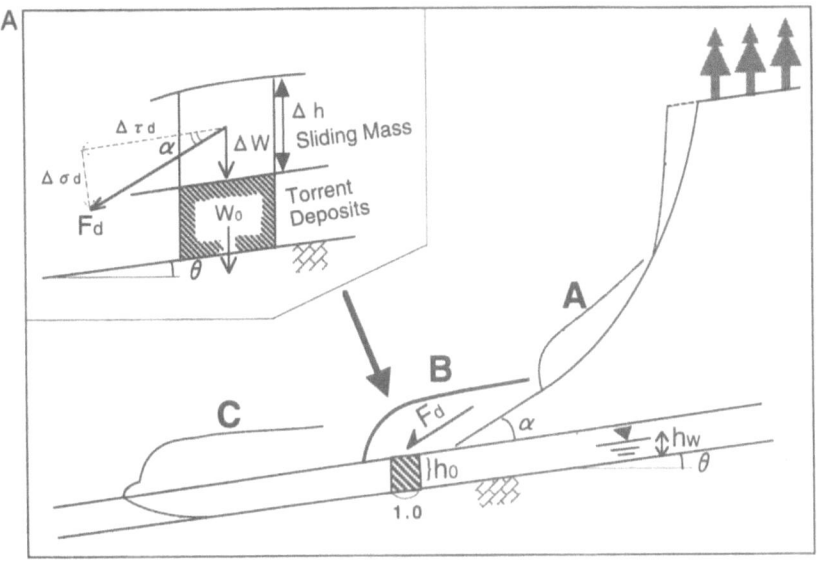

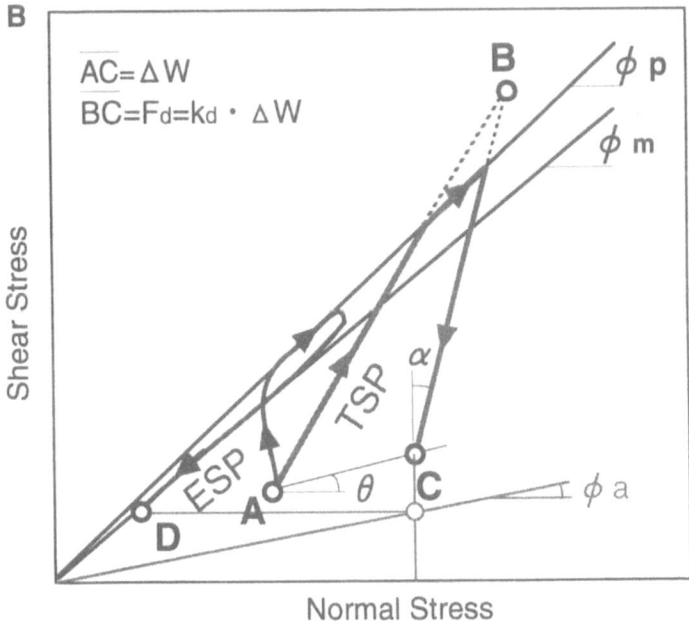

Figure 8 Model of the Gamahara type of the landslide triggered debris flow.
A: illustration of the model; B: stress path of the torrent deposit during loading.
α : Angle of thrust between the slope and the torrent bed; F_d : Dynamic stress;
kd : Dynamic coefficient ($F_d/\Delta W$).

investigation strongly indicates that an unsaturated failed mass moved onto the saturated torrent deposit (at least its bottom).

3.3 Model of the Gamahara Type of Landslide-Triggered-Debris Flow

A model of the Gamahara type of landslide-triggered debris flow is schematically shown in Figure 8-A. A landslide mass (not liquefied, not yet debris flow) occurs at a mountain slope. In the case of Gamahara debris flow, the initial landslide was a retrogressive slide at the previous landslide scarp. The mass moves along the slope (A), and it loads onto a torrent deposit at the foot of the slope (B). When a torrent stream or subsurface flow exists and some part of deposits is saturated, the torrent deposit can be sheared by undrained loading and transported together with the sliding mass(C). Let us consider a column of a unit width inside the torrent deposit. In the position (A) of the sliding mass, the self-weight of column (W_0) is working. When the sliding mass rides on the torrent deposit (B) with a certain velocity, it gives a dynamic loading on the column. Here, we assume the loaded stress as the sum of the static stress of ΔW (load due to the self weight of the sliding mass) and the dynamic (impact) stress working in the moving direction of the sliding mass (F_d).

The stress working on the bottom of column is presented in Figure 8-B. The initial stress is expressed by the point "A" which corresponds to the position (A) of sliding mass in Figure 8-A. When no excess pore pressure is generated during loading, the stress point moves to the point (C) by adding the static stress (ΔW) to the initial stress. And also by adding the dynamic stress (F_d) to this, the stress moves to the point (B). Therefore, the stress path in the real case tends to moves from the point A to the point B. However, when the stress reaches the failure line, the stress path moves along the failure line as seen in the figure. When the dynamic stress is lost, the total stress moves to the stress point (C), namely the sum of W_0 and ΔW. Denoting the angle of thrust at collision to the torrent deposit as α and the dynamic stress as F_d using a dynamic coefficient $k_d =$

($F_d/\Delta W$), dynamic shear stress and normal stress are expressed as,

$$F_d \cdot \cos \alpha = \tau_d, \qquad F_d \cdot \sin \alpha = \sigma_d \qquad (2)$$

The stress path from A-B-C is the total stress path in case of no pore pressure generation. However, excess pore pressure is likely to be generated during loading and also during shearing after failure. In this case, the effective stress path will be deviated from this total stress path like a curved line from A to D.

3.4 Ring Shear Test for Mechanical Simulation

To examine the model concept of Figure 8, I tried to take sample of the torrent deposit downstream of B in Figure 7, and the debris was sampled from the Jurassic sedimentary rocks at the valley side slope (marked as ◎ in Figure 6)

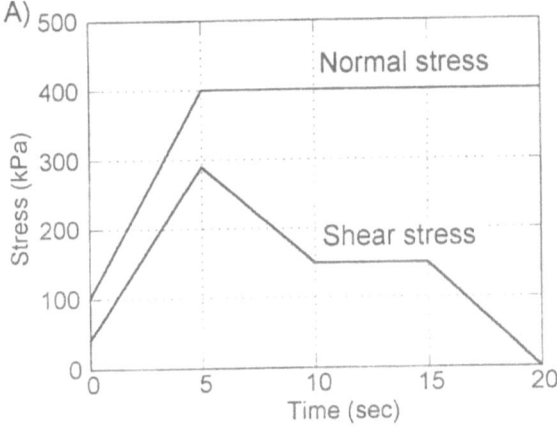

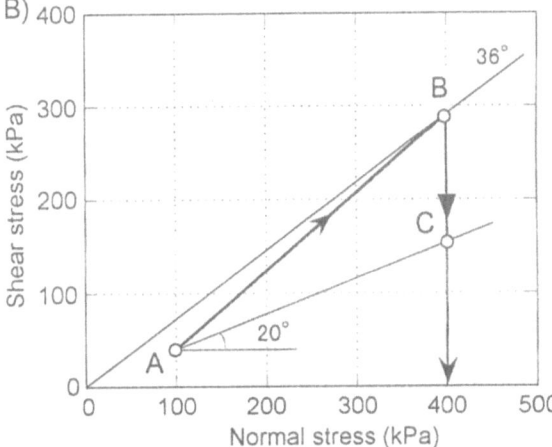

Figure 9 Control signals for the mechanical simulation test.

A: Time series variation of the signal; B: Stress path of the signal.

because most original deposits inside the valley floor were scraped out. Using this sample, the situation of the model of Fig. 8 was simulated by the dynamic loading undrained ring-shear apparatus.

We adopted the test condition corresponding to the field situation that the landslide mass of ① and ② loads on the torrent deposit in the down-slope of B in the Figure 7 . Namely parameters in Figure 8 are; $\Delta h = 18$ m, $h_0 = 6$ m, $hw = 1$ m, $\theta = 20°$, $\alpha = 0°$ because the slope is very smooth. The total unit weight of soils is $\gamma = 2.0$ gf/cm^3.

We assumed the dynamic coefficient $k_d = 0.42$ so that the stress path should reach the failure line of this material. The control signals given from the computer to the servomotor and the servo-oil piston are shown in Figure 9. The line segment of A-C is the static stress increment due to the self-weight of loaded debris of about 18 m. The line segment of B-C is the dynamic (impact) stress increment component. (For the practical purpose, the normal stress after loading was set to be 400 kPa instead of 406 kPa as the round number). The real stress increment at impact is the stress path from A to B. The loading and unloading speed in the test was set to be 5 sec for loading time, 5 sec for unloading time, 5 sec for constant, 5 sec for decreasing shear stress to zero as seen in Figure 9. Pore pressure coefficient ($B_D = \Delta u / \Delta \sigma$) was measured to check the degree of saturation after consolidation.

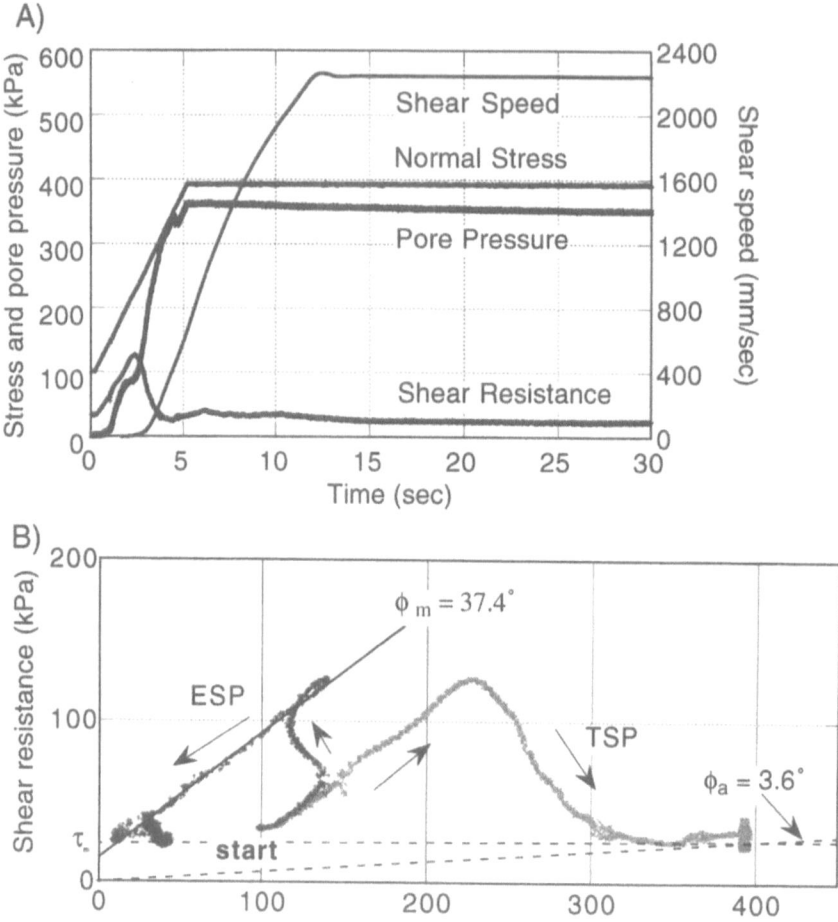

Figure 10 Test results for the Gamahara debris flow.
A: Time series data; B: Stress paths.
$B_D = \Delta u/\Delta \sigma = 0.79$, $e = 0.67$,
ESP: effective stress path; TSP: total stress path.

The upper one of Figure 10 presents variation of normal stress, pore pressure, shear resistance and shear speed. After failure, pore pressure is much increased and shear resistance became very low. Shear velocity at the center of sample reached to the maximum speed of this apparatus of 224 cm/sec. The lower one of Figure 10 shows the effective stress path (ESP) and the total stress path (TSP) monitored for a test. The effective stress path reached a failure line, and then it decreased to a low stress point along a failure line. It is a typical stress path of sliding surface liquefaction (Sassa 1996, Sassa et al 1996). And the apparent friction angle in the steady state is 3.6° which is obtained by the shear resistance in the steady state divided by the total normal stress. Therefore, a rapid motion will be continued in the steeper torrent bed than 3.6°.

In the position B of debris in Figure 8, the sliding surface liquefaction will take place at the bottom of debris. Then, the debris together with the torrent deposit moves to the position C. The position C in the Gamahara torrent is similar to the position B because the average gradient of torrent to the main river is 18° and the angle of thrust $\alpha = 0$. Therefore, the debris should continue to move until the alluvial fan in the Himekawa river because of low apparent friction angle of 3.6°, though the moving debris could be changed to a liquefied debris flow if enough water was supplied and the debris was disturbed during the long travel course.

From the stress at failure in this test, the dynamic coefficient necessary to cause failure can be estimated. The shear stress at failure was 125 kPa from Figure 10. The static shear stress after loading is $(W_0 + \Delta W) \sin \theta = 151$ kPa. Because the static shear stress after loading is greater than the peak shear resistance in the undrained loading, the dynamic stress (impact force) is not necessary to cause shear failure of the torrent deposit.

This test result could support the model concept illustrated in Figure 8. The initial landslide itself was not a liquefied landslide because almost no rainfall on the day of debris flow. Not much, but 49 mm of rainfall was observed one day before the debris flow. So there should be a saturated part inside the torrent deposits. In that case, a failed mass from the slope could cause shear failure of the torrent deposit. The large stress drop after failure indicates the sliding surface liquefaction phenomenon. The sliding surface liquefaction is similar to the Nikawa landslide induced by the Hyogoken-Nanbu earthquake.

Earthquake loading in the Nikawa landslide and the loading by the failed mass in the Otari debris flow is rather similar, because one is cyclic loading and another is a half cycle of loading, but the basic concept is common in the sense of dynamic loading.

Figure 11 Photo of the Harihara landslide triggered debris flow.

512

4. TYPE OF HARIHARA DEBRIS FLOW

4.1 General Situation of the Harihara Landslide-Debris Flow

During the rainy season in July 1997, heavy rainfall attacked in the Western Japan. This rainfall caused many landslides and debris flows in many prefectures. The most catastrophic disaster was a rapid landslide-debris flow in Izumi City, Kagoshima Pre-fecture at about mid-night on 9 July. Figure 11 shows aerial oblique view of the landslide-debris flow. A landslide of about $13 \times 10^4 \, m^3$ moved into the Harihara torrent and spread over the alluvial fan over the almost completed check dam. The debris mass destroyed 19 houses and 13 non-residential buildings on the alluvial fan, and 21 residents were killed and 13 were injured (Shimokawa et al. 1998). The total precipitation from 7 to 10 July in Izumi City was 613.5 mm and the maximum hourly rainfall was 62 mm(Meteorological Agency, Japan).   After a heavy rainfall stopped, a debris flow attacked 3.5 hours later without rainfall.   People in the village noticed a heavy sound at about 23 hours on 9 July and checked the torrent, but they did not evacuate because rainfall already stopped and the water level in the torrent was decreasing. However, the debris flow attacked about one hour after at 0:40 hrs on 10 July (Hirano et al.1997).

Figure 12 shows the longitudinal central section along this Harihara landslide-debris flow path. The mobilized apparent friction angle was 10.9°. The source area was a landslide, and the situation in the alluvial fan after the check dam was clearly debris flow.   Shimokawa et al. investigated the field in details, including the detailed observation of 5-6 m deep sections of the landslide deposit. They concluded the landslide mass moved until the check dam without major disturbance. Some standing trees were observed in the upstream of the check dam. The almost saturated landslide mass (due to long and heavy rainfall) was probably liquefied to a debris flow state. And it covered the village (Shimokawa et al. 1998).

4.2   Mechanical Simulation Test of the Harihara Landslide-Debris Flow

Sassa et al investigated this landslide-debris flow. The initial landslide and also

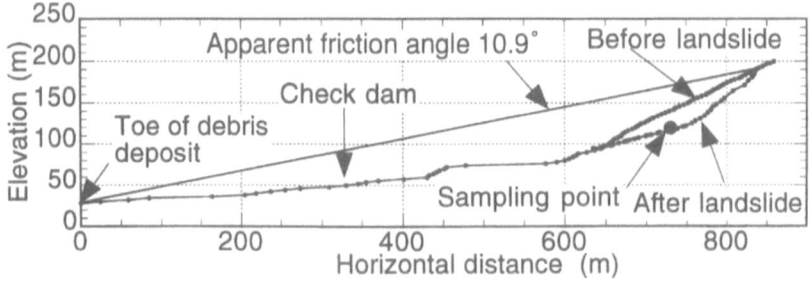

Figure 12 Longitudinal section of the whole travel course.

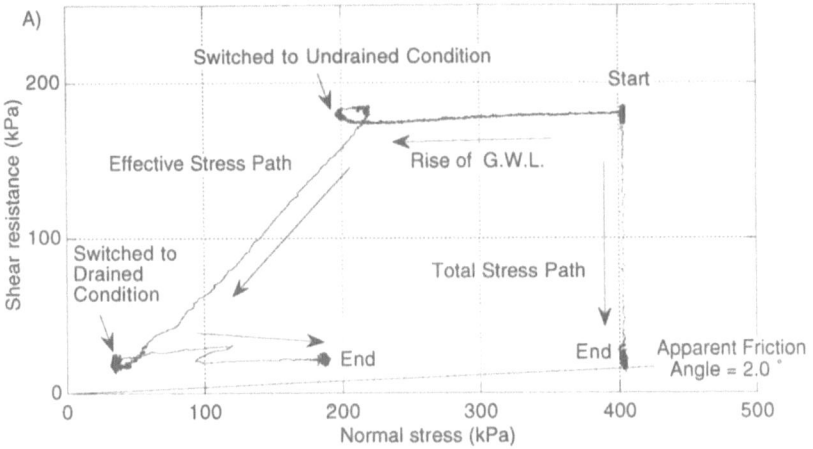

A)

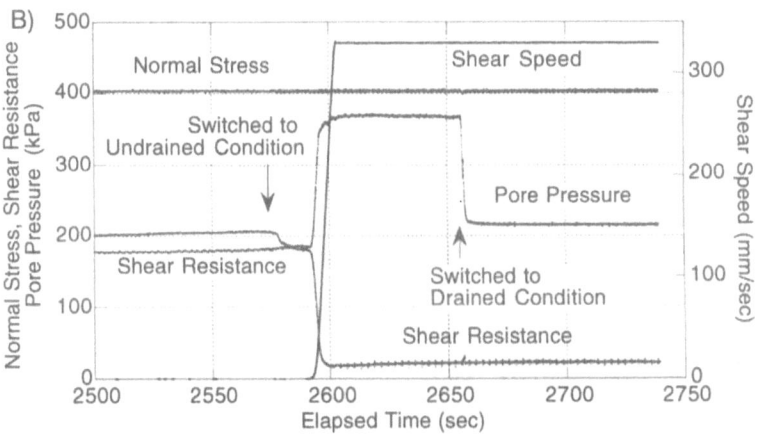

B)

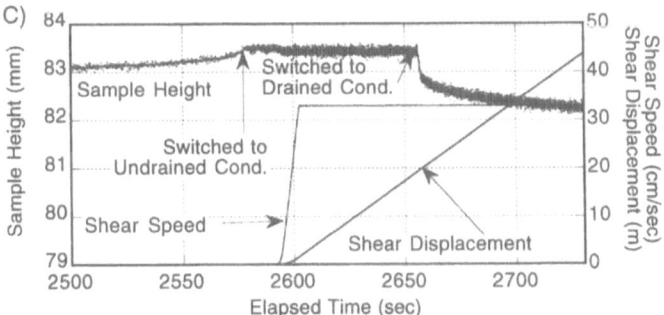

C)

Figure 13 Mechanical simulation test of the Harihara landslide initiation.
The shear box was switched from the drained state to the undrained state immediately after the initiation of failure, and returned to the drained state after the steady state of shearing. $B_D = \Delta u / \Delta \sigma = 0.96$.
a)Stress path; b)Time series data of stress; c)Time series data of displacement.

the undrained state soon after the failure and shearing starts; in order to measure the effective stress path. 4) After reaching the steady state motion, the shear box is changed to the drained state again to examine the speed of pore pressure dissipation.

Figure 13 shows the stress path (a) and the time series data (b, c) of the test. "Start" in (a) shows the initial stress corresponding to the stress along the potential sliding surface (25 m) in the slope (24°). The horizontal shift of stress path is due to pore pressure increase corresponding to the rise of ground water table. When the stress path reached to the failure line and shear displacement started to increase, the shear box was changed to the undrained state. It was written in the figures. The effect of closing the pore pressure supply tube appeared as a little decrease of pore pressure, because the average pore water pressure in the shear box was a bit smaller than the supplied pressure. Thereafter, excess pore pressure was accelerated to generate with the progress of shear displacement. Then, a rapid shearing started with a rapid drop of shear resistance as shown in (b, c). And the mobilized apparent friction angle in the steady state was only 2.0° as shown in (a). It is a very typical sliding surface liquefaction resulted from grain crushing of sandy soils in the shear zone. After enough shearing at the steady state, the shear box was changed again to the drained state. The stress path showed the excess pore pressure dissipated. However, the mobilized shear resistance remained at a very low level. It means that the pore pressure was maintained in the shear zone, even though drainage from the shear zone should have occurred. When pore pressure dissipates and the effective stress tends to increase, further grain crushing and comminution should take place, and pore pressure must be generated to reach a certain critical effective normal stress under which grain crushing does not occur. The volume reduction after the change to the drained condition found in (c) supports this interpretation.

To check the heavy grain crushing and the resulting decrease of permeability in the shear zone of this andesitic debris, the author performed another test shown in Figure 14. In this test, pore water pressure wad gradually increased from the initial stress of A. And the stress reached the failure line at B. And the shear displacement started to increase. However, the drained state of shear box, where the drainage valve in the top of shear box is open, was still maintained after the initiation of post-failure shear displacement (B-C), and the extent of decrease of apparent friction angle after failure was observed. As seen in the stress path, the apparent friction angle decreased to 5.3°. And the valve was closed to measure the excess pore water pressure after the point C at the relatively stable point. The stress path reached the failure line at D. The excess pore pressure generated in the shear zone seemed to be distributed in the whole sample box. Because the effective stress should be the point E corresponding to the total stress point C. This interpretation will need further examination. Then, shear zone was exposed and carefully observed after the test. Figure 15 shows the photo. The shear zone was quite silty while the sample above/below this layer is in the original sandy state. Permeability was measured before and after the test by supplying water from the bottom of shear box and monitoring the water discharge from the top of

A)

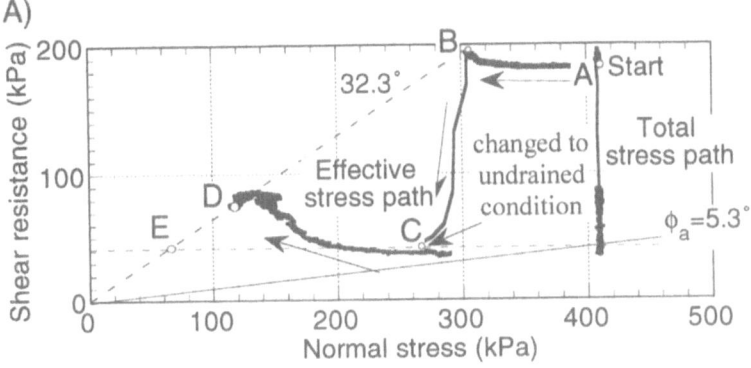

B)

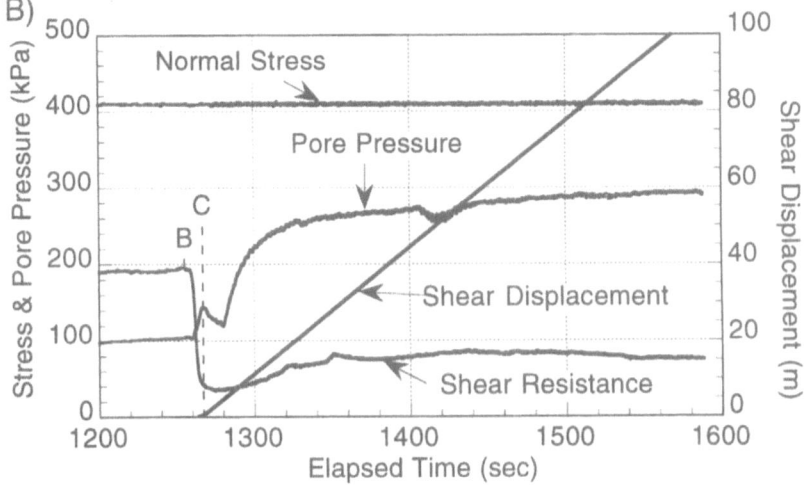

Figure 14 Mechanical simulation test in the drained state after failure.
The shear box was maintained in the drained state after the initiation of failure. It was changed to the undrained state after the shear resistance had almost stabilized. $B_D = \Delta u / \Delta \sigma = 0.96$.
a) Stress path; b) Time series data.

the torrent deposit of the Harihara River were composed of andesitic debris. We took sample the andesitic debris from the side-wall of initial landslide. Using this sample mechanical simulation was carried by the dynamic loading undrained ring shear apparatus.

The high gear was used for the Gamahara torrent debris flows, but for this case the medium gear was used (the maximum speed is 33 cm/sec) because a higher normal stress was needed. To simulate the Harihara landslide by the ring shear apparatus, the following procedure was taken. 1)Apply the initial shear stress and normal stress corresponding to the stress in the slope to the sample. 2) Increase pore water pressure at a slow constant speed (100 kPa/1300sec) corresponding to the rise of ground water level. 3) Change the shear box from the drained state to

516

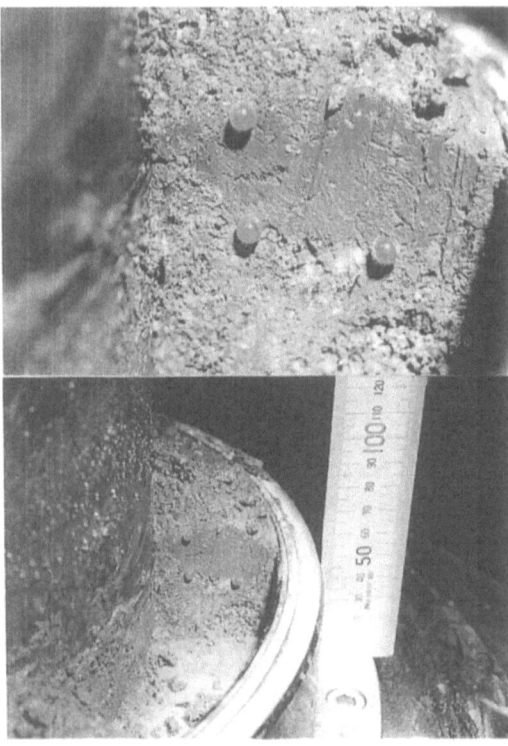

Figure 15 Photo of section through the shear zone after the test. Pins show the border between the shear zone and less-disturbed zones. The shear zone was very silty.

shear box. Therefore, it is the average value of permeability including the shear zone. The value changed from 1.0×10^{-4} cm/sec before the shear test to 4.8×10^{-7} cm/sec after the test. These test results indicate that this material mobilizes a very low apparent friction angle of $2 \sim 6°$ after the initiation of shear failure under this level of normal stress, because these two values were obtained from one very undrained condition and another very drained condition.

Figure 16 illustrates the model of Harihara landslide triggered debris flow. Heavy rainstorm initiates a landslide by rising pore water pressure inside the slope. The sliding surface liquefaction occurs by shearing and grain crushing in the shear. The rapid motion is continued by the sliding surface liquefaction in the saturated layer of the moving landslide mass or in the saturated torrent deposits or

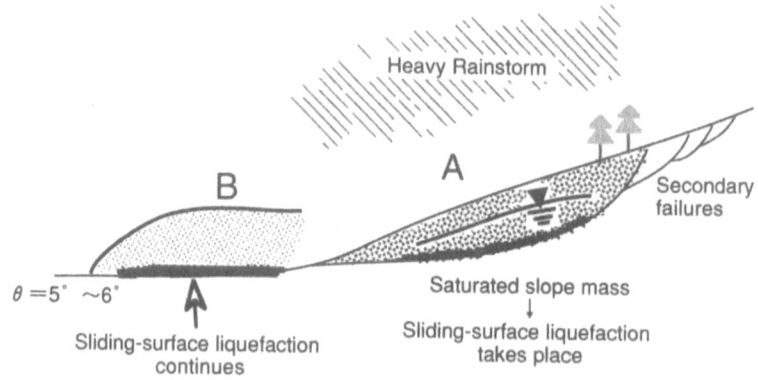

Figure 16 Model of the Harihara type of landslide triggered debris flow.

in the border of both layers. Due to the low friction in the shear zone, the landslide mass can move without many disturbances if the travel course is smooth. In the case of Harihara debris flow, there was a check dam in the course. Therefore, the wet (rather saturated) landslide mass was much disturbed and almost liquefied by the collision to the dam. The flowing situation after the dam was same with a typical debris flow.

CONCLUSION

1) Two debris flows in Gamhara and Harihara in Japan were triggered by landslides. The initiations of landslides are not always controlled by heavy rainfall. Some slopes such as a head scarp of previous landslides are very critical without rainfall. They can cause landslides with no or relatively small rainfall.

2) There is a mechanism of rapid and long run-out motion without enough water to saturate landslide mass. That is the sliding surface liquefaction. Shearing can cause grain crushing in some materials under a certain pressure in the shear zone, and it results in volume reduction and causes excess high pore water pressure when a shear zone is saturated. The sliding surface liquefaction phenomenon was found by Sassa 1996, by investigation of the Nikawa landslide triggered by the Hyogoken Nanbu earthquake. It occurred in very dry season, but the landslide mobility was very high. The mobilized friction during motion was similar to debris flows.

3) The recent two landslide triggered debris flows in Japan presented good examples of two types; The Gamahara type is that not-saturated landslide mass loads onto a saturated torrent deposit (at the bottom at least). A shear surface is formed in the saturated part of the torrent deposit by undrained loading and possibly the sliding surface liquefaction. Then, the landslide mass together with the torrent deposits starts to move at a very low friction and continues its process along the torrent. The Harihara type is that the initial landslide, itself, causes the sliding surface liquefaction in its saturated shear zone soon after its motion, and the landslide mass moves rapidly onto a torrent. The sliding surface liquefaction continues in the bottom of saturated part of landslide mass or in the saturated torrent deposits, possibly in the border of both. Accordingly it keeps its high mobility until it comes to a gentle slope less than the apparent friction angle.

4) This research focuses the initiation process of landslide triggered debris flows. The moving debris can be much disturbed and include more water during its long travel distance. In this case the debris possibly shifts to a fully liquefied debris flow. The effective stress is zero or very low in the fully liquefied mass. The sliding surface liquefaction will not occur because grain crushing does not proceed under a very low effective stress.

ACKNOWLEDGEMENT

The author acknowledges Dr. H. Fukuoka, Assoc. Professor, Mr. Wang Fawu and other postgraduate students of Landslide Section of the Disaster Prevention Research Institute, Kyoto University for their cooperation.

REFERENCES

Hirano, M. M. Hikita & H. Hashimoto (1997): On the Debris Flow Disaster in the Harihara River, Izumi City, July 1997. Proc. 16th Japanese Society for Natural Disaster Science, pp.101-102 (in Japanese).

Marui, H., O. Sato & N. Watanabe (1997): Gamahara torrent debris flow on 6 December 1996, Japan. Landslide News, No.10, pp.4-6.

Nikkei Construction (1997): Nagano debris flow disaster in the blind spot of risk assessment, 1997.1.24 issue, pp.46-49 (in Japanese).

Sassa, K. (1995): Keynote lecture: Access to the dynamics of landslides during earthquakes by a new cyclic loading high-speed ring shear apparatus. Proc.6th International Symposium on Landslides, "Landslides", Balkema, Vol.3, pp.1919-1939.

Sassa, K. (1996): Prediction of earthquake induced landslides. Proc. 7th International Symposium on Landslides, "Landslides", Vol.1, Balkema Ltd., pp.115-132.

Sassa, K. H. Fukuoka, G. Scarascia-Mugnozza & S. Evans (1996): Earthquake-induced-landslides: distribution, motion and mechanism. Special Issue of Soils and Foundations. pp. 53-64.

Sassa, K., Fukuoka, H. & F.W. Wang (1997): Mechanism and risk assessment of landslide-triggered-debris flows: Lesson from the 1996.12.6 Otari debris flow disaster, Nagano, Japan. Landslide Risk Assessment, editors: Cruden and Fell, Balkema Ltd., pp.347-356.

Sassa, K., Fukuoka, H. & F.W. Wang (1998): Mechanism of rapid long run-out motion in the Sumikawa reactivated landslide in Akita Prefecture and the Harihara landslide-debris flow in Kagoshima Prefecture, 1997, Japan. J. of the Japan Landslide Society, to be published in Vol.35, No.2 (in Japanese).

Shimokawa, E., Jitouzono, T. & S. Ogawa (1998): The Harihara debris flow disaster in Izumi city. Report of the 1997.7 Heavy Rainfall Disaster in the Western Japan (Scientific Grant, Ministry of Education, Science, Culture and Sports, Representative: E. Shimokawa, No.09600003), pp.19-30 (in Japanese).

Takahashi, T. (1978): Mechanical characteristics of debris flow. J. Hydraulic. Div. ASCE, Vol.104, No.HY8, pp.1153-1169.

Takei, A. Editor (1993): Criteria of precaution and evacuation, Sabo Engineering, Bunei-do Publisher, pp.252-262 (in Japanese).

The Japan Society of Erosion Control Engineering (1992): Series of Sabo Text No. 6-1 "Prevention of sediment disasters", Sankai-do Publisher, pp. 143-166 (in Japanese).

Wieczorek, G.F., Wilson, R.C., Mark, R.K., Keefer, D.K., Harp, E.L., Ellen, S.D., Brown III, W.M. & P. Rice (1990): Landslides warning system in the San Francisco Bay Region, California. Landslide News, No.4, pp.5-8.

Assessment of Failure and Success of Preventing Damages of Debris Flows Caused by Landslide in Laogan Ravine, Yunnan, China

Tianchi LI

International Center for Integrated Mountain Development (ICIMOD)
P. O. Box: 3226, Kathmandu, Nepal

ABSTRACT

Laogan Ravine is a tributary of Xiaojiang River which is a branch of the Yangtse River in North-eastern Yunnan province, China. The railway from Dongchuan to Kunming, a highway and Tuanjie irrigation canal pass through the lower reach of the Laogan Ravine. During 70's through 80's, debris flows occurred frequently in the Laogan Ravine and created great damages to the railway, highway and the irrigation canal, total economic losses was 19.5 million Yuan. On July 26, 1985 a large debris flow overturned a bus and killed 12 passengers, silted up the open railway tunnel and highway bed with total 410m long. The Road Department and Railway Department of Dongchuan city used many methods separately to prevent debris flow damages but they were all not in success. Therefore, development and enlargement of debris flow in Laogan Ravine became a key blocking section of traffic not only for Dongchuan city but also for north-eastern Yunnan province therefore Laogan Ravine debris flow control was a great issue for the development of Dongchuan city. In 1991, area approach was used to control debris flow in Laogan Ravine. The area approach consisted of 2 main dams with total height of 17.5m, 1080 m draining channel, 60 long catch drain, 4 fixing-bed dams, 155 m long protection dyke, 8 check dams and afforestation of 160 ha. This debris flow control project was completed in 1994 with total investment of 1.372 million Yuan. Since then, debris flow has not occurred in the Laogan Ravine.

1.   Introduction

Laogan Ravine, located at 103^0 20' 30" E to 103^0 23' 00" E and 26^0 2' 20" N to 26^0 6' 00" N, 11 Km south of Dongchuan city, Yunnan province, China, is a tributary on the right valley slope in the middle reach of the Xiaojiang River, which is a tributary of upper Yangtze River. The catchment area of the Laogan Ravine is 7.7 Km^2 and the altitude ranges from 1300 m asl to 2600 m asl. The main stream is 5.7 Km long with a gradient of 17-25%. There are several villages in the catchment area, namely Zhiga, Heishan, Xiaoxincun and Xiaomaidi. The railway from Dongchuan to Kunming, a highway and a irrigation canal, which all were built in late 50's, pass through the lower reach of the Laogan Ravine (Fig. 1).

Above the altitude of 1900 m asl, bed rocks are mainly composed of limestone, dolomite of Devonian and Carboni ferous and lower Permian. Below the altitude of

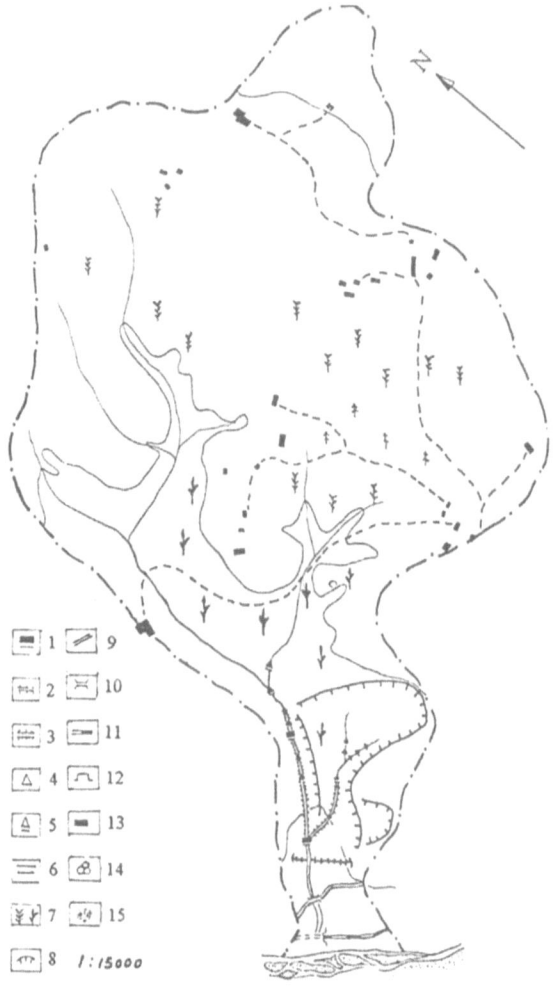

<table>
<tr><td>1</td><td>Catch dam</td><td>6</td><td>Anti-scouring rib</td><td>11</td><td>Railway</td></tr>
<tr><td>2</td><td>Catch drain</td><td>7</td><td>Forest area</td><td>12</td><td>Open tunnel</td></tr>
<tr><td>3</td><td>Diversion flume</td><td>8</td><td>Collapse area</td><td>13</td><td>House</td></tr>
<tr><td>4</td><td>Fix-bed dam</td><td>9</td><td>Road</td><td>14</td><td>Mortar masonry</td></tr>
<tr><td>5</td><td>Check dam group</td><td>10</td><td>Bridge</td><td>15</td><td>Masonry</td></tr>
</table>

Fig. 1   Map of Laogan ravine showing landslides and the line infrastructure and control measures

1900 m asl, mainly Emeishan basalt of Permian. The deep active Xiaojiang Fault pass through the middle reach of the catchment. The bed rocks are highly weathered and fractured due to the activities of the Xiaojiang fault. The outcrop area of bedrock is relatively small, most of the catchment area is covered by landslide and slope wash deposits. Semi-weathered basalt can be seen on the north side of the middle and lower sections of the mainstream.

Climatically, the catchment is under monsoon and has distinct vertical climatic zones. The main elements of climatic conditions are present in Table 1.

Table 1   Main climatic elements of Laogan Ravine, Dongchuan, Yunnan, China

Altitute (m)	Average annual temp.	Average annual rainfall	Average annual evaporation
1300 - 1600	20^{0}C	800 mm	3700 mm
1600 - 2100	17.5^{0}C	900 mm	1706 mm
2100 - 2600	14.0^{0}C	1040 mm	1200 mm

Most of precipitation is concentrated in June through August with high intensity of rainstorms. According to local rainfall record, maximum one hour rainfall was 40.4 mm in the lower valley area, 36.7 mm in lower mountain area and 61.2 mm in mountain area. Maximum 24 hour rainfall was 83.7 mm in the lower valley area, 117.2 mm in lower mountain area and 100.4 mm in mountain area. Concentrated rainfall and local rainstorm with high intensity were the trigger of the debris flow in the Ravine.

2.   Debris Flows in the Catchment Area

As mentioned above, the bedrocks, in particular the basalt rocks were strongly weathered and fractured due to the influence of tectonic movement of the Xiaojiang fault. The historic earthquakes of great magnitude and effects in the Xiaojiang river is presented in Table 2.

The strong earthquake of February 5, 1996 at Xiniu mountain in Dongchuan area reactivated a huge landslide in the lower catchment. The landslide covered a area of 2.44 Km^2 with total loose earth material of 13 million m^3. The slidden mass pushed the mainstream course to the right side.

The total loose solid materials from small landslides and rockfalls in the lower reache were estimated to be only 40 X 104 m^3 before 1962, and about 200 x 10^4 m^3 before the strong earthquake in 1966. After the strong earthquake in 1996 through 1989, the total loose solid materials were 13,653 million m^3. And the area of landslide increased from 0.27 Km^2 in 1963 to 0.505 Km^2 in 1989.

Table 2   **Historic earthquakes of great magnitude and effects in the Xiaojiang River region**

Date	Approximate Epicenter	Magnitude	Effects
1. February 26, 1713	Xundian	6 ½	Ground fissures, many slumps and shallow slides in upper watershed of Xiaojiang river.
2. August 2, 1733	Dongchuan	6 ¾	Fault rupture 100 Km long. Many landslides; one large landslide buried village on the Daqiao tributary killing 40 people.
3. September 6, 1833	Songming	8	Ground fissures, liquefaction (sand boils), many landslides in upper and middle watershed of Xiaojiang River.
4. March 15, 1927	Xundian	5 ½	Slumps and shallow slides in upper watershed of Xiaojiang River.
5. February 5, 1966	Dongchuan	6.5	Fault rupture, ground fissures and liquefaction (sand boils).
6. February 13, 1966	Dongchuan	6.2	Many rock falls, slumps, slides; reactivated ancient landslides in middle Xiaojiang River watershed. Combined effects of both February, 1966 earthquakes.

There are two tributaries in the upper reach of the catchment. The mainstream course is 5.7 Km with bed gradient of 17% in the middle and upper reaches above the confluence of the two tributaries, and 25% in the low reach, where the ravine is narrow and deep cut with side slope angel of 40^0-50^0. In the low reach, mainstream runs by the right side due to slidden material. Large runoff created by high intensity short time rainfall, eroded the toe part of landslide mass, forming debris flow in the mainstream valley of lower reach. For example, on June 26, 1985 a flood flow with discharge of 30 m^3/s, eroded slidden material of landslide toe, forming a viscous debris flow with a discharge of 91.5 m^3/3 and velocity of 6.5 m/s and 14.5 x 10^4 m^3 loose solid materials were transported and deposited on the deposit fan in the confluence area with the Xiaojiang river.

3.   The Damages and Failures of Debris Flow Prevention Works

The railway, road and irrigation canal pass parallely through the lower reach of the Laogan Ravine as shown in Fig. 1. As the railway runs at the lowest location, it had received more damages than the highway and the irrigation canal.

3.1   The Damages of the Railway

In 1959, the railway crossed the ravine with middle-scale bridge of 3 span of 5 x 12 m. One debris flow silted up the bridge and washed away large section of the railway closeby the bridge and formed a new course of the ravine in the summer of the same year.

In 1960, a large scale bridge with 4 span of 7 x 12 m was constructed to cross the ravine. This large bridge was gradually silted up from 1961 through 1970. The rate of deposition was 70 cm per year in average.

In 1971, an open tunnel with more than 100 m long and a protection dyke of 150 m long was built above the silted bridge.

In 1985, a large scale debris flow buried the open tunnel, then an artificial tunnel was built, above which debris flow could pass down into the Xiaojiang river.

For preventing debris flow, damages to the railway, the Railway Department spent 13 million Yuan during 30 years period from 1959 to 1989. During the same period, the ravine course bed silted up about 30 m from 1305 m asl to 1335 m asl with a rate of 1 m deposition per year.

3.2   The Damages of Highway

The highway runs across the ravine about 80 m higher than the railway upstream. At beginning, a small bridge was constructed across the ravine course, soon later the small bridge was destroyed and a pavement was constructed for allowing debris flow down. The cost for repairing and maintaining the section of highway was 15 - 20 x 10^4 yuan annually. Adding the cost of a new bridge the total cost was more than 5.4 million yuan.

3.3   The Damages of Irrigation Canal

The Tuanjia irrigation canal, constructed in 1958, is alocated above the highway about 50 meters upstream. Two very large scale debris flows in 1985 and 1987 destroyed the cover and foundation of the tunnel of irrigation canal, causing economic loss of more than 40 x 10^4 yuan (Plate 1).

Apart from the damages of infrastructure, one disastrous debris flow of June 26, 1985, killed 12 persons and seriously injured 3 persons in a running bus which was washed away by debris flow and destroyed in the Xiajiang river causing a great social impact in the Dongchuan area.

During the 30 years period from 1958 through 1988, the debris flows in the Laogan Ravine caused 19.5 million yuan in economic losses, apart from killed 11 persons. Although the Department of Railway and Highway used different structures to prevent debris flow in the ravine, but no one's structure was in success. With the development and enlargement of the debris flows in the ravine, debris flow effective control became an important issue for transportation and irrigation. As all the preventing measures constructed by the Departments of Road and Railways were failured for saving the transportation, there was a need to find out another alternative method.

4. Debris Flow Control by Area Approach

In 1988, the Institute of Debris Flow Protection (IDFP), Dongchuan city, was approached to investigate this debris flow ravine and design effective control measures with assistance from the Chengdu Institute of Mountain Hazards and Environment, CAS. It was found out that the formulation of the debris flow in Laogan Ravine was mainly associated with the landslide on the south slope of the up-lower reach of the ravine. The railway and highway could not be saved without controlling the landslide. Based on detailed field investigation and analysis of the landslide and the process of debris flow, an integrated area approach was prepared and implemented during the period of 1991 through 1994. The control measures can simply summarised as below.

4.1 Landslide Stabilisation

Two medium scale sediment-trapped dams with total height of 17.5 m were constructed respectively at the toe area of the main landslide to control down cutting and toe erosion of the main landslide on the south slope of the up-lower reach of the ravine. And a 155 m long protective dyke and 8 small check dams were also constructed on the landslide area to control surface erosion.

4.2 Chanalisation of the Main Stream Course

For protecting down cutting and side erosion of the main stream course, a total 1080 m long stream course was chanalised with one side wall in the case of that one bank side was of loose materials and two side walls in the case of that two bank sides were composed of loose materials. The stream course was chanalised also with 4 small fixed-bed-dams which will make the chanalised course more stronger (Plate 2).

4.3 Afforestation in the Catchment Area

The afforestation activity was undertaken at the same time period. From 1991, trees were selected according to the altitudes and soil types and were planted on the bare mountain slopes in the middle and upper reaches as well as on the mass of landslides in the lower reach. In the middle reach with elevations from 1930 to 2400 m asl, the planting area was 130 ha. The planted trees are robinis pseudoacacia, albizia mollis, cupressus funebris, pinus yunnanensis, pinus armandii etc. In the landslide area of the upper-lower reach, albizia mollis, coris sinica etc. were planted in the area of 29 ha.

Plate 1   Debris flow of 28 July 1987 destroyed highway and irrigation canal in
           the lower reach of Laogan ravine

Plate 2   View of Laogan ravine looking upstream, two years after completion
           of the project in 1994

5. Impact Assessment of Area Approach

5.1 Engineering Control works

The two medium scale sediment-trapped laws arrested about 42,000 m3 sediment behind the dams and stabilised the main landslide by raising the ravine bed level above the toe of the landslide. The 1,080 m long artificial diverson flume with 78 scouring-proof ribs and 4 fixing-bed dams confined the flood within the flume and also prevented the side erosion of the ravine, then production of load sediment were greatly reduced. Before the measures taken place, about 150,000 m3 loose solid materials were transported in the form of debris flow down to the Xiaojiang river, and after completion of the engineering works, there was only sediment laden flow and light erosion in the low reach of the ravine.

5.2 Afforestation Activity

The tree plantation area in the middle and upper reaches is 130 ha. With active participation of local people, the living rate of planted tree was 90.4%. Taking 20 years as a period, the profit is 4800 yuan/ha year and the total income for the local people is 624,000 yuan. In the landslide area, the trees of albizia mollis and coris sinica etc. were planted in the area of 29 ha. Taking 5 years as a period, the profit is 750 yuan/ha. Year, then the total income is estimated to be 21750 yuan. Apart from the income to the local people, the vegetation will plays an important role in soil erosion control and shallow small landslide stabilisation in the middle and upper reaches of the ravine.

The debris flow in the Laogan river caused 19.5 million yuan in economic losses in the 30 years from 1958 to 1988, the cost of area approach control project was only 1.37 million yuan. The debris flow hazard was controlled and great potential income will be generated by the project. The ration of benefit/cost is valuable.

5.3 Main Lessons Learnt

Debris flow is a major mountain hazard, sometimes it may cause catastrophic disasters when a debris flow destroy houses, roads and other properties of individuals and human community.
A debris flow watershed can be devided generally into three zones, namely formation zone, transportation zone and deposition zone. In most cases, landslides located in the upper and middle reaches are the formation zone of debris flow.

In a debris flow watershed, debris flow cause disasters mainly in the transportation and deposition zones when houses and infrastructures are located in these zones.

Alignment of line infrastructures such as road and irrigation canal should avoid to pass through the deposition zone of debris flow.

Measures to mitigate damages from debris flow to highway, railway located in deposition zone of debris flow should be emphasised to control slope instabilities in middle and upper reaches and to improve the environment conditions of all the catchment area. Integrated area approach has been recognised as a success one to tackle this problem.

6. ICIMOD's Programme on Landslide Hazard Mitigation

With financial support from the Government of Japan, ICIMOD has been developing and implementing four-year (1995 to 1998) programme on landslide hazard management and control in the Hindu Kush-Himalayan region. One of the goals of this programme is to develop better understanding of natural and human processes causing landslide hazards and identify approaches for their control and management. Under this programme, there are several sub-programmes, including the state-of-the art review on landslide management and control in some selected countries, regional training programme, preparation of a Climatic-Hydrological Atlas of Nepal, and national training course, etc.

Through the activities of the landslide hazard management and control programme, ICIMOD facilitated the exchange and disseminated the relevant information about hazard mitigation among the regional countries. The long-term output of the landslide hazard management and control programme will undoubtedly improve the overall level and skill of landslide hazard mitigation in the Hindu Kush-Himalayan region.

REFERENCE

Chen Xunqian, Yang Wenke, Niu Xiankun, Kang Zhicheng, Zhang Shucheng, and Chen Jingwu, 1981: Debris Flows along the Siaojiang River, Dongchuan, Yunnan, in Collected papers on Debris Flows (No. 1), Chengdu Institute of Geography, Chengdu, China pp. 13-19, 1981.

Institute of Debris Flow Protection of Dongchuan City, Planning and Design Reports of Debris Flow Control in Laogan River, Yunnan, China, pp. 45, 1989.

Li Tianchi, 1980: A preliminary study of landslides in the Dongchuan Region, Yunnan, unpublished paper from Chengdu Institute of Geography, Chengdu, China pp. 1 - 5, 1980.

Wieczorek G. F; Wu Jishan and Li Tianchi, 1987: Deforestation and Landslides in Yunnan, China, Proceedings of Conference XVIII International Erosion Control Association, February 26 - 27, 1987, Reno, Nevada, USA

The Estimation of the Hazard Potency of Debris Flows and The Step to Step Method

Gernot FIEBIGER

Director, Austrian Forest-technical Service in Avalanche, Torrent & Erosion Control, District Hallein & Salzburg, Paracelsusstr. 4, A-5027 Salzburg, Austria

ABSTRACT During the longtime experience in Hazard Mapping of Torrents and Debris Flows it was shown up that a systematical way is necessary to shorten the time of work and fasten the development of Hazard Maps. It was seen that the **Step to Step Method** is the most efficient methodology and grants small time spent on the development of the Hazard Map. At least six (6) steps are neccesary. **Step one (1)** is the study of the basic conditions of the watershed, the reconnaissance of the catchment area. **Step two (2)** ist the learning of the hydrology of the watershed and of the channel hydraulics. It is the definition of the design event of the debris flow and/or flood potency of the watershed. **Step three (3)** leads us on site in the watershed for field work. The result of the outdoor work will be the description of the geomorphologic features and the quantification of the debris balance and debris load and/or bedload. **Step four (4)** is the important step for the quantification of the design event by combination of the results of step two and step three. **Step five (5)**, again field work, is the distribution of the design event upon the area that will be mapped, respectively is determinded by land use planning. **Step six (6)** is the integration of the single scenarios of disasters showed up in step five. The final **step seven (7)** brings the sipervision in scientific and legal sense and the confrontation with the afflicted people.

Keywords: Hazard potency, hazard mapping, process.

INTRODUCTION

Since World War II the Alps developed with increasing intensity. Trafic, transit, tourism and settlement burst in some regions. Also the natural regions and the areas of disaster reasons and areas of disaster impact were developed and the menace of these reaches collided with the trends of anthropogenous development. Natural disasters like floods, debris flows, avalanches, rockslides and landslides increased in quality and quantity.

That development lead also to the increase of control measures it was known soon that in future the financing of this kind of control will be impossible. This reconnaisance was the reason to enforce passive methods in Austria. The most important of these methods is hazard mapping. To carry out a hazard map scientific steps and administrative processes with parallel supervising. The process of hazard mapping is shown by figure 1.

Table 1: Step to Step Method to carry out a hazard map in torrents and debris flows areas.

Tab. 1: Stufenweise Erarbeitung eines Gefahrenzonenplanes für murfähige Wildbäche

STEP 1	BASIC RECONNAISSANCE Maps, Topography, Geology,Torrent Survey, Quantification and qualification of the catchment; Chronicle; Debris flow Yes/No;
STEP 2	HYDROLOGY Calculation of the flood potency regarding to a specific return period (150 years) Quantification, Hydraulics
STEP 3	FIELD SURVEY, Debris/Bedload Balance; Quantification; Geomorphology; Accumulations and deposits
STEP 4	QUANTIFICATION OF THE DESIGN EVENT - How much bedload/debris (step 3) could be mobilized by runoff and discharge (step 2) and causes the debris flow; ----Design Debris Flow
STEP 5	DANGEROUS AREA MAPPING Distribution of the design debris flow according to various scenarios; Delimination of danger and hazards; Mapping of endangered areas (zones)
STEP 6	HAZARD MAPPING - INTEGRATION OF POSSIBLE DANGEROUS AREAS AND MAPPING OF THEM - Hazard zones - Reservation area - Reference area - Special reaches

531

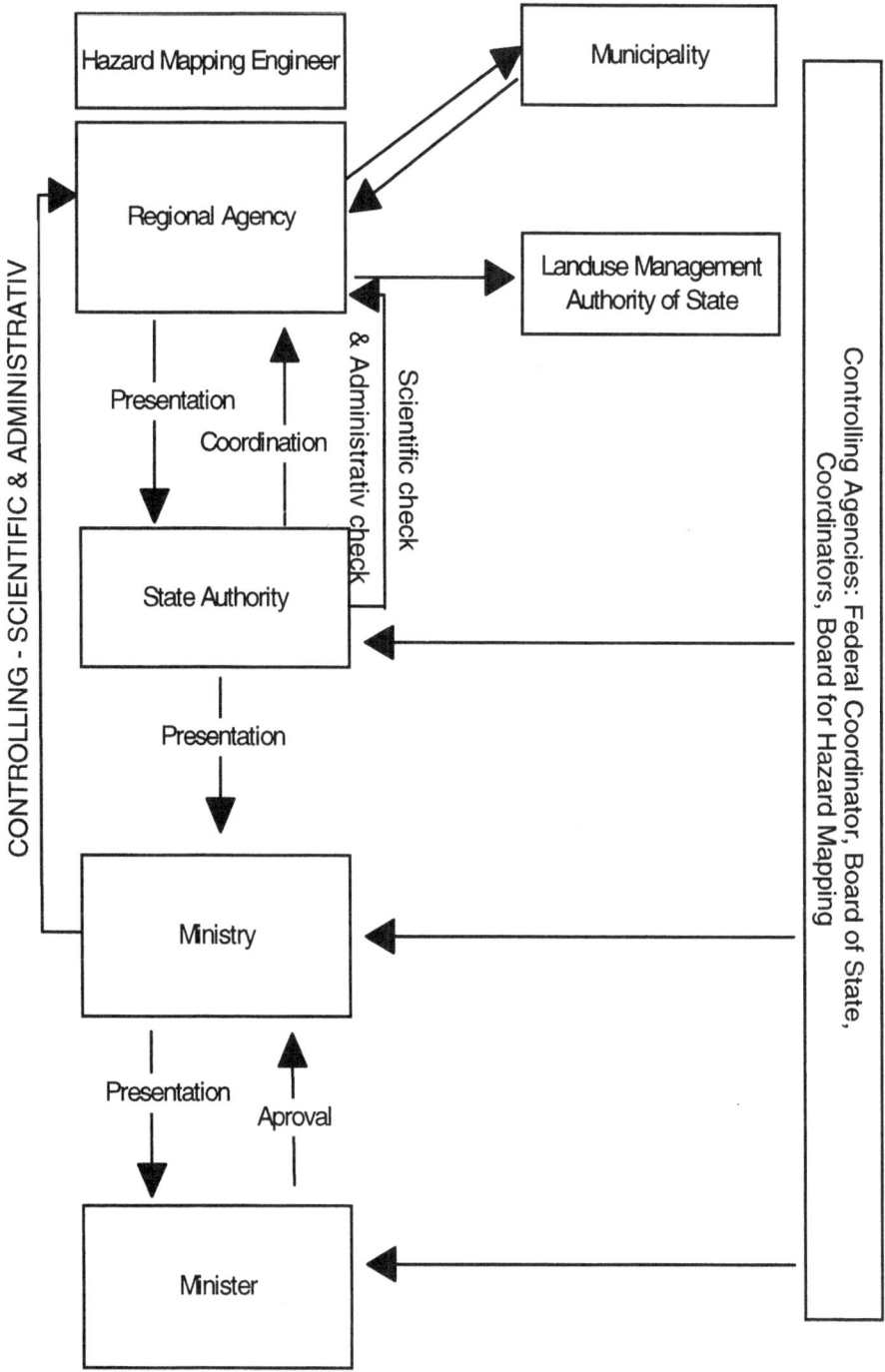

Figure 1 : The process of hazard mapping
(SCHEURINGER 1998)

THE STEP TO STEP METHOD IN HAZARD MAPPING

The step to step method in Hazard Mapping is parted in two parts a scientific part and an administrativ controlling part. This paper deals only with the scientific part the development of the hazard map. This part has six steps and starts with the basic reconnaissance of the watershed, leads form hydrologic investigations and field work to the definition of the design event. The last steps are the mapping of the hazardous events and the integration of various scenarious in the hazard map. Table one (1) gives an overview of the various steps of the method and their content. The single steps are descript in the following part.

Step one (1) - Basic reconnaissance & studies

The first step is occupied with the general discription of the torrential watershed, its meteorological situation and conditions, the geologic and geomorphologic features and the vegetation cover of the catchment. One of the most important investigations are the kinds of debris sources and and the definition of the potency of disasters & character of the torrential watershed respectively debris flow catchment. Also a description of the anthropogenous influences and existing control measures and their influences is neccessary. The study of the Torrent Chronicle if it is available also is very important.

The Torrent Survey informs first on the name of the torrent and defines the political situation the municipality district, subwatershed, water management cadaster and cipher of torrent and cipher of cadastre.

The general discription of the torrent describes the site, the area of catchment (watershed), the shape and exposures of the watershed, the elevation of the summit and the mouth above sealevel, the length of the water course and the difference in height and the avarage slope. The length of the water course should also be determined in upper, middle and lower course. Important tributaries should also be mentioned and registered.

The meteorogical situation and condition describes the climatic zone, the characteristic precipitation, maximum one day precipitation, maximum and mean annual precipitation. Recorded and extrapolated amounts should be reported. The source of the basis data should be confirmed by the name of station its elevation and the time of recording duration. These informations are rounded up by the critical weather situations the main wind direction and the known track of storms (typhoons, blizzards, hurricans etc.) In geology the description of the tectonics and geology and petrologic strata in the various parts of the watershed and also should include the bedrock sediments and deposits for the upper middle and lower course.

The vegetation should be characterized by the main phytosociological reaches the actual and potential vegetaton cover and the important patterns like the percentage of forests, the actual and potential timber line etc.. The influences of management on the watershed and its areas (Pasturage, Waste land, Barren land) should be shown up special the anthropogenous influences (runoff increasing and decreasing factors) on the upper, middle and lower course.

In the section kind of debris sources the superficial deposits, loos material, the bulk of debris sources, actual and possibel sliding areas (potential landslides) the kind of bedload and its relation with disasters are descript.

The potency of disasters and character of torrent is the record of the combined results of step two (2),(Hydrology) and step four (4) quantifiying the design event. The floods with and without bedload with various return periods respectively various frequencies and the amount of debris flow should be reported. The possibilities of jaming along the course and the possibility of shifting specially on the apex of the debris cone should be mentioned. Silent tests, the transport capacity of the recipient and last but not least the character of the potential hazard event rounds up the character of the torrential watershed.

Anthropogenous influences and the administration of the watershed, existing control measures and/or proposals for future control measures rounds up the torrent survey. Last not least the torrent chronicle gives an overview of historical recorded reports on floods and debris flows and reported statments of local persons.

Step two (2) - Hydrology
The calculation of hydrological data of the torrential watersheds depends on the type and the character of the torrent. It is a wide spread between the debris flow and the flood with low bedload. The calculation of the hydrology of a torrential watershed is only possible for the discharge of floods for various return periods and frequencies. The calculated flood is the base to combine the debris or bedload potency to the design event. For the calculation of the flood with a return period of 150 years we developed in Austria a procedure of hydrograph calculation based on the USSCS (United States Soil Conservation Service) method, modified by MALZER and KREPS (OFNER 1987).

For the calculation one needs the following input, the area of the watershed [km^2], the amount of the design precipation [mm/m^2] and its duration [h], the length of the course [km], the slope [%], the roughness coefficient, Strickler coefficient [m$^{1/3}$/s], the percentage of forestation [%], the type of soil [A, B or C], the soil moisture before the rainfall or storm, the number of the characteristic curve of the precipitation and the return period for the design event.

The result of the calculation brings the runoff originating precipitation, the runoff coefficient, the initiative periode, the load of the flood and the peak flood and the plotted hydrograph.

Step three (3) - Reconnaissance of the torrent catchment

The estimation of the debris input from sources in the upper course respectively in the headwater and the estimation of the bedload potency at the apex of the debris cone are one of the most difficult problems for the planner of the hazardous region. There are only few supports to solve this problem. The geomorphology of a torrent watershed as the reason for hazardous floods and debris flow cannot be generalized. A good knowledge of the dynamics and causes of mass movements and geomorphological features is necessary.

Hochwasserganglinie
EINÖDGRABEN
nach US-Soil/Malzer/Kreps

Fläche :	3,1	km²
Niederschlagssumme :	70	mm/m²
Niederschlagsdauer :	0,5	Std.
Lauflänge :	5	km
Gefälle :	26	%
Rauhigkeitsbeiwert :	20	$m^{1/3}/s$
Bewaldung :	50	%
Bodentyp :	C	
Vorbefeuchtung :	mittel	
Kurvennummer :	77	
Jährlichkeit :	150	

abflußwirks. Niederschl.:	23,00	mm/m²
Abflußfaktor :	0,33	
Anlaufzeit :	0,44	Std.
Hochwasserfracht :	71.300,00	m³
Hochwasserspitze :	27,60	m³/s

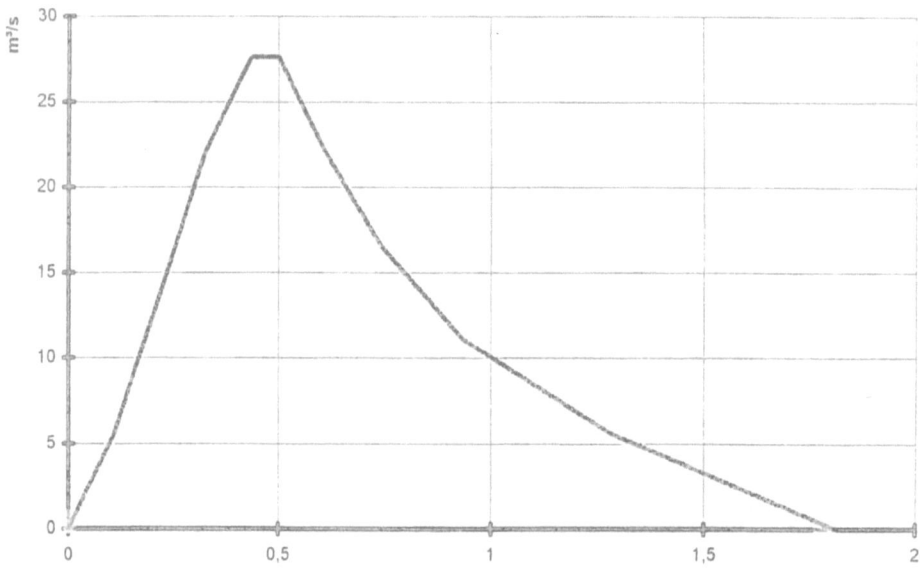

Figure 2: Hydrograph of 2 torrent watershed (EINÖDGRABEN)

During the inspection of the upper course a lot of relevant torrent reaches are seen. There are reaches of cleaning out the cross section form bedload changing with such where massmovements like landslides and sackungen are present. In bedrock courses the amount of bedload can decrease to zero. Last not least the planner has to decide immediatively on the inspection in the field. In upper courses with large mass movements it is often impossible to calculate the amount of debris and/or bedload of the design event. In sensitive instable reaches collapses of the banks are possible up to 100 m. The activated debris can be transported during the event to the debris cone and cause there the disasters.

One of the important criteries for the assessment of mass movements and the amount of bedload is the assessment of the slopes after long precipitation periods. Additional to the estimation of geomorphological instabilities one can do various investigations like drilling and digging the determination of clay minerals and their percentage and some more investigations. During field investigations the planner has in general not the equipment to do the above investigations.

Relating to the techniques of the inspection and survey of a torrent watershed it is more convenient to do the inspection from the watershed and spring to the debris cone. In that way it is possible to assess the origin of debris and bedload from its sources and roots and solves the problem by summarizing up and balancing the debris capacity. The decisive bedload at the apex of the debris cone is the result of the mechanism of influences and effects along the upper course. The more or less corectness of the results is granted by the upon descript survey method.

For a possible duplication it is necessary to document the assessment. To that reason forms for a debris/bedoad balance are developed (Fig. 3). The upper course is parted in various reaches and the according debris and/or bedload is integrated within.

Step four (4) - Quantification of design event

The important decision of the design event by combination of the result of the hydrological calculation and the debris/bedload balance.
It is necessary to determine the hydrograph in regard to its load and peak discharge and its possibilities to activate debris and bedload balance potencies are the design event.

Last not least it is of importance to assess the discharge and flood as the transporting medium. Flood events with more or less bedload have an other behavier in development and fluid mechanics than debris flows where the influence of peak discharge decreases compared to other parameters like woody debris, coarse and fine parts of the bedload and so on.

Step five (5) - Dangerous (disasterous) area mapping

Again field work is requested but now on the debris cone. The design event, the debris flow or the flood with bedload, has to be distributed under various conditions and scenarios on the debris cone starting at the apex.

The criterias for the zoning process of endangered areas are according to the guidlines of the Austrian Ministry of Agriculture and Foresty, Engineering Service in Avalanche & Torrent & Erosion Control (Table 2)

DEBRIS/BEDLOAD BALANCE

Name of torrent/watershed: EINÖDGRABEN

Elevation a.sl.	running meter	m³/rm	Σ m³	remarks
1070	300	3,0	900	gorge, local small landslides
1150	250	5,0	1250	right bank sackung; accumulation of woddy debris month
1150	400	2,0	800	tributary right bank in loose material
1280	500	7,0	3500	Increasing landslides from the right bank; high sliding reaches; woody debris
1360	400	10,0	4000	moraine and mylonitic phyllite on the right bank; woody debris; local sliding also from the left bank
1330			300	Crossing of forest road
1460	400	12,0	4800	Sackung; high moraine layers on the right bank
1550	400	20,0	8000	Sliding in moraine layers; woody debris jaming
1650	350	8,0	2800	Decreasing landsliding; lower slope
1700	200	7,0	1400	Forest road construction
tributary	500	3,0	1500	Torrent partly on bedrock
			29250	Sum

DEBRIS/BEDLOAD CAPACITY 30.000 m³

Figure 3: Debris/Bedload Balance of EINÖDGRABEN/RAURIS

(SCHEURINGER 1988)

Table 2: Criterias for the mapping of disastrous areas endangered by flood and/or debris flow.

TR = Torrent Red   TY = Torrent Yellow (FIEBIGER 1997)

Criteria		Hazard Zone	Design event 150 year return period	Frequent event 10 year return period
1)	stagnant water	TR	water depth > 1,5 m	Border line HQ10 > 50cm HQ 1 > 20 cm
		TY	water depth < 1,5 m	Border line HQ10 < 50 cm HQ1 < 20 cm
2)	running water	TR	height of energy line > 1,5 m	height of energy line HQ 10 > 0,25 m
		TY	height of energy line < 1,5 m	height of energy line HQ10 < 0,25 m
3)	erosion (rill-erosion)	TR	depth > 1,5 m	rill erosion possible
		TY	depth < 1,5 m	runoff without rill erosion
4)	bedload deposit	TR	height of deposit > 0,7 m	Bedload deposit possible
		TY	height of deposit < 0,7 m	no bedload deposit
5)	Bank collapse due to depth erosion & undercut	TR	upperedge of the cross section & slope	---------------
		TY	Safety	---------------
6)	Debris Flow & Soil creeping	TR	Boarder of debris flow deposit	---------------
		TY	Boarder of debris flow deposit	---------------
7)	Regressing erosion	TR	possible size	no assessment
		TY	take notice of 3) & 5)	

remarks:

1): Bogs, small lakes, small tronghs, wells, pools & ponds are not represented

2): Foundation for the broad of the safety strip for the individual case

Step six (6) - Hazard Mapping

In Hazard mapping endangered areas, Torrent (Debris Flow) Red Zone & Torrent (Debris Flow) Yellow Zone and Reference areas, Brown Reference Area and Violett Reference area and a Blue Reservation Area are mapped.

The hazard zones, the reference and reservation areas are defined as follows:

Red hazard zone:

The Red hazard zone encloses areas endangered by torrents, debris flows and/or avalanches to such an extend that their permanent use isnot possible for settlement and/or traffic purposes and infrastructures because of the expected damages by the recurrent design event" or frequency of danger or only can be ensured by excessively high efforts. Another definition says; Red Zone includes areas which are endangered by torrential floods or debris flows to such an extend so that permanent utilisation for settlements and infrastructures is not possible because of enormous danger for human being, traffic and infrastructures.

Buildings and infrastructures are not allowed to be implemented in the zone because of the great danger and amount of natural hazard.

Yellow hazard zone:

The Yellow hazard zone encloses all other areas endangered by torrents and/or avalanches the permanent use of which for settlement and/or traffic purpose because of that danger is impeded.

Yellow Zone recovers areas with reduced danger. Buildings and infrastructures may be protected by reinforcements and special architectural designing. People within buildings are safe but outside endangered nevertheless. Danger is decreasing with distance from Red Zone.

In areas which are already settled, an expert has to be consulted by the authorities in new buildings and infrastructures are intentioned. These expert opinion has to be observed in the decisionof the authorities. Such terms may be - reinforced walls, no doors and windows at the flood or avalanche impact side, reinforced windows, anchring of roof construction, etc..

In unsettled areas flood or avalanche danger have to be eliminated by technical control works before their dedication as developing area - public funds are not available for this purpose.

Buildings and infrastructures with gathering of people without possibility of short term evacuation in case of danger are not allowed ot be estabilshed in the Yellow Zone (schools, stations of cable railways, camping sites etc.)

Beside these two hazard zones are blue reservation and brown reference reaches exhibited.

Blue reservation area:

These are areas which have to be kept free for protective measures (including protective forests). Reference to technical measures, forest-technical, forest-biological and soil-bioengineering measures, respectively special management to ensure protectional effects.

Brown refence area:

These are areas which are not endangerend by torrents and/or avalanche but a danger by rockfall, landslide, water-logging etc. is recognized or must be assumed.

Reference to rockfall, landslide other natural hazards.

Violett reference area:

These are areas of protective topographic effects which have be preserved for this purpose.

Reference to special topographic, conditions, state of soil, state of terrain.

Step seven (7)- Supervision; scientific & administrative (legal)

The draft of the hazard zone map has to be sumitted to the mayor of the concerning community and layed out there for public inspection during four weeks. Everybody is entitled to give his written opinion on the hazard zone map. After this public announcement the draft of the hazard map has to be checked by a commission of four qualified and competent experts:
- representative of Ministry of Agriculture and Forestry
- ferderal service for torrent and avalanche control
- administration at the provincial level
- province and concerning community (municipality)

This commission has to consider the comments delivered by the people during public announcement.

The Commission makes it decisions by simple majority vote, in case of parity the vote of the representative of Ministry if Agriculture is deciding. Finally the reviewed plan has to be approved officially by th Federal Minister of Agriculture and Forestry and handed over to the relevant authorities (municiipality and district administration). The original remains at the relevant agency office for public inspection and use

Land use planning:

Beyond this regulations according to federal laws, executive rules for hazard zones are hold in provincial laws for land use planing. In concerning laws of provinces is stated generally, that areas which are endangered by floods, avalanches, mudflows, rockfalls and landslides are not allowed to be defined as developing areas. As the local authorities (municipalities and rural communities, controlled by the provincial governments) are responsible for land use and building planning - concerning authority of the mayor- hazard zone maps have to be observed strictly in relevant decisions. In addition Ministry of Agriculture and Forestry has decided, that in case of non observans of hazard zone maps public funds for flood and avalanche control works are not available further. Respectively money already used must be reimbursed in case of disregarding hazard zones.

Indication for construction authorities and work stations

Red hazard zone: In this zone the development and installation of buildings is not allowed.
Yellow hazard zone: In this zone the expert opinions of the federal Service for torrent and avalanche control has to be demanded by the construction authority and work station. In case of a permission for constructions and/or other infrastructures the following points have to be considered according to the provincial laws :
a)   Areas in the Yellow zone can only be dedicated as developing areas if the are situated within an existing, connected and settled area or in the immediate vicinity of it and if the settled area due to this will not be extended towards the

existing natural danger. Protective measures for the planned buildings and infrastructures have to be prescribed by the cinstruction authority and/or work station.

b) Buildings and installations with the possibility of crowds of people in times of hazard are not allowed to be situated in parts of the yellow zones where human lives are endangered outside to these building if they cannot be evacuated and closed within short time (f.e. schools, stations of cable cars, camping sites, sport grounds, centers of public events, etc.).

Blue reservation area: This area has to be kept free of any development.

Brown reference area: In case of development of such an area expert opinion of geologist, soil mechanic or concerning special fields has to be used by the construction authority.

Violett reference area: An expert opinion of the Ferderal Service for torrent and avalanche control has to be observed before any change of site conditions can be considered or allowed.

White (unplanned)reaches within the treated area: An expert opinion of the Federal Engineering Service for torrent and avalanche control is not required.

White (unplanned) reaches outside the treated area: If it seems necessary an expert opinion of the Federal Engineering Service for torrent and avalanche control has to be demanded.

REFERENCES:

FIEBIGER, G., (1997): Le zonage des risques naturels en Autriche. France-Autriche conference en restauration du terrain en montagne. Grenoble, France. Wildbach- und Lawinenverbau No. 134,Jg 61,1997,pp.155-164. (Hazard Mapping in Austria; Journal of Torrent, Avalanche, Landslide and Rockfall Engineering No.134, Vol.61,1997,pp.153-164; also in English and German)

Ofner, G., (1987): Praxisorientierte Erstellung von Ganglinien als Grundlage für Retentions-maßnahmen. Wildbach- und Lawinenverbau No.106,Jg.51, h1987, pp.163-173. (The calculation of hydrographs as base to retaining measures - a contribution oriented on practice. Journal of Torrent, Avalanche, Landslide and Rockfall Engineering No. 106, Vol.51, pp. 163-173)

Scheuringer, E., (1988): Ermittlung der maßgeblichen Geschiebefracht an Wildbach-Oberläufen. Wildbach- und Lawinenverbau, Jg. 52., 1988, pp.87-95. (Finding out of the dicisive debris load from torrent upper courses. Journal of Torrent, Avalanche, Landslide and Rockfall Engineering No.109,Vol.52., 1988, pp. 87-95)

Scheuringer, E., (1998): Grundlagen und Grundsätze der Gefahrenzonenausweisung der Wildbach- und Lawinenverbauung. Der Alm- und Bergbauer, Jg. 48, 3, pp. 58-61, 1998. (Base and principles of HazardMapping in Austria. The Mountain FarmerVol.48; 3; pp.58-61; 1998)

Statistical Analysis on the Planimetry of Debris Flow Fans

Xilin LIU

Professor of the Institute of Mountain Disasters and Environment, Chinese Academy of Sciences, P. O. Box 417, Chengdu, Sichuan 610041, China

ABSTRACT   The planimetry of debris flow fans could be represented by the combination of geometric figures. Two planimetric models of debris flow fans are presented in this paper based on aerial photograph observation and statistical analysis of 52 debris flow fans in Xiaojiang Valley of southwestern China and in Alpine valleys of northeastern Italy. The combination of an isosceles triangle with a semicircle best fits the fans in Xiaojiang Valley and a circular sector is best for the fans in northeastern Italian Alps. By using statistical analysis, two groups of morphometric relationships between debris flow fans and debris flow basins are established. The basin area and basin relief are identified as the significant factors influencing the plan shapes and the dimensions of debris flow fans. The statistical analysis on geomorphic parameters of debris flow drainage basin could provide an empirical method to predict the debris flow depositional extent.
Key words: statistical analysis, planimetry, debris flow fan

1. INTRODUCTION

Debris flow fans are common geomorphic features in mountainous areas. Their favourable slopes and proximity to water make them very suitable for various land uses such as agriculture, transportation and urban development. Debris flows have been studied in detail in recent years because of their potential for devastation.

Debris flows have their greatest potential to cause destruction in debris flow fan areas because human activities are concentrated on fan surfaces. In order to sustain development in mountainous regions, and to mitigate the damaging impact of debris flow hazards, geomorphologists, engineering, geologists, governmental institutions and international organizations have developed various ways to reduce the consequences of debris flows (Hungr et al., 1987; Jackson et al., 1987; Kellerhals and Church, 1990).

Debris flow fans are the deposits which result from debris flows. The determination of the fan dimensions (area, length, width) could provide important information for land use planners, disaster insurers and others in concerning the   optimal land use in the fan area. If a basin is identified as an area of active debris flows, the associated fan dimensions may be obtained from aerial photographs and topographic maps. However, if a basin is identified only as having the potential for debris flow activity, an alternative mathematical method for estimating fan dimensions may be required in order to predict the depositional area of debris flows and establish zone of potential hazard. If the factors which control the process of debris flow generation and fan deposition can be estibalished, then a relationship between basin slope and fan morphology may be established.

2. STUDY AREAS

Xiaojiang Valley in southwestern China

The development of Xiaojiang Valley in Yunnan Province (Fig. 1) has been controlled by an old and active structural fault formed in the Proterozoic era.The main fault surface has a steep dip of approximately 55 degrees. The width of the Valley which follows the fault ranges from hundreds to thousands of meter. The dominant rock types are Proterozoic slate, quartzite and shale, Cambrian dolomite and mudstone, Permian basalt and pelitic limestone, and Triassic sandstone. The highest peak is 4344m above sea level (a.s.l.) and most of other peaks are also more than 4000 m a.s.l.. The local base level of the rivers is 1100 m a.s.l..

The climate is typically semiarid subtropical, with an annual rainfall of about 700 mm and potential evaporation of about 3700 mm. The annual average temperatuer is about 20 degree centigrade, with a maximum of more than 40 degree centigrade and a minimum of -2 degree centigrade. Vegetation cover in the Valley is sparse and was estimated at only 8.88% in 1983. It consists mainly of scattered trees with bushes and grasses.

There are totally 107 areas of gullies and hillslopes where annually active debris flows occur. Many gullies are simultaneously active in their discharge of debris flows. Also debris flows occur repeatedly in single gullies. In Jiangjia Ravine, a permanent Debris Flow Observation and Research Station was established by Chinese Academy of Sciences, and many rainstorm-induced debris flow events were recorded. The large numbers and frequent activity of debris flows in Xiaojiang Valley make it become an ideal site for scientific research on debris flow fans.

Alpine valleys in northeastern Italy

The region studied in Italy (Fig. 2) is characterized by general uplift and tilting during the Cretaceous-Paleogene and the Pliocene-Quaternary (Zanferrari et al., 1982). Thus, neotectonic activity, which is mainly expressed by vertical differential movements along both preexisting tectonic discontinuities and recently fomed faults, is more intense towards

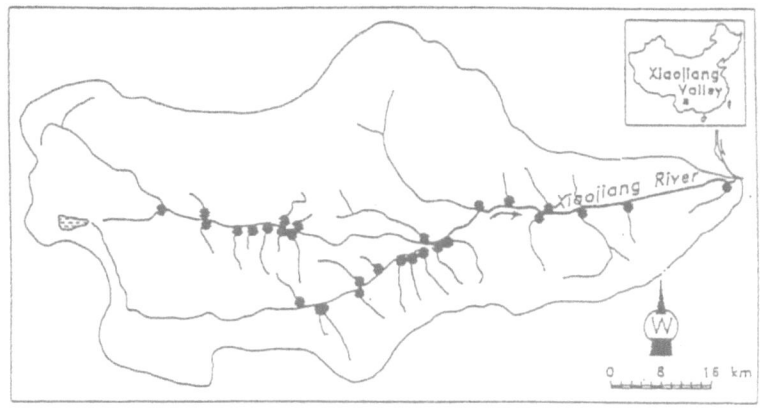

Fig. 1 Sketch map of the study area-Xiaojiang Valley in southwestern China

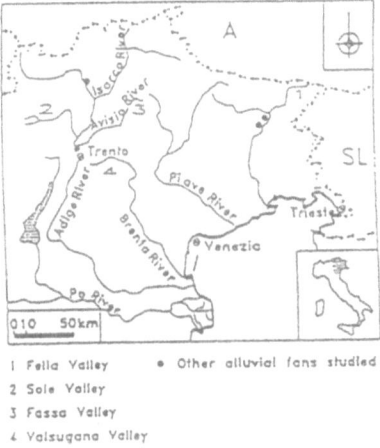

1 Fella Valley	● Other alluvial fans studied
2 Sole Valley	
3 Fassa Valley	
4 Valsugana Valley	

Fig. 2 Location map of the study area-Alpine valleys in northeastern Italy

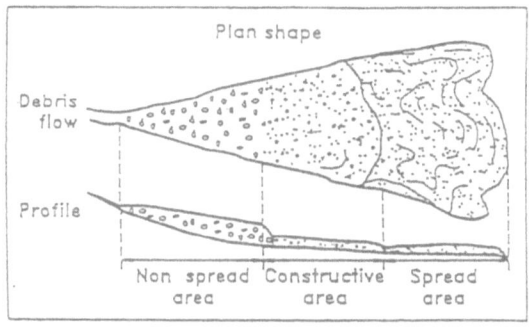

Fig. 3 Conceptual model of ideal debris flow fan

the eastern part of this area where seismicity is still severe. It has resulted in a general rejuvenation and erosion of the drainage basins, as a consequence of the local lowering of the base levels of streams and the increase in relief (Marchi and Tecca, 1995).

The average annual precipitation shows a wide range between different valleys. Precipitation is 800-1100 mm in Sole Valley, 950-1300 mm in Fassa Valley, 1000-1500 mm in Valsugna Valley and 1600-1800 mm in Canale Valley (Marchi et al., 1993). Land use in valley bottoms is devoted mainly to agriculture, roads and urban settlements. Forests (mostly coniferous) are widespread on lower valley slopes, up to an elevation of about 1800-2000 m a.s.l.. Barren rock and debris prevail at the higher elevations with a sparse vegetation cover of shrubs and alpine meadows.

The generation of debris flow fans in this region began in the end of last glacial stage (approximately 10000 years ago). Recent and historical records of flow events indicate that most of these fans are still active. The latest investigation by Marchi et al. (1993) identified 37 debris flow fans in this area. Debris flow processes cause serious property damage and, sometimes, loss of lives.

3. THE MODEL OF IDEAL DEBRIS FLOW FAN

A debris flow fan is usually divided into three parts (Fig. 3) (Bull, 1968): a non-spread area or fan head located at the top of fan and having a narrow lingual shape; a constructive area or main fan body situated in the middle of fan; and a spread area or fan toe at the bottom of fan.

The shape and growth of a debris flow fan is affected by its depositional environment. In piedmonts and intermontane basins, the fan may expand freely if there are no restrictions of local landforms. In mountainous valleys, if the erosion of receiving river is relatively weak, or if the fan is on inside of a convex bend of the river, it may also freely develop because the sedimentation from the river can create enough space for expansion of the fan. If the erosion of receiving river is strong, or if the fan is on outside of a concave river bend, the expansion of the fan is restricted. In both of the study areas, there are parts of the valleys with narrow floors which confine the fan dimension (Tang et al., 1991; Marchi and Tecca, 1993). Also, human activities may alter the areal extent of the fans.

This study aims at those debris flow fans which are of the following three characteristics:

(1) generated by debris flows which have integrated drainage basin; thus morphometric parameters of which may be measured easily;
(2) created by rainstorm-induced debris flows;
(3) laterally unconstrained (not strongly influenced by artificial

words), for which the natural morphometric relationship between the basins and the fans may be fully analysed.

Fan shape is measured by length and width. Morphometric properties of the basins may be defined by parameters such as drainage basin area, drainage basin relief, main channel length, main channel gradient, drainage basin perimeter, basin shape ratio, basin relief ratio, drainage density (Patton, 1988), Strahler's ruggedness number (Strahler, 1964) and melton's ruggedness number (Melton, 1965). Some of these parameters are interdependent, which can not be called as independent variates. Previous researches (Bull, 1964; Church and Mark, 1980; Kostaschuk et al., 1986; Kochel, 1990; Liu, 1990; Liu et al., 1992; Oguchi and Ohmori, 1994), have given priority to study of those specific variates, and to the selection of the following morphometric parameters of fans and basins:

(1) Fan area (km^2) - the plan area of each fan.

(2) Maximum fan length (km) - the distance from the apex to the toe along the fan bisector.

(3) Maximum fan width (km^2) - the maximum distance perpendicular to maximum fan length. Drainage basin area (km) - the plan area of the basin.

(4) Drainage basin relief (km) - the difference in elevation between the highest point in the drainage system and the fan apex.

(5) Main channel length (km) - the plan length of the main channel.

(6) Main channel gradient (%) - the difference in elevation between the channel head and the fan apex, divided by main channel length.

29 debris flow fans in Xiaojiang Valley and 23 debirs flow fans in the northeastern Italian Alps study area were selected for statistical analysis. Morphometric parameters were measured from topographic maps at scales of 1:50000 in China and 1:10000 or 25000 in Italy, and from aerial photographs at scales of 1:38000 or 42000 in China and 1:25000 or 40000 in Italy.

4. STATISTICAL ANALYSIS OF DEBRIS FLOW FAN PLANIMETRY

Aerial photographic analysis of 52 debris flow fans indicated that nonspread area and the constructive area of fan can be simulated by an isosceles triangle in both Xiaojiang Valley and the northeastern Italian Alps study area (Fig. 4), while the spread area of the fan in Xiaojiang Valley is best approximated by a semi-circle, and the spread area of the fan in the northeastern Italian alps study area is best represented by a

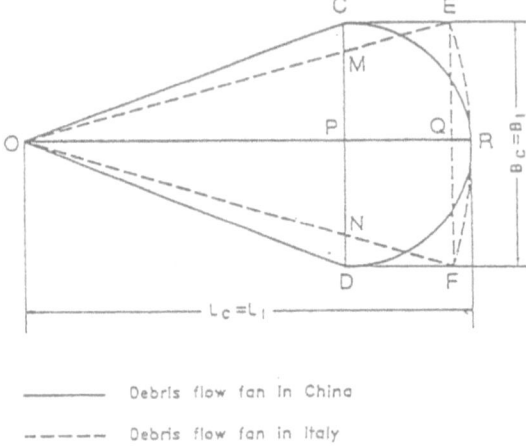

Fig. 4   Generalized planimetry of debris flow fans in Xiaojiang Valley of
southwestern China and in the alpine valleys of northeastern Italy

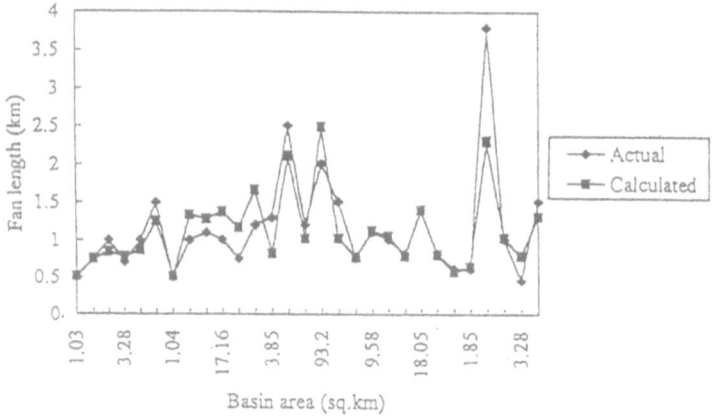

Fig. 5   Actual and calculated values of the fan length in Xiaojiang Valley
of southwestern China

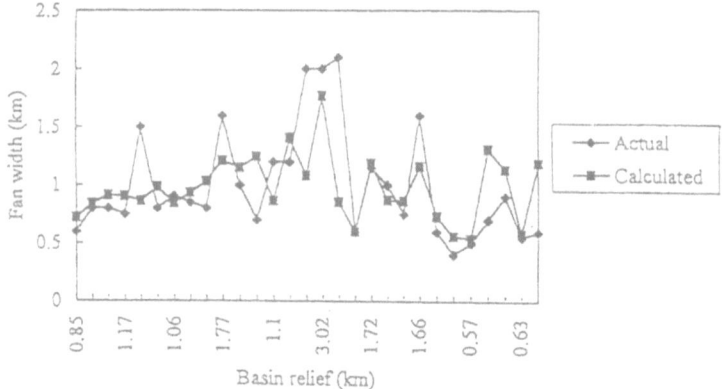

Fig. 6   Actual and calculated values of the fan width in Xiaojiang Valley
of southwestern China

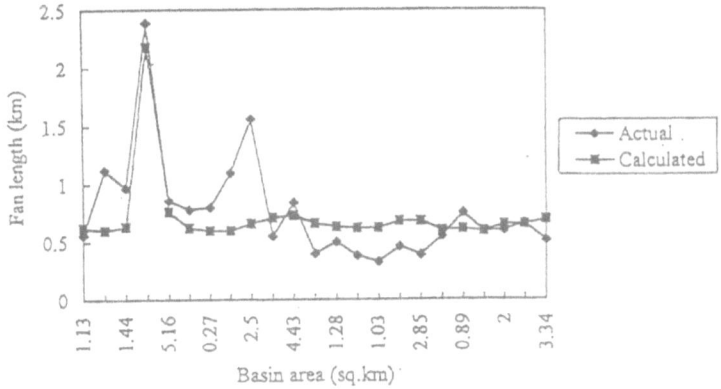

Fig. 7   Actual and calculated values of the fan length in the northeastern
Italian Alps study area

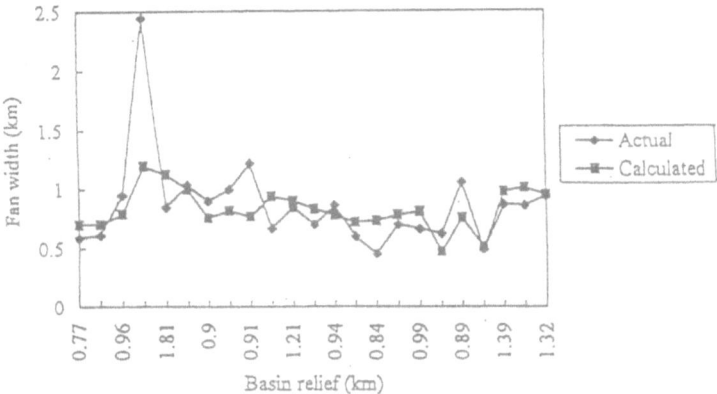

Fig. 8   Actual and calculated values of the fan width in the northeastern
Italian Alps study area

segment of a circle.

In Xiaojiang Valley, the area of debris flow fan (S_c) can be mathematically deduced as follows (fig. 4):

$S_{triangle}$ COD = CD x OP/2

CD = B_c

OP = OR - PR = OR-CP = L_c - B_c/2

$S_{triangle}$ COD = $B_c(L_c - B_c/2)/2$

that is,

$S_{triangle}$ COD = $L_c B_c/2 - B_c^2/4$

$S_{semi-circle}$ CRD = $CP^2/2$

CP = $B_2/2$

$S_{semi-circle}$ CRD = $B_c^2/8$

S_c = $S_{triangle}$ COD + $S_{semi-circle}$ CRD

therefore,

$$S_c = L_c B_c/2 + (/2-1)B_c^2/4 \quad (1)$$

Where S_c is the debris flow fan area (km²); L_c is maximum fan length (km); B_c is maximum fan width (km).

In Xaojiang Valley of southwestern China, regression analysis indicates that the optimum regression equations between L_c, B_c and the basin morphometry are non-linear power functions. The morphometric relationships may be expressed as follows:

$$L_c = 0.5044A_c^{0.3508} \quad (r=0.8622) \quad (2)$$

Where A_c is the drainage basin area (km²). The actual and calculated values are shown in Fig. 5.

$$B_c = 0.8046H_c^{0.7152} \quad (r=0.6481) \quad (3)$$

Where H_c is the drainage basin relief (km). The actual and calculated values are shown in Fig. 6.

In the northeastern Italian Alps study area, the debris flow fan area (S_i) can be calculated from the following mathematical deduction (Fig. 4):

S_{sector} EOF = OR^2 (EOF/360)

EOF = 2EOQ = 2arc sin(EQ/OE)

EQ = $B_i/2$

OE = OR = L_i

EOF = 2arc sin($B_i/2L_i$)

S_{sector}EOF = L_i^2 arc sin ($B_i/2L_i$)/180

Therefore,

$$S_i = L_i^2 \text{ arc sin } (B_i/2L_i)/180 \quad (4)$$

where S_i is debris flow fan area (km²); L_i is maximum fan length (km); B_i is maximum fan width (km).

Statistical analysis implies that, in the northeastern Italian Alps study area, the optimum regression equations between L_i, B_i and the basin morphometry are also non-linear functions. The morphometric relationships may be expressed as follows:

$$L_i = 0.5884e^{0.0486A_i} \quad (r=0.5528) \quad (5)$$

Where A_i is drainage basin area (km²). The actual and calculated values are shown in Fig. 7.

$$B_i = 0.81H_i^{0.5636} \quad (r=0.6212) \quad (6)$$

Where H_i is the drainage basin relief (km). The actual and calculated values are shown in Fig. 8.

5. DISCUSSIONS

Several researchers have examined relations between the area of an alluvial fan (S) and that of its drainage basin (A). Bull (1964) indroduced a power function $S=cA^b$, and Church and Mark (1980) have shown that c varies locally but b has a constant value of 0.9 under comparable environments for fan development. Latter studies recognized that debris flow processes are different from bedload transport and hyperconcentrated flows, and have been classfied into debris flow fans, entirely fluvial fans and mixed fans (Kostaschuk et al., 1986; Costa, 1988; Kochel, 1990; Liu 1990; Marchi et al., 1993; Zarn and Davies, 1994). For the debris flow fans in subhumid, glaciated Bow Valley (Alberta, Canada), b is 0.48 (S = $0.17^{0.18}$) (Kostaschuk et al., 1986), and in the arid loessial conditions (Gansu, northwestern China), b is 0.83 (S = $0.06A^{0.38}$) (Liu, 1990). This study found that, in Xiaojiang Valley, b is 0.50 ($S_c = 0.24A_c^{0.50}$, 29 samples), and for the northeastern Italian Alps study area, b is 0.28 ($S_i = 0.28A_i^{0.28}$, 23 samples). In individual valley of the northeastern Italian Alps: b is 0.56 in Tagliamento Valley (6 samples); b is 0.55 in the Valsugana Valley (7 samples); b is 1.36 in Venosta Valley (7 samples) (Marchi et al., 1993).

It is clear from these examples that not only c but b varies from location to location. Besides the different geological and tectonic background, several factors probably contribute to the variability of b. First, different numbers of samples may influence the regression coefficients (c and b) and second, different erosion processes in the debris flow basins which include both sheet wash and gravity erosion (collapses, avalanches, landslides, etc.) result in debris availability being related, not only to basin area, but also to the erosion type and intensity which may cause uncertainity in value of b. Also fan area may be affected by the main channel gradient and the drainage basin relief (Kostaschuk et al., 1986). Costa (1984) noted that, the mechanical strength of a debris flow will exert some control on the areal extent of the associated deposit; lower strength flows result in a more extensive deposit than higher stength flows. Kostaschuk et al. (1986) inferred that, strength may be related to basin slope; thus the more steep sloping basins could produce a flow of lower strength and proportional larger fan area than a gentler basin. Statistical results in this study support their inferences.

The planimetry of debris flow fans in Xiaojiang Valley can be generally represented by a combination of an isosceles triangle with a semi-circle, and the planimetry of the fans in the northeastern Italian Alps study area is generalized by a combination of an isosceles triangle with a circular segment. Therefore, for the fans with same length and width, fan area in Xiaojiang Valley is always larger than that in the northeastern Italian Alps stusy area (Fig. 4), and for the fans with same area, the fans in Xiaojiang Valley are more wider in their main bodies and have more prominent toes, while the fans in the northeastern Italian Alps study area are relatively narrow in their main bodies and have relatively flat toes. This is because much more debris is available for the transverse and longitudinal expansion of the fans in Xiaojiang Valley than that in the northeastern Italian Alps study area within a similar depositional space.

6. CONCLUSIONS

This study into the planimetry of debris flow fans has led to the followintg conclusions:

(1)The maximum fan length is generally positively related to drainage basin area in both of study areas, and the maximum fan width is generally represented as a power function of drainage basin relief in both of study areas;

(2)In both study areas, drainage basin area and drainage basin relief are the most significant factors affecting the shapes and dimensions of debris flow fans.

(3)Simultaneous equations (1), (2) and (3) could be used to estimate the length, width and area of debris flow fans which develop in the environment similar to Xiaojiang Valley of southwestern China; simultaneou equations (4), (5) and (6)could be used to estimate the extent of debris flow fans that develop in then environment similar to the studied Alpine valleys of northeastern Italy.

ACKNOWLEDGEMENTS

This research was funded by the National Natural Science Foundation of China (Grant 49000011). The draft of this paper was completed when the author was in the Institute for Prevention of Hydrological and Geological Hazards, National Research Council, Padova, Italy in 1994-1995 as a Senior Visiting Scholar supported by an Overseas Studies Scholarship from Chinese Academy of Sciences. Original data from Italy were supplied by Dr. L. Marchi and P.R. Tecca. Many thanks go to Dr. L. Marchi for his helpful discussions and Dr.R.A. Kostaschuk for his initial review. Their valuable comments have greatly improved the manuscript. Mr. G. peruzzo drafted some of the figures.

REFERENCES

Bull, W. B. (1964) : Geomorphology of segmented alluvial fans in western Fresno County,California. United States Geological Survey Professional Paper, 352E: 89-129

Bull, W.B. (1986) : Alluvial fans. Journal of Geologic Education, 16:101-106. Costa, J.E., 1984. Physical geomorphology of debris flow. In: J.E. Costa and P.J. Fleis (Editors), Developments and Applications of Geomorphology. Springer-verlag, Berlin, pp. 268-317.

Costa, J.E. (1988) : Rheologic, geomorphic, and sedimentologic differentiation of water floods, hyperconcentrated floods, and debris flows. In: V.R. Baker, R.C.Kochel and P.C.Patton (Editors), Flood Geomorphology. Hohn wiley & Sons Ltd, Toronto, pp. 113-122.

Church, M.A. and Mark, D.M. (1980) : On size and scale in geomorphology. Progress in Physical Geography, 4:342-390.

Hungr O., Morgan C.G., Vandine F.D. and Lister R.D. (1987) : Debris flow defenses in British Columbia. Geological Society of America, Reviews in Engineering Geology, 7:201-222.

Jackson, L.E.Jr., Kostaschuk, R.A. and MacDonald, G.M. (1987) : Identification of debris flow hazard on alluvial fans in the Canadian Rocky Mountains. Geological Society of America, Reviews in Engineering Geology, 7:115-124.

Kellerhals, R. and Church, M. (1990) : Hazard management of fans, with examples from British Columbia. In: A.H. Rachocki and M. Church (Editors), Alluvial Fans: A Field Approach. John Wiley & Sons Ltd., Chichester, pp. 335-354.

Kochel, R.C. (1990) : Humid fans of the Appalachian Mountains. In: A.H. Rachocki and M. Church (Editors), alluvial Fans: A Field Approach. John wiley & Sons Ltd.,Chichester, pp. 109-129.

Kostaschuk, R.A., MacDonald, G.M. and Putnam, P.E. (1986) : Depositional process and alluvial fan-drainage basin morphometric relationships near Banff, Alberta, Canada. Earth Surface Processes and Landforms, 11:471-484.

Liu, Xilin (1990) : Debris flow fan area. Journal of Catastrophology, 5:278-281 (in Chinese).

Liu, Xilin, Tang, C., Zhu, J. and Zhang, S. (1992) : Forecast on depositional area of debris flow from drainage basin background. Journal of Natural Disasters, 1:312-323 (in Chinese).

Marchi, L. and Tecca, P.R. (1995) : All uvial fans of the Eastern Italian Alps: morphometry and depositional processes. Geodinamica Acta,

8:20-27.

Marchi, L., Pasuto, A. and Tceea, P.R. (1993) : Flow processes on alluvial fans in the Eastern Italian Alps. Annals of Geomorphology, 37:447-458.

Melton, M.A. (1965) : The geomorphic and paleoclimatic significance of alluvial deposits in Southern Arizona. Journal of Geology, 66:1-38.

Oguchi, T. and Ohmori, H. (1994) : Analysis of relationships among alluvial fan area, source basin area, basin slope, and sediment yield. Annals of Geomorphology, 38:405-42.

Patton, P.C. (1988) : Drainage basin morphometry and floods. In: V.R. Baker, R.C. Kochel and P.C. Patton (Editors), Flood Geomorphology. John wiley & Sons Ltd., Toronto, pp. 51-64.

Strahler, A.N. (1964) : Quantitative geomorphology of drainage basins and channel neworks.

In: V.T. Chow (Editor), Handbook of Applied Hydrology. McGraw- Hill, New York, pp. 4.40-4.74.

Tang, Chuan, Zhu, J., Duan, J. and Du, R. (1991) : Research on debris flow fans in Xiaojiang Valley, Yunnan Province. Mountain Research, 9:179-184 (in Chinese).

Webster, R. and Yaalon, D.H. (1994) : The research paper: An informal guide for authors. Catena, 21:3-11.

Zanferrari, A., Bollettinari, G., Carobene, L., Carton, A., Carulli, G.B., Castaldini, D., Carallin, A., Panizza., M., Pellegrini, G.B., Pianetti, F. and Sauro,V. (1982) : Evoluzione neotettonica dell Italia Nord-Orientale. Memorie di Scienze Geologiche, 35:355-376.

Zarn, B. and Davies. T.R.H. (1994) : The significance of processes on alluvial fans to hazard assessment. Annals of Geomorphology, 38: 487-500.

The Largest Debris Flow in the World, Seimareh Landslide, Western Iran

Zieaoddin SHOAEI and Jafar GHAYOUMIAN

Soil Conservation and Watershed Management Research Center
Tehran P. O. Box 13445-1136, Iran

ABSTRACT: Landslides and debris flow are the most prevailed events in western Iran. The largest one is Seimareh landslide occurred in prehistoric time as the largest in the eastern hemisphere.
About 30 km^3 of material detached from the north flank of Kabirkuh anticline left a scarp of about 16km long and 5km wide. It seems that an earthquake that based on carbon-dating method occurred 10370±120 years BP must be the main triggering factor. The large angular bloke with the diameter of about 60m that has traveled on the low angle of slopes of the plain (<20°) to the distance of about 15km suggests a complex mechanism of motion in this landslide.
Keyword: The largest, Landslide, Debris flow

1. INTRODUCTION

The review of history of Zagros range in Iran with the trend of northwest southeast confirms the high seismicity of this area. Frequently some big earthquakes have reported in Kabirkuh anticline that could be the most effective factor of large scale and disastrous landslides. Some evidence such as river terraces at high elevation, anomaly of underground water table and very thick river sediment confirm the high tectonic activities of this area.

Due to the geomorphology, geology and tectonic, the stratigraphy, Kabirkuh is divided into two series of high and low strength of formations. The high strength material that has formed the scarps consists of limestone (Mid.-Cret.), lime (Oligo.-Mio.) and the low strength materials are marl and shale (upper Cret.-Eoc.), marl, lime, sand and gypsum (Mio.-Plio.).

The sequences of impermeable marl and shale layers within well-fractured limestone, active tectonic, local steep slopes caused high susceptibility of sliding that resulted to many falls, slides and turbulent geological structures.

Most of the lakes within the mountainous area of Lorestan, Kordestan and other western provinces of Iran could be due to the occurrence of landslide (Shoaei & Ghayoumian, 1997). Among hundreds of gravitational and structural control landslide in Zagros, the "Seimareh landslide" is one of the largest in the eastern hemisphere. In this study, field investigation of this landslide that firstly reported by Harison and Falcon (1935) was implemented.

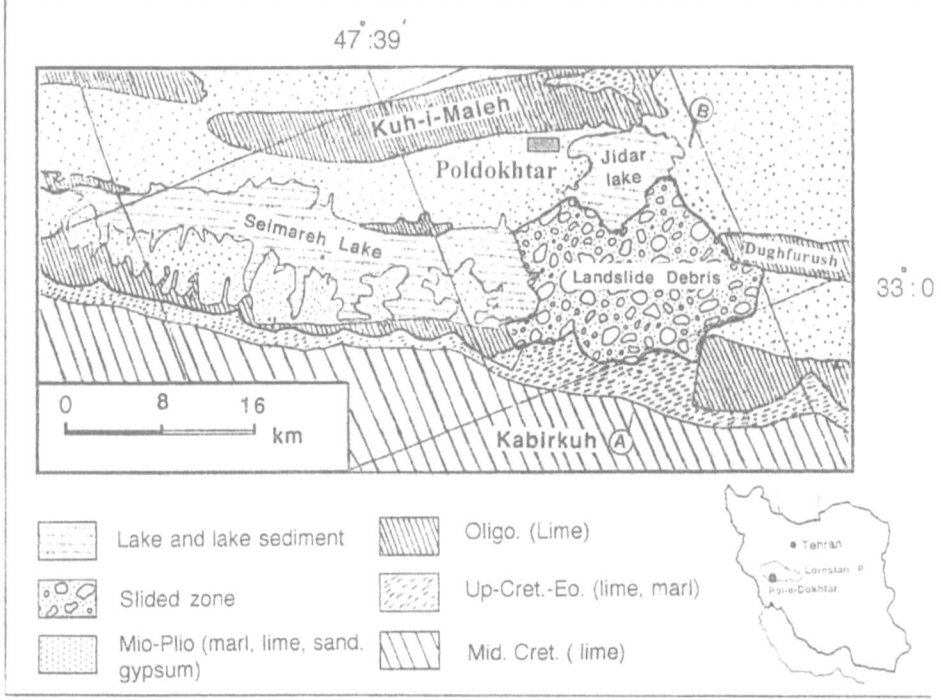

Figure 1. Geological map of Seimareh landslide.

2. STUDY AREA

The study area is located at 33°: 0′ to 33°: 15′ latitude and 40°: 30′ to 47° : 40′ longitude, 15km southeast of Poldokhtar city (Fig. 1 & I in Fig 2) on the border of Lorestan province. This basin is a large syncline surrounded by Malehkuh and Kabirkuh anticlines at north and south (Fig. 1).

At present Kashkan and Seimareh rivers are passing through this basin and along these two anticlines (Fig. 2-F and G, respectively).

3. GEOLOGY AND GEOMORPHOLOGY

The geomorphology of Zagros range in western mountainous provinces of Iran with the elevation up to about 3000m is closely controlled by simple and almost symmetrical folds in which the limestone bodies being strong and the resistant feature-making rocks. The principal chain in Lorestan province is called Kabirkuh (Fig. 1, Fig. 2-B) and is an anticline feature in middle Cretaceous limestone forming whale-back mountain. On its flank lies a zone of layers in which the upper Cretaceous-Eocene marly body has been more rapidly weathered away (Fig. 2-H). Then, come an imposing sierra formed by the Asmari limestone (Oligo.) which has weathered out. The valley on the northeast is controlled by this formation upon which a relatively thin skin of Fars series (Mio. Plio., marl, sand, gypsum) have remained after a course of differential erosion.

At the eastern end, the syncline is about 9km wide, and the Asmari limestone of Kuh-i-Dughfurush (Fig. 2-C) forms the northeastern wall, but, owing to the northwestward plunge of this structure, the syncline widens out in a westerly direction where, the Asmari limestone of Kuh-i-Maleh forms the north eastern slope of the syncline.

Upon this simple geology and geomorphology foundation, a series of uncommon surface deposits is superimposed. The most conspicuous is a broad apron spread out from south to north completely covering the valleys with a wild confusion of disorderly limestone blocks. This apron of debris has an area of about 166km^2. Large blocks of limestone were mixed with fragment of smaller sizes down to that of dust. All the pieces are angular. The surface morphology is very rough with irregular waves. Some ponds still remain.

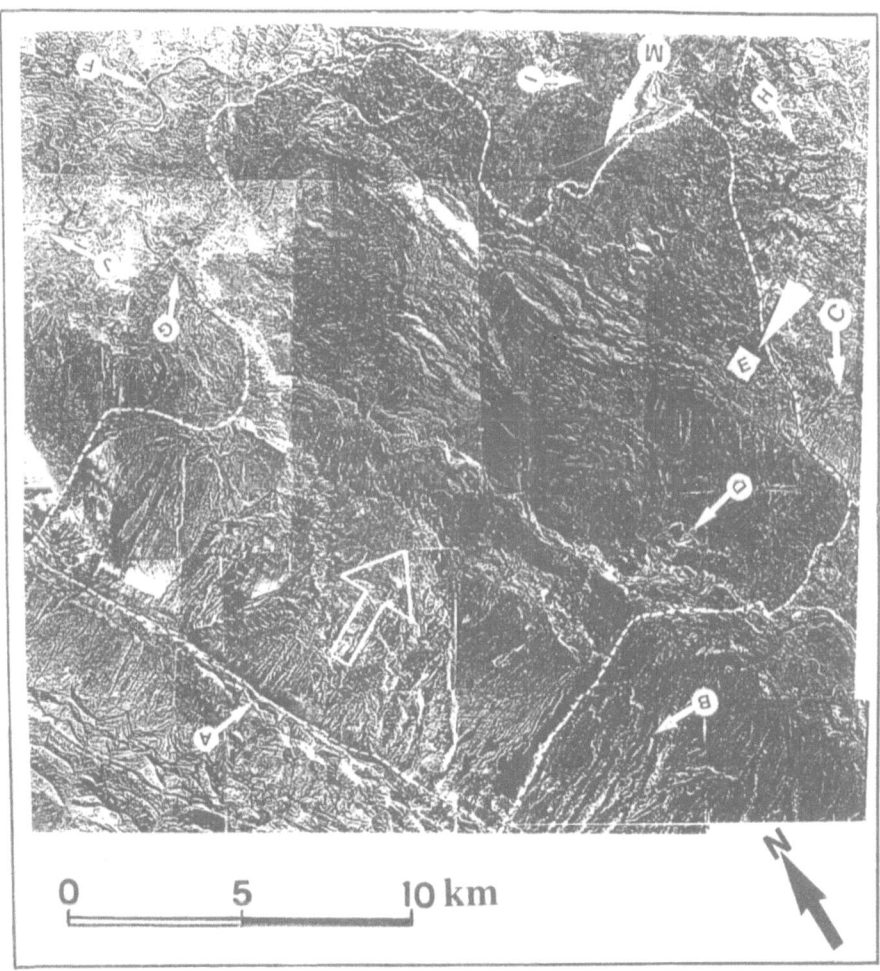

Figure 2. Aerial photograph of Seimareh landslide (taken 24 Aug. 1955)

The flanking sierra of Kabirkuh before the landslide, was made up of a regularly dipping sheet of limestone inclined toward the valley at a dip of less than 20°. On most hills a coherent sheet of limestone at this attitude must normally be in stable condition. Talus on the sides of these limestone mountains is still encountered with slopes up to 30°.

Maleh anticline is located at the north of historical Poldokhtar city and along the north margin of Seimareh basin. This anticline consists of a series of marl and shale formations, but its highest scarps with the elevation of 2000m consists of Asmari limestone.

Kashkan river is one of the large rivers in the Lorestan province. This river originates from the northeast mountain (Fig. 2-F). This river causes deep erosion through river terraces and passing the deep valley of Maleh anticline extends to the northeast and enters to the Seimareh basin. At present, this river joins to the Seimareh river at a location called Babakhrazm. Seimareh river is one of the most water-rich rivers in Lorestan provinces. This river originates from the northern mountains of Lorestan province (Fig. 2-G). Seimareh river southeast of Poldokhtar joins to Kashkan river and after passing historical Gavmishan bridge extends to the north flank of Kabirkuh and after passing a narrow and deep valley of Kabirkuh anticline enters to the Khusestan province to form Karkheh river.

4. SEIMAREH LANDSLIDE

Seimareh landslide in one of the largest and rare landslide events in the world. This landslide has occurred at the northeast flank of Kabirkuh anticline (Fig. 1, 2-B). A huge mass with an average thickness of about 400m of Asmari limestone and upper parts of Pabdeh formation (marl and marly limestone) with the total volume of about $30km^3$ moved down from the 1880m elevation to the maximum travel distance of about 20km toward Seimareh valley. Deposited material has covered $166km^2$ of the area of Seimareh basin.

After the occurrence of Seimareh landslide, due to the direction of displaced material to the north, part of the Seimareh syncline, the old valley of Seimareh and Kashkan rivers were clogged and formed many lakes, but have since been drained, either by seepage or as their overflow cut. Two largest were the Jidar and Seimareh Lakes. The silt and sand sediments of these lakes with the thickness of about 30m to 100m covered the weathered unconformity on Gachsaran formation (Mio.-Plio.) northeast and north west of the displaced material in Seimareh basin (Fig. 2-J).

The area of Jidar and Seimareh lakes' sediments are $35km^2$ and $85km^2$, respectively. Probably some small ponds were also formed and some of them are still visible as sinkholes and ponds (Fig. 2-D). The sediment of Seimareh lake shows different sedimentary structures such as cross-bedding, lamination, graded-bedding, convoluted structure, nodule, biogenic structures such as root-casts and other plant remains, and also, gastropod shells.

The scarp produced by the downfall of this material is bounded by two Asmari limestone cliffs. This scar has a length of about 16km (Fig. 2-A) with the average width about 5km (Fig. 3).

Figure 3. Bird's eye view of Seimareh landslide. The direction of photography is shown in Fig. 2-M.

A minimum thickness of 300m of rock has been removed to form the Asmari limestone part of the moved material and while the thickness of 200m of under-lying shale and marl form the matrix of displaced debris. The volume of rocks involved is therefore about 24 to 32km^3 (16×5×(0.30 to 0.40km)).

4.1 Mechanism of Motion

Once a mass was set in motion, the rules of static and dynamic friction and ideal limiting value such as the strength of moved material come to apply.

The region is prone to large magnitude earthquakes and seismic shock may have been the main triggering factor for the first detachment of many landslides. It is very probable that the Seimareh landslide in Zagros, which has a debris volume of about 30 millions cubic meters (30km^3) perhaps the largest in the eastern hemisphere, was triggered by a large-magnitude earthquake.

A slab of Tertiary Asmari limestone with the dimension of 16km in length, average 5km in width and 300m to 400m thickness slid off the northern flank of Kabirkuh. The moved material crossed two valleys and an intervening ridge and extended 20km from the source (Figs. 1& 2). Taking for this material the specific gravity of 2.5, it is merely a question of arithmetic to show that approximately 80,000 millions tons of rocks (16000×5000×400×2.5) block have been involved in the catastrophe.

The matrix is composed of un-graded angular fragments down to the size of powder. The great cubes are also found resting high upon the nose from which the gradient is so low that an angular mass could not have traveled down it. One of these near the 1250m high point on the nose of Kuh-i-Dughfurush measured about 60×30×30m (point E in Fig. 2). The possibility of this block having originated from the flank of Kuh-i-Dughfurush has been considered, but the sides of that mountain are smooth and regular and reveal no scar, change of nature or shape where the boulder sheet could be detached from it. Taking the specific gravity of 2.5, the total weight of this block is 135000 tons that traveled to a distance about 15km from the source without any progress in its roundness

The Eocene in which gliding took place is composed of thin bedded and marly limestone and marls that are normally well compacted and not liable to form a liquefied debris when saturated.

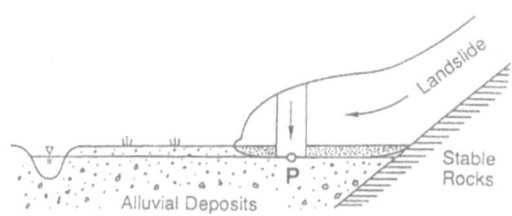

Figure 4. Schematic profile of a landslide inducing rapid loading onto alluvial deposits (Sassa et al., 1994).

Still some other effective factors to reduce the strength of detached material appear to be needed to allow the slipping to commence upon an inclined plane with a dip of less than 20°. The composition of Seimareh landslide mass is uniform in its heterogeneity, great rectangular masses jostling with particle of dust from the riverside to the farthest lobe of about 20km away. Only a fluid turbulence could yield such a wide distribution with no tendency to grading of constituent parts in the process or to rounding of their sharp edges.

Some different mechanisms of landslides motion were suggested (Sassa et. al. 1994, Shoaei & Sassa 1997). Sassa (Sassa et al., 1994) suggested that large scale landslide usually move over alluvial or un-consolidated material found at the toe of slopes as schematically shown in Figure 4.

When the high porous and semi to well-saturated deposits is overload by landslide material, the collapse in the structure of grains aggregates will occur, relevantly, pore-pressure must be generated in the saturated layers due to the rapid loading that will be resulted to an undrained condition. So that, the apparent friction angle mobilized in the alluvial or high porous material can be much smaller than the effective angle of material. In this case, a sliding surface will be formed on the saturated layer and the landslide mass can travel long.

In the case of Seimareh landslide, there is no doubt that the toe deposit and Fars series (Fig. 5) before the landslide was completely saturated. Some possibilities can be considered, a): the occurrence of landslide was during the rainy seasons that has resulted to the high saturation of the material at the toe. It is reasonable to suppose that the rainfall was the same order as that as present, which is about 400mm to 700mm annually, but, varies greatly from year to year. In some cases, the rain amounts few 10mm in a day. Such rain would at least tend to provide well saturation in downward deposit, b): the landslide occurred in two stages, first, due to the high erosion of Fars series, the limestone slab probably had previously been undercut by Seimareh river that caused a narrow and deep valley. This resulted to the instability of the lower part of the mountain which presumably came down first (Fig. 5-B).

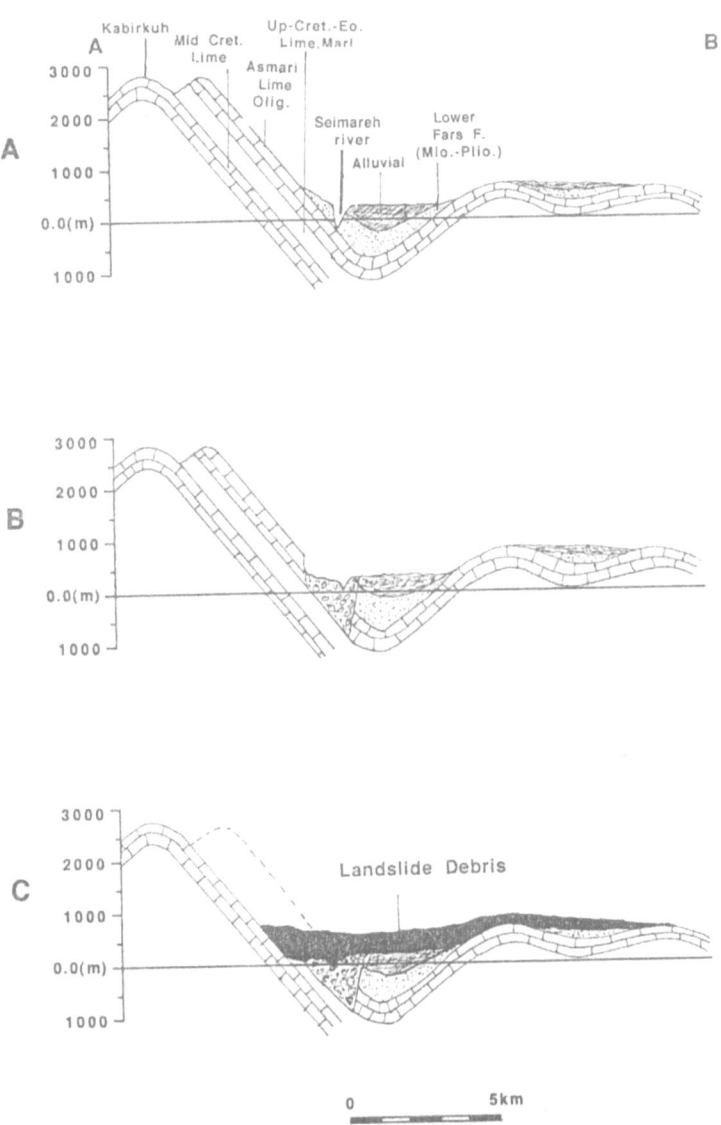

Figure 5 Section of Seimareh area to show the possible mechanism of motion, (the direction of section is shown in Figure 1).

Then, the Seimareh river was dammed and the water rising caused a complete saturation in Fars series and its overlaid deposits. At the second stage, the main landslide was triggered by a big earthquake. The displaced material which overloaded on the downward saturated material, resulted to the generation of high pore water pressure and completely liquefied layer. Under such condition, it is possible to interpret the motion of debris for long travel distance on nearly very low angle slopes (Fig. 5-C)

4.2 History of Motion

The landslide took place recently from geological point of view, for the debris has moved across the old terraces and the plain of the lower Seimareh river. It was prehistoric because a ruined town built in pre-Sassanian time (1800 BP) stand upon the silts of the Seimareh lake. Considering some other historical evidences, it can be concluded that the landslide took place much more than two thousands years ago.

In some works it has reported that the basal sediment of the lake has a radiocarbon date of 10370±120 years BP (Stein 1940, Oberlander 1965; Watson & Wright, Jr. 1969, van Zeist & Bottema 1977).

The Seimareh earthquake is considered to be the main triggering factor. It was a "Zagros-Type hidden earthquake" (Date: 11000BP, Macroseismic Epicenter: 33°N-47°E, Ms=7.0, (Berberian, M. 1994, Amberseys & Melville 1982) possibly under the southwestern fold limb of the Kabirkuh anticline. The length of this anticline is more than 200km.

ACKNOWLEDGMENT

The Soil Conservation and Watershed Management Research Center (SCWMRC), Iran, is gratefully acknowledge for financial support for field investigation. The authors wish to thank Mr. Amir Sarreshtehdari of SCWMRC and Mr. Taher Farhadi Nejad of Natural Resource and Animal husbandry Research Center of Lorestan province for their contribution to this paper.

REFERENCES

Berberian, M. 1994. Natural Hazards and the First Earthquake Catalogue of Iran, Historical Hazards in Iran Prior to 1900. International Institute of Earthquake Engineering and Seismology (IIEES) press. Vol. 1, 603 pages.

Harrison, J. V. & N. L. Falcon & P. B. Mailing 1935. The Geology of Lorestan from a reconnaissance, NIOC report No. 490.

Oberlander, T. 1965. The Zagros Stream: a New Interpretation of Transverse Drainage in an Orogenic Zone. Syracuse Univ. Press, Syracuse Geography Ser.. 168 pages.

Watson, R. A. & Jr. H. E. Wright 1969. The Saidmarreh Landslide, Iran. Geol. Soc. Am. Inc., Sp. Pap.. 123, 115-1387.

Van Zeist, W. & S. Bottema 1977. Palynological Investigation in Western Iran. Palaeohistoria, XIX. 20-85.

Amberseys, N. N. & C. P. Melville 1982. A History of Persian Earthquakes. Cambridge University Press: London. 219 pages.

Sassa, K. & H. Fukuoka & J. H. Lee & Z. Shoaei & Zhang, Z. & Z. Xie & S. Zeng & B. Cao 1994. Prediction of Landslide Motion Based on the Measurement of Geotechnical Parameter. Proc., International Workshop on Prediction of Rapid Landslide Motion, 26 Jul. 1997, Kyoto, Japan. 13-43.

Shoaei, Z. & K. Sassa 1997. Slope Stability due to the Different Mechanism of Pore Pressure during Earthquake on Saturated Soil. Proc. International Symposium on Engineering Geology and the Environment, IAEG, Athens, Greece, 23-27 June, 1997. 1049-1053.

Shoaei, Z. & J. Ghayoumian 1997. Landslide Dam in Iran, Mechanism and History. Proc. International Symposium on Engineering Geology and the Environment, IAEG, Athens, Greece, 23-27 June, 1997. 1043-1047.

Assessment of Hazard Potential of Debris Flows in Relation to the Reclamation of Forests on the Foot of Volcanoes

Kazuo OKUNISHI and Hiroshi SUWA

Disaster Prevention Research Institute, Kyoto University, Kyoto 611-0011, Japan

ABSTRACT   This paper focuses on the hazards due to debris flows in the alluvial fans on the foot of volcanoes. A case study is carried out to assess the hazards associated with the transformation of forests to golf courses. A close examination of geomorphic evidence gives the history of occurrence of the debris flows of varied ages in the past and the estimate of their recurrence. To generalize this result, hazard potential is defined, from which hazard and risk are derived according to the time span, location and details of the reclamation.
Keywords: Hazard potential, Reclamation, forest, debris flow, environment.

1. INTRODUCTION

Forests have been severely destroyed until the modern times when the value of forests was re-evaluated. Another wave of forest reclamation occurred recently where the forest management had an economical difficulty. This paper proposes a method for assessing the potential hazards of the reclaimed forestlands, on the basis of a case study at the southern foot of the Yatsugatake volcanoes. The field investigation is focused on the alluvial fan where topographic changes due to deposition of debris flows and succeeding dissection have been marked and hazards due to debris flows can be evaluated through the observation of the depositional features.

2. EXISTING RESEARCHES

The volcanoes that constitute the mountain body of Yatsugatake erupted in varied ages. Since each of the eruption stimulated mass movements, the lava layers of different ages are intercalated with deposits of debris flows and pyroclastic flows. Kochi (1977) has compiled the geological map of Yatsugatake and its environs. Other physiographic conditions related to natural hazards have been compiled by Fujimi Town (1985). Stream net and the location of major springs as investigated by Marui *et al.* (1993) are shown in Fig. 1. Water and sediment discharges at the reaches of the stream that are not fed by major springs is unstable and debris flows occur there frequently. Past occurrence of debris flows has been investigated by Nagano Nagano National Forest Management Bureau (1984), and it was found that the latest debris flow occurred between 1961 and 1966. Deposition of older and larger debris flows was found at several sites. More systematic investigation of the past occurrence of debris flows was carried out by Kosaka (1992) who recognized three ages of the occurrence of the debris flows. Accordingly, the old-age debris flows covered entire surface of the upper half of the Kikkakezawa fan. Middle-age debris flows caused inundation of a width of about 200m around the current course of the Kikkakezawa gully. Most of the new-age debris flows did not go more than 50 meters off its current course except a course which followed the middle-age debris flows.

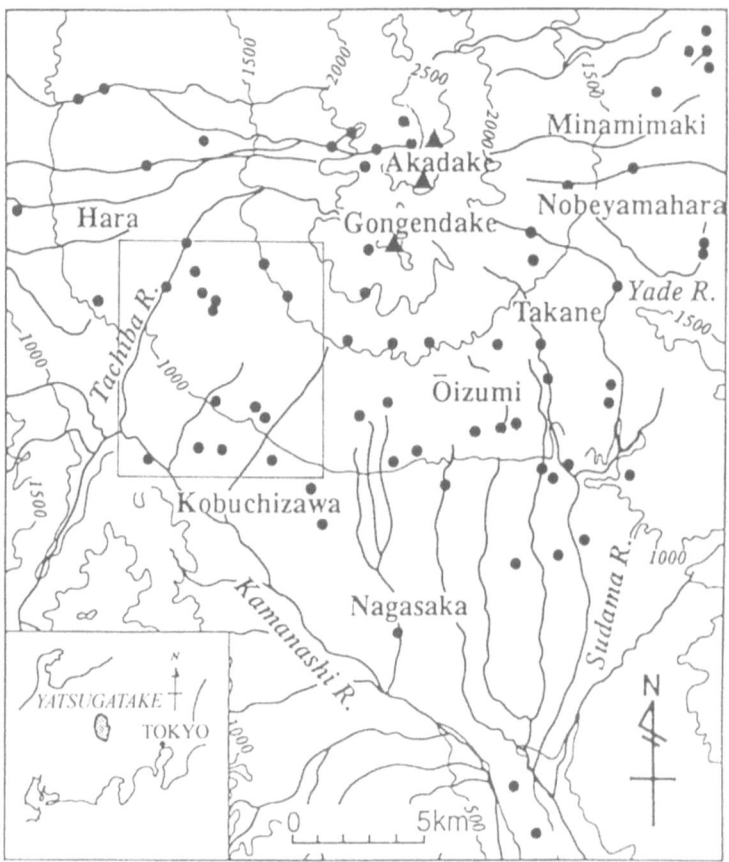

Fig. 1 Drainage network and springs on the southern foot of the Yatsugatake Volcanoes (after Marui *et al.*, 1993). The rectangle shows the extent of Fig. 2.

3. LAND USE AND THE PLANNED RECLAMATION

Current land use in the Kikkakezawa fan and its environ is shown in Fig. 2. Among the streams in this area, R. Tatsuba is the largest and the most dissecting. The Kikkakezawa gully has an alluvial fan (hatched with vertical lines and marked with "Do". in Fig. 2) covered by debris flow deposits, as well as the neighboring gullies. A Flood plain on the valley floor of the R. Tatsuba and the areas of the alluvial fans below the spring line (Y) of an altitude of about 1,100m (see Fig. 1) have constituted farmlands and settlements (*S*). Railways (RW), national road (R20) and highway (H) pass these areas. The line with the mark "Y" thus represents the front of development. The area above it (*K*) have been used as forests (largely private) in the past, although recent construction of the road (PR) is stimulating the development of residential areas, recreational areas (*R*) including golf courses (*G*). Thus the development front has currently been shifted to the line "Z". It is migrating further upslope. An expansion of the golf course (*G*) was planned at the upslope part of the Kikkakezawa fan (*Gp*).

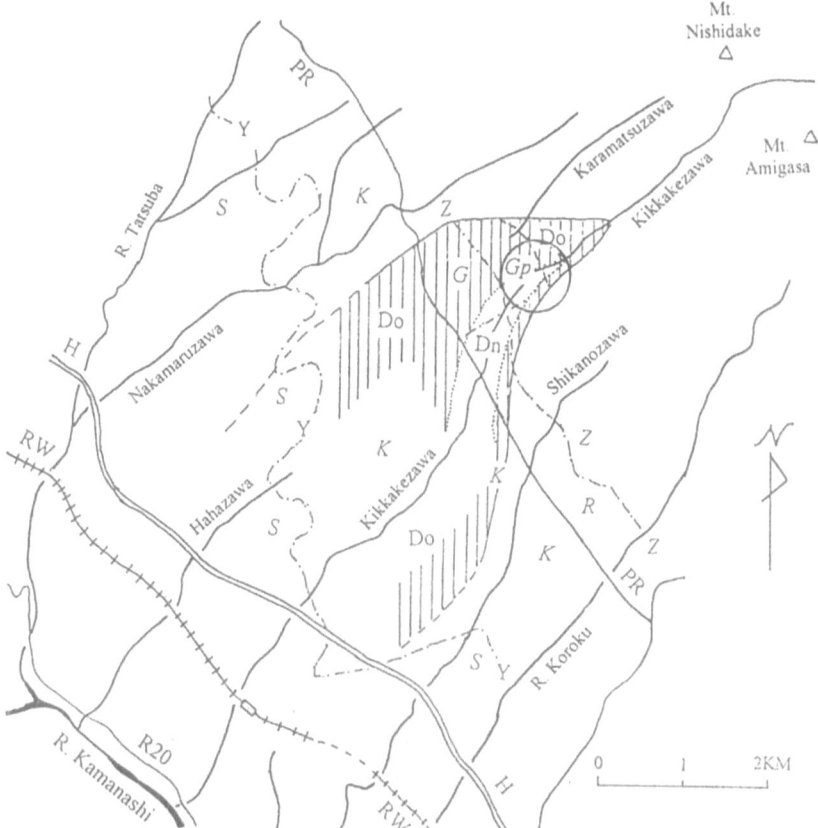

Fig. 2 Land use of the Kikkakezawa Fan and environs. *S*: settlement, Y: Major spring line, *K*: forest zone: Do: extent of the old-age debris flows in the Kikkakezawa Fan, Dn: depositional areas of the middle- and new-age debris flows, *R*: Sport grounds and resort residences, *G*: existing golf courses, *Gp*: planned extension of the Golf courses, PR: newly built road, H: highway, RW: railway, R20: national road. The circle shows the location of Fig. 3.

4. FIELD INVESTIGATION BY THE AUTHORS

A detailed investigation of the debris flow deposits was carried out by Okunishi (1994) to assess the risk of natural hazard associated to the expansion plan of the golf courses, and reviewed by Okunishi and Suwa (unpublished). The depositional features in a representative part of the alluvial fan of the Kikkakezawa gully are illustrated in Fig. 3. The feature marked with ① is a partially eroded natural levee. Two matrix-free freeze deposits of the new-age debris flow were found between this feature and the current channel. The one marked with ② was dated by Nagano National Forest Management Bureau (1984) between 1961 and 1966 as stated above. Another one marked with ③ seems older, since larches planted after its occurrence are now grown-up. Depositional features of middle-age debris flows are located behind the natural levee on the left side of the current channel of the Kikkakezawa gully. Among them, some present a succession of a long and slender deposit of boulders and a massive mound of boulders. It is a typical deposition of debris flow, where the head part consisting of matrix-free boulders freezes at a point leaving a boulder mound, and the following part of higher fluidity continues to flow down leaving a slender deposit. Suwa and Okuda (1983) have closely observed such depositional processes on the eastern slope of Mt. Yakedake. Another group of the deposits of middle-age debris flows is a combination of continuous lines of heaped boulders. They are thought to have constituted the lateral deposits of repeated debris flows (Suwa and Okuda, 1984). This type of deposits is also found on the right side of the current channel of the Kikkakezawa gully (④). It is suggested that the middle-age debris flows deviated from the current channel from around the point ⑤. Thus, about a half of the area of planned golf course (*Gp*) was inundated by the new- and middle-age debris flows.

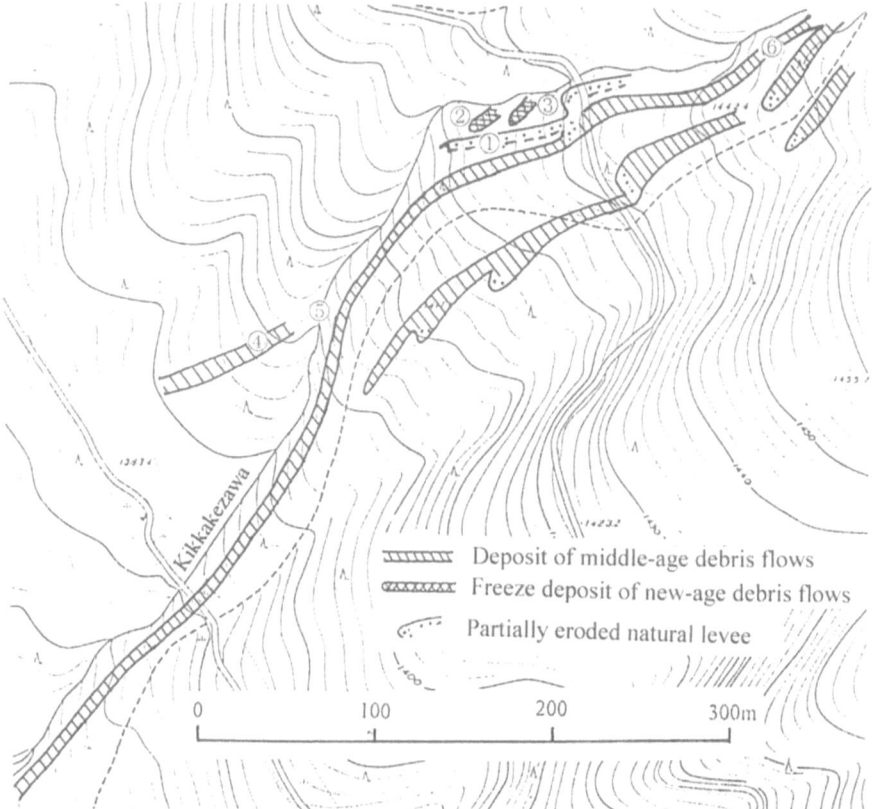

Fig. 3. Major results of the reconnaissance of the deposits of debris flows along the Kikkakezawa gully.

5. RISK, HAZARD AND HAZARD POTENTIAL

A schematic transverse cross section of the Kikkakezawa fan at an altitude of

1,430m is shown in Fig. 4. The topographic conditions of the debris flows of different ages are easily derived from this illustration. The old-age debris flows entirely covered the alluvial fan. Since the depth of these deposits is less than the relative height of their surface from the current channel bed (about 30m), their recurrence needs in advance a marked aggradation of the Kikkakezawa gully, and a continuation of raised sediment production in the headwaters. Therefore, it can be said that these debris flows will not recur unless the sediment production becomes very much activated. The conditions are similar for the middle-age debris flows. Compared with the old-age debris flows, however, they were of smaller scale and covered smaller areas of the fan. The geomorphic thresholds for their recurrence is lower. Moreover, recent aggradation of about 10 meters of the channel of the Kikkakezawa gully around the point ⑤ in Fig. 3 has made it easy for a debris flows to inundate to the fan surface from such a point. It is not easy to estimate their recurrence interval in future, but Okunishi (1994) has postulated it between 100 years and 1,000 years on the basis of the possible dates of the middle-age debris flows.

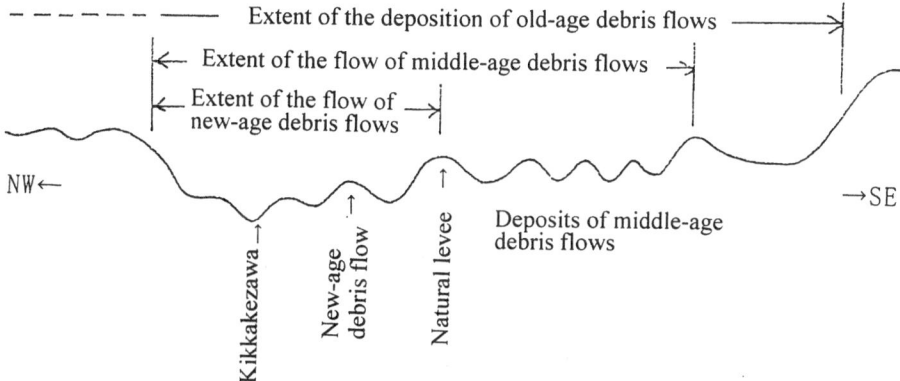

Fig. 4 A schematic transverse cross section of the Kikkakczawa fan at an altitude of 1,430m and the depositional features.

The recurrence interval of the new-age debris flows are thought to be around 30 years, on the basis of the date of the latest one. It is frequent enough and their magnitude is enough to break a levee and other conventional barriers.

Back to Fig. 2, the newly built road (PR) crosses the upper parts of the alluvial fans of the Kikkakezawa gully and neighboring gullies. The areas on the upper side of this road commonly have a high hazard potential of debris flows, since they are covered with large boulders which constituted the head parts of the debris flows, which is very destructive. Qualitatively, such areas can be said to be very risky, but a quantitative assessment of the risk needs some procedure.

It is, thus, desirable to define the hazard potential, hazard and risk separately. Hazard potential is defined as the possible hazards for any land use at any part of a given area for infinite time span in future. It is practically the maximum possible hazard of any frequency, and it reflects all hazardous phenomena in the past, not only in historical period but also in the geological history. Forest can preserve the

568

evidence of such phenomena better than any other form of land use. It is, however, not easy to find such an evidence at any site in forests for a given plan of reclamation. Hence, a comprehensive assessment of hazard potential for a wide area is advantageous. In Fig. 2, for example, hazard potential in an alluvial fan can be approximated as a function of the distance from the fan head, irrespective of the distance from the stream channel.

The hazard can be defined as statistically expected hazards for any reclamation plan at a given site (Dunne, 1991). It is expressed in terms of physical magnitude of phenomena. It depends on the time span of the planned land use, as well as on hazard potential, since the phenomena of a recurrence interval greater than this time span can be ignored. It also depends on the distance and the relative height from the channel, since the extent of inundation of the debris flows of a given recurrence interval should be limited. For example, possible inundation area of debris flows in a time span of several decades is limited to the area of new-age debris flows (Dn) in Fig. 2.

The risk of hazards is, on the contrary, assessed in terms of loss of human lives and property. As well as it depends on the hazard, it depends on the degree of the removal of trees, topographic transformation, construction of buildings and other structures, style of human activity, and physical and social competence to the natural hazards.

REFERENCES
Dunne, T. (1991) Stochastic aspects of the relations between climate, hydrology and landform evolution, Transactions of Japanese Geomorphological Union , **12**, 1-24.
Fujimi Town (1985) Report of the fundamental research on systematic improvement of rivers, 143p.+appendix (in Japanese).
Marui, A., Yasuhara, M. Kazahaya, K., Suzuki, Y., Shimano, Y. and Takayama, S. (1993) Hydrological environment in southern parts of Mts. Yatsugatake, J. Jap. Assoc. Hydrological Sciences (in Japanese with English abstract).
Kochi, S. (1977) Geology of the Yatsugatake Area, 92p.+geological map (in Japanese).
Kosaka, T. (1992) Geology of the Southwestern Foot of Mt. Nishidake and Mt. Amigasa in the Yatsugatake Volcanoes, 11p. + appendix (in Japanese).
Nagano National Forest Management Bureau (1984) Report of Research on Transformation of the Hirohara National Forests, 395p. (in Japanese)
Okunishi, K. (1994) Geomorphic Environment and Hazard Potential of the Alluvial Fan of the Kikkakezawa Gully in Fujimi Town, Nagano Prefecture, 11p. +11figures+21photos (in Japanese).
Okunishi, K. and Suwa, H. (unpublished) Assessment of hazard potential at volcanic alluvial fans, submitted to J. Jap. Soc. Natural Disaster Sci. (in Japanese with English abstract)

Motion and Fluidization of a Hariharagawa Landslide, South Japan

Hiromu MORIWAKI and Teruko SATO

National Research Institute for Earth Science and Disaster Prevention
Tsukuba 305-0006, Japan

ABSTRACT   On July 10, 1997 a debris flow disaster occurred in the Harihara district, Izumi City in Kagoshima Prefecture, causing 34 casualties (21 fatalities) and completely destroying 29 houses. The debris flow was initiated by a landslide on a hillside by the Hariharagawa River which was triggered by heavy rainfall. A kinematic analysis and field surveys showed that the landslide fell a long distance at high speed, and concluded the high mobility was possibly caused by fluidization at failure.
Keywords: Landslide, Debris flow, Simulation, Mobility, Fluidization.

INTRODUCTION

Izumi City in the northeast of Kagoshima Prefecture, south Japan had severe rainfall for three days between July 7 and 9, 1997. The recorded rainfall was 349 mm for the 3 days. The intermittent rainfall on July 9 was particularly intensive, so that the recorded rainfall, 275 mm was a record for a single day, the maximum hourly rainfall recorded at, 62 mm is the third record since 1979. Then, a landslide which was generated on the flank in the upper stream of the Hariharagawa River, flew into the torrent and directly hit the settlement (Harihara district; 86 households, 262 persons) on the small alluvial fan 500 m downstream at 0:44 a.m. on July 10 (Fig.1). As a result it claimed the lives of 21 persons, injured 13 more and completely destroyed 29 houses. Human damage was created by the failure to evacuate: the residents had no experience of debris flow and the disaster occurred at midnight.

This landslide was characterized by high mobility and high speed although it occurred on a gentle slope. In this paper, the mechanism and motion of the debris were examined.

1. LANDSLIDE AND SITUATION

The landslide occurred on a gentle hillside (about 26 degrees) facing westerly direction. The slope was composed of weathered andesite rock and soil. The

Fig.1   Location

main scarp at the source area is 195 m above sea level. The landslide was about 200 m in length, about 80 m in width, about 27 m at maximum depth and about 16.5 ten thousand cubic meters in volume. The surface shape was oval and the slip surface was circular. The majority of the debris seemed to have flown from the source area and the residual soil was little.

A check dam with a height of 14 m, a length of 85 m and a sand capacity of 2.2 ten thousand cubic meters, was completed near the mouth of the valley at the end of June, 1997. Hence the dam was unoccupied. The average river-bed gradient of the dam upstream before the disaster is about 17 degrees and that downstream is 5 degrees since a small alluvial fan was formed at the piedmont. In addition, there was a farm reservoir of about 1,600 square meters between the dam and the source area. The effective capacity was estimated to be approximately 5 thousand cubic meters·

2. DOWNWARD FLOW AND SEDIMENTATION OF THE DEBRIS

Fig.2 shows the location of the flow and sedimentation area of the debris, including the source area. Fig.3 indicates a longitudinal profile along the debris path before and after the disaster. The horizontal runout distance to the toe of debris sediment from the main scarp of the source area is about 960 m, and the horizontal distance is about 175 m.

The debris flew into the Hariharagawa River from the collapse source and ran down the river bed like a bobsled course (Fig.4). Part of the debris on the left side of the flow ran up the jutting slope in front of the dam and stopped, passing the reservoir (340 m downstream from the main scarp). Carcasses of fish which lived

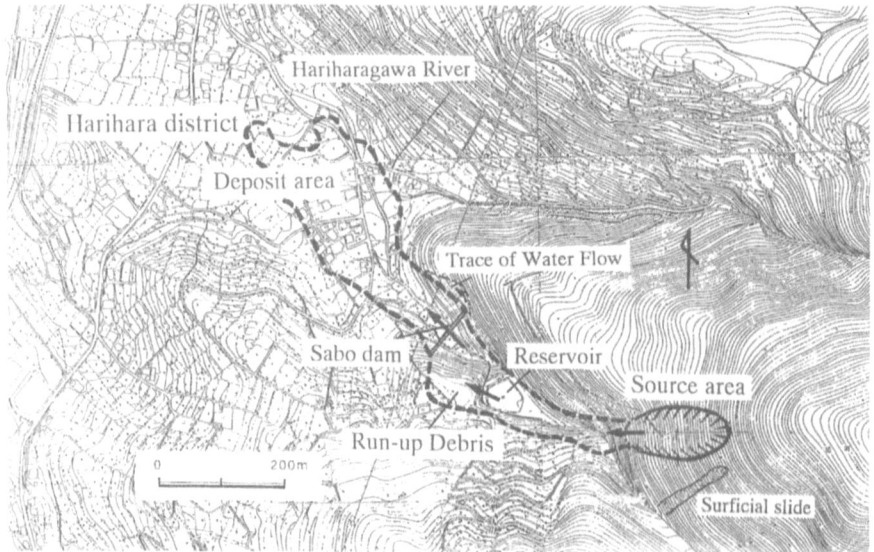

Fig.2 Landslide and flow area

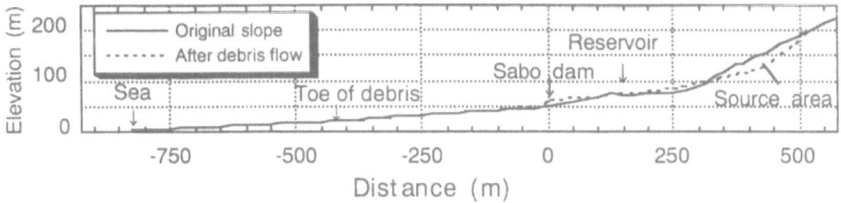

Fig.3   Longitudinal profile (before and after debris flow)

in the reservoir were found near the top of the debris riding up on the jutting slope which means that the debris pushed the water out of the reservoir.

The main part of the debris excluding the left part, approached the dam and collided with it. Though part of the debris was stopped by the dam, the majority flowed over the top. This destroyed the right end concrete section of the dam and washed it away on the side of the settlement. In the meantime, a trace by water running over the slope from the right end concrete of the dam was observed.   Neither an eroding mark nor residual soil on the trace existed. It seems that this trace was made by the flow of the water in

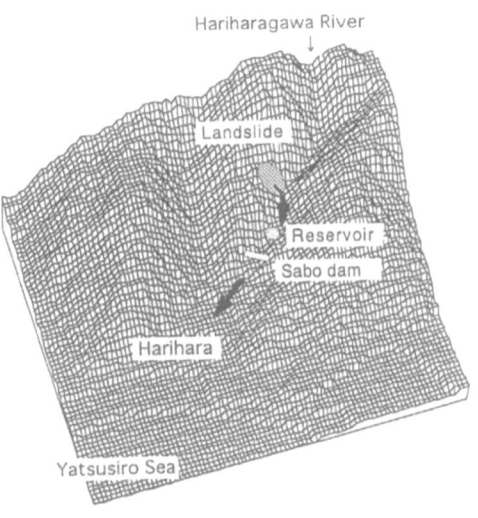

Fig.4   Bird's view of Harihara district

the reservoir and the dam, which was pushed out by the flowing debris, because the vegetation on the slope was knocked over, in the direction of the flow. The alluvial fan originally inclined from the stream line to the right hand side because the ridge of the left side slope covered with old debris flow, extended. Accordingly the deposit was spread and piled up from the stream line of the dam on the right side, going straight. The sedimentation depth on the right hand side of the alluvial fan was also greater than that of the center in the fan. The sedimentation gradient of a debris flow going straight through the dam was 4 - 5 degrees.

3. MOBILITY OF DEBRIS

Scheidegger (1973) dealt with mobility of debris statistically. He showed that a landslide ran down a greater distance as the soil volume increased, using the relationship between H/L and landslide volume for large-scale landslides. Here H is a vertical runout height and L a horizontal runout distance. Fig. 5 revised the relationship in Scheidegger's data (white circles), adding the case of the Hariharagawa landslide. The H/L of the present case is 0.20. The H/L was obtained from the position of the debris which ran up

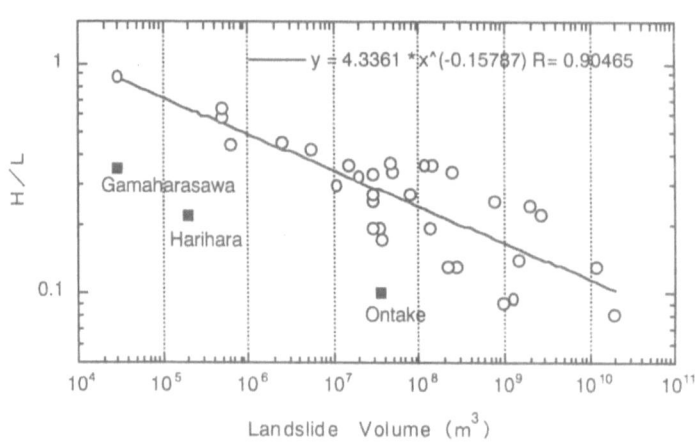

Fig.5 Relationship between H/L and volume

the slope on the left side just in front of the dam. The value of H/L in the present case is small, compared with Shceidegger's tendency. That is, the fluidity of this debris is higher than in a usual collapse. Fig.5 includes another two cases; (1)1996-Gamaharasawa debris flow, Nagano Prefecture (Moriwaki, 1997), and (2)1984-Mt. Ontake debris avalanche, Nagano Prefecture (Moriwaki, 1985). Both cases were also originated by landslides. All three cases show similar fluidity when viewed in terms of the scale of the collapse. The points common to the above three cases were all landslide-induced type and flowed in the torrent.

With regard to the long distance fluidity of the flow in Harihara's case, there are some causes as follows; the characteristics of the debris itself, with high mobility and geomorphological conditions which allowed easy flow into the torrent (the confluence direction). The confluence angle to the torrent was small, and it was in the direction in which the resistance to flow was also comparatively little. For example, the debris will run onto the opposite slope and stop in a short distance if the downward of the flow direction is right-angled to the opposite slope. In addition, the river bed was gently curved so as to allow the debris to flow easily. However, the slope jutting into the torrent and the dam, played the role of diminishing the energy of the sliding mass. It partially checked the flowing debris. The debris would have run down further without the presence of the jutting slope and the dam.

4. NUMERICAL CALCULATION

This chapter deals with the movement of the debris from the source area to the dam. With regard to the downward flow from the dam, accurate kinematic analysis from start to stop is difficult because the attenuation of kinetic energy created by the collision with the dam and physical property changes generated by falling debris from the dam (soil structure destruction) etc. The inevitable value for the calculation is the coefficient of dynamic friction. The first calculation was executed on the movement of debris on the left side of the flow which rode on to the slope in front of the dam and stopped, in order to get the value. Next, calculation on the movement to the dam was

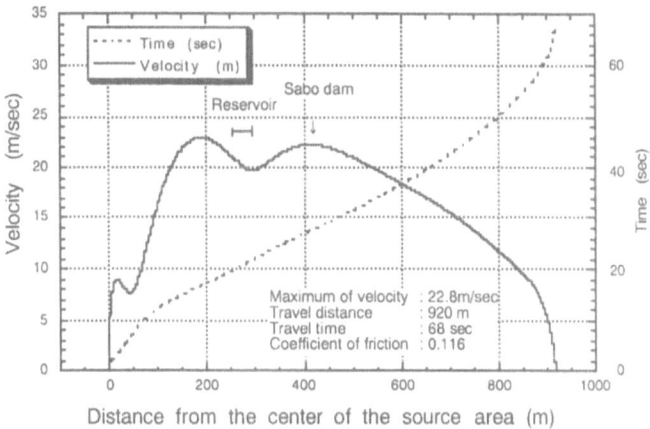

Fig.6 Velocity and running time calculated

attempted using the value of the coefficient of dynamic friction calculated. In the calculation, the lumped mass model was used which regards the sliding debris as approximately one mass (Moriwaki, 1985). In this model, resistance in proportion to the dead weight worked on the sliding body. The initial velocity was zero. Increase and decrease of the debris mass and the effect of the water in the reservoir at the mid-point have not been included. The starting point of the model was 124 m above sea level. Serration width in time t in the calculation is 0.1 sec.

<First calculation>: The debris flowed 335 m (horizontal distance) and stopped at 80 m above sea level on the jutting slope. The coefficient of dynamic friction was 0.116 in first circulation.

<Second calculation>: Using the coefficient value of dynamic friction obtained, calculation of the motion of the sliding mass which continued to the dam was conducted. The result is shown in Fig. 6. The motion downstream from the dam is calculated under the assumption that the dam did not exist. In Fig.6, the decreases in the velocity behind the peak curve to the dam appear twice between the source area and the dam. The first is caused by the motion which slips on a circular arc slip plane in the source area. It gradually becomes a low-gradient while the resistance increases and the sliding force also decreases. The second decrease of velocity was generated because the reservoir on route set it levelly in calculation. The toe of the debris reached the dam 23 seconds after the collapse, when the debris is supposed to keep its original length. The velocity at this time was about 20.2 m/sec (73 km/hr). The maximum velocity occurred at 22.6 mm/sec (82 km/hr) at the point approximately 190 m in horizontal distance from the source area. This distance corresponds to the distance from where the toe of the debris slightly entered the reservoir in a similar assumption. The time-of-arrival at the houses in the settlement (about 150 m-200 m from the dam), which is close to the dam, is estimated to be about 30-32 second after the collapse, when the debris was supposed to have passed over the dam at the same velocity.

574

5. MECHANISM OF DEBRIS FLUIDIZATION

On-site surveys clarified that the majority of the debris did not stay in the source area and that the coefficient of dynamic friction required from kinematic analysis was remarkably small. In addition, the geomorphological index of mobility, the *H/L* was also small. The matters above mentioned mean that a landslide with a high water content was fluidized in the collapse and flowed at high speed. In order to be fluid in a collapse, the slope needs to have both a high water moisture content and be comprised of loose soil.

A slope with a gradient of 26 degrees is generally stable. It does not fail unless the pore water pressure rises considerably. It has been confirmed that the inside of the deposit in the sedimentation area was in the mud state with a high water content, while the surface was dry. The negative volume change (rearrangement of the soil particle structure) was caused during failure when the surface layer was loose, the pore water pressure in the soil rises further, and the shear resistance (coefficient of dynamic friction) will decrease further. The soil was found to be considerably porous after boring exploration in the vicinity of the collapse slope. In short, the local conditions in which the soil layer density is low, created a high possibility of producing fluidization in a collapse.

Sassa et al. (1997) carried out the ring shear test using the soil in the source area, and found generation of the rapid excess pore water pressure and a lowering of shear resistance lowered (apparent angle of friction in the steady state is 2 degrees). They estimated that the ground water level rose to the limit of the equilibrium of the slope stability, and it generated liquefaction along the slip plane when the shear started. Their examination agrees with the hypothesis described in this paper.

From the above considerations, it is concluded that the present landslide became fluid in a collapse and descended into the torrent at a high speed.

REFERENCES

Moriwaki, H., N.Oyagi. and S. Yazaki (1985）: A Gigantic Debris Avalanche and its Dynamics at Mt. Ontake Caused by Naganoken-seibu Earthquake, 1985. Proc. of IVth International Conference and Field Workshop on Landslides, 1985, Tokyo, pp.359-364.
Moriwaki, H. and Yazaki S.(1997): On the Motion of Landslide-induced Debris Flow in Gamaharasawa Torrent, Otari Village, Nagano Pref. Proc. of 36th Annual Conference of Japan landslide Society, pp.91-92. (in Japanese)
Sassa, K. et al. (1997) : Study on the Initiation Mechanism of the Flowslide-Debris Flow in Izumi, Kagoshima Pref., July 1997. Proc. 16th Annual Conf. of Japan Society for Natural Disaster Science, pp.107-108. (in Japanese)
Scheidegger, A.E.(1973): On the Prediction of the Reach and Velocity of Catastrophic Landslides. Rock Mechanics 5, pp.231-236.

Cyclic-Loading Ring-Shear Tests to Study High-Mobility of Earthquake-Induced-Landslides

Fawu WANG[*1], Kyoji SASSA[*2] and Hiroshi FUKUOKA[*2]

*1 Graduate School of Science, Kyoto University, Kyoto, Japan
*2 Disaster Prevention Research Institute, Kyoto University, Kyoto 611-0011, Japan

ABSTRACT   To study the mechanical behavior of sandy soils under cyclic-loading, a series of undrained cyclic-loading ring-shear tests were conducted. The stress condition of the tests simulated the stress condition of a soil element subjected to a horizontal seismic force in the sliding zone of an infinitely long slope. Osaka group coarse, sandy soil, silica sand and Toyoura standard sand were used. The sliding surface liquefaction behavior of different soils shows a relation dependent on grain crushing susceptibility. The effects of the initial stress state and frequency of cyclic loading on the high-mobility were also analyzed.
Keywords: cyclic-loading, ring-shear test, high mobility, sandy soil.

1. INTRODUCTION

Earthquake-triggered-landslide becomes a serious geo-hazard in mountainous area with the progress of regional development in high seismic area. Those with high-mobility often cause great damage and loss to the society (Sassa et al. 1996, Sassa et al. 1997). Study on the mechanism of high-mobility is very important for disaster mitigation. A concept of "Sliding Surface Liquefaction" was proposed by Sassa et al. (Sassa et al. 1996, Sassa 1996), in studies of some landslides triggered by the 1995.1.17 Hyogoken-Nanbu earthquake through the undrained ring-shear tests. Sliding surface liquefaction takes place along a sliding surface, need not destruction of soil structure that is required for (mass) liquefaction. Sliding surface liquefaction can take place even in medium or dense soil structure because grains crushing along the sliding surface results in volume shrinkage and pore pressure generation (Sassa et al. 1996).

Wang et al. (1998a) conducted a series of ring-shear tests to study the relation between grain crushing (by drained shear-speed-controlled ring-shear test) and susceptibility of sliding surface liquefaction (by undrained shear-speed-controlled ring-shear test) on three types of samples. The samples were Osaka group coarse, sandy soil, silica sand and Toyoura standard sand. It was found that, the soil susceptible to grain crushing is prone to generate sliding surface liquefaction. The excess pore pressure behavior of these samples in the undrained ring-shear was similar to the volume change behavior in drained ring-shear test.

The purpose of this study is to analyze the mechanism of high-mobility of earthquake-triggered-landslide through cyclic-loading ring-shear test, and discuss the effects of initial conditions of slopes and frequencies of cyclic-loading on the high mobility. The undrained ring-shear apparatus DPRI-5 (Sassa 1997, Wang et al. 1997) was used.

2. SAMPLE PROPERTIES AND TEST PROGRAM

The samples used by Wang et al. (1998a, b) were also used in this study. They are Toyoura standard sand, silica sand and Osaka group coarse, sandy soil.

Toyoura standard sand (T-sample) is predominantly a uniform sub-angular to rounded quartz fine sand with approximately 90% of quartz and 4% of chert. Silica sand (S-sample) is construction material for industrial use. It is made of weathered silica sand, and has a uniform grain size distribution. The grain is almost angular. It consists of 92 ~ 98% of quartz, and a little amount of feldspar. In this study, the silica sand no.7, with a mean grain size of 0.16 mm was used. Osaka group coarse, sandy soil (O-sample) was taken from the source area of the Takarazuka landslide, which was triggered by 1995.1.17 Hyogoken-Nanbu earthquake (Sassa et al. 1996). Osaka group is a limnetic and marine deposit of Pliocene to Mid-Pleistocene distributed around Osaka area. It comes from weathered granite and is composed of 77

Table 1 Classification property of samples

Classification property	T	S	O
Mean grain size, D_{50} (mm)	0.18	0.16	0.80
Effective grain size, D_{10} (mm)	0.11	0.09	0.28
Coefficient of uniformity, Uc	1.7	2.1	4.7
Maximum void ratio, e_{max}	0.98	1.30	1.17
Minimum void ratio, e_{min}	0.61	0.71	0.66
Specific gravity, Gs	2.64	2.64	2.61

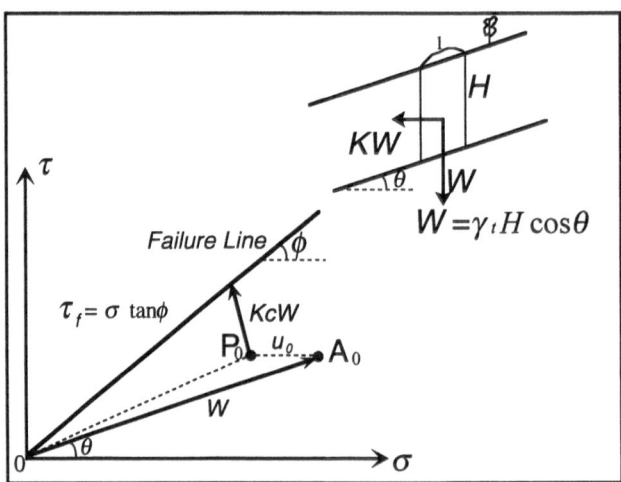

Figure 1. Schematic diagram of the undrained cyclic loading tests. A_0: Initial stress state in dry condition. $\sigma_0 = W\cos\theta$, $\tau_0 = W\sin\theta$; P_0: Initial stress state when initial pore pressure, $u_0 = \gamma_w h_w \cos^2\theta$ exists.

percent of quartz and 23 percent of feldspar. The grain is angular. In this study, due to the need of apparatus, the grains larger than 4.75 mm were sieved out. Wang et al. (1998a) found that O-sample and S-sample are easily to be crushed, while T-sample is very difficult to be crushed. Some classification and index properties of the employed samples are listed in Table 1.

The consolidated-undrained cyclic-loading test (CU-C) was planned. It simulates the stress condition of a soil element subjected to a horizontal seismic force in the sliding zone of an infinitely long slope (Figure 1). To represent an "ideal" slope, depth of potential sliding surface was assumed to be 30 m, slope angle to be 25°, and initial pore pressure $u_0 = \gamma_w h_w \cos^2\theta$ to be 20 kPa (h_w was assumed to be 2.43 m). $\gamma_t = 20$ kN/m^3 was taken as the unit weight of all the samples. The initial normal stress $\sigma_0 = 482.9$ kPa, and initial shear stress $\tau_0 = 225.2$ kPa. A condition of 15° of slope angle ($\sigma_0 = 548.6$ kPa, and $\tau_0 = 147.0$ kPa) was also involved to analyze the effect of initial stress state. The seismic coefficient ratio K/Kc is defined as the cyclic stress amplitude parameter. Where, K is horizontal seismic coefficient, and Kc is the critical seismic coefficient for the stress path to reach the failure line by adding KcW to the initial stress point "P_0" in Figure 1. Kc can be given by Equation (1).

$$Kc = \tan(\phi - \theta) - [\tan\theta + \tan(\phi - \theta)]u_0\sin\theta/W \qquad (1)$$

Where: ϕ is the angle of internal friction of sample. Thus, the critical increments of normal stress and shear stress causing slope to failure are calculated as: $\Delta\sigma_c = -KcW\sin\theta$, $\Delta\tau_c = KcW\cos\theta$. To study the post-failure behavior of soils, $K/Kc = 1.5$ was used. The amplitude of normal stress increment is $\Delta\sigma = K/Kc\Delta\sigma_c$ and shear stress increment is $\Delta\tau = K/Kc\Delta\tau_c$.

Cyclic-loading of constant amplitude lasted for 10 cycles with different frequencies of 0.02 Hz, 0.1 Hz and 0.5 Hz (Fig. 2). The CU-C test was named in the order of soil name-cyclic-loading (frequency) -slope angle. For example, O-C(0.1)-25° means that the sample is O-sample, frequency of loading is 0.1 Hz, and the angle of ideal slope is 25°.

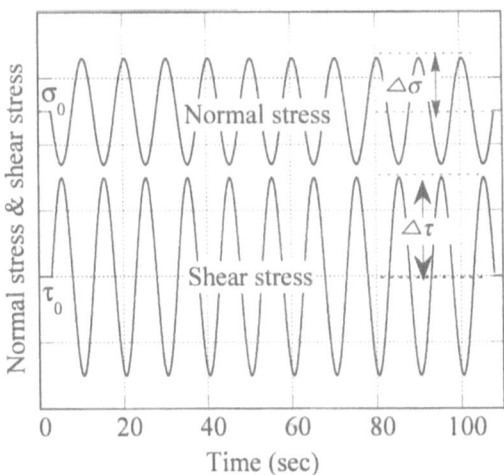

Figure 2. Stress pattern used in the undrained cyclic-loading tests (a case of frequency = 0.1 Hz).

3. TEST PROCEDURE

Sample was set in the shear box with natural falling method. Before any load was loaded, the initial relative density of the sample is D_{ri}. Then CO_2 was supplied from the lower part of the

Table 2 Sample relative densities and amplitude of cyclic-loading.

Test No.	Dri(%)	Drc(%)	ϕ (°)	Kc	$\Delta\sigma$ (kPa)	$\Delta\tau$ (kPa)
T-C(0.1)-25°	32.7	43.2	32.8	0.128	48.5	104.1
S-C(0.1)-25°	51.6	67.8	32.8	0.128	48.5	104.1
O-C(0.1)-25°	22.8	111.1	34.5	0.157	59.7	128.1
O-C(0.5)-25°	22.8	117.6	34.5	0.157	59.7	128.1
O-C(0.02)-25°	22.8	115.4	34.5	0.157	59.7	128.1
O-C(0.1)-15°	22.8	115.8	34.5	0.348	76.7	286.3

shear box to replace the air in the sample. De-aired water was infiltrated to saturate the sample. After sample was consolidated at a normal stress about 50 kPa, the degree of saturation was confirmed by B_D (=$\Delta u / \Delta\sigma$) value, a ratio of response pore pressure increment to the loaded normal stress increment under undrained condition (Sassa, 1985). Only these samples with B_D value higher than 0.95 were conducted cyclic-loading ring-shear tests. After sample was normally consolidated under the initial normal stress σ_0, the initial pore pressure u_0 and the initial shear stress τ_0 were applied in the drained condition. The relative density of samples after consolidation, D_{rc}, as well as D_{ri}, are listed in Table 2. The residual angle of internal friction, ϕ, based on drained ring shear tests (Wang et al. 1998b), were used to calculated Kc, and normal stress increment $\Delta\sigma$ and shear stress increment $\Delta\tau$. After changing sample to the undrained condition, the normal stress increment ($\Delta\sigma$) and shear stress increment ($\Delta\tau$) were applied simultaneously.

4. TEST RESULTS AND ANALYSES

4.1 Results of Different Samples in CU-C Tests

At first, CU-C tests were conducted for O-sample, T-sample and S-sample with cyclic frequency = 0.1 Hz (see Table 2). Figure 3 is time-series data for O-C(0.1)-25°. Normal stress showed a way as it was loaded. Shear resistance decreased from the first cycle, it means that the sample failed, because shear resistance was smaller than the shear stress loaded. After about two cycles, shear resistance reduced to a low steady value. The sample reached the steady state. The shear displacement

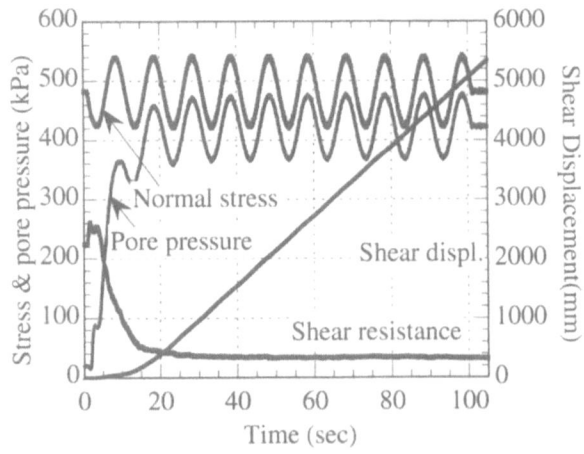

Figure 3. Time-series data of test O-C(0.1)-25° .

showed an accelerated process during the first two cycles, then increased at a constant speed (the maximum speed of DPRI-5). Pore pressure generated rapidly at the first two cycles. After the steady state was reached, the pore pressure did not increased any more, but changed upper and down with the cyclic normal stress.

Figure 4(a) shows the effective stress path (ESP) and total stress path (TSP) of O-C(0.1)-25°. After the start point, ESP went to failure line soon. Then, ESP went down along the failure line rapidly during the first two cycles (it can be checked by TSP), and sample reached a steady state. The apparent friction angle ϕ_a (tan ϕ_a = shear resistance / total normal stress) at the end of cyclic loading is 3.6°. The ESP showed a typical pattern of high-mobility and sliding surface liquefaction. That is, after reached the failure line, because of

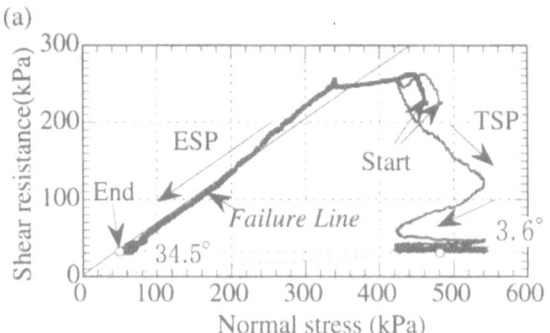

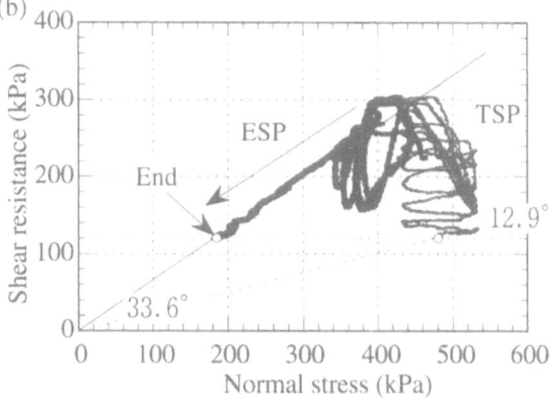

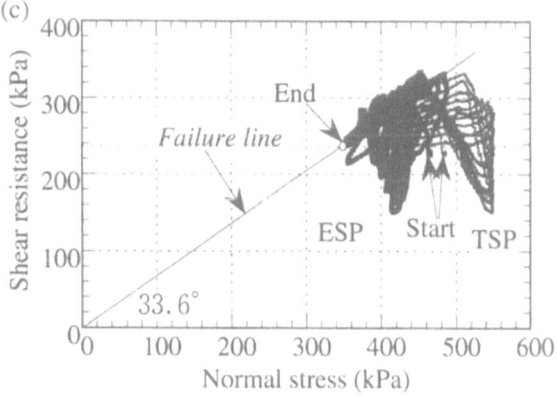

Figure 4. Stress paths of undrained cyclic-loading test.
(a): O-C(0.1)-25°, (b): S-C(0.1)-25° and (c): T-C(0.1)-25°.
TSP: Total stress path; ESP: Effective stress path.

the grain crushing along the sliding surface resulting in volume shrinkage and generation of pore pressure, the stress level reduced to very low.

Figure 4(b) shows the stress paths of S-C(0.1)-25°. A similar stress path style as that in O-C(0.1)-25° was observed. The difference is that the apparent friction angle ϕ_a for S-sample is 12.9° at the end of cyclic loading.

Figure 4(c) shows the stress paths of T-C(0.1)-25°. Comparig them with those of O-sample and S-sample, it is found that most of the stress points distributed around the start point. Although the ESP showed a little bit decrease along the failure line, the end of TSP was located above the start point. The T-sample did not show any mobility.

Under cyclic-loading, the measured pore pressure can be divided into two components: fluctuating component and residual component (Wang et al. 1998a). The fluctuating component is mainly caused by variation of normal stress increment. It is almost equal to the increment of normal stress under fully saturated condition. The residual component is mainly due to plastic shear displacement. The residual component was refined from the total pore pressure. Then, excess pore pressure ratio, expressed as excess pore pressure, the residual component with respect to the initial effective normal stress, was obtained and used to evaluate the degree of mobility. The resultant excess pore pressure ratios of these samples are shown in Figure 5. After shear displacement reached 10 mm, curves of excess pore pressure ratio for all of the sample show increasing tendency, while before that, O-sample shows no increase, and T-sample and S-sample shows a result of dilatancy. In O-C(0.1)-25°, pore pressure generated rapidly from 10 mm, and reached the steady state, while, in T-C(0.1)-25° and S-C(0.1)-25°, pore pressure generated relatively slow. After all of the ten cycles cyclic-loading, the excess pore pressure ratio is still smaller than 0.2 and 0.6, respectively. It means that O-C(0.1)-25° shows high mobility, T-C(0.1)-25° did not show mobility, while S-C(0.1)-25° shows a trend similar to O-C(0.1)-25°, but the rate of pore pressure generation is a bit larger.

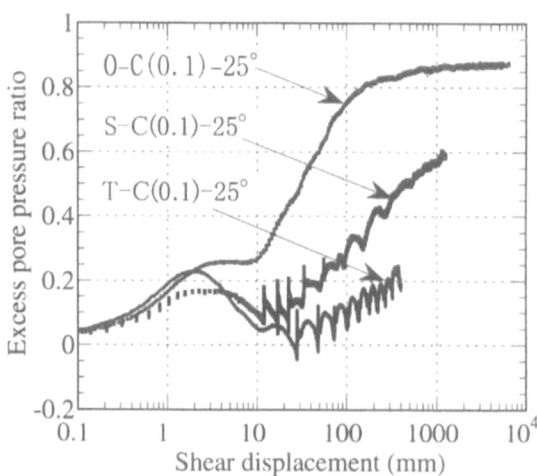

Figure 5. The residual pore pressure ratio of various samples in the undrained cyclic-loading tests.

The difference of excess pore pressure ratio among the three samples shows an internal connection with grain crushing. In sample susceptible to grain crushing, excess pore pressure generate easily, and the sample is prone to sliding surface liquefaction (like O-sample), while, in a sample difficult to grain crushing, it is difficult to generate high-mobility and sliding surface liquefaction, because excess pore pressure generates difficulty (like T-sample). It is indicated that, in cyclic-loading, grain crushing also strongly affected the excess pore pressure behavior of sandy soils.

4.2 Results of O-samples in Different Initial States

This series of tests is to compare the pore pressure behavior at different slope angle of 15° and 25°. Figure 6 is the test results of O-C(0.1)-25° and O-C(0.1)-15°. Under the same value of K/Kc (= 1.5), the pore pressure responses of O-sample, can be compared. Both curves show a slight decrease or at least no increase from 4 mm to 10 mm shear displacement. This is caused by the dilative behavior when the soils begin to be sheared. Thereafter, the excess pore pressure ratio increased rapidly as the shear displacement increased, and then

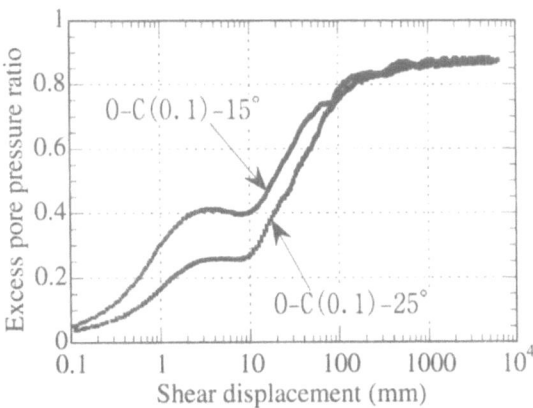

Figure 6. The residual pore pressure ratio of O-samples in the undrained cyclic-loading tests at different initial slope angle.

caused high-mobility. The difference is that the pore pressure response is higher in 15° slope than that in 25° slope. It is indicated that, under the same K/Kc value, high-mobility phenomenon takes place easier in gentler slope. This is because that, when K/Kc values are kept the same, the shear stress increment acting in the gentler slope is larger than that acting in the steeper one.

4.3 Results of O-samples at Different Frequencies

Frequency is an important index of seismic wave. To investigate how the frequency affects the pore pressure behavior under the undrained condition, another two frequencies of 0.02Hz and 0.5 Hz were used in the C-CU tests on O-sample. The results were compared with that of O-C(0.1)-25°. As shown in Figure 7, O-sample generated high mobility

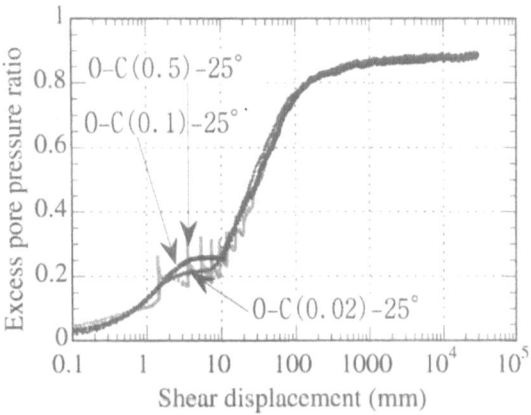

Figure 7. The residual pore pressure ratio of O-samples in the undrained cyclic-loading tests with different frequencies.

shear displacement for all tests, and before that, although test data of 0.5 Hz can not be clarified because of the noise, the excess pore pressure ratio of 0.1 Hz is larger than that of 0.02 Hz. There are two possible reasons for this phenomenon. (1) For the same shear displacement, greater number of cycles is loaded in the higher frequency. Pore pressure generation may be affected by the cycle number of cyclic loading; (2) Higher frequency causes faster pore pressure generation. Faster pore pressure generation will effectively build up excess pore pressure in the shear zone, because excess pore pressure will dissipate upwards and downwards from the shear zone in the shear box.

5. CONCLUSION

The following conclusion may be drawn from the results of the undrained cyclic-loading ring-shear tests.
(1) The high-mobility is prone to generate in the sandy soil susceptible to grain crushing, because of the rapid generation of excess pore pressure resulting from volume shrinkage.
(2) For the same shear displacement, at different initial slope angles for a soil, the high-mobility generated easier at the gentler slope; and, higher frequency causes excess pore pressure generate more quickly.

REFERENCES

Sassa, K. (1985):The mechanism of debris flows. Proc. of XI Int'l Conf. on Soil Mechanics and Foundation Engineering, San Francisco, Vol. 3, pp.1173-1176.

Sassa, K. (1996): Prediction of earthquake induced landslides. Proc. of 7th Int'l Symposium on Landslides, Rotterdam: Balkema, Vol. 1, pp.115-132.

Sassa K., Fukuoka H., Scarascia-Mugnozza G. and S. Evans (1996): Earthquake induced landslides. Soils and Foundations, Special Issue, pp. 53-64.

Sassa, K. (1997): A new intelligent type dynamic loading ring shear apparatus. Landslide News. No. 10, pp. 33.

Sassa, K., Fukuoka, H. and F. W. Wang (1997): Mechanism and risk assessment of landslide-triggered-debris flows: Lesson from the 1996.12.6 Otari debris flow disaster, Nagano, Japan. Proc. of the Int'l Workshop on Landslide Risk Assessment, Rotterdam: Balkema, pp. 347-356.

Wang, F. W. and K. Sassa (1997): Shear displacement behavior of sandy soils in different states of saturation and the seismic coefficient in cyclic-loading ring-shear tests. Natural Disaster Science, J., Vol. 19, No. 1, pp.31-45.

Wang, F. W. and K. Sassa (1998a): Experimental study on the factors affecting high-mobility of landslides by ring-shear tests. Proc. of 8th Int'l. IAEG Congress, Rotterdam: Balkema, in press.

Wang, F. W. and K. Sassa (1998b): Ring-shear tests on sliding surface liquefaction behavior of sandy soils. Annuals of DPRI, Kyoto University. in press.

Initiation of Rapid and Slow Landslides in Experimental Model

Tayoko KUBOTA*[1] and Yasuo TAKEDA*[2]

*1 Forestry and Forest Products Research Institute, Ibaraki, Japan
*2 School of Agricultural Sciences, Nagoya University, Nagoya, Japan

ABSTRACT The initiation of liquefied landslides in an experimental flume model and its mechanism was studied, so far it has been studied generally in laboratory strength tests. Loose and compacted fine silica sands were employed, and for their soil masses, stresses and displacement were logged minutely with 0.03 sec intervals during the failure. Loose soil mass failed suddenly and its mechanism was confirmed to a sudden subsidence accompanied with a rapid generation of pore water pressure leading to a reduction of shear stress and consequent acceleration of the failure mass. While, in compacted soil mass, it was found that those failures could not be reproduced.
Keywords: landslide, liquefaction, model experiments, pore pressure, velocity.

1. INTRODUCTION

The mechanism of failure in saturated sandy slope that fails suddenly and moves to a long distance like a fluid, have been reported by Sassa, 1984 and Hiura, 1991. It is called as liquefied landslide same as liquefaction in a saturated deposit of loose sands.

The liquefaction is the term describing a transformation of solid state to a liquid state as a consequence of pore water pressure generation that will reduce the effective stress. Many studies of liquefaction have been implemented, since a lot of catastrophic liquefaction was triggered by Niigata earthquake in 1964. Early studies have been conducted under seismic loading conditions using cyclic triaxial tests. Castro (1969) found a possibility that liquefaction is also triggered by monotonic loading. And the mechanism of liquefaction triggered by monotonic loading has explained the mechanism of liquefied landslide.

As mentioned above, the mechanism of the initiation of liquefied landslides has been fairly understood by triaxial tests. In the laboratory strength tests, however, actual condition of failure in the slope deposits can not be reproduced, also, the motion after the failure can not be observed. Eckersley (1990) induced liquefied landslides in small instrumented coking coal stockpiles and found that the undrained condition may not be a prerequisite to initiate liquefied landslides, although most previous laboratory studies have involved undrained loading. However, only a few works that have reproduced liquefied landslides were conducted. Therefore, it is desirable to confirm the initiation and the motion of

583

584

liquefied landslides in conditions more similar to actual slopes. In this study, the mechanism of initiation of liquefied landslides and also the acceleration of liquefied landslides were considered in a small flume test.

2. OVERVIEW OF EXPERIMENTS

2.1. Test Apparatus

The test apparatus is shown schematically in Fig.1. This apparatus consists of transparent acrylic flume with the size of 1800 mm length, 240 mm width and 150 mm depth. There is no roughness on the floor of the flume that its slope angle is changeable easily. A manometer is placed under the flume bed at the location of 450 mm from the upper end; a pressure transducer for measuring pore water pressure is fixed to the manometer. A set of small pressure transducers for measuring total and shear stresses are placed on the flume bed near the manometer. A two cm diameter styrene foam ball is placed on the lower side of these transducers that is connected to a potentiometer. When a failure takes place, this ball moves with the failure mass. Measuring the movement of this styrene foam ball by the potentiometer, a downward displacement of the failure mass can be measured. Though the potentiometer used in this experiment has a 160gf wire tension, this tension is offset by a 160gf weight connected to the styrene foam ball. A laser displacement transducer is placed on the flume. A target as shown in Fig.1 is buried in the sample (2 cm above the flume bed), and moves up and down with the movement of the soil mass. Measuring this movement, a vertical displacement around the shear zone can be measured.

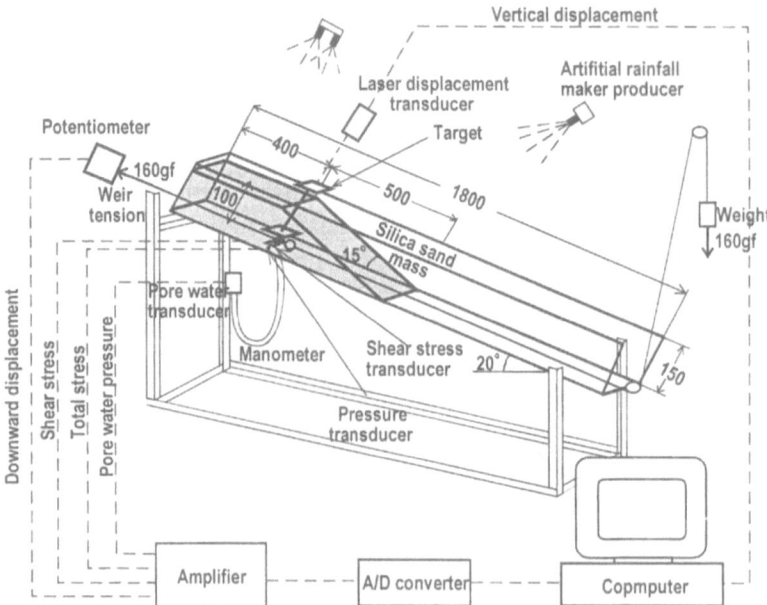

Fig.1 Over view of the test apparatus and logging system (all dimension is in mm)

2.2. Material

A soil mass was constructed using fine silica sand (d_{50} = 0.076 mm) in the flume as shown in Fig.1. It has 10 cm thickness in 40 cm length and has 15° dip in 50 cm length of total 90 cm length. The grain size distribution of used silica sand was shown in Fig.2. The used silica sand has the same particle sizes of potentially liquefiable soil in the nature.

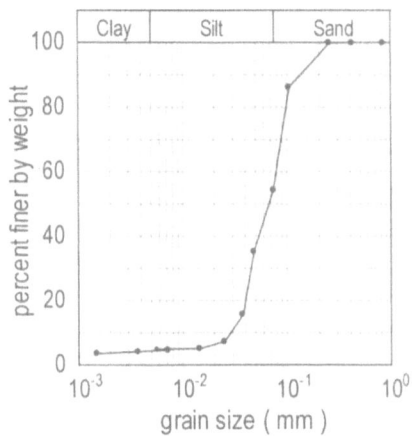

Fig.2 Grain size of distribution of silica sand

2.3. Method of experiment

The landslide was triggered by supplying steady intensity artificial rainfall (1.39 mm/min) to the soil surface. Preliminary test was conducted in a horizontal flume. Then the experiment for 20° slope was conducted. Two types of soil masses were made in each cases. One of them was very loose (bulk density ≒ 0.9 g/cm³) and another was a little compacted (dry density ≒ 1.3 g/cm³). During the experiments, pore water pressure, total normal stress, shear stress, downward displacement and vertical displacement were logged with approximately 0.03 sec intervals, respectively. Video camera was focused on the soil mass section to record the displacement patterns of failure initiation and final landslides. The surface displacement in the preliminary tests was measured by video camera records.

3. TEST RESULTS

3.1. Preliminary Test

Stress and displacement records of the loose and compacted sample are shown in Fig.3a and b, respectively. The loose soil mass subsided suddenly by about 1.5 cm at 22'30". The inter vertical displacement also shows that contraction took place around the shear zone. Consequence of the subsidence, the pore water pressure increased rapidly. The total stress increased slowly before failure and fluctuated complexly, then reduced after the subsidence. Photos.1a, 1b and 1c show that a sudden subsidence of the loose soil mass accompanied with rapid pore water pressure generation within a short time. The photo corresponds to Fig.3(a).

586

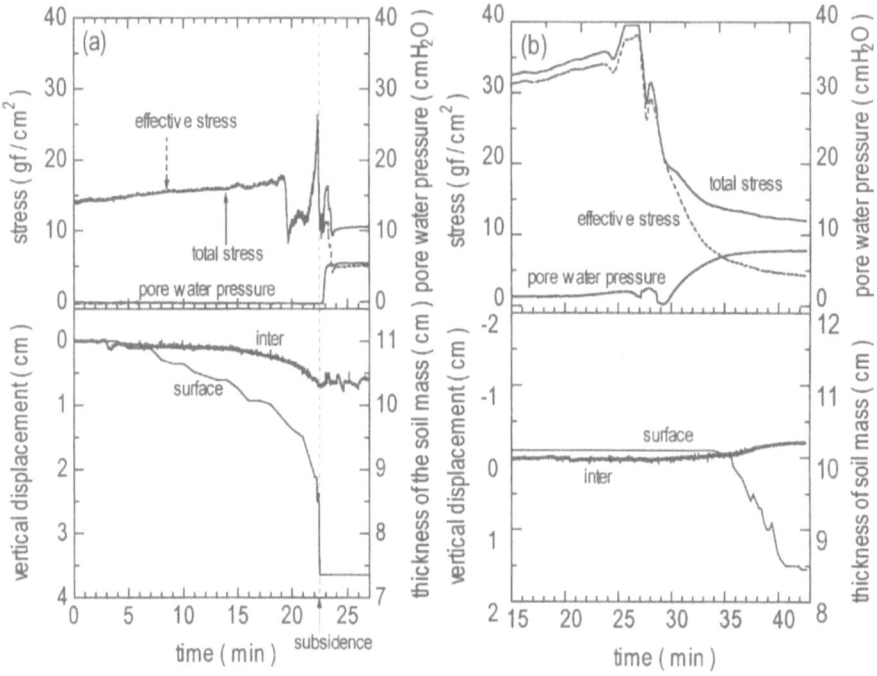

Fig.3 Stress and displacement measurements in the horizontal flume test.
(a) loose silica sand mass; bulk density = 0.9 g / cm³, relative density = -5.1 %.
(b) compacted silica sand mass; dry density =1.33 g / cm³, relative density = 79 %.

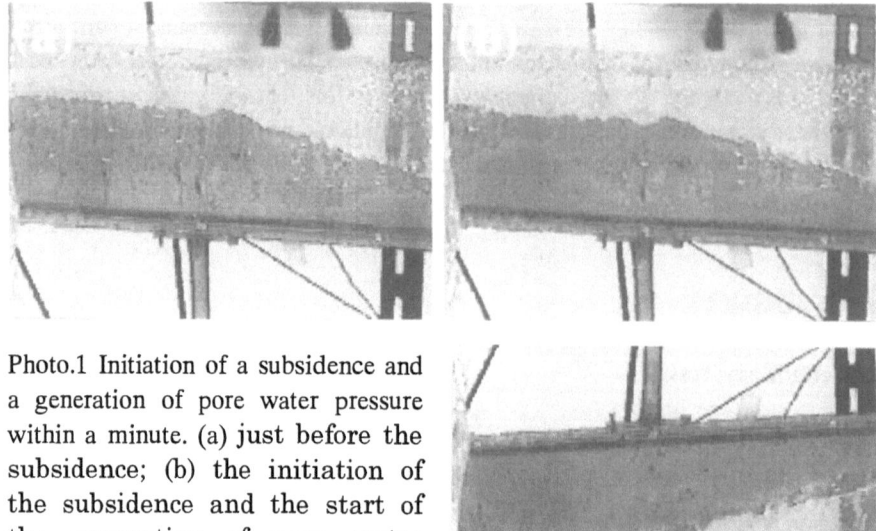

Photo.1 Initiation of a subsidence and a generation of pore water pressure within a minute. (a) just before the subsidence; (b) the initiation of the subsidence and the start of the generation of pore water pressure; (c) the rise of pore water pressure

In the case of the compacted sand, only a part of the soil mass at which the slope angle changes (Fig.1) failed slowly. Different from the case of the loose sample, sudden initiation of a subsidence did not take place and the soil mass moved slowly as the surface vertical displacement shows. While, the inter vertical displacement shows that the expansion, which was not observed in the loose soil mass, took place around the shear zone. The pore water pressure began to increase gradually after artificial rainfall supplied from the soil surface reached flume bed. The rapid fluctuation of pore water pressure and total stress were not observed.

3.2. Test for Slopes

Stress and displacement records for the loose and compacted sands are shown in Fig.4a and b, respectively. The vertical displacement, however, was not recorded after the target fell down with the movement of the soil mass. The soil mass of loosely packed silica sand failed suddenly, accelerated rapidly and nearly flowed. Firstly, the soil mass subsided at 10.68 min, as it is shown in vertical displacement record; the pore water pressure increased sharply and transiently. Then, the soil mass started to move rapidly, as it is shown in downward displacement record. The pore water pressure dissipated after the failure because the soil mass on the transducers moved away. The total normal stress and the

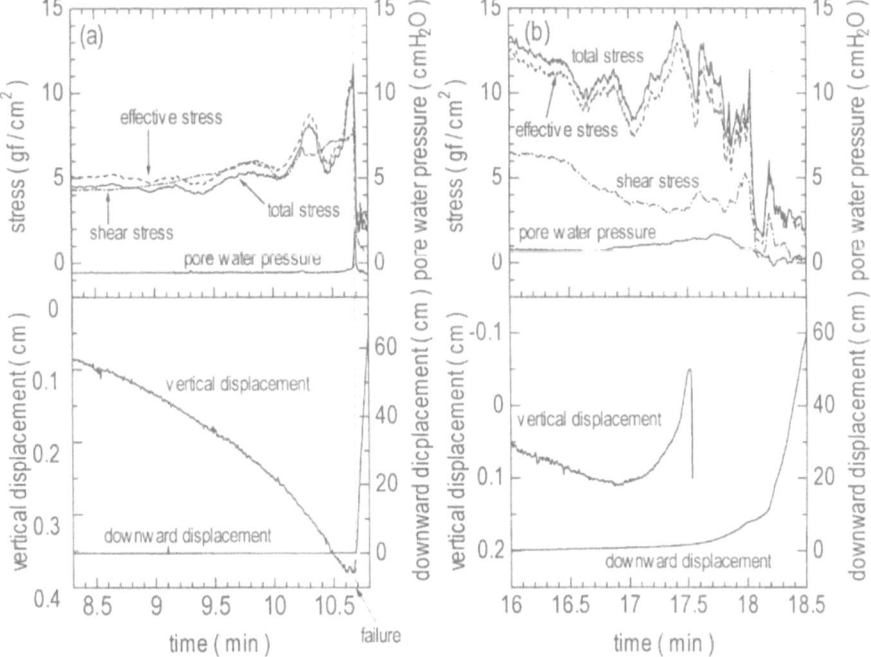

Fig.4 Stress and displacement measurements in the test for 20 ° slope. (a) very loose silica sand mass; bulk density = 0.9 g / cm³, relative density = -5.1 %. (b) compacted silica sand mass; dry density =1.30 g / cm³, relative density = 73 %.

588

shear stress increased gradually and reduced rapidly after the subsidence. However the reduction of the total and shear stress were also affected by the reduction of the amount of the soil mass due to moving away.

On the other hand, the compacted soil mass failed very slowly. A crack occurred on the soil mass, then, a soil block slide down slowly along the slide plane. Another crack occurred, then the soil block slide down similarly. These retrogressive failures took place repeatedly. The pore water pressure increased gradually and dissipated due to the downward displacement. It seems that the soil mass around the shear zone began to expand since about 35 sec before the failure (16.9 min) as it is shown in the vertical displacement (Fig.4b). The dissipation of pore water pressure as a consequence of the expansion of the soil mass, which is observed in the strength tests, was not observed. In this test also, the rapid increase in pore water pressure was not observed but it fluctuated slowly.

4. VELOCITY OF LANDSLIDES

4.1.Comparison of the velocities in two types landslides

In the case of the loose and compacted soil mass, the failure mode and the velocity of the failure mass were very different. The difference of the velocity accompanied with the movement between there two landslides was compared that is shown in Fig.5. The velocity was calculated from downward displacement records shown in Fig.4a and b. The loose soil mass accelerated and slide down rapidly. The highest speed of the failure mass was 11.3 cm/sec. Then,

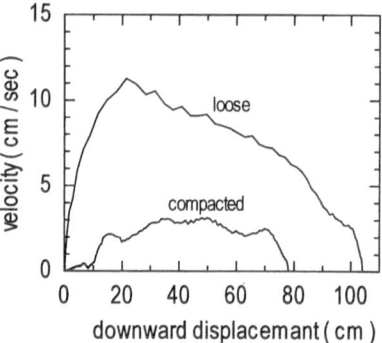

Fig.5 The comparison of the failure mass velocties of the loose and compacted soil mass

it decelerated gradually and stopped. On the other hand, the compacted soil mass did not accelerate and moved at nearly steady velocity.

4.2. Comparison of the time to landslides

The time to failure is predicted generally using the velocity of surface displacement applying the features of creep transformation. Several methods have been proposed. One of them is the method using the relationship between the inverse of velocity of surface displacement (1/v) and the time from the beginning of the movement (Fukuzono, 1985). The relationship between 1/v and time is approximately linear, and the point of the intersection between this approximate line and the time axis gives time of a start of the landslide. By using this method, in this study, the times to landslide in the loose and compacted soil masses were compared to illustrate the difference of initiation and the mechanism of landslides. Fig.6 compares the difference of time to final landslide in both cases. The inverses of the velocity were calculated severally from the displacement records during the movement like a creep. The loose soil masses changed to a landslide in only a few seconds. This shows that liquefied landslides take place very suddenly and unexpectedly. While, the compacted soil masses failed on the slide plane required much more time to change to a final landslide.

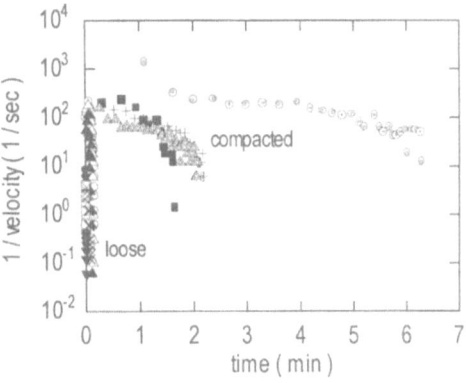

Fig.6 Comparison of the difference of times to final landslide from the start of movement between the loose and compacted soil mass

5. SUMMARY AND CONCLUSIONS

Generally the mechanism of liquefied landslides is explained in terms of soil mechanics as follows. Firstly, saturated cohesionless soil liquefies, which extends to a generation of excess pore water pressure. Then the strength of soil reduces suddenly, leading to an unbalanced driving force, then the failure mass accelerates.

In the experiment for a horizontal plain, it was observed that the very loose soil mass could subsided suddenly accompanied with the rapid generation of pore water pressure. The loss of total stress was also confirmed. While, in the compacted soil mass, the pore water pressure and the total stress fluctuated gently. Similar fluctuations of pore water pressure and stresses were also observed in the tests for slope. In the slope tests, the loose soil mass accelerated after the subsidence, therefore, a failure like a liquefied landslide occurred. As the result of stress and displacement measurements, the mechanism of the initiation and the acceleration of liquefied landslides are considered to be the result of the subsidence leading to a rapid generation of pore water pressure. While, in the compacted soil mass, it was found that liquefied landslides did not take place and the pore water pressure fluctuated slowly during the failure. In this series of experiments, the mechanism of liquefied landslides which has been so far studied in laboratory strength tests, was examined in a flume test that showed a good agreement with laboratory results.

The relationship between the inverse of velocity and time shows that fluid like landslides take place suddenly and unexpectedly, while, landslides which move on the slide plane started to move slowly. Although acceleration of the failure mass can not be observed in the laboratory strength tests, by means of flume experiments, the velocity of landslides could be calculated actually and also the sudden initiation and acceleration of liquefied landslides could be confirmed.

6. ACKNOWLEDGEMENTS

The author acknowledges Prof. Takeda, Nagoya University for his cooperation through this work. The author is also most grateful to Prof. Sassa, Kyoto University for his support. Other cooperators in Nagoya University, Assoc. Prof. Fukuoka, Kyoto University and Assoc. Prof. Kaibori, Hiroshima University for their advice are also acknowledged. This study was conducted for a master's thesis at Nagoya University.

REFERENCES

Eckersley, D. J. (1990) : Instrumented Laboratory Flowslides, Geotechnique, Vol. 40, No. 3, pp. 489-502.

Hiura, H., K.Sassa, K.Ohte and K. Kaibori (1991) : Case Study of the Nakaba Liquefied Landslide, Journal of Japan Landslide Society, Vol. 27, No. 4, pp.26-32.

Fukuzono, T. (1985) : A Method to Predict the Time of Slope Failure Caused by Rainfall Using the Inverse of Velosity of Surface Displacement, Journal of Japan Landslide Society, Vol. 22, No. 2, pp.8-13.

Sassa,K. (1984) : The Mechanism Starting Liquefied Landslides and Debris Flows, Proc. 4th Int. Sym.on Landslide, Toronto, Canada, Vol. 2, pp. 349-354.

An Experimental Study on the Rainfall-Induced-Flowslides

Gonghui WANG[*1] and Kyoji SASSA[*2]

*1 Graduate School of Science, Kyoto University, Kyoto, Japan
*2 Disaster Prevention Research Institute, Kyoto University, Kyoto, Japan

ABSTRACT By using a small flume, a series of tests were conducted on silica sands to initiate rainfall-induced-flowslides. Based on the monitoring of sliding distance, and pore pressures, excess pore pressure generation and flowslide motion were discussed. By changing the initial dry density of samples, the effect of initial void ratio on the excess pore pressure generation and the initiation of flowslides were examined.
Keywords: failure, laboratory test, flowslide, excess pore pressure, and rainfall.

1 INTRODUCTION

Flowslides refer to those failures that are characterized by rapid movement, fluid-like motion, very flat final slopes and correspondingly large run-out distance. They are mostly induced by rainfall, melted snow or some other factors. Hence, the mountain area in a sub-aqueous environment suffers more from the dangers of failure and subsequent flowsliding. For example, the Gamahara torrent debris flow happened on 6 December 1996 in Japan, in a relatively high temperature snow-melting season, killed 23 people who were constructing a check dam for debris flow. The Haribara debris flow on 10 July 1997 in Japan, initiated by heavy rainfall resulted in a loss of 21 lives and huge loss of properties. Both of these two catastrophic flowslides are considered to be the result of liquefaction (Sassa 1998), a process by which a soil mass suddenly loses a large proportion of its shear resistance and flows in a manner resembling a liquid until the shear stress acting on the mass become as low as the reduced shear resistance.

By now, significant research effort has been devoted to the cause of liquefaction. There is so much knowledge about it has been built up. However, many of them are based on the testing of small specimens under idealized conditions. The fabric effects, stress anisotropy, entrapped air, deformation of thin shear zones and some currently unknown processes, which may also have significant influences on the development of liquefaction, have not been thoroughly studied (Eckersley 1990). Along this line of thought, Eckersley (1990) triggered liquefaction and flowslides in coking coal stockpiles by raising the water table, and showed that excess pore pressures were generated during rather than before movement. However, in this work, the sliding displacement was only recorded by video cameras, and the observation of the deformation within the pile were not carried out. Kubota (1997) performed a series of tests on silica sand no.8 (Table 1) and loess to study the mechanism of liquefied landslides. It is concluded that the generation of pore pressure was a result of sudden initiation of subsidence. Another laboratory flowslide study was conducted on loose saturated fine quartz sands, and the motion of liquefied sands and pore pressures during moving were analyzed by Spence & Guymer (1997). However, how the excess pore pressure generates, and how the liquefaction and slide motion are affected by grain size, void ratio, and some other factors, are still poorly understood. Therefore, in the present work, by changing the initial void ratio, a series of laboratory experiments was performed on silica sand no.8 (Table 1) in a flume under artificial raining. Based on the results, the effects of initial void ratio on the generation of excess pore pressure and commenced flowslide were examined.

2 PROPERTIES OF THE SAMPLE

Silica sand no.8 (S8) was used in this research. Some main characteristics of this soil are listed in Table 1. The permeability coefficients range from 0.001 to 0.01(cm/s) when the void ratios vary from 0.9 to 1.3.

Table 1.Description of S8 used in the tests.

Property	D_{10} (mm)	D_{60} (mm)	D_{50} (mm)	ρ_{dmin} (g/cm^3)	ρ_{dmax} (g/cm^3)	Gs
Value	0.0060	0.061	0.057	0.99	1.42	2.63

3 ARRANGEMENT OF EXPERIMENTAL APPARATUS

The experimental apparatus was illustrated in Figure 1. The flume with transparent side was 180 cm long with 24 cm in width and 15 cm in height. To make the same friction between sand and the base of the flume with that in sands, silica sand was stuck on the surface of the flume base. At the places of longitudinal distance of $x_1 = 45$ cm, $x_2 = 90$ cm and $x_3 = 140$ cm from the upstream of the flume, three identical holes with diameter of 1.5 cm lined along the central line were dug as seen in Figure 1. Vinyl tubes were inserted into these holes with one end being flush with the flume bed (end A), and the other end (end B) connecting with pressure transducers, so that the pore water pressures could be measured. On the side of the hole at place of x_1, a normal stress transducer and a shear stress transducer were also installed to measure the stresses in the soils. Right down the place of x_1 a styrene foam ball with 2 cm in diameter connecting with a linear displacement transducer through a wire was laid there. During the test, the ball buried in the sample moves with the sample together, and then the time series of moving distance could be monitored. Because the resolution of this linear displacement transducer with large capacity was low, the displacement smaller than 0.8 mm can not be obtained. Therefore, a laser displacement sensor with resolution of 0.015 mm and capacity of 15 mm was used by fixing a target on the wire and shining a laser beam upon the target to get the small displacement. A weight of 160g was attached to the other end of the wire to balance the pulling

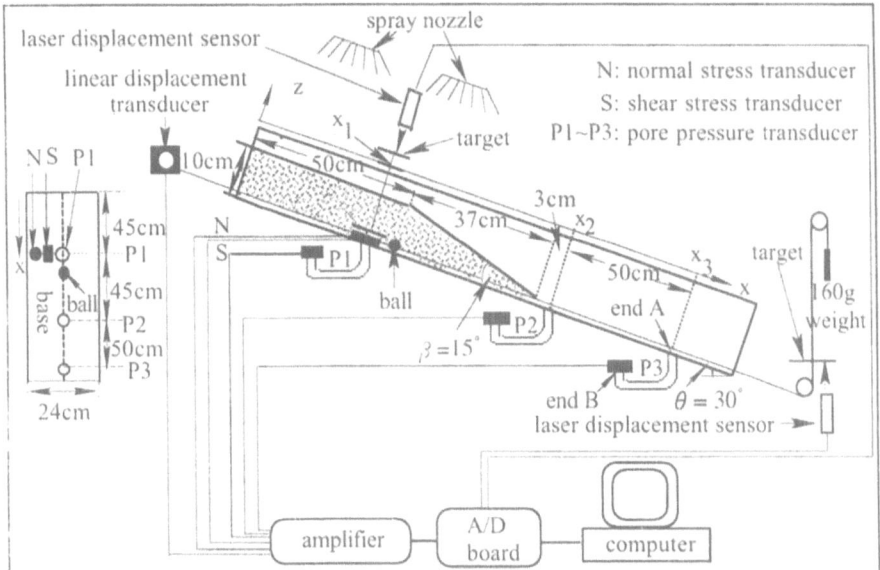

Fig.1 Arrangement of experimental apparatus

resistance of linear displacement transducer. At the place of 3 cm above the base at the place of x_1, a plate (3 cm$\times$3 cm) with many holes was laid parallel to the base. The plate was connected with another target outside of the soils, and then using another laser displacement sensor could monitor the normal displacement of the soils in the zone near the base. Above the flume two spray-nozzles were placed to form a uniform artificial rainfall. A video camera was used to monitor the whole test process from one side of the flume.

4 TEST PROCEDURE AND TEST CONDITION

4.1 Test procedure

The tests were conducted as follows.

(1) Installation of the transducers: Normal stress transducer and shear stress transducer were fixed flush with the surface of the flume base. The pore pressure transducers were installed with end B being 1 cm lower than end A and end B being ensured to be de-aired.

(2) Measurement of initial value: Because end B is lower than end A, there existed an initial water pressure acting on the pore pressure transducers before testing.

(3) Preparation and placement of sample: To get loose samples with different void ratio, water was added firstly to make the initial water content rise up to 5%, and then the sand was stirred evenly. After that, the sample was packed into the flume. To make the sample uniform, while packing, the sample was placed in a

series of layers of 2cm thickness parallel to the flume base, and then each layer was tamped. Finally, the superfluous parts were removed and the shape was made to be just like that shown in Figure 1. Initial dry density was determined from the oven-dried weight of used mass and sample's volume.

(4) Sprinkling water and recording: When the sprinkling began, the data logging system and video camera started to record. To get the quick change of pore pressure during quick failure, each of the instruments was logged at 0.05 second intervals. The start point was treated as the zero point of time.

4.2 Test condition

Table 2 presents the adopted test conditions and some test data. In all the tests, the flume angle was $\theta = 30°$. Density was expressed by density index I_d, $I_d=(e_{max}-e)/(e_{max} - e_{min})$, where e_{max} and e_{min} are the maximum void ratio and minimum void ratio of dry sand respectively. In all these tests the given rainfall intensity was 1.7 mm/min. In Table 2, e: Initial void ratio; I_d: Initial density index; Δu: Excess pore pressure; V_{max}: the maximum speed that flowslide had even reached; ΔS: Run-out distance before reached the sliding in very small velocity (Explained in detail in Section 5.2).

Table 2. Summary of flowslide test data.

Test	e	I_d	Δu	ΔS	V_{max}
1	1.29	0.46	0.10	13.4	4.1
2	1.41	0.31	0.48	33.9	8.6
3	1.46	0.25	0.53	*	*
4	1.46	0.25	0.62	*	*
5	1.48	0.22	1.01	81.2	36.8
6	1.50	0.20	1.02	79.0	35.0
7	1.51	0.19	0.99	29.6	8.5
8	1.58	0.10	0.64	45.6	29.8
9	1.68	-0.03	0.51	58.6	39.6
10	1.70	-0.05	0.13	11.8	16.1
11	1.71	-0.06	0.33	24.0	31.1
12	1.77	-0.14	0.20	39.3	18.1

*: Sliding distance did not obtain.

5 TEST PHENOMENON AND DISCUSSIONS

The detailed history of failure varied with each experiment. In general, the whole process required 20 to 30 minutes for slope wetting, and a few seconds for failure. After failure, excess pore pressure was generated and flowslide was initiated.

5.1 Excess pore pressure

Precise monitor of transient pore pressures in relation to slope movement demonstrated that excess pore pressures in the shear zone did not exist prior to the initiation of failure (Fig. 2). As shown, there was just very small pore pressure, u_b, generated immediately before the failure. After failure, pore pressure rose up suddenly, showed a sharp transient, thereafter fell down accompanying the moving because of the decreasing of mass height overlaying the base and the dissipation of pore pressures. It should be noted that the maximum value of pore pressure (u_m) measured was much greater than u_b.

Figure 2 presents the same fact that excess pore pressures were generated during rather than before the quick failure of slope, just as what had been pointed out by Eckersley (1990) and Wang et al. (1998). To analyze the relationship between excess pore pressure and failure motion, the excess pore pressure of each experiment was calculated. For simplification, excess pore pressure (Δu) was treated as the difference between u_m and u_b (Fig.2).

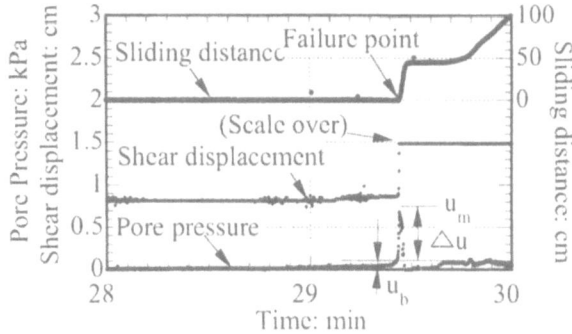

Fig.2 Time series of sliding distance and pore pressure (Test 8).

5.2 Flowslide motion

The temporal time series of shear displacement, sliding distance and sliding velocity before and after the failure is shown in Fig.3. As there existed relative moving between the soil layers during flowing (reviewed by video records), here the velocity refers to the velocity along the flume base because the sliding distance was got along the base. As shown, shear displacement along the base increased very slowly before the quick failure. After failure, the mass flowed downwards quickly, showed a rapid movement and large run-out distance.

Figure 4 shows the variation of velocity during moving in relation to the sliding distance. It is indicated that the moving process involved three stages.

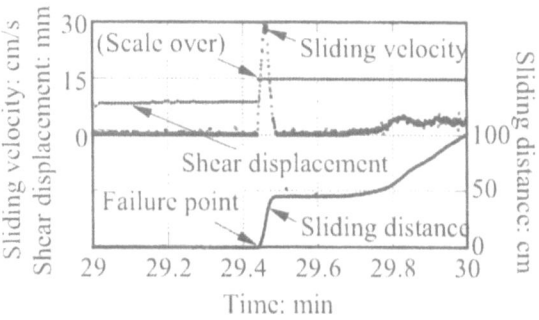

Fig. 3 Variation of shear displacement, sliding distance and sliding velocity (Test 8).

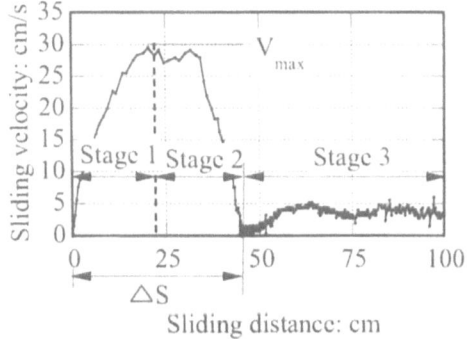

Fig.4 Variation of sliding velocity in relative to sliding distance (Test 8).

(1) Stage 1 (Accelerating process): After failure, the mass accelerated to slide until reached a certain velocity, which was the maximum value during all the sliding process. (2) Stage 2 (Decelerating process): After velocity reached its maximum value, the material decelerated to slide, with increase of sliding distance, the velocity reduced to a very small value. (3) Stage 3 (Slow sliding period): After a certain distance (ΔS) of rapid movement, the mass flowed downward finally at very small velocity, which kept almost a constant value.

This variation of flowslide velocity can be interpreted in relation to the generation of excess pore pressure. At the first stage, because excess pore pressure was generated, shear resistance was reduced and unbalanced driving force was resulted in. Then the accelerating of sliding was led to. With increase of slid distance, pore pressure dissipated and deceleration was resulted in, finally the driving force and shear resistance reached the balance and then the sliding velocity tend to the same value.

6 EFFECTS OF INITIAL VOID RATIO

As pointed out, void ratio plays an important role in the liquefaction (Castro 1969), special attention was therefore paid on the effects of initial void ratio on the generation of excess pore pressure during failure and commenced flowslide.

6.1 The effect on excess pore pressure

The test results show that the generated excess pore pressure for each test greatly depended on the initial density index. The excess pore pressures versus density indexes were plotted in Figure 5. As presented, with the increase of density index, the increment of pore pressure increased till a certain value of I_d (the I_d value at which excess

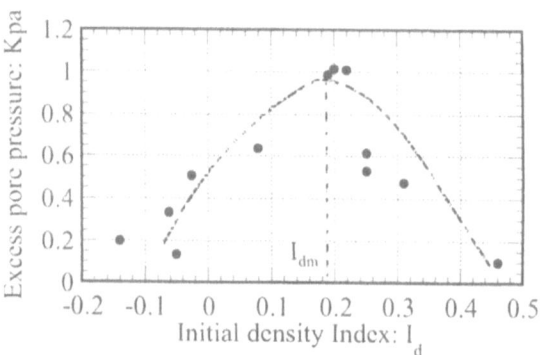

Fig. 5 Relationship between excess pore pressure and initial density index.

pore pressure reached its peak value is denoted as I_{dm}), and thereafter it became to decrease. Although there are many factors involved in affecting the initiation of excess pore pressure, in this situation, the main reason could be as follows. For the tests that I_d is smaller than I_{dm}, the high permeability is greater, and then, the dissipation will be quicker, it will be difficult for the pore pressure to build up. For the tests that I_d is greater than I_{dm}, because of the small void ratio, the volume contractant will be small, and then, there will be less pore water pressure initiated accompanying the failure. Therefore, according to this figure, it is concluded that there is an optimization density index for excess pore pressure generation. This density index could be different with the changing of grain size, sample's initial height, slope angle and some other factors.

6.2 The effect on the flowslide motion

The variations of sliding velocity in relative to sliding distance for some tests are shown in figure 6. As shown, after failure, the flowslides experienced the accelerating period, decelerating period and slow moving period, just as what had been described in Section 5.2. During the quick moving period (accelerating period and decelerating period), the maximum velocities and run-out distances are different for the samples with different initial density index. During the final slow moving period, there appeared no obvious differences between their moving velocities.

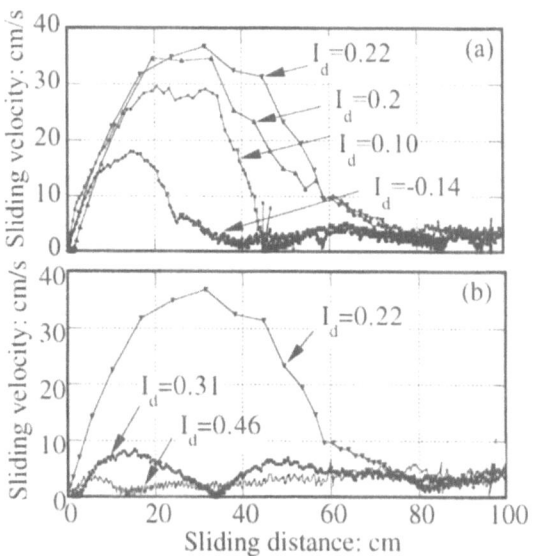

Fig. 6 Variation of sliding velocity in relative to sliding distance for sample with different initial density index. (a): for these tests with initial density index smaller than I_{dm} (Test 5, 6, 8, 12); (b): for these tests with initial density index greater than I_{dm} (Test 1, 2, 5).

Although there were some exception (Test 9, 10) because of the difficulties with test sample preparation and low precision in measurement, the maximum velocity and run-out distance of a flowslide became greater with increase of the initial density index (Fig.6a). There existed a turn point at which both of the maximum velocity and run-out distance reached their peak value. After that, with increase of initial density index, they became to decrease (Fig.6b). This variation tendency showed a good accordance with the change tendency of excess pore pressure generation in relation to initial density index (Figs.5, 6). As shown, the test with greater excess pore pressure generation usually had greater maximum velocity and larger run-out distance. Therefore it was concluded that great excess pore pressure generated during failure are mainly responsible for rapid movement and large run-out distance of commenced flowslide.

7 CONCLUSION

By using a small flume, a series of tests was conducted on silica sand (S8) to study the generation of excess pore pressure and flowslide motion under the condition of raining. By changing the initial dry density, the effects of initial void ratio on the excess pore pressure generation and flowslide motion were assessed. The main findings are summarized as follows.

(1) Excess pore pressures were generated during movement. Excess pore pressure appeared to be very sensitive to the initial density index. There is an optimization density index for excess pore pressure generation.

(2) The flowslide motions were different for the samples with different initial density index. The maximum velocity (V_{max}) and run-out distance (ΔS) of the flowslide during moving depend greatly on the initial density index. The relationship between these two values and initial density indexes showed a good accordance with that between excess pore pressure generation and the initial density index.

ACKNOWLEDGEMENTS

The authors want to thank Dr. Fukuoka, assistant professor of DPRI, Kyoto University, Mr.Wang Fawu, graduated student of Kyoto University, for their important suggestion and discussion throughout this research. Miss T. Kubota is to be thanked for offering the apparatus. The authors also wish to thank T. Sammori, H. Moriwaki, and I. Miyoshi for their warm help in this research.

REFFERENCE

Castro, G. (1969): Liquefaction of sands. Ph.D. Thesis, Harvard University, Mass.

Eckersley, J. D. (1990): Instrumented Laboratory Flowslides. Géotechnique Vol.40, No.3, PP.489-502.

G.H. Wang, K. Sassa, H. Fukuoka & F. W. Wang. (1998): Study on the excess pore pressure generation in laboratory-Induced-landslides. Proc. 8th Congress of the International Association of Engineering Geology. (In press).

Kubota, T. (1997): An experimental study on the mechanism of liquefied landslides. M.S. Thesis, Nagoya University, Japan.

Sassa, K. (1996): Prediction of earthquake induced landslides. Special Lecture of 7th International Symposium on Landslides, "Landslides", Vol.1. Rotterdam: Balkema. PP.115-132.

Sassa, K. (1998): Recent urban landslide disasters in Japan and their mechanisms. Proc. 2nd International Conference on Environmental Management, "Environmental Management", Vol.1. Rotterdam: Balkema. PP.47-58

Spence, K. J. & Guymer, I. (1997): Small-scale laboratory flowslides. Géotechnique Vol.47, No.5, PP.915-932.

Computed Powder Avalanche Impact Pressures on a Tunnel-bridge in Auβerfern-Tirol

Lambert RAMMER*[1], Horst SCHAFFHAUSER*[1] and Peter SAMPL*[2]

*1 Institut für Lawinen- und Wildbachforschung Forstliche
   Bundesversuchsanstalt, Hofburg-Rennweg 1, A-6020 Innsbruck, Austria
*2 AVL LIST GesmbH, Kleiststr. 48, A-8020 Graz, Austria

ABSTRACT A tunnel-bridge has been constructed 40 meters above the bottom of the Gröber avalanche path to improve the traffic safety along the Bschlaber access-road (Außerfern, Tirol). A maximum permissible static load of 20 kPa was allowed in the design criteria. The results of the three dimensional numerical simulation of the powder avalanche in the following are used as a basis for similar projects for the safety of railroads and highwaysystems in the Austrian Alps.
Keywords: Avalanche dynamic, dynamic pressure, tunnel bridge.

1. OBJECTIVIES AND METHODOLOGY

A powder avalanche model has been developed in cooperation with the Austrian Torrent and Avalanche Control (WLV), the AVL (Research Institute für Internal Combusting Engines) in Graz, Austria and the Institute for Avalanche and Torrent Research at the Austrian Federal Forest Research Station (FBVA) in Innsbruck. The results of the model are in good agreement with mapped field studies of natural disasters (H. J. Hufnagl, 1986). This mathematical model is used to forecast the run-out distance of the avalanche, as well as the vertical distribution of densities, velocities and total pressures. For the last two years this model has been used as a tool to solve engineering problems in cable-way constructions such as like the influence of designes relating to the behaviour of powder avalanches.

The AVL powder avalanche model is based on the fundamental differential equation-system (three-dimensional Reynolds-averaged Navier-Stokes equation), which governs the conservation of mass, momentum, snow concentration and the effects of turbulence are taken into a two equation turbulence model (Brandstätter, 1992; Sampl, 1993).

By means of fixed 3 D-Eulerian grid which is fitted by a coordinate transformation to the surface of the avalanche catchment and the SIMPLE algorithm (Patankar, 1980) the before mentioned equation are solved.

For calculating powder avalanches by the AVL/FIRE software there are USER-functions applied (s. fig. 1.) . The snow concentration is represented by using

600

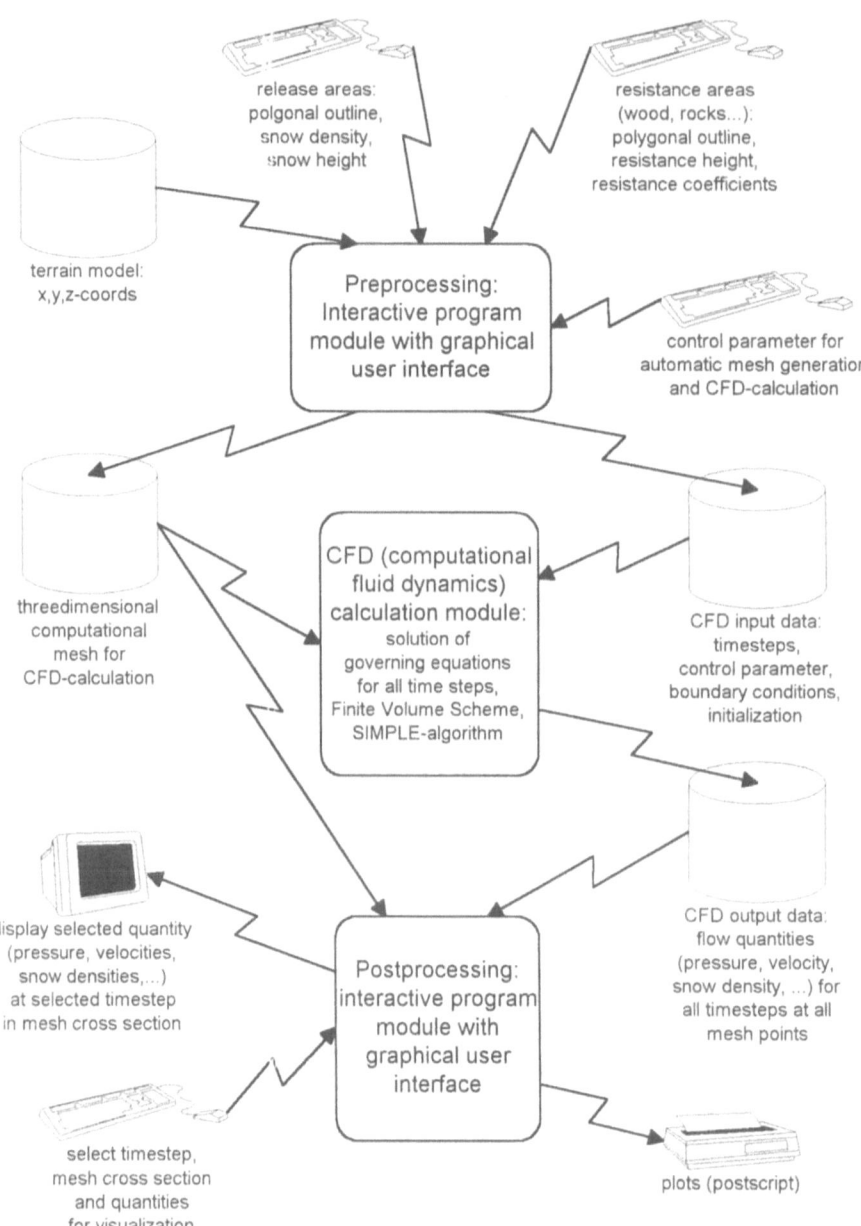

release areas:
polgonal outline,
snow density,
snow height

resistance areas
(wood, rocks...):
polygonal outline,
resistance height,
resistance coefficients

terrain model:
x,y,z-coords

Preprocessing:
Interactive program
module with graphical
user interface

control parameter for
automatic mesh generation
and CFD-calculation

threedimensional
computational
mesh for
CFD-calculation

CFD (computational
fluid dynamics)
calculation module:
solution of
governing equations
for all time steps,
Finite Volume Scheme,
SIMPLE-algorithm

CFD input data:
timesteps,
control parameter,
boundary conditions,
initialization

display selected quantity
(pressure, velocities,
snow densities,...)
at selected timestep
in mesh cross section

Postprocessing:
interactive program
module with
graphical user
interface

CFD output data:
flow quantities
(pressure, velocity,
snow density, ...) for
all timesteps at all
mesh points

select timestep,
mesh cross section
and quantities
for visualization

plots (postscript)

Fig. 1: Avalanche simulation
with AVL-FIRE

the passive scalar and the local density is calculated via URSDEN-function depending on the passive scalar. The motion of the snow-air mixture is caused by the .interaction of gravitation and the density differences. With a density of 30kg/m^3 in a USRINI-function is the avalanche started. At the bottom a fixed wall-boundary condition with wall-roughness is used. For the other forces (top, lateral) of the computational domain is set constant to one bar $(= P_{STAT} + P_{dyn})$.

Bulk Continuity

$$\frac{\partial \bar{\rho}}{\partial t} + \frac{\partial \left(\bar{\rho} \bar{u}_j \right)}{\partial x_j} = 0$$

Conservation of Snow Fraction

$$\frac{\partial \left(\rho_2 \bar{c} \right)}{\partial t} + \frac{\partial \left(\rho_2 \bar{c} \bar{u}_j \right)}{\partial x_j} = \frac{\partial \left(\rho_2 \overline{c' u'_j} \right)}{\partial x_j}$$

t	time
x_i	co-ordinates (i=1,2,3)
u_i	velocity (i=1,2,3)
p	pressure
ρ	density
τ_{ij}	stress tensor (i,j=1,2,3)
η	viscosity
c	snow volume fraction
η_t	turbulent viscosity
k	turbulent energy (from k-equation)

overbar indicates averaged quantities,
primed quantities represent turbulent fluctuation parts;
indices: 1=air, 2=snow

Bulk Momentum

$$\frac{\partial \left(\bar{\rho} \bar{u}_i \right)}{\partial t} + \frac{\partial}{\partial x_j} \left(\bar{\rho} \bar{u}_i \bar{u}_j \right) = -\frac{\partial \bar{p}}{\partial x_i} + \frac{\partial}{\partial x_j} \left(\bar{\tau}_{ij} - \bar{\rho} \overline{u'_i u'_j} \right) + \bar{\rho} g_i$$

$$\bar{\tau}_{ij} = \eta \left(\frac{\partial \bar{u}_i}{\partial x_j} + \frac{\partial \bar{u}_j}{\partial x_i} \right) - \frac{2}{3} \eta \frac{\partial \bar{u}_k}{\partial x_k} \delta_{ij}$$

Bulk Density

$$\bar{\rho} = (1 - \bar{c}) \cdot \rho_1 + \bar{c} \cdot \rho_2$$

Eddy Viscosity Concept

$$-\overline{\rho u'_i u'_j} = \eta_t \left(\frac{\partial \bar{u}_i}{\partial x_j} + \frac{\partial \bar{u}_j}{\partial x_i} \right) - \frac{2}{3} \left(\bar{\rho} k + \eta_t \frac{\partial \bar{u}_k}{\partial x_k} \right) \delta_{ij}$$

Eddy Diffusity Concept

$$-\rho_2 \overline{c' u'_i} = \frac{\rho_2}{\bar{\rho}} \frac{\eta_t}{\sigma_f} \frac{\partial \bar{c}}{\partial x_i}$$

The tunnel-bridge is an excellent mounting support for measuerment devices for the calibration of the AVL – Model. In the middle of the bridge a load panel of 1m² has been installed for recording of the impact pressures of the powder-portion. Measurements can be done of forces as well perpendicular to the panel as the shear forces to the panel. Four 4 wire full bridge load cells are used for normal load, and two load cells are used as well for the shear force in vertical direction as in horizontel direktion. On the opposite side of the bridge there are in three different levels pressure transducers to measure the suction caused by the flow of the snow-air mixture.

2,5m above and 2,5m below the bridge airspeed indicators are fixed on special mounts. Each of these instruments have two pressure transducers for recording the dynamic pressure and the atmospheric pressure of the powder component of the avalanche blowing across the bridge. Taking the difference between the pressures and estimating the density of the snow-air mixture, we can calculate the velocity of the powder component of the avalanche.

The behaviour of the dense flow part passing under the bridge is checked by radar (H. U. Gubler 1991) from SFISAR, Davos.

Control of measurements and data storage is done by programmable dataloggers. The sample rate for every measured value is 8 per second. Normally the loggers are in standby state. In this case data are allocated to a storage area with predefined size, so that 20 seconds of measurement can be recorded. This area is a ring memory, the newest data are always written over the oldest data. When an avalanche becomes apparent, i.e. when the normal load exceeds a predefined value, data storage in another storage area is triggered. This memory is configured as fill and stop and can hold about 6 minutes of measurement data. Controlling the measurements this way, all data from 20 seconds before and 6 minutes after the trigger event are recorded.

For safety of data a backup is done to a battey buffered memory module. In case of a breakdown of the power supply of the dataloggers, data will still be available in the memory module.

In the release area of the avalanche an automatic weather station has been installed to record all avalanche relevant snow and weather parameters. Data are transmitted by radio to the bridge.

Periodical checks of all the installation and data retrieval is done by modems and phone line.

Because there has been a lack of snow in the last winters, only two very small avalanches have been recorded yet. The first in Feb. 1993, when a maximum normal load to the load panel of 0,45kN/m² and a maximum speed of the dense flow part of 140km/h were recorded. Those data could not be used for model verification, but the reliability of the installation was shown. In the second case, in Jan. 1995 only a dense flow part of the avalanche was recorded with a maximum speed of about 100km/h.

2. DISCUSSION OF THE RESULTS

With the exception of one case the load cell devices have not been activated in the last winter periods as there has been no avalanches with a large powder component, to compare the results of the powder avalanche simulation with the experimental data.

In the release area of the Gröbertal avalanche the slope angles are between 25° to 48° and the track is inclined at >40°. The whole catchment is exposed to SE and SW directions, with a release area of 1.4 km². The geometrical boundary conditions (slope-morphological model) is automatically generated and the calculation grid is filled up with 60.000 mesh cells at which the calculation is timed in time steps of 0.1 seconds .

With a snow mass of 23.000 ts the avalanche simulation is started. During the impact phase-forty seconds after releasing and forty meters above the bottom - the powder part attains a maximum speed of 70 m sec^{-1} with a total maximum pressure of 17.5 kPa (s. fig.2).   Because of difficulties in the software development it was not possible to install the tunnel-bridge in the calculation grid so in this case the flow is calculated without it.

3. CONCLUSION

Before the AVL powder avalanche model has been developed in Austria impact forces of the powder component were calculated by means of the VOELLMY formula (1955), modified by the graphic method of LEYS (1975). The expertice for the permissible static load (20 kPa) was elaborated by the local head of the Austrian Torrent and Avalanche Control (F. DRAGOSITS 1988). The important goal within the development of the avalanche modelling at the FBVA in Innsbruck is the validation of the AVL-model. For the calibration of the numerical simulation model exist only a few exactly mapped catastrophic avalanches. So this led to the consideration of installing measurement devices like radar or load cells which are located in the avalanche path. This gives us the possibility by the aid of artificially released avalanches to measure avalanches dynamics paramter for the calibration of the model.

Except for the optimizing of hazard zoning and mapping should the three dimensional numerical AVL powder snow simulation model be a tool for the solution of engineering problems relating to the avalanche control.

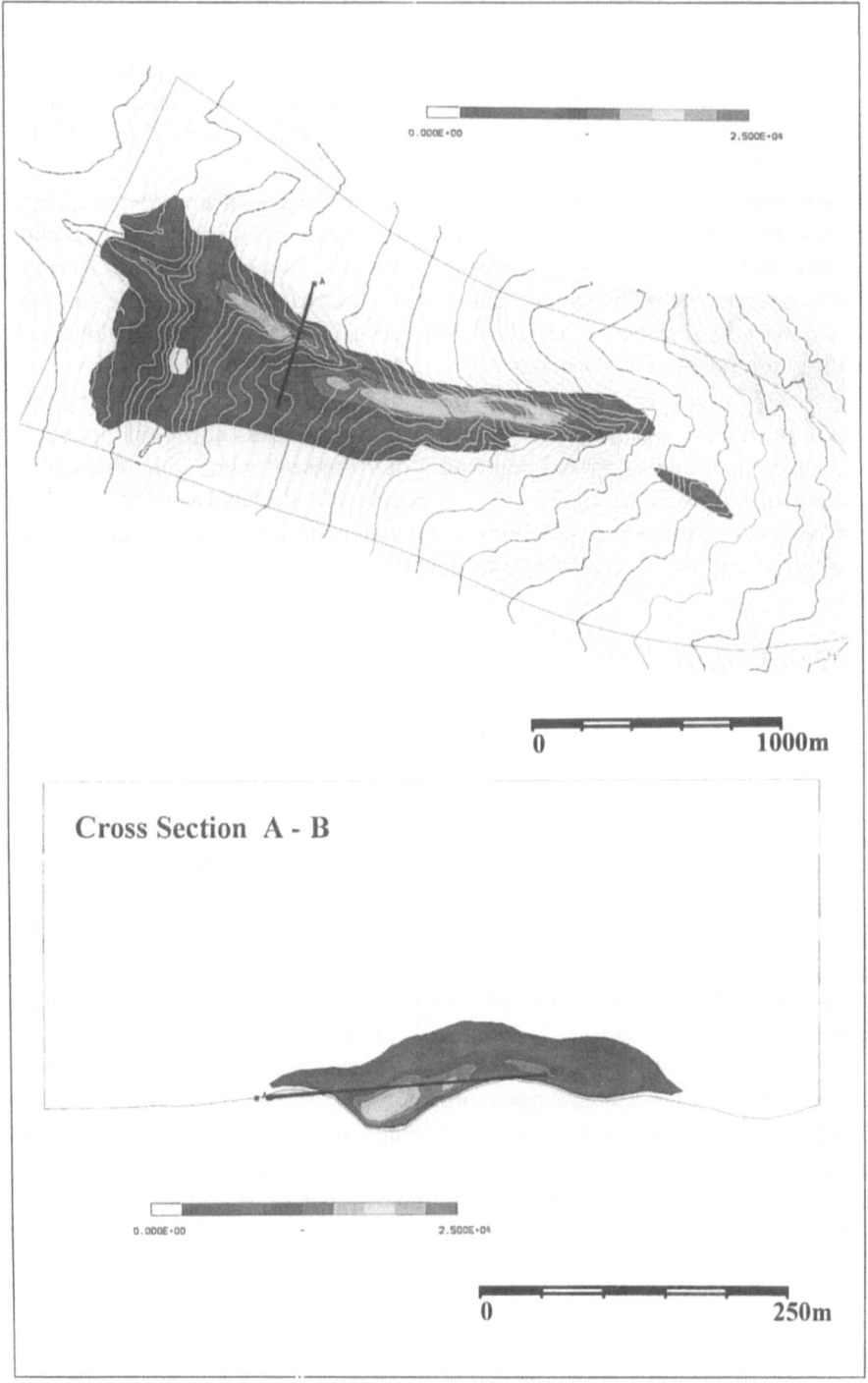

Fig. 2: Gröbertal Avalanche (A-B=Tunnelbridge) :
Dynamic Peak Pressure (0-25kPa) 40m above ground,
Cross Section : Dynamic Peak Pressure (0-25kPa).

REFERENCES

Brandstätter, W., F. Hagen, P. Sampl and H. Schaffhauser. 1992. Three-dimensional simulation of powder snow avalanches in complex terrain. *Zeitschrift des Vereins der Dipl. Ingenieure der WLV Österreichs.* Wildbach- und Lawinenverbauung . 56 (120), 107-137.

Leys, E. 1975. Einige Grundlagen für die Lawinenzonenplanung.
In: Tagungspublikation Interpraevent 1975. Klagenfurt Vol. 1, der Forschungsgesellschaft für vorbeugende Hochwasserbekämpfung, 375-390.

Patankar, S. V. 1980. Numerical Heat Transfer and Fluid Flow. MC Graw-Hill Book Company. New York. 1980.

Rammer L. 1992. Dynamic Measurements on Powder Avalanches. In: Tagungspublikation Interpraevent 1992, Bem. Vol 2, der Forschungsgesellschaft für vorbeugende Hochwasserbekämpfung, 421-431.

Sampl, P. 1993. Current Status of the AVL Avalanche Simulation Model - Numerical Simulation of Dry Snow Avalanches. In: Proceedings of the „Pierre Beghin" international workshop on rapid gravitational mass movements. 1993. Grenoble, CEMAGREF, 169-282.

Voellmy, A. 1995. Über die Zerstörkraft von Lawinen. *Schweiz. Bauzeitung.* **73**, 3-25.

Evaluation of Slope-failure-debris Mass Using a Digital Elevation Model with Stereo Pair Aerophotographs

Satoshi TSUCHIYA

Faculty of Agriculture, Shizuoka University, Shizuoka 422-8529, Japan

ABSTRACT   In order to measure the land mass of slope failure, terrain elevations were derived from stereo pair aerophotographs, and three-dimensional perspective terrain data of the slope failures that occurred in the northwestern Nagano Prefecture of Japan were created. Based on these data, the debris mass of slope failures was evaluated and the retrogressive failure process was considered.
Keywords: aerophotograph, slope failure, debris flow, retrogressive failure.

1. INTRODUCTION

A debris flow occurred and left 14 people dead on December 6 1996 at the Gamaharasawa Torrent on a branch of the Himekawa river in the northwestern region of the Nagano Prefecture in Japan. This debris disaster was triggered by a slope failure occurring at an altitude of 1300 m in the upper stream of the Gamaharasawa Torrent (Kawakami,1997). In July of the previous year (1995) a similarly large slope failure had occurred as the result of heavy rains (49 mm of precipitation in an hour, and a total of 395 mm in the Himekawa region). And, in April of 1997, the slope repeatedly failed as a result of moderate rains and melting snow. In this paper, we estimate the land mass of slope failures using three-dimensional terrain data automatically generated from stereo pair aerophotographs, and we consider the failure of this slope based on the events of 1995 to 1997.

Slope failure occurred
on 6 Dec., 1996

Gamahara Torrent

Fig. 1 Study site and surround (aerophotograph taken on December 7, 1996).

2. STUDY AREA AND STEREO CORRELATION PROCEDURE

Three stereo pair aerophotographs, taken on July 19, 1995 (1:10000 scale), on December 7, 1996 (the day after the debris disaster, 1:6000 scale, about 1900 m of flight altitude), and on April 26, 1997 (1:12000 scale, about 2800 m of flight altitude) were prepared. A stereo pair of gray level images was scanned with an accuracy of 150 dots per inch using each of these aerophotographs and recorded in digital format. Figure 1 shows a gray level image of the study area taken on December 7, 1997, centering on the slope failure scar and debris flow, an area of approximately 0.18 km^2.

Stereo correlation requires matching pixel locations in the left (L) and right (R) images of the stereo pair (Mori, 1981,1985). As shown in Fig. 2, this is accomplished by centering a correlation window on a pixel location (L_{xy}) in the left image and, assuming no relief displacement, defining a larger search window centered at a corresponding pixel location ($R_{xy'}$) in the right image. The correlation window is then systematically moved to all pixel locations within the search window, and the correlation between the pixels in the left and right images is computed. The pixel location at which the maximum correlation is recorded defines the match between L_{xy} and R_{xy}. A difference in pixel location in the x-direction between L_{xy} and $R_{xy'}$ is the parallax difference dx. The dx value is proportional to the elevation between the measured point and a standard plane (Welch, R et al., 1992).

For purpose of analysis, it is necessary to determine the size of the correlation window on the image. Generally, an increase in the size of the correlation window decrease in a close roughness of relief and tree canopy. On the other hand, if we assume that the window is reduced in size, the probability of mismatching is increased. In this study, experiments were conducted with correlation windows ranging from 15-by-15 to 45-by-45 pixels in 3 pixel steps, with the optimum results obtained using a 21-by-21 correlation window. To obtain the coefficient of correlation, we moved the correlation window of the right image along a horizontal line in a right-hand direction while the left window was fixed, and repeated this movement a total of three times.

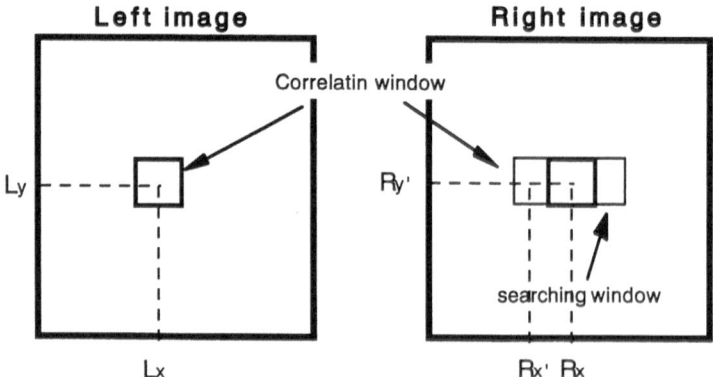

Fig. 2 Generalized stereo correlation point matching procedure.

The location of the pixel on the right image that correspond to that of the left pixel was determined at the maximum value of the coefficient of correlation. However, in the case of apparent mismatching, the correlation coefficient was calculated using an average brightness value of 3-by-3 pixels. Pass points for the orientation were selected using a map of 1:5000 scale, however, the accuracy of the orientation may have been affected by forest or snow-cover. Consequently the analysis accuracy of the obtained three-dimensional coordinates may be within the range of 3 to 5 m. 3.

3. ANALYSIS RESULTS

Figure 3 shows a three-dimensional perspective display derived from aerophotographs taken in 1995, 1996 and 1997 at the dip of a 20 degree angle. As shown in this figure, the heads and circumference of the slope failure originally were steep slope. There were two relatively large-scale slope failures, previous to the 1996 slope failure, on the each of the left and right bank of the Gamaharasawa Torrent at an altitude of 1250 to 1300 m above sea level. It appears that the slope failure which occurred on the right bank triggered the debris flow, although the decline in residual mass following the 1995 failure was also a factor. The black colored portion of 1996 image may be attributable to melting snow and rockfall scarring. There was the flow scar evidence in the middle slope, probably ground water was sure to spout out on the steep cliff and its underneath. In addition, the 1997 image confirms that the head mass largely collapsed in the left side and flowed down into the Gamaharasawa Torrent. It follows from this that the July 1995 slope failure was caused by the heavy rains at an altitude of 1300 m above sea level and, that it failed mainly at the right-side heads in December 1996, then again at the left-side heads in April 1997.

Next, we consider the slope failure process in terms of the generated terrain map. Figure 4 shows the contour map based on the obtained three-dimensional data. The measured elevation indicates the surface of the forest canopy which we could only restrict to a rang of 10 to 20 m in height due to the thick forest covering in the July 1995 aerophotograph, other aerophotograph featured snow cover a top of fallen leaves, and in these the elevation might have corresponded to the snow-surface elevation. In both these cases, therefore the measured elevation is closer to the actual grand surface elevation. However, on the aerophotograph taken on December 7, 1996, it is difficult to distinguish the terrain texture in detail due to the high reflection on the photograph caused by the fresh snow cover. Because of this, there is little difference in the shape of the contour lines among the three images.

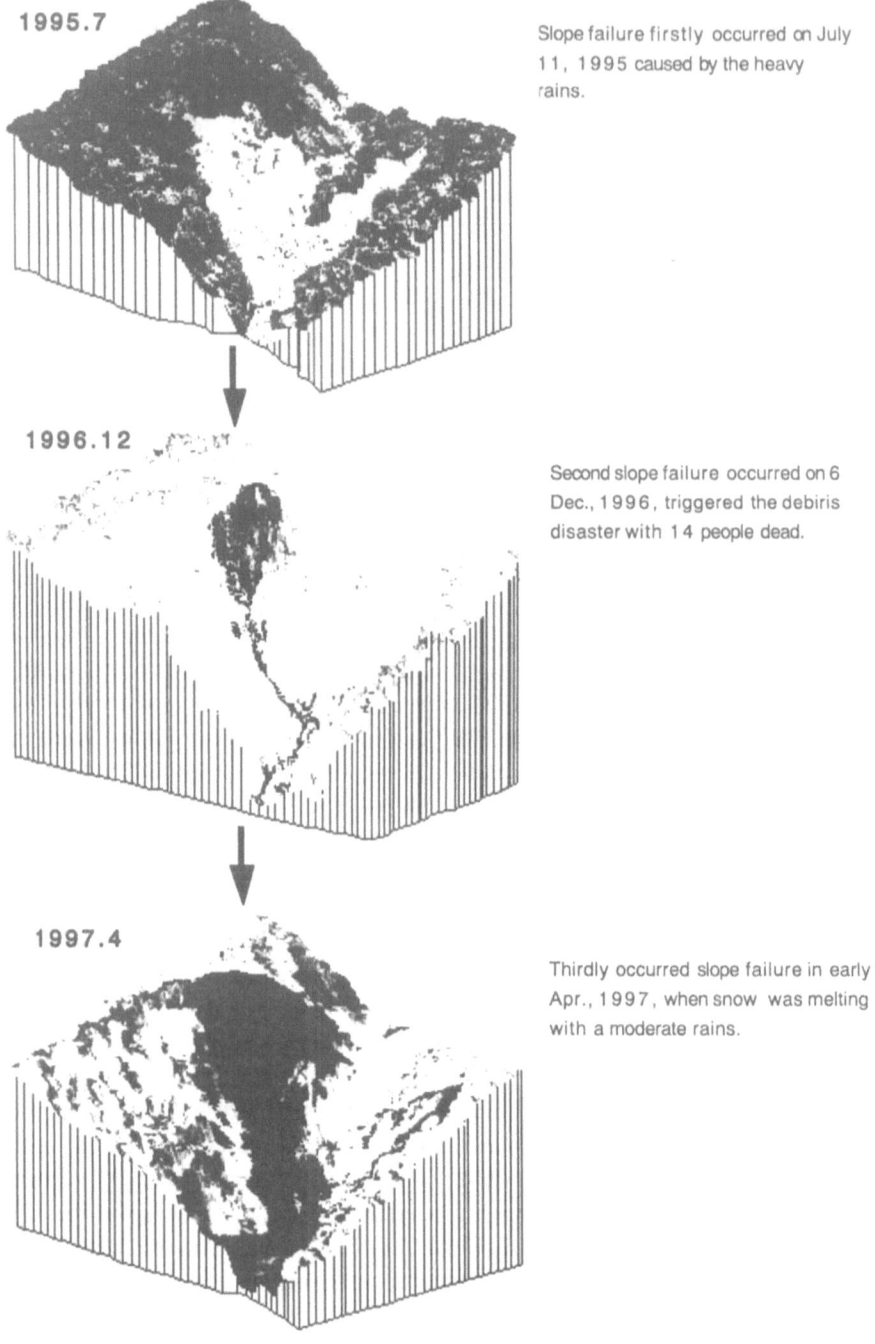

1995.7

Slope failure firstly occurred on July 11, 1995 caused by the heavy rains.

1996.12

Second slope failure occurred on 6 Dec., 1996, triggered the debiris disaster with 14 people dead.

1997.4

Thirdly occurred slope failure in early Apr., 1997, when snow was melting with a moderate rains.

Fig. 3 Perspective viewing of slope failures derived from stereo aerophotographs.

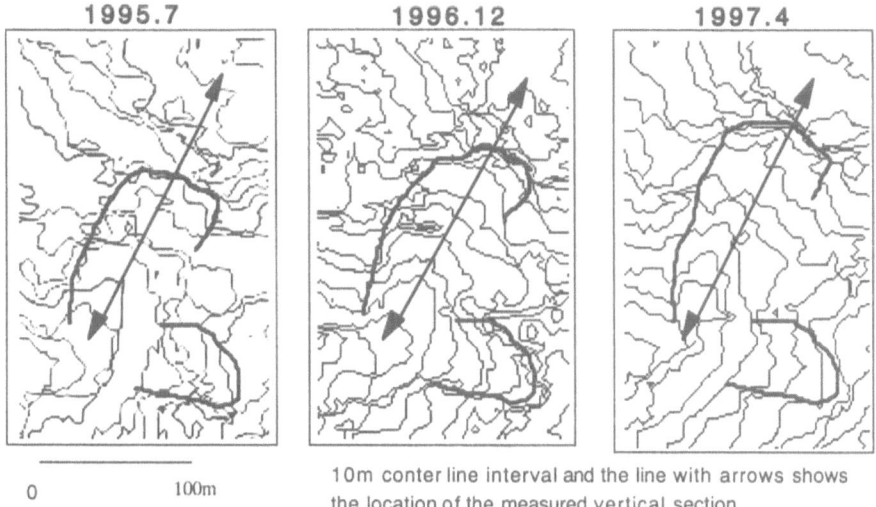

0   100m

10m conter line interval and the line with arrows shows
the location of the measured vertical section

Fig. 4 Contour map obtained by the analysis (marked curves show the boundary of the slope failure).

We marked the boundary between the collapsed and un-collapsed area based on the degree of curvature of contours in Fig. 4. Comparing the collapsed area with 3 times, it appears that the slope failure occurred on December 6, 1996, that failed and enlarged on the right side head area of the 1995 slope failure. And then the 1997 slope failure failed enlarge at the heads of 1996 slope failure, repeatedly.

Figure 5 shows vertical sections of the three slope failures, the location of which are located in Fig. 4 with arrows. The scale of the slope failure occurring in July, 1995 was measured as about 110 m in horizontal, 50 to 60 m in width and about 20 m in average depth. The failed mass was evaluated as being at least 50,000 m³. This volume was 50 to 100 times larger than that of shallow slope failures commonly occurring in Japan (Tsuchiya S. et al., 1996). And comparing the vertical sections of 1995 and 1996, the latter head scar was located at a horizontal distance of about 20 m to the back side.

We may conclude that most of the debris from this slope flowed down into the Gamaharasawa Torrent in 1997. In 1996, the mass of 25,000 to 30,000 m³ debris was derived from the retrogressive section area in Fig. 5 and the enlarged plain area in Fig. 4. The 1997 slope failure was determined to have an approximately 20,000 m³ debris mass using the same method.

4. SITE MEASUREMENTS

On July 27, 1997 when the snow first began to melt, we surveyed the terrain using a laser measurement system. As shown in Fig. 6, snow remained in places but debris flow scar and groundwater outlets were also clearly visible. The debris flow scar appeared in early April, and the groundwater discharge which seeped out through the volcanic rocks at the bottom of the steep cliff, centered on the collapse and flowed down. There is no doubt that the slope failure occurrence was closely associated with ground water seepage. In site surveying of the terrain, in order to measure the scale of the slope failure occurring in early April 1997, we surveyed a vertical section along the center of the collapse and horizontal sections.

Figure 6 shows the results of the comparison between the vertical and horizontal sections. Based on Fig. 7, the height of the failed cliff was about 45 m, the elevation between the junction of the Gamaharasawa Torrent and the bottom of cliff was about 80 m, and the horizontal distance was 150 m. The average slope angle ranged from 25 to 30 degrees, and the valley was typically V shaped. The width at the bottom of the cliff was 90 m, the width at the section located at a horizontal distance 120 m up from the conjunction was 70 m, the width in the area of the conjunction was 40 m.

It might be concluded that the slope failure formed in the shape of an unfolded fan, result in progressing and enlarging to the upward retrogressively. The slope failure mass occurring in early April 1997 was determined to range from 17,000 to 20,000 m^3 because the retrogressive distance between the December 1996 failure and the April 1997 failure was in the range of 15 to 20 m (Fig. 7), the average cliff height was about 25 m, and the width of the collapse was about 45 m as a half of width at the bottom of the cliff in Fig. 7.

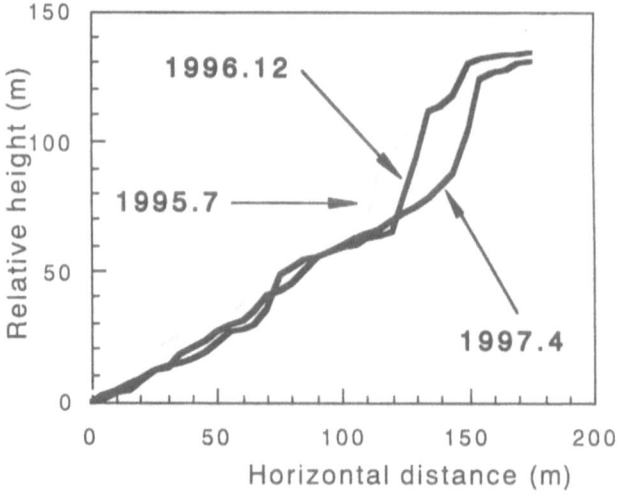

Fig. 5 Comparison of the terrain profile derived from stereo aerophotographs.

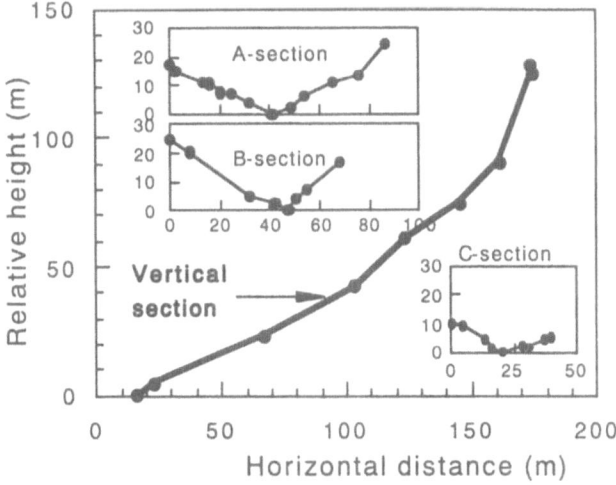

Fig. 6 Measured terrain profiles (A-section: at the bottom of the failure cliff; B-section: at the junction point; C-section: at the end of collapse).

Vertical cross section

A : Criff   B : ground water seepage   C : under ground water flow
D : Kuruma bed   E : scar of debris flow

Fig. 7 The view of the collapse (photo taken on April 4, 1997). The height of the cliff was about 45 m, the elevation between the junction of the Gamaharasawa Torrent and the bottom of the cliff was about 80 m, the horizontal distance was 150 m, and the average slope angle was about 25 to 30 degrees.

5. CONCLUSIONS

In creating three-dimensional terrain elevations from stereo pair aerophotographs, the orientation of the photographs is a very important factor on the accuracy of the results. However, it is doubtful that the stereo pair aerophotographs taken on December 7, 1996 just after the debris disaster were oriented with a high degree of accuracy, due to fresh snow cover. Because of this, we used average brightness values of 3-by-3 pixels in the correlation window on the images. Consequently, the accuracy of the analysis may be in a range of 3 to 5 m on the ground. We conclude the following in relation to the slope failure that occurred at 1300 m above sea level in the Gamaharasawa Torrent:

1) The July 11, 1995 slope failure was caused by heavy rains had a mass of at least 50,000 m^3.

2) The December 6, 1996 slope failure had a mass of 25,000 to 30,000 m^3, and the early April, 1997 failure had that of 17,000 to 20,000 m^3.

3) The slope has been failing and enlarging retrogressively in an upstream direction.

4) The slope failures that occurred at about 1300 m in altitude formed the shape of an unfolded fan, probably due to the rising of groundwater.

REFERENCES

Kawakami, H. (1997): The study of debris disaster occurring in Otari village of Nagano Prefecture in 1997, Report of the scientific grant of the Ministry Education, 6.1-8 (in Japanese).

Mori, C., Hattor, S., Uchida, O. (1985): Stereo plotting from aerial photographs with patch-by-patch correlation, Remote sensing and aerial photograph, Vol. 24, No. 1, pp. 13-22 (in Japanese with English abstract).

Mori, C., Hattori, S., Imai, K., Ogawa, I. (1981): Automatic contour production using image correlation aerial stereo Photographs, Remote sensing and aerial photograph, vol. 20, No. 4, pp. 4-13 (in Japanese with English abstract).

Tsuchiya, S., Oosaka, O., Tamaki, S. (1996): On the shape of the vertical section and soil properties in shallow landslide, Symposium on the natural disasters in Chubu district. pp. 101-102 (in Japanese).

Welch, R. & Papacharalampos, D. (1992): Three-dimensional terrain visualization on personal computers, Photogrammetric engineering & remote sensing, vol. 58, No. 1, pp. 71-75.

Energy Approach to Evaluation of Grain Crushing

Dimitri A. VANKOV[*1] and Kyoji SASSA[*2]

*1 Graduate School of Science, Kyoto University, Kyoto, Japan
*2 Disaster Prevention Research Institute, Kyoto University, Kyoto 611-0011, Japan

ABSTRACT.   This paper describes the results of investigation of particle crushing of two kinds of sandy sample, performed by ring shear apparatus. For interpretation of obtained data the energy approach was applied. Both samples show that particle breakage depends on dissipated energy during testing. Character of this dependence is different for samples with different initial grain size distribution. No correlation with the type of loading (cyclic or monotone) was observed. Pattern of dependence of density of the sample on   particle breakage suggests that type of loading could have significant influence. Using of dissipation energy as criteria for describing particle breakage behavior of soils seems to be reliable and useful method
Keywords: grain crushing, Marsal's breakage factor, energy approach, ring shear apparatus,

1. INTRODUCTION

The majority of important engineering properties of sandy soils such as strength behavior, volume change, pore pressure developments and permeability depend on the integrity of soil particles or ability of particles to crush under applied load.

In landslide mechanism of aggregate or particle crushing can play decisive role. The recently proposed theory of sliding surface liquefaction (Sassa 1996), is based on the assumption, that pore pressure could be generated because of negative dilatancy due to particle crushing and comminuting along the sliding zone. This process takes place in medium and dense soils, which usually are not considered as liquefaction prone deposits. Therefore, the task of identification and quantification of particle crushing is very important.

As it was stated by Lade et al. (1996) that virtually all investigations involving soil testing above normal engineering geological pressure have resulted in considerable particle breakage. However, particle crushing may occur in relatively low pressures.

Among factors that affect the amount of particle breakage in a soil, are follows (Lade et al. 1996):

- stress level and stress magnitude; larger amount of particles generated when stress level are higher; the quantity of particle breakage is also a function of time; even under constant stress of sufficient magnitude particle breakage continues with time, but at decreasing rates;
- particle size; as the particle size increases, particle crushing is also increases;
- increasing the particle angularity increases particle breakage;
- well-graded soils do not break down as easily as uniform soils;

- as the relative density increases, the amount of particle breakage decreases;
- mineral composition; increasing the mineral hardness decreases the amount of particle crushing.

The most popular device for investigation of grain crushing phenomena is high pressure (up to 50 MPa) triaxial testing device. In case of triaxial testing only relatively small sample deformations is possible. Consequently, for producing sufficient amount of crushed material researcher is forced to use high pressure. However, the ring shear apparatus is very suitable equipment for this matter due to its ability to limitless shear displacement. Since the sample in the ring shear apparatus subjected to direct shear condition, the shearing proceeds along well-determined shear zone. Also important, that is very easy to collect the sample for grain size distribution analysis after ring shear test.

2. EXPERIMENTAL OUTLINES

2.1 Sample Preparation

Two types of samples were used in this investigation. The first type

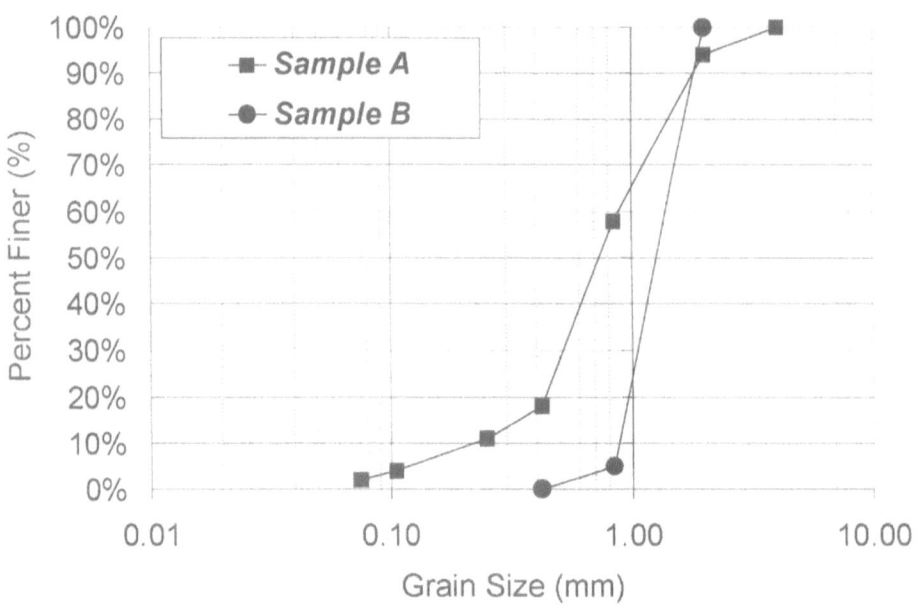

Fig. 1. The initial grain size distribution curves

(Sample A) is soil belongs to Plio-Pleistocene Osaka widely distributed in Osaka-Kobe area. Particles have angular shape. In the mineral composition quartz and albite are predominant minerals. The procedure of preparation of samples to the tests were follows. Because of Osaka formation includes some clay layers, at first, sandy soil was separated from the clayey blocks. Then sandy soil was dried at the temperature of 105°C during 48 hours and cooled. After that soil was disintegrated by means of rubber hammer and sieved through the sieve with diameter 4.00 mm. The sieved soil assumed to be ready for the test. Samples were kept in hermetically closed metal boxes for preserving them from hygroscope moisturizing.

The second type of sample (Sample B) was prepared from the Osaka formation soil by choosing of faction with restricted grain size distribution. Dried soil was sieved through column of sieves with opening diameters 2.00 mm, 0.84mm and 0.42 mm. The fraction from 0.84mm sieve was chosen as sample. **The grain size distribution curves are represented in Fig. 1.**

2.2 Apparatus

The cyclic loading ring shear apparatus (DPRI-4) was used in this research, which was modified from the cyclic loading ring shear apparatus (DPRI-3), developed by Sassa et al. in 1991. The most important parameters of apparatus are plotted in Tab.1. In the ring shear apparatus, (an annular donut-like specimen (Fig. 2), subjected to normal stress is confined laterally, and ultimately

Tab. 1. Technical parameters of ring shear apparatus (DPRI-4)

Parameter	Value
Shear box:	
-inner diameter (mm)	210
- outer diameter (mm)	290
Maximal normal stress (kPa)	2500
Maximum shear speed (cm/sec)	18.33
Resolution of gap control system (mm)	0.001
Maximum data acquisition rate (readings/second)	200
Maximum frequency of loading (Hz)	5

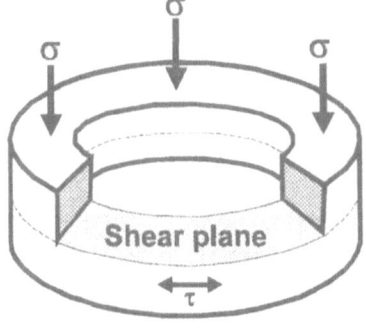

Fig. 2. The ring shear sample

caused to rupture on a plane of relative rotary motion, so-called shear plane. The total normal stress and shear stress being transmitted through the soil across the shear plane are precisely known. The inner diameter of the sample is sufficiently large to the outer diameter for uncertainties arising from a assumed non-uniform stress distribution across the shear plane to be reduced to an acceptable level. The lower half of the sample is carried on a rotating table driven by shear motor. The upper half of the sample is reacted via arm against one or pair of fixed proving rings (transducers) that measuring the shear resistance value.

The shear torque could be applied as cyclic or in monotone loading. In both type of loading shear speed control mode and shear torque control mode are provided.

Computer with data acquisition rate of 200 readings per second registers all test data. The values of normal stress, shear torque, shear resistance, pore

618

pressure, vertical displacement of loading plates and shear displacement are recorded. The structure of ring shear apparatus is described in details by Vankov, 1996.

3. TEST PROCEDURE

Once sample was settled and the shear box assembled consolidation pressure was applied. Only dry sandy soil was used in investigation.

Tab. 2. Sample A; Test parameters

№	Normal Stress (kPa)	Shear Speed (cm/sec)	Shear Displacement (cm)	Dissipated Energy (kJ/m²)	Initial Density (g/cm³)	Final Density (g/cm³)	Marsal's Factor (%)
1	197	0.66	10.21	0.964	1.36	1.39	5
2	197	0.59	100.27	11.533	1.38	1.44	5
3	197	2.6	1037.38	123.211	1.38	1.48	15
4	197	2.71	13250.01	1687.973	1.38	1.54	18
5	197	4.5	101733.63	13195.947	1.39	1.58	22
6	980	3.06	10706.57	6961.296	1.43	1.63	37

Tab. 3. Sample B; Test parameters for monotone loading

№	Normal Stress (kPa)	Shear Speed (cm/sec)	Shear Displacement (cm)	Dissipated Energy (kJ/m²)	Initial Density (g/cm³)	Final Density (g/cm³)	Marsal's Factor (%)
7	500	0.23	584.233	1976.84	1.44	1.54	54
8	740	0.22	609.237	3161.51	1.45	1.56	62
9	250	0.24	573.641	864.46	1.42	1.48	39
10	1240	0.22	84.727	593.83	1.47	1.59	60
11	500	0.31	71.804	230.72	1.42	1.47	42
12	250	0.31	109.134	162.93	1.42	1.44	34

Tab. 4. Sample B; Test parameters for cyclic loading

№	Normal Stress (kPa)	Shear Stress Amplitude (kPa)	№ of Cycles Applied	Loading Frequency (Hz)	Dissipated Energy (kJ/m²)	Initial Density (g/cm³)	Final Density (g/cm³)	Marsal's Factor (%)
13	500	400	100	0.50	107.30	1.45	1.64	14
14	500	400	100	0.40	104.46	1.44	1.59	11
15	500	400	100	0.25	94.11	1.49	1.59	12
16	500	400	100	0.20	105.80	1.42	1.55	10
17	500	320	100	0.10	1354.68	1.44	1.64	46
18	500	400	100	0.20	96.84	1.44	1.56	6
19	500	390	50	0.10	53.91	1.44	1.55	5
20	500	400	25	0.10	48.72	1.44	1.56	3

Consolidation of such type of soil proceeds very fast, in our research it was 30 minutes. All samples were normally consolidated. After that shear stress was applied accordingly to test program. For Sample A all tests were done in monotone loading mode. For the Sample B both monotone and cyclic loading tests were conducted. All test parameters are summarized in Tabs. 2, 3 and 4. Totally 20 tests were carried out on both samples, with different type of loading, number of cycles and frequency of loading (for cyclic tests), normal stress and shear displacement. Monotone tests were conducted with constant shear speed, however the value of the shear speed was vary from test to test. Fukuoka (1991) demonstrated that in dry granular material variation of the friction angle during shear is much affected by the shear displacement and also by the normal stress, but not by the shear speed value.

After shearing was finished, the shear zone was removed for grain-size distribution analysis. The column of the sieves (with opening diameters 4.00, 2.00, 0.84, 0.42, 0.25, 0.105, and 0.075 mm) was shaken during 35-40 minutes.

4. TEST RESULTS AND DISCUSSION

In this research breakage factor defined by Marsal 1967, was used for quantification of particle crushing. His method involves the change in individual particle sizes between the initial and final grain size distributions. The differences in the percentage retained are computed for each sieve size. This difference will be either positive or negative. Marsal's breakage factor, B, is the sum of the differences having the same sign as presented in Fig.3. This difference could be either positive or negative. The lower limit of Marsal's index is zero percent, and has a theoretical upper limit of 100%. All obtained values of this breakage factor are presen-ted in Tab. 2, 3 and 4.

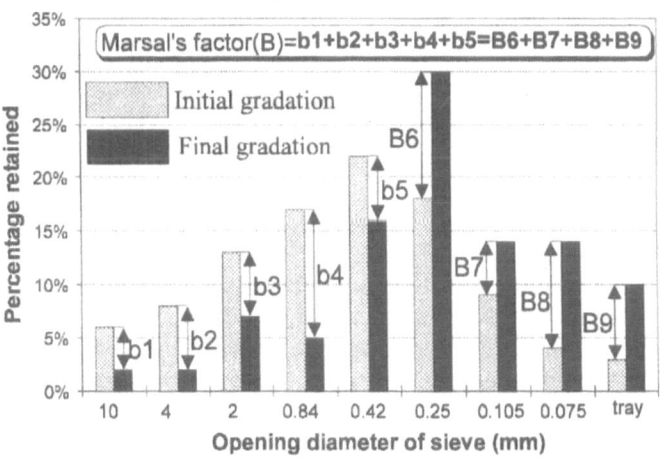

Fig.3 Definition of Marsal's breakage factor

Lade et al. (1996) pointed out that correlations between particle crushing and effective mean normal stress (using the triaxial testing de-vice), at failure as well as void ratio at failure were not fully satisfactory. Most soil mechanics parameters derived from the analysis of test data will not effectively correlate with the amount of particle breakage observed in various types of tests. However, it has been demonstrated that increases in confining pressure as well as shearing of the soil to larger stress and strain levels increase the amount of particle breakage.

Considering this, the magnitude of total energy input per unit volume of specimen during the test seem to be a more appropriate parameter to use for correlation with particle breakage. The reason for this is based on the fact that computations of energy incorporate the magnitudes of both stress and strains, both of which greatly affect the amount of particle crushing. A form of this type of correlations was proposed earlier (Miura&Ohara 1979). However, they proposed using plastic work instead of total input energy. They extracted the elastic energy from the total energy to obtain the plastic work. The amount of elastic energy present in specimens in which particle crushing is significant is very small when compared with to the magnitude of plastic work. Consequently, there is a little difference between the total work and plastic work, and the use of total energy simplifies the computational procedure considerably.

All mentioned above is true for our case with the only difference that in present research the ring shear apparatus was employed. Since the shearing proceeds along predetermined shear zone, and shear displacement has dimension of length but not dimensionless as strain is, the total input energy is related to this shear zone. It should be noted, that in the ring shear apparatus, the total input energy is not equal the energy applied to the soil, because certain friction between upper and lower confining rings exists. This friction could be measured with appreciable preciseness before test and eliminated during calculations after.

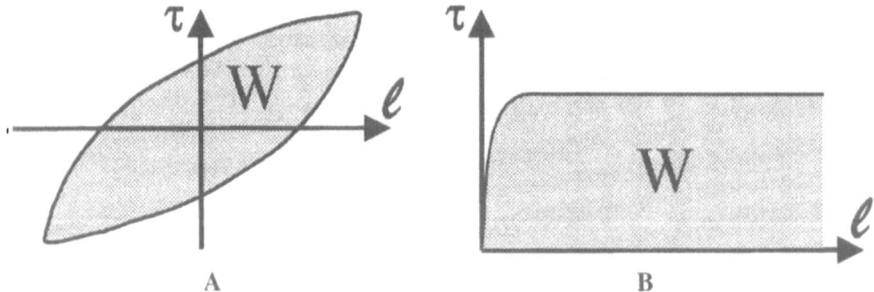

Fig. 4. Dissipated energy for cyclic (A) and monotone (B) loading

However, the term "input energy" to be not very correct and in this research the term "dissipated energy" will be used instead. In case of cyclic loading the energy dissipated in the soil during one cycle of loading is represented by the area of hysteresis loop, and for monotone loading it is area under shear resistance line, as it shown in (Fig. 4). The total dissipated energy could be calculated by means of simple formula (Figueroa 1994) with minor modification to direct shear conditions as follows:

$$W = \sum_{i=1}^{n-1} \frac{1}{2}\left(\tau_i + \tau_{i+1}\right)\left(l_{i+1} - l_i\right) \qquad (1)$$

where τ is shear resistance; l is shear displacement; n number of points recorded. The Equation (1) is differ from that proposed by Figueroa by substitution shear strain on the shear displacement. Obtained energy, in such case, will be dissipated energy per unit area of shear zone, not input energy per unit volume. Results of calculations of the total dissipated energy during tests are presented in Tabs. 2, 3 and 4.

In Fig. 5 the Marsal's particle breakage factor is plotted versus total dissipated energy. The tendency of grain crushing for the natural soil sample (Sample A) is very different from that of artificially prepared sample (Sample B).

For Sample A, the value of particle breakage factor is gradually increases with total dissipated energy increase. This is because the natural sample contains particles of different diameter, and probably different shape. The weakest particles are crushed within the low level of dissipated energy. With increasing of amount of total dissipated energy more and more particles are involved into process of breakage and this ultimately lead to increase of Marsal's breakage factor.

However, the dependence of breakage factor on the total dissipated energy for Sample B is absolutely different. It looks like that until the total dissipated energy will reach the certain threshold value (in our case it does appear to be about 50-60 kJ/m^2) there is no grain crushing at all. When the amount of total dissipated energy overcomes this threshold value grain crushing started and Marsal's particle breakage factor rises up to significant value (34%) within rather narrow range (48-162 kJ/m^2) of total dissipated energy. Then, the rate of dependence of grain crushing on dissipated energy substantially decreases. For instance, changes in the Marsal's breakage from 34% to 62% correspond to range of the total dissipated energy from 162 kJ/m^2 to 3161 kJ/m^2. This behavior is the same for both cyclic and monotone loading. Such phenomenon proved the suggestion that grains with certain size, and probably, shape could be crushed only within relatively narrow range of total dissipated energy.

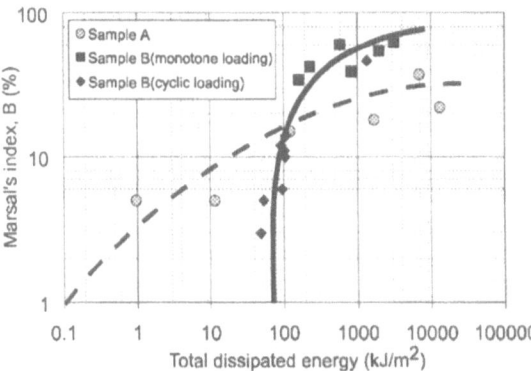

Fig. 5. Marsal's breakage factor related to total dissipated energy

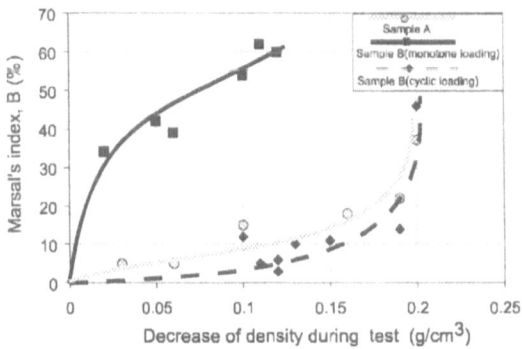

Fig. 6. Marsal's breakage factor related to decrease of density

In Fig. 6 Marsal 's breakage index is plotted versus decrease of density during the test, obtained by difference between initial density of sample before shearing and final density of sample after shearing. It is interesting, that densification of Sample A is very similar to densification of Sample B during cyclic loading. Both of them demonstrate rapid increase of particle breakage after reaching certain level of density. This could be explained that during cyclic loading, shear displacement is small and particle oscillates almost on the same place. This lead to optimum distribution of soil particles within given volume. In case of Sample A, the voids are filled by the particles of the lesser diameter, while shearing. In both cases grain crushing starts when all voids are filled and further redistribution of soil particles are impossible.

On contrary, Sample B being tested under monotone loading conditions shows that within very narrow range of the total dissipated energy the particle breakage increases significantly (up to 34%) and then the rate of the rate of dependence of grain crushing changes in density decreases. It should be added, that the highest particle breakage value (62%) correspond to decrease of the density of 0.11 g/cm^3. In Sample B all particles has almost the same size and similar, probably angular, shape. In this case when shearing started, grain crushing should start immediately due to shape of the particles. However, after certain amount of particles has been crushed, shear zone became more homogeneous and dependence of particle crushing on the density proceeds steadily.

5. CONCLUSIONS

After series of ring shear test was performed on the dry samples, the following conclusions could be drawn.

- Both samples show that particle breakage depends on dissipated energy during testing.
- Character of this dependence is different for samples with different initial grain size distribution. No correlation with the type of loading (cyclic or monotone) was observed.
- Pattern of dependence of density of the sample on particle breakage suggests that type of loading could have significant influence.
- Using of dissipation energy as criteria for describing particle breakage behavior of soils seems to be reliable and useful method.

REFERENCES

Figueroa, J.L. et al. (1994): Evaluation of Soil Liquefaction by Energy Principles. Journal of Geotechnical Engineering ASCE :Vol. 120: No 9: 1554-1569.

Fukuoka H., (1991): Variation of the Friction Angle of Granular Material in the High-Speed High-Stress Ring Shear Apparatus; Bulletin of the Disaster Prevention Research Institute, Kyoto University, Vol. 41, pp. 243-279.

Lade, P. V., Yamamuro, J. A., Bopp, P. A. (1996): Significance of Particle Crushing in Granular Materials; Journal of Geotechnical Engineering, Vol. 122, No. 4, pp. 309-316.

Miura, N and Ohara, S. (1979): particle Crushing of the Decomposed Granite Soil Under Shear Stresses, Soils and Foundations, !9(3), pp. 1-14.

Sassa, K.(1996): Prediction of Earthquake Induced Landslides; Special Lecture, Proc. 7th Int. Symp. on Landslides, BALKEMA, Vol.1, 115-132

Vankov, D. A. (1997): Behavior of Dry and Saturated Sandy Soil Subjected to Different Frequencies of Loading. Master Thesis, Fac. of Science, Kyoto University.

Comparison of Shear Behaviour of sandy soils by Ring-Shear Test with Conventional Shear Tests

Yasuhiko OKADA[*1], Kyoji SASSA[*2] and Hiroshi FUKUOKA[*2]

*1 Graduate School of Science, Kyoto University, Kyoto, Japan
*2 Disaster Prevention Research Institute, Kyoto University, Kyoto 611-0011, Japan

ABSTRACT   The relation of shear behaviour of dry Toyoura standard sand between the ring-shear test, triaxial compression test and two types of shear-box tests was compared. Test results shows that the peak internal friction angles from the ring-shear test fitted those from shear-box tests, while there was a certain difference from triaxial compression test. Then, the liquefaction potential of saturated Osaka-group coarse sandy soil by ring-shear test and triaxial compression test was examined. The liquefaction was observed only in the loose state by the triaxial compression test, while it appeared in all three tests by the ring-shear test. The stress paths of liquefaction in the medium and dense state by the ring-shear test is different from liquefaction in the loose state by both tests.
Keywords: ring-shear apparatus, peak friction angle, liquefaction.

1. INTRODUCTION

Shear tests to measure shear strength of soils play a quite important role in the practical engineering. Shear-box test and triaxial compression test have been primarily used in the present investigation because of its simplicity. And now it is known that each of shear apparatuses have considerable varieties of peak internal friction angle over ranges of void ratio, but the relation between ring-shear apparatus and other shearing apparatuses has not been investigated. Shear-box test and ring-shear test simulate or reproduce the stress condition along a shear plane, while triaxial compression test simulates the relation between stress and strain within a homogeneous sample. Thus, these tests have different aims respectively. However, shear strength parameters of c and ϕ obtained by both tests are used for analysis of landslides. Therefore, to compare the exact shear characteristics by both tests is important. The soil movement like landslide is regarded phenomenologically as the shearing, internal friction angle of soil is crucial factor for risk assessment. The difference of internal friction angles from different shearing apparatuses and the appropriateness or position of ring-shear apparatus with great advantage of limitless shear displacement over other shearing apparatuses must be investigated.

Within some types of landslide, such as sliding, liquefaction and creep, liquefaction-induced landslide can be most hazardous, however most research into liquefaction in landslides have been conducted using triaxial test (Seed 1968).

Recently, a concept of "Sliding Surface Liquefaction" was newly proposed by Sassa et al. (Sassa 1996, Sassa et al. 1996) through undrained ring-shear tests. But this is only from ring-shear test and it has never been performed by triaxial compression test. As for this new concept which can explain rapid landslides, it is important to perform some experiments from the viewpoint of which condition or which type of shear test can produce it and compare the test results. In this paper, explanation of ring-shear test is presented first, next is comparison of peak internal friction angles from the ring-shear test, the triaxial compression test and the shear-box tests. Finally shear behaviour of saturated sandy soils from the undrained ring-shear test and undrained triaxial compression test are examined.

2. Experimental Setup

2.1 Apparatus

The fifth and sixth version in a family of ring-shear apparatus developed by Disaster Prevention Research Institute, Kyoto University, in 1996 (DPRI-5 and DPRI-6) were employed. These had been designated as DPRI-5 for a smaller shear box with 120 mm (inside) and 180 mm (outside) diameters, and DPRI-6 for a larger shear box with 250 mm (inside) and 350 mm (outside) diameters (Sassa 1998). For the triaxial compression apparatus, the cylindrical sample has 100 mm diameter and 200 mm height before consolidation. Minor principal stress σ_3 is kept constant by air servo. Constant strain in sample is given by the uplift of the bottom plate of the screw jack.

2.2 Test Condition and Procedure

Shear-speed-controlled drained ring-shear tests on the dry Toyoura standard sand and undrained ring-shear tests on the saturated Osaka-Group coarse sandy soil were conducted. Likewise, strain-controlled drained and undrained triaxial compression tests on dry and saturated samples were also performed to compare the shear behaviour. Shear speed for the ring-shear test was 0.3 mm/sec up to 1.0 m shear displacement and 3.0 mm/sec up to 10.0 m with constant normal stress σ=200 kPa, and compression speed for triaxial test was 0.05 mm/sec with constant minor principal stress σ_3=200 kPa. Tests with constant minor principal stress σ_3=100 kPa were added to get failure envelope lines for dry test.

Tests on the dry Toyoura standard sand with three different void ratio (loose, medium and dense) by drained ring-shear tests and triaxial compression tests are referred to as Test-R-TL, R-TM, R-TD, T-TL1, T-TL2 and so forth. In the same way, tests on saturated Osaka-group coarse sandy soil by undrained ring-shear test and triaxial compression test are Test-R-OL, T-OD and so on, they are all listed in Table 1.

Toyoura standard sand has a mean diameter of D_{50}=0.17 mm, a uniformity coefficient of Uc=1.7 and specific gravity of 2.64. Osaka-Group coarse, sandy soil

Table 1 Test No. and conditions

Test No.	Apparatus	Dry or saturated	e_0	σ or σ_3 (kPa)	stress history
R-TL	Ring	Dry	0.89	σ=200	1.5
R-TM	Ring	Dry	0.72	σ=200	4
R-TD	Ring	Dry	0.65	σ=200	2
T-TL1	Triaxial	Dry	0.91	σ_3=200	1
T-TL2	Triaxial	Dry	0.89	σ_3=100	2
T-TM1	Triaxial	Dry	0.73	σ_3=200	1
T-TM2	Triaxial	Dry	0.74	σ_3=100	2
T-TD1	Triaxial	Dry	0.63	σ_3=200	1
T-TD2	Triaxial	Dry	0.64	σ_3=100	2
R-OL	Ring	Saturated	0.75	σ=200	1
R-OM	Ring	Saturated	0.61	σ=200	1
R-OD	Ring	Saturated	0.55	σ=200	2.5
T-OL	Triaxial	Saturated	0.76	σ_3=200	1
T-OM	Triaxial	Saturated	0.60	σ_3=200	1
T-OD	Triaxial	Saturated	0.55	σ_3=200	1

Where stress history means the ratio of maximum normal stress σ (minor principal stress σ_3) divided by those (σ, σ_3) of initial value on testing respectively. Ring means Ring-shear apparatus and Triaxial means Triaxial compression apparatus.

has a mean diameter of D_{50}=0.9 mm, a uniformity coefficient of Uc=5.2 and specific gravity of 2.61.

Experimental procedures are as following;

(1) pluviate the dry-ovened sample in the shear box or cylindrical membrane. To make medium and dense sample, some tamping was conducted. (2) pass CO_2 gas through the sample from the lower drainage valve to the upper one for around 2 hours for undrained test. (3) infiltrate de-aired water into the sample in the same way for CO_2 for undrained test. (4) consolidate the sample at 50 kPa normal stress or minor principal stress, and then, measure the B_D (=Δu / $\Delta \sigma$) or B (=Δu / $\Delta \sigma_3$) value to confirm the degree of saturation for undrained test. Tests on samples with these values higher than 0.95 were used in this paper. (5) consolidate under certain value to adjust the void ratio (loose, medium and dense). (6) consolidate under the initial normal stress σ=200kPa and minor principal stress σ_3 (200 or 100 kPa). (7) start shearing.

3. Check of Undrained Ring-Shear Test

In order to investigate pore water pressure generation before and after failure the capability of undrained testing should be produced. To check whether ring-shear apparatus can keep undrained condition during long shearing, the preliminary ring-shear test on water instead of soil was conducted. The structure of undrained shear box was drawn in Fig. 1. To prevent of water leakage from

626

contact face between upper ring and rubber edge on the lower ring, the contact force between them is controlled by a servo oil piston with a high precision of 0.001mm. If no water leakage was permitted during shearing, no volume change would happen with pore water pressure keeping constant value. The result of this preliminary test on water which was conducted under 150 kPa normal stress, 10.0 mm/sec shear speed and up to 10^4 mm shear displacement is presented in Fig. 2. Vertical displacement is the change of sample height, thus the increase of this value indicates the volume shrinkage, namely, water leakage takes place. The increment of vertical displacement in this figure is almost zero during long shearing, so undrained condition could be kept in this test. In ring-shear test, the friction between upper ring and rubber edge

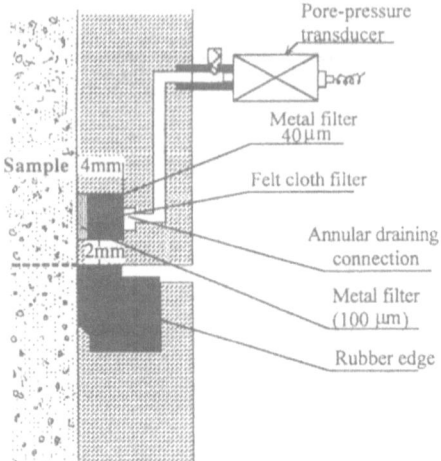

Fig. 1 The structure of undrained shear box (DPRI-5, DPRI-6). (Sassa 1997)

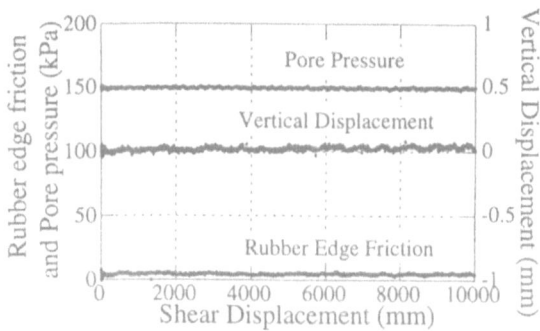

Fig. 2 Relation between shear displacement vs. rubber edge friction, pore pressure and vertical displacement on water (DPRI-5). Normal stress =150 kPa. Shear speed=10.0mm/sec.

definitely exists, and this value should be subtracted from the shear resistance of the soil measured by ring-shear test, it was 5.0kPa from this preliminary test result.

4. Comparison of Peak Internal Friction Angles from Drained Ring-Shear Test and Triaxial Compression Test and shear-box test on Dry Toyoura Standard Sand

The comparison of the internal friction angles of Toyoura Standard sand from triaxial compression test and some kinds of direct shear test was conducted. The results from Mikasa-Type (Japanese Geotechnical Society 1979) constant stress shear-box tests and Mikasa-Type constant volume shear-box tests had considerable difference of peak internal friction angles over the ranges of denser condition, even though little difference was obtained over looser condition. These differences are affected by the influence of side friction between sample and shear box when sample tends to dilate in dense condition. Even if normal stress was kept constant, the actual normal stress on the shear plane should be larger than applied normal stress because of the influence of side friction in dense condition. It causes higher peak internal friction angles for constant stress shear-box tests. On the

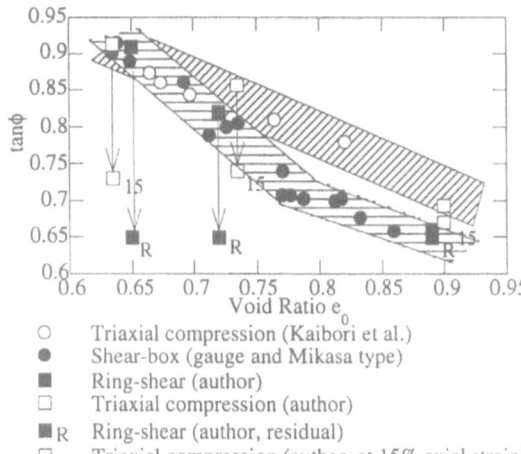

○	Triaxial compression (Kaibori et al.)
●	Shear-box (gauge and Mikasa type)
■	Ring-shear (author)
□	Triaxial compression (author)
■R	Ring-shear (author, residual)
□15	Triaxial compression (author, at 15% axial strain)

Fig. 3 $\tan\phi$ of Toyoura standard sand by some types of shearing apparatuses vs e_0.

contrary, no volume change during shearing happens for constant volume shear-box tests. So the influence of side friction is quite little for this tests (Sassa 1982). To remove side friction effect, improved shear-box test, gauge-type shear-box test, was developed by Sassa (1982). This type of shear-box test can produce constant normal stress applied on the shear plane in practice. It makes it possible to avoid the increment of normal stress in dense condition by the influence of side friction during shearing. Obtained peak internal friction angles from ring-shear tests and triaxial compression tests by author added to those from Mikasa-type constant volume shear-box test, gauge-type constant stress and volume shear-box tests and triaxial compression test (Kaibori 1980) are plotted in Fig. 3. Shear-box test is representative of Mikasa-type shear-box test and Gauge-type shear-box test in Fig.3. Ring-shear apparatuses (DPRI-5 and DPRI-6) are designed for removing the influence of side friction between sample and shear box. As guessed readily, peak internal friction angles from ring-shear tests showed a close consistency with those from Mikasa-type constant-volume, gauge-type shear-box tests. Namely, ring-shear test produces the same peak internal friction angles by above-mentioned shear-box test which removes the influence of side friction. The internal friction angles from triaxial compression tests by author were calculated by the least square method with an assumption of zero cohesion from the each two samples with almost same void ratio (loose, medium and dense), for example Test-T-TL1 and T-TL2. These do not have considerable difference between other results from triaxial compression tests by Kaibori et al. In loose condition the higher peak friction angles are obtained by triaxial compression tests than those from other shear tests and in dense condition nearly equal. This fact seems to be affected by the relation between the diameter of shear box and the height of sample of shear-box test (Inoue 1964) and ring-shear test, more detailed information are necessary.

Internal friction angles from ring-shear tests after shearing up to 10^4mm and from triaxial compression tests at 15% axial strain (30mm axial displacement) are also plotted in Fig. 3. Friction angles from ring-shear tests converged to around 33 degrees (tanϕ=0.65) with long shearing, and samples reached the same residual state. On the other hand, those from triaxial compression tests did not converge and sample did not reach a residual state. Residual friction angle is the main issue for reactivated landslides. Ring-shear test which can only produce the residual friction angle is one of the most practical shearing apparatuses and plays especially important role in landslide research.

5. Examination of Shear Behaviour between Undrained Ring-Shear Test and undrained Triaxial Compression Test on the Saturated Osaka-Group Coarse Sandy Soil

In ring-shear test simulating the Hyogo-ken Nanbu Earthquake (1995) on the Osaka-group coarse sandy soil, the stress path went downward along the failure line and very low apparent friction angle was obtained. It is called sliding surface liquefaction (Sassa 1996), that is explained later (Fig. 4). The investigation into sliding surface liquefaction on loess with different void ratio by different over consolidation was conducted by Zhang (1996). But up to the present, the effect of the difference of void ratio of sandy soils, such as Osaka-group coarse sandy soil, on how sliding surface liquefaction would occur, have not been clarified. And it is not also clear which type of shear apparatus can produce sliding surface liquefaction. The Osaka-group coarse sandy soil was selected as sample, undrained ring-shear test and triaxial compression test on samples with different void ratio (loose, medium and dense) were performed respectively. For triaxial compression test, the stress state over 15% axial strain may not give correct value. However, up to 26.2% axial strain (52.3mm axial displacement) which is the maximum of linear transducer, tests were continued and searched as a reference in these tests.

5.1. Mass Liquefaction and Sliding Surface Liquefaction

The difference of conventional mass liquefaction and sliding surface liquefaction is illustrated in Fig. 4 (Sassa 1996). Mass liquefaction is normally caused by destruction of the meta-stable structure for a loose saturated soil mass. Grain crushing is not necessary. The stress path usually fails before reaching the failure line. On the contrary, sliding surface liquefaction takes place only along the sliding surface. The destruction of loose structure is not required. In the latter case, excess pore pressure is generated by grain crushing and comminution. So as long as grains are crushed and comminuted along the sliding surface and it causes volume shrinkage, the sliding surface liquefaction is considered to take place with any void ratio. Stress path in the sliding surface liquefaction necessarily reaches the failure line, thereafter, it moves downward. Stress paths of ring-shear test on loose sample (Test-R-OL) and triaxial compression test on loose sample (Test-T-OL) are shown in Fig. 5(a) and (b) respectively. In both tests, pore pressure monotonically continued to increase without decrease, and very low effective normal stress or effective mean stress were obtained. The notable difference is that failure line was observed from Test-R-OL but not from Test-T-OL. The stress path of Test-T-OL is typical one of conventional mass liquefaction, destruction of meta-stable soil structure produced such as the state that the soil particles float in the water. It behaves something like liquid. Sliding and subsequent grain crushing is not necessary. As for Test-R-OL, the stress path reached the failure line at 5.78 mm shear displacement, thereafter, grain crushing took place due to sliding. Finally, the stress path came to a standstill when effective normal stress got as little as not causing any grain crushing.

To the samples with almost same void ratio,

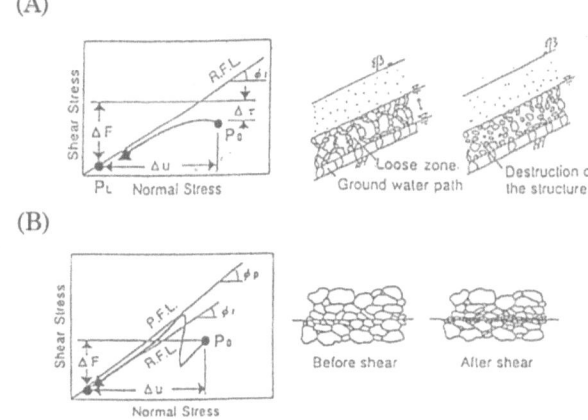

Fig.4 Stress path and illustration of mass liquefaction (A) and sliding surface liquefaction (B). (Sassa 1996)
RFL: residual failure line, PFL: peak failure line

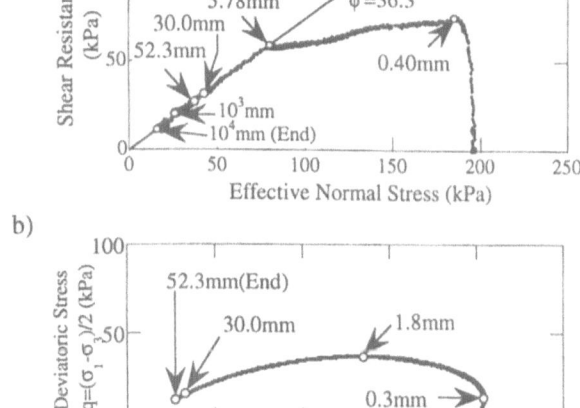

Fig. 5 Stress path of undrained tests in the loose Osaka-group coars sandy soil.
a) by ring-shear test (Test-R-OL), shear speed 0.3~3.0 mm/sec, e_0=0.75, σ=200 kPa. b) by triaxial compression test (Test-T-OL), axial compression speed 0.05mm/sec, e_0=0.76, σ_3=200 kPa

triaxial compression test produced mass liquefaction on the one hand while ring-shear test did sliding surface liquefaction on the other hand. This difference might depend on the types of shearing apparatus. Homogeneous deformation was given by triaxial compression test, shearing deformation was concentrated in the predetermined shearing zone for ring-shear test. Much grain crushing was produced by ring-shear test than triaxial compression test and destruction of meta-stable soil structure was harder to take place in ring-shear test. However, it is very sensitive problem what happened in soils especially under loose condition, much more experiments are required.

5.2 Undrained Shear Strengh

The undrained shear strength of saturated Osaka-group coarse sandy soil from the series of ring-shear tests and triaxial compression tests are presented in Fig. 6 (a) and (b) respectively. Abscissa (shear displacement) of Fig. 6(a) is set at 52.3 mm which is equal to that (axial displacement) of Fig. 6(b) to compare each other. It is the same that the denser sample (R-OD and T-OD) showed higher undrained shear strength for both ring-shear tests and triaxial compression tests. All three tests from ring-shear apparatus (Test-R-OL, R-OM and R-OD) exhibited the similar tendency that undrained shear strengths were decreasing gradually after showing peak value at around 2, 10 and 6 mm respectively. On the other hand, different trends from triaxial compression tests observed, that is, Test-T-OL was something like Test-R-OL and R-OD, Test-T-OM showed increasing and Test-T-OD showed dramatic decreasing after reached quite large undrained shear strength. The decreasing of undrained shear strength of Test-R-OL, R-OM and R-OD were due to

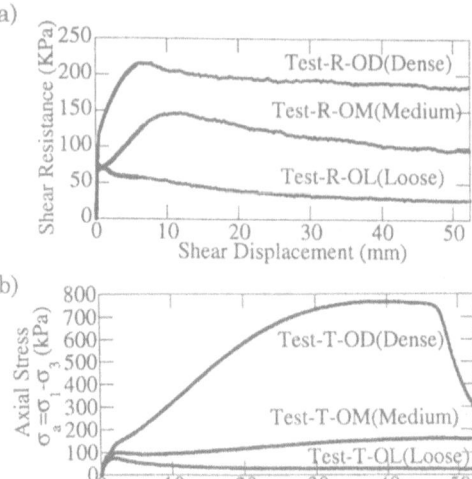

Fig. 6 Undrained shear strength of Osaka-group coarse sandy soil.
a) by ring-shear tests, e_0=0.75 (loose), e_0=0.61 (medium) and e_0=0.55 (dense). b) by triaxial compression tests, e_0=0.76 (loose), e_0=0.60 (dedium) and e_0=0.55 (dense)

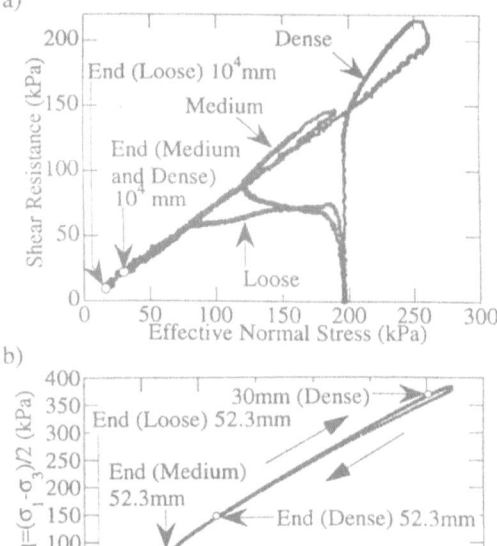

Fig. 7 Effective stress paths of the undrained Osaka-group coarse sandy soil.
a) by rng-shear tests (loose, medium, dense), shear speed 0.3~3.0 mm/sec. b) by triaxial compression tests (loose, medium, dense), axial compression speed 0.05mm/sec

excess pore pressure generation caused by grain crushing because the stress paths went along the failure line, they are finally reached the very low undrained shear strength after long shearing (10^4mm) (Fig. 7). Apart from the results of Test-T-OL which generated mass liquefaction without reaching failure line, distinct decreasing of undrained shear strength was not observed from Test-T-OM and T-OD up to 46 mm axial

Photo 1 Shear plane in cylindrical sample by triaxial compression test (Test-T-OD) at around 50 mm axial displacement. e_0=0.55, σ_3=200 kPa, axial compression speed = 0.05mm/sec

displacement. This means almost zero grain crushing was generated from triaxial compression tests, and this is the crucial difference from ring-shear test.

The dramatic decreasing of undrained shear strength from Test-T-OD after 46 mm axial displacement are examined here. In the previous study (Poulos et al. 1981), there is a unique relationship between effective mean stress p' and void ratio in the steady state. A dense sample shows a large effective mean stress p' in the steady state during the undrained tests, accordingly it mobilizes a large shear resistance. But the undrained shear strength of Test-T-OD was decreasing rapidly after it was considered to be reached steady state at around 40mm axial displacement. Even though the acceptable axial displacement is 30mm (15% axial strain) by standard, this dramatic undrained shear strength decreasing is worth discussing. Photo 1 is of the cylindrical sample at about 50mm axial displacement before finishing test. Clear shear plane was observed, grain crushing must take place along it. As Fig. 6 shows, the rate of decreasing of undrained shear strength of Test-T-OD was quite larger than those from ring-shear tests, other possible reasons might exist.

The peak shear resistance were inversely proportional to the void ratio from ring-shear tests, and then all of the stress paths of Test-R-OL, R-OM and R-OD went left-downward along the failure line and reached very low shear resistance because of grain crushing (Fig. 7(a)). The peak shear resistance from triaxial compression tests were also inverse proportion to the void ratio. However, the stress path of Test-T-OM and T-OD did not go left-downward along the failure line up to the standard axial displacement, that is 30 mm axial displacement (Fig. 7(b)), it fits the previous study. But after that, the stress path of Test-T-OD abruptly went left-downward along the failure line. This phenomenon had not been obtained by triaxial compression test, further interpretation of this phenomenon might be necessary.

6. CONCLUSION

1) Peak friction angles of the Toyoura standard sand from ring-shear test fitted those from the Mikasa-type constant-volume shear-box test and gauge-type shear- box test which remove the side friction effect. But the difference of those between triaxial compression apparatus and other shearing apparatuses exists especially in the range of void ratio from 0.75 to 0.85, while in denser condition they fitted each other.

2) After long shearing (10^4mm), same value of internal friction angle (residual friction angle) was obtained from ring-shear tests on the Toyoura standard sand with different void ratio, while different internal friction angles at 15% axial strain were measured by triaxial compression tests, this apparatus could not produce residual state because of a very limited deformation.

3) Ring-shear tests produce the sliding surface liquefaction to the Osaka-group coarse sandy soil over ranges of void ratio, and final stress points reached a very low shear resistance. On the contrary, liquefaction phenomenon was not observed except for loose condition from triaxial compression test. Although a stress path similar to the sliding surface liquefaction appeared in the dense state after 40 mm axial displacement over the capacity of this apparatus.

REFERENCES

Japanese Geotechnical Society (1979): Mechanical tests on soil (II). Standards on soil tests, pp.423-593. (in Japanese)

Inoue, H. (1964): Research on direct shear test of sand. Proceedings of the Japan Society of Civil Engineering. No.101, pp.15-24. (in Japanese)

Kaibori, M. (1980): Direct Shear Test on Saturated Toyoura Standard Sand. Master Thesis for Graduate School of Kyoto University, pages 93.

Poulos, S.J. (1981): The steady state of deformation. Geotechnical Engineering Division, ASCE, Vol.107, No.5, pp.553-561.

Sassa, K. (1982): Portable direct shear test. Proc. Workshop of the Kansai Branch of the Japanese Society of Soil Mechanics and Foundation Engineering. (in Japanese)

Sassa, K. (1996): Prediction of earthquake induced landslides. Proc. of 7th Int'l Symp. on Landslides, Rotterdam:Balkema. Vol.1:115-132.

Sassa, K., Fukuoka, H., Scarascia-Mugnozza G., and S. Evans (1996): Earthquake induced landslides. Soils and Foundations, Special Issue, pp.53-64.

Sassa, K. (1997): The development of reproducing apparatus for earthquake-induced landslide, Proc. of No.36 Annual Congress of Japanese Landslide Society, pp.223-224. (in Japanese)

Sassa, K. (1998): Mechanisms of landslide triggered deblis flows. Evvironmental Forest Science, Kluwer Academic Pub. (this book)

Seed H. B. (1968): Landslides during earthquakes due to soil liquefaction. Journal of the Soil Mechanics and Foundations Division, ASCE, Vol.94, pp.1055-1122.

Zhang, D.X. (1996): A study on the mechanism of loess landslides induced by earthquakes. Journal of Natural Disaster Science, Vol.18, No.1, pp.27-41.

Soil Bioengineering— an Environmental Alternative for Erosion and Torrent Control

Christoph GERSTGRASER

Department of Soil Bioengineering and Landscape Construction, University of "Bodenkultur", Vienna, Austria

Keywords: soil Bioengineering ,Erosion Control, Slope Stabilization, Riverbank Protection

1. INTRODUCTION

In torrent catchment areas, heavy rainfall very often causes slope failure and erosion of soil and riverbanks which are major environmental hazards. As a consequence, agricultural or forest land is lost and new debris source areas are created. Such areas are potential sources of danger because debris flow could occur which might threaten or destroy adjoining properties, buildings, or roads on the alluvial fan.

The risk of slope failures and erosion is enhanced when the vegetation cover is removed (Rickson/Morgan, 1995). Therefore, soil bioengineering is a suitable method for erosion control, slope stabilization and riverbank protection, because soil bioengineering uses only plants or native vegetation combined with structural elements made of wood, stone, or other auxiliary materials. A few weeks after their installation, the plants sprout and vegetation covers the ground. Soil bioengineering can be regarded as an environmental alternative to hard engineering works because the use of forest products like bushes and trees is environmentally cost effective and sustainable. Additionally, the visual impact of civil engineering works is reduced and such enhances the quality of landscape.

For torrent control, different methods of soil bioengineering are available. Using the right method depends on the erosion or failure processes. Considering this, the following soil bioengineering methods give an overview which of these methods are mainly used in the alpine area of South Tyrol (Italy) and Austria (Florineth/Gerstgraser, 1997).

2. SEEDING TECHNIQUES

The most common method of soil conservation is seeding of a grass and herb mixture. A dense cover of grass and herbs provides very good surface protection against rainfall and wind. The effect of a dense grass cover is that the vegetation absorbs rainfall energy and, therefore, prevents soil detachment by raindrops (Gray/Sotir, 1996). Furthermore, the roots restrain soil particles and increase the water storage capacity of the soil which delays the onset of runoff. Seeding techniques are used for hill and slope stabilisation and are the only soil bioengineering technique for successful erosion control above the timber line.

Hydroseeding is mainly used on steep slopes with a smooth surface and mild climate. Seed (25 g/m^2), organic fertilizer (100 g/m^2), mulch (e.g. cellulose, straw 80 g/m^2) and an algal product as glue (100 g/m^2) are mixed in a special barrel with water and pumped out on to the slopes (2 l/m^2). On very steep slopes, it is advisable to fasten a jute mesh on the slope because it fixes the hydroseed.

Bitumen Straw Seeding according to SCHIECHTL is the best method on exposed areas and areas mainly above the forest line. In a 3-4 cm thick layer of straw (700 g/m^2), seed (25 g/m^2), and fertilizer (100 g/m^2) are spread out and covered with an instable bitumen emulsion. The straw ensures mechanical protection and the bitumen emulsion absorbs the warmth, which is necessary above the forest line, and functions like a greenhouse. On very steep slopes and on failure edges, a stable jute or coconut mesh instead of the bitumen emulsion covers the straw. On areas where rockfall could happen or rocks could break out of the soil it is useful to nail an iron mesh into the ground covering the straw layer.

3.   SLOPE STABILIZATION METHODS

There are different types of failure mechanisms. Generally, slope failure occurs when the shear stress exceeds the shear resistance in a slope. Normally, slopes fail along a critical sliding surface and, therefore, the depth which can be stabilized with plants is limited by the depth of the roots. Only if the roots can penetrate the sliding surface, effective soil reinforcement is achieved. The greatest trigger of landslides is water infiltration, for that reason, water drainage is very important for efficient slope stabilization.

The beneficial effect of vegetation is that the soil is mechanically reinforced by the root system and that plant transpiration reduces soil moisture and pore pressure which enhance the stability of the slopes. Except for the drain fascines all of the applied soil bioengineering methods are linear systems. To achieve a larger surface protection these soil stabilization methods are supplemented by seeding techniques. Some of the methods described could be used for the stabilization of cut or fill slopes of roads as well.

3.1. DRAIN FASCINES

Drain fascines are one of the best methods to stabilize and drain wet slopes. They consist of live branches of willows tied together with wire. The lowest third should also contain dead branch material which channels the water unhindered. The tip end of the branches always points in flow direction. Usually the fascines are 30 to 60 cm thick and have no length limit. They are put into ditches and after placement, the fascines must be covered with soil so that all the branches are embedded and can take root and grow. To prevent the fascines from sliding or washing away, they are fixed every 2 m with wooden poles.

3.2. BRUSHLAYERING

Brush layers are one of the best methods to stabilize loose rock slopes. Fresh, green willow cuttings and rooted plants are layered in parallel on 1-1.5 m wide berms on the face of the slope. The berms should have an angle of inclination (sloping up) of 10-20 percent to the outside. Wherever there is a risk of slope failure, ditches should only be dug in short segments. The plants are placed closely together and the tips or the leafy ends should protrude 15 - 20 cm beyond surface. The vegetation used should be resistant to rock fall and rubble and must have the ability to produce adventitious root systems. Vertical spacing between layers is dictated by the erosion potential of the slope and may be between 1 or 2 meters.

3.3. VEGETATED PILE WALLS

On sites where there is light rock fall, shallow soil movement, and where erosion has not been stopped completely, vegetated pile walls could be used. They are applied instead of wattle fences on slopes, because wattle fences are poorly covered with soil which causes a short lifetime and no protection against soil movement. In order to build a vegetated pile wall, iron or wood piles are hammered into the ground on which a larch log is attached and deciduous trees are placed on the logs and covered by soil. Additionally, due to the stabilizing function of the plant, the logs prop up the soil material too. It is necessary to use stump sprout deciduous trees since they are sufficiently durable to resist mechanical damage by rock or rubble.

3.4. VEGETATED CRIBWALL

Vegetated cribwalls, normally made of wood, are one of the best methods to immediately secure endangered parts of slopes and toes of slopes. They are built from logs and anchor logs held together with nails or bolts. For higher stability against sliding, cribwalls should not be placed horizontally on the slope but at an angle of 10-15% toward the slope. Additionally, the whole construction is secured with 2-2.5 m long iron poles which are hammered into the ground. The space between the logs is filled with soil material. Plants should not stick out of the wall more than a quarter of their length. To reach vegetation quickly, it is advantageous to use green willow branches and strongly rooted pioneer plants. In order to achieve good plant development, the face wall should have a ratio of at least 2:1. If this is not the case the plants will not be exposed to light sufficiently.

3.5. VEGETATED STONE WALLS AND GABIONS

Vegetated stone walls and gabions are useful for the stabilization of slope toes or steep slope cuts because they are flexible, permeable, and durable. During the building, live plants are placed into the joints between the stones so that they reach into the soil behind the construction. The joints must be filled with soil material to ensure plant growth. Green willow branches and rooted plants which have the ability to produce adventitious root systems should be used as vegetation. The branches should not protrude from the wall more than 10 cm to prevent desiccation.

4.   RIVERBANK PROTECTION

Flowing water creates hydrodynamic forces which affect the whole surface of a river bank. The use of soil bioengineering methods depends on various factors but mainly on the intensity of the hydrodynamic forces. There will be always situations where hard structures are appropriate but up to a shear stress of 300 N/m^2 soil bioengineering is an alternative. A pre-condition for applying soil bioengineering methods in torrents is river bed protection, because if the river bed is eroded its banks will also be underminded.

The best bank protection is ensured by dense, flexible vegetation covers. Bank vegetation causes resistance to flow and decreases flow velocity. This leads to a reduction of the forces on the bank surfaces and creates protection against bank erosion. In addition, the bank is reinforced by the roots of the trees. Rigid and solitary plants are vulnerable and less effective. They cause localized increases in velocity and turbulence that may generate local scour and can become a locus of serious channel instability (Thorne et al, 1998). To ensure long-term bank protection, maintenance of the vegetation is required.

4.1   BRUSH MATTRESS WITH WILLOWS

A brush mattress of willows is one of the best method to protect river banks. They can resist a bed shear stress up to 300 N/m^2 (Florineth, 1995). Fresh, green willow branches with a thickness of 3-8 cm are closely placed cross the flow direction on the bank surface. The butt end of the branches must reach into the water. The aim is to obtain a very dense layer of branches which is fastened down on wooden poles with a coconut string, wire or logs. The toe of the brush mattress is the most endangered part, therefore, it is protected by a wooden crib wall, rough coniferous trees, or blockstones. To enhance the diversity of plants some deciduous trees may be placed between the willow branches in addition. Finally, the branches are covered by a 3-5 cm thick layer of soil to protect them against desiccation. The advantage of the brush mattress is that the butt ends reach into the water and this ensures that the upper ends of the branches are supplied with water. For that reason a very dense vegetation forms out quickly

4.2   FASCINES

Live branches of willows are tied together with wire in long, 30 to 50 cm thick bundles, their length is limited by the handling. The fascines are placed on the river bank and the tip end of the branches must always show in flow direction. After the fascines are secured every 1-1.5 m with wooden poles, they are dumped and covered with soil so that the branches are embedded. Placing one fascine behind the other ensures a very efficient bank protection, especially, shortly after installation when the fascines have not yet formed out dense sprouts. They are used to protect bank toes and shallow riverbanks with a slope angle ratio of 1 : 3 because on steeper slopes the fascines could easily get dry.

4.3   WATTLE FENCE

Wattle fences could be used for bank protection on small rivers and gullies. Wooden piles are hammered at a distance of 1-1.2 m into the ground. Between these piles, elastic 3-8 cm thick live branches of willows are alternated woven. Also, the branches could be woven diagonally with the butt end in the ground which guarantees a better water supply of the branches. The wattle fence is a vertical construction and its height should not exceed 60 cm. Otherwise, there is a risk that the upper sprouts shadow the lower branches which prevent them from growing and finally, these branches die and rot.

REFERENCES

Florineth, F. (1995): Weidenspreitlage als Weg zur schnellen Uferbepflanzung und –sicherung. Gesells. f. Ingenieurbiologie, Mitt. 4, Aachen, pp. 51-67.

Florineth, F. / Gerstgraser, Ch. (1997): Ingenieurbiologie – Studienblätter zur Vorlesung. Herausgeb. Arbeitsbereich Ingenieurbiologie u. Landschaftsbau, Universität für Bodenkultur, Wien, 146p.

Gray, D.H. / Sotir, R.B. (1996): Biotechnical and Soil Bioengineering Slope Stabilization. John Wiley and Sons, New York, 365p.

Morgan, R.P.C. / Rickson, R.J. (1995): Slope Stabilization and Erosion Control: A Bioengineering Approach. Chapman and Hall, London, 274p.

Thorne, C., Amarasinghe, I., Gardiner, J., Perala-Gardiner, C., Sellin, R., Greaves, M., Newmann, J. (1997): Bank Protection using Vegetation with Special Reference to Willows. EPSRC/EA Scoping Study-Project Record, University of Nottingham.

Attachment : Methods of Soil Bioengineering

Drain Fascine

Cross section

Top View

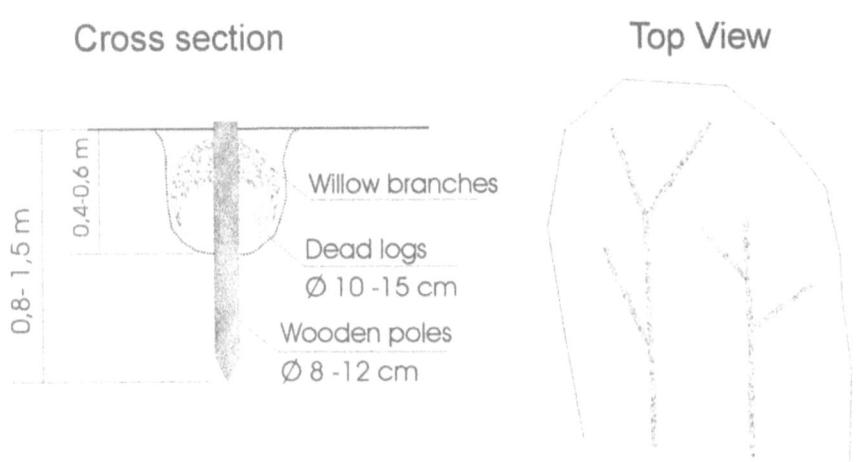

Willow branches

Dead logs
Ø 10 -15 cm

Wooden poles
Ø 8 -12 cm

0,8- 1,5 m

0.4-0.6 m

Brush Layer

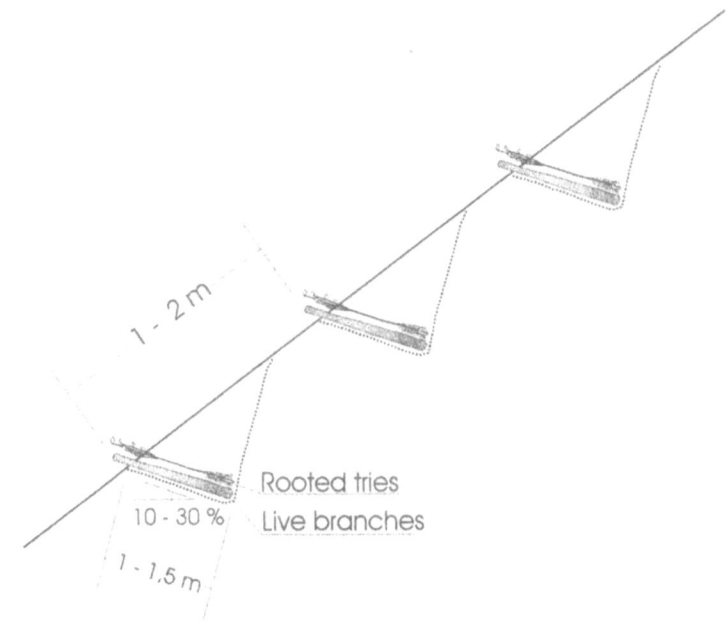

1 - 2 m

Rooted tries
10 - 30 %   Live branches
1 - 1,5 m

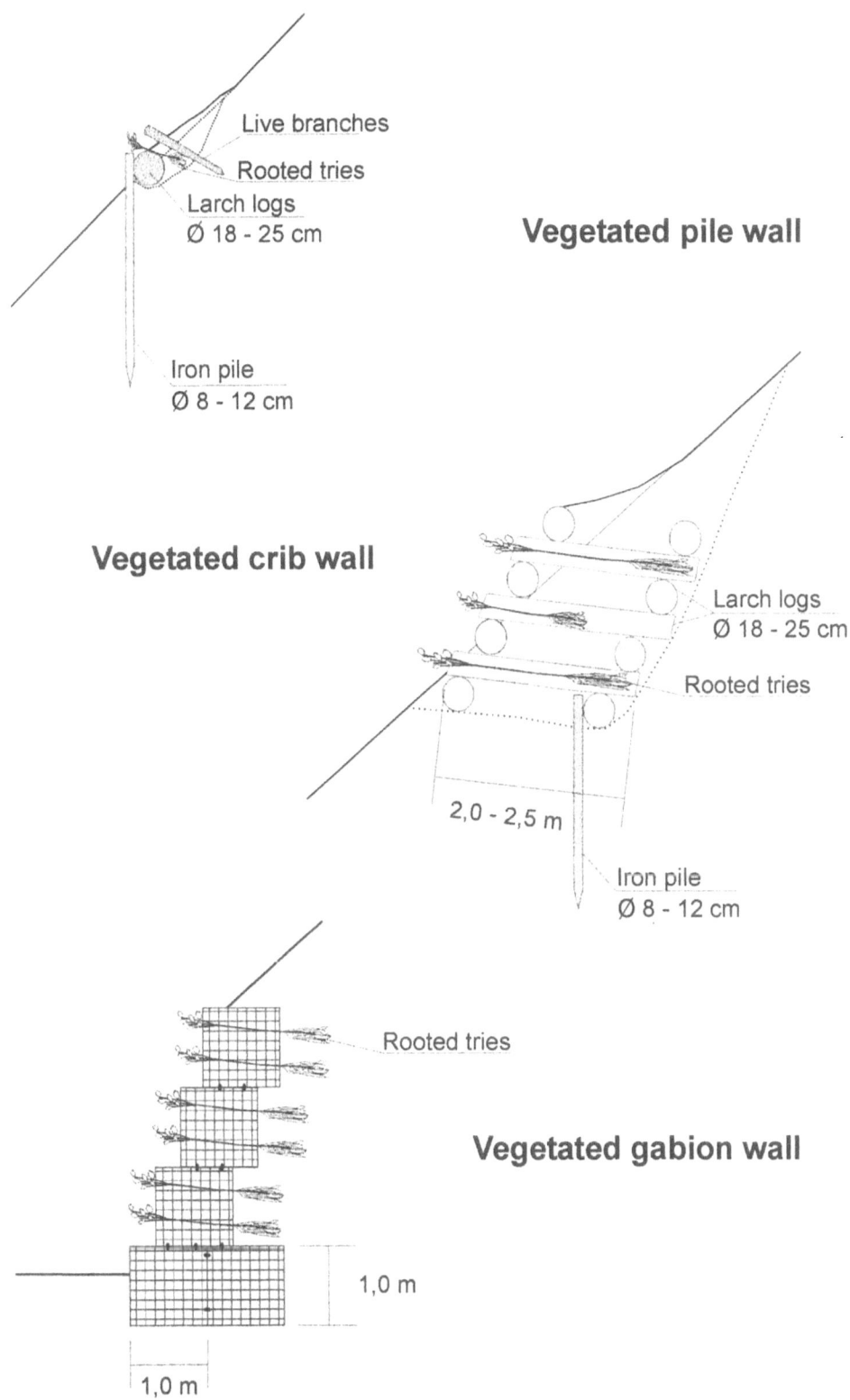

Live branches

Rooted tries

Larch logs
Ø 18 - 25 cm

Vegetated pile wall

Iron pile
Ø 8 - 12 cm

Vegetated crib wall

Larch logs
Ø 18 - 25 cm

Rooted tries

2,0 - 2,5 m

Iron pile
Ø 8 - 12 cm

Rooted tries

Vegetated gabion wall

1,0 m

1,0 m

Brush mattres§

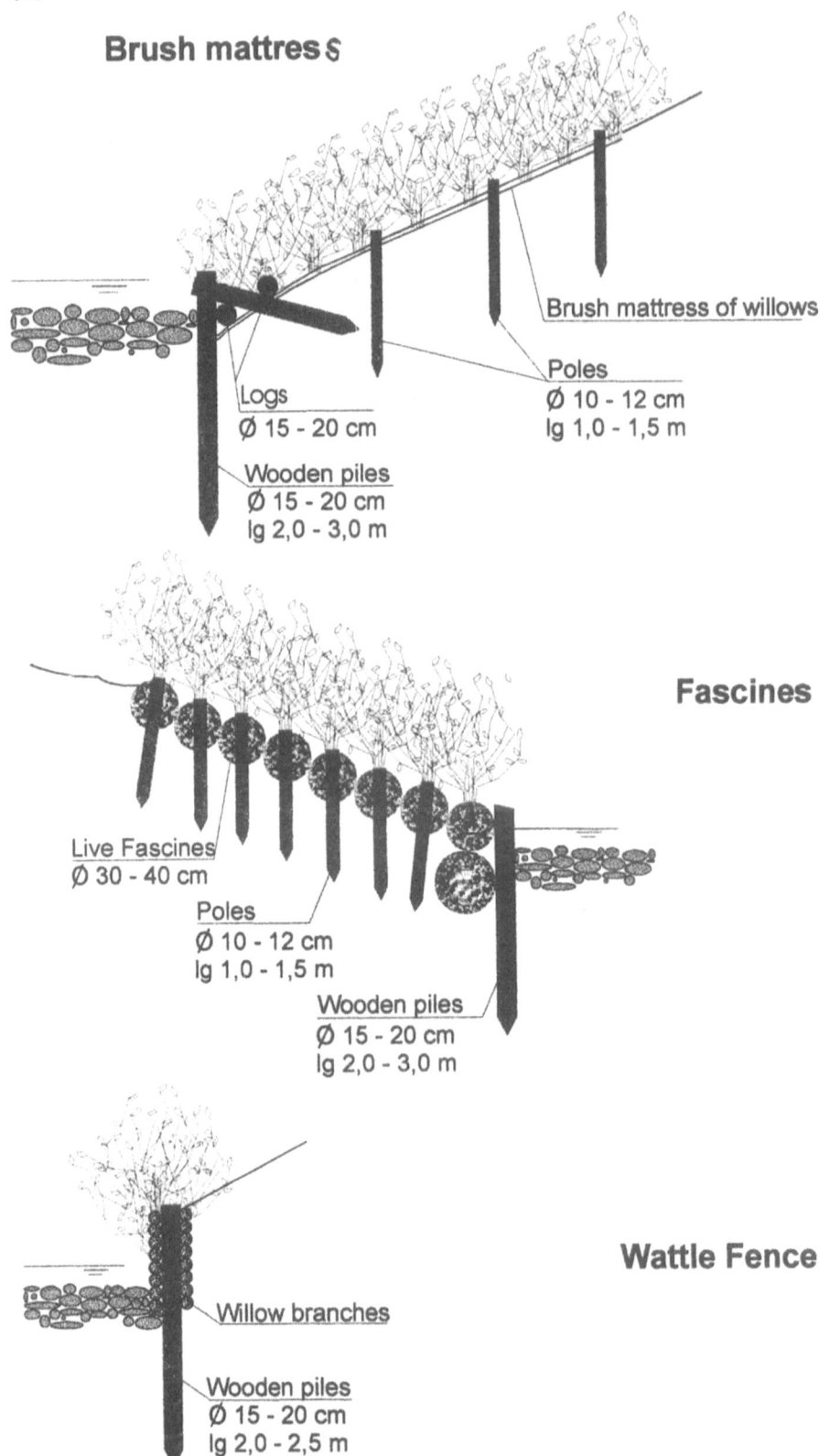

Brush mattress of willows

Poles
Ø 10 - 12 cm
lg 1,0 - 1,5 m

Logs
Ø 15 - 20 cm

Wooden piles
Ø 15 - 20 cm
lg 2,0 - 3,0 m

Fascines

Live Fascines
Ø 30 - 40 cm

Poles
Ø 10 - 12 cm
lg 1,0 - 1,5 m

Wooden piles
Ø 15 - 20 cm
lg 2,0 - 3,0 m

Wattle Fence

Willow branches

Wooden piles
Ø 15 - 20 cm
lg 2,0 - 2,5 m

Forest Fire Prevention Through Prescribed Burning in *Acacia Mangium* Plantation in South Sumatra, Indonesia

Bambang Hero SAHARJO[1,2] and Hiroyuki WATANABE[2]

*1 Laboratory of Forest Protection, Division of Forest Management, Faculty of Forestry, Bogor Agricultural University, Indonesia
*2 Laboratory of Tropical Forest Resources and Environments, Division of Forest and Biomaterials Science, Graduate School of Agriculture, Kyoto University, Kyoto 606-8502, Japan

ABSTRACT A 15 m x 10 m plot was established at 3 ,4, 5, and 6-years of plantation age of *A.mangium* located in one forest concession area in South Sumatra, Indonesia. In each plot, available fuel, fuel moisture content, seed storage on and under the soil surface, tree height, and diameter are measured before burning. 3 ton/ha of fuel loads and 8 cm fuel bed depths were used as treatment. Flame length and flame temperature using Tempilaq (temperature indicating liquid) is measured during burning. Fires were set on the sides all sides of the plot at the same time. After burning, fuel remaining, natural seedling regeneration in the forest floor, injured trees, crown scorch, and shrub invasion were calculated.
Keywords: *Acacia mangium*, flame length, fuel bed depth, fuel load, prescribed burning

1. INTRODUCTION

One of the reasons why industrial forest plantations are valuable is the high demand for raw materials, especially for pulp and paper, which has increased year by year. For pulp, the 1997 projected pulp capacity of Indonesia increased 4.6 million tons/year. In 1998, the estimate is 5.2 million tons/year, and in the year 2010 capacity will increase to 11 million tons/year. To guarantee sustainability of the raw materials, a million hectares of industrial forest plantation must be established. Without this, the natural rain forest will surely be sacrificed. Many factors, however, affect the success of forest planting, one of those being forest fires (Saharjo 1997).

Fire is still a critical factor in managing industrial forest plantation, especially for Acacia (*Acacia mangium*). The daily increase of fire damaged areas, especially

641

in the dry seasons, with the disappearance of young *A.mangium* plantations does not have happen if fuel load in the plantation is not so high and tree maintenance is improved (Saharjo and Watanabe 1996). If nothing is done, entire plantations are quickly destroyed. One possible solution to solve this problem is to conduct prescribed burning.

Prescribed burning is carried out now for many purposes (Gill and Bradstock 1996). It may be used to protect wood products, water supplies, animal production and heritage items, control weeds, maintain biodiversity and improve human safety, eliminate undesirable species, and more (Wheelan 1995). The alternative chosen is influenced by the values threatened by fire. The need for controlling or eliminating fire risk increases as fuel hazards and values increase. It becomes a truism to the fire control planner that high fire risk must not be permitted in any area which has both high fuel hazards and high destructible values (Brown and Davis 1973).

The objective of this research is to clarify the feasibility of prescribed burning being conducted in the plantation and it effects on tree mortality and available fuel after fire.

2. STUDY SITE

Research was carried out in newly established *A. mangium* plantation stands aged 3, 4, 5, and 6 years belonging to one forest concession holder , at Subanjeriji forest block, South Sumatra, Indonesia, from August 1997 to February 1998.

The mean rainfall is approximately 2,800 mm and monthly rainfall ranges from 92 mm in July to 278 mm in February. According to Schmidt and Fergusson (1951), the climate of this area belongs to rainfall type A ($0<Q<0.143$). Mean maximum air temperature in this area is 32.8°C in August, mean minimum air temperature is 22.3°C in December and mean relative humidity is about 85 % annually.

The plantation soil is Red Yellow Podsolic and soil classes (USDA) are: Haplaquox, Dystropepts, Kandiudults and Hapludox. This area has the following characteristics (Hikmatullah et al. 1990): acid tuff plain, acid tuff and fine felsic sedimentary rocks, flat to undulating (slopes<8 %), slightly dissected, high weathering parent material, terrain flat to undulating, low amplitude with low variability, drainage varying from well drained to imperfectly drained, 5-125 m altitude (15-50 m range), soil depth 101-150 cm, total nitrogen was 0.11-0.20 %, organic carbon 1.09-3.51 %, available phosphorus 0-15 ppm., K-exchange 0-0.05 me/100 g, cation exchange capacity 0-16 me/100 g, base saturation 5.6-21.8 5, free salinity 0-3.9 mmnos/cm, pH 6.5-5.5, Al saturation 61-80 %, Al toxicity (Al content is very high), organic matter 3.1-16.1 %, and soil fertility based on Indonesian criteria is low. Organic layer varies from 0.1-0.5 m.

3. MATERIALS AND METHODS

3.1. Field experiment

Four quadrates 150 m^2 (10 m x 15 m) in area were set up for the four different stand ages. In this plot, seed storage in the soil surface and under the soil surface, fuel load, and fuel moisture samples were collected and measured before burning. Then, fuel loads up to 45 kg (based on a fuel load 3 ton/ha permitted to remain in the forest floor) were delivered to the plot, and fuel bed depth was set at 8 cm. Fires were set on the sides of the plot at the same time. After burning, remaining fuel was measured.

3.1.1. Seed storage

3.1.1.1. On the soil surface (0 m)

Four of 1 m x 1 m subplots were established in the forest floor in each plantation year in order to measure seed storage. The number of seeds discovered was collected and measured.

3.1.1.2. Under the soil surface

Seed storage was measured using a ring sample of 400 cm^3 in volume (surface area was 100 cm^2 and ring height was 4 cm). This ring was used to take seed at 0-5 cm and 6-10 cm under the soil surface for different plantation years. Five samples were taken in every plot. The number of seeds discovered was collected, measured and brought to the laboratory to determine germination percentage (viability).

3.2. Experimental burning procedure

In each plot, five subplots of 1 m^2 (1 m x 1 m) were chosen for calculation of available fuel and moisture content. Available fuel was determine by collecting all the materials dead and a live in the subplot. These materials were then brought to the laboratory for measuring weight and moisture content. Fuel moisture content was measured based on dry weight calculation, after placement in an oven for 24 hours at 105°C.

Available fuel in the plots was then set aside in order to clean the plot. Based on maximum fuel load allowed to remain in the plantation, 3 ton/ha with a fuel bed depth of 8 cm (Saharjo and Watanabe 1997), 45 kg of this fuel load was separated in the plot. The plot was then burned using matches in all locations and the fire was allowed to propagate naturally. Flame temperature at 0 m and 1 cm under the soil surface was estimated by using "Tempilaq" (temperature indicating liquid), which melts at a specified temperature and provides estimates of maximum temperatures. Each liquid was set in an aluminum pipe 2 cm in diameter and 30 cm long (Saharjo 1995). The temperature sensors were placed in the vegetation at two locations.

Rate of the spread of fire was estimated by measuring the average distance perpendicular of the moving flame front per minute. A stopwatch and measurement tape were used. Crown scorch was estimated by measuring the percentage of canopy whose leaves were dark. Fire intensity was calculated by using Byram's equations (Wheelan 1995), FI: 273 (L) $^{2.17}$, where FI is fire intensity (kW/m) and L is flame length (m).

After the fire, all fuel left in the plot was collected and measured to determine how much fuel remained after burning.

3.3. Fire effects

The effects of fire on the trees and the characteristics of the forest floor fuel load is monitored monthly to determine how many trees die and how much of fuel load increases.

4. STATISTICAL ANALYSIS

A completely random design of variance was used to test for differences among tree growth parameter, diameter, and height and fire behavior parameter, flame length, rate of the spread of fire, and fire intensity, based on the following model (Steel and Torrie 1981):

$$Yjk = U + Tj + Ejk$$

Where, Yjk = Tree growth and fire behavior for plantation j in the k replication

U= Mean of the treatment population sampled

J = Plantation year

k = replication

Ejk = random component

To detect significant differences in tree growth and fire behavior parameters (p $\leq$ 0.05), the Duncan test (Steel and Torrie 1981) was applied.

5. RESULTS AND DISCUSSION

Diameter of trees (Tab.1) in the plot ranged from 10.3 cm to 15.8 cm, and height from 10.7 m to 20.0 m.

Table 1. Plantation year, diameter and height of trees in the plot

Plantation year	N/ha	Diameter (cm)	Height (m)
1991/1992	1250	15.8 ± 4.2 b	19.4 ± 3.6 c
1992/1993	1250	11.7 ± 2.6 ab	14.8 ± 2.3 b
1993/1994	1250	10.3 ± 2.0 a	12.4 ± 0.4 a
1994/1995	1250	11.2 ± 2.1 a	10.7 ± 1.6 a

* Means are significantly different when standard error is followed by different letters ($p \leq 0.05$)

Seed storage (Tab.2) in the forest floor varied from 16.5 seeds/m^2 in the 1993/94 plantation to 215.3 in 1991/1992. Seed storage under the soil surface varied from 90 to 850 seeds/m^2 at 0-5 cm, and 20 to 700 seeds/ m^2 at 6-10 cm. Germination percentage of seeds taken both from the soil surface and under the soil surface was quiet high, ranging from 80 – 90 %.

Table 2. Seed storage in the forest floor and under the soil surface

Plantation year	Seed storage in the forest floor (per m^2)	Seed storage under the soil surface (per m^2)	
		0 – 5 cm	6 – 10 cm
1991/1992	215.3 ± 135.0 b	850.0 ±1 0.0 c	700.0 ± 400.0 c
1992/1993	31.3 ± 18.0 a	300.0 ± 280.0 b	500.0 ± 300.0 b
1993/1994	16.5 ± 10.6 a	190.0 ± 120.0 a	20.0 ± 20.0 a
1994/1995	5.0 ± 3.0 a	90.0 ± 10.0 a	0.0 ± 0.0 a

* Means are significantly different when standard error is followed by different letters ($p \leq 0.05$)

Available fuel (Tab. 3) varied from 13.7 ton/ha in 1993/94 to 25.1 ton/ha. Available fuel in 1993/94 was lower than in other areas because this plot was monitored and maintained by the R & D section. If all the available fuel burn directly, it is estimated that all the trees will die, as happened in 1994 (Saharjo and Watanabe 1996).

Table 3. Weather condition and fire behavior when burning was being conducted

	Plantation year			
	1991/1992	1992/1993	1993/1994	1994/1995
Weather conditions				
Air temperature (°C)	30	33	35	33
Relative humidity (%)	80	80	75	80
Wind speed (m/s.)	1.7	1.6	1.7	1.8
Fire behavior				
Fuel moisture (%)	5.5 - 30.0	12.2 - 43.9	11.6 - 31.2	3.5 - 49.5
Available fuel (ton/ha)	20.2±0.5 b	23.2±0.3 c	13.7±0.4 a	25.1±0.3 d
Pre-fire fuel (kg)	45.0	45.0	45.0	45.0
Post-fire fuel (kg)	1.9	2.5	1.2	0.5
Fuel bed depth (cm)	8.0	8.0	8.0	8.0
Rate of the spread of fire (m/min.)	1.4 ± 0.2 a	1.5 ± 0.2 a	1.6 ± 0.2 a	1.7 ± 0.2 a
Flame length (m)	0.74±0.06 a	0.78±0.04 a	0.85±0.04 b	0.93±0.03 c
Flame temperature (°C)				
0 m (soil surface)	149-159	139-149	149-159	149-170
1 cm(under soil surface)	< 38	< 38	< 38	< 38
Crown scorch height(m)	8 – 10	9 – 10	5 – 6	8 – 10
Fire intensity (kW/m)	136.9±23.9 a	151.1±16.5 a	183.1±21.1 b	224.3±15.8 c

* Means are significantly different when standard error is followed by different letters (p≤0.05)

Flame length as an indicator of fire intensity was not high, ranging from 0.74 m in 1991/1992 to 0.93 m in 1994/1995. The highest fire intensity was in 1994/1995, with 224.3 kW/m and a maximum temperature of 170°C. This flame heat had different effects on trees canopy. Crown scorch height varied from 5-6 m in 1993/94 to 9-10 m in 1992/1993. This temperature did not have any effect on seed in the soil surface, because the heat penetration is less than 38°C, and the seed is still dormant.

Six months after the fire, no trees were dead and no natural regeneration was evident in the forest floor. Fuel load again to cover in the plantation, however, three months after fire ranging from 2.6 to 5.3 ton/ha (Tab. 4).

Burning consumed at least 85-90 % of fuel load in the plot. This means such an area will become safe for a while, until fuel load increases again. High consumption of this fuel load was caused by a low fuel moisture content, and because the load was dominated by leaves and dead fuel. The composition of fuel load (dead and live fuel) obviously affects the rate of the spread of fire and flame temperature. The highest maximum flame temperature (1994/95) of 170°C was caused by a composition with more live fuel than in other plantations. This means that more heat energy is needed to burn it. The resulting fire intensity reached in this plantation, 224.3 kW/m, was the highest achieved. At first, the rate of spread of fire is not high, but when heat energy increases, speed also increases. In addition, air temperature and relative humidity also affect burning.

Table 4. The effect of fire on tree mortality and available fuel 6 months after fire

	Plantation year			
	1991/1992	1992/1993	1993/1994	1994/1995
Tree mortality	0.0	0.0	0.0	0.0
Available fuel (ton/ha)				
3 months after fire	5.3 ± 0.2 b	2.6 ± 0.1 a	2.7 ± 0.2 a	5.3 ± 0.2 b
6 months after fire	8.2 ± 1.7 b	4.6 ± 0.3 a	7.6 ± 1.4 b	9.5 ± 1.7 c

* Means are significantly different when standard error is followed by different letters ($p \leq 0.05$)

One of the reasons for prescribed burning in forest plantations is to reduce the high available fuel stored in the plantations. This results from poor maintenance activities and low rates of material decomposition. Table 3 shows that all the plantations have more available fuel than should be permitted (3 ton/ha). This means that all the plantations have a high risk of fire invasion, and trees will easily be killed. This happened in 1994 (Saharjo and Watanabe 1996), where 20,000 ha of *A. mangium* plantation (of 120,000) ha planted burned in only three months. The risk is also increased by the potential for arson around the plantation.

After prescribed burning, no trees were damaged and no natural regeneration was found in the forest floor. This means that the procedure for burning can be based on the previously stated rule. 3 months after the fire, fuel load increases again in the plot. It seems that this fuel is dominated by falling leaves and old branches.

6. CONCLUSION

The results of the research show that prescribed burning can be used to reduce high available fuel in the plantation without any damage to the trees. Burning consumed at least 85-90 % of fuel load. Six months after fire there is no natural regeneration on the forest floor .

If prescribed burning is conducted, it must be following the rule: Maximum fuel load at 3 ton/ha, fuel bed depth of 8 cm, fuel moisture content in the range of 5 % - 20 % and dominated by dead fuel, air temperature around 30 – 35 °C, relative humidity at 75 – 80 %, wind speed of 1.6 – 1.8 m/min., sunny day, in the afternoon between 13.00 and 16.00 p.m. and fire should be ignited on the side of the site and following the wind direction.

REFERENCES

Brown, A. A. and K. P. Davis. (1973). Forest Fire: Control and Use. 2nd edition. McGraw-Hill Book Company, New York. 658 p.

Gill,A. M. and R. A. Bradstock. (1996). Prescribed Burning: Patterns and Strategies. Paper Presented at the 13th Conference on Fire and Forest Meteorology, Lorne, Australia, October 1996. 6 p.

Saharjo,B. H. (1995). The Changes in Soil Chemical Properties Following Burning in a Shifting Cultivation Area in South Sumatra, Indonesia. Wallaceana 75: 23-26

Saharjo,B. H. and H. Watanabe. (1996). Fire Threaten Industrial Forest Plantation: The Case Study in South Sumatra, Indonesia. Paper presented at the 13th Conference on Fire and Forest Meteorology, Lorne, Australia, October 1996. 22 p.

Saharjo, B. H. and H. Watanabe. (1997). Tree Spacing Minimizing Fuel Load in *Acacia mangium* Plantation: A Case Study in South Sumatra, Indonesia. Proceeding of the 108th of the Japanese Forestry Society Annual Meeting. pp: 119.

Saharjo, B. H. (1997). The Role of Industrial Forest Plantation in Supporting Pulp and Paper Industries: A Case Study in South Sumatra, Indonesia. Paper Presented at the 6th Annual International Workshop BIO-REFOR, Brisbane, Australia, December 1997. 8 p.

Schmidt, F. H. A. and J. H. S. Fergusson. (1951). Rainfall Type Based on Wet and Dry Periods of Ratios from Indonesia With Western New Guinea. Verhandelingen No.42. Directorate Meteorology and Geophysics. Jakarta, Indonesia. 77 p.

Seabright, D. (1995). Meeting With the Minister. Asian Timber 14 (9): 28-31

Steel, R. G. D. and J. H. Torrie. (1981). Principles and Procedures of Statistics: A Biometrical Approach. 2nd edition. McGraw-Hill International Book Company. 633 p.

Wheelan, R. J. (1995). The Ecology of Fire. Cambridge University Press. 344 p.

Demands on a Nature Orientated Flood Control

Hansjoerg HUFNAGL

Forest-technical Service for Torrent and Avalanche Control, Villach, Austria

ABSTRACT

Ecological criteria (water and landscape) have, until now, not played an important role in flood control measures. Nowadays, protection should cover more requirements at once. Flood control measures should only be carried out when all possibilities to protect the stream have been exhausted. Therefore, the ecological requirements are increasingly relevant and have to be taken into account whenever flood control measures are necessary. The aesthetics of the landscape, as well as the social economical aspects are increasingly important, too ; however, it is essential to give preference to the ecological requirements over others. In the present paper, above all, the ecological requirements are discussed; in addition to it the demands concerning the aesthetics, the image and the character of a landscape along with the social economical aspects are also taken into account. Finally, the aims of projects, when planning flood control measures, are discussed.

Keywords: ecological requirements, nature orientated flood control, torrent control

1. ECOLOGICAL REQUIREMENTS

In natural flowing waters the dynamic of discharge, which is depending on geology and geomorphology of the landscape, creates the stream´s structure and its diversity. This diversity of bed structure and current depends on a high temporal dynamic. It can be the starter for different ecological processes which, through their succession, can lead to an incredibly big variety of species and a large amount of individuals in and around the stream.

Biological organisms find their habitat not only in the sedimentary surface but also in the anorganic structure lying underneath. This structure is called "hyporheic habitat". It is extremely important, not only because it provides habitats but also because it is a potential shelter for the under water fauna in case of flooding or other disturbance in the ecological system.

Because of the high diversity of bed structure and current natural flowing waters have a high "resilienz", which means, that the ecosystem is able to overcome passing disturbances (e.g. debris flows) and achieve their characteristic ecological balance again rapidly.

Bearing in mind that the structure of the flowing water´s abiotical parameters is the basis of the complex biotical or ecological system, there should be considered following requirements in principle when talking about nature orientated flood control :

1) The diversity of bed structure and current of flowing waters and its underlying dynamic should never be permanently reduced in the reach of low and mean water discharge and only when strictly necessary in the reach of flood flow discharge.

2) An intervention within the diversity of bed structure and current of the ecosystem "stream" is only permitted if there is a guarantee that the stream is able to restore its original phenotype of bed structure and current diversity in the low and mean water channel by itself after finishing defence works.

3) The "continuity of streams" in the aggrading reach of the lower part of a watercourse and in the reach with aggrading and eroding processes (section of "furkation") has to be preserved in its totality.

As far as the erosional section of a stream´s upper course is concerned, it is also necessary, whenever it is possible, to preserve the continuity. In case of a division of the ecological coherence, for instance through "drop structures" (dams) in this part of the stream, appropriate technical measures should be taken to contravene (e.g. leave the foundation drain open).

Criteria that have to be respected when planing and realizing protection measures (Honsig/Hufnagl et al. 1996):

- Morphological dynamic of the stream, type of stream morphology
- diversity of structures depending on stream morphology
- ecological balancing the surface
- stream´s centerline
- open ground, structure of bed, grain size distribution
- appearance of the currents and its distribution
- migration
- association of biotopes, integration with the immediate environment
- vegetation typical for specific waters
- local building material
- rare species and habitats

1.1. Morphological dynamic of the stream, type of stream morphology

Dynamical transformations, like erosion and sedimentation, must be possible, even after human interventions, to preserve the ecological system´s stability. A stable way of building does not induce a "stable" ecological system. Every technical intervention within natural flowing waters changes the dynamic and very often reduces it. Indeed, the highly dynamical flowing waters´ fauna and flora depends on this dynamic and is therefore stable.

When planing and realising protection measures for torrents, one should give preference to alternatives which respect the naturally morphological dynamic of the stream to a greatest extend.

Indeed, flowing waters´ dynamic is mainly characterised by erosion, sedimentation, aggradation processes and the formation of meanders. These processes can be separated temporally and spatially, can happen in concurrence and can even merge together. (Weinmeister 1994)

These permanently changing phenomena determine the torrents´ typical runoff behaviour, like the so-called "hyperconcentrated flow" and the "debris flow" as well as the "turbulent flood discharge".

In the stream`s sections of erosion, but also in the sections where all different processes like erosion, sedimentation and aggradation take place ("furkation"), the natural morphological dynamic leads to the development of multiple "step-pool-systems".

In order to preserve the morphological dynamic, it is essential, in course of the realisation of the project, to give the stream enough room to move, so that those diverse "step-pool-systems" can develop again naturally. Therefore, it is important that a certain amount of bedload transport is still possible after the realisation of torrent control measures.

For this reason it is essential, too, to organize a specific "stream management" at the erosion areas.

In gently steep middle reaches and especially in lower reaches one should also keep an eye on the development of natural stream` widths. Revetment works in these sections are therefore only to be carried out when absolutely necessary. Depending on the types of streams, one should choose bioengineering methods or control measures combining the use of rocks and wood.

1.2. Diversity of structure depending on stream morphology

You need morphological dynamic to have varied stream bed structures. Protection measures usually lead to a qualitative and quantitative loss of precious structures like flat banks, scours, unprotected steep banks, heaps of rubble, heaps of dead wood.

As a result, there is a loss of precious habitats. Biocoenosis are considerably threatened by a structure that is not varied.

Consequently, it is essential to respect the original diversity of bed structures.

If the natural dynamic of the stream is not acceptable at all because of security reasons, the diversity of structures should be recreated artificially.

When artificaly built bed structures are necessary "step-pool" areas, meanders and areas of aggradation should be planed depending on the original stream morphology. This can be one way to avoid the entire loss of habitats.

1.3. Ecological balancing the surface

Control measures often lead to a drastic reduction of that area which is usually used and affected by torrents. As a consequence, many different habitats as well as biocoenosis disappear.

When protection structures are necessary, the surface should, at least, be balanced with regard to their different ecological values. The surfaces affected by mean, low or high water discharge are very different in quality and each of them is a uniqueness in its biotic structure and functionalism.

The areas affected by low water, for example where river moss or cardamine grow, or splash water biocoenosis near drops, have, as a consequence, the same ecological value as sloppy meadows or a pioneer forest of alders in the deposit area of a debris stream.

Therefore, control measures should only be carried out when similar alternative areas are provided as compensation for destroyed biological surfaces and for the multiple combination of species, which is affected.

The aim is not to create artificial biological areas, but to let nature take its course in chosen areas so that it is able to recreate new biotopes without human influence. Even when measures are taken on apparently unproductive surfaces, like washouts and steeply (vertically) bare banks, one has to organise alternative surfaces because these are very often areas with a high dynamic of succession.

Concerning the ecological balance of surface it should also be refered to the possibilities offered by the means of hazard-zoning. Especially the naturally existing flood retention plains should get protected and measures that would reduce the water retention on these surfaces should restrictly be banned.

The influence, that the planning of hazard zones carries on the regional and municipal development planning, should be used and extended when a ecological balancing the surface is necessary.

When measures have to be carried out and whenever negotiating with land owners is necessary, it is important, regardless of the conceptual thoughts, to try to include the largest possible tract of land for the realisation of control measures.

Because the most important demand for nature orientated flood control says "nature needs space".

1.4. The stream´s centreline

When defence works are placed, alterations of the stream`s natural centreline are made in many cases.

Channel shiftings and meanderings are mostly discarded for a straighter "line".

Very often, as a consequence, the stream is shortened; this causes a significant increase of the stream`s velocity, when bed and bank roughness stay similar.

In order to conform with the requirements regarding ecology and landscape aesthetics, it is important to try to respect or reconstitute the characteristic centerline of the stream.

The centreline expresses the stream`s typical morphological type. Therefore it is necessary to conform the stream`s centreline to the particular physical processes of each part of the stream (erosion, aggradation, "furkation") when planning an intervention.

1.5. Open ground, structure of bed, grain size distribution

Defence works in the upper course or modifications of the centreline in the lower course, mostly lead to changes in the structure of streams` substratum. The grain size distribution can significantly transmute into a smaller structure. This can clog up the stream bed, affects vertical ("hyporheic") habitats adversely and makes many benthos living community absolutely impossible.

The habitable surface and its biological energy, which are responsible for the ecological systems` stability, are reduced.

The logical minimum requirement, when any type of defence works is necessary : no sealing of stream beds.

The maintenance of natural beds` structures goes together with the maintenance of the natural conditions of current and stream bed gradient.

1.6. Appearance of the current and its distribution

These factors are connected together with the structure of substratum and therefore also with the stream`s morphological diversity of structure. An intervention within the stream`s current leads to a reduction of the diversity of structure and would have a direct influence on the pattern of the current which is distributed irregular over the stream`s bed.

Currents that are irregular formed stand out for quick changes of velocity, which result of the irregularity of the step-pool-system or of the meandering and accumulation forms of lower reaches. In this case the flowing water, which is concentrated in the depth lines of the structured stream`s bottom, hits areas of slight water concentration and widened water runoff.

The irregularity of the currents makes it possible, especially in the case of low water runoff, for different sorts of aquatic beings to survive.

When defence works are necessary, it is important to preserve or reconstitute the original pattern of currents.

1.7. Migration

Control measures, especially transverse structures, can reduce or stop migration possibilities for fish and macro zoo benthos. This can significantly weaken the genetical exchange between populations and therefore lead to a loss in the genetical variety.

The very important compensation migrations, especially after flood events are reduced if not stopped.

Compensation flights are not hindered by transverse structures, but depend on the sort and extend of bank`s vegetation.

The preservation of the flowing water`s continuity and other compensation mechanisms (for example the possibility of migration from tributaries) is essential therefore.

In many cases (especially in gentle flowing waters) the migration possibilities can also be measured against the passable way for fish.

1.8. Association of biotopes, integration with the immediate environment

Biotope connection is an important ecological factor, especially as far as civilised land is concerned. Torrents, in particular, with their different types of attendant vegetation, have an important role to play as a landscape integrating factor. Because of the many different types of human interventions in the landscape, the immediate environment of streams is one of the ultimate refuge area for rare organisms, which cannot find their habitat within cultured areas.

Defence works, for instance longitudinal structures, can lead to a reduction or disappearance of the biotope's combining function and biotope dividing barriers can be established in that way..

The type of bank vegetation is dependant mainly on the type of stream. Its minimum width should amount two or three times the stream bed's width. This width is essential for a well balanced ecological system.

Maintenance, as well as planting new vegetation, shall establish a "protection buffer" against other cultural practice, for instance agriculture.

1.9. Vegetation typical for specific waters

The bank vegetation and the stream itself are in intimate coherence concerning the insertion of organic substances and the shade the vegetation confers. At the same time, the vegetation has an influence on erosion and sedimentation processes, especially within the area of the bank.

However, the bank's ecological value is not only important for the water itself. It also represents protection, shelter and habitat for all sorts of animal and biological species. Indeed, this stream bank vegetation has a special influence on certain bird types. Thus, the bird population increases significantly when the stream bank's vegetation widens up to 10 meter (Kartaus 1990).

Consequently, it is essential to opt for a balanced bank vegetation that respects the type of stream.

However, one shouldn't quickly re-plant the stream bank after defence works are finished, but give nature the possibility to take control again over those surfaces and to "reconstruct" the landscape with new well adapting structures.

To revegetate a stream bank hastily after finishing the control measures in order to improve the aesthetics or the image of the landscape is not ecologically efficient in most cases; indeed, the rivalry of "foreign" roots only hinders the development of plants that would grow naturally.

Keeping the root stocks of trees excavated before and incorporating them upside down in the bank slope helps stabilising it immediately and at the same time creates essential ecological niches and shelters for different small animals and fish.

1.10. Local building material

When defence works have to be carried out, it is important to avoid using construction material that isn`t appropriate for the site. For instance, the use of limestone for riprapping would not conform ecological requirements, when the substratum of the catchment area consists of granite or gneiss.

Furthermore, it is important not to plan bedded rockfills in meandering streams and in streams either with little or without bedload transport and incorporate wooden defence works in streams having a high level of bedload transport.

Concrete is never considered as typical for the site, but in some cases it is impossible to substitute it with local building material. Therefore, the necessary defence works should be carried out in that way, that it helps to increase the ecological value of the structure (for example: leave the foundation drain open, lowering of slit drains under the stream bed, etc.)

Criteria to follow when using construction material in torrents:
- Coarse stones in torrents with a high sediment discharge and debris flow level ("bedload torrents" and "debris flood/flow" torrents)
- Limestone in limestone streams, granite stone in granite stone streams, etc.
- Bioengineering construction material in gently steep streams with low bedload transport and for revetment works along inside bends.
- Wood in torrents with low sediment discharge.
- Wood in the case of gently sloped lower courses and meandering streams
- Combination of wood and stones in case of gently inclined middle reaches.
- Coarse stones in gently steep or steep courses.
- Utilisation of stream`s own rounded stones (boulders).
- Utilisation of local cut and untreated wood.

1.11. Rare species and habitats

Even the most nature orientated constructions aren`t ecologically acceptable, when regional peculiarities get lost. These peculiarities can be ecological precious upland marshes as well as rare and endangered animal species like river crayfish, etc..

Therefore, the habitats of rare species and rare living spaces in general must not be involved in building works at all.

2. REQUIREMENTS REGARDING THE AESTHETICS, THE IMAGE AND THE CHARACTER OF A LANDSCAPE

Streams not only have ecological functions but also play a significant landscape designing role in the scenery.

Especially in tourist regions flowing waters represent an increasingly important social value. The physical good health and the quality of life of human beings is also ruled by the landscape formed environment. In an highly technological environment the value of a natural and harmonic landscape rises.

When flood control measures are necessary, it is therefore not only essential to take in account the ecological aspects but there is an increased demand also for aesthetic solutions.

The idea of the aesthetics of a landscape is mainly based on marks of the image and the character of the scenery.

The image of the landscape should be judged by the extent of its naturalness and unaffectedness.

The character of the landscape in question depends on is special trait and oddity, which is defined by the "typical composition" of landscape elements and their "gearing together".

Therefore the requirements regarding the aesthetics of a landscape change depending on the countryside that is included.

3. REQUIREMENTS REGARDING THE SOCIAL ECONOMICAL ASPECTS

Especially in settlement areas, water plays a social economical role and contributes to the quality of life of human beings.

Water is attractive and soothing (fresh, humid air, regular babble) and can lead to a multitude of sporting activities.

When planing flood control measures one has to take in account beside the ecological and aesthetic aspects also additional criteria which deal with the necessities of humans concerning the versatility of the medium "water".

Social economical demands :

- Water as a playground for children
- Water for different kinds of aquatic sports (fishing, rafting, swimming, boating, etc.)
- Water as a recreational area
- water used for the extinction of fires, etc.

Conflicts between the requirements regarding the ecological, aesthetical and social economical aspects can occur. This is especially the case when human needs are given the preference over ecological demands.

Nature can very quickly be over used by humans, therefore it is essential to give preference to the ecological requirements over the social and human ones.

4. REQUIREMENTS REGARDING THE AIMS OF THE PROJECTS, WHEN PLANNING FLOOD CONTROL MEASURES

The prime condition when defence works have to be carried out should be the respect of the ecological environment. This means that the different threats shouldn`t be adjusted to the claims of human utilisation, but to adjust the claims of human utilisation to the potential natural hazards.

The means to reduce natural danger should be bound to <u>disaster prevention</u> and technical measurements are to submit the possibilities of landscape planing.

In case of planing torrent control measures considerations of land use planing aspects, as follows, should be taken into account:

- preservation and protection of the natural inundation areas and debris flow deposition areas still available (flood and debris retarding)
- organisation of an alarm and evacuation system for potential endangered areas
- compulsory transfer from endangered areas
- constitution of rules and regulations for agricultural and forest management in chosen areas
- preservation of particular highly dynamic eroded reaches (to preserve the "ecological stability")
- protection of the association of biotopes
- organisation of protection zones for specific species of animals and plants

These measures aim to give nature more space and therefore to restore a greater ecological independence. Consequently, one can improve the compatibility of "hard" constructions that are absolutely necessary.

Nevertheless, all interventions should respect the ecological criteria shown in (1) additionally. The bed structure of naturally flowing streams and the untouched elements of landscape should be taken as models (ecological "Leitbild").

All details within the planning of an ecological flood control management are bound to the limits of ecology and nature's capacity of regeneration. If security measure has gone beyond this ecological limit (this should be exceptional), alternative or compensation measures have to be organised.

FINAL COMMENT

When flood control measures have to be carried out, the question is not "which quantity of flood a channel can take", but rather "how much space is left for flowing waters in our civilisation, settlement and industrial society" (FRITSCH et al. 1996).

It should be also considered to give priority to the question of the sensitivity and acceptance of damage instead of avoiding damages at any price.

Thus, flood control - originally a technical problem - has become a social process which reflects the change of opinion in ecological tendency of our society and its ability to take risks.

REFERENCES

Fritsch M., Gelzer S., Gunzenreiner U., Willi H., Zahno M. (1996): Differenzierte und bodennutzungsabhängige Hochwasserschutzkonzepte als Instrumente zur integralen Kulturraum- und Ressourcenbewirtschaftung, Intern. Symposion Interpraevent 1996, Tagungspublikation, Vol. 4, p. 231-242

Honsig W., Hufnagl H., Patek M., Wagenleitner H., Heumader J., Scheiber T. (1996): Ökologische Planungsinhalte und Kriterien bei Projekten der Wildbachverbauung / Ecological planing contents in torrent control projects, Positionspapier der Arbeitsgruppe „Wildbachverbauung und Ökologie", Journal of Torrent, Avalanche, Landslide and Rock Fall Engineering, Vol. 60, No. 131, 1996

Hufnagl H. (1994): Ökologische Gestaltungsmöglichkeiten – Erfahrungen aus Kärnten / Ecological design possibilities – experiences in Carinthia, Journal of Torrent, Avalanche, Landslide and Rock Fall Engineering, Vol. 58, No. 126, 1994

Weinmeister H.W. (1994): Machbarkeit und Grenzen der WLV – physikalische und ökologische Aspekte / Feasibility and limits of torrent control – physical and ecological aspects, Journal of Torrent, Avalanche, Landslide and Rock Fall Engineering, Vol. 58, No. 126, 1994

FORESTRY SCIENCES

21. T.C. Hennessey, P.M. Dougherty, S.V. Kossuth and J.D. Johnson (eds.): *Stress Physiology and Forest Productivity.* Proceedings of the Physiology Working Group, Technical Session, Society of American Foresters National Convention (Fort Collins, Colorado, USA, 1985). 1986   ISBN 90-247-3359-6

22. K.R. Shepherd: *Plantation Silviculture.* 1986   ISBN 90-247-3379-0

23. S. Sohlberg and V.E. Sokolov (eds.): *Practical Application of Remote Sensing in Forestry.* Proceedings of a Seminar on the Practical Application of Remote Sensing in Forestry (Jönköping, Sweden, 1985). 1986   ISBN 90-247-3392-8

24. J.M. Bonga and D.J. Durzan (eds.): *Cell and Tissue Culure in Forestry.* Volume 1: General Principles and Biotechnology. 1987   ISBN 90-247-3430-4

25. J.M. Bonga and D.J. Durzan (eds.): *Cell and Tissue Culure in Forestry.* Volume 2: Specific Principles and Methods: Growth and Development. 1987   ISBN 90-247-3431-2

26. J.M. Bonga and D.J. Durzan (eds.): *Cell and Tissue Culure in Forestry.* Volume 3: Case Histories: Gymnosperms, Angiosperms and Palms. 1987   ISBN 90-247-3432-0
                                            Set ISBN (Volumes 24–26) 90-247-3433-9

27. E.G. Richards (ed.): *Forestry and the Forest Industries: Past and Future.* Major Developments in the Forest and Forest Industries Sector Since 1947 in Europe, the USSR and North America. In Commemoration of the 40th Anniversary of the Timber Committee of the UNECE. 1987
                                            ISBN 90-247-3592-0

28. S.V. Kossuth and S.D. Ross (eds.): *Hormonal Control of Tree Growth.* Proceedings of the Physiology Working Group, Technical Session, Society of American Foresters National Convention (Birmingham, Alabama, USA, 1986). 1987   ISBN 90-247-3621-8

29. U. Sundberg and C.R. Silversides: *Operational Efficiency in Forestry.*
    Volume 1: Analysis. 1988   ISBN 90-247-3683-8

30. M.R. Ahuja (ed.): *Somatic Cell Genetics of Woody Plants.* Proceedings of the IUFRO Working Party S2.04-07 Somatic Cell Genetics (Grosshansdorf, Germany, 1987). 1988
                                            ISBN 90-247-3728-1

31. P.K.R. Nair (ed.): *Agroforestry Systems in the Tropics.* 1989   ISBN 90-247-3790-7

32. C.R. Silversides and U. Sundberg: *Operational Efficiency in Forestry.*
    Volume 2: Practice. 1989   ISBN 0-7923-0063-7
                                            Set ISBN (Volumes 29 and 32) 90-247-3684-6

33. T.L. White and G.R. Hodge (eds.): *Predicting Breeding Values with Applications in Forest Tree Improvement.* 1989   ISBN 0-7923-0460-8

34. H.J. Welch: *The Conifer Manual.* Volume 1. 1991   ISBN 0-7923-0616-3

35. P.K.R. Nair, H.L. Gholz, M.L. Duryea (eds.): *Agroforestry Education and Training.* Present and Future. 1990   ISBN 0-7923-0864-6

36. M.L. Duryea and P.M. Dougherty (eds.): *Forest Regeneration Manual.* 1991
                                            ISBN 0-7923-0960-X

37. J.J.A. Janssen: *Mechanical Properties of Bamboo.* 1991   ISBN 0-7923-1260-0

38. J.M. Bonga and P. Von Aderkas: *In Vitro Culture of Trees.* 1992   ISBN 0-7923-1540-5

FORESTRY SCIENCES

39. L. Fins, S.T. Friedman and J.V. Brotschol (eds.): *Handbook of Quantitative Forest Genetics.* 1992   ISBN 0-7923-1568-5

40. M.J. Kelty, B.C. Larson and C.D. Oliver (eds.): *The Ecology and Silviculture of Mixed-Species Forests. A Festschrift for David M. Smith.* 1992   ISBN 0-7923-1643-6

41. M.R. Ahuja (ed.): *Micropropagation of Woody Plants.* 1992   ISBN 0-7923-1807-2

42. W.T. Adams, S.H. Strauss, D.L. Copes and A.R. Griffin (eds.): *Population Genetics of Forest Trees.* Proceedings of an International Symposium (Corvallis, Oregon, USA, 1990). 1992   ISBN 0-7923-1857-9

43. R.T. Prinsley (ed.): *The Role of Trees in Sustainable Agriculture.* 1993   ISBN 0-7923-2030-1

44. S.M. Jain, P.K. Gupta and R.J. Newton (eds.): *Somatic Embryogenesis in Woody Plants,* Volume. 3: Gymnosperms. 1995   ISBN 0-7923-2938-4

45. S.M. Jain, P.K. Gupta and R.J. Newton (eds.): *Somatic Embryogenesis in Woody Plants,* Volume 1: History, Molecular and Biochemical Aspects, and Applications. 1995   ISBN 0-7923-3035-8

46. S.M. Jain, P.K Gupta and R.J. Newton (eds.): *Somatic Embryogenesis in Woody Plants,* Volume 2: Angiosperms. 1995   ISBN 0-7923-3070-6
     Set ISBN (Volumes 44–46) 0-7923-2939-2

47. F.L. Sinclair (ed.): *Agroforestry: Science, Policy and Practice.* Selected Papers from the Agroforestry Sessions of the IUFRO 20th World Congress (Tampere, Finland, 6–12 August 1995). 1995   ISBN 0-7923-3696-8

48. J.H. Goldammer and V.V. Furyaev (eds.): *Fire in Ecosystems of Boreal Eurasia.* 1996   ISBN 0-7923-4137-6

49. M.R. Ahuja, W. Boerjan and D.B. Neale (eds.): *Somatic Cell Genetics and Molecular Genetics of Trees.* 1996   ISBN 0-7923-4179-1

50. H.L. Gholz, K. Nakane and H. Shimoda (eds.): *The Use of Remote Sensing in the Modeling of Forest Productivity.* 1996   ISBN 0-7923-4278-X

51. P. Bachmann, M. Köhl and R. Päivinen (eds.): *Assessment of Biodiversity for Improved Forest Planning.* Proceedings of the Conference on Assessment of Biodiversity for Improved Planning (Monte Verità, Switzerland, 1996). 1998   ISBN 0-7923-4872-9

52. G.M.J. Mohren, K. Kramer and S. Sabaté (eds.): *Impacts of Global Change on Tree Physiology and Forest Ecosystems.* Proceedings of the International Conference on Impacts of Global Change on Tree Physiology and Forest Ecosystems (Wageningen, The Netherlands, 26–29 November 1996). 1997   ISBN 0-7923-4921-0

53. P.K.R. Nair and C.R. Latt (eds.): *Directions in Tropical Agroforestry Research.* Selected Papers from the Symposium on Tropical Agroforestry organized in connection with the annual meeting of USA (Indianapolis, U.S.A., 5 November 1996). 1998   ISBN 0-7923-5035-9

54. K. Sassa (ed.): *Environmental Forest Science.* Proceedings of the IUFRO Division 8 Conference Environmental Forest Science. (Kyoto, Japan, 19–23 October 1998). 1998   ISBN 0-7923-5280-7

KLUWER ACADEMIC PUBLISHERS – DORDRECHT / BOSTON / LONDON